International Conference on Computing and Applied Informatics 2016

Journal of Physics: Conference Series Volume 801

Medan, Indonesia
14 – 15 December 2016

ISBN: 978-1-5108-3844-4
ISSN: 1742-6588

Printed from e-media with permission by:

Curran Associates, Inc.
57 Morehouse Lane
Red Hook, NY 12571

Printed by Curran Associates, Inc. (2017)

For permission requests, please contact the Institute of Physics
at the address below.

Institute of Physics
Dirac House, Temple Back
Bristol BS1 6BE UK

Phone: 44 1 17 929 7481
Fax: 44 1 17 920 0979

techtracking@iop.org

Additional copies of this publication are available from:

Curran Associates, Inc.
57 Morehouse Lane
Red Hook, NY 12571 USA
Phone: 845-758-0400
Fax: 845-758-2633
Email: curran@proceedings.com
Web: www.proceedings.com

Table of contents

Volume 801

International Conference on Computing and Applied Informatics 2016
14–15 December 2016, Medan, Indonesia

Accepted papers received: 3 January 2017
Published online: 27 March 2017

Preface

Papers

Artificial Intelligence

Bioinformatics

Data Mining

E-Government

E-Health

Machine Learning

Natural Languange Processing

Networking

Smart City

Software in Science and Engineering

IOP Conf. Series: Journal of Physics: Conf. Series **801** (2017) 011001 doi:10.1088/1742-6596/801/1/011001

INTRODUCTION

On behalf of the Organizing Committee, I am delighted to welcome all participants to the International Conference on Computing and Applied Informatics (ICCAI) 2016 which took place from December 14 to December 15. This is an inaugural conference in the field of Computing and Applied Informatics which was held in Medan, Indonesia. With a very exciting program, it had become a premier forum for discussion, dissemination and achievement of research in the related fields.

Nearly 300 academics, researchers, practitioners and presenters from 7 countries have gathered and 215 papers have been submitted in this year's conference. Each paper was reviewed with tight criteria from our invited and voluntary multi-national reviewers. Based on the review results, 100 papers have been accepted, which lead to an acceptance rate of 46.51%. All accepted papers were orally presented in 3 technical sessions and was published in the conference proceeding.

The conference is a prestigious event that would not been successful without extensive effort from many volunteers. I would like to express my sincere appreciation and thanks to all participants and supporters, especially the members of Faculty of Computer Science and Information Technology (Fasilkom-TI), Faculty of Mathematics and Natural Science (FMIPA), Faculty of Engineering (FT) and The International Affairs Office of University of Sumatera Utara. I would also like to thank the keynote speakers, Prof. Hiromi Homma, Prof. Madya, Dr. Rahmita Wirza O.K. Rahmat, who were willing to meet the ICCAI invitation. My sincere gratitude to Rector of University of Sumatera Utara, Prof. Dr. Runtung Sitepu, M,Hum and Dean of Faculty of Computer Science and Information Technology, Prof. Dr. Opim Salim Sitompul, M.Sc..

I would also like to give my appreciation to many others who are involved in the success of this event, including the reviewers, student volunteers, and staffs at the University of Sumatera Utara.

Lastly, on behalf of the organizing committee, I would like to welcome you to our first international conference on computing and applied informatics, ICCAI 2016. Our hope is that you have an enjoyable and inspiring moment during our conference. Thank you for your participation on ICCAI 2016.

Dr. Eng. Himsar Ambarita
Chair, ICCAI 2016

IOP Conf. Series: Journal of Physics: Conf. Series **801** (2017) 011001 doi:10.1088/1742-6596/801/1/011001

IOP Conf. Series: Journal of Physics: Conf. Series **801** (2017) 011002 doi:10.1088/1742-6596/801/1/011002

Conference Information

Dates	: December 14 - December 15, 2016
Organizer	: Faculty of Computer Science and Information Technology University of Sumatera Utara
Venue	: Hotel GranDhika Setiabudi Medan Jalan Dr. Mansyur no. 169 Medan, Sumatera Utara Indonesia
Phone	: +62-61-42081234
Fax	: +62-61-42081234
Email	: reservation@grandhika-hotel.com
Website	: www.grandhika-hotel.com
Official Language	: English
Secretariat	: Faculty of Computer Science and Information Technology University of Sumatera Utara Jalan Universitas No. 9A, Kampus USU Medan, 20155 Sumatera Utara, Indonesia
Email	: iccai@usu.ac.id
Conference Website	: http://iccai.usu.ac.id

IOP Conf. Series: Journal of Physics: Conf. Series **801** (2017) 011002 doi:10.1088/1742-6596/801/1/011002

Committee

Honorary Chair
Prof. Dr. Runtung Sitepu, S.H, M.Hum (Rector – University of Sumatera Utara)

Steering Committee
Prof. Hiromi Homma (Toyohashi University of Technology, Japan)
Prof. Dr. Masashi Daimaruya (Muroran Institute of Technology, Japan)
Prof. Dr. Rahmat Budiarto (Al Baha University, Saudi Arabia)
Muhammad Fermi Pasha, B.Comp.Sc., M.Sc., Ph.D (Monash University, Malaysia Campus)
Prof. Dr. Hideki Kawai (Muroran Institute of Technology, Japan)
Prof. Zainal A.Hasibuan, Ir., MLS, PhD (University of Indonesia, Indonesia)
Prof. Dr. Opim Salim Sitompul (University of Sumatera Utara, Indonesia)
Drs. Mahyuddin K. M. Nasution, M.IT, PhD (University of Sumatera Utara, Indonesia)
Prof. Dr. Herman Mawengkang (University of Sumatera Utara, Indonesia)
Prof. Dr. Saib Suwilo (University of Sumatera Utara, Indonesia)
Dr. Ir. Rinaldi Munir, MT (Bandung Institute of Technology, Indonesia)
Prof. Dr. rer.nat. Dedi Rosadi, M.Sc. (University of Gadjah Mada, Indonesia)
Benny B. Nasution, Ph.D (Politeknik Negeri Medan, Indonesia)
Dr. Fitri Arnia, M.Eng.Sc (Syiah Kuala University, Indonesia)

Editorial Board
Dr. Eng. Himsar Ambarita (University of Sumatera Utara, Indonesia)
Drs. Mahyuddin K. M. Nasution, M.IT, PhD (University of Sumatera Utara, Indonesia)
Romi Fadilah Rahmat, B.Comp.Sc, M.Sc (University of Sumatera Utara, Indonesia)
Prof. Dr. Masashi Daimaruya (Muroran Institute of Technology, Japan)
Prof. Dr. Rahmat Budiarto (Al Baha University, Saudi Arabia)
Muhammad Fermi Pasha, B.Comp.Sc., M.Sc., Ph.D (Monash University, Malaysia Campus)
Dani Gunawan, S.T, M.T (University of Sumatera Utara, Indonesia)
Cheng-Yuan Chang (Chung Yuan Christian University, Taiwan)

IOP Conf. Series: Journal of Physics: Conf. Series **801** (2017) 011002 doi:10.1088/1742-6596/801/1/011002

Chair
Dr. Eng. Himsar Ambarita (University of Sumatera Utara, Indonesia)

Co. Chair
Romi Fadilah Rahmat, B.Comp.Sc., M.Sc (University of Sumatera Utara, Indonesia)
Suherman, PhD (University of Sumatera Utara, Indonesia)

Executive Members
Dr. Erna Budhiarti Nababan, M.IT (University of Sumatera Utara, Indonesia)
Emerson Sinulingga, PhD (University of Sumatera Utara, Indonesia)
Dr. Fahmi F (University of Sumatera Utara, Indonesia)
Mohammad Fadly Syah Putra, M.Sc (University of Sumatera Utara, Indonesia)
Muhammad Anggia Muchtar, ST, MMIT (University of Sumatera Utara, Indonesia)
Dani Gunawan, S.T, M.T (University of Sumatera Utara, Indonesia)
Dr. Eng Taufiq Bin Nur, ST, M.Eng.Sc (University of Sumatera Utara, Indonesia)

Local Members
Prof. Dr. Muhammad Zarlis (University of Sumatera Utara, Indonesia)
Dr. Poltak Sihombing, M.Kom (University of Sumatera Utara, Indonesia)
Dr. Syahril Effendi, M.IT (University of Sumatera Utara, Indonesia)
Dr. Sawaluddin, M.IT (Univeristy of Sumatera Utara, Indonesia)
Dr. Elviawaty Muisa Zamzami, S.T, M.T, M.M (University of Sumatera Utara, Indonesia)
Dr. Maya Silvi Lydia, M.Sc (University of Sumatera Utara, Indonesia)
Amalia, S.T, M.T (University of Sumatera Utara, Indonesia)
Ainul Hizriadi, S.Kom, M.Sc (University of Sumatera Utara, Indonesia)
Indra Aulia, S.TI, MKom (University of Sumatera Utara, Indonesia)
Dedy Arisandi, S.T, M.Kom (University of Sumatera Utara, Indonesia)
Dian Rachmawati, S.Si, M.Kom (University of Sumatera Utara, Indonesia)
Baihaqi Siregar, S.Si, M.T (University of Sumatera Utara, Indonesia)
Ivan Jaya, S.Si, M.Kom (University of Sumatera Utara, Indonesia)
Jos Timanta Tarigan, S.Kom, M.Sc (University of Sumatera Utara, Indonesia)
Sarah Purnawati, S.T, M.Sc (University of Sumatera Utara, Indonesia)
Mohammad Andri Budiman, S.T, M.Comp.Sc, M.E.M (University of Sumatera Utara, Indonesia)
Ulfi Andayani, S.Kom, M.Kom (University of Sumatera Utara, Indonesia)
Sri Melvani Hardi, S.Kom, M.Kom (University of Sumatera Utara, Indonesia)
Amer Sharif, S.Si, M.Kom (University of Sumatera Utara, Indonesia)
Seniman, S.Kom, M.Kom (University of Sumatera Utara, Indonesia)

IOP Conf. Series: Journal of Physics: Conf. Series **801** (2017) 011002 doi:10.1088/1742-6596/801/1/011002

Venue Map

Hotel GranDhika Setiabudi Medan
Jalan Dr. Mansyur no. 169
Medan, Sumatera Utara
Indonesia
Phone : +62-61-42081234
Fax : +62-61-42081234

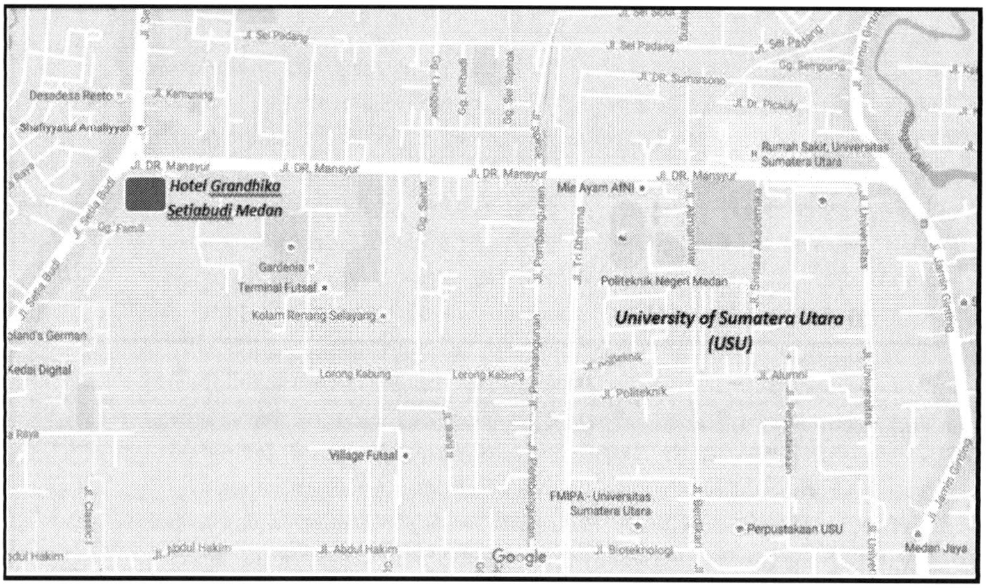

ICCAI 2016 Venue

Floor Plan

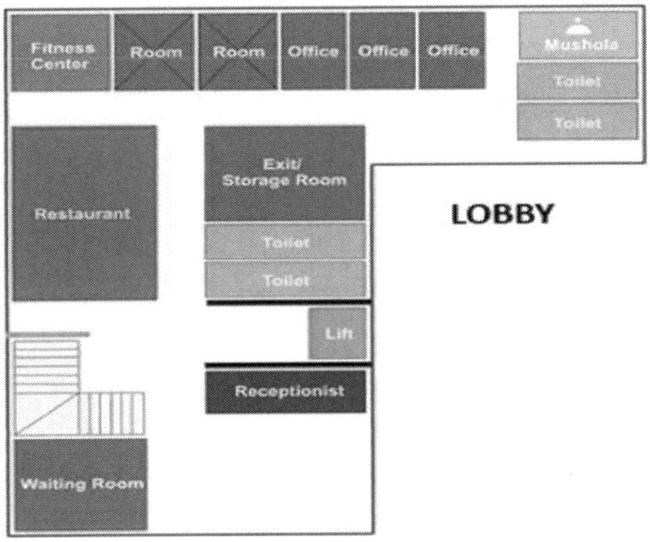
Fitness Center
Room
Room
Office
Office
Office
Mushola
Toilet
Toilet
Restaurant
Exit/ Storage Room
LOBBY
Toilet
Toilet
Lift
Receptionist
Waiting Room

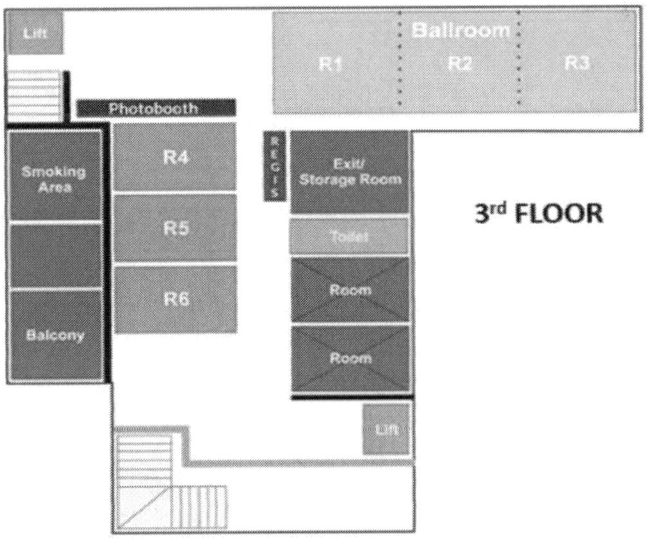
Lift
Ballroom
R1
R2
R3
Photobooth
Smoking Area
R4
R5
R6
REGIS
Exit/ Storage Room
3rd FLOOR
Toilet
Room
Room
Balcony
Lift

The 1st International Conference on Computing And Applied Informatics (ICCAI) 2016

Applied Informatics toward Smart Environment, People, and Society

CALL FOR PAPERS

December 14th-15th, 2016 | Medan, Indonesia

DESCRIPTION

The 1st International Conference on Computing and Applied Informatics is an international event that gathers researchers, scientists, engineers, scholars, experts, academics, and students who are interested to disseminate, report and share their experiences, innovations, ideas, research results in the field of computing and informatics

TOPICS

Artifical Intelligence | Augmented Reality | Smart City | Information Security | Big Data | Bioinformatics | Socio Informatics | Natural Language Processing | Computational Fluid Dynamics | Data Mining | Language Generation | Software Engineering | Machine Learning | GIS | E-Government | E-Commerce | E-Health | Human Computer Interaction | Image Vision | Computer Graphics | Networking | Computer-assisted Survey and Research | Computational Mechanics | Numerical Analysis Science | Computational Modeling | Computer in Medicine and Biology | Computer in Social Sciences | Numerical Analysis Engineering | Numerical Methods in Engineering | Computer Based Researches

IMPORTANT DATES

Full Paper Submission
November 14th, 2016

Notification of Acceptance
November 21st ,2016

Early Bird
November 28th, 2016

Camera Ready
December 4th, 2016

Further Information

iccai@usu.ac.id | http://iccai.usu.ac.id
ICCAIUSU | +62878-6894-9412
ICCAIUSU | ICCAI USU

KEYNOTE SPEAKERS

Prof. Hiromi Homma
(Toyohashi University of Technology, Japan)

Prof. Madya Dr. Rahmita Wirza O.K. Rahmat
(Universiti Putra Malaysia)

PAYMENT

EARLY BIRD

Author
Domestic Student (IDR 1.750.000) - Domestic Non Student (IDR 2.500.000)
International Student (USD 175) - International Non Student (USD 250)

Participant
Domestic Student (IDR 650.000) - Domestic Non Student (IDR 850.000)
International Student (USD 65) - International Non Student (USD 85)

REGULAR

Author
Domestic Student (IDR 2.000.000) - Domestic Non Student (IDR 2.750.000)
International Student (USD 200) - International Non Student (USD 275)

Participant
Domestic Student (IDR 750.000) - Domestic Non Student (IDR 1.000.000)
International Student (USD 75) - International Non Student (USD 100)

Additional Page Charge (USD 20 / Rp. 200.000)

Proceeding Publisher :
IOP Conference Series conferenceseries.iop.org IOP Publishing

Accepted Papers Will be Published in Journal of Physics: Conference Series Indexed by :
Scopus THOMSON REUTERS
IET Inspec INIS SPIRES Compendex
VINITI Abstract Journal NASA Astrophysics Data System

IOP Conf. Series: Journal of Physics: Conf. Series **801** (2017) 011004 doi:10.1088/1742-6596/801/1/011004

Peer review statement

All papers published in this volume of *Journal of Physics: Conference Series* have been peer reviewed through processes administered by the proceedings Editors. Reviews were conducted by expert referees to the professional and scientific standards expected of a proceedings journal published by IOP Publishing.

IOP Conf. Series: Journal of Physics: Conf. Series **801** (2017) 012001 doi:10.1088/1742-6596/801/1/012001

Fuzzy linear programming for bulb production

I Siregar[1], H Suantio[1], Y Hanifiah[1], M A Muchtar[2] and T H Nasution[3]

[1]Department of Industrial Engineering, University of Sumatera Utara, Almamater Street, University of Sumatera Utara, Medan – North Sumatera, 20155
[2]Departement of Information Technology, University of Sumatera Utara, Universitas Street, University of Sumatera Utara – North Sumatera, 20155
[3]Departement of Electrical Engineering, University of Sumatera Utara, Almamater Street, University of Sumatera Utara, Medan – North Sumatera, 20155

Email: ikhsan.siregar@usu.ac.id

Abstract. The research was conducted at a bulb company. This company has a high market demand. The increasing of the market demand has caused the company's production could not fulfill the demand due to production planning is not optimal. Bulb production planning is researched with the aim to enable the company to fulfill the market demand in accordance with the limited resources available. From the data, it is known that the company cannot reach the market demand in the production of the Type A and Type B bulb. In other hands, the Type C bulb is produced exceeds market demand. By using fuzzy linear programming, then obtained the optimal production plans and to reach market demand. Completion of the simple method is done by using software LINGO 13. Application of fuzzy linear programming is being able to increase profits amounted to 7.39% of the ordinary concept of linear programming.

1. Introduction

In a production planning, there are several constraints that limit the production of a company. These constraints can be the engine capacity, the availability of working time, and the availability of raw materials. One of the factors that influence the success of a company is a good production planning. A production planning is said to be good if those planning could reach the market demand by using the minimum cost to get optimal results.

In the previous research results, Linear Programming Method was used to minimize the cost [1]. In addition, linear programming can also be used in telecommunication products [2]. The use of linear programming method is used on land planning [3]. In the previous research, linear programming methods are closely related to mathematics and computer [4]. Meanwhile, for the methods used in this research, the Fuzzy method also applied in previous studies discusses the integration with Fuzzy TOPSIS method [5]. In addition, fuzzy method is used in the information transmission as well [6]. In industrial engineering, Fuzzy methods used in determining the criteria for suppliers that therefore selected by suppliers which are the worthiest to be chosen [7]. Fuzzy methods are similarly used in determining the design of an alga based eco-industrial park that influenced by the variability of the price of the product [8].

Results of previous studies can be seen that there are no studies that discuss the formulation of product demand, especially bulb products by using Fuzzy method. Thus through this research, the optimal value of the company can be calculated and observed.

To fulfill the market demands by using fuzzy logic approach to optimal the resource allocation and determine the value of production interval constraints. From the interval value is then searched the most optimal value. By using Fuzzy Linear Programming, researchers can look for the optimum amount of light bulbs manufactured products to meet market demand and in accordance with the limited production resources.

2. Methods

The data collection was conducted in one of the companies that producing light bulbs in North Sumatra. This research is a quantitative research; the quantitative research is the systematic and empirical studies of quantitative attributes and relationships. As the object of study is the light bulb brand Dai-Ichi G40 and Type A and Type B. The research was carried out by conducting a preliminary study to determine the condition of the company, production processes, and data collection. Data taken in this study is a form of production capacity data, the timing of engine damage, the availability of working time, usage and inventory of raw materials, product selling prices, cost of production, and the cycle time of each work station. Cycle time data is taken by using a stopwatch.

Production planning method that used in this research is Fuzzy Linear Programming (FLP). FLP methods used to maximize the value of Z obtained from Linear Programming by using fuzzy numbers.

2.1 Determining the objective function and constraint

In this study, the decision variable are symbolized by X_i . The decision variables are:

X_1 = Dai-Ichi Type G40 Bulb
X_2 = Stanless Star Tipe A Bulb
X_3 = Stanless Star Tipe B Bulb

The objective function of the model is to maximize the marginal returns (Z) of each bulb produced. The calculation of the marginal benefit of the product by using the formula:

$$\textit{Marginal Benefit} = \textit{Selling Price/unit} - \textit{Production Cost/unit} \tag{1}$$

The mathematical formulation for the objective function can be written as follows:

$$MAX\ Z = \sum_{i=1}^{m} HX_i \tag{2}$$

In this study, which became the first constraint of the function is an installed production capacity of each machine. The mathematically formulations of the first constraint functions to the installed production capacity are:

$$\sum_{i=1}^{m} X_i \leq \sum_{j=1}^{n} \sum_{k=1}^{o} Y_{jk} \tag{3}$$

The second constraint function is a function of working hours availability constraints. It is used to look at the relationship between the productions times with the amount of product produced. The formulations used to formulate the second constraint functions are:

$$\sum_{i=1}^{m} a_i X_i \leq \sum_{j=1}^{n} R_j \tag{4}$$

The third constraint function is a function of raw material usage constraints. It is used to look at the relationship between the use and availability of raw materials to the amount of product produced. The formulations used to formulate the third constraint functions are:

$$\sum_{l=1}^{p} \sum_{i=1}^{m} c_l X_i \leq \sum_{l=1}^{p} T_l \tag{5}$$

IOP Conf. Series: Journal of Physics: Conf. Series **801** (2017) 012001 doi:10.1088/1742-6596/801/1/012001

2.2 Establishment of fuzzy linear programming model

Before the formulation of the fuzzy linear programming models, the p_0 value must be determined. Mathematically the p_0 value is calculated by:

$$p_o = Z_{j\,(t=1)} - Z_{j\,(t=0)} \tag{6}$$

Fuzzy value is symbolized by λ. General formulation of fuzzy linear programming models, namely:

Max : λ

With constraint : $\lambda p_o + B_o X_i \leq d_o + p_o$.

3. Research Result

3.1 Fuzzy concept linier programming model

The concept of fuzzy logic provides the interval linear programming models. The concept of fuzzy logic t = 0 means that all the functions of the constraints that have been established are not using the limit values of tolerance interval. The concept of fuzzy logic t = 1 means that all the functions of the constraints that have been formed by using the limit values of tolerance interval. Objective function and constraints function are expressed in linear programming models can be seen as follows:

$$\begin{array}{llllllllllll}
\text{MAX Z} & 90\,X_1 & + & 110\,X_2 & + & 140\,X_3 & & & & & \\
& X_1 & + & X_2 & + & X_3 & \leq & 5280000 & - & 167904 & t \\
& X_1 & + & X_2 & + & X_3 & \leq & 4400000 & - & 191840 & t \\
& X_1 & + & X_2 & + & X_3 & \leq & 7920000 & - & 303336 & t \\
& X_1 & + & X_2 & + & X_3 & \leq & 5280000 & - & 256080 & t \\
& X_1 & + & X_2 & + & X_3 & \leq & 7040000 & - & 209792 & t \\
& X_1 & + & X_2 & + & X_3 & \leq & 6160000 & - & 138600 & t \\
& X_1 & + & X_2 & + & X_3 & \leq & 5280000 & - & 229152 & t \\
& X_1 & + & X_2 & + & X_3 & \leq & 5280000 & - & 110352 & t \\
& 14{,}01\,X_1 & + & 14{,}99\,X_2 & + & 15{,}67\,X_3 & \leq & 50688000 & + & 12672000 & t \\
& 7{,}20\,X_1 & + & 8{,}30\,X_2 & + & 10{,}80\,X_3 & \leq & 25000000 & + & 1000000 & t \\
& 0{,}67\,X_1 & + & 0{,}77\,X_2 & + & X_3 & \leq & 2000000 & + & 500000 & t \\
& X_1 & + & X_2 & + & X_3 & \leq & 2000000 & + & 500000 & t \\
& X_1 & + & X_2 & + & X_3 & \leq & 2500000 & + & 500000 & t \\
& X_1 & + & X_2 & & & \leq & 2000000 & + & 100000 & t \\
& & & & & X_3 & \leq & 400000 & + & 100000 & t \\
& 0{,}07\,X_1 & + & 0{,}08\,X_2 & + & 0{,}10\,X_3 & \leq & 250000 & + & 20000 & t \\
& & & 0{,}84\,X_2 & + & 1{,}09\,X_3 & \leq & 1750000 & + & 250000 & t \\
& 0{,}04\,X_1 & + & 0{,}04\,X_2 & + & 0{,}05\,X_3 & \leq & 150000 & + & 10000 & t \\
& X_1 & & & & & \geq & 316817 & & & \\
& & & X_2 & & & \geq & 1669172 & & & \\
& & & & & X_3 & \geq & 400173 & & & \\
& & & & & X_1, X_2, X_3 & \geq & 0 & & &
\end{array}$$

From the calculation by using the software LINGO 13 obtained the calculation results in Table 1 and Table 2.

Table 1. Linear programming result t=0

Month	Type C	Type A	Type B	Profit (Rupiah)
August	316.817 unit	1.669.172 unit	400.173 unit	258.067.240
September	315.339 unit	1.698.055 unit	443.051 unit	274.081.080
October	313.205 unit	1.725.947 unit	489.697 unit	274.042.480

Table 2. Linear programming result t=1

Month	Type C	Type A	Type B	Profit (Rupiah)
August	367.857 unit	1.732.143 unit	500.000 unit	293.642.860
September	315.339 unit	1.698.055 unit	620.818 unit	302.081.080
October	313.205 unit	1.725.947 unit	599.999 unit	302.042.500

3.2. Fuzzy linear programming

The fuzzy linear programming model formulation for company was as follows:

$$
\begin{array}{lrcrcrcrcl}
\text{MAX Z} & \lambda & & & & & & & & \\
 & -35575620\,\lambda & + & \text{s90}\,X_1 & + & 110\,X_2 & + & 140\,X_3 & \geq & 258067240 \\
 & -167904\,\lambda & + & X_1 & + & X_2 & + & X_3 & \leq & 5112096 \\
 & -191840\,\lambda & + & X_1 & + & X_2 & + & X_3 & \leq & 4208160 \\
 & -303336\,\lambda & + & X_1 & + & X_2 & + & X_3 & \leq & 7616664 \\
 & -256080\,\lambda & + & X_1 & + & X_2 & + & X_3 & \leq & 5023920 \\
 & -209792\,\lambda & + & X_1 & + & X_2 & + & X_3 & \leq & 6830208 \\
 & -138600\,\lambda & + & X_1 & + & X_2 & + & X_3 & \leq & 6021400 \\
 & -229152\,\lambda & + & X_1 & + & X_2 & + & X_3 & \leq & 5050848 \\
 & -110352\,\lambda & + & X_1 & + & X_2 & + & X_3 & \leq & 5169648 \\
 & 12672000\,\lambda & + & 14{,}01\,X_1 & + & 14{,}99\,X_2 & + & 15{,}67\,X_3 & \leq & 63360000 \\
 & 1000000\,\lambda & + & 7{,}20\,X_1 & + & 8{,}30\,X_2 & + & 10{,}80\,X_3 & \leq & 26000000 \\
 & 500000\,\lambda & + & 0{,}67\,X_1 & + & 0{,}77\,X_2 & + & X_3 & \leq & 2500000 \\
 & 500000\,\lambda & + & X_1 & + & X_2 & + & X_3 & \leq & 3000000 \\
 & 500000\,\lambda & + & X_1 & + & X_2 & + & X_3 & \leq & 3000000 \\
 & 100000\,\lambda & + & X_1 & + & X_2 & & & \leq & 2100000 \\
 & 100000\,\lambda & + & & & & & X_3 & \leq & 500000 \\
 & 20000\,\lambda & + & 0{,}07\,X_1 & + & 0{,}08\,X_2 & + & 0{,}10\,X_3 & \leq & 270000 \\
 & 250000\,\lambda & + & & & 0{,}84\,X_2 & + & 1{,}09\,X_3 & \leq & 2000000 \\
 & 10000\,\lambda & + & 0{,}04\,X_1 & + & 0{,}04\,X_2 & + & 0{,}05\,X_3 & \leq & 160000 \\
 & & & X_1 & & & & & \geq & 316817 \\
 & & & & & X_2 & & & \geq & 1669172 \\
 & & & & & & & X_3 & \geq & 400173 \\
 & & & & & & & X_1,X_2,X_3 & \geq & 0
\end{array}
$$

From the calculation using the software LINGO 13 obtained calculation results in Table 3.

Table 3. Fuzzy linear programming results

Month	λ	Product			Profit (Rupiah)
		Type C	Type A	Type B	
August	0,536	377.219 unit	1.669.172 unit	425.572 unit	Rp. 277.138.710
September	0,500	315.339 unit	1.698.055 unit	520.818 unit	Rp. 288.081.080
October	0,499	313.205 unit	1.725.947 unit	499.999 unit	Rp. 288.042.480

By using the concept of fuzzy linear programming, it can be analyzed that the market demand has been reached. The profits generated for the month of August, September, and October respectively is Rp. 277.138.710, Rp.288.081.080 and Rp.288.042.480. Companies can also define necessary intervals additions by using the value λ 0.536, 0.500, and 0.499. In other words, the largest scale value t = 1 to 0.536 = 0.464. Of the value of t, the company only needs 327 878 seconds as the overtime or equal to 1 hour.

4. Conclusion

Application of fuzzy linear programming is used to increase profits. The increased profits amounted to 7.39%, 4.67% and 4.65% of the ordinary concept of linear programming. Suryo research results obtained the increase of 10.6% from the usual linear programming. From these two studies, it can be concluded that the FLP can optimize the number of production companies. FLP combines the usual linear programming model and the concept of fuzzy logic as a means of decision-making in determining the optimum number of products taking into resource constraints of production. In a subsequent study can be added more than one objective function so that it can be modeled as Multi-Objective Fuzzy Linear Programming.

Acknowledgment

This research was supported by University of Sumatera Utara. We thank our colleagues from Information System Center who provided insight and expertise that greatly assisted the research. All the faculty, staff members and lab technicians of Industrial Engineering Department, whose services turned my research a success. Dr. Himsar Ambarita, my mentor, whose reminders and constant motivation encouraged me to meet the deadlines.

My Wife, Kartika Widya Astuty and children, Omaar Al Faridzi Siregar and Alisha Rizki Siregar, family members and friends, without whom I was nothing; they not only assisted me financially but also extended their support morally and emotionally.

References

[1] Al-Shihabi S,. A hybridofmax–min ant system and linear programming for the k-covering problem, *Computers & Operations Research* 76(2016) pp 1–11
[2] Fortz B, Compact mixed integer linear programming models to the minimum weighted tree reconstruction problem, *European Journal of Operational Research* 256 (2017) pp 242–251
[3] Kumar P, Mixed integer linear programming approaches for land use planning that limit urban sprawl, *Computers & Industrial Engineering* 102 (2016) pp 33–43
[4] Touil I, A feasible primal–dual interior point method for linear semidefinite programming, *Journal of Computational and Applied Mathematics* 312 (2017) pp 216–230.

[5] Chen SM, Multicriteria decision making based on the TOPSIS method and similarity measures between intuitionistic fuzzy values, *Information Sciences* 367–368 (2016) pp 279–295.
[6] Fu L, A fuzzy-theory-based method for studying the effect of information transmission on nonlinear crowd dispersion dynamics, *Commun Nonlinear Sci Numer Simulat* 42 (2017) pp 682–698
[7] Ghorabaee and Mehdi K, 2016 Multi-criteria evaluation of green suppliers using an extended WASPAS method with interval type-2 fuzzy sets, *Journal of Cleaner Production* vol 137 pp 213-229
[8] Ubando, Aristotle T, 2016 Fuzzy mixed integer non-linear programming model for the design of an algae-based eco-industrial park with prospective selection of support tenants under product price variability, *Journal of Cleaner Production* vol 136 pp 183-196

System for Detecting Potential Lost Person based on Conditional Random Field

R S Kusuma [1], G A P Saptawati[2]

Schools of Electrical Engineering and Informatics, Bandung Institute of Technology, Bandung, Indonesia

Email: [1]riadysastrak@gmail.com, [2]putri@informatika.org

Abstract. Global Positioning System (GPS) technology has been used widely in transsportation industry to help company in managing taxis. The most popular GPS utilization for taxi company is to identify the position of taxis and monitor theirs the mobility. Nowdays, data collected from GPS tracker is combined with data from taxi meter are analyzed to provide region information regarding potential passengers. Zicheng Liao's proposed a system based on GPS taxi data to detect anomalous area/region which was then interpreted as region with to predict rare passengers. The system was developed based on conditional random field (CRF) method and features position, velocity, passenger loading information. Our research was aimed to develop tool based on GPS data to detect potential lost person. We motivated by Liao research and modified the algorithms and features of CRF. Our experiments showed that the system has precision of 98.86% and recall of 87.478%.

1. Introduction

Nowdays, data collected from GPS can be analyzed to detect the position of a person to know the whereabouts of a person so that if the person is experiencing problems, supervisors are informed of his current position and able to provide help. However, with the limitations of GPS, supervisors do not know when the supervised experience problems. This issue has been addressed by a GPS tracker device. GPS tracker now has an emergency button that the supervised can press to tell the supervisor that he was in danger so that supervisors can immediately go to the scene to deal with the problem. However, this is still not effective regarding children or old men.

A case in point is the case of children and old man with Alzheimer. Old men with Alzheimer often wander aimlessly and reach unfamiliar areas unconsciously. Kids are also prone to getting lost when they find things that make them curious. This is very dangerous situation because they might be kidnapped. When they wander around initially, they do not realize that they are in danger, so they do not press the emergency button on the GPS tracker.

Therefore, old men with Alzheimer and kids need supervision constantly. It is impossible for supervisors to supervise them all the time since they may have other matters to attend to. To handle such cases, we propose a monitoring systems to detect potential lost person based on his/her daily habits.

2. Related Works

Zicheng Liao developed a system for detecting anomalies in GPS data from visual analysis [1]. The system was built based on machine learning with Conditional Random Fields (CRF) method for detecting anomalies in GPS data. In addition to building models for machine learning, the system has a function to display maps with the locations of potential anomalies detected. This system has been tested at a taxi company and the results are quite good since company are able to know where the passengers are rare in each region.

The system displays a map along with the potential anomaly in each region. The system required a model created by the GPS data collection on all taxi. In modelling, it takes labeled training data. The labeling process itself is carried out by the supervisory monitoring system that displays the entire taxi service. Supervisors will provide labels on certain areas in accordance to his knowledge about urban traffic monitoring whether or not the area is an anomaly.

GPS anomaly detection system requires a taxi GPS data. GPS data will be used by the system to analyze behavior patterns by observing taxis' movements. The model is based on training data that has been labeled by the supervisor. It was of a general nature, which means that the model uses the data of all existing taxis and that the model will also be used to detect anomalous areas for all taxis. It was made using CRF method with features that represents the characteristics movement patterns of taxis in general. The features cover speed, time, location, and passenger loading information.

3. Proposed System for Detecting Potential Lost Person

This section describes the development of potential lost person detection system.

3.1. Training data labelling

Training data is needed to build a model that will be used to classify whether or not a person is potentially get lost. Each user of the system (in this case, they could be kids or old man with Alzheimer has their own individual training data to build model based on their own habits and behavior.

Training data were prepared by labeling the GPS data which has already been collected using density-based outlier detection methods. Density-based outlier detection method can detect outliers in spatial data, so that it can be used to classify normal data and outlier data based on density. The normal data will be labeled as normal, while the outlier data will be labeled as an anomaly.

Density-based outlier detection is an anomaly detection based on density data. Density-based outlier detection is an approach to detect the anomaly data based on local data density around the data. Figure 1 illustrates how density-based outlier detection works. In Figure 1, there are two sets of points, C1 and C2, which have different densities. Density-based outlier detection detects the points on C1 and C2 as normal because the points in C1 and C2 have relatively uniform distance to the points around it, creating two areas with a certain density. On the contrary, point o1 and o2 are detected as anomalies, since they have different distances with the points around it.

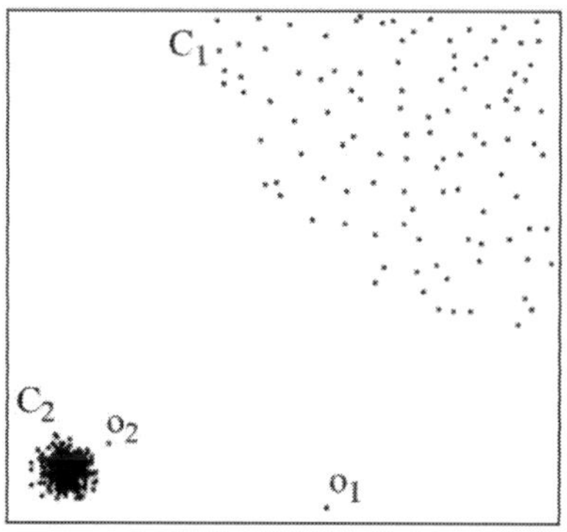

Figure 1 Illustration of density-based outlier detection

Density-based outlier detection clustered data based on the similarity of local reachable density (LRD) of any data. LRD can be determined by calculating the distance of a point with some of the closest points in the dataset using K-Nearest Neighbors (KNN). Below are the main steps for density-based outlier detection:

- Find the K-nearest neighbors of each point in dataset
- Call the maximum distance to K-nearest points that is found in previous step as K-distance. For example, if K=3 and the three nearest neighbors have distances of 1.2, 2.5 and 6.4. Then, the K-distance for this point will be 6.4.
- Next, for certain number of points (MinPts), the reach-distance is calculated.
- Then the LRD of each point is calculated using below formula:
- Finally, LOF scores is calculated using below formula:

3.2. CRF on lost person detection
CRF method is used to detect the anomaly of a pattern. Zicheng Liao used CRF method for GPS anomaly detection. CRF receives sequential data as the input. In lost person detection, CRF receives GPS data streams. The data should be split into smaller sequential data so that it can be effectively processed by CRF.

CRF method requires some feature function to conduct training and classification process on GPS data. Feature function must be defined in accordance with the needs of lost person habits so CRF method can work effectively to detect lost person. Feature function receives the input sequence of data (s), the position of the data in the sequence of data (i), the data label at position i (li), and the data label at position i-1 (li-1). It then produces a real number as the output, although the output is usually either 0 or 1. Output 1 indicates that the label at that position is correct and the output of 0 signifies that the label in that position is wrong. Features that are used to detect the lost will be discussed in the following sections:

3.2.1. Position feature. In GPS anomaly detection system, there is a position feature. If the position is often found in a certain area, the system will detect the data as normal and conclude that person is not lost. The positions on the GPS data sequences is processed by IQR (interquartile range) outlier detection methods that can detect outliers in the input dataset. Outliers are detected by calculating the IQR on the latitude and longitude separately. If the data on the current position is not an outlier in its latitude or longitude, then this feature detects that the person is not lost. But if the data on the current position is an outlier in its latitude or longitude, then this feature detects that the person is lost. Data on the current position can also be determined from the previous label. If the data on the current position is not an outlier in its latitude or longitude, then the position feature states that the current label is same as the previous label. But if the data on the current position is an outlier, then this feature states the current label is opposite to the previous label.

3.2.2. Direction feature The direction of movement can determine whether someone is getting lost or not. Direction is obtained from the difference between current GPS data and the previous GPS data with an assumption that the people who are lost have a tendency to walk with an uncertain direction. If the directions on the GPS data sequences have nearly the same amount and shows a uniform distribution, that means the person does not walk at straight directions and that the person is being lost.

3.2.3. Speed feature The speed feature is a supporting feature in this system. Walking faster usually suggests that the person is outside his usual area. In addition, faster movement also increases the emergency of receiving treatment if the person is lost. If the current data is observed to have a greater speed than the average, then this feature will detect that the person is being lost. Otherwise, if the current data shows a speed less than the average, then this feature will state that the current label is same as the previous label

3.3. System Implementation

Lost person detector use density-based outlier detection and CRF as the main method to detect whether the person is being lost or not. Lost person detector have 2 main phases: training phase and classification phase. In training phase, lost person detector obtain sequences of GPS data collected previously and then label GPS data with as normal or anomaly with density-based outlier detection method to build training dataset. Training datasets are used by CRF method to build a classification model. In classification phase, lost person detector receives sequences of GPS data and then label it accordingly using the model produced from training phase. The system flow can be seen in Figure 2

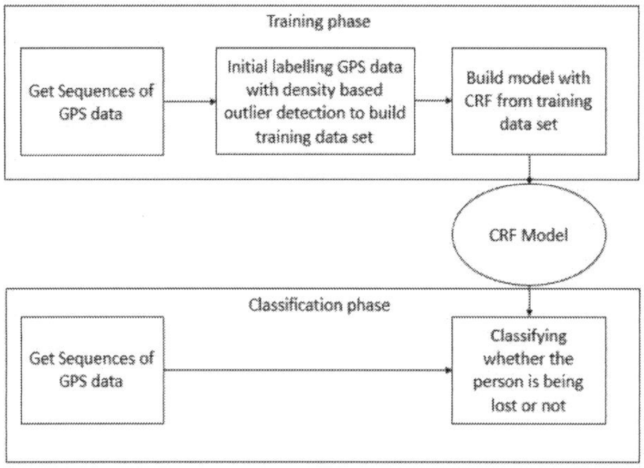

Figure 2 Lost person detector system flow

Figure 3 shows a GPS data visualization of the normal area of a supervised person's usual activity in the past 6 days. Figure 4 shows GPS data visualization of the area where the supervised person is being lost. Lost person detector use GPS data in Figure 3 and Figure 4 as training data. Figure 5 is the output of the classification phase using 10-fold cross-validation. The red markers show normal GPS data, while the purple markers show that supervised person is being lost in that area.

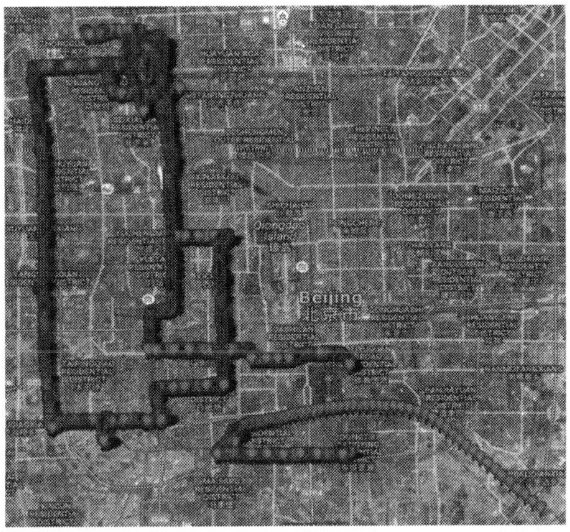

Figure 3. Normal area of supervised person's usual activity.

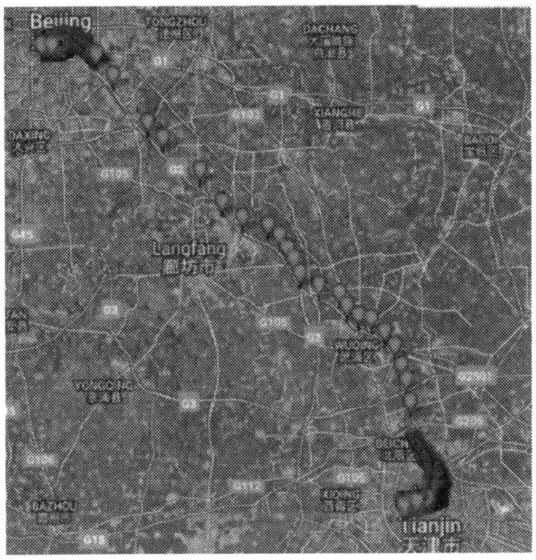

Figure 4. Area that supervised person is being lost.

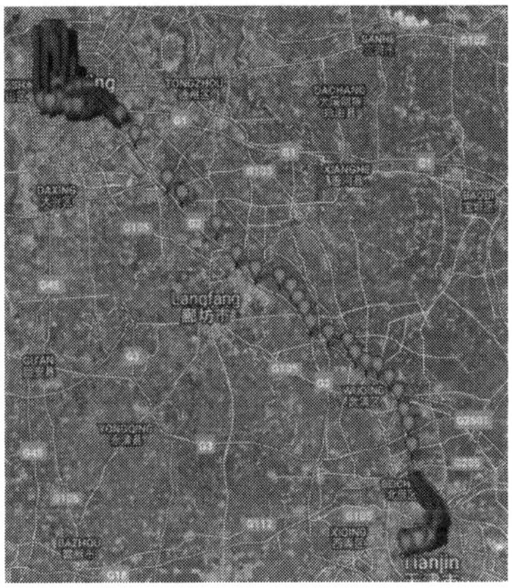

Figure 5. Output of lost person detector.

4. Experimental Result

GPS data used for experiment obtained from the collected data by Zheng et al in 2009 [9]. There are four scenarios to perform this evaluation. For each scenario, the data is collected from the same person. For scenarios 1 and 2, the normal GPS data being used is a collection of GPS data in a duration of six days. For scenarios 3 and 4, the normal GPS data being used is a collection of GPS data in a duration of three days. Anomalies that are used for each scenario are the GPS data retrieved when the person was being lost in one day.

Table 1 Lost detector experiment

Data	Training data set labelling	Precision	Recall
GPS Data collected in 6 days	Manual labelling	98.86786%	87.47855%
	Density- based outlier detection labelling	94.93670%	77.18696%
GPS Data collected in 3 days	Manual labelling	98.74367%	89.526421%
	Density-based outlier detection labelling	81.50087%	76.80921%

In comparison, the precision and recall of the classification process with training data labeled using outlier detection method is smaller than those with manually labeled training data. This is because the manually labeled training data incorporates expert insights which makes it more accurate than labeling training data using outlier detection method.

However, the difference is insignificant, and it is still considered effective to detect lost people using outlier detection method in labeling the training data since the recall is higher that 75%. The only major disadvantage is that the notification might be delayed for a while and there is slightly more probability of false alarms, where the system detects a person as lost while actually he is not.

When comparing the experiment results with different duration of training dataset, very little differences in precision and recall are seen. When outlier detection is used, the precision resulted from using three days GPS data decreased more greatly compared to that obtained from using six days GPS data. This is because the models have not saved a lot of movement patterns and habits so there are more false alarms. However, when the data that is labeled manually, the recall when using three days GPS data is more than that when using six days GPS data. This is because the three days GPS data has already formed a good enough model of the behavior patterns of the supervised person. If the training dataset contains too many normal data than necessary, over time CRF models will grow to be inflexible.

5. Conclusion

It can be concluded that the adjustments to the GPS anomaly detection system to detect lost person should be done by making changes to the training data labeling method and a change of model building. Labeling use density-based outlier detection method changes raw GPS to labeled training data that will be used to create the model. The model is created specifically for each person using the CRF. CRF method uses three features which are position, speed and direction of movement

References

[1] Liao Z 2010 Anomaly Detection in GPS Data Based on Visual Analytics *IEEE Symp. on Visual Analytics Science and Technology* p 51 (Salt Lake City: IEEE)

[2] Breunig M M, Kriegel H-P, Ng R T and Sander J 2000 LOF: Identifying Density-Based Local Outliers. SIGMOD '00 *Proceedings of the 2000 ACM SIGMOD intl. conf. on Management of data* pp 93-104 (New York: ACM)

[3] Dunning T and Friedman E 2014 Practical Machine Learning A New Look at Anomaly Detection (Sebastopol: O'Reilly Media)
[4] Ester M, Kriegel H-P, Sander J and Wu X 1996 A Density-Based Algorithm for Discovering Clusters *KDD-96 Proc*. pp 226231 (Munich: University of Munich)
[5] Kriegel H-P 2010 Outlier Detection Techniques *16th ACM SIGKDD Conf. on Knowledge Discovery and Data Mining*
[6] Lafferty J, McCallum A and Pereira F 2001 Conditional Random Fields: Probabilistic Models for Segmenting and Labeling Sequence Data *Intl. Conf. on Machine Learning* pp 282-289 (Pittsburgh: Carnegie Mellon University)
[7] National Academy of Public Administration 1995 The Global Positioning System: A Shared National Asset (Washington D.C: National Academy Press)
[8] Upton G and Cook I 1996 Understanding Statistics (New York: Oxford University Press)
[9] Zheng Y, Zhang L, Xie X and Ma W-Y 2009 Mining interesting locations and travel sequences from GPS trajectories World Wild Web pp 791-800 (Madrid: ACM Press)

IOP Conf. Series: Journal of Physics: Conf. Series **801** (2017) 012003 doi:10.1088/1742-6596/801/1/012003

Distributed Multi Robot Simultaneous Localization and Mapping with Consensus Particle Filtering

F Hidayat, B R Trilaksono and H Hindersah

Institut Teknologi Bandung Jl. Ganeca No. 10 Bandung

Email: fadhil.2persen@gmail.com, briyanto@lskk.ee.itb.ac.id, hilwadi@stei.itb.ac.id

Abstract. These paper present distributed computations of join probabilities SLAM with consensus particle filtering algorithm. We consider groups of robot observe an unknown environment and build the global maps. In this paper, every local map is a global map. Global maps build by using its own information and the information obtained from the other robots. We use particles likelihood as transfer parameters. The information can be transferred to other robot if reach global agreement on particles weight. The global agreement can be obtained by using the Consensus Particle Filtering algorithm, computation is done locally on each robot. After reaching an agreement, then the global map can be built and information of the global map is owned on each robot.

1. Introduction

Simultaneous localization and mapping (SLAM) is a method to build a map (mapping) and determine location on the map (localization) simultaneously using single robot or a group of robots. SLAM is part of three main issue in robot navigation that is mapping (mapping), localization (localization), and path planning (path planning) [1]. Mapping is a problem of integrating environmental information obtained from sensor installed on the robot into map representation. The map representation is formed depending on type of sensor used in mapping. On a feature-based SLAM, a map is a collection of landmarks location. Problem of mapping requires the robot localization. Localization is a problem of determining location of the robot on map. Problems of SLAM are a chicken and egg problem. When estimating location the robot requires map information in other words localization is relative to the map. Otherwise map representations can be formed by knowing location of the robot. Problems of localization and mapping is further complicated because there is no initial location information and no landmarks location information.

Researches on SLAM play an important role in the development of robot technology. Solutions for SLAM using single robot has been developed in various forms. SLAM on a single robot can be solved by the probabilistic approach called probabilistic SLAM. The probabilistic SLAM try to find distribution of robots location and map given observation information obtained from sensors, control input, and robots initial starting position. Extended Kalman Filter (EKF) is an algorithm that can be used to find a solution of single-robot probabilistic SLAM [2,3,4,5]. However, solution of EKF SLAM has a problem on the convergence of maps location, computational complexity, and non-linearity [2,6]. This problem is caused by sample-space SLAM is growing with the increasing of environmental observation results. The problem of EKF SLAM can be resolved with a Rao-Blackwellized SLAM (RB SLAM) [2]. Rao-Blackwellized theorem solves the problem by factoring the SLAM problem into two forms. First, calculate estimated robots location based on observation results, input control and robots initial position. Next is to build maps based on the robots location and maps that have been formed previously. Mapping can be solved analytically while robot localization solved by sampling. This SLAM's factorization solutions resolve the problem that appears on the EKF SLAM by reducing sample-space of SLAM problem.

Algorithms that implement RB SLAM is FastSLAM 1.0 [7] and FastSLAM 2.0 [8]. FastSLAM algorithm applies Particle Filter algorithm to estimate robots location and EKF algorithm to update map using measurement result obtained from sensor. This algorithm represents localization and mapping in a set of weighted particles [9]. Construction of a map based on information represented by the particle with the highest weight.

SLAM now widely implemented in various fields such as robot mapping for indoor or outdoor environment, unmanned aerial vehicle (UAV) mapping and observe certain areas, unmanned underwater vehicle (UUV) to map sea floor, underground robot to map underground, and space robot to map conditions in outer space. However, single robot SLAM has limitation in mapping capabilities only for small environment. The next challenge of SLAM is to map larger environment. The main idea for SLAM with larger environment is to use more than one robot doing SLAM or also called multi-robot SLAM. Challenges arise in multi-robot SLAM are perform coordination between robots, integrate information obtained from different robots into a single consistent map, perform communication among robots [10], and measure relative position between group robots [11].

Research on multi-robot SLAM has been done by Howard (2006) [11] and A. León et.al (2009) [12] in centralized form. Centralized multi robot SLAM requires a robot that acts central processing while non-center robot provide measurement information from its own sensor. Centralized multi robot SLAM has problems in vulnerable to system failure at the robot center and large transfer parameter. This problem can be resolved by applying the distributed multi-robot SLAM. Distributed multi-robot SLAM is a method to distribute computation of SLAM to all robot members without center processing. Research on distributed multi-robot SLAM has been done in [13,14,15] by accumulating local map obtained from each robot member. Transformations of local maps to global map are done by looking at overlap of local maps; this method is implemented in grid-based map. Implementation of distributed multi-robot SLAM to build feature-based SLAM has been done in [16]. This method requires synchronization of the time. Then, Jafri et.al (2011) [17] and Aragues et.al (2011) [18] propose the method that does not need time synchronization.

All of these previous works are focused on merging the map from member of robot groups. In this paper, we propose distributed multi-robot SLAM using global agreement of information that is transferred among the robot member. We apply the Consensus Particle Filter algorithm [19,20] from multi agent target tracking to reach the global agreement of transfer parameter. We use the particle weight representation as transfer parameter and consensus algorithm is computed to reach the global agreement to particles weight. The tutorial of consensus algorithm can be seen at [21].

In the rest of this paper, formulation of the single-robot SLAM and multi robot SLAM is described in section 2. Consensus particle filtering using particle weight parameter is given in section 3. Section 4 provides the simulation results. Finally, conclusions and discussion are summarized in section 5.

2. Methodology

Our approaches to distributed multi robot SLAM shown in algorithm 1. This section provides problem formulation of single-robot SLAM and multi-robot SLAM. The SLAM approach is feature-based; therefore the map is represented by a set of 2D landmarks position.

Algorithm 1 Distributed Multi Robot SLAM with Consensus Particle Filtering Algorithm

1. ***DistributedMultiRobotSLAMAlgorithm***( ):
2.   ***if*** *robot k* ***is connected*** *with robot k'* $\rightarrow$ ***false***
3.   *SingleRobotSLAM*( )
4.   ***else if*** *robot k* ***is connected*** *with robot k'* $\rightarrow$ ***true***

5. *DistributedMultiRobotSLAM*( )
6. ***end if***

2.1. Single Robot SLAM

Single robot SLAM is implemented when there is no communication or without having transfer data from other robots. Probabilistic SLAM can be denoted as the following equation (1) and (2).

$$p(x_t, m | z_{0:t}, u_{0:t}) \quad (1)$$

$$p(x_{0:t}, m | z_{0:t}, u_{0:t}) = p(m | x_{0:t}, z_{0:t}) p(x_{0:t} | z_{0:t}, u_{0:t}) \quad (2)$$

Where x_t is vector of robot position at time t, m is map mad up from location of landmarks, $z_{0:t}$ is sensor observation, and $u_{0:t}$ is input control given to the robot. Prediction stage is estimate pose robot based on the previous step, denoted as equation (3).

$$p(x_t, m | z_{0:t}, u_{0:t}, x_0) = \int p(x_t | x_{t-1}, u_t). p(x_{t-1}, m | z_{0:t-1}, u_{0:t-1}, x_0) dx_{t-1} \quad (3)$$

Update phase make corrections to prediction result using the results of observations, denoted as equation (4).

$$p(x_t, m | z_{0:t}, u_{0:t}, x_0) = \frac{p(z_t | x_t, m). p(x_t, m | z_{0:t-1}, u_{0:t}, x_0)}{p(z_t | z_{0:t-1}, u_{0:t})} \overset{Bayes}{=} \eta p(z_t | x_t, m). p(x_t, m | z_{0:t-1}, u_{0:t}, x_0) \quad (4)$$

In FastSLAM 1.0 algorithms [7], robot trajectory represented using weighted sample and map computed analytically. By using samples weight w, then the joint posterior distribution SLAM represented as equation (5) follows.

$$\{w_t^i, x_{0:t}^i, p(m | x_{0:t}^i, z_{0:t})\}_i^N \quad (5)$$

In equation (5), N is the number of sample particle, and i is index of the particle. Prediction phase can be done by spread the sample particle according to proposal distribution (6).

$$x_t^i \sim \pi(x_t | x_{0:t-1}^i, u_t) \quad (6)$$

Map building can be written as equation (7).

$$p(m | x_{0:t}^i, z_{0:t}) = \prod_t^T p(m_k | x_{0:t}^i, z_{0:t}) \quad (7)$$

Particle weight can be computed by using equation (8).

$$w_t^i = w_{t-1}^i p(z_t | x_{0:t}^i, z_{0:t-1}) \left(\frac{p(x_t^i | x_{t-1}^i, u_t)}{\pi(x_t^i | x_{0:t-1}^i, u_t)} \right) \quad (8)$$

Data association problem can be solved by applying equation (9).

$$\hat{n}_t^i = \underset{n_t}{\operatorname{argmax}} \, w_t^i(n_t) \quad (9)$$

2.2. Multi Robot SLAM

Algorithms for distributed multi robot SLAM with consensus particle filtering is shown in algorithm 2.

Algorithm 2 Distributed Multi Robot SLAM with Consensus Particle Filtering Algorithm

1. ***DitributedMultiRobotSLAM Algorithm***( ):
2. *Prediction* $x_t^i \sim \pi(x_t | x_{0:t-1}^i, z_{0:t}, u_t)$
3. *Check Connection with Robot k'*
4. *Measure Relative Position*
5. *Transfer and get information from robot k'*
6. *Consensus Particle Filtering on Particles Weight*
7. *Built global map using approximation of global weight*

To perform multi-robot SLAM, each robot needs to measure relative position of other robot k' using its own position k. Method to perform the relative position measurement can be seen at [14]. Assume that there are two robot i and j in multi-robot SLAM system communicate each other and transferring the information $d^{j0}_{j,1:t^j_{i,k}}$ and $d^{i0}_{i,1:t^i_{j,k}}$ respectively. Superscript $j0$ and $i0$ means that the information is form in local frame reference of recipient robot. The objective of measuring relative position of the robot is to find the global frame reference using roto-translation method written as equation (10).

$$X^{i0}_{j0} = X^{i0}_{i} \oplus X^{i}_{j} \ominus X^{j0}_{i} \tag{10}$$

Each robot compute local particle weight equation (8) locally and transfer the local weight to other robots. Information obtained from robot i can be written as equation (11) and (12).

$$d^{i0}_{i,1:t} = \langle w^{i0}_{i,1:t}, x^{i0}_{i,1:t}, p(z^{i0}_{t} | x^{i0}_{i,0:t}, z^{i0}_{i,0:t-1}) \rangle^{N}_{i=1} \tag{11}$$

$$p(z^{i0}_{i,t} | x^{i0}_{i,0:t}, z^{i0}_{i,0:t-1}) = \int p(z^{i0}_{i,t} | x^{i0}_{i,0:t}, m^{i0}) p(m^{i0} | x^{i0}_{i,0:t-1}, z^{i0}_{i,0:t-1}) dm \tag{12}$$

After receiving the information from other robot, consensus particle filtering is executed to calculate global agreement of particle weight called global weights. When reaching consensus then compute importance weights by using global weight. Hence, equation (8) can be rewritten by equation (13).

$$w^{i}_{k,t} = w^{i}_{t-1} \prod_{k=1}^{K} p\left(z^{i0}_{k,t} | x^{(j)i0}_{k,0:t}, z^{i0}_{k,0:t-1}\right) \frac{p\left(x^{(j)}_{k,t} | x^{(j)}_{k,t-1}, u_{k,t}\right)}{\pi\left(x^{(j)}_{k,t} | x^{(j)}_{k,0:t-1}, u_{k}\right)} \tag{13}$$

Data association problem in equation (9) can be rewritten as equation (14).

$$\hat{n}^{i}_{t} = \underset{n_t}{\operatorname{argmax}}\, w^{i}_{k,t}(n_t) \tag{14}$$

3. Consensus Particle Filtering

This paper use consensus particle filtering with consensus particle weight [19] adopted from target tracking problem. Assume there are K number of agent in agent network and each member of K having its own scalar value s_k and its own weight variable $\omega^{(i)}_{k,k}$. General consensus algorithm described at algorithm 3.

Algorithm 3 General Consensus Algorithm

1. ***Consensus Algorithm***$(s_k, s_{k'})$:
2. $\xi^{(0)}_{k} = s_k$
3. $for\ i = 1,2,\dots,\infty\ then$
4. $\xi^{(i)}_{k} = u\left(\xi^{(i-1)}_{k}, \left\{\xi^{(i-1)}_{k'}\right\}_{k' \epsilon Nk}\right)$ /* $\left\{\xi^{(i-1)}_{k'}\right\}_{k' \epsilon Nk}$: s_k *neighbooring agents**/
5. ***broadcast***$(\xi^{(i)}_{k})$
6. if(***consensus***) $then$ ***end for***.

We use averaging consensus algorithms for the update function $u(.)$ written as equation (15).

$$u\left(\xi^{(i)}_{k}, \left\{\xi^{(i)}_{k'}\right\}_{k' \epsilon Nk}\right) = \omega^{(i)}_{k,k} \xi^{(i)}_{k} + \sum\nolimits_{k' \epsilon Nk} \omega^{(i)}_{k,k'} \xi^{(i)}_{k'} \tag{15}$$

Global weight w^{i}_{t} of particle with index i at time t is proportional to product of local likelihood, described in equation (13). The averaging consensus (15) performed for each particle to calculate approximation to average of all particles weight.

$$w^{(j)}_{t} \triangleq w^{(j)}_{t-1} \frac{1}{K} \sum_{k=1}^{K} w^{(j)}_{k,t}, \qquad j \in \{1,2,\dots,N\} \tag{16}$$

4. Simulation

In order to show the performance of the algorithm, we built simulation of multi-robot SLAM by modified FastSLAM simulation obtained from [22]. The simulation consists of 4 robots exploring 50 x 50 m^2 environment with 29 landmarks. We assume that configuration of each robot are equal. Each robot use 50 particles for prediction phase. The simulation use velocity and steering angle as control variable. Initial noise to velocity is 0.3 m/s and initial noise to steering angle is $3\pi/180$ radian. Maximum observation range is 30 m and initial noises to observation are 0.1 m for range and $1\pi/180$

radian for bearing. Time interval between control signals is 0.025 second and time interval between observations is 0.2 second. Maximum distance of communication range between robots is 30 meter. Number of consensus iteration is 5 iterations. The simulation configuration is shown in figure 1.

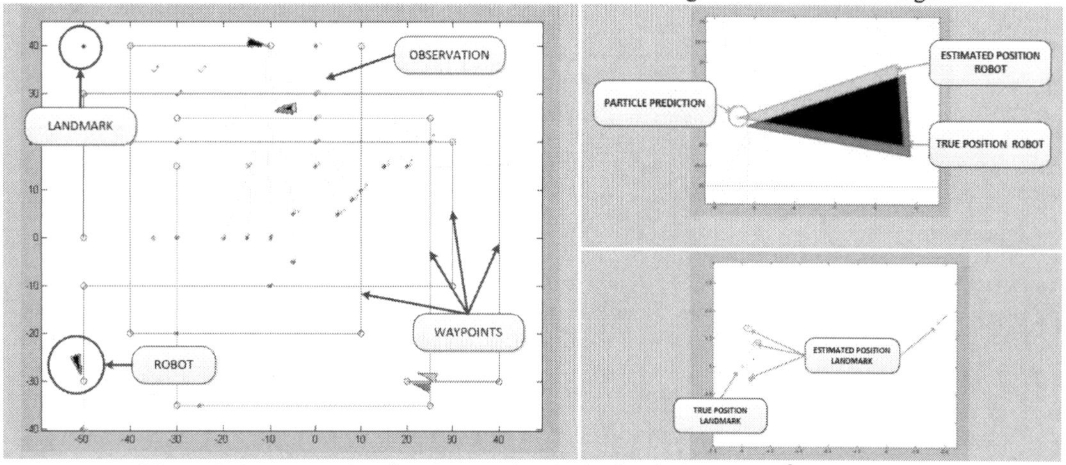

Figure 1. A team consist of 4 robots explore 30 landmarks in 50x50 m^2 environment.

We compare result of the simulation with simulation of single robot SLAM. The comparasion result shown in table 1.

Table 1. Simulation result.

Description	Loops	Run Time	Landmarks	Map RMSE	Localization RMSE	
Single Robot	**32.412**	**60.197 s**	**29 Total**	**Location**	**Position**	**Orientation**
Robot 1	32.412	± 0,464 s / loop	19 Observed	1,229 m	2,209 m	0,035 π/180 rad
Robot 2	32.412	± 0, 464 s / loop	29 Observed	0,160 m	0,015 m	0,003 π/180 rad
Robot 3	32.412	± 0, 464 s / loop	27 Observed	0,071 m	0,250 m	0,004 π/180 rad
Robot 4	32.412	± 0, 464 s / loop	27 Observed	1,483 m	0,230 m	0,003 π/180 rad
Multi Robot	**32.412**	**72.317 s**	**29 Total**	**Location**	**Position**	**Orientation**
Robot 1	32.412	± 0,558 s / loop	29 Observed	0,785 m	1,758 m	0,051 π/180 rad
Robot 2	32.412	± 0,558 s / loop	29 Observed	0,179 m	0,139 m	0,005 π/180 rad
Robot 3	32.412	± 0,558 s / loop	29 Observed	0,139 m	0,336 m	0,006 π/180 rad
Robot 4	32.412	± 0,558 s / loop	29 Observed	0,139 m	1,253 m	0,022 π/180 rad

Number of communication between robots is shown in table 2.

Table 2. Number of communication between robots.

Comunication	Robot 1	Robot 2	Robot 3	Robot 4
Robot 1	-	0	88	196
Robot 2	0	-	134	171
Robot 3	88	134	-	256
Robot 4	196	171	256	-

Regarding to the simulation result, multi-robot SLAM with consensus particle filtering improve map estimation. This method also merge the observation between robots.

5. Conclusion

We have presented the algorithm for multi-robot SLAM with consensus particle filtering. These methods improve map estimation and merge the observation between robots. However, increasing of computing time cannot be avoided. The increase in computing time depends on the number of iterations used in consensus algorithm.

Acknowledgements

This work was funded by Ministry of Research, Technology and Higher Education of the Republic of Indonesia. The authors gratefully acknowledge them for providing the funding.

References

[1] C. Stachniss, *Robotic Mapping and Exploration*, B. Siciliano, O. Khatib, and F. Groen, Eds. Berlin, Germany: Springer, 2009.

[2] Hugh Durrant-Whyte and Tim Bailey, "Simultaneous Localisation and Mapping," *Robotics & Automation Magazine, IEEE*, vol. 13, no. 2, pp. 99-110, June 2006.

[3] M.W.M.G. Dissanayake, P. Newman, S. Clark, H.F. Durrant-Whyte, and M. Csorba, "A Solution to the Simultaneous Localization and Map Building (SLAM) Problem," *IEEE TRANSACTIONS ON ROBOTICS AND AUTOMATION*, vol. 17, no. 3, pp. 229 - 241, Agustus 2002.

[4] Jiang Yan, Liu Guorong, Luo Shenghua, and Zhou Lian, "A Review on Localization and Mapping Algorithm Based on Extended Kalman Filtering," in *International Forum on Information Technology and Application*, vol. 2, Chengdu, 2009, pp. 435-440.

[5] G. Dissanayake, S. Huang, Z. Wang, and R. Ranasinghe, "A Review of Recent Developments in Simultaneous Localization and Mapping," in *IEEE International Conference on Industrial and Information Systems*, 2011, pp. 477 - 482.

[6] Tim Bailey and Hugh Durrant-Whyte, "Tutorial Simultaneous Localization and Mapping (SLAM): Part II," *IEEE Robotics & Automation Magazine*, pp. 108-117, September 2006.

[7] M. Montemerlo, S. Thrun, D. Koller, and B. Wegbreit, "FastSLAM: A Factored Solution to the Simultaneous Localization and Mapping Problem," in *AAAI National Conference on Artificial Intelligence*, 2002, pp. 593-598.

[8] M. Montemerlo, S. Thrun, D. Koller, and B. Wegbreit, "FastSLAM 2.0: An Improved Particle Filtering Algorithm for Simultaneous Localization and Mapping that Provably Converges," in *IJCAI International Joint Conference on Artificial Intelligence*, 2003, pp. 1151-1156.

[9] M. Sanjeev Arulampalam, Simon Maskell, Neil Gordon, and Tim Clapp, "A Tutorial on Particle Filter for Online Nonlinear/Non-Gaussian Bayessian Tracking," *IEEE Transactions on Signal Processing*, vol. 50, pp. 174-188, February 2002.

[10] D. Fox et al., "Distributed Multirobot Exploration and Mapping," in *Proceedings of the IEEE*, vol. 94, July 2006, pp. 1325-1339.

[11] Andrew Howard, "Multi-Robot Simultaneous Localization and Mapping using Particle Filter," *ACM Journal International of Robotics Research*, vol. 25, no. 12, pp. 1243-1256, December 2006.

[12] A. León, R. Barea, L.M. Bergasa, M. Ocana E. Lopez, and D. Schleicher, "SLAM and Map Merging," *Journal of Physical Agents*, vol. 3, no. 1, pp. 11-23, January 2009.

[13] S. Carpin, "Fast and Accurate Map Merging for Multi-Robot Systems," *Autonomous Robots*, vol. 25, no. 3, pp. 305-316, 2008.

[14] L. Carlone, M. K. Ng, J. Du, B. Bona, and M. Indri, "Rao-Blackwellized Particle Filters Multi Robot SLAM with Unknown Initial Correspondences and Limited Communication," in *IEEE Internation Conference on Robotics and Automation Anchorage Convention District*, Anchorage, 2010, pp. 243 - 249.

[15] Luca Carlone, Miguel Kaouk Ng, Jingjing Du, Basilio Bona, and Marina Indri, "Simultaneous Localization and Mapping Using Rao-Blackwellized Particle Filters in Multi Robot Systems," *Journal of Intelligent & Robotic Systems*, vol. 63, no. 2, pp. 283 - 307, August 2011.

IOP Conf. Series: Journal of Physics: Conf. Series **801** (2017) 012003 doi:10.1088/1742-6596/801/1/012003

[16] Arturo Gil, Oscal Reinoso, Monica Ballesta, and Miguel Julia, "Multi-Robot Visual SLAM Using a Rao-Blackwellized Particle Filter," *Elsevier Robotics and Autonomous System*, vol. 58, pp. 68-80, 2010.

[17] Syed Riaz un Nabi Jafri, Luca Brayda, and Ryad Chellali, "Distributed Feature Based RBPF Multi Robot SLAM," in *IEEE Conference on Robotics and Biomedics*, Phuket, Thailand, 2011, pp. 7-11.

[18] R. Aragues, J. Cortes, and C. Sagues, "Distributed Consensus Algorithms for Merging Feature-Based Maps with Limited Communication," *Elsevier Robotics and Autonomous Systems*, vol. 59, pp. 163-180, 2011.

[19] O. Hlinka, F. Hlawatsch, and P.M. Djuric, "Distributed particle filtering in agent networks: A survey, classification, and comparison," *Signal Processing Magazine*, vol. 30, no. 1, pp. 61 - 81, December 2012.

[20] O. Hlinka, O. Sluciak, and F. Hlawatsch, "Likelihood Consensus and Its Application to Distributed Particle Filtering," *IEEE Transactions on Signal Processing*, vol. 60, no. 8, pp. 4334-4349, August 2012.

[21] R. Olfati-Saber, A. Fax, and R. M. Murray, "Consensus and Cooperation in Networked Multi-Agent System," in *IEEE Proeedings*, 2006.

[22] Tim Bailey. https://openslam.org/. [Online]. https://openslam.org/

IOP Conf. Series: Journal of Physics: Conf. Series **801** (2017) 012004 doi:10.1088/1742-6596/801/1/012004

Comparison of Exponential Smoothing Methods in Forecasting Palm Oil Real Production

B Siregar[1], I A Butar-Butar[1], RF Rahmat[1], U Andayani[1] and F Fahmi[2]

[1]Dept. of Information and Technology, Faculty of Computer Science and Information Technology, University of Sumatera Utara Jl. Dr. Mansur 9 Medan Indonesia
[2]Dept. of Electrical Engg, Faculty of Engineering, University of Sumatera Utara Jl. Dr. Mansur 9 Medan Indonesia

Email: fahmimn@usu.ac.id

Abstract— Palm oil has important role for the plantation subsector. Forecasting of the real palm oil production in certain period is needed by plantation companies to maintain their strategic management. This study compared several methods based on exponential smoothing (ES) technique such as single ES, double exponential smoothing holt, triple exponential smoothing, triple exponential smoothing additive and multiplicative to predict the palm oil production. We examined the accuracy of forecasting models of production data and analyzed the characteristics of the models. Programming language R was used with selected constants for double ES (α and β) and triple ES (α, β, and γ) evaluated by the technique of minimizing the root mean squared prediction error (RMSE). Our result showed that triple ES additives had lowest error rate compared to the other models with RMSE of 0.10 with a combination of parameters $\alpha = 0.6$, $\beta = 0.02$, and $\gamma = 0.02$.

1. Introduction

Palm oil is one of plantation crops that have an important role for the plantation subsector. Industrial development of palm oil benefit in increasing the income of farmers and society, production of the raw material that create added value for the country, the export of CPO which generate foreign exchange, and it provides job opportunities for more than two million workers in the various subsystems.

Palm oil plantations in Indonesia are managed in the form of small plantations and large farms. Large estate consists of a country estate as a Perkebunan Nusantara (PTPN) and private estates [1].

For each plantation, there will be monthly evaluation processing regarding quality objectives. This was done to look at the performance of the production for the month whether it has reached a production target or not. In order to achieve the production target, the availability of information about real production in previous years plays a very important role. Target production in the future can be expected only if the previous year's production data is available and of good quality. In the plantation production targets are calculated by looking at the production of previous years done by manual calculations. Therefore we need a method that can provide information on production forecasts as a reference in determining the company plans more realistic production targets.

Research on Palm oil production forecasts have been done by several methods. Methods existed included the use methods of causal and time-series. Research conducted in 2013 using two methods to compare the production forecasts Palm oil is used in the form of multiple regression methods of causal

IOP Conf. Series: Journal of Physics: Conf. Series **801** (2017) 012004 doi:10.1088/1742-6596/801/1/012004

and time-series methods such as exponential smoothing. In the test, the value of Mean Absolute Percentage Error (MAPE) obtained using regression is 21:08% while the value of MAPE obtained using exponential smoothing method is 12.78% per year [2].

In another study, a research was conducted on the production prediction using neural networks with back propagation algorithm. Usually neural network was utilized in image based analysis such as meter reading [3] or image recognition [4]. The study used seven data as a variable based on the quality of the land as well as rainfall, altitude, and slope, age of the plant, rocks, solium, and the acidity of the soil. Experiment with several layers to get the best parameters: 3 layer, 4 layer and 5 layer. The best results obtained in experiment with 3 layer iteration to 30,000, with the learning rate by 0.9. Training results obtained with R2 = 0.9998 and RMSE = 0.0709 and test results with R2 = 0.8901 and RMSE = 2.2196 [5].

Another research in 2015 predicted Palm oil production using RBF neural network. The best results for Palm oil production forecasts obtained through a combination of parameters: the number of variables = 1, the number of input nodes = 5, the value of learning rate = 0.75 and the maximum epoch value = 3000 with the results of error MAPE = 11.75% [6].

Another research in 2011 utilized exponential smoothing to forecast sea surface temperatures, the researchers predicted the next year based on the results of the model were compared to analyze the characteristics of the model and also comparing the values obtained using three methods: Single Exponential Smoothing, Double exponential smoothing Holt and Holt -winters. Of the three methods produced the best performance on the method and Holt-Winters exponential smoothing Double Holt [7].

In this study we compared several exponential smooothing methods used to predict the actual production of palm oil such as single ES, double exponential smoothing holt, triple exponential smoothing, triple exponential smoothing additive and multiplicative and analyze the performance.

2. Material and Methods

The general architecture of this study can be seen in Figure 1.

2.1 Data Collection

This section shows the retrieval of data, then the data transformation process to the creation of a data plot. The dataset used in this study is a document from Palm oil production. The data used was secondary data, i.e data Palm oil plantation PTPN III Plantations Sei Meranti which have been collected from 2010-2014. The data was stored in the form of csv and tables that no normalization of data until the data can be included in a database or data in comma-separated values extension * .csv.

Collecting data in this study was using survey techniques, performed directly in the office Factory Palm oil (MCC) Sei Plantations Meranti. The data consisted of data from the production of fresh fruit bunches (FFB) Palm oil, palm oil production data and core data obtained from FFB oil that had been processed.

The next stage was the transformation of data, where the data was normalized so that data values can be processed easily. These data were converted into a value in the range of 0.1 up to 0.9 by using equation 1 [7].

$$x' = \frac{0.8\,(x-a)}{b-a} + 0.1 \tag{1}$$

The next step was to create a data plot. Plot Palm oil production data in 2010-2014 in Plantations Sei Meranti which had been transformed can be seen in Figure 2. Based on Figure 2, Palm oil production starting from the year 2010 to 2014 were likely to increase. Only in early 2014 Palm oil production declined. Based on information from the management of the farm in 2014, there was a drastic climate change, so that not only the plantations Sei Meranti are decreased, but also almost all the plantations that exist in PTPN III experienced the same thing.

Overall it can be concluded that the production of oil in the plantation Palm Sei Meranti has a positive trend (increasing).

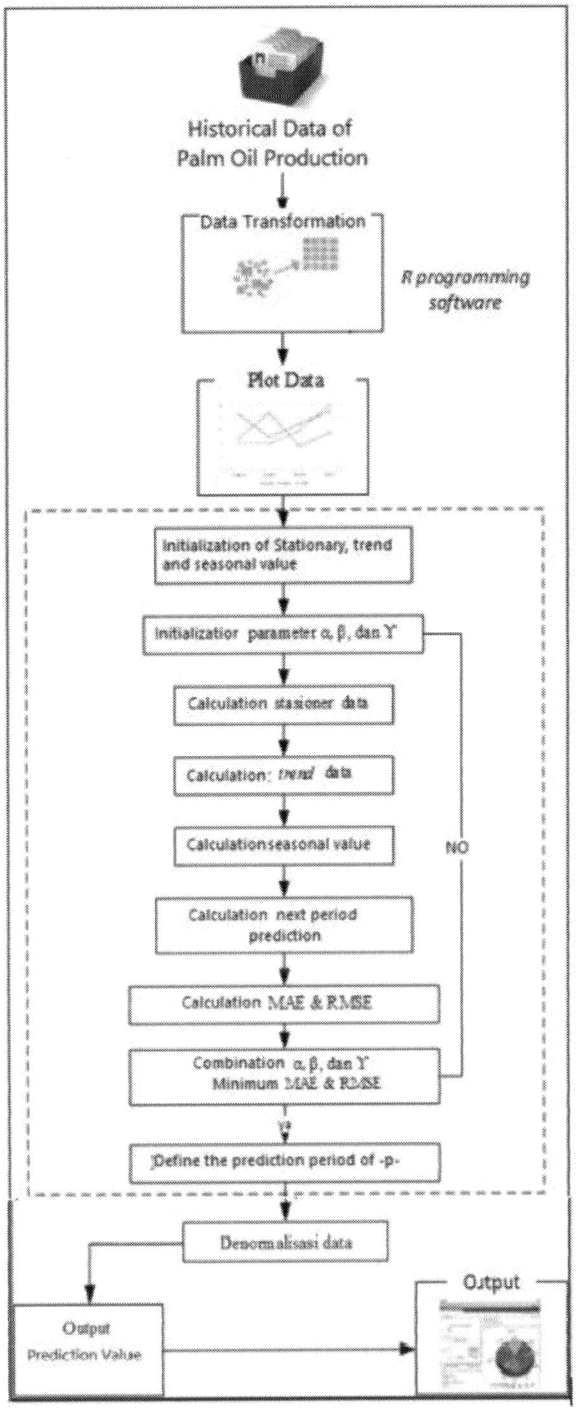

Figure 1. General Architecture

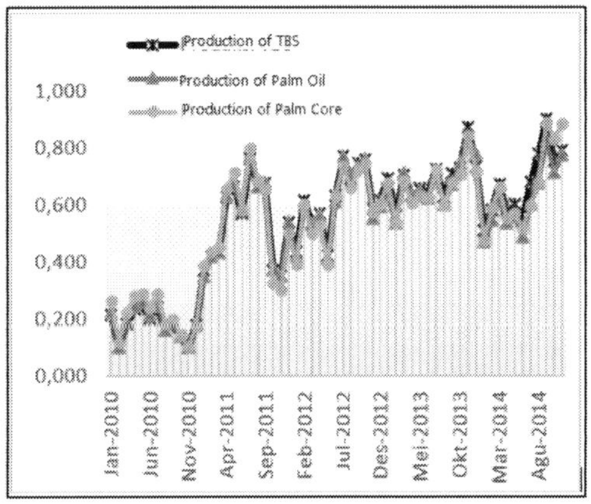

Figure 2. Graph production of Palm oil PTPN III Sei Meranti

2.2 Exponential Smoothing Method

2.2.1. Double exponential smoothing (DES) Holt. In DES Holt method, the smoothed trend component separately using different parameters, namely α and β. In this technique the value of the trend can be smoothed by using different weights. However, these two parameters need to be optimized so the search for the best combination of parameters is more complicated than using only one parameter.
In addition, the components of season in this technique are not taken into account. Here is the process undertaken Double exponential smoothing holt Holt.

Step 1: Determine the initial initialization value. Value is stationary, and the trend is beginning to use equation 2 to equation 3 [8].

$$S_L = X_L \text{ atau } S_L = \frac{1}{L} + (X_1 + X_2 + ... + X_L) \quad (2)$$

$$T_L = \frac{1}{L}\left\{\frac{(X_{L+1} - X_1)}{L} + \frac{(X_{L+2} - X_2)}{L} + ... + \frac{(X_{L+L} - X_L)}{L}\right\} \quad (3)$$

Step 2: Initialize the value of the parameters a and b where each range is between 0 - 1. In this study, we used the method of trial and error to produce the value of α and β are optimal based on the value Root Mean Square Error (RMSE) of the most minimum. Values α and β are optimal will be determined directly by the application program that has been designed.

Step 3: Calculation of stationary of data can be done using Equation 4 [8].

$$S_t = \alpha Xt + (1 - \alpha)(S_{t-1} + b_{t-1}) \quad (4)$$

\

Step 4: Calculation of trend data can be done using the equation 5 [8].

$$T_t = \beta(S_t + S_{t-1}) + (1 - \beta)b_{t-1} \quad (5)$$

Step 5: Once the data is stationary and trendy value has been established, then the forecasts can be found using equation 6 [8].

$$F_{t+m} = S_t + b_t m \quad (6)$$

Step 6: Then calculate the forecast errors using RMSE and MAPE. If MAPE produced not the smallest then repeat step two and so on until we got the smallest error value.
Step 7: Finding the forecasts for the next few periods, for example, for 12 months. By using the S parameter (level), and T (trend) of the last data specify the length of the forecast period p. By re-using Equation 6, set the value of m = 1, to find the forecasts one period or one month following the next. Suppose for 1 year ahead or equal to 12 periods ahead then set the value of m with m = 1, m = 2 and m = 3 and so on up to m = 12.

2.2.2 *Triple exponential smoothing Additive and Multiplicative.* From the analysis that has been done before, it was found that the amount of oil production Palm Plantations Sei Meranti from the first months of 2010 until the final months of 2014 was influenced by trends and seasonal factors. It is seen from Figure 2, in the seventh month the number of production Palm oil spiked higher then month-a month after the demand will decrease and then going back upward as it approached the seasonal period in the next year. This situation repeated every year, which means influenced by seasonal factors. Because the data is affected by seasonal factors, the other methods that can be implemented are Triple exponential smoothing method.
The following processes are carried out in Triple exponential smoothing (TES) Additive and Multiplicative.

Step 1: Determine the initial initialization value. ie stationary value, trend and seasonally initial value using equations 2, 3 and 7 [8].

$$I_L = \frac{X_k}{S_L}, k = 1,2,...,L \tag{7}$$

Step 2: Initialize the value of the parameters α, β and γ where each range is between 0-1.

Step 3: Calculation of TES data is stationary with additive and multiplicative method can be performed using equations 8 and 9 [8].

Aditive : $S_t = \alpha(X_t - I_{t-L}) + (1-\alpha)(S_{t-1} + T_{t-1})$ (8)

Multiplicative: $S_t = \alpha \frac{X_t}{I_{t-L}} + (1-\alpha)(S_{t-1} + T_{t-1})$ (9)

Step 4: Calculation of trend data can be performed using equations 10 and 11 [8].

Aditive : $T_t = \beta(S_t - S_{t-1}) + (1-\beta)T_{t-1}$ (10)

Multiplicative: $T_t = \beta(S_t + S_{t-1}) + (1-\beta)T_{t-1}$ (11)

Step 5: For the calculation of the initial value I, in one cycle of the first season (12 first period), can be performed using Equation 12 and 13 [8].

Aditive : $I_t = \gamma(X_t - S_t) + (1-\gamma)I_{t-L}$ (12)

Multiplicative: $I_t = \gamma \frac{X_t}{S_t} + (1-\gamma)I_{t-L}$ (13)

Step 6: With all three parameter values that have been obtained, then the forecasts can be found by Equation 14 [8].

$$F_{t+m} = (S_t + T_t m)I_{t-L+m} \tag{14}$$

Step 7: Then calculate the forecast errors using RMSE and MAPE. If MAPE produced not the smallest then repeat step two and so on until you get the smallest error value.

Step 8: Finding the forecasts for the next few periods, for example, for 12 months. By using the S parameter (level), T (trend), and I (seasonal) of the last data specify the length of the forecast period p by re-using Equation 14.

2.3 Output

After obtaining the approximate value of the desired period, the next step is to normalize the data, which returns the data into the form of the initial unit using Equation 15.

$$x = \frac{(x'-0,1)(d_{maks}-d_{min})}{0,8+ d_{min}} \tag{15}$$

Then the results of normalize will be processed into the form of plots and tables to facilitate the user in observing the forecast. Implementation of this method is done using the software R Programming.

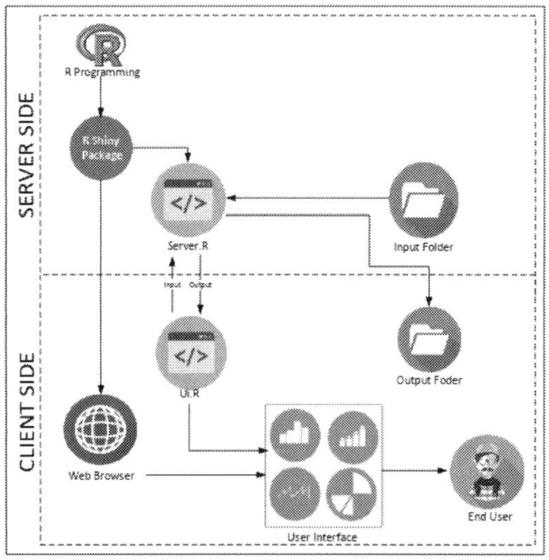

Figure 3. Process in R Programming

Based on Figure 3, there are two parts in the process of forecasting system using the R software, the server side and client side. The entire core code of the implementation process of exponential smoothing is stored in a file named server.R. The process of data input using csv file extension stored in the input folder, which is then called directly by server.R.

While user interface.R handle the process of how the output is displayed to the user. For the output of this system, the R language provides an easy way to integrate server.R and ui.R use shiny library. The results of exponential smoothing configuration can be accessed directly by the web browser by the user.

3. Results

Several tests were conducted to determine the performance of the system in forecasting realization Palm oil production using exponential smoothing method. Based on the draft research steps, then each data will be processed using three methods already mentioned. Testing parameters for all three methods with different combinations of parameters were used. Besides being able to manually select a combination of parameters, the system can also look for optimizing the parameters that produce the smallest error.

By executing the equations exponential smoothing models with the help of R software, a comparison was made between the results obtained form the model with the results of observations for each method. The data on production of fresh fruit bunches (FFB), a constant value of α, β and γ are determined by the system are respectively $\alpha = 0.7$ and $\beta = 0.01$ for Holt DES method, $\alpha = 0.6$, $\beta = 0 , 02$ and $\gamma = 0.02$ for TES additive method and $\alpha = 0.5$, $\beta = 0.02$ and $\gamma = 0.02$ for TES multiplicative method. The parameter optimization results using these three methods can be seen in Figure 4.

Figure 4 shows that the model DES holt, TES additives and multipikatif was very precise, especially for some of the initial data. Forecast data generated by additive and multiplicative method of time January 2010 until December 2014 was very precise compared observation data.

Data FFB production is a seasonal cycle of data will be repeated the following year, the multiplicative and additive models try to predict the repetition of the production cycle.

The analysis was conducted based on the accuracy of the model DES, TES TES additive and multiplicative FFB production data using statistical instruments such as the Root Mean Square Error (RMSE), Mean Absolute Error (MAE), Mean Absolute Percentage Error (MAPE) [8]. RMSE, MAE and MAPE refers to the magnitude of the error rate (errors) of an estimate, the smaller the value of these instruments, the better the forecasts were made. The results of the evaluation of the forecasting models each dataset with all three methods mentioned can be seen in Table 1.

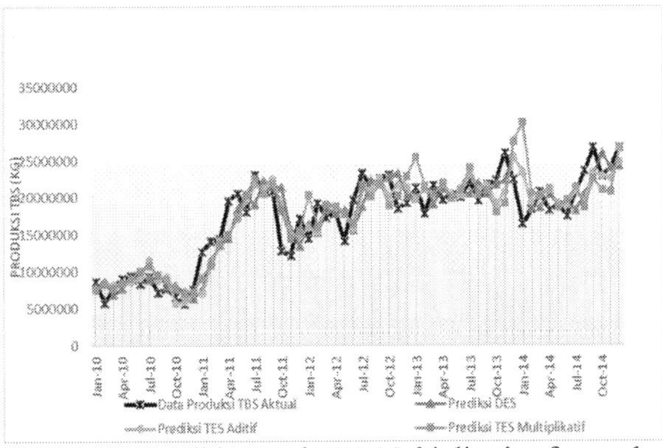

Figure 4. DES Holt, TES Aditive, and TES Multiplicative for production 2010-2014

Table 1 Error estimation for different forecasting models

Dataset	*Exponential smoothing*	Estimated Error		
		MAE	RMSE	MAPE
Production of FFB	Holt (α=0.7;β=0.01)	0.08	0.11	19.37
	Aditive (α=0.6;β=0.02; γ=0.02)	0.07	0.09	17.18
	Multiplicative (α=0.58;β=0.02 ; γ=0.02)	0.09	0.12	20.04
Production of Palm Oil	Holt (α=0.74;β=0.01)	0.09	0.11	20.79
	Aditive (α=0.7;β=0.01 ; γ=0.03)	0.07	0.10	18.06
	Multiplicative (α=0.64;β=0.02 ; γ=0.03)	0.09	0.12	21.31
Production of Palm Kernel	Holt (α=0.75;β=0.01)	0.09	0.11	21.9
	Aditive (α=0.68;β=0.01 ; γ=0.03)	0.08	0.10	18.8
	Multiplicative (α=0.72;β=0.01 ; γ=0.21)	0.10	0.13	22.9

An instrument of error such as MAE, RMSE, and MAPE (%) of the additive model is smaller than Holt and multiplicative models for data in seasonal cycles on the production data Plantations Sei Meranti.

Model triple exponential smoothing (Holt-Winters additive and multiplicative) has good precision, especially the additive method for data that are seasonal cycle such as production data FFB, palm oil, and palm kernel as shown in Table 4.

From Table 4 the exponential smoothing models, which is the method TES additive is the best method which produces the smallest error in forecasting for the data that is periodic with each value RMSE = 0:09 to production data FFB, RMSE = 0.10 for the data production of palm oil, and RMSE = 0.10 for the data production of palm kernel.

After the comparison of the forecasts of the results of field observations from the period January 2010 to December 2014 described Figure 4 and Table 1, then we made further forecasts. In the forecasts for the next two years ie from January 2015 to December 2016 for all three methods of exponential smoothing was provided. Advanced forecasting results can be seen in Figure 5 (a) to (c).

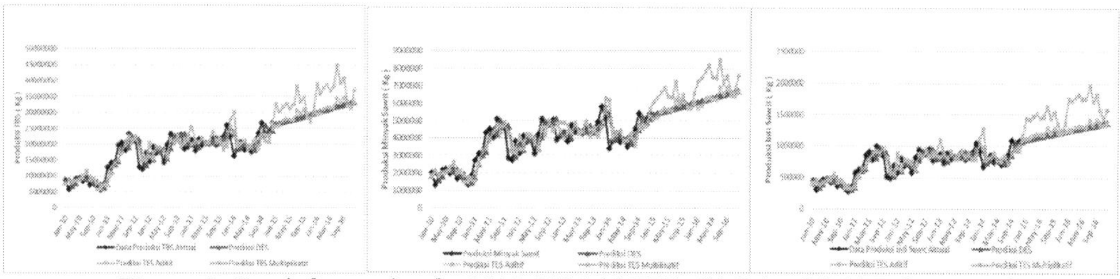

Figure 5. Graph forecasting for 2010-2016 (a) FFB (b) Palm Oil (c) Palm Kernel

4. Discussions

Based on the evaluation of system forecasting result it was obtained that the best results to forecast FFB production, palm oil, and palm kernel produced by the method of triple exponential smoothing additive through a combination of parameters in a row is α= 0.6, β = 0:02, and γ = 0:02 to FFB production with the results of error RMSE = 0:09, α = 0.7, β = 0:01, and γ = 0:03 to palm oil production with the results of error RMSE = 0:10, and α= 0.68, β = 0:01, and γ = 0:03 to palm kernel production with the result of error RMSE = 0.10.

From the test results it is shown that each of the data output has the characteristics of time series data differently that each of the data output forecasts have different parameters.

The use of double exponential smoothing method holt to predict what kind of data Palm oil production is not appropriate because the observation data has a seasonal component. Forecasts for the data containing seasonal components should be solved by using the method of triple exponential smoothing.

For further development, the study suggested the same case can use other methods that produce results forecast by the smaller error value. The forecasting is a reference to be considered in setting production targets that required high accuracy.

5. Conclusion

Compared to other exponential smoothing methods, Triple exponential smoothing additive can provide better forecasting of palm oil real production with suggested parameters as the result from this study.

References

[1] Batubara EA 2015 Implementasi algoritma learning vector quantization pada prediksi produksi Palm oil di PT. Perkebunan Nusantara I Pulau Tiga *under publication*

[2] Kacaribu RA 2013 Aplikasi Peramalan Produksi Palm oil dengan Metode Regresi Ganda dan *Exponential Smoothing. under publication*

[3] Triantoro T, Batubara FR and Fahmi F 2014 Image based water gauge reading developed with ANN Kohonen Proceedings of 2014 *International Conference on Electrical Engineering and Computer Science, ICEECS*

[4] Siswanto A, Tarigan P and Fahmi F 2013 Design of contactless hand biometric system with relative geometric parameters *Proc. of 2013 3rd Int. Conf. on Instrumentation, Communications, Information Technol., and Biomedical Engineering: Science and Technol. for Improvement of Health, Safety, and Environ., ICICI-BME*

[5] Hermantoro and Purnawan RA 2009 Prediksi produksi Palm oil berdasarkan kualitas lahan menggunakan model Artificial Neural Network (ANN) *Agroteknose* **4(2)**: 1-6.

IOP Conf. Series: Journal of Physics: Conf. Series **801** (2017) 012004 doi:10.1088/1742-6596/801/1/012004

[6] Jannati R 2015 Prediksi produksi panen Palm oil menggunakan jaringan saraf *radial basis function* (RBF) *under publication*
[7] Salamena GG 2011 Pengujian model peramalan deret waktu *sea surface temperature* (SST) teluk ambon luar dengan metode exponential smoothing *Oseanologi dan Limnologi di Indonesia* **37(1)**: 43-55.
[8] Maranata L 2013 Peramalan jumlah produksi kakao di sumatera utara dan konsumsi kakao di Indonesia dengan pemulusan eksponensial ganda metode linier satu parameter dari brown. *under publication*
[9] Siang JJ 2009 Jaringan saraf tiruan & pemrogramannya menggunakan MATLAB. *Penerbit ANDI*: Yogyakarta.
[10] Makridakis S, Wheelwright SC and McGee VE 1999 Metode dan Aplikasi Peramalan *Erlangga*: Jakarta.

D&T: An Euclidean Distance Optimization based Intelligent Donation System Model for Solving the Community's Problem

D N Utama[1], Fitroh[2], Nuryasin[2], E Rustamaji[2], Nurbojatmiko[2], I Qoyim[2]

[1]Laboratory Optimization Models and Systems for Decision Support, Information Systems Department, UIN Syarif Hidayatullah, Jakarta

[2]Information Systems Department, UIN Syarif Hidayatullah, Jakarta

Email: dit.utama@uinjkt.ac.id

Abstract. The trust is a main difficulty to propose a donation system to the community. A specific information system is scientifically estimated able to escalate the trust level of one community in donating; where, their donation can reinforce them to solve the socioeconomic problem in one region. The concept of fuzzy-logic has been practically embedded in measuring an inequality index of socioeconomic aspect, particularly for health and education sectors. Moreover, the concept of the Euclidean distance measurement is operated to measure the distance value of two parameters (geographical location and inequality). The hill-climbing optimization method that can recommend the most recommended donation recipient is embedded into system model to meet donor and recipient of donation. Here the intelligent donation system model is scientifically constructed. The proposed system model undoubtedly can solve the socioeconomic problem in one community. In this study, the urban village Sawah, Ciputat, Indonesia was taken as an object of the research where the empirical data coming from.

1. Introduction

Commonly in urban village in Indonesia, exclusively in health and education sectors, the socioeconomic inequality is still wide-ranging. It touches the inequality index value 0.65 approximately [1], where the value 1.00 illustrates the poorest condition of socioeconomic gap. Several government programs have been implemented to answer the problem, however much effort needs to be conducted well.

One way to decrease the socioeconomic inequality is to increase the power of community donation role. Nevertheless, the other challenge appears. The trust is a main dilemma to encourage the community to donate. The research problem to answer "how to increase the community trust to donate" is an interesting question to be responded.

Several researches surrounding donation term, by this time, have been conducted in many countries. [2] conducted a research regarding the analysis and expectation of negative impact of food insecurity. They analyzed the uncertainty connected with in-kind food donation. The uncertainty term here was extended to the donor, product, and supply chain structure. They also developed the model to predict the quantity of in-kind donations. [3] was doing a study to identify the factors affecting consumer donation in U.S. retail stores. The conceptual framework was developed, where the factors were derived from parameters marketing, consumer behavior, and psychology literatures. The concept probably can assist for future empirical studies. Furthermore, in phycology aspect, [4] conducted a study related to an "inspiration-helping" hypothesis that can inspire people to donate. They combined a positive and negative emotion to observe the effectivity of people's appeal in donating. [5] examined how scope impacts consumer donation behavior. Concluded here, that consumers are more unresponsive to scope when donating if valuations are based on emotions.

This paper is explaining the proposed donation system model based on the measurement concept of Euclidean distance of parameters geographical location and inequality index that is combined with optimization process. It can facilitate the donor and recipient of donation intelligently. To deliver it, the introduction section is followed by sections research methodology, results and discussion, and conclusion and further works.

2. Research Methodology

The concept of fuzzy-logic method [6] that combined with Williamson idea [7] particularly used to develop the model of inequality index measurement [1]. Where here, it becomes a part of the proposed system model. It was combined with Euclidean distance notion then to find the objective distance between donor and recipient of donation. Indeed, inequality index of [1] portrays the index socioeconomic gap in two sectors, health and education. It (called fuzzy-Williamson index) is an extended version of Williamson index that only consider socioeconomic gap in general.

Furthermore, the concept of hill-climbing optimization [8] used to find the recipient's shortest distance among lots alternatives to be recommended to get the first donation before others. The recommendation used by donor to act. The hill-climbing itself is a heuristic optimization method to find the parameters that give a near-best value of objective function. And the final one, the object oriented method coming from [9] operated to configure the proposed system model. Basically, two diagrams usecase and class are functioned to depict the relation between system and actors and the relationship among classes in the system model respectively. The steps of the research is universally depicted in Figure 1.

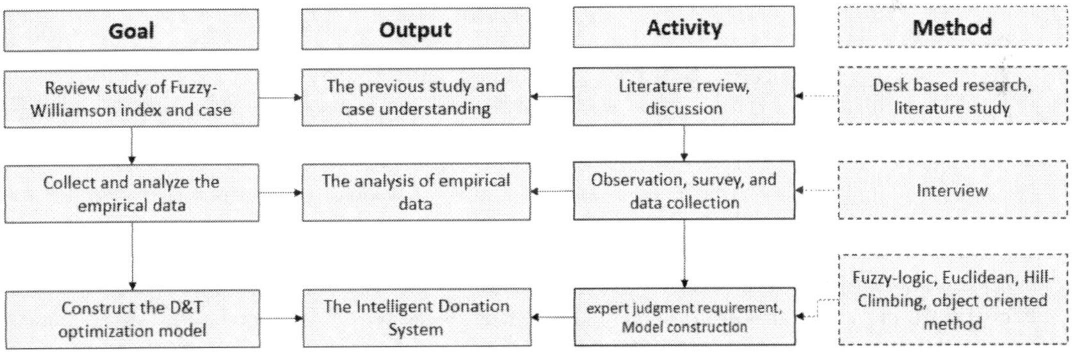

Figure 1. Research Framework

3. Results and Discussion

3.1. The Constructed System Model

Fundamentally, the constructed system consist of three human actors; `Admin`, `Agent`, and `Donor`. The whole system is managed by administrator (`Admin`), where practically the administrator validates (usecase `Validating`) all register process of agent and openly access the report (usecase `Reporting`). Furthermore, the agent is practically placed in urban village. The actor is responsible to register all recipients (usecase `Registering`), validate all donors' donation (usecase `Validating`), and can access the report (usecase `Reporting`) as well. The complete usecase diagram of the constructed system is described in Figure 2.

The actor `Donor` has the important role here. Besides the actor can register directly (via usecase `Registering`, and then validated by administrator), he / she can select the recommended recipients (that have been processed through usecase `Optimizing`) to give donation (usecase `Donating`) and trace the donation (usecase `Tracing`) directly, and also see the report. In detail, the actor `Donor` can trace by communicating directly with the recipients regarding the donation.

Generally, all entity relationships in the constructed system is configured by class diagram (Figure 3). There are fifteen classes fundamentally. Specifically for geographical location, in Indonesia, one location is divided into four levels: province, district, sub-district, and urban village. The urban village itself can have more specific attributes i.e. hamlets and neighborhood.

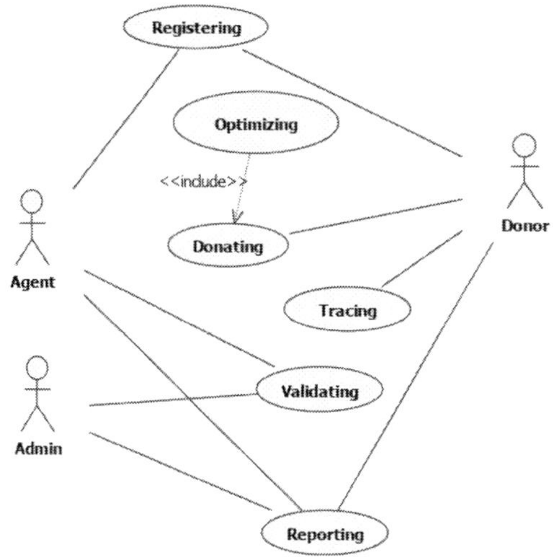

Figure 2. Schematic View of the Constructed System

Moreover, the recommended recipient is proposed by process of hill-climbing optimizing based on the computation process of real geographical location and inequality index distance combination. Where purposely, the optimization model is described in Figure 4. The objective function is determined by considering the real geographical location distance and socioeconomic gap distance (both health and education parameters). In general, it is described in equation (1); where $\boldsymbol{x}$ and $\boldsymbol{y}$ are parameter examples, and $\boldsymbol{D_x}$ and $\boldsymbol{D_y}$ are distance of parameters $\boldsymbol{x}$ and $\boldsymbol{y}$ respectively. All distances are calculated from zero point. This formula is coming from the Euclidean distance measurement [10], generally it is formulated in equation (2).

To normalize the Euclidean distance value for the real geographical location distance ($\boldsymbol{D_{LN}}$), the equation (3) is used. It is the equation to normalize a value in the condition when the highest value is representing the best. On the other hand, to normalize Euclidean distance value for the parameter inequality index ($\boldsymbol{D_{IN}}$), the equation (4) is used. The equation (4) is used to find the normal value when the most minimal value describes the best [11]. Thus, explicitly, the objective function for the constructed optimization model in this study is defined in equation (5).

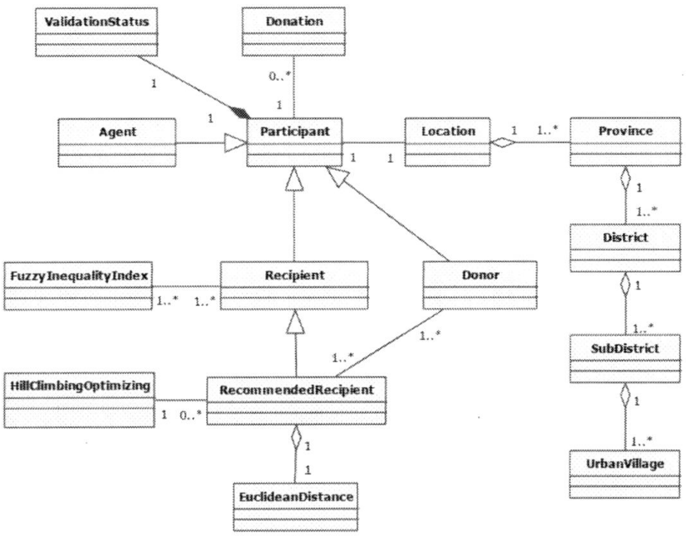

Figure 3. The Class Diagram of the Constructed System

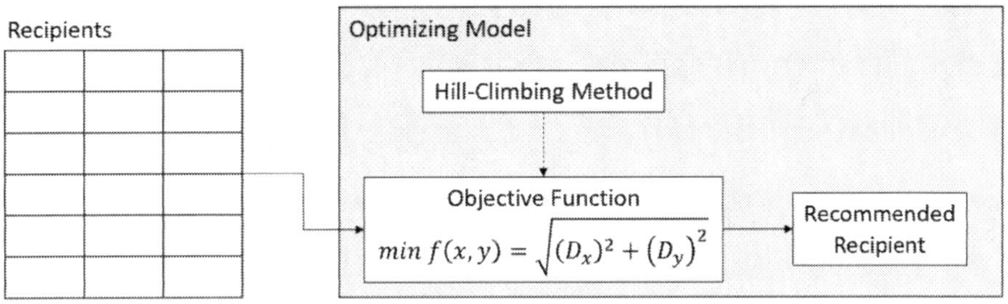

Figure 4. Schematic View of Optimization Model

$$\boldsymbol{min\ f(x,y) = \sqrt{(D_x)^2 + (D_y)^2}} \quad (1)$$

$$\boldsymbol{|x-y| = \sqrt{(x-y)^2}} \quad (2)$$

$$\boldsymbol{D_{LN} = \frac{D_C}{\max(D_1..D_n)}} \quad (3)$$

$$\boldsymbol{D_{IN} = \frac{\min(D_1..D_n)}{D_C}} \quad (4)$$

$$\boldsymbol{min\ f(D_{LN}, D_{IN}) = \sqrt{(D_{LN})^2 + (D_{IN})^2}} \quad (5)$$

3.2. Empirical Measurement and Optimization Experiments

In the urban village Sawah, there are 385 family cards for empirical sample size. The value of Euclidean distance for each family card (based on the distance of both real geographical location and inequality index) to the donor is drawn in hill-valley graph (Figure 5). The graph was arranged by sorting all

IOP Conf. Series: Journal of Physics: Conf. Series **801** (2017) 012005 doi:10.1088/1742-6596/801/1/012005

numbers of family card. Thus, the graph is a representation of all families' distance (from both parameters geographical location and gap).

In addition, for the purpose of the laboratory experiment, the data example of the geographical location distance is practically randomized based on one randomized virtual location (as a donor point example). Afterward, by using the hill-climbing method with ten experiments, the recommended recipient who get the donation firstly is the recipient with Euclidean distance value 0.03 (the lowest one, see Figure 6). It means the donor is suggested to donate that family (the nearest recipient candidate) first.

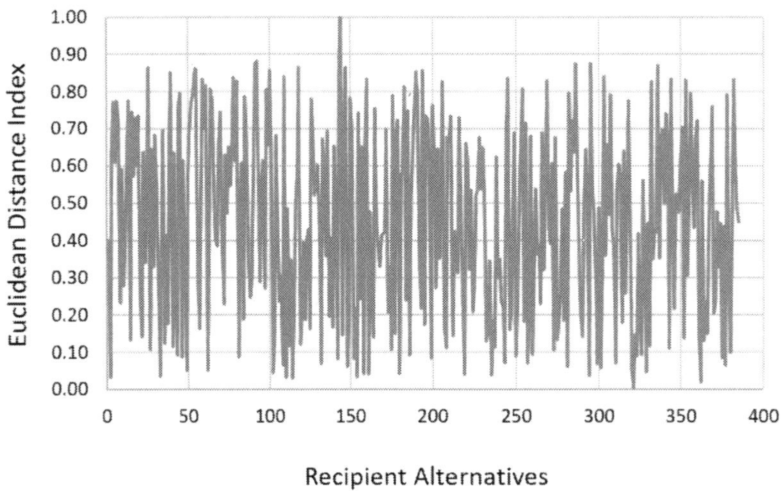

Figure 5. The alternatives of Recipient based on Empirical Data

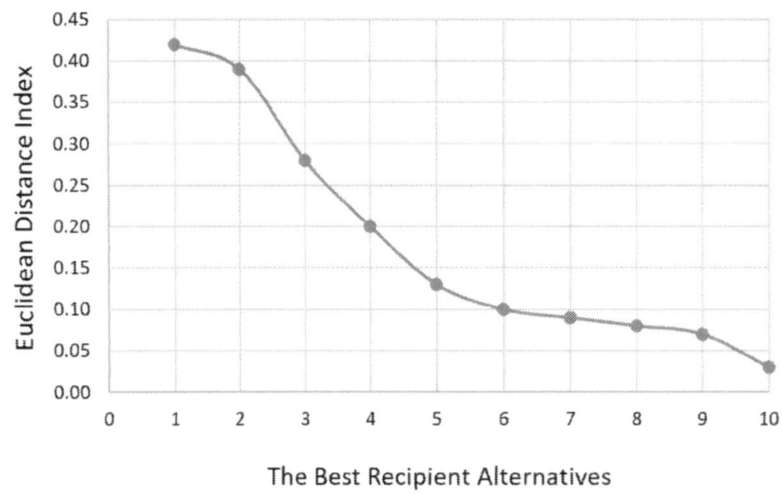

Figure 6. The Most Recommended Recipient Alternatives based on Hill-Climbing Optimizing

4. Acknowledgment
We particularly would like to thank Lembaga Penelitian dan Pengabdian kepada Masyarakat (LPPM), UIN Syarif Hidayatullah, Jakarta, who has supported our works.

5. Concussion and Further Works
The constructed system model is able to recommend the most recommended donation recipient from more than 350 family cards (based on empirical sample) by using the concept of hill-climbing optimization. The recommendation is proposed by minimizing the combination of both parameters geographical location (between donor and recipient) and the inequality index (that merges the health and education gaps). The optimization model of donation recipient is embedded into donations system.

Other aspects can be taken into account to make the recommendation of donation recipient more objective; such as family condition. Or the potency of the donation power also can be examined, both regular or irregular donation.

References
[1] Utama, D.N., Fitroh, Nuryasin, Rustamaji, E., Nurbojatmiko, Qoyim, I. in press. A measurement model based on fuzzy weighted index to examine the socioeconomic inequality (case study: urban village Sawah, Ciputat, Indonesia). *Journal of Business and Economic Policy*, vol.3 No.4.

[2] Davis, L.B., Jiang, S.X., Morgan, S.D., Nuamah, I.A., Terry, J.R. 2016. Analysis and prediction of food donation behavior for domestic hunger relief organization. *International Journal of Production Economics*, vol.182, pp. 26-37.

[3] Savas, S. 2016. Factors affecting donations in U.S. retail stores: a conceptual framework. *Journal of Retailing and Consumer Services*, vol.33, pp. 178-185.

[4] Liang, J., Chen, Z., Lei, J. 2016, Inspire me to donate: the use of strength emotion in donation appeals. *Journal of Consumer Psychology*, vol.26 no.2, pp. 283-288.

[5] Hasford, J., Farmer, A., Waites, S.F. 2015. Thinking, feeling, and giving: the effects of scope and valuation on consumer donations. *International Journal of Research in Marketing*, vol.32 no.4, pp. 435-438.

[6] Zadeh, L.A. 1996. Fuzzy logic = computing with words. *IEEE Transactions of Fuzzy Systems*, vol.4 no.2, pp. 103-111.

[7] Williamson, J. 1965. Regional inequality and the process of national development. *Economic Development and Cultural Change*, vol. 14, pp. 3-45.

[8] Utama, D.N. 2015. *The Optimization of the 3-d Structure of Plants, Using Functional-Structural Plant Models. Case Study of Rice (Oryza sativa L.) in Indonesia*, PhD Thesis, Georg-August Universität Göttingen.

[9] Mathiassen L, Munk-Madsen A, Nielsen P A and Stage J 2000 *Object-Oriented Analysis and Design* Marko Publishing ApS Aalborg Denmark

[10] Deza, E., Deza, M.M. 2009. *Encyclopedia of Distances*. Springer

[11] Utama D N, Saputra M D, Wafiroh L N, Putra M A A and Lestari P 2016 F-multicriteria based decision support system for road repair and maintenance (case study: three areas in Tangerang Selatan, Province Banten, Indonesia) *Proceeding of Academics World 39th International Conference* pp 27-31

Design Support System for Open Distance Learning Student Teamwork

A Putranto[1], Y D Pradipto [2]

[1]School of Computer Science, Doctor of Computer Science, Bina Nusantara University, Indonesia
[2]Psychology Department, Faculty of Humanities, Bina Nusantara University, Indonesia

Email: aputra@binus.edu, ypradipto@binus.edu

Abstract. Open distance learning students in doing team assignment, they seldom face to some problems such as student fell unfair in numbers of workload contribution, instructors also do not know which students do more work than others. So there are some questions ie: how to connect between instructor, team members, and working documents. Methods will be used are first, analyzing current condition and last by designing systems to connect between instructor, team members, and document. The expected result is support systems for open distance learning student teamwork.

Introduction

Background

Indonesia is a country in Asia that have more than 14.000 islands. A lot of islands are not supported by the internet, online learning in Indonesia is necessary for improving higher education [1]. Education Institutions who running e-learning have Learning Management Systems to help instructors and students can connect each other's [2] [3]. In Learning Management Systems provide subject content materials, assignments, and discussion forums [2].

To support Open Distance Learning need a small group discussion, students will have active learning habits, and students will be more understandable in learning [4][5]. Doing teamwork must have aesthetic rules. From Buddha way, there are some rules i.e. talking right, doing right, working right, and trying right [6]. Students also do not want to be considered telling untrue about their job from his team members. Falsification and astray thought are also not considered right for teamwork [7].

Informal conversation on Social Media such as Whatapps, Line, Twitter etc. can provide student learning information [8].

In doing assignment especially in a small group, students are seldom faced problems i.e.: student have teammates who get only a few work contributions, and instructor confused to get information which student have more work.

Previous Research

Open and Distance Learning Education

Open and Distance Learning Education is an alternative education method that geographical between instructors and students are apart, and need a medium tools to send messages or study content material from instructors to students [9].

E-Learning tools
In running distance learning education, institutions must have e-learning tools. And some institution educations are not prepared deploying e-learning tools [10][1]. Education Institutions who running e-learning have Learning Management Systems to help instructors and students can connect each other's [2] [3]. In Learning Management Systems provide subject content materials, assignments, and discussion forums [2].

Distance Learning Characteristic
Distance Learning characteristic is a separation geographic between instructors, learners and also instruction system which need some medium to work as a tool that can be used to send messages or subject learning between learners and instructors [9].

Higher Education Distance Learning Medium
Medium in Distance Learning Higher Education that can be used are i.e.: printed medium (such as book, module, or learning rules), digital medium (such as audio, video, radio broadcast, television broadcast, interactive video, computer, VCD) [9].

Group
Group definition is two or more people interaction who have specific goals. There are 2 kinds of group i.e. formal group and informal group or people often call social group [11].

Group development phase
Group development can be passed in 5 (five) steps [11]: 1. Forming phase. There are two forming steps : i. The first step, people join in a group, ii. Next step, the decision to a directed goal, structure, and leadership. 2. Storming phase. In this phase, there are some clashes, and toward an agreement. 3. Norming phase. Norming phase is a phase that a relation and cohesion of a group be one. In this phase, there will be constructed a same expectation set or norm on team member behavior. 4. Performing phase. A phase where group structure accepted by team members. Energy also toward to understanding of each work and each member work on group goals.

Challenge Group Management
Challenge on manage group is drawbacks and benefits [11]. Drawbacks or retreats : do not like team member, do not trust team member, stereotyping team member, communication problems, stresses and pressures. Benefits : differ and diffuse idea, limitation group thought, improve focus or understand from other ideas or views.

Methodology
Field study research.Researcher will find institution which run distance education in Indonesia, The analysis and design method that will be used are Object Oriented Analysis and Design (OOAD). On analysis method show about how is student doing team assignment in current condition. On design are shown use case diagram, activity diagram and user interface.

Result and Discussion
Current Condition Analysis
There are two types of assignments, the first type is personal assignments, and the second type is team assignments. In next, the paper only discusses team assignments.

Assume students can be from difference islands in Indonesia, i.e. student A from Java Island, student B from Borneo Island, and C from Bali Island, and the instructor from Jakarta, in Java Island. After they finish their assignment, one of them sent their assignment to current learning management systems. And the instructors take their assignment through learning management systems.

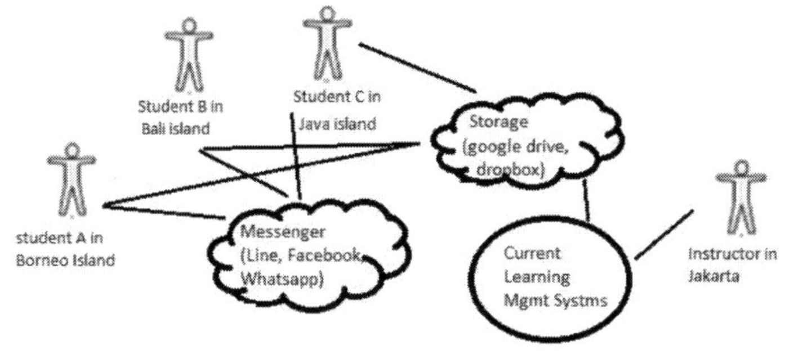

Figure 1. Teamwork Distance Learning Education Systems

Students in doing team assignment, will follow steps such as: a. Students join in a group. They determine their team members b. Students accept / take team assignment c. Students share their workload d. Students do their own workload team assignment e. One member of team submits their team assignment and send to instructor through learning management systems

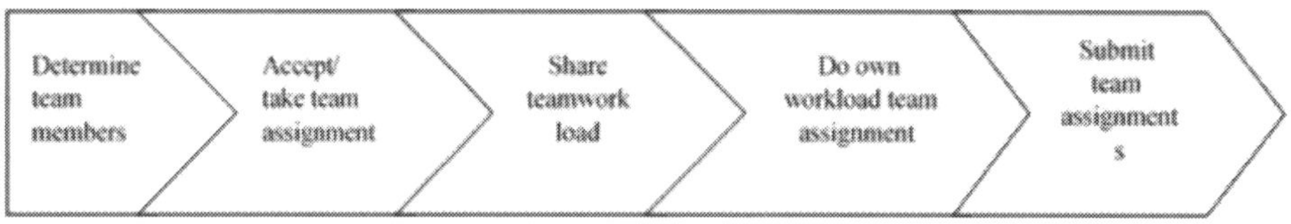

Figure 2. Current team assignment flow

Students do their own work in their places that separate geographically. After every team member finished their own assignments, then they combine to be one and submit to their instructors.

Some problems occur: a. students do their workload but do only a part of their assignments, b. students do their work more than others, c. students submit to team member late, over deadline set by team. Those problems cause: a. students feel his workload more than others, b. students feel grade given by their instructors are not fair, c. instructors only know and give grade according to submitted assignments.

Student Teamwork Support System Design

To support teamwork process, there needs Team Leader menu, to choose a leader of the team. The team leader will determine types of teamwork will be done. 2 (two) types of working process: (1) Relay assignment or (2) Parallel assignment. Relay assignment to support a kind of assignment that will be run look likes relay race that passes over a token. Parallel assignment to support the assignment that has parallel work.

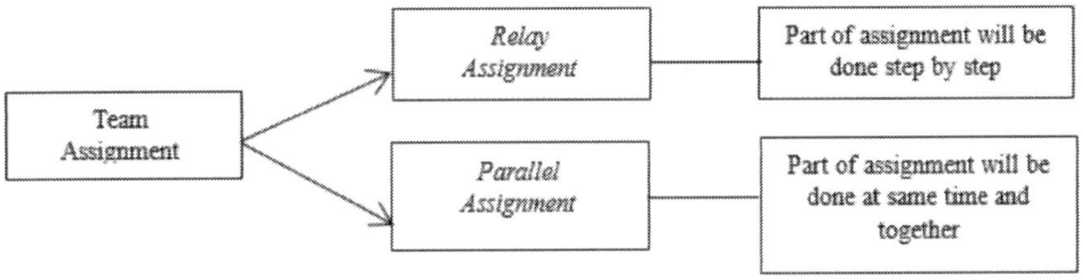

Figure 3. Types of team assignment

There are some features to support the systems: 1. A menu that provides types of team assignment : relay and parallel assignment. 2. A file storage feature, a place to store their teamwork files. 3. A social media feature, a place where a social media will be used to communicate. 4. A contribution feature, where instructors can see how much their team members contribution in doing their assignments.

Team leader creates team and enters the name of team members. Leader choose the way assignment will be done, by relay or by parallel. Leader create the name of working file. Leader choose storage to put the working file. Leader defines an application of social media. Members will have discussion in social media. After team submits the assignment, instructor see their work contribution. Use case diagram will be shown on figure below.

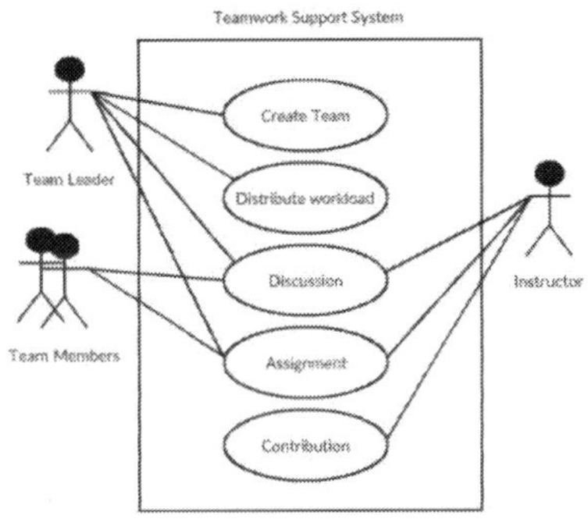

Figure 4. Teamwork Support Systems Use Case Diagram

Team members choose one of team members as leader. The leader entry team leader name and its members. Leader choose kind of assignment whether relay or parallel. Team leader entry storage media where to put working files. Leader create working files, and give name on the working files. Leader divide assignment workload to all member of the team. Leader entry the name of social media that will be used for communication among them. Every member does their job according workload that has been divided by leader. Every member has chance to review their mate jobs. One of team member submit the assignment to instructor. Instructor can see the score of member's contribution.

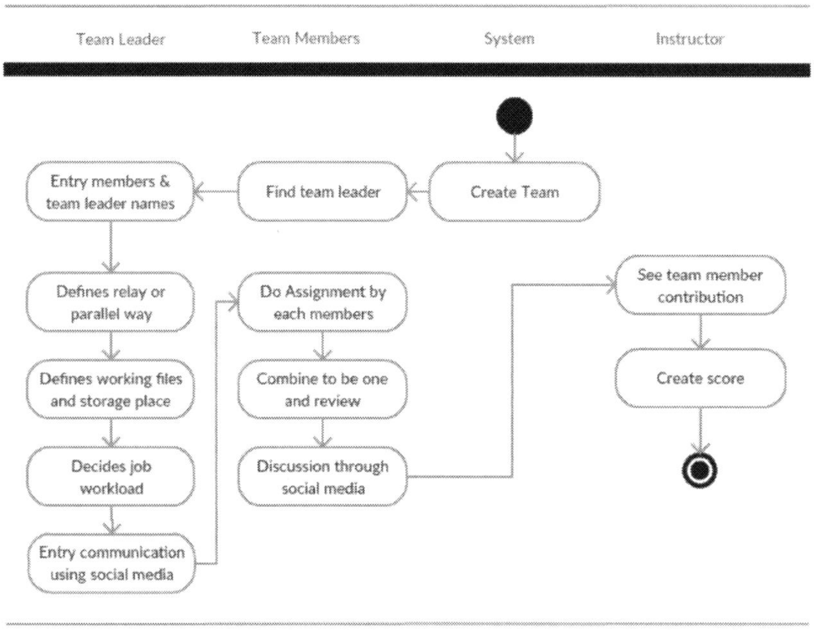

Figure 5. Proposed Teamwork Support Systems Activity Diagram

User interfaces are created from proposed activity diagram "contribution". Instructor can see student activity, and score can be get from it.

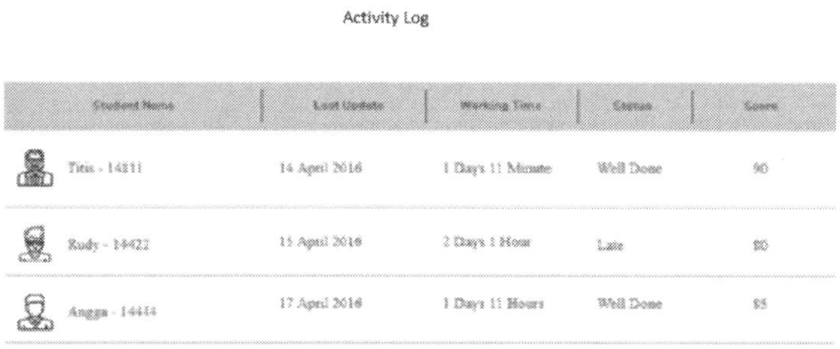

Figure 6. Student Activity Log User Interface

Algorithm to get the score:

If submission date > submission dateline then

$$\text{Score} = \text{Score} - \sum_{i=1}^{n} L$$

If x > 0 then

$$\text{Score} = \text{Score} + \sum_{i=1}^{x} C$$

endif

end

Legend:

$_n$ = number date of late submission | $_x$ = number of contribution
L = score addition for late | C = score addition for contribution

Conclusion

The student teamwork support systems for ODL will solved the problems: 1. instructor will know addition scores to their assignment from number day of late submission and number of contribution, 2. team members know who work and submit the assignment earliest and on schedule given by team leader. 3. The contribution score can be added by number of contribution (C) and subtracted by number of late submission date (L).

References

[1] Diki D 2013 International Collaboration of Distance Learning Universities for Online Learning in Indonesia *LUX* vol 2 no 1 pp 1–8
[2] Hasibuan Z A and Santoso H B 2005 The use of E-learning towards new learning paradigm: Case study student centered E-learning environment at Faculty of Computer Science University of Indonesia *Proc. - 5th IEEE Int. Conf. Adv. Learn. Technol. ICALT 2005* vol 2005 pp 1026–1030
[3] Zuhairi A, Julaeha S and Sinar T E A 2013 Support System for Large Diverse Groups of Open and Distance Learning Students in Indonesia *Infonomics-Society* vol 4 no 3 pp 1197–1204
[4] Ferreri S P and O'Connor S K 2013 Redesign of a Large Lecture Course Into a Small-Group Learning Course *Am. J. Pharm. Educ.* vol 77 no 1 p 13
[5] Laru J, Näykki P and Järvelä S 2012 Supporting small-group learning using multiple Web 2.0 tools: A case study in the higher education context *Internet High. Educ.* vol 15 no 1 pp 29–38
[6] Jalaludin A H and Idi H 2011 *Filsafat Pendidikan : Manusia, Filsafat, dan Pendidikan.* Jakarta: PT RajaGrafindo Persada
[7] Mukthar L 2014 *Filsafat Ilmu* Jakarta: Kencana
[8] Chen X, Vorvoreanu M and Madhavan K P C 2014 Mining Social Media Data for Understanding Students' Learning Experiences *IEEE Trans. Learn. Technol.* vol 7 no 3 pp 246–259
[9] Pannen P D V, Mustofa D, Lukman M, Nurhajati S 2008 *The Utilization of Information and Technology by Student in Universitas Terbuka.* SEAMEO SEAMOLEC
[10] Suryadi A D 2015 Pengukuran Tingkat Kesiapan Penerapan e-Learning Menggunakan TRI (Technology Readiness Index), Studi Kasus : UIN Suska Riau *Open Access J. Inf. Syst.*, pp 237–241
[11] Robbin S P 2012 *Management.* England: Pearson Education Limited

IOP Conf. Series: Journal of Physics: Conf. Series **801** (2017) 012007 doi:10.1088/1742-6596/801/1/012007

E-Learning Application of Tarsier with Virtual Reality using Android Platform

H N Oroh[1], R Munir[2] and D Paseru[1]

[1] Program Studi Teknik Informatika, Fakultas Teknik Universitas Katolik De La Salle Manado, Kombos, Kairagi 1, Telp.(0431): 877512

[2] Sekolah Teknik Elektro dan Informatika, Institut Teknologi Bandung, Jl. Ganesha No. 10, Kota Bandung

E-mail: harlanoroh@gmail.com, rinaldi@informatika.org, dpaseru@unikadelasalle.ac.id

Abstract. Spectral Tarsier is a primitive primate that can only be found in the province of North Sulawesi. To study these primates can be used an e-learning application with Augmented Reality technology that uses a marker to confronted the camera computer to interact with three dimensions Tarsier object. But that application only shows tarsier object in three dimensions without habitat and requires a lot of resources because it runs on a Personal Computer. The same technology can be shown three dimensions' objects is Virtual Reality to excess can make the user like venturing into the virtual world with Android platform that requires fewer resources. So, put on Virtual Reality technology using the Android platform that can make users not only to view and interact with the tarsiers but also the habitat. The results of this research indicate that the user can learn the Tarsier and habitat with good. Thus, the use of Virtual Reality technology in the e-learning application of tarsiers can help people to see, know, and learn about Spectral Tarsier.

1. Introduction

Tarsier is one of the smallest primates found on several islands in Indonesia. Tarsier has several types namely, Bancanus Tarsier, Spectral Tarsius, Pygmy Tarsier, and others. Spectral Tarsier can be found in the provinces of North Sulawesi. Their habitat is a secondary forest with trees that are small and medium. They can be found in tree cavities and voids are formed ushered dense bamboo trees. Tarsier is carnivores by eating insects.

Virtual Reality is a technology that allows users to interact with the artificial environment simulated by computer. It can give a real impression. It is widely used in the fields of healthcare, automotive, education, and others. One example application in the field of education is Sea World VR2 created by eiNpictures. This application allows users to be a diver in a sea that full of marine creatures and make the user know about a sea creature [4]. Tools to develop the Virtual Reality-based application is Unity and Blender. Virtual Reality can be run on Android smartphone supported by Gyroscope Censor and Magnetic Field Censor. Users may use additional tools such as VR Headset and Gamepad.

E-learning of Tarsier can be done by using existing application. *Aplikasi Pembelajaran Satwa Langka* (Endangered Species Learning Application) is one of e-learning application of tarsier with

Augmented Reality technology [11]. The Application displays in three dimensions' object by using marker faced the camera computer to interact so three dimensions' tarsier object superimposed onto the real world. The weakness of the application is not shown the habitat tarsier and need a special marker to interact. The same technology can be shown three dimensions' objects is Virtual Reality. The Virtual Reality advantages than the Augmented Reality is making the user like venturing into the virtual world. To overcome the weakness of e-learning application that will be constructed e-learning application with Virtual Reality using Android Platform. This application not only shows the Tarsier but also its habitat and can interact with the virtual environment so users seemed to get into the virtual world.

Based on the description above, that can be seen on an opportunity to develop e-learning application of Tarsier with Virtual Reality using Android platform to introduce and provide information about Tarsier. Android Platform was chosen because it provides the resources needed. This application can help people to see, know, and learn about Spectral Tarsier.

The formulated research question is "how to develop e-learning application of Tarsier with Virtual Reality using Android platform that can help people to see, know, and learn about Spectral Tarsier?". In this regard, the research objectives are as follows:

1. Provide information and knowledge to the public about Spectral Tarsier and its habitat.
2. To develop an e-learning application of Tarsier with Virtual Reality using Android Platform.

2. Literature Review

2.1. Tarsier

Tarsiers first appeared in the Eocene era and known as "living fossils". This primate can rotate the head at an angle of 180° and jumps. Generally, family's tarsier has a characteristic body covered with thick hair and smooth, big eyes and big ears. For Tarsier found in Sulawesi, have distinctive characteristics that their white hair behind the ears, while the lid hair gray [7].

2.2. Virtual Reality

Virtual Reality is the computer-generated simulation of a 3D model environment, which seems very real to the person experiencing it, using the special electronic equipment. The objective is to achieve a strong sense of being present in the virtual Environment [8]. VR environment generally present a visual experience that is displayed on a computer screen or through a stereoscopic viewer, but some simulations include additional information by sensing results [11]. The user can interact with the Virtual Reality environment through the use of standard input tools such as a keyboard, mouse, and others.

Augmented Reality (AR) is a variation of Virtual Environments (VE), or known as Virtual Reality (VR). Augmented reality is an attempt to combine the real world and the virtual world in real-time in 3D model created by the computer so that the boundary between two worlds become thin [6].

Virtual Reality and Augmented Reality has a fundamental difference that is from the input to the console being used. At the Virtual Reality all the input data are pre-programmed, whereas, in Augmented Reality, the input is derived from the surrounding environment on the console [1]. Another difference is the Virtual Reality allows users joined in a virtual environment as a whole, whereas Augmented Reality allows users see the real environment with virtual objects which added to or incorporated into the real environment [6].

2.2.1. Component of Virtual Reality. A VR system is made up of 2 major subsystems, the hardware and software. The hardware can be further divided into computer or VR engine and I/O devices, while the software can be divided into application software and database as illustrated below [2].

2.2.2. Virtual Reality Works. Virtual Reality basically works in two ways namely stereoscopic 3D viewing and head tracking [8]. Stereoscopic 3D viewing is separating display screen into two parts with two dimensions' image of each for the left eye and right eye so as to give the illusion of three dimensions view. Head Tracking is a process to detect a pose and head movement.

2.2.3. Types of Virtual Reality. Types of Virtual Reality can be classified into three main categories namely non-immersive, immersive, and semi-immersive [2]. Immersive System is the type that will be used. This type can provide the highest level of immersion so that user will feel like be part of the virtual environment. This uses Head Mounted Display (HMD) as the main tool and complementary tools are tracking devices, data gloves, and other.

2.2.4. Methods of Virtual Reality. Methods of Virtual Reality can be divided into four namely simulation, avatar, projectors, and desktop [9]. Simulation-based Virtual Reality is the methods that will be used. This method provides a simulation to the user with virtual reality.

2.3. Tools to Develop Virtual Reality
There are two software tools to develop Virtual Reality. Those tools are Blender and Unity. Blender is an application that supports the creation of modelling, rigging, animation, and other [3]. Unity is a game engine that can manage multiple data such as three dimensions' objects, sound, texture, and other [10].

2.4. Similar Application
The following are e-learning application using Virtual Reality technology. *Aplikasi Pembelajaran Satwa Langka* (Endangered Species Learning Application) is an application that uses a type of semi-immersive system and runs on the computer by using a marker that faced the camera computer to interact with an object [11]. Sea World VR2 is a mobile application that uses a type of immersive system to allow the user to be a diver in a sea that full of marine creatures [4]. VR Safari is a mobile application and uses a type of immersive system that delivers an environment of safari park to the user as in the car [5].

3. Proposed Virtual Reality Design

3.1. Analysis of Similar Application
The following table will only show the comparison between Virtual Reality application by eiNpictures, Tumboimbela, and Chai.

Table 1. The Comparison of *Aplikasi Pembelajaran Satwa Langka* (Endangered Species Learning Application), Sea World VR2, and VR Safari [11] [4] [5].

	Endangered Species learning Application	**Sea World VR2**	**VR Safari**
I. Type of *Virtual Reality*	Semi-immersive (Augmented Reality)	Immersive	Immersive
II. Operating System	Windows	Android	Android
III. Language	Indonesian	English	English
IV. Helping Tools	Webcam and marker	Headset VR and Gamepad	Headset VR and Gamepad
V. Feature			
a. 3D Object	√	√	√
b. Virtual environment	√	√	√
c. Animation	√	√	√
d. Information	√	-	-
e. Feed the object	-	-	-

f. Change time condition	-	-	-
g. Explore the world	-	√	√

3.2. Requirement Specification

The following are the features that will be developed on the e-learning application of tarsier.

1. The Application can show Spectral Tarsier and its habitat in 3D model along with information.
2. The User can explore the virtual world with a Gamepad.
3. The User can change time condition of virtual world so that the application can show Spectral Tarsier activity based on time of day or night.
4. The Application can provide a button to feed the Spectral Tarsier.
5. The User can control the Tarsier objects such as jumping, eating, sleeping, and a voice with buttons on the Gamepad.

3.3. Design Modeling

There are four use cases of e-learning application of Tarsier. Figure 3 depicts those use case in use case diagram. There is only one actor that is a user.

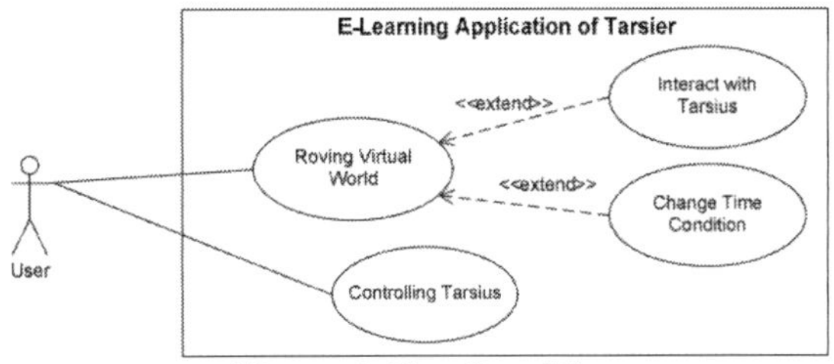

Figure 1. Use Case Diagram

3.4. Function for the Application Development

To develop the application require a lot of function. The following is a list of function that is most needed.

Table 2. Function Needed For Application Development.

Function Name	Information
Head Tracking	Detecting head movements and shows the object in the direction of the angle from the head.
Input Gamepad	Set the input buttons are derived from gamepad devices.
Stereoscopic	Setting the display screen Android smartphone into two parts.
Change Condition	Changing the conditions of the time from the virtual environment to daytime into a night and vice versa.
Tarsier Activities	Determining the activities that can do of Tarsier object based on the time condition.
Avoid User	Regulate the movement of Tarsier object to the user.
Feed	Regulate the movement of Tarsier object when user feeds the Tarsier.
User Movement	Regulate the movement or displacement of the user when the user is walking in a virtual environment.
Tarsier Animation	Set the animation of objects Tarsier based on input by the gamepad.

3.5. Animation Controller and Design Map of Virtual World

The following are animation diagram of Tarsier object and virtual world design map for the future will be explored by the user. Figure 4 depict Tarsier object can do in the virtual world. Figure 6 is a design map of Virtual World.

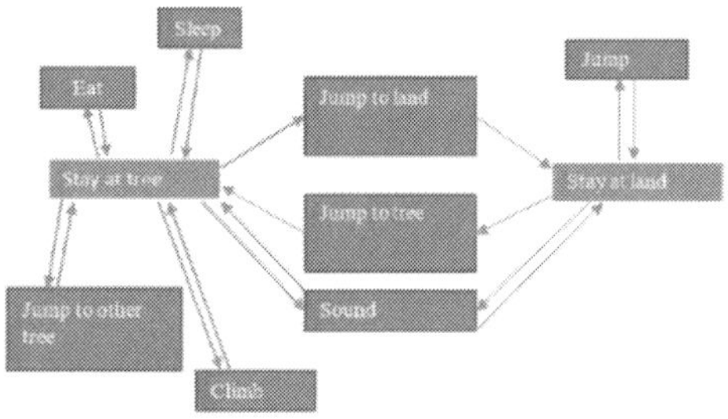

Figure 2. Animation Controller

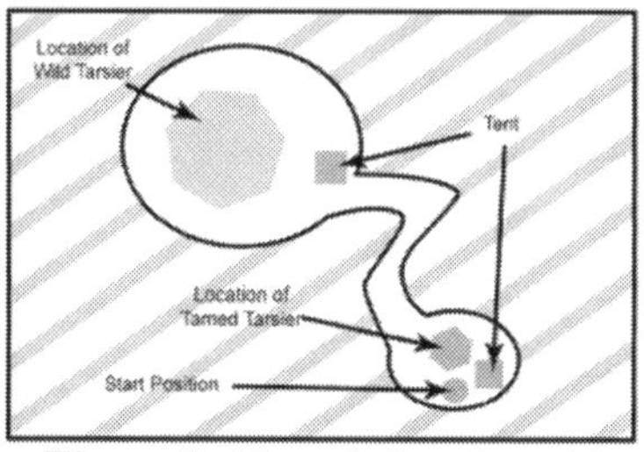

Figure 3. Map of virtual world

4. Implementation and Testing

4.1. Technology for the Application Development
Supporting software for this application is as follows.

Table 3. Software Required For Application Development.

Programming	Unity 5.3.3
3D Object	Blender v2.71
Animation	Blender v2.71
Image Editing	Adobe Photoshop CS6
Modeling	Microsoft Office Visio 2007

4.2. Implementation of Application
The following will depict the application interfaces.

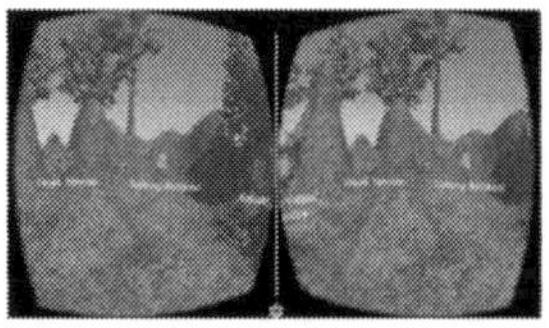

Figure 4. Main Menu Display

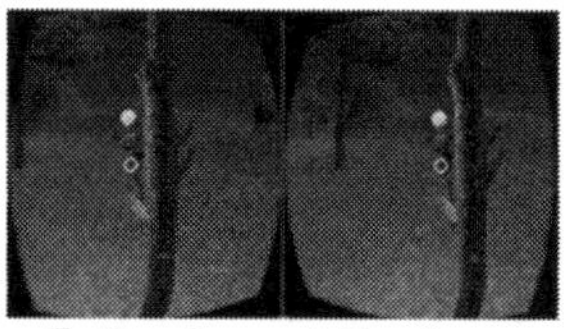

Figure 5. Tarsier and Environment Display

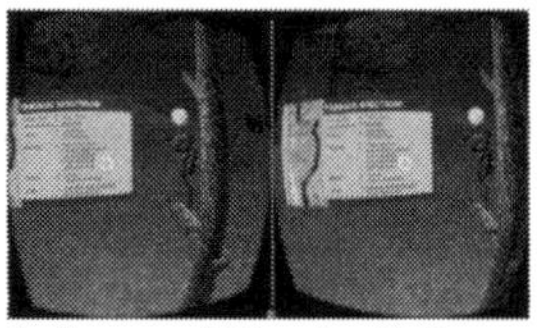

Figure 6. Spectral Tarsier Information Display

Figure 7. Menu of Tarsier Object Display

Before running this application, connect the Gamepad to the smartphone first. The following is an explanation of the use of the application after running.

1. Enter the smartphone into a VR Headset and user wearing VR Headset.
2. The menu selection is done by looking at the menu item and pressing the A button on the Gamepad or button provided by VR Headset.
3. For virtual world tarsier menu, Buttons used in Gamepad are the A button to interact with objects and Analog button to walk.
4. Before entering into the virtual world tarsier, the application will provide information about objects that can interact with the user.
5. For tarsier object menu, the buttons used in Gamepad are the A button to sleep, B button to eat, X button to jump, Y button to sound, and Start button to return to the main menu.

4.3. Testing

The following are the specifications used in the test.

1. Smartphone with Intel Atom Z3560 processor, 2 GB RAM, and Android 5.0 operating system.
2. Headset VR-type RITECH REIM 2 and Gamepad-type Ipega PG-9028.

The testing process was based on requirements specification. The following points were the tests performed by the user on the application.

1. Application showing Spectral Tarsier and its habitat in 3D model along with information.
2. User exploring the virtual world.
3. User changed time condition be during the day or night.
4. User fed the Tarsier by using the feed menu displayed when the user pressed the A button on an object.
5. User controlled the Tarsier object like jumping (X button), eating (B button), sleeping (A button), and making sound (Y button).

5. Conclusion

After the development processes that has been done, the developed application is able to help the people who want to see, know, and learn Spectral Tarsier with criteria:

1. Showing Spectral Tarsier and its surroundings in 3D model with information.
2. Interacting with Tarsier (e.g. feeding) and virtual world (e.g. changing time condition, exploring).
3. Making the users feel as if they are on the real habitat of Tarsier.
4. Running well on an Android smartphone that has gyroscope and magnetic field censor.
5. Receiving inputs from the Gamepad such as walking, choosing navigation, and controlling Tarsier.

References

[1] Asfari U, Setiawan B, and Sani N A, 2012 Pembuatan Aplikasi Tata Ruang Tiga Dimensi Gedung Serba Guna Menggunakan Teknologi Virtual Reality Studi Kasus Graha ITS Surabaya. Jurnal Teknik ITS Vol. 1, No. 1, Fakultas Teknologi Informasi Jurusan Sistem Informasi Institut Teknologi Sepuluh Nopember.

[2] Bamodu, O., Ye, X. 2013 Virtual Reality System Components. Paris : Atlantis Press.

[3] Blender (n.d.). About Blender. Available at: https://www.blender.org/about/ [22 April 2016]

[4] Einpictures 2016 Sea World VR. Available at: http://www.einpictures.co.kr/?c=29/34 [4 April 2016]

[5] Google Play (2016). Aplikasi Virtual Reality. Available at: https://play.google.com/store/search?q=Aplikasi%20VR&c=apps [20 April 2016]

[6] Indrawati Y, Ichwan M, and Putra W, 2013 Media Pembelajaran Interaktif Pengenalan Anatomi Manusia Menggunakan Metode Augmanted Reality. Jurnal Informatika Vol.4, No.2, Jurusan Teknik Informatika Institut Teknologi Nasional Bandung

[7] Kurniawan M B, and Pratama B, 2010 Mengenal Hewan & Tumbuhan Asli Indonesia. Jakarta: Cikal Aksara (Imprin Argomedia Pustaka).

[8] Linowes, J. (2015). Unity Virtual Reality Projects. Birmingham: Packt Publishing.

[9] Nesamalar K E, and Ganesan G, 2012 An Introduction do Virtual Reality Techniques and Its Applications. International Journal of Computing Algorithm Vol 01, Issue 02, Department of Information Technology Sri Kaliswari College, Sivakasi

[10] Sihite B, Samopa F, and Sani N A, 2013 Pembuatan Aplikasi 3D Viewer Mobile dengan Menggunakan Teknologi Virtual Reality Studi Kasus Perobekan Bendera Belanda di Hotel Majapahit. Jurnal Teknik Pomits Vol. 2, No. 2, Fakultas Teknologi Informasi Jurusan Sistem Informasi Institut Teknologi Sepuluh Nopember

[11] Tumboimbela K, 2012. Aplikasi Pembelajaran Satwa Langka Tarsius Berbasis Augmented Reality. Unpublished theses. Fakultas Teknik Jurusan Teknik Informatika Universitas Katolik De La Salle Manado. Manado

Augmented Reality Implementation in Watch Catalog as e-Marketing Based on Mobile Aplication

D Adrianto, F A Luwinda and V Yesmaya

Computer Science Department Bina Nusantara University, Indonesia

Email: dadrianto@binus.edu, fluwinda@binus.edu, vyesmaya@binus.edu

Abstract. Augmented Reality is one of important methods to provide user with a better interactive user interface. In this research, Augmented Reality in Mobile Application will be applied to provide user with useful information related with Watch Catalogue. This research will be focused on design and implementation an application using Augmented Reality. The process model used in this research is Extreme Programming. Extreme Programming have a several steps which are planning, design, coding, and testing. The result of this research is Augmented Reality application based on Android. This research will be conclude that implementation of Augmented Reality based on Android in Watch Catalogue will help customer to collect the useful information related to the specific object of watch.

1. Introduction

1.1. Augmented Reality

Augmented Reality is a technique to provide more information than already exist in the real environment by overlaying virtual objects generated by computer. The virtual object created by computer may be comes in two or three dimensional object. Augmented reality works by analyze the real objects which capture by camera as an input and then create additional virtual objects combined with the real objects and shown as an output.

Augmented reality is able to track down the real objects by camera movement and able to manipulate the output by adding additional virtual objects combined. To be able to track down the object, augmented reality will depends on object recognition by edge detection [1].

Augmented reality manipulate user perspective of the physical world by adding additional 3D information that produced by computer. This technique is used in Boeing (Airplane System) to generate the wiring system design by showing wiring path on engineer helmet [2].

1.2. Extreme Programming

Extreme Programming is developing technique that focusing on object-oriented projects where the programmers are located in the same location without any effect of how many programmers hired [3].

Extreme Programming is an agile development technique which the practices based on Pair Programming technique, collective code ownership, and frequent meetings are inherently collaborated. This technique required the programmer needs to be in the same location with the people who fully understand about the business process of the system. In order to have a better improvement using this technique, teamwork becomes one of the important keys [4].

The basic activities of Extreme Programming technique are planning and managing, designing, coding and testing.

1.3. E-Marketing
E-Marketing is an implementation of the application doing the commercial activity which facilitates information of products or services offered to customer [6].

E-Marketing is more efficient than traditional marketing which provides the opportunity as advertisement which can be access anywhere and anytime, increases overall potential of advertisement, increases the income of the companies, and decreasing pollution and energy consumption [7].

2. Literature Review

2.1. Vuforia
Vuforia is Software Development Kit (SDK) which provided augmented reality in mobile application. Vuforia used computer vision method to detect object in the real world (input). To detect the objects, Vuforia implements edge detection algorithm. Vuforia able to detect object not only in 2D but also in 3D objects. Vuforia also able to detect markerless object. Markerless object is an object without matrices marker.

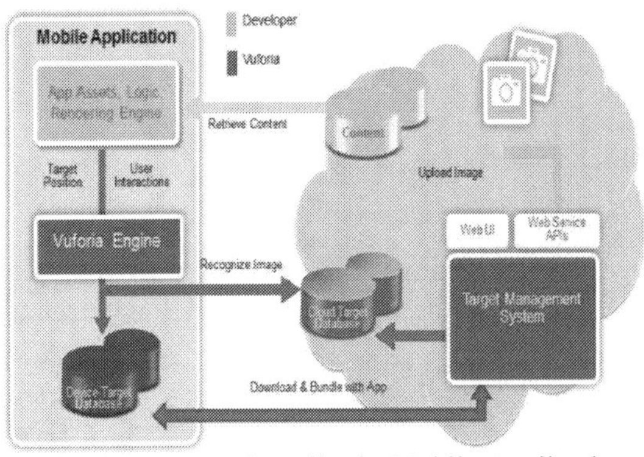

Figure 1. Process of Augmented Reality in Mobile Application using Vuforia

In figure 1 shown the process how Vuforia works in mobile application from detecting object until shown the augmented scene (output which already manipulated by computer which adding virtual objects and combined with the real objects) on mobile side.

Target in Vuforia is an object in real world which capture by camera as an input. Vuforia divided the target into several types:

i. Image targets, such as: Photo, Board Game, Paper Magazine, Cover Book, Product Packaging, Poster, and Greeting Cards. This type of target shows a simple image of Augmented Object.
ii. Frame markers, is 2D image with specific pattern which is able to be used as the pieces of game in board game.
iii. Multi-target, such product packaging in square shape. This type of target shows a simple image of 3D Augmented Object.
iv. Virtual-buttons, is an object which shown as a button.

3. Methodology
As the development method, in this research, Extreme Programming detail process will be described by figure 2 below.

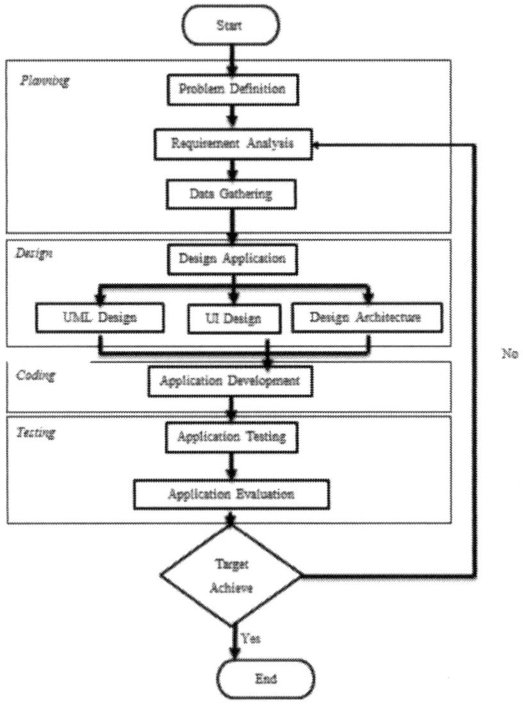

Figure 2. Development Stage

The development process divided into four main steps which are Planning, Design, Coding, and Testing. In planning phase, the problem needs to be defined in detail. In design phase, the structure of the system needs to be design first which describe in "UML Design" followed by UI Design to create the design of the application and Design Architecture as the framework used in code. In coding phase, the system will be developed. In testing phase, the application will be tested using Measurable Human Factors and Eight Golden Rules to evaluate the application whether the application is already align with the purpose of this research.

4. Result

Below are the specifications of the mobile devices suggested:

i. Processor : Dual-Core 1.2 GHz
ii. RAM : 1 GB
iii. Available Memory : 100 MB
iv. Camera : Available (No Specific Resolution Required)

Below are the displays of the User Interface of the main feature of the application which shown the augmented reality implementation.

Figure 3. Main Menu Screen

Figure 4. Catalogue's Menu Screen

Figure 5. Bracelet's Menu Screen

Figure 3 shown a screen which will be shown after login process is done. In this screen, there are 4 others menu which are Catalogue, Bracelet, Order, and Guide. Figure 4 shown a screen which will be shown if the Catalogue's Menu on Main Menu Screen is clicked. On this screen, product information along with the product's image will be shown. Figure 5 shown a screen which will be shown if the Bracelet's Menu on Main Menu Screen is clicked. On this screen, the augmented reality process is happen. When the camera is targeting the hand of the user then the application will run the augmented reality to detect the real object, in this case the object is human hand, and then manipulate the display by adding virtual objects which is the watch that already chosen in Catalogue's Screen combined with the hand of the user.

So, the augmented image which shown to the user seems like the user hands wearing the watch chosen on Catalogue's Screen.

Below are the evaluation result based on Eight Golden Rules:

i. Consistency

Figure 5 shows the consistency of EGHYPT.

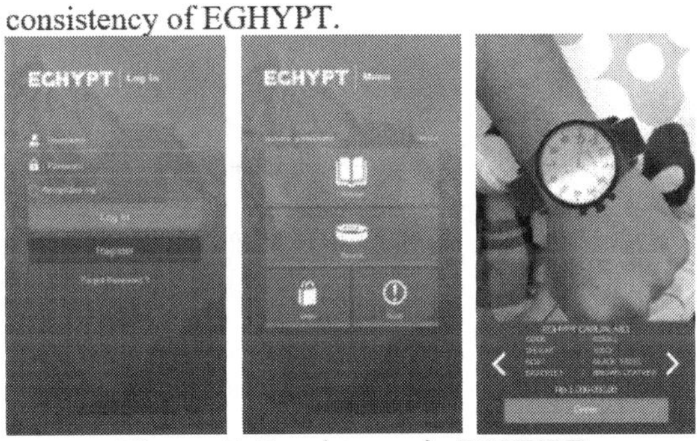

Figure 5. Consistency in EGHYPT

ii. Cater for Universal Usability

Figure 6 shows a "Guide Page" to help user to understand about EGHYPT.

iii. Offer Informative Feedback

EGHYPT already provided with an informative feedback. Figure 7 shows a sample of informative feedback in EGHYPT.

iv. Design Dialogs to Yield Closure

Figure 8 shows a sample of Closure in EGHYPT's Registration Page when user already done the registration process there will be a "Closure Message".

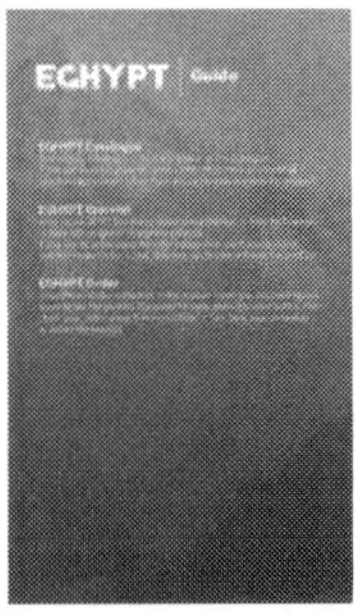

Figure 6. Cater for Universal Usability in EGHYPT

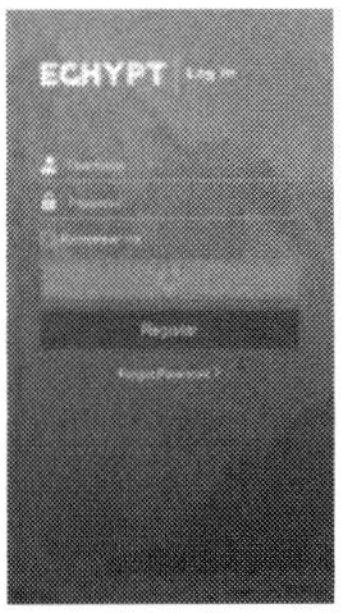

Figure 7. Informative Feedback in EGHYPT

Figure 8. Closure Message in EGHYPT.

v. Offer Error Prevention and Simple Error Handling
Figure 9 shows a sample of Error Prevention in EGHYPT's login page when users try to log in using invalid data.

vi. Permit Easy Reversal of Actions
Figure 10 shows a sample of Easy Reversal in EGHYPT's home page when users click back button, application will shows a confirmation dialog.

vii. Support Internal Locus of Control
Figure 11 show that EGHYPT already provided users with full control of navigation in home page.

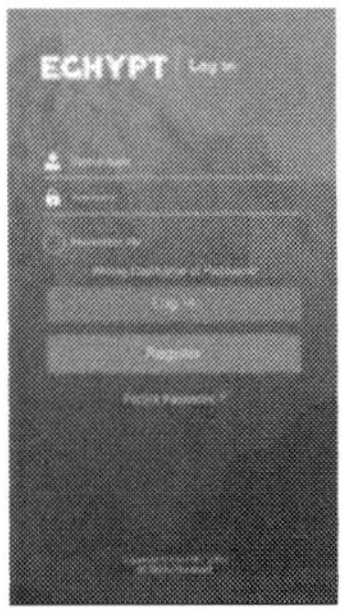

Figure 9. Error Prevention in EGHYPT

Figure 10. Easy Reversal Action in EGHYPT

Figure 11. Internal Locus of Control in EGHYPT

viii. Reduce Short Term Memory Load
Figure 12 show that EGHYPT already provided users with interactive design which able to reduce short term memory load of users. Such as the use of specific image for a button.

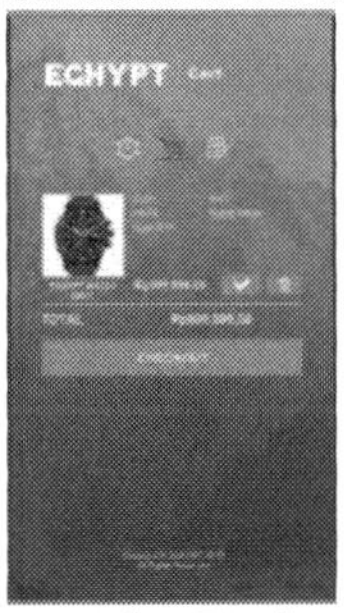

Figure 12. Interactive Design in EGHYPT

Below are the Measurable Human Factors:

i. Speed to Learn
According to the users' survey found that 53.33% of respondents agree that EGHYPT is easy to use and 43.33% of respondents agree that EGHYPT is very easy to use.

ii. Recall at Later Time
According to the users' survey found that 66.67% of respondents agree that EGHYPT is provided with specific icon that easy to remember.

iii. Performance Speed
According to the users' survey found that 53.33% of respondents agree that EGHYPT is responsive and 30% of respondents agree that EGHYPT is very responsive.

iv. Accuracy (Minimize Error)
According to the users' survey found that 56.67% of respondents agree that EGHYPT is provided users with clear error information and support and 30% of respondents agree that EGHYPT is provided users with very clear error information and support.

v. Enjoyment
According to the users' survey found that 30% of respondents agree that EGHYPT is interest of using EGHYPT and 66.67% of respondents agree that EGHYPT is very interest of using EGHYPT.

5. Conclusion

In this research, Vuforia SDK is used to develop the augmented reality application. The result of this research is an e-marketing application uses augmented reality in mobile application to provide information of product to customer by targeting the object product using this application.

References

[1] N.R.Raajan, A Vinodhini, N Vidhyalakshmi 2013 Augmented Reality Based Assistance *International Journal of Engineering and Technology (IJET), 5*(2), 1652-1657

[2] Peng H 2015 Application Research on Face Detection Technology based on OpenCV in Mobile Augmented Reality *International Journal of Signal Processing, Image Processing and Pattern Recognition, 8*(2), 249-256

[3] Lucas Layman, Laurie Williams and Lynn Cunningham 2004 Exploring Extreme Programming in Context: An Industrial Case Study *Agile Development Conference IEEE*

[4] Stephen Wood, George Michaelides and Chris Thomson 2013 Successful extreme programming: Fidelity to the methodology or good teamworking? *Information and Software Technology, 55*, 660-672

[5] Fojtik R 2011 Extreme Programming in development of specific software *Procedia Computer Science, 3*, 1464-1468

[6] New trends of Intelligent E-Marketing based on Web Mining for e-shops 2014 *International Conference on Strategic Innovative Marketing, 175*, 75-83

[7] Azam Hamidi, Maryam Safabakhsh 2011 The impact of information technology on E. Marketing. *Procedia Computer Science, 3*, 365-368

Virtual Application of Darul Arif Palace from Serdang Sultanate using Virtual Reality

M F Syahputra[1], T Annisa[1], R F Rahmat[1], M A Muchtar[1]

[1] *Department of Information Technology, Faculty of Computer Science and Information Technology, University of Sumatera Utara, Medan, Indonesia*

Email: ncafadly@usu.ac.id, triannisa92@students.usu.ac.id, romi.fadillah@usu.ac.id, anggi.muchtar@usu.ac.id

Abstract. Serdang Sultanate is one of Malay Sultanate in Sumatera Utara. In the 18th century, many Malay Aristocrats have developed in Sumatera Utara. Social revolution has happened in 1946, many sultanates were overthrown and member of PKI (Communist Party of Indonesia) did mass killing on members of the sultanate families. As the results of this incident, many cultural and historical heritage destroyed. The integration of heritage preservation and the digital technology has become recent trend. The digital technology is not only able to record, preserve detailed documents and information of heritage completely, but also effectively bring the value-added. In this research, polygonal modelling techniques from 3D modelling technology is used to reconstruct Darul Arif Palace of Serdang Sultanate. After modelling the palace, it will be combined with virtual reality technology to allow user to explore the palace and the environment around the palace. Virtual technology is simulation of real objects in virtual world. The results in this research is that virtual reality application can run using Head-Mounted Display.

1. Introduction

Some of cultural heritages have been destroyed or extinct. It makes people be difficult to know about some places that ever exist in the past. Such as Darul Arif Palace from Serdang Sultanate which is one of Melayu civilization in East Sumatera. Sultan Sulaiman Shariful Alamshah, the fifth Sultan of Serdang Sultanate, built Darul Arif Palace in 1889. When social revolution happened in 1946, sultan family and relatives abandoned the palace [1].

The digital research of historical relics is currently focused in the area of archeological studies and computer application [2]. Virtual reconstructions are based on 3D modelling techniques. The virtual model is a valid cognitive tool and a fundamental medium, so that user can interact with 3D models and agents in a virtual environment. This technology can be applied to the world of cultural heritage as a media for preservation, reconstruction, documentation, research and promotion [3].

Applied Virtual Reality Technology allows user to explore ancient culture and places that no longer exist or that might be too dangerous or expensive to visit. The expection from Virtual Reality Technology is a virtual environment that allows user to get experience like in the real world.

In this research, we will reconstruct Darul Arif Palace using 3D modelling techniques and make it available for virtual exploration. This application will be able to run using head mounted display. User will know about the description of Darul Arif Palace in past, such as the layout of palace, detailed ornament, interior and exterior palace.

Duan, et al. [2] introduced a 3D reconstruction and exploration system of Anyang Yinxu. Yinxu was the capital city of Shang Dynasty. Shang was the earliest ancient capital in the history of China. This area has preserved many ruin such palace, ancestor temple, emperor's mausoleum, and household utensils used in Shang Dynasty. The applied techniques are 3D modelling technique based on stereo photo-pairs, virtual display technique based on CAVE, explore technique based on virtual animation, and scene interaction control technique based on VRML nodes.

Bruno et al suggested some guidelines for development of VR system for digital archaeological exhibition [3]. It illustrated a complete methodology to create a virtual exhibition system, based on realistic high-quality 3D model. Calabria in 18th and 9th BC was the subject of this research. Using a low cost multimedia stereoscopic system called MNEME which allows user to interact in easy way.

Saputra et. al [4], developed the Virtual Tour application of Darul Aman Palace from Langkat Sultanate. Using virtual reality technology in buildings that have become extinct by using image-based modelling technology, image-based modelling is a method using a set of image to produce a 3D object. Modelling is done by using photographs of the palace in the past and create a virtual environment so users can explore the palace and surrounding.

2. Research methodology

There are several steps on our proposed method, namely collecting data, modelling 3D object, optimizing 3D object, materializing and texturing and then setting in virtual reality environment including lighting. Figure 1 shows the general architecture of the proposed method.

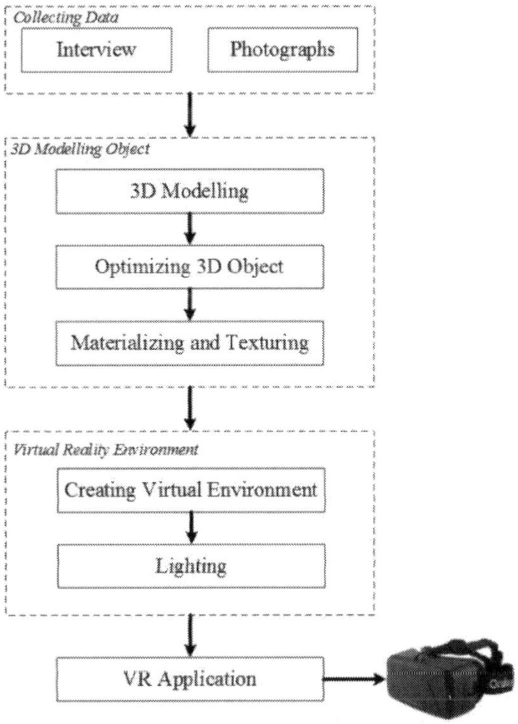

Figure 1. General Architecture

2.1. Collecting Data
The data used in this research consists of three types, namely photographs, mockups of palace and video description about the layout of the rooms in the palace. The photographs taken in the period of 1928-1937 archived digitally. The images are obtained from two sources, namely Tengku Lukman Sinar Library archives and The Koninklijk Instituut voor Taal-Land en Volkenkunde (KITLV)/Royal Netherlands Institute of Southeast Asian and Caribbean Studies.

2.2. 3D Modelling Object
There are three basic primitives or fundamental ways to create a 3D shape, namely points, curves, and polygons. Points are the simplest primitive; a 3D shape can be composed of a large amount of points to cover the whole 3D surface, where each point can be represented by a 3D vertex. A more practical and compact way of representing 3D shapes is using curves approximated by lines or points. Curves are used to describe the complex 3D surfaces [5]. In 3D modelling, there are three components of the object, namely vertex, edge and face. Vertex are points. Edge connects vertex to vertex. The series of edge forms polygon. Face is surface that cover the object. Face has two sides, front and back. In 3D space, front face called normal. In computer graphics, normal is used to determine surface direction toward the light source. There are two used techniques for 3D modelling, namely polygonal modelling and NURBS modelling.

2.2.1. Polygonal Modelling. Polygonal Modelling is the most common 3D modelling technique. A model defines as polygon called mesh. Polygon has size, shape and position so that the polygon forms the surface [6]. The advantage of polygonal modelling is simple process on computer for displaying and manipulating the object, flexible shape and having a resolution mesh that varies according to application. Polygonal modeling is used to generate low poly objects with low resolution.

2.2.2. NURBS Modelling. NURBS (Non-Uniform Rational B-Splines) is 3D modelling technique using curves. NURBS has become standard in 3D modelling object, especially to create objects with curved shapes that have a high level of complexity. This technique is different from the Polygonal. In NURBS, an object consists of curves. Transformation of curves will affect the shape of the object. NURBS is the most popular method because NURBS curve can be formed only by three point. NURBS has control point (CV) which makes it easier to control the curves. One point of CV can control one area [7].

2.3. Optimization 3D Object
When modeling 3D object, vertices and face count on the object is not efficient. Therefore, modelling requires plenty of vertices in order to get a smooth surface. Vertices reduction is useful for rendering process, make the object become low poly and simplify the UV mapping on the object.

It is not usually used on meshes created by modeling carefully and economically where all vertices and faces are necessary to correctly define the shape. But if the mesh is the result of complex modeling, sculpting or applied subsurface modifiers, the Decimate modifier can be used to reduce the polygon count for a performance increase, or simply remove unnecessary vertices and edges [8]. The Decimate modifier is a quick and easy way of reducing the polygon count of a mesh. However, using decimate modifier will change the face position.

2.4. Materializing and Texturing
With UV mapping, we can tell the software to place textures at an exact location. When we create a UV map, the computer has all the instructions to place the textures where the artist wants on a face.

UV texturing permits polygons that make up a 3D object to be painted with color from an image. The image is called a UV texture map, but it is just an ordinary image [9]. The UV mapping process involves assigning pixels in the image to surface mappings on the polygon [10]. UV is the alternative to XY. It only maps into a texture space rather than into the geometric space of the object. But the rendering computation uses the UV texture coordinates to determine how to paint the three-dimensional surface.

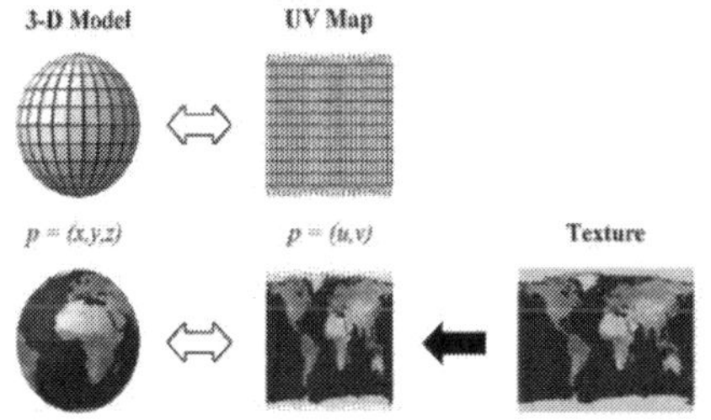

Figure 2. UV mapping dan texturing process

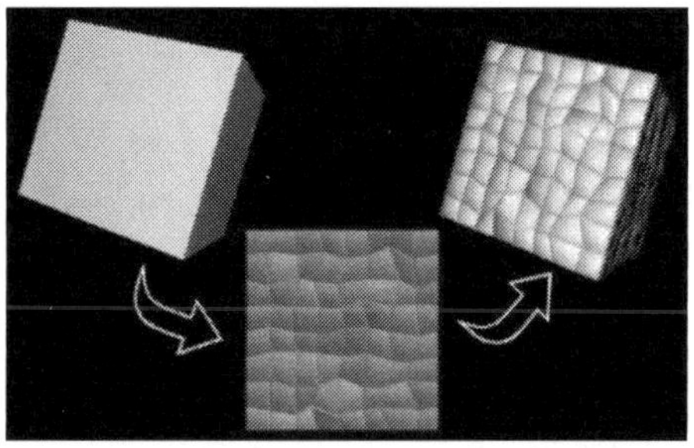

Figure 3. Normal mapping on low poly object

In a few decades, innovations in rendering and texture mapping techniques had been done to generate realistic images such as height mapping, bump mapping, normal mapping and displacement mapping. In this research, the mapping technique used is the normal mapping.

Normal mapping is a technique used for faking the lighting of bumps and dents. It is used to add details without using more polygons. A common use of this technique is to enhance the appearance and details of a low poly model. Low poly object has a few vertexes.

Normal maps are commonly stored as regular RGB images where the RGB components correspond to the X, Y, and Z coordinates, respectively, of the surface normal.

2.5. Virtual Environment

Using terrain engine on the Unity3D, we can make a ground and desired height and add entities to produce a landscape.

Landscape for Istana Darul Arif based on the site plan of the palace complex as shown in figure 4. In this research, the landscape will be made only the environment around the site of the palace.

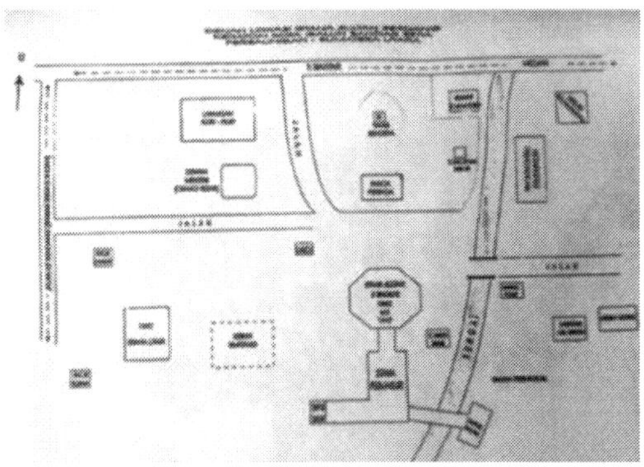

Figure 4. Site plane of Darul Arif Palace

2.6. Lighting

The placement of light sources, the generation of shadows and incorporation of solar illumination contribute greatly to the overall interpretation image. The interplay of light with surface, reflections, the contrast between foreground and background and the way light refract though transparent object. All of these affect to the realism of an image. Most of higher end visualization and modelling packages have an array of light types added to a model as a part of rendering [11].

In this research, we use a directional light to visualize the solar illumination and point light to visualize the light from the bulb. Over exposure can cause an object as if illuminate light.

3. Result and discussion

3.1. Usage of Data
The interviews with descendant of the Serdang Sultanate, obtained the data as shown in table 1.

Table 1. Data about Darul Arif Palace.

Type of Data	Amount	Source
Photos of exterior the palace in the past	20	TBM Tengku Luckman Sinar archives and KITLV archives
Photos of interior the palace in the past	11	TBM Tengku Luckman Sinar archives
Photos of mock up	19	TBM Tengku Luckman Sinar archives
Video about description of the palace	1	TBM Tengku Luckman Sinar archives

3.2. Modelling Result
Figure 5 shows the exterior of Darul Arif palace. Using polygonal modelling to create simple geometry shape such as wall, stairs and pillar. Boolean operation is used to get easier way when modelling hole on the wall. Array operation is used to repeat the same form from the object.

The other technique is NURBS for modelling ornament that consist complex curves. Then, the objects are converted to mesh, so the curves from NURBS modelling become vertices. Also, extrusion technique is used to form 3D object from the mesh.

Figure 5. The Result from 3D modelling

Figure 6. Ornaments on a wooden trellis window

Figure 7. 3D object of Darul Arif Palace

3.3. Material and Texturing Result

3D object texturing is the process adding details of the object, determining the material characteristic from the object. Object material can be obtained from the specific properties from the application, for example, reflectivity, transparency and refraction. Texture can be used to make a variety of colors, patterns, smoothness/roughness of a layer object in more detail.

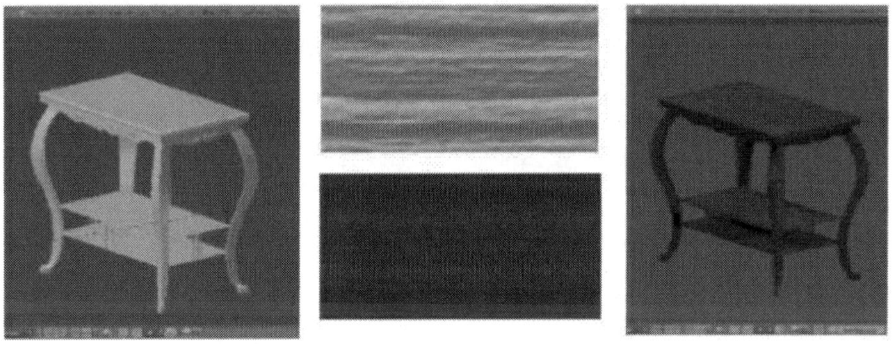

Figure 8. Texturing a table using normal mapping technique

3.4. Create Virtual Environment

Modelling landscape surrounding the palace show in figure 9 and figure 10. There are river on the right side of the palace, the hedges in front of palace and tree around the palace.

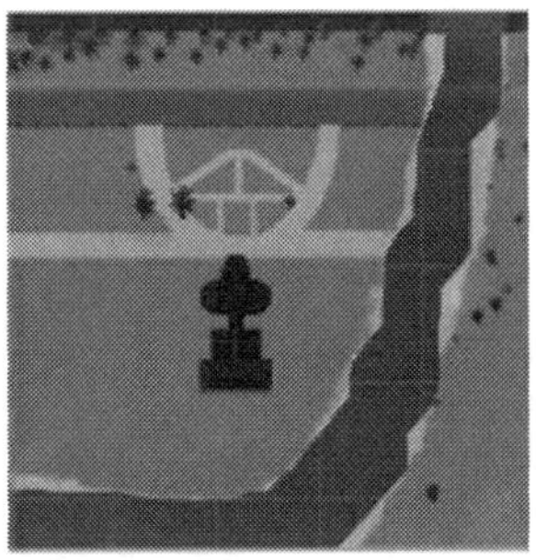

Figure 9. Top view the landscape

Figure 10. Left view of the palace

Collision detection is used to detect the intersection between two or more object when a object is given component of collider, it is like put some invisble net around the object. Collider is used on walls, chairs, tree and the others object.

Green lines as shown in figure 11 are the collider that used on the object. Using collider to restrict available area for the player. On the object, collider is used to make the object more realistic because by using collider, player can walk through the object.

Figure 11. Collider at the environment around the palace and on the palace objects

Figure 12. Balaiurung Room in Darul Arif Palace

3.5. Virtual Application of Darul Arif Palace

Virtual Application of Darul Arif Palace is able to run using HMD from Oculus Rift. Oculus Rift will project two display for each user eyes. Then, user brain will combine it automatically. It will produce three-dimensional sensation, as if the world and the objects in game truly present in front of anyone who wears the device.

The application has several menus, namely Play Immersive and Non-Immersive. As shown in figure 12, it explores the palace using HMD. The difference between Play Immersive and Non-Immersive is the prefab player from the unity3D. Play Immersive have to run using HMD because HMD run the user sight and user rotation.

3.6. World User Interface (UI)

In non-VR projects, UI is frequently overlaid on top of the screen to show things like health, score, etc. as what we often refer to as a HUD (Heads Up Display), known as non-diegetic UI that does not exist within the world, but makes sense for the player in the context of viewing the game. We generally need to position our UI within the environment itself using World Space Canvas render mode. This will allow our eyes to focus on the UI, known as Spatial UI. Placement of the UI within the world also needs some consideration. If it is too close to the user, it can cause eyestrain, and if it is too far, it can feel like focussing on the horizon. In this research, using spatial UI makes additional information UI for the users.

Figure 13. World UI Space used on application

Figure 14. UI Screen Overlay used on application

IOP Conf. Series: Journal of Physics: Conf. Series **801** (2017) 012009 doi:10.1088/1742-6596/801/1/012009

4. Conclusion

Based on the discussion and evaluation of the previous chapters, there are several conclusions, namely:

- The application is capable of displaying 3D objects palace and made it into a virtual reality environment using Unity 3D tool.
- Presentation of virtual reality devices such as head-mounted display can provide a better virtual experience and greatly affect the immersive level in virtual reality applications compared with using only a monitor media device.
- Virtual Reality, allows the user to gain new experiences and new knowledge. Merging the virtual and interaction in the surrounding environment. Immersive from Virtual Reality encourage people to see and learn, useful for research and conservation of cultural heritage.
- The system does not perform as expected, due to required hardware still low quality to run the Oculus Rift and optimizations that do not maximized to achieve better application performance.

5. References

[1] Basarshah-II and Sinar T L 2003 *Kronik Mahkota Kesultanan Serdang* Yandira Agung Medan

[2] Duan X, Gu B and Shi C 2007 Virtual exploration application of cultural heritage for Anyang Yinxu *2nd Workshop on Digital Media and its Application in Museum & Heritages* pp 237-241 IEEE

[3] Bruno F, Bruno S, Sensi G D, Luchi M L, Mancuso S and Muzzupappa M 2010 From 3D reconstruction to virtual reality: A complete methodology for digital archaeological exhibition *J. Cultural Heritage* **11** 42-49

[4] Syaputra M F, Siregar B and Purnamawati S 2014 Desain Aplikasi Virtual Reality Istana Darulssalam, Tanjungppura. *Senarai 2014* pp 249-253

[5] Gutiérrez , Mario A, Vexo, Frédéric and Thalmann D 2008 Stepping into Virtual Reality Springer : London

[6] Kisworo T and Wicaksono A 2011 Simulasi Kecelakaan Kereta Api dengan Menggunakan Teknik Polygonal Modelling Thesis STMIK Amikom Yogyakarta

[7] Flemming B 1999 3D *Modelling and Surfacing* (Morgan Kauffman : Burlington)

[8] Veniero F 2013 http://www.3dtotal.com/tutorial/1729-how-to-engrave-glass-in-blender-part-1-blender-misc-by-filippo-veniero-object-drink-model

[9] Mullen T 2009 *Mastering Blender* (Indianapolis Indiana: Wiley Publishing)

[10] Murdock K L 2008 *3ds Max 2009 Bible* (Indianapolis Indiana: Wiley Publishing)

[11] Ogley C L 2007 The "Truthlikeness" of Virtual Reality Reconstruction of Architecture Heritage: Concepts and Metadata. *Proc. 2nd ISPRS Int. Workshop 3D-ARCH 2007 3D Virtual Reconstruction and Visualization of Complex Architectures* (ISPR)

Augmented Reality for Searching Potential Assets in Medan using GPS based Tracking

M A Muchtar[1], M F Syahputra[1], N Syahputra[1], S Ashrafia[2] and R F Rahmat[1]

[1]*Department of Information Technology, Faculty of Computer Science and Information Technology, University of Sumatera Utara, Medan, Indonesia*

[2]*Faculty of Sciene, University of Amsterdam, Netherland*

Email: anggi.muchtar@usu.ac.id, nca.fadly@usu.ac.id, nugroho.syahputra@students.usu.ac.id, shahnaz.ashrafia@gmail.com, romi.fadillah@usu.ac.id

Abstract. Every city is required to introduce its variety of potential assets so that the people know how to utilize or to develop their area. Potential assets include infrastructure, facilities, people, communities, organizations, customs that affects the characteristics and the way of life in Medan. Due to lack of socialization and information, most of people in Medan only know a few parts of the assets. Recently, so many mobile apps provide search and mapping locations used to find the location of potential assets in user's area. However, the available information, such as text and digital maps, sometimes do not much help the user clearly and dynamically. Therefore, Augmented Reality technology able to display information in real world vision is implemented in this research so that the information can be more interactive and easily understood by user. This technology will be implemented in mobile apps using GPS based tracking and define the coordinate of user's smartphone as a marker so that it can help people dynamically and easily find the location of potential assets in the nearest area based on the direction of user's view on camera.

1. Introduction

In order to the development of region, every city is required to introduce its variety of potential assets. Medan, capital of North Sumatera Province, has variety of potential assets which has not been widely known by the public because of lack of socialization and information. Meanwhile, potential assets which includes infrastructure, facilities, people, communities, organizations and customs, can affects the social, economic and cultural aspects in Medan.

Nowadays, development of technology has proved able to help humans in their daily activities, including to get any required information. The most widely used technology is smartphone supported with many features such as GPS, accelerometer sensor and digital compass, which can allow users to get detailed information simpler and easier. Therefore, we need to create a mobile application which can help people easily find any information of potential assets in Medan.

There are so many mobile apps which provide search and mapping locations which can be used to find the location of potential assets in user's nearest area. However, the available information, such as text and digital maps, sometimes do not much help the user clearly and dynamically.

A new technology called Augmented Reality (AR) is technology that combines what is in reality and what is generated from computer by enhancing what we see, hear, smell and feel [1]. This technology is mostly implemented to give users real-time, clearer and more interactive information.

In previous research, AR had been implemented in education [2-3], health [4-5], military [6], advertising [7-8], entertainment [9], and navigation field, such as Layar, Wikitude and Metaio. The most implementation of augmented reality technology is to find and to get information of nearby places from user. In 2012, Shiddiqi and Hendrianto [10] implemented augmented reality technology in android based mobile app and combined the integrated sensors, such as accelerometer, compass and GPS for locating a mosque. The detected nearest mosque will be labelled with some information of the mosque, such as name, description and the distance between the mosque and user. The information retrieved from the server mosque then stored in internal storage media for mobile phones.

AR technology was also implemented for developing cultural heritage sites [11]. They used video images from outdoor as an input by using object panorama, SURF and GPS to augments 3D model visualization system of cultural heritage sites. This research was aimed to ensure efficient management of cultural heritage sites being left in traces and to have public in Korea watch their cultural heritage sites with realistic sense through 3D model visualization system of cultural heritage sites. The 3D model was augmented in cultural sites by using smartphone camera and detailed information of cultural sites would be displayed without difficulty at any outdoor place as well as indoor. In 2014, indoor augmented reality was also conducted for finding any nearby places, users can find a location when being indoor by using QRCode and RFID as markers of augmented reality [12].

An innovation in AR technology done by Sundaram, et al. in 2015. They integrated Cloud, which can act as virtual storage space, with Augmented Reality technology for finding nearby places of interest. The Cloud was used to store and access the location based augmented reality mobile application. This application aimed to help in the field of tourism and navigation [13].

Another research for augmented reality was conducted in 2011 for treating cockroach phobia [14]. They implemented augmented reality technology in a serious game in mobile phone. Experiment was conducted with a 25-year-old woman with cockroach phobia participated in the research. They concluded that based on testing results, the use of mobile game was able to reduce her level of fear effectively.

In this research, we implement augmented reality technology in mobile application that aims to help user easily find the location of potential assets in user's nearest area, especially in Medan, by giving more real, interactive and detail information using GPS based tracking. Section II describes the proposed method. Experimental results and discussion are presented on Section III. Section VI describes the conclusion and suggestions for future research.

2. Research methodology

In this section, we will discuss about our proposed method for building an augmented reality mobile app based location aimed to give user real-time information about the potential assets in his nearest area by displaying virtual 2D/3D content in real world vision. The name of this app is Medan KulinAR which can find the potential assets especially for Medan. Figure 1 shows the architecture of two-tier system implemented in our proposed method. The two-tier system consist of several components, such as mobile application (frontend), web admin, web service, database server (backend).

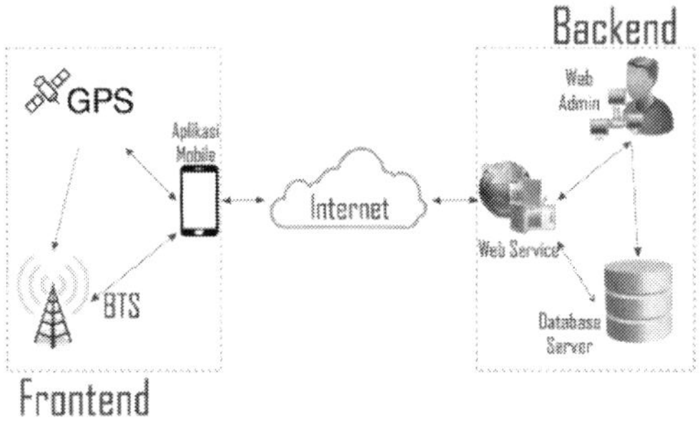

Figure 1. Architecture of Two-Tier System

The installed Medan KulinAR android mobile app is used as an interface between the user and the system in frontend side. The interface is built using Eclipse IDE, Android SDK, Mixare library and Java programming for the augmented reality.

The server side (backend) is the supporting application that serves as data source for the frontend application [15]. This application consists of three main parts, namely web administrator, web service and database server. Web administrator built using PHP programming serves as processor of data resources stored in database server which is managed by admin. Web service built using the combination of REST API and JSON (JavaScript Object Notation) serves as exchange data media between database server and mobile app. MySQL database server stores data resources for the application. The mobile network will connect the frontend application and web service and GPS receiver in mobile device will be connected to GPS satellite that serves to determine user's position, so that the information of potential assets locations will be relevant with user's position. The general architecture of our proposed method for frontend application is shown in figure 2.

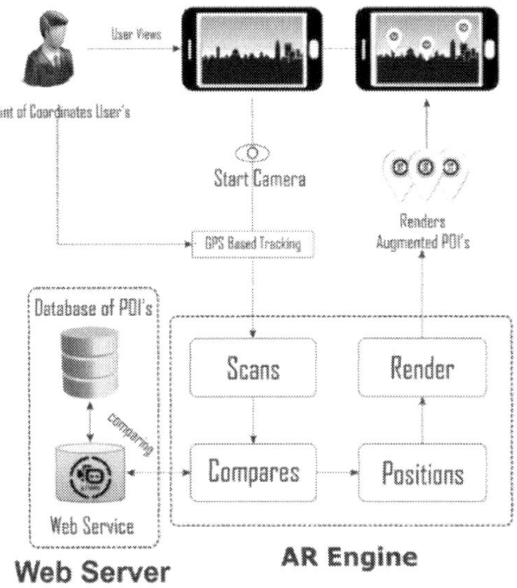

Figure 2. General Architecture (Frontend)

Frontend system begins with the camera capturing the real condition which is directed by the user, and then GPS detects coordinates of the user's location used as a target, and the detected coordinates will be matched with the coordinates of nearest location stored in the database system. The matching process will be repeated until the system finds the coordinates of nearest location. After matching process is completed, the system will proceed positioning and orientation. It aims to determine the location and position of virtual content added to the reality through the camera screen of user's smartphone. While positioning, system calibrates the digital compass with augmented video to match the orientation and position of the camera to the direction of the user's view. Then the last process is rendering to display the virtual content in accordance with the position and orientation set by the system. Virtual content displayed on this application is information of 2D object location which is visible in reality.

3. Result and discussion

The used input data in this application is the coordinate point of location detected by GPS in user's smartphone which is stored in web server and defined as a marker of augmented reality in Medan KulinAR app.

In this research, experiments are conducted based on category of selected marker content and difference in shape of marker that can change according to the distance between user and the location. There are several categories of potential asset able to be selected by user, namely ATM machine, hospital, culinary, supermarket, etc. Figure 3 and figure 4 show the testing results for a few selected categories.

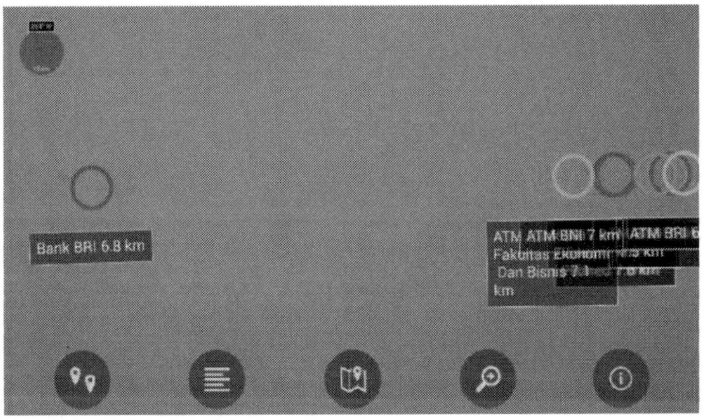

Figure 3. Testing Result for ATM

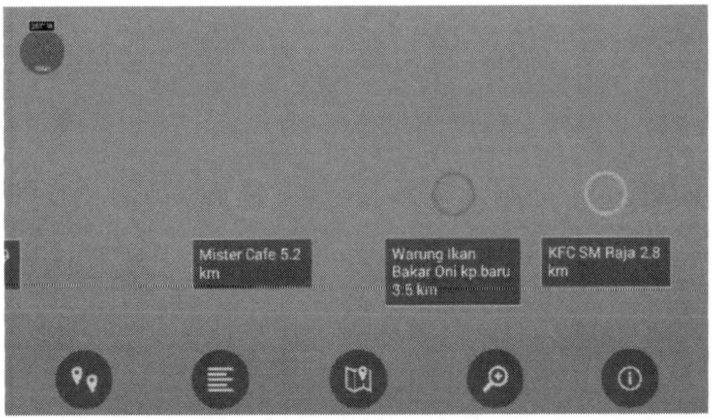

Figure 4. Testing Result for Culinary

IOP Conf. Series: Journal of Physics: Conf. Series **801** (2017) 012010 doi:10.1088/1742-6596/801/1/012010

For the testing based on distance, the shape of marker can change according to the distance between user and the location. The farther the distance, the smaller the circle and vice versa. If the user is right at the intended position, then the marker will be inverted triangle. Figure 5 and figure 6 show the testing results based on the distance.

Figure 5. Testing Result for Distance > 1 km

Figure 6. Testing Result for Distance < 100 m

4. Conclusion and Future Work

Based on the implementation of our proposed method, GPS based tracking, there are several conclusions obtained in this research. First, the implementation of augmented reality technology can help people easily find the potential assets in their nearest area in Medan. This application can give user detailed information of the potential assets, such as its address or direction, dynamically and more real according to its actual position based on user's view on camera.

Future work for augmented reality research can be done by adding other platforms that support the mobile app. Another experiment possibly done by another researcher is adding the information of locations which is not limited in a city without reducing the performance of the app.

5. References

[1] Agrawal M, Kulkarni A, Joshi S and Tiku N 2015 Augmented Reality *Int. J. Advance Research in Computer Science and Management Studies* **3** 114-122

[2] Serio Á D, Ibáñez M B and Kloos C D 2013 Impact of an augmented reality system on students' motivation for a visual art course *Computers & Education* **68** 586-596

[3] Cheng K H and Tsai C C 2013 Affordances of augmented reality in science learning: Suggestions for future research *J. Science Education and Technology* **22** 449-462

[4] Hansen C, Wieferich J, Ritter F, Rieder C and Peitgen H O 2010 Illustrative visualization of 3D planning models for augmented reality in liver surgery. *Int. J. Computer Assisted Radiology and Surgery* **5** 133-141

[5] Nicolau S, Soler L, Mutter D and Marescaux J 2011 Augmented reality in laparoscopic surgical oncology *Surgical oncology* **20** 189-201

[6] Graham M, Zook M and Boulton A 2013 Augmented reality in urban places: contested content and the duplicity of code *Transactions of the Institute of British Geographers* **38** 464-479

[7] Scholz J and Smith A N 2016 Augmented reality: Designing immersive experiences that maximize consumer engagement *Business Horizons* **59** 149-161

[8] Kim Y G and Kim W J 2014 Implementation of augmented reality system for smartphone advertisements *Int. J. Multimedia and Ubiquitous Engineering* **9** 385-392

[9] Lv Z, Halawani A, Feng S, Réhman S U and Li H 2015 Touch-less interactive augmented reality game on vision-based wearable device *Personal and Ubiquitous Computing* **19** 551-567

[10] Shiddiqi A M and Hendrianto D 2012 Augmented reality on android operating system based device; case study: mosque finder *Jurnal Ilmiah Kursor* **6** 197-204

[11] Han J G, Park K W, Ban K J and Kim E K 2013 Cultural heritage sites visualization system based on outdoor augmented reality *Proc. 4th Int. Conf. on Intelligent Systems and Control (AASRI'13)* pp 64-71 (AASRI Procedia)

[12] Leeladevi B, Rahul R and Tolety S 2014 Indoor location identification with better outlook through augmented reality *Proc. the IEEE Int. Conf. on Advanced Communication Control and Computing Technologies (ICACCCT)* pp. 398 - 401. IEEE

[13] Sundaram M V, Vasudevan S K, Ritesh A and Santhosh C 2015 An innovative app with for location finding with augmented reality using cloud *Proc. 2nd Int. Symp. on Big Data and Cloud Computing (ISBCC'15)* pp 585-589 (Procedia Computer Science)

[14] Botella C, Breton-Lopez J, Quero S, Baños R M, Garcia-Palacios A, Zaragoza I and Alcaniz M 2011 Treating cockroach phobia using a serious game on a mobile phone and augmented reality exposure: A single case study *Computers in Human Behavior* **27** 217-227

[15] Savitri W P and Ananta M I 2004 Perbandingan model two-tier dengan three-tier dalam arsitektur client/server untuk mengolah perintah query pada aplikasi database *Seminar Nasional Aplikasi Teknologi Informasi* **1** 1-8.

IOP Conf. Series: Journal of Physics: Conf. Series **801** (2017) 012011 doi:10.1088/1742-6596/801/1/012011

Bioinformatics analysis of the oxidosqualene cyclase gene and the amino acid sequence in mangrove plants

M Basyuni*, **R Wati**

Department of Forestry, Faculty of Forestry, University of Sumatera Utara, Jl. Tri Dharma Ujung No. 1 Medan, North Sumatera 20155, Indonesia

*Email: m.basyuni@usu.ac.id

Abstract. This study described the bioinformatics methods to analyze seven oxidosqualene cyclase (OSC) genes from mangrove plants on DDBJ/EMBL/GenBank as well as predicted the structure, composition, similarity, subcellular localization and phylogenetic. The physical and chemical properties of seven mangrove OSC showed variation among the genes. The percentage of the secondary structure of seven mangrove OSC genes followed the order of α helix > random coil > extended chain structure. The values of chloroplast or signal peptide were too low, indicated that no chloroplast transit peptide or signal peptide of secretion pathway in mangrove OSC genes. The target peptide value of mitochondria varied from 0.163 to 0.430, indicated it was possible to exist. These results suggested the importance of understanding the diversity and functional of properties of the different amino acids in mangrove OSC genes. To clarify the relationship among the mangrove OSC gene, a phylogenetic tree was constructed. The phylogenetic tree shows that there are three clusters, *Kandelia KcMS* join with *Bruguiera BgLUS*, *Rhizophora RsM1* was close to *Bruguiera BgbAS*, and *Rhizophora RcCAS* join with *Kandelia KcCAS*. The present study, therefore, supported the previous results that plant OSC genes form distinct clusters in the tree.

1. Introduction

Mangrove plants are halophytes that are distributed in the intertidal zone of tropical and sub-tropical regions where they exist in conditions of high salinity, extreme tides, strong winds, high temperature and muddy, anaerobic soils [1]. Mangrove plants have long been well known a source of phytochemical compounds or biologically active compounds [2]. Mangroves are also prosperous sources of pentacyclic triterpenoids and phytosterols [3].

The biosynthesis of polycyclic phytosterol and pentacyclic terpenoids shares the common pathway up to the branching point of 2,3-oxidesqualene. A variety of triterpenoids and phytosterols widely distributed in plants are biosynthesized from a common precursor 2,3-oxidosqualene by the enzyme oxidosqualene cyclases (OSCs). 2,3-Oxidosqualene, therefore, locates at the branching point of isoprenoid pathway toward phytosterols or terpenes biosynthesis [4]. The cyclization of oxidosqualene into triterpenoids and phytosterols are one of a fascinating reactions found in nature and their biosynthetic pathway has been shown to be complicated and divergent. The enzymes, which perform these reactions, belong to OSCs family. In higher plants, OSC family member cycloartenol synthase and lanosterol synthase are responsible for sterol biosynthesis, and other OSCs are involved for triterpenoid synthesis.

A number of studies have been suggested that OSCs are regulatory steps in the isoprenoids pathway orienting the biosynthetic flux towards either pentacyclic triterpenes or phytosterols, which are the precursors of large variety physiologically active compounds [5-6]. Several successful cloning of triterpenoid synthases from mangrove plants has proved the presence of multiple OSCs as distinct proteins even in one plant species. Taken together, this hypothesis suggested that the cyclization reactions were carried out by different enzymes [7-9].

In view of the diversity of the triterpenoid product formed in mangrove plants [3], it became interesting to determine how each triterpene synthase controls the product specificity. In this context, the importance of OSCs is as the first committed enzyme in triterpenoid biosynthesis. However, the bioinformatics analysis of OSC genes are rarely reported, especially from mangrove plants. Here we used the bioinformatics methods to analyze seven mangrove OSC genes as well as we predicted the structure, composition, similarity, subcellular localization and phylogenetic to provide references for the OSC genes and to further understand the diversity and functional of properties of amino acids in mangrove OSC genes.

2. Materials and method

2.1. *Materials*

A total of seven mangrove oxidosqualene cyclase genes deposited in the DDBJ/EMBL/GenBank were collected. The DDBJ/GenBank/EMBL accession numbers of the DNA sequence and amino acid sequence of used this analysis are as follows: AB289585, BAF80443 (*BgbAS*), AB289586, BAF80444 (*BgLUS*), AB292609, BAF73930 (*KcCAS*), AB257507, BAF35580 (*KcMS*), AB292608, BAF73929 (*RsCAS*), AB263203, BAF80441 (*RsM1*), AB263204, BAF80442 (*RsM2*).

2.2. *Physicochemical properties and secondary structure of the OSC gene*

The composition, physical and chemical properties of DNA and amino acid sequences of OSC were analyzed using Protparam online analysis (web.expasy.org/protparam/). The computed parameters include the molecular weight, theoretical isoelectric point values, amino acid composition, atomic composition, extinction coefficient, estimated half-life, instability index, fat coefficient, and grand average hydrophilicity. HNN Secondary structure prediction method online (https://npsa-prabi.ibcp.fr) was used to analyze the secondary structure of OSC genes.

2.3. *Potential transit of peptide and subcellular localization of OSC gene*

The targetP 1.1 Server online (www.cbs.dtu.dk/services) was used for transit peptide prediction. The location assignment is based on the predicted presence of any of the *N*-terminal pre-sequences chloroplast transit peptide (cTP), mitochondrial targeting peptide (mTP) and secretory pathway signal peptide (SP). PSORT Prediction online (psort.hgc.jp/form.html) was used to analyze the subcellular localization of OSC genes.

2.4. *Similarity and phylogenetic analysis of OSC gene*

The amino acid sequences were aligned and similarity scores were obtained using the FASTA ver. 3.4t26 [10] of the DNA Data Bank of Japan (Mishima, Shizuoka, Japan). The best score of results is shown in Table 5. Phylogenetic analysis of deduced amino acid alignment from mangrove actin genes was conducted with CLUSTAL W ver. 1.83 [11] of the DNA Data Bank of Japan (Mishima, Shizuoka, Japan) followed by drawing with TreeView, ver. 1.6.6 [12] based on a neighbor-joining method. Bootstrap analysis with 1000 replications was used to assess the strength of the nodes in the tree [13]. The DDBJ/GenBank/EMBL accession numbers of the DNA and amino acid sequence of used this analysis is described in the Materials subsection.

3. Results and Discussions

The results will be discussed in three subsections; they are physicochemical properties and secondary structure of the OSC gene, the potential transit of peptide and subcellular localization of OSC gene, and similarity and phylogenetic analysis of OSC gene.

3.1. Physical and chemical properties and secondary structure of the OSC gene

Table 1 shows the initiation codon of OSC genes was ATG, where the stop codons were TGA and TAG. The open reading frame length was varying from 2277 bp to 2316 bp. A number of encoded amino acids were 758 to 771 with no negatively or positively charged residues. *RsCAS* had the smallest relative molecular mass, *RsM2* had the largest one. Mangrove OSC genes had similar theoretical isoelectric point value (0.90 to 0.92). *BgbAS* and *RsCAS* had a minimum total number of atoms, *RsM2* had a maximum total number of atoms. The half-life period was accurately same among OSC gene. Based on stability coefficients, *BgbAS* and *RsCAS* were stable proteins, the remaining were non-stable proteins. These results suggested the importance of understanding the diversity and functional of properties of the different amino acids in mangrove OSC genes [4].

Tabel 1. Physical and chemical properties of the OSC genes in mangrove

Nucleotide accession number	***BgBAS***	***BgLUS***	***KcCAS***	***KcMS***	***RsCAS***	***RsM1***	***RsM2***
Length of genes/bp	2280	2286	2277	2286	2277	2280	2316
Open reading frame length/bp	2280	2286	2277	2286	2277	2280	2316
Start site and codon	1ATG	1ATG	1ATG	1ATG	1ATG	1ATG	1ATG
Stop site and codon	2280TGA	2286TGA	2277TGA	2286TGA	2277TGA	2280TGA	2316TAG
Number of encoded amino acids	759	761	758	761	758	759	771
Relative molecular mass	187857	187868	187950	188205	187836	188203	190134
Theoretical isoelectric point values	4.91	4.92	4.91	4.92	4.91	4.91	4.90
Positively charged residues	0	0	0	0	0	0	0
Negatively charged residues	0	0	0	0	0	0	0
Total number of atoms	24010	24073	24010	24127	24010	24050	24234
Extinction coefficient	0.160	0.155	0.161	0.155	0.160	0.161	0.167
Half life period	4.4h	4.4h	4.4h	4.4h	4.4h	4.4h	4.4h
Instability coefficient	39.58	44.04	42.00	41.25	39.34	41.68	44.45
Fat coefficient	26.58	28.04	26.17	28.26	26.26	26.67	27.63
Overall average hydrophilicity	0.716	0.729	0.712	0.735	0.707	0.724	0.769

Table 2. Secondary structure analysis of mangrove OSC gene

Protein accession number	**α helix number**	**α helix ratio (%)**	**Extended chain structure number**	**Extended chain structure ratio (%)**	**Random coil number**	**Random coil ratio (%)**
BAF80443	341	44.93	90	11.86	328	43.21
BAF80444	337	44.28	100	13.14	324	42.58
BAF73930	335	44.2	100	13.19	323	42.61
BAF35580	354	46.52	88	11.56	319	41.92
BAF73929	352	46.44	97	12.8	309	40.77
BAF80441	362	47.69	87	11.46	310	40.84
BAF80442	323	41.89	114	14.79	334	43.32

The secondary structure of OSC genes consist of α helix, extended chain structure, and random coil with their relative proportion was depicted in Table 2. Table 2 shows that the percentage of the secondary structure of seven mangrove OSC genes followed the order of α helix > random coil > extended chain structure. It has been reported that there are five signature QW motifs located in OSC genes [8-9], these QW motifs connect surface α helix and are though to stabilize OSC by tightening the protein structure and avoiding structural damage during the cyclization [14]. This study, therefore, suggested the prediction of secondary structure was performed for basic structural information.

3.2. Potential transit of peptide and subcellular localization of OSC gene

Table 3 shows the possibility of the potential transit peptide in mangrove OSC genes. There are three possibilities: chloroplast transit peptide, mitochondrial target peptide and signal peptide of secretory pathway along with the prediction probability. The values of chloroplast or signal peptide were too low, indicated that no chloroplast transit peptide or signal peptide of secretion pathway in seven OSC genes. It is interesting to note that target peptide value of mitochondria varied from 0.163 to 0.430, indicating that it was possible to exist. The highest mitochondrial target peptide was *KcCAS, K. candel* cycloartenol synthase with the product of 100% cycloartenol [9] and the lowest mitochondrial peptide was *RsM1, R. stylosa* multifunctional triterpene synthase with the product of germanicol, β-amyrin, and lupeol in the proportions 63:33:4[8].

Tabel 3. Possibility of the potential transit peptide in mangrove OSC gene

Protein accession number	Reliability			
	Chloroplast transit peptide	**Mitochondrial target peptide**	**Signal peptide of secretory pathway**	**Reliability prediction**
BAF80443	0.042	0.163	0.111	2
BAF80444	0.052	0.285	0.049	3
BAF73930	0.040	0.430	0.028	4
BAF35580	0.070	0.159	0.091	2
BAF73929	0.049	0.429	0.028	4
BAF80441	0.059	0.147	0.091	2
BAF80442	0.045	0.426	0.070	4

Tabel 4. Subcellular localization of mangrove OSC gene

Protein accession number	Degree of certainty							
	Ctm	**Gb**	**Pm**	**Er**	**Mb**	**N**	**Mps**	**Mim**
BAF80443	nd	0.400	0.600	0.300	0.300	nd	nd	nd
BAF80444	0.100	nd	nd	nd	0.748	0.300	0.100	nd
BAF73930	0.179	nd	nd	0.600	0.545	nd	nd	0.100
BAF35580	0.100	nd	nd	nd	0.748	0.300	0.100	nd
BAF73929	0.179	nd	nd	0.600	0.685	nd	nd	0.100
BAF80441	nd	0.400	0.600	0.300	0.300	nd	nd	nd
BAF80442	0.100	nd	nd	0.100	0.748	nd	0.100	nd

nd= not detected, Ctm= Chloroplast thylakoid membrane, Gb= Golgi bodies, Pm= Plasma membrane, Er= Endoplasmic reticulum, Mb=Microbody, N= Nucleus, Mps= Mitochondrial matrix space Mim= Mitochondrial inner membrane

Table 4 shows subcellular localization of OSC genes in mangroves. The subcellular localization of these genes was mostly in microbody, chloroplast thylakoid, and endoplasmic reticulum. Recently, it has been reported that the expression of two triterpenoid synthases, *BgbAS* and *RsM1* under *GAL1* promoter in GIL77 increased the triterpenoid content of both whole cell body and plasma membrane fractions [15]. In the present data (Table 4), both gene *BgbAS* and *RsM1* were located at the plasma membrane, supported previous results on their subcellular localization located in the plasma membrane [15].

3.3. Similarity and phylogenetic analysis of OSC gene

The deduced amino acid sequence of OSC genes from mangrove exhibited 53-93% between themselves (Table 5). The deduced amino acid sequences of *BgbAS* showed significant similarity (93 %) to *RsM1*. *KcMS* is close to *BgLUS* and showed high similarity (90%), on the other hand, *KcCAS* depicted 93% sequence homology to *RsCAS* (Table 5). Both genes are cycloartenol syntheses. *KcMS* produced 50% of lupeol along with β-amyrin (25%) and α-amyrin (25%) [7], therefore shared high sequence homology with *BgLUS* [8], which produced 100% lupeol.

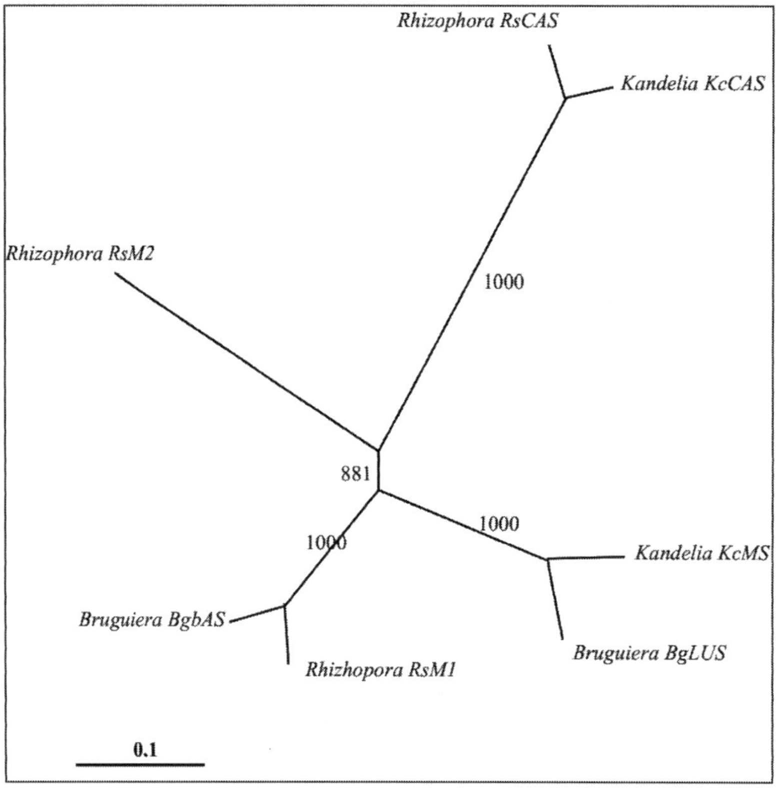

Figure 1. Phylogenetic tree of mangrove OSC genes. Phylogenetic tree of deduced amino acid sequences was constructed with the neighbor-joining method of the CLUSTAL W [11]. The indicated scale represents 0.1 amino acid substitutions per site. Numbers indicate bootstrap value from 1000 replicates. The DDBJ/GenBank/EMBL accession numbers of the amino acid sequence of used this analysis are indicated in the Materials subsection.

To clarify the relationship among the mangrove OSC gene, a phylogenetic tree was constructed on the basis of their amino acid sequences (Figure 1). The phylogenetic tree shows that there are three clusters, *Kandelia KcMS* join with *Bruguiera BgLUS*, *Rhizophora RsM1* is close to *Bruguiera BgbAS*, and *Rhizophora RcCAS* join with *Kandelia KcCAS*. Of seven OSC genes, *RsM2* form a cluster that is close to cycloartenol synthase. This multifunctional triterpenoid synthase are proposed to be at an evolutionally transient states between one product specific OSC to anther OSC, on the other hand, they are distributed between the monofunctional β-amyrin synthase (bAS), lupeol synthase (LUS), and cycloartenol synthase (CAS) [8, 16]. The enzymatic reaction products of *RsM1* and *RsM2* differed from those of their neighboring genes in the tree [8]. It has been reported that plant OSCs are evolutionarily descended from CAS [8, 16], the present study, therefore, supported the previous results that plant CAS, bAS, and LUS genes form distinct clusters in the tree [7-9, 16].

Tabel 5. The similarity of amino acid sequence between OSC genes in mangrove

Genes	1	2	3	4	5	6	7
1.BgbAS	100						
2.BgLUS	71	100					
3.KcCAS	58	55	100				
4.KcMS	71	90	55	100			
5.RsCAS	58	55	93	55	100		
6.RsM1	93	71	58	71	58	100	
7.RsM2	66	62	53	63	53	66	100

4. Conclusions
The present study demonstrated the importance of understanding the diversity and functional of properties of the different amino acids in mangrove OSC genes. Furthermore, our data supported previous results on the subcellular localization of *BgbAS* and *RsM1* genes located in the plasma membrane and in which the plant OSC genes form distinct clusters in the tree.

Acknowledgment
This study was partly supported by an International Research Collaboration and Scientific Publication Grant (No. 117/SP2H/LT/DRPM/II/2016 to MB) from the Directorate for Research and Community Service, Ministry of Research, Technology and Higher Education, Republic of Indonesia.

References
[1] Tomlinson P B 1986 the Botany of Mangroves Cambridge University Press London.
[2] Bandaranayake W M 2002 Bioactivities, bioactive compounds and chemical constituents of mangrove plants *Wetl. Ecol. Manag.* **10**:421–452.
[3] Basyuni M, Oku H, Baba S, Takara K, Iwasaki H 2007 Isoprenoids of Okinawan mangroves as lipid input into estuarine ecosystem *J Oceanogr.* **63**:601–608.
[4] Abe I, Rohmer M, Prestwich GD 1993 Enzymatic cyclization of squalene and oxidosqualene to sterols and triterpens *Chem. Rev.* **93**: 2189–2208.
[5] Henry M, Rahier A, Taton M 1992 Effect of gypsogenin 3,0-glucuronide pretreatment of Gypsophila paniculata and Saponaria officinalis cell suspension cultures on the activities of microsomal 2,3-oxidosqualene cycloartenol and amyrin cyclases *Phytochemistry* **31**: 3855–3859.
[6] Taton M, Ceruti M, Cattel L, Rahier A 1996 Inhibition of higher plant 2,3-oxidosqualene cyclases by nitrogen-containing oxidosqualene analogues *Phytochemistry* **43**:75–81.
[7] Basyuni M, Oku H, Inafuku M, Baba S, Iwasaki H, Oshiro K, Okabe T, Shibuya M, Ebizuka Y 2006 Molecular cloning and functional expression of a multifunctional triterpene synthase cDNA from a mangrove species *Kandelia candel* (L.) Druce *Phytochemistry* **67**: 2517–2524.

[8] Basyuni M, Oku H, Tsujimoto E, Kinjo K, Baba S, Takara K 2007 Triterpene synthases from the Okinawan mangrove tribe, Rhizophoraceae *FEBS J.* **274**:5028–5042.

[9] Basyuni M, Oku H, Tsujimoto E, Baba S 2007. Cloning and functional expression of cycloartenol synthases from mangrove species *Rhizophora stylosa* Griff. and *Kandelia candel* (L.) Druce *Biosci. Biotechnol. Biochem.* **71**: 1788–1792.

[10] Pearson W R, Lipman D J 1988 Improved tools for biological sequence comparison *Proc. Natl. Acad. Sci. USA* 85: 2444–2448.

[11] Thompson J D, Higgins D G, Gibson TJ 1994 CLUSTAL W: Improving the sensitivity of progressive multiple sequence alignment through sequence weighting, position-specific gap penalties, and weight matrix choice *Nucleic Acid Res.* 22: 4673–4680.

[12] Page R D 1996 TreeView: an application to display phylogenetic trees on personal computers *Comput. Appl. Biosci.* **12**: 357–358.

[13] Felsenstein J 1985 Confidence limits on phylogenies: an approach using the bootstrap *Evolution* **39**: 783–791.

[14] Wendt K U, Poralla K, Schultz GE Structure and function of a squalene cyclase *Nature* **277**: 1811–1815.

[15] Inafuku M, Basyuni M, Oku H 2016 Triterpenoid modulates the salt tolerance of lanosterol synthase deficient *Saccharomyces cerevisiae,* GIL77 *Saudi J. Biol. Sci.* http://dx.doi.org/10.1016/j.sjbj.2016.10.009.

[16] Shibuya M, Zhang H, Endo A, Shishikura K, Kushiro T, Ebizuka Y 1999. Two branches of the lupeol synthase gene in the molecular evolution of plant oxidosqualene cyclases. *Eur. J. Biochem.* **266**: 302–307.

Bioinformatics approach of salt tolerance gene in mangrove plant *Rhizophora stylosa*

M Basyuni[1*], Sumardi[2]

[1]Department of Forestry, Faculty of Forestry, University of Sumatera Utara, Jl. Tri Dharma Ujung No. 1 Medan, North Sumatera 20155, Indonesia
[2]Faculty of Pharmacy, Tjut Nyak Dhien University, Jl. Rasmi No. 28 Medan 20123, Indonesia

*Email: m.basyuni@usu.ac.id

Abstract. This study describes bioinformatics approach on the analyze of the salt tolerance genes in mangrove plant, *Rhizophora stylosa* on DDBJ/EMBL/GenBank as well as similarity, phylogenetic, potential peptide, and subcellular localization. The DNA sequence between salt tolerance gene from R. stylosa exhibited 42-77% between themselves The target peptide value of mitochondria varied from 0.163 to 0.430, indicated it was possible to exist. These results suggested the importance of understanding the diversity and functional of properties of the different amino acids in mangrove OSC genes. To clarify the relationship among the salt-tolerant genes in *R. stylosa*, a phylogenetic tree was constructed. The phylogenetic tree shows that there are three clusters, first branch of *Cu/Zn SOD* and *reverse transcriptase* genes, the second branch consists of the majority genes and the last group was MAP3K alpha protein kinase only. The present study, therefore, suggested that salt tolerance genes form distinct clusters in the tree.

1. Introduction

Mangroves are woody plants that grow at the interface between land and sea in the tropical and subtropical area. The growth of the plants is impeded by the salt stress, due to water deficit, ion imbalance, and ion toxicity. Although the stress-tolerance mechanisms of the plants appeared to be complex and divergent, the mechanisms of plant cells to cope with the salt stress and the molecular strategies are well documented. Mangrove plants thus overcomes the salt stress by combination of several type mechanisms: (1) adjustment of the osmotic pressure by accumulation of small molecules of osmolytes such as glycine-betaine or sugar alcohols [1]; (2) salt extrusion across the plasma membrane using ion transporters [2]; (3) accumulation of salt in vacuole using tonoplast transporter [3]; (4) induction of stress-related proteins [4]. (5) salt stress induced changes in triterpenoid content and triterpenoid synthase gene [5-8].

Mangroves are salt tolerant and have mechanisms to take up water against strong osmotic potentials [9]. It has been reported that *Rhizophora stylosa* is distributed in the more coastal area compared to the habitat of another member of Rhizophoraceae [10]. This study suggested that *R. stylosa* is more tolerant to salinity than the other species. Mangroves are therefore an ideal model for studying the molecular mechanisms of salinity tolerance at the cellular level and study the salt tolerance mechanism of individual genera or species [10].

Recently it has been reported that triterpenoid synthase modulates the salt tolerance of lanosterol synthase deficient *Saccharomyces cerevisiae,* GIL77 [11]. This study suggested that salt tolerance genes (*BgbAS* and *RsM1*), strongly inhibited the growth of GIL77 cells under stressed salt conditions. In this context, it is interesting to analyze the salt tolerance genes from mangrove using bioinformatics method. However, the bioinformatics analysis of salt tolerance genes are rarely reported, especially from mangrove plants. Here we used the bioinformatics methods to analyze 15 salt tolerance genes from *R. stylosa* [10] on DDBJ/EMBL/GenBank as well as similarity, phylogenetic, potential transit peptide, and subcellular localization and to provide references for the genes and to further understand the diversity and functional of salt tolerance genes.

2. Materials and method

2.1. Sample collection

A total of 15 salt tolerance gene of *Rhizophora stylosa* deposited officially in the DDBJ/EMBL-/GenBank were colleted. The DDBJ/EMBL/GenBank accession numbers of the DNA sequence used in this analysis are as follows: FS997281 (Rs125/Serin rich protein), FS997186 (Rs30/Protein di-sulfide isomerase), FS997171 (Rs15/β-amyrin synthase), FS997172 (RS16/Multifunctional triterpene synthase), FS997173 (Rs17/Lupeol synthase), FS997267 (Rs111/Isoprene synthase), FS997310 (Rs154/Myrcene synthase), FS997159 (Rs3/Cu/Zn SOD), FS7202 (Rs46/Peroxidase), FS997262 (Ear-ly responsive to dehydration 4), FS997210 (Rs53/Mitochondrial F1 ATP synthase), FS997165 (Rs 7165/MAP3K α protein kinase), FS997177 (Rs21/MADS domain transcription factor), FS997255 (Rs99 AP2/EREBP transcription factor), and FS997175 (Rs19/Reverse transcriptase).

2.2. Similarity and phylogenetic analysis of salt tolerance genes

The DNA sequences were aligned, and similarity scores were obtained using the FASTA ver. 3.4t26 [12] of the DNA Data Bank of Japan (Mishima, Shizuoka, Japan). The best score of results is shown in Table 1. Phylogenetic analysis of deduced amino acid alignment from mangrove actin genes was conducted with CLUSTAL W ver. 1.83 [13] of the DNA Data Bank of Japan (Mishima, Shizuoka, Japan) followed by drawing with TreeView, ver. 1.6.6 [14] based on a neighbor-joining method. Bootstrap analysis with 1000 replications was used to assess the strength of the nodes in the tree [15]. The DDBJ/GenBank/EMBL accession numbers of the DNA and amino acid sequence of used this analysis is described in the sample subsection.

2.3. Possibility of the potential peptide and subcellular localization in salt tolerance gene

The targetP 1.1 Server online (www.cbs.dtu.dk/services) was used for transit peptide prediction. The location assignment is based on the predicted presence of any of the N-terminal pre-sequences chloroplast transit peptide (cTP), mitochondrial targeting peptide (mTP) and secretory pathway signal peptide (SP). PSORT Prediction online (psort.hgc.jp/form.html) was used to analyze the subcellular localization of salt tolerance gene from *R. stylosa*.

3. Results and Discussions

The results will be discussed in two subsections; they are similarity and phylogenetic analysis of salt tolerance genes and possibility of the potential peptide and subcellular localization in salt tolerance gene

3.1. Similarity and phylogenetic analysis of salt tolerance genes

The DNA sequence between salt tolerance gene from *R. stylosa* exhibited 42-77% between themselves (Table 1). The DNA sequences of clone Rs154 (myrcene synthase) showed significant similarity (77 %) to clone Rs54 (mitochondrial F1 ATP synthase). Clone Rs16 (multifunctional triterpene synthase) is close to clone Rs17 (lupeol synthase) and showed relatively high similarity (75%), on the other hand, clone Rs16 Among the salt tolerant genes analyzed, clone RS15 (multifunctional triterpene synthase) and clone Rs17 (lupeol synthase) showed the lowest similarity (42%) DNA sequence homology to Rs46 (peroxidase) and Rs3 (Cu/Zn SOD), respectively (Table 1).

To clarify the relationship among the salt-tolerant genes in *R. stylosa*, a phylogenetic tree was constructed. The phylogenetic tree shows that there are three clusters in the tree, the first branch of Cu/Zn SOD and reverse transcriptase genes, the second branch consists of the majority genes and last group contained MAP3K alpha protein kinase only. The enzymatic reaction products of salt tolerance genes differed from those of their neighbor clones in the tree. This result suggests that the relationships in the phylogenetic tree have limited significance in predicting the product profile of the genes.

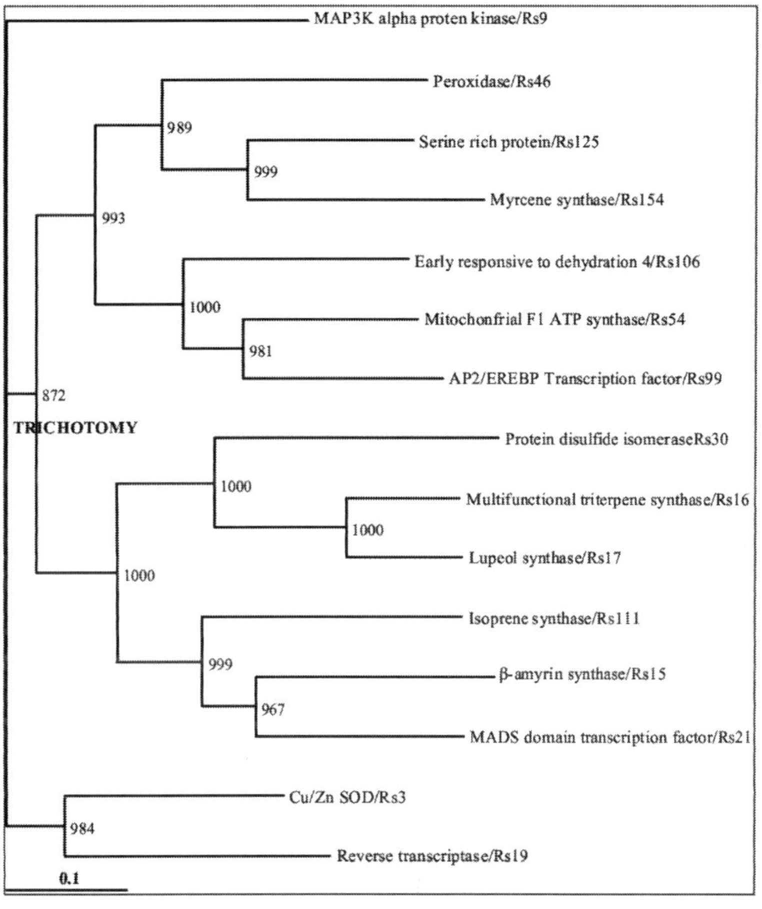

Figure 1. Phylogenetic tree of salt tolerance genes in *R. stylosa*. Phylogenetic DNA sequences were constructed with the neighbor-joining method of the CLUSTAL W [13]. The indicated scale represents 0.1 amino acid substitutions per site. Numbers indicate bootstrap value from 1000 replicates. The DDBJ/GenBank/EMBL accession numbers of the DNA sequence of used this analysis are described in the Materials subsection.

Tabel 1. DNA sequence similarity between salt tolerance gene in *R. stylosa*

Clone ID	1	2	3	4	5	6	7	8	9	10	11	12	13	14	15
1. Rs125	100														
2. Rs30	55	100													
3. Rs15	57	54	100												
4. Rs 16	45	56	54	100											
5. Rs17	43	57	59	75	100										
6. Rs111	54	59	64	56	58	100									
7. Rs154	72	68	61	65	70	74	100								
8. Rs3	48	54	58	52	42	57	70	100							
9. Rs46	51	48	59	42	50	54	67	47	100						
10. Rs106	51	48	54	50	51	59	66	50	51	100					
11. Rs54	53	53	64	56	57	47	77	48	53	57	100				
12. Rs9	52	53	56	45	50	51	66	51	47	49	44	100			
13. Rs21	45	51	62	53	51	58	62	45	45	51	57	58	100		
14. Rs99	60	52	49	68	56	72	48	60	60	64	57	57	58	100	
15. Rs19	56	53	49	65	56	55	53	60	54	56	58	57	60	51	100

3.2. The potential peptide and subcellular localization in salt tolerance gene

Table 2. Possibility of the potential transit peptide in salt tolerance gene

Clone ID	Reliability			
	Chlroplast transit peptide	**Mitochondrial target peptide**	**Signal peptde of secretory pathway**	**Reliability prediction**
Rs125	0.169	0.129	0.031	4
Rs30	0.252	0.145	0.027	4
Rs15	0.113	0.214	0.055	3
Rs 16	0.138	0.129	0.106	4
Rs17	0.151	0.051	0.136	4
Rs111	0.146	0.168	0.034	3
Rs154	0.322	0.078	0.025	4
Rs3	0.298	0.150	0.040	4
Rs46	0.281	0.054	0.134	5
Rs106	0.675	0.169	0.045	3
Rs54	0.257	0.073	0.306	5
Rs9	0.169	0.015	0.443	4
Rs21	0.167	0.156	0.047	3
Rs99	0.679	0.135	0.008	3
Rs19	0.178	0.131	0.194	5

Table 2 shows the possibility of the potential transit peptide in salt tolerance genes. There are three options: chloroplast transit peptide, mitochondrial target peptide and signal peptide of secretory pathway along with the prediction probability. The target chloroplast varied from 0.146 to 0.679, with the highest values of chloroplast belongs to clone Rs99 (0.679), indicated that chloroplast transit peptide present in salt tolerance gene. It is interesting to note that target peptide value of mitochondria was less compared with chloroplast transit or mitochondrial peptide. The highest signal peptide of the secretory pathway was Rs54 (mitochondrial F1 ATP synthase).

Tables 3 shows subcellular localization of salt tolerance genes in *R. stylosa*. The subcellular localization of these genes was mostly in the endoplasmic reticulum (membrane, lumen), plasma membrane, outside and Golgi body. Recently, it has been reported that the expression of two triterpenoid synthases, *BgbAS* and *RsM1* under GAL1 promoter in GIL77 increased the triterpenoid content of both whole cell body and plasma membrane fractions [11]. Table 3 shows that Rs15 and Rs16, with the highest degree, Rs21 were located at the plasma membrane, supported previous results on their subcellular localization of triterpenoid genes located in the plasma membrane [11]. There have been several views that the plasma membrane is the first line of defense against changes in physicochemical variables of the thriving environment [16-17]. It has also been suggested that the plasma membrane might be a primary site of salt injury [18]. Therefore, the integrity and functionality of the membrane is a critical factor for salt tolerance [19].

Table 3. Subcellular localization of salt tolerance gene in *R. stylosa*

Clone ID	Degree of certainty				
	Endoplasmic reticulum (membrane)	**Plasma membrane**	**Endoplasmic reticulum (lumen)**	**Outside**	**Golgi body**
Rs125	0.550	nd	0.100	0.100	0.100
Rs30	0.100	nd	0.100	0.370	0.100
Rs15	0.932	0.496	0.100	nd	0.190
Rs 16	0.550	0.190	0.100	0.100	nd
Rs17	0.100	na	0.100	0.380	0.100
Rs111	0.100	0.100	0.100	0.370	0.100
Rs154	0.100	0.658	0.100	0.490	nd
Rs3	0.550	nd	0.100	0.100	0.100
Rs46	nd	nd	0.100	nd	0.100
Rs106	0.550	0.190	0.100	0.100	nd
Rs54	0.550	0.685	0.100	0.100	nd
Rs9	0.100	nd	0.100	0.499	0.100
Rs21	0.550	0.717	0.100	0.100	nd
Rs99	0.550	0.190	0.100	0.100	nd
Rs19	0.550	0.190	0.100	0.100	nd

nd= not detected

4. Conclusions

The present study demonstrated the importance of understanding the diversity and functional of properties of the different DNA sequences in salt tolerance genes in *R. stylosa*. Two salt tolerance of triterpenoid genes were located in the plasma membrane and in which the plant salt tolerance genes form distinct clusters in the tree.

Acknowledgment
This study was partly supported by an International Research Collaboration and Scientific Publication Grant (No. 117/SP2H/LT/DRPM/II/2016 to MB) from the Directorate for Research and Community Service, Ministry of Research, Technology and Higher Education, Republic of Indonesia.

References

[1] Sakamoto A, Murata N 2000 Genetic engineering of glycine betaine synthesis in plants: current status and implication for enhancement of stress tolerance *J. Exp. Bot.* **51**: 81–88.

[2] Kura-Hotta M, Mimura M, Tsujimura M, Washitani-Nemoto S, Mimura T 2001 High salt-treatment-induced Na+ extrusion and low salt-treatment-induced Na+ accumulation in suspension-cultured cells of the mangrove plant, *Bruguiera sexangula Plant Cell Environ.* **24**: 1105–1112.

[3] Mimura T, Kura-Hotta M, Tsujimura T, Ohnishi M, Miura M, Okazaki Y, Mimura M, Maeshima M, Washitani-Nemoto S 2003 Rapid increase of vascular volume in response to salt stress *Planta* **216**: 397–402.

[4] Ashihara H, Wakahara S, Suzuki M, Kato A, Sasamoto H, Baba S 2003 Comparison of adenosine metabolism in leaves of several mangrove plants and a poplar species *Plant Physiol. Biochem.* **41**: 133–139.

[5] Oku H, Baba S, Koga H, Takara K, Iwasaki H 2003 Lipid composition of mangroves and its relevance to salt tolerance *J. Plant Res.* **116**: 37–45.

[6] Basyuni M, Baba S, Inafuku M, Iwasaki H, Kinjo K, Oku H 2009 Expression of terpenoid synthase mRNA and terpenoid content in salt-stressed mangrove *J. Plant Physiol.* **166**: 1786–1800.

[7] Basyuni M, Baba S, Kinjo Y, Oku H 2012 Salinity increases the triterpenoid content of a salt secretor and a non-salt secretor mangrove *Aquat. Bot.* **97**:17–23.

[8] Basyuni M, Baba S, Kinjo Y, Putri LAP, Hakim L, Oku H 2012 Salt-dependent increase in triterpenoids is reversible upon transfer to fresh water in mangrove plants *Kandelia candel* and *Bruguiera gymnorrhiza J. Plant Physiol.* **169**: 1903–1908.

[9] Tomlinson, P.B., 1986. The Botany of Mangroves. Cambridge University Press, London. 419 pp.

[10] Basyuni M, Kinjo Y, Baba S, Shinzato N, Iwasaki H, Siregar EBM, Oku H 2011 Isolation of salt stress tolerance genes from roots of mangrove plant, *Rhizophora stylosa* Griff., using PCR-based suppression subtractive hybridization *Plant Mol. Biol. Rep.* **29**: 533–543.

[11] Inafuku M, Basyuni M, Oku H 2016 Triterpenoid modulates the salt tolerance of lanosterol synthase deficient *Saccharomyces cerevisiae,* GIL77 *Saudi J. Biol. Sci.* http://dx.doi.org/10.1016/j.sjbj.2016.10.009.

[12] Pearson WR, Lipman DJ 1988 Improved tools for biological sequence comparison *Proc. Natl. Acad. Sci. USA* 85: 2444–2448.

[13] Thompson JD, Higgins DG, Gibson TJ 1994 CLUSTAL W: Improving the sensitivity of progressive multiple sequence alignment through sequence weighting, position-specific gap penalties, and weight matrix choice *Nucleic Acid Res.* 22: 4673–4680.

[14] Page RD 1996 TreeView: an application to display phylogenetic trees on personal computers *Comput. Appl. Biosci.* **12**: 357–358.

[15] Felsenstein J 1985 Confidence limits on phylogenies: an approach using the bootstrap *Evolution* **39**: 783–791.

[16] Liu J, Zhu Y, Du G, Zhou J, Chen J 2013 Exogenous ergosterol protects *Saccharomyces cerevisiae* from D-limonene stress *J. Appl. Microbiol.* **114**: 482–491.

[17] Turk M, Plemenitas A, Gunde-Cimerman N 2011. Extremophilic yeasts: plasma-membrane fluidity as determinant of stress tolerance. *Fungal Biol.* **115:** 950–958.

[18] Mansour MMF, Salama KHA 2004. Cellular basis of salinity tolerance in plants. *Environ. Exp. Bot.* **52**: 113–122.

[19] Chen SL, Li JK, Wang SS, Huttermann A, Altman A 2001. Salt, nutrient uptake, and transport, and ABA of *Populus euphratica*; a hybrid in response to increasing soil NaCl. *Trees Struct. Funct.* **15**: 186–194.

Bioinformatics study of the mangrove actin genes

M Basyuni[1*], M Wasilah[1], Sumardi[2]

[1]Department of Forestry, Faculty of Forestry, University of Sumatera Utara, Jl. Tri Dharma Ujung No. 1 Medan, North Sumatera 20155, Indonesia
[2]Faculty of Pharmacy, Tjut Nyak Dhien University, Jl. Rasmi No. 28 Medan 20123, Indonesia

[*]Email: m.basyuni@usu.ac.id

Abstract. This study describes the bioinformatics methods to analyze eight actin genes from mangrove plants on DDBJ/EMBL/GenBank as well as predicted the structure, composition, subcellular localization, similarity, and phylogenetic. The physical and chemical properties of eight mangroves showed variation among the genes. The percentage of the secondary structure of eight mangrove actin genes followed the order of α helix > random coil > extended chain structure for *BgAct1, KcAct1, RsAct1*, and *A. corniculatum Act.* In contrast to this observation, the remaining actin genes were random coil > extended chain structure > α helix. This study, therefore, shown the prediction of secondary structure was performed for necessary structural information. The values of chloroplast or signal peptide or mitochondrial target were too small, indicated that no chloroplast or mitochondrial transit peptide or signal peptide of secretion pathway in mangrove actin genes. These results suggested the importance of understanding the diversity and functional of properties of the different amino acids in mangrove actin genes. To clarify the relationship among the mangrove actin gene, a phylogenetic tree was constructed. Three groups of mangrove actin genes were formed, the first group contains *B. gymnorrhiza BgAct* and *R. stylosa RsAct1*. The second cluster which consists of 5 actin genes, the largest group, and the last branch consist of one gene, *B. sexagula* Act. The present study, therefore, supported the previous results that plant actin genes form distinct clusters in the tree.

1. Introduction

Higher plants including mangrove plants contain families of actin-encoding genes which are divergent and differently expressed and constitutively involved in basic housekeeping function required for cell maintenance [1]. They are used as endogenous internal references for normalizing gene expression studies. Actin gene, therefore, plays a significant role in simple eukaryotic cellular processes such as motility, cell growth regulation, cell differentiation and stability structure [2].

The identification of plant actin genes will help the researcher to choose the appropriate internal standard with stable expression under particular conditions, either biotic or abiotic stress [1]. The actin genes of dicot and monocot plants are well conserved and more distinct from each other. Therefore, actin genes are more strongly linked than duplications, represents that actin gene is purely from gene family of ancient times [1].

Mangrove is halophytes that are defined ecologically by their location in upper inter-tidal zones of tropical and sub-tropical climates and physiologically by their ability to withstand high concentrations of salt or low levels of soil aeration. A number cDNA cloning of actin genes have been reported from mangrove plants, *Bruguiera gymnorrhiza* [3], *Kandelia candel* [3], *Rhizophora stylosa* [4], *Aegiceras corniculatum* [5], several others have been deposited in the DDBJ/GenBank/EMBL. This information can be an excellent molecular tool and allow us used the bioinformatics methods to analyze eight mangrove actin genes as well as predicted the structure, composition, similarity, subcellular localization and phylogenetic to provide references for the actin genes and to further understand the diversity and functional of properties of amino acids in mangrove actin genes.

2. Materials and method

2.1. Materials collection
A total of eight mangrove actin genes officially deposited in the DDBJ/EMBL/GenBank were collected. The DDBJ/GenBank/EMBL accession numbers of the DNA sequence and amino acid sequence of used this analysis are as follows: AB491932, BAI39620 (*Bruguiera gymnorrhiza BgAct1*), AB491931, BAI39619 (*Kandelia candel KcAct1*), AB573024, BAK52268 (*Rhizophora stylosa RsAct1*), DQ831135, ABI144668 (*Aegiceras corniculatum Act*), JX679722, AFV91359 (*B. sexangula Act*), JX679721, AFV91358 (*B. gymnorrhiza Act*), JX679720, AFV91357 (*B. cylindrica Act*), and JX679723, AFV91360 (*K. candel Act*)

2.2. Physicochemical properties and secondary structure of the actin gene
The composition, physical and chemical properties of DNA and amino acid sequences of actin genes were analyzed using Protparam online analysis (web.expasy.org/protparam/). The computed parameters include the molecular weight, theoretical isoelectric point values, amino acid composition, atomic composition, extinction coefficient, estimated half-life, instability index, fat coefficient, and grand average hydrophilicity. HNN Secondary structure prediction method online (https://npsa-prabi.ibcp.fr) was used to analyze the secondary structure of mangrove actin genes.

2.3. Potential transit peptide in the mangrove actin gene
The targetP 1.1 Server online (www.cbs.dtu.dk/services) was used for transit peptide prediction of mangrove actin genes. The location assignment is based on the predicted presence of any of the N-terminal pre-sequences chloroplast transit peptide (cTP), mitochondrial targeting peptide (mTP) and secretory pathway signal peptide (SP).

2.4. Similarity and phylogenetic analysis of mangrove actin gene
The amino acid sequences were aligned, and similarity scores were obtained using the FASTA ver. 3.4t26 [6] of the DNA Data Bank of Japan (Mishima, Shizuoka, Japan). The best score of results is shown in Table 4. Phylogenetic analysis of deduced amino acid alignment from 8 mangrove actin genes was conducted with CLUSTAL W ver. 1.83 [7] of the DNA Data Bank of Japan (Mishima, Shizuoka, Japan) followed by drawing with TreeView, ver. 1.6.6 [8] based on a neighbor-joining method. Bootstrap analysis with 1000 replications was used to assess the strength of the nodes in the tree [9]. The DDBJ/GenBank/EMBL accession numbers of the DNA sequence and amino acid sequence of used this analysis is described in subsection Materials collection.

3. Results and Discussions
The results will be discussed in three subsections; physicochemical properties and secondary structure of the actin gene, potential transit peptide in the salt tolerance gene, and similarity and phylogenetic analysis of mangrove actin gene.

3.1. Physicochemical properties and secondary structure of the actin gene
Table 1 shows the physical and chemical properties of mangrove actin genes. The initiation codon ATG of actin gene only found in *A. corniculatum* actin (DQ 831135), where the stop codons were TAA, AAG, and GTA. No open reading frame length found, all are partial actin genes. Some encoded amino acids were 33 to 297. *K. candel Act* had the smallest relative molecular mass; *K. candel KcAct1*

had the largest one. Mangrove actin genes had variation theoretical isoelectric point value (5 to 10). *K. candel Act* had a minimum total number of atoms; *B. gymnorrhiza BgAct1* had a maximum total number of atoms. The half-life period varied among actin gene. Based on stability coefficients, *BgAct1*, *KcAct1*, *RsAct1*, and *A. corniculatum Act* were stable proteins, the remaining were non-stable proteins. These results suggested the importance of understanding the diversity and functional of properties of the different amino acids in mangrove actin genes [1].

Tabel 1. Physical and chemical properties of the actin genes cDNA in mangrove plants

Nucleotide accession number	AB491932	AB491931	AB573024	DQ831135	JX679722	JX679721	JX679720	JX679723
Length of genes/bp	894	894	894	930	221	209	203	204
Open reading frame length/bp	nd	nd	nd	nd	nd	nd	nd	nd
Start site and codon	nd	nd	nd	55ATG	nd	nd	nd	nd
Stop site and codon	894 TAA	894 TAA	894 TAA	930 AAG	99 TAA	87 TAA	90 AAG	84 GTA
Number of encoded amino acids	297	297	297	292	33	29	30	28
Relative molecular mass	33377.30	33459.35	33274.23	32331.91	3606.24	3203.78	3334.97	3075.60
Theoretical isoelectric point values	5.37	5.27	5.27	5.17	10.01	10.00	10.00	9.70
Positively charged residues	31	30	30	30	4	4	4	3
Negatively charged residues	40	40	40	42	1	1	1	1
Total number of atoms	4684	4681	4670	4529	520	467	484	446
Extinction coefficient	1.288	1.285	1.247	0.897	1.938	2.182	2.096	2.273
Half life period	2.8 h	2.8 h	2.8 h	30 h	20 h	1.3 h	30 h	20 h
Instability coeffiient	37.81	35.85	38.64	30.08	49.74	41.94	40.88	46.11
Fat coeffiient	86.73	83.43	87.37	85.89	91.52	104.14	100.67	107.86
Overall average hydrophilicity	-0.179	-0.207	-0.146	-0.213	0.061	0.114	0.173	0.257

The secondary structure of OSC genes consist of α helix, extended chain structure, and random coil with their relative proportion was depicted in Table 2. Table 2 shows that the percentage of the secondary structure of eight mangrove actin genes followed the order of α helix > random coil > extended chain structure for *BgAct1, KcAct1, RsAct1*, and *A. corniculatum Act*. In contrast to this observation, the remaining actin genes were random coil> extended chain structure > α helix. This study, therefore, indicated the prediction of secondary structure was performed for necessary structural information. This study also suggested the diversity of secondary structure among the mangrove actin genes.

Table 2. Secondary structure analysis of mangrove actin gene

Protein registry number	α helix number	α helix ratio/ %	Extended chain structure number	Extended chain structures ratio/ %	Random coil number	Random coil ratio/ %
BAI39620	121	40.74	56	18.86	120	40.40
BAI39619	120	40.40	58	19.53	119	40.07
BAK52268	122	41.08	57	19.19	118	39.73
ABI14668	105	35.96	61	20.89	126	43.15
AFV91359	5	15.15	12	36.36	16	48.48
AFV91358	5	17.24	10	34.48	14	48.28
AFV91357	5	16.67	11	36.67	14	46.67
AFV91360	5	17.86	9	32.14	14	50.00

3.2. Potential transit peptide in the salt tolerance gene
Table 3 shows the possibility of the potential transit peptide in mangrove actin genes. There are three possibilities: chloroplast transit peptide, mitochondrial target peptide and signal peptide of secretory pathway along with the prediction probability. The values of chloroplast or signal peptide or signal

peptide secretory pathway were too small, indicated that no chloroplast transit peptide or mitochondrial peptide or signal peptide of secretion pathway in eight actin genes.

At primary structural level, transit peptide sequence are highly divergent, showing that the transit peptide contains multiple domains that provide either distinct or overlapping functions [10]. Therefore it will provide an update on the limited structural information of some transit peptide and the evolution [10]

Table 3. Possibility of the potential transit peptide in salt tolerance gene

Protein accession number	Reliability			
	Chloroplast transit peptide	Mitochondrial target peptide	Signal peptide of secretory pathway	Reliability prediction
BAI39620	0.111	0.023	0.103	2
BAI39619	0.136	0.021	0.087	2
BAK52268	0.124	0.022	0.092	2
ABI14668	0.173	0.109	0.097	2
AFV91359	0.127	0.154	0.173	3
AFV91358	0.092	0.229	0.139	3
AFV91357	0.080	0.238	0.156	3
AFV91360	0.100	0.212	0.147	3
ABI51282	0.177	0.040	0.083	2

3.3. Similarity and phylogenetic analysis of mangrove actin gene

Eight mangrove actin sequences in this study were aligned each other as displayed in Table 4. The highest similarity (100%) was shown by *B. gymnorrhiza* BgAct1, *K. candel* KcAct1 and *R. stylosa* RsAct1 with four short DNA sequences of *B. sexangula* Act, *B. gymnoorhiza* Act, *B. cylindrica* Act, and *K. candel* Act (221-209 bp). Among of these short sequences also show identical sequences (100% similarity). The clustering of the mangrove actin genes suggests that these sequences may have involved in the evolution of plant dicots and they are functionally conserved [1]. This study makes the actin gene useful for phylogenetic analysis, especially using neighbor joining.

Tabel 4. The similarity of amino acid sequences between mangrove actin gene

Clones	1	2	3	4	5	6	7	8
1. *B. gymnorhiza* BgAct1	100							
2. *K. candel* KcAct1	97	100						
3. *R. stylosa* RsAct1	98	96	100					
4. *A. corniculatum* Act	69	68	69	100				
5. *B. sexangula* Act	100	100	100	18	100			
6. *B. gymnorhiza* Act	100	100	100	20	100	100		
7. *B. cylindrica* Act	100	100	100	20	100	100	100	
8. *K. candel* Act	100	100	100	21	100	100	100	100

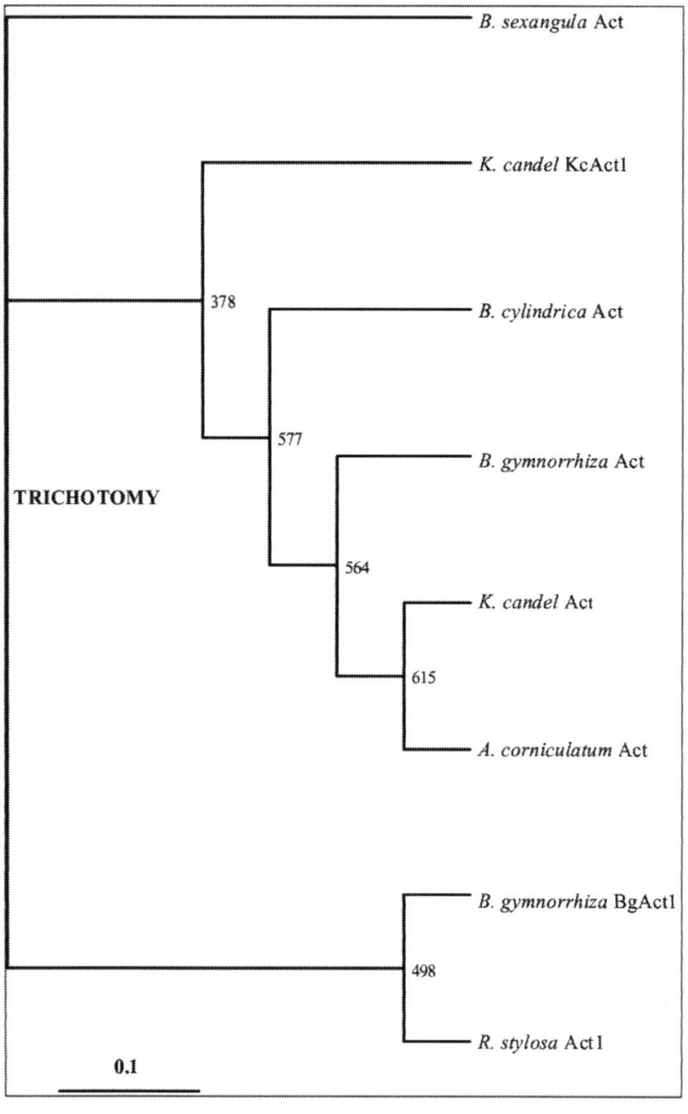

Figure 1. Phylogenetic tree of mangrove actin genes. Phylogenetic tree of deduced amino acid sequences was constructed with the neighbor-joining method of the CLUSTAL W [7]. The indicated scale represents 0.1 amino acid substitutions per site. Numbers indicate bootstrap value from 1000 replicates. The DDBJ/GenBank/EMBL accession numbers of the amino acid sequence of used this analysis are reported in the Materials subsection.

Figure 1 shows there are three groups of mangrove actin genes; the first group contains *B. gymnorrhiza BgAct* and *R. stylosa RsAct1*. The second cluster which consists of 5 actin genes, the largest group and the last branch consist of one gene, *B. sexagula Act.* The clustering of the mangrove actin genes suggests that these sequences may have involved in the evolution of plant dicots and they are functionally conserved

4. Conclusions

The present study demonstrated the importance of understanding the diversity and functional of properties of the different amino acids in eight mangrove actin genes. Furthermore, our data improve our understanding of the structure of actin genes in the mangrove.

Acknowledgment

This study was partly supported by an International Research Collaboration and Scientific Publication Grant (No. 117/SP2H/LT/DRPM/II/2016 to MB) from the Directorate for Research and Community Service, Ministry of Research, Technology and Higher Education, Republic of Indonesia.

References

[1] Slajcherova K, Fiserova J, Fischer L, Schwarzerova K 2012 Multiple actin isotypes in plants: diverse genes for diverse roles? *Frontiers Plant Sci.* **3**:1–8.

[2] Wery J, Dalderup MJM, Linde JT, Boekhout T, van Ooyen AJJ 1996 Structural and phylogenetic analysis of the actin gene from the yeast *Phaffia rhodozyma Yeast* **12**: 641–651.

[3] Basyuni M, Baba S, Inafuku M, Iwasaki H, Kinjo K, Oku H 2009 Expression of terpenoid synthase mRNA and terpenoid content in salt-stressed mangrove *J. Plant Physiol.* **166**: 1786–1800.

[4] Basyuni M, Kinjo Y, Baba S, Shinzato N, Iwasaki H, Siregar EBM, Oku H 2011 Isolation of salt stress tolerance genes from roots of mangrove plant, *Rhizophora stylosa* Griff., using PCR-based suppression subtractive hybridization *Plant Mol. Biol. Rep.* **29**: 533–543.

[5] Ya-Lan Peng, You-Shao Wang, Ji-Dong Gu 2015 Identification of suitable reference genes in mangrove *Aegiceras corniculatum* under abiotic stresses. Ecotoxicology (2015) 24:1714–1721

[6] Pearson WR, Lipman DJ 1988 Improved tools for biological sequence comparison *Proc. Natl. Acad. Sci. USA* 85: 2444–2448.

[7] Thompson JD, Higgins DG, Gibson TJ 1994 CLUSTAL W: Improving the sensitivity of progressive multiple sequence alignment through sequence weighting, position-specific gap penalties and weight matrix choice *Nucleic Acid Res.* 22: 4673–4680.

[8] Page RD 1996 TreeView: an application to display phylogenetic trees on personal computers *Comput. Appl. Biosci.* **12**: 357–358.

[9] Felsenstein J 1985 Confidence limits on phylogenies: an approach using the bootstrap *Evolution* **39**: 783–791.

[10] Bruce BD 2000 Chloroplast transit peptides: structure, function and evolution *Trends Cell Biol* **10**: 440–447.

IOP Conf. Series: Journal of Physics: Conf. Series **801** (2017) 012014 doi:10.1088/1742-6596/801/1/012014

Utilization starch of jackfruit seed (*Artocarpus heterophyllus*) as raw material for bioplastics manufacturing using sorbitol as plasticizer and chitosan as filler

M Lubis[1], M B Harahap[2], A Manullang[1], Alfarodo[1], M H S Ginting[1] and M Sartika[1]

[1]Departement of Chemical Engineering, University of Sumatera Utara, Almamater Kampus USU, Medan, 20155, Indonesia

[2]Departement of Physics, Universitas Negeri Medan, Williem Iskandar, 20221, Medan, Indonesia

Email: maulida70@gmail.com

Abstract. Starch is a natural polymer that can be used for the production of bioplastics because its source is abundant, renewable and easily degraded. Jackfruit seeds can be used as raw material for bioplastics because its contains starch. The aim of this study to determine the characteristics of jackfruit seeds and determine the effect of chitosan and sorbitol on the physicochemical properties of bioplastics from jackfruit seeds. Starch is extracted from jackfruit seeds were then characterized to determine its chemical composition. In the manufacture of bioplastics starch composition jackfruit seeds - chitosan used was 7: 3, 8: 2 and 9: 1 (g/g), while the concentration of sorbitol used was 20%, 25%, 30%, 35%, and 40% by weight dry ingredients. From the analysis of jackfruit seed starch obtained water content of 6.04%, ash content of 1.08%, the starch content of 70.22%, 16.39% amylose content, amylopectin content of 53.83%, 4.68% protein content, fat content 0.54%. The best conditions of starch bioplastics jackfruit seeds obtained at a ratio of starch: chitosan (w/w) = 8: 2 and the concentration of plasticizer sorbitol 25% with tensile strength 13.524 MPa. From the results of FT-IR analysis indicated an increase for the OH group and the group NH on bioplastics due to the addition of chitosan and sorbitol. The results of mechanical tests is further supported by analysis of scanning electron microscopy (SEM) showing jackfruit seed starch has a small granule size with the size of 7.6 μm and in bioplastics with chitosan filler and plasticizer sorbitol their fracture surface is smooth and slightly hollow compared bioplastics without fillers chitosan and plasticizer sorbitol.

1. Introduction

Plastics as we know are used as packaging materials for various products including food. Plastic chosen as the packaging material because it is safe, robust (waterproof, light, and heat) and the price is cheap [1]. However, synthetic plastic are derived from non-renewable resources that can caused environmental damage. Most plastics are made from petroleum. Therefore, the plastic is not considered environmentally friendly because it is not biologically degraded lands and will certainly contaminate the soil [2]. Due to its shortcomings, it has been necessary to develop an ecological solution to overcome these problems, one of which is to develop biodegradable plastics (bioplastics). This means that these plastics can be decomposed by microorganisms naturally. Development of biodegradable plastic materials using renewable natural materials (renewable resources) is highly expected [3].

One of the main component of bioplastics is starch. Starch is widely used in the form of biodegradable films in varied applications because it is a renewable, abundant and inexpensive material [4]. In a jackfruit contained approximately 100 to 500 jackfruit seeds or about 8-15% by weight of jackfruit itself with high starch content [5]. The addition of plasticizer serves to improve the flexibility bioplastic. Sorbitol is a plasticizer which is good enough to reduce internal hydrogen bonds that will improve the intermolecular distance [6]. Chitosan is used as a filler in the manufacture of bioplastics. One way to reduce the hydrophilic nature (less resistant to water) bioplastics is by mixing starch with biopolymers hydrophobic (water resistant) such as cellulose and chitosan [3].

The aim of this paper is to report a study of the utilization of jackfruit seed (*Artocarpus Heterophyllus*) as raw material for bioplastics manufacturing using sorbitol as Plasticizer and chitosan as filler which can be overcome environmental problems due to use of syntetic plastics.

2. The Material and Method

2.1. Materials
Jackfruit seed was obtained from fruit merchants located at Traditional Market Tanjung Rejo, Medan, Indonesia. Sorbitol, aquadest, and acetic acid were obtained from Laboratory of Chemical Process Industries, Chemical Engineering, University of Sumatera Utara.

2.2. Starch Extraction of Jackfruit Seed Starch
Jackfruit seeds (100 gr) were peeled and washed witch clean water. Seeds were cut with a size of approximately 1 cm^2 and then blended with 100 ml water. Starch slurry was filtered using a plastic sieve to obtain liquid filtrate (starch suspension). The resulting suspension were then deposited for 24-48 hours until the starch settles perfectly. Starch sediment was filtered using Whatman filter paper no. 1 to obtain a wet starch. Starch was dried using oven on temperature of 70° C for 30 minutes. Then starch was sieved with strainer 100 mesh.

2.3. Film Preparation
Number of starch and Chitosan mass wanted was 10 grams. Starch solution made with a ratio of starch : distilled water is 1 : 20 as much as 100 grams in 500 ml beaker glass. Then chitosan solution made in accordance with the volume that has been calculated on a beaker glass. Then the starch solution was heated while stirred with velocity of 400 rpm and temperature of 88,82 °C for 25 minutes using hotplate. After 25 minutes, sorbitol was added with variation 20%, 25%, 30%, 35% dan 40% on starch solution. After stirred for 15 minutes, the solution was cooled down and cast onto flat and dried with temperature 60 °C for 24 hours. After bioplastic dried, it was removed from the flat and saved in the desicator and ready to be analyzed.

2.4. Characterization of Starch
The procedure of water content and ash content of jackfruit seed starch is based on AOAC standard. Analysis of starch, amylose, amylopectin, fat and protein was observed in Jasa Uji Ilmu, Fakultas Teknologi Industri Pertanian, Universitas Padjajaran.

2.5. Characterization of Bioplastic
FTIR analysis of bioplastic was observed in Laboratorium Penelitian, Fakultas Farmasi, University of Sumatera Utara. Tensile strength test is based on ASTM standard D638. SEM analysis was observed in Laboratorium Fisika, Universitas Negeri Medan.

3. Results and Discussion

3.1. Characterization of jackfruit seed starch

Based on sedimentation method [7] from 100 grams of jackfruit seed could produce jackfruit seed starch 26,67 grams or percentage of starch was 26,67 %.

Characteristics of jackfruit seed starch was aimed to know the percentage of each component contained in the starch produced, include starch (amylum), water, ash, fat, protein, amylose and amylopectin content so that known the quality of starch produced [8]. The result of jackfruit seed starch was provided inside the Table 1.

Table 1. Results of jackfruit seed starch

Components of Starch	Percentage (%)	Standard of Indonesia Industry (%) [9]
Starch (amylum)	70,22	Min 75
Amylopectin	16,39	-
Amylose	58,83	-
Water	6,04	Max 14
Ash	1,08	Max 1,5
Fat	0,54	-
Protein	4,68	-

3.2. *Fourier-transform infrared spectrometry*

The FTIR result of starch, chitosan,bioplastic without adding filler and plasticizer was provided in figure 2 below.

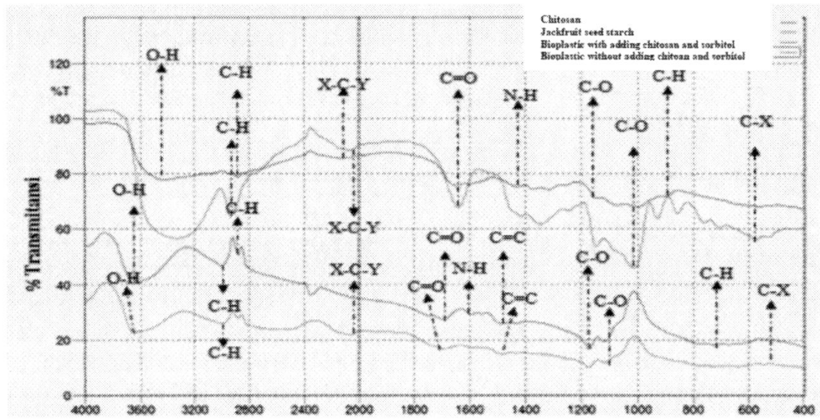

Figure 2. The result of *Fourier Transform Infra Red* (FTIR)

From the analyzing result of FTIR can be seen that jackfruit seed starch which owns O-H Alcohol (H-bonded), C-H alkanes (stretch) and C-H aldehydes, C-O amide, S=O sulfates, sulfonamides, sulfones and C-O ester. The presence of OH group binded with hydrogen, stretching of C-H alkanes, stretching of C=O amide and stretching of C-O ether had represented the content of jackfruit seed starch which is consisted of amylose and amylopectin and reducing glucose ($C_6H_{10}O_5$)n [10].

From the analyzing result of FTIR chitosan which owns the hydrogen bonding groups O-H, C-H alkanes (stretch), N-H amide, C=O amide, C-N amine, ether groups and C-O ester. Chitosan has a hydroxyl group (OH) which is negatively charged and the amine group (NH_2) positively charged, so that chitosan is able to form a strong ionic bonds [11]. The existing of those groups had shown the characteristic of chitosan which is characterized by the presence of amino group at wave number of absorption peak 1315,45 cm^{-1} and 1145,72 cm^{-1}, and the presence of hydroxil group at wave number 3433,29 cm^{-1}.

IOP Conf. Series: Journal of Physics: Conf. Series **801** (2017) 012014 doi:10.1088/1742-6596/801/1/012014

From the analyzing result of FTIR bioplastic without adding chitosan and sorbitol can be seen the presence of alcohol group O-H, alkanes group, aldehydes, C-H alkenes, amide C=O, aromatic group C=C and C-O ester.

From figure 2 also can be seen the FTIR result of bioplastic with adding chitosan as filler and sorbitol as plasticizer showed the groups of O-H alcohol, alkenes, aldehydes, C-H alkenes, C=O aromatic, the primer and secondary groups of amine, C=C aromatic and ester C-O. The differences result between bioplastic without adding filler chitosan and plasticizer sorbitol and bioplastic with adding filler chitosan and plasticizer sorbitol could be seen from the emergence of the N-H amino group of a functional group which is owned by chitosan with wave number 1593,20 cm^{-1}.

From the result, can be seen that bioplastic with adding chitosan as filler and sorbitol as plasticizer had increasing wave number of OH group in which the increase of O-H group is due to the increasing of hydrogen bonds formed on bioplastics. The hydrogen bonds consist of bonds between chains of amylose-amylose, amylose-amylopectin, chitosan-chitosan, and amylose-chitosan-amylopectin [8].

3.3. *Tensile strength of Bioplastic*

The following graphs shows the effect of chitosan and sorbitol addition on tensile strength bioplastic.

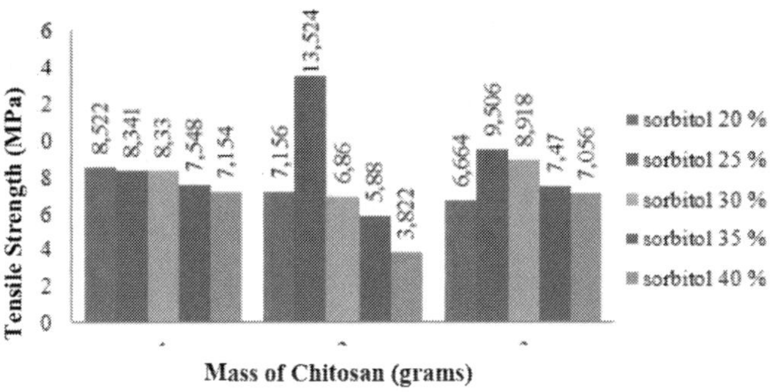

Figure 3. The effect of chitosan and sorbitol addition on tensile strength bioplastic from jackfruit seed starch

Figure 3 shows bioplastic with chitosan content 2 grams and 25% sorbitol provided the maximum tensile strength for 13,52 MPa. While, bioplastic with chitosan content 2 grams and 40% sorbitol provided the minimum tensile strength for 3,82%.

From figure 3 can be seen that with increasing mass of chitosan cause the incresing value of tensile strength. The addition of chitosan affects the chemical bonds constituent of bioplastic thus improved its tensile strength. With increasing concentrations of chitosan, there will be more hydrogen bonds contained in the bioplastic so that the chemical bonds of bioplastics will be stronger and difficult to break up, because it requires a large energy to break the bond. This is caused by the physics changes of bioplastic particles, so that plastics are increasingly homogeneous and has a dense structure [12].

A drop tensile strength value is shown for bioplastic with chitosan content 3 grams and concentration of sorbitol 25%. Buzarovska et al (2008) reported that a decrease in the results of tensile strength is caused also by uncompletely distribution of each constituent of bioplastic. Plasticizer has two OH groups that are used to interact with the starch molecule and interacts with the amine group of chitosan that will form a new hydrogen bonds [13].

Meanwhile, figure 3 also shows that with increasing concentrations of sorbitol, cause the value of tensile strength of bioplastics decreased. This case shows the addition of sorbitol inversely proportional to the tensile strength of bioplastics produced. When sorbitol is added, the molecules of the plasticizer in the solution lies between biopolymer bonding chain and be able to interact by forming hydrogen bonds in the bond between the polymer chains resulting in weak interactions between molecules biopolymers [14].

3.4. *Scanning Electron Microscopy (SEM)*

Characteristic of SEM jackfruit seed starch, bioplastic without chitosan as filler and sorbitol as plasticizer and bioplastic with adding chitosan as filler and sorbitol as plasticizer are shown on figure 4 below.

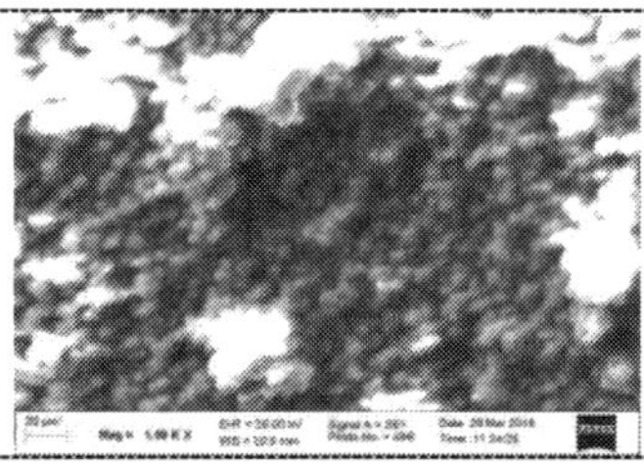

Figure 4. Jackfruit seed starch with magnification 1000x

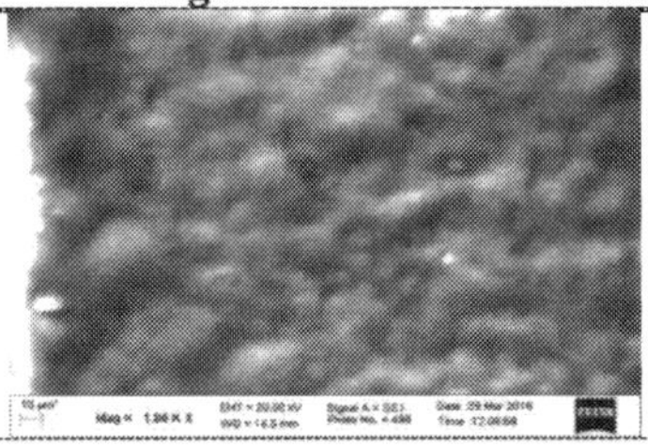

Figure 5. Bioplastic without adding chitosan as filler and sorbitol as plasticizer with magnification 1000x

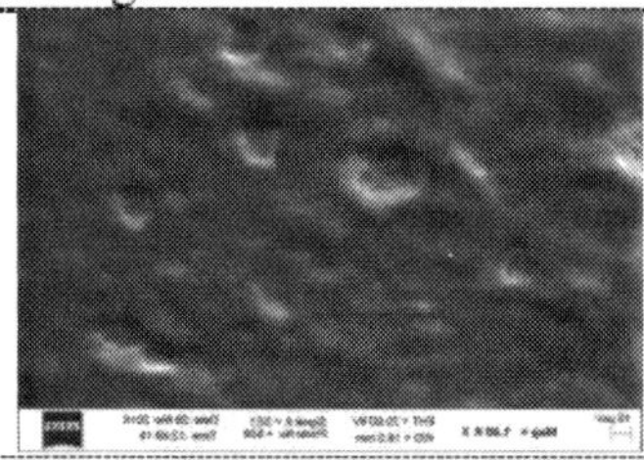

Figure 6. Bioplastic with chitosan content 2 grams and sorbitol 25% with magnification 1000 x

Figure 4 shows the result of analyzing SEM on jackfruit seed starch. The morphological jackfruit seed starch didn't have a uniform particle size. This is due to sieving process carried out at a size of 100 mesh, so that particles with a size smaller than 100 mesh also passed the sifting process. The surface morphology of starch varies depending on the type of starch source. SEM analysis results of jackfruit seed starch had a semi-round shape granules. Marta et al., (2014) showed the similar results with observations using 1000x magnification electron micrograph. The jackfruit seed starch had a spherical shape with an average size of granules 6-13 μm [15].

The SEM analysis results of jackfruit seed starch shows that starch granules had a size of 7,6 μm. In the classification of starch, there are two types of starch granule size, ie small granule size (5-10 m) and large granule size (25-40 m) [16]. Based on the classification, jackfruit seed starch had a small granule size.

From figure 5 can be seen the presence of white granules of jackfruit seed starch of bioplastic without adding chitosan as filler and sorbitol as plasticizer which indicates that jackfruit seed starch do not evenly mixed in the bioplastics produced. However, there are no void fill found in bioplastic without adding chitosan as filler and sorbitol as plasticizer. Sorbitol in bioplastics will form a large cavity of bubbles. This is because sorbitol is hydrophilic so that the bioplastics is easy to fragile. This proves the above analysis of tensile strength, with increasing concentrations of sorbitol causing decreased tensile strength bioplastic [12].

From figure 6 shows that jackfruit seed starch and chitosan completely dissolved in bioplastic solution and produced a homogeneous bioplastics. From the analysis of SEM is also can be seen the void fill on bioplastics produced. This is due to the the void that is trapped in bioplastics when mixing. Figure 6 shows the surface of bioplastic looked smooth and more compact because the existing of adding plasticizer and a little bit found space for the existing of adding chitosan as filler so when the tensile strength was observed, the product will have better capability [8].

4. Conclusion
Based on the results of research can be concluded that the analysis of jackfruit seed starch obtained moisture content 6,04 %, ash content 1,08 %, starch content 70,22 %, amylose content 16,39 %, amylopectin content 53,83 %, protein content 4,68 %, and fat content 0,54 %. The best condition of bioplastic from jackfruit seed starch obtained at comparison of starch : chitosan (w/w) = 8:2 and concentration of sorbitol 25 % with tensile strength 13,524 MPa.

References
[1] A. W. Utomo, B. D. Argo, and M. B. Hermanto, "Effect of Temperature and Drying Duration toward Psychochemical Characteristic of Biodegradable Plastic from Starch Composite of Aloevera–Chitosan," *Bioprocess J. Trop. Commod.*, vol. 1, no. 1, 2013.
[2] D. Kurniawan, F. Arifan, and N. Izzah, "Preparation Preliminary Study of Biodegradable Plastics Based of Cassava Compounds with Additives Limonene Extraction of Orange Leather," in *The 2nd International Seminar on New Paradigm and Innovation on Natural Sciences and Its Application*, 2012.
[3] Y. Darni and H. Utami, "Preparation and Characteristic Study of Mechanical Properties and Hydrophobicity Bioplastics from Sorghum Starch," *J. Chem. Environtmental Eng.*, vol. 7, no. 4, pp. 88–93, 2010.
[4] J. S. Alves, K. C. Reis, E. G. T. Menezes, F. V Pereira, and J. Pereira, "Effect of Cellulose Nanocrystals and Gelatin in Corn Starch Plasticized Films," *Carbohydr. Polym.*, 2014.
[5] M. A. and K. Sajjaanantakul, "Physico-chemical Properties of Fflour and Starch from Jackfruit Seeds (Artocarpus heterophyllus Lam.) Compared with Modified Starches," *Int. J. Food Sci. Technol.*, vol. 39, pp. 271–276, 2004.
[6] T. H. McHugh and K. J.M., "Sorbitol vs Glycerol Plasticed Whey Protein Edible Film : Integrated Oxygen Permeability and Tensite Property Evaluation," *J.Agic food Chem*, vol. 2, no. 4, pp. 841–845, 1994.
[7] Z. Anita, F. Akbar, H. Harahap, D. T. Kimia, F. Teknik, and U. S. Utara, "Effect of Glycerol Addition on Mechanical Properties of Biodegradabilty Film from Cassava Peel Starch," *J. Chem. Eng. USU*, vol. 2, no. 2, pp. 37–41, 2013.
[8] M. H. S. Ginting and M. F. R. Tarigan, "□ Effect of Gelatinization Temperature and Chitosan on Mechanical Properties of Bioplastics from Avocado Seed Starch (Persea americana mill) with Plasticizer Glycerol," *Int. J. Eng. Sci.*, vol. 4, no. 12, pp. 36–43, 2015.
[9] W. Rahmawati, Y. A. Kusumastuti, and N. Aryanti, "Characterization of Taro Starch (Colocasia esculenta (L.) schott) as Alternative Sources of Starch Industry in Indonesia," *J. Chem. Technol. Ind.*, vol. 1, no. 1, pp. 347–351, 2012.
[10] A. Irwan, Sunardi, and A. Syabatini, "Acrylamide-based Superabsorbent Polymer (AAM) Grafted BananaTuber Starch (Musa paradisiaca)," *Semirata FMIPA Univ. Lampung Proceeding*, pp. 1–5, 2013.
[11] D. Listiyaningsih, "Manufacture and Characterization of Dioscorea esculenta Starch-Chitosan with Plasticizer Polyvinyl Alcohol (PVA)," Universitas Negeri Semarang, Semarang, 2013.
[12] P. Coniwanti, L. Laila, and M. R. Alfira, "Development of Biodegradable Film Plastic from Corn Starch with Chitosan and Glycerol Addition," *J. Chem. Eng.*, vol. 20, no. 4, pp. 22–30, 2014.
[13] A. Buzarovska, A. Grozdanov, M. Avella, G. Gentile, and M. Errico, "Potential Use of Rice Straw as Filler in Eco-Composite Materials," *Aust. J. Crop Sci.*, vol. 1, no. 2, pp. 37–42, 2008.
[14] A. Purwanti, "Tensile Strength and Elongation Analysis of Chitosan Plastic with Plasticizer Sorbitol," *J. Technol.*, vol. 3, no. 2, pp. 99–106, 2010.
[15] M. S. Madruga, F. S. M. de Albuequerque, I. R. A. Silva, D. S. do Amaral, M. Magnani, and V. Q. Neto, "Chemical, Morphological and Functional Properties of Brazilian Jackfruit (Artocarpus Heterophyllus) Seeds Starch," *Elsevier Food Chem.*, vol. 143, pp. 440–445, 2014.
[16] A. Jading, E. Tethool, P. Payung, and S. Gultom, "Physicochemical Characteristics of Sago Starch Results of Fluidized Drying Using Cross Flow Fluidized Bed Dryer Powered Solar and Biomass," *Reactor*, vol. 13, no. 3, pp. 155–164, 2011.

Designing Knowledge of The PPC with Semantic Network

Ari Fajar Santoso[1], Iping Supriana[2], and Kridanto Surendro[2],

[1] School of Industrial Engineering and Information System,
Telkom University, Indonesia
arifajar@student.itb.ac.id, arifajar@telkomuniversity.ac.id, arifajar2012@gmail.com
[2] School of Electrical Engineering and Information Technology,
Bandung Institute of Technology, Indonesia
iping@informatika.org, endro@informatika.org

Abstract. A manufacturing process is a very complex activity, which involves a variety of knowledge, such as the product design and description, manufacturing operations, tools, machines, and the relationships between these entities. Production planning control (PPC) as dominant part of manufacturing process has a very important role in determining the measures taken by the management of company. The PPC construction is an interactive process that requires the collaboration of both ICT and manufacture experts. Building PPC process can be used as a basis for making knowledge in manufacturing domain that will be used in manufacturing intelligent system. This paper describes about designing knowledge with semantic network in the production scheduling.

Keywords : PPC; semantic network; knowledge; intelligence.

1. Introduction

In manufacturing companies, parts production into one of the core processes of large companies. The production department in which there are process planning, control, execution and evaluation of the production of a product accurately and systematically. Production data recording is done in principle production process. One part of the manufacturing is a control system and material requirements commonly, production, planning, control and inventory in called Production Planning Control (PPC) became the dominant, crucial and complex in the production process. In the domain of informatics, knowledge production mainly PPC becomes the center of attention for a major role in the company. Production scheduling problems become urgent because it involves many units are inter-related and operational very rigid in its implementation, so that the records and information to be important in supporting the implementation of the PPC.

PPC systems have a number of fundamental flaws. Manufacturing planning and control systems a recurrently dominated by systems based upon material, control and planning. A potential alternative to PPC systems is suggested after research into the economic batch scheduling problem. An alternative approach to manufacturing planning and control is developed, Based on the ideas of economic batch scheduling, and enhanced through artificial intelligence techniques [1].

Applications of Artificial Intelligence is made up of two main parts, the knowledge base and the inference mechanism . Knowledge base contains facts about objects in a particular domain and mutual relationship with each other. The knowledge base can also contain thought, theory, practical procedures and mutual relations. Knowledge is a concept that has gained importance in recent years as interest in knowledge management [2]. Inference mechanism is a series of procedures used to test basic knowledge systematically while answering questions, solving problems or making decisions in the domain that have been determined. Schematic representation generally categorized as declarative knowledge or procedural declarative scheme is used to explain the facts and statements (statement). This method including logic, ontology, scripts, and frames, semantic networks.

Semantic (associative) network can be seen (Sowa, 2000) as a directed graph consisting of nodes and edges. Nodes represent individual objects of described world and edges connecting these nodes represent relationships between these objects. Very often we can meet with following types of the relations [3]:

- Is-Instance-Of (ISA) relationship is used to express the fact that a particular object (an instance of a particular class) belongs to the specified class.
- A-Kind-Of (AKO) relationship can express that a class is a subclass of another class.
- A-Part-Of (APO) relationship serves to express that a certain class of objects is composed of certain parts.

Semantic networks are knowledge representation schemes involving nodes and links (arcs or arrows) between nodes. The nodes represent objects or concepts and the links represent relations between nodes. The links are directed and labeled; thus, a semantic network is a directed graph. In print, the nodes are usually represented by circles or boxes and the links are drawn as arrows between the circles. This represents the simplest form of a semantic network, a collection of undifferentiated objects and arrows. The structure of the network defines its meaning. The meanings are merely which node has a pointer to which other node. The network defines a set of binary relations on a set of nodes. Most semantic networks are cognitively based. They also consist of arcs and nodes which can be organized into a taxonomic hierarchy. Semantic networks contributed ideas of spreading activation, inheritance, and nodes as proto-objects [4].

Semantic network have been used successfully to representation knowledge domain well-establish taxonomies to simplify problem solve. Some of advantages of semantic network are :1. Flexibility in adding, modifying, or deleting new nodes and arcs, 2.Ability to inherit relationships from other nodes, 3. Ease of drawing inferences about inheritance hierarchy [5].

Semantic model allows users to ask questions about what is happening in the system being modeled with a more natural way. Therefore, semantic models can be used to connect the physical world, to complicate matters reference information and aggregation, data points of interest can be managed through a database, file, application, or service components each have an interface method itself and naming conventions for access the data [2]. When talking about integration-oriented architecture-based operating system information, in this context, we are really referring to utilize the semantic model as the core functional applications to provide a model to navigate the data and associated relationships that represent knowledge in the target domain.

The basic advantage of the semantic network is that it contains information similarly as information is stored in human memory, and it is machine-understandable. This means that it can be machine-processed - is it possible to analysis facts and information included in the semantic network and to acquire new knowledge about represented
Facts [6].

The purpose of this paper is to map domain entities as a frame or basic knowledge of scheduling of material requirements planning in the production process. This knowledge base can be as the basis for designing the next production intelligence system.

The recommended approach is based on a semantic network that can capture the basic elements of the frame information and reciprocal relationship. Semantic network is defined as a representation of knowledge that includes linkages between topics. The second approach to Lisp. Lisp is a convenient language for representing any graph structure, including semantic nets. Lists provide the ability to create objects of arbitrary complexity and these objects may be bound to names, allowing for easy reference and the definition of relationships between them.

2. Production Planning Control Cycle

The production planning control process is a quite complex domain which requires precision in building knowledge about it. And it's a part of the management production that have functionality to record, process and make some reports of production transactions. Manufacturing is an art and science that involves recording transactions in a particular manner. Today there are many production users apart from the business organizations themselves like plants, shareholders, suppliers, academicians, students, researchers etc. and it becomes utmost important for them to know the exact meaning and relation of the production terms [7]. In some organization, there are several users of production process who do not know well about the concept of management production.

The production cycle starts on the order of marketing. Marketing needs analyzed recorded then made plans for production. Preliminary calculations based on a rough estimate. The production cycle begins on the evidence of process order. Analyze each order and event from source documents, then based on the evidence of production process, such as order marketing, bill of material (BOM), purchasing receipts, inventory order, were recorded in the form of document production as master production schedule. A typical manufacturing planning and control system consists of three process : front end process , engine process, and back end process[8].

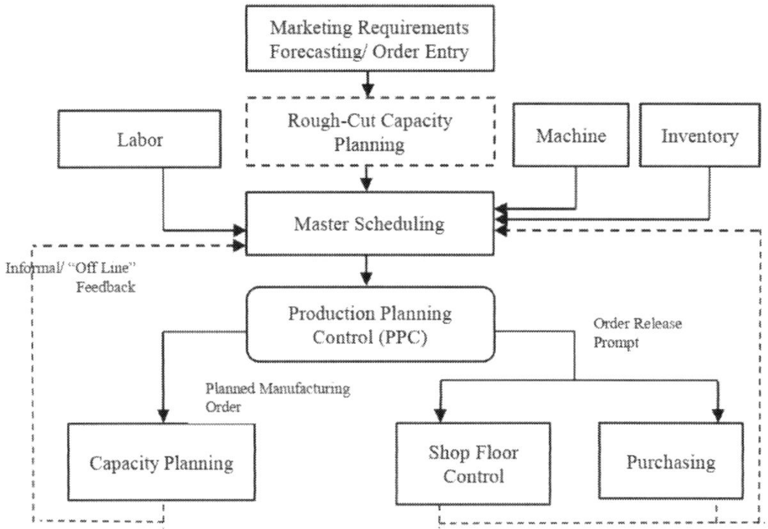

Figure 1. Production Planning Control System Architecture

Model Production Planning System architecture is an extension of the Ronald S. Dattero , the PPC system architecture, as in Figure 1. The development of production processes by considering the machine elements, human and inventory. As the figure shows, the system takes a schedule of marketing requirements as a major input and produces two major outputs: a schedule of planned manufacturing orders, and a set of order release prompts to the shop floor and to Purchasing, include requirement labor and machine. Those elements are a factor in making production scheduling constraints. The main problem in production scheduling, the constraints of each element. Further optimizing the feasible area.

The PPC system divides the manufacturing task into subtasks such as master scheduling, shop floor control, and inventory planning. Subtasks that are fairly standard , in the installations we have seen, are denoted with solid boxes in Figure 1. Other subtasks (modules) such as maintenance planning, purchasing, and tool planning are often present as well. A rough-cut capacity planning module, used to aid the development of master schedules, is also available in many implementations.

3. Related Work

From his research Yingzhong proposed about Knowledge of unit processes, knowledge can be considered as the theoretical or practical understanding of a subject, which generally includes facts, information, descriptions, or skills acquired through experiences or education. With the practice and development of manufacturing industries and technologies, a lot of unit process's domain knowledge has been created, which mainly includes the following three aspects: concepts of unit processes and their taxonomies, manufacturing capabilities of a unit process and process parameters of a unit process [9].

Abhinav also proposed a semantic network based approach is being presented for natural language processing of qualitative information available from industrial systems in the form of textual descriptions. Focus of this research in compared to other current automated methods to manipulate text messages which are computationally expensive, this technique takes advantage of the semi structured nature of the text and domain limited vocabulary in industrial environments in order to create an architecture that processes textual information efficiently and effectively [10].

Garcia-Crespo proposed about the Industrial manufacturing processes representation is a key challenge for leveraging interoperability among business partners. The Semantic representation of information enables the creation of intelligent systems, which can interpret and understand potentially automated tasks, harnessing added-value decision-making processes. A. Garcia-Crespo and B. Ruiz-Mezcua said that The model is based on different situations a problem might yield and the correspondent behavioural responses which should be generated. Using the concept of "Situation" as the conceptual corner-stone and building block of descriptions, we discuss how semantics provides a natural knowledge representation strategy, which eases the resource-intensive process of acquiring knowledge [7].

Antonio also proposed about the modeling of manufacturing processes independently of the structure and composition of the industrial plant. The rationale for the proposal is in the incorporation of knowledge, supported by ontologies regarding manufacturing processes, industrial machinery and the industrial plants involved. With this knowledge, process engineers can focus their efforts exclusively to define the activities to be performed on the raw material (abstract model). To refine the proposal, a case study is put forward in which the management system was able to infer the way for the raw material to be processed according to the abstract model and automatically composed the activities necessary for achieving it. Finally, an implementation of this management system for processes is presented together with the design of two models of industrial plant that demonstrate the validity of our proposal [11].

Sormmaz DN and Khoshnevis B, 1997, propose about the Knowledge representation for manufacture processes In general, the manufacturing activities are guided by process planning, which is a pivotal link between design and manufacturing. Process planning involves the selection of necessary manufacturing processes and the decision of their sequences to transform a designers idea (namely the designed part) into a physical component economically and competitively. The selection and decision depend on the manufacturing knowledge heavily [12].

4. Designing the Semantic Network in The PPC Process

A manufacturing process is a very complex activity, which involves a variety of knowledge, such as manufacturing operations ,the product design and description, machines, tools, and the relationships between these entities. Yingzhong Zhang, and Xiaofang said that Knowledge representation for manufacture process is the design approach to manufacturing processes unit. Manufacturing consists of several major processes. A unit process is a fundamental part of the product manufacturing process. In the face of a very large number of units existing and emerging processes, how to form a unit of knowledge representation process is very difficult. So do the two approaches: Taxonomy of unit processes and unit process capabilities. Taxonomy of unit processes is a primary knowledge representation work for organizing unit according to reviews their processes some common characteristics. While the process unit capabilities are

Referred to as the limitations or performance constraints and the range of applicability of a given unit manufacturing process [9].

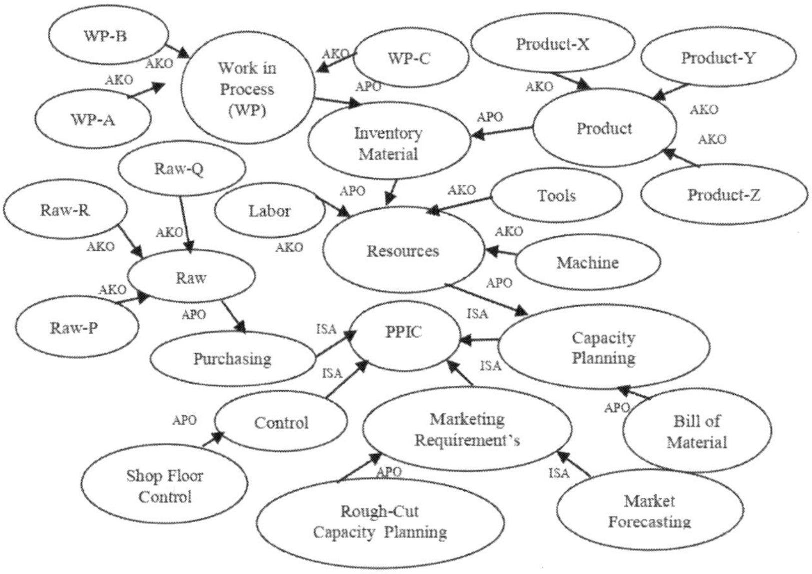

Figure 2. Production Planning's Semantic Network of Manufacturing

Proposed approach is illustrated in the example of Production Planning Control environment. This case study shows how semantic networks and flow PPC process help to frame the Capacity Planning 's, Shop Floor Control's, Purchasing's and Marketing Requirements 's preferences. Thus it helps to decide what information should be shown on the master scheduling. Semantic networks may serve as a supporting tool in the management production strategy.

Four object semantic networks of major process scheduling were identified :

- Semantic network of the Process capacity planning's frame - refers to the basic features and links of what affects the capacity planning process in selecting the information about his PPC.
- Semantic network of the process shop flow control's frame - refers to the basic features and links of what influences the shop flow control in create information process.
- Semantic network of the process purchasing's frame - refers to the basic features and links of what influences purchasing in the selection of published information about the PPC.
- Semantic network of the process marketing requirement's frame - refers to the basic features and links of what influences marketing requirement's in the selection of published information about the PPC.

With the practice and development of manufacturing industries and technologies, a lot of unit process's domain knowledge has been created, which mainly includes the following three aspects: concepts of unit processes and their taxonomies, manufacturing capabilities of a unit process and process parameters of a unit process [9].

Semantic network of the capacity planning's frame shows what affects the decision about information published on bill of material (BOM), capacity planning process, resources. Figure 2 shows that production planning's semantic network of manufacturing. The shop flow control is the semantic network representation control of production, taking into account the resources of companies which consists of labor, machine, tools. Control also noticed how much inventory material, such as work in process product, and the final product that is so. The process of purchasing's frame to explain how much raw material required in the production process and raw material supplies, rested on raw-p, q and a raw-raw-z. The process of marketing requirement's semantic network shows the need to assess the market of rough-cut capacity planning and forecasting required demand of the market potential. The PPIC's semantic network shows sub-class consisting of capacity planning, control, marketing requirements, purchasing of material needs.

5 Conclusion

Designing production process with semantic network can be used as a basis for making knowledge in production planning control domain that will be used in production intelligent system. The Production Planning Control (PPC) process is a quite complex domain which requires precision in building knowledge about it. And it's a part of the introduction domain that have functionality to record, process and make some reports of production transactions. The PPC in manufacturing firms has a very important role in determining the measures taken by the management of company.

Production Planning Control cycle must be understood by the users of production data. It can improve the understanding of knowledge in process production. Knowing about classifications of production planning control transaction is used for designing production and planning control transaction semantic network.

References

1. Ronald S. Dattero. : Enhancing Manufacturing Planning and Control Systems Through Artificial Intelligence Techniques University of Dayton eCommons MIS/OM/DS Faculty Publications Department of Management Information Systems, Operations Management, and Decision Sciences. Missouri State University - Springfield John J. Kanet University of Dayton, jkanet1@udayton.edu Edna M. White (1989)
2. Tim Hanis., Dave Noller .: The role of semantic models in smarter industrial operations,.hanistt@us.ibm.com, Chief Architect, IBM Integrated Information Core, IBM, nollerd@us.ibm.com, Senior Architect, Industrial Sector Solutions
IBM (2012)
3. Sowa, J. F.: Knowledge Representation: Logical, Philosophical, and Computational Foundations, Brooks/Cole Publishing Co., Pacific Grove, CA, ISBN 0534949657 (2000)
4. Sormmaz DN., Khoshnevis B. : Process planning knowledge representation using an object-oriented data model. Int J ComputIntegr Manuf 10(1–4):92–104 (1997)
5. Adedeji B. Badiru., John Y. Cheung.: Fuzzy Engineering Expert Systems with Neural Network Applications, ISBN: 978-0-471-29331-6,. John Wiley & Sons, New York (2002)
6. Steyvers, M., Tenenbaum, J. B. : The Large-Scale Structure of Semantic Networks: Statistical Analyses and a Model of Semantic Growth, Cognitive Science, Vol. 29, Iss. 1, p. 41–78. ISSN:0364-0213 (2005)
7. A. Garcia-Crespo., B. Ruiz-Mezcua., J.L. Lopez-Cuadrado., J.M. Gomez-Berbis.: Conceptual model for semantic representation of industrial manufacturing processes Computer Science Department, Universidad Carlos III de Madrid, Av. de la Universidad 30, 28911 Legane´s, Madrid, Spain. Computers in Industry 61, 595–612 (2010)
8. Vollmann , T. E., Berry, W. L. ,Whybark, D.C.: Manufacturing Planning and Control Systems (Homewood, IL: Irwin) (1984)
9. Yingzhong Zhang., Xiaofang Luo., Hong Zhang., John W. Sutherland.: A knowledge representation for unit manufacturing processes, Published online: 11 May 2014# Springer-Verlag London (2014)
10. Abhinav Saxena., George Vachtsevanos.: Processing Textual Information from Industrial, Systems Using Semantic Networks, Georgia Institute of Technology, Atlanta GA 30332, {asaxena,gjv}@ece.gatech.edu. IICAI (2005)
11. Antonio Ferrándiz-Colmeiro., Virgilio Gilart-Iglesias., Francisco Maciá-Pérez.: Semantic Processes Modelling Independent of Manufacturing Infrastructures, , Computer Science Department, University of Alicante, {aferrandiz, vgilart, pmacia}@dtic.ua.es. 978-1-4244-6850-8/10. IEEE (2010)
12. Sormmaz DN., Khoshnevis B. : Process planning knowledge representation using an object-oriented data model. Int J ComputIntegr Manuf 10(1–4):92–104 (1997)

FWFA Optimization based Decision Support System for Road Traffic Engineering

D N Utama, F A Zaki, I J Munjeri, N U Putri

Laboratory Optimization Models and Systems for Decision Support, Information Systems Department, UIN Syarif Hidayatullah, Jakarta

dit.utama@uinjkt.ac.id

Abstract. Several ways and efforts have been already conducted to formally solve the road traffic congestion. However, the objective strategy type of road traffic engineering could not be proven truly. Try and error is one inefficient way in road traffic engineering to degrade the level of congestion. The combination between fuzzy-logic and water flow algorithm methods (called FWFA) was used as a main method to construct the decision support system (DSS) for selecting the objective strategy in road traffic engineering. The proposed DSS can suggest the most optimal strategy decision in road traffic engineering. Here, a main traffic road of Juanda in area Ciputat, Tangerang Selatan, province Banten, Indonesia; was selected as a research object in this study. The constructed DSS for road traffic engineering was structurally delivered in this paper.

1. Introduction

Road traffic engineering is one way to reduce the traffic congestion level. Where, the congestion is an ordinary problem that needs an extraordinary way to solve [1], particularly in big cities' main road network [2]. However, the implementation of traffic engineering practically is the other problem. The high cost effect of the implementation will appear. Thus, the engineering strategy has to be selected objectively and firstly executed in a model domain before in a real implementation.

Conceptually, many approaches have been offered to answer the congestion problem thru research projects conducted in several countries. [3] proposed an approach to mitigate the motorway traffic congestion by installing an auxiliary lane. They develop a model to simulate the implementation of solution. [4] implemented the expert system for dynamic traffic light control to improve the condition of road traffic congestion. Furthermore, [5] developed the quantitative system analytic model for traffic controlling. The model was predicted can improve the air quality and solve the problem of traffic congestion.

In addition, [6] constructed an intelligent vehicle system to reduce the traffic congestion. The constructed system aimed to maintain the homogenous vehicle movements in minimalizing the congestion. The system was based on mobile device practically. Also [7] proposed the solution strategy to mitigate a traffic congestion with diversion and lane restrictions. Finally, [8] used the binary logit model to reduce road traffic congestion and air pollutant. Can be concluded that by encouraging the students and employees to use the public transport, the congestion can be conceivably downgraded.

This paper presents the construction of decision support system (DSS) based on fuzzy water flow algorithm (FWFA) that is used to mitigate the congestion via selecting the road traffic engineering strategy. The strategy in engineering a road traffic in one main traffic road in Indonesia was recommended. To deliver it structurally, the section introduction is followed by sections related works, research methodology, results and discussion, and conclusion and further works.

2. Related Works

Numerous researchers have previously used water flow algorithm (WFA) as their research main method, exclusively in computer modeling area. [9] started to use WFA to optimize bin packing problem. Compared to genetic algorithm (GA) method, WFA was more proven optimal. [10] implemented WFA algorithm in segmenting text and detecting reference text line. The WFA method was modified version, especially in expansion values of water flow specified angle and unwetted image frames function enlargement. [11] also did their research in manufacturing field. They used WFA to solve the problem in building the layout of manufacture. WFA was ever operated as well to solve the problem of travelling sales problem [12] and finding the shortest path [13]. And lastly, WFA was functioned as main method of optimization model in answering the road traffic congestion [1].

3. Research Methodology

This study (where the activities are described clearly in Figure 1) is expanded version of [1] research. It was conducted by using main traffic road in area Tangerang Selatan (province Banten, Indonesia) as an object of the research. Through using [14] method, the empirical data in pick time (6 a.m. to 10 a.m.) of five working days (Monday to Friday) was extracted becoming data source of the research.

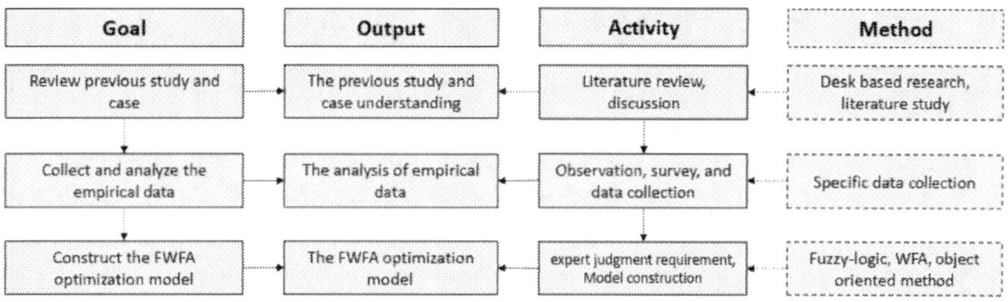

Figure 1. Research Activities

The main method used in developing the system is the combination of WFA [9] [1] and the concept of fuzzy-logic [15]. The WFA operated to model the traffic condition and also to seek the best decision alternative (via optimization process). While, the concept of fuzzy-logic used to parameterize and prioritize all involved parameters in the model. Here, we found a judgment for seeing the urgency of decision alternatives coming from experts' point of view. We required the multi expert judgment and convert it to crisp-output value then. The value used as the coefficient of decision alternatives.

The object oriented method [16] was used to document the constructed system. Two diagrams explicitly used in this study is usecase and class diagrams. The usecase diagram is used to describe the communication pattern between system and actors. Where, the actors theoretically could be human and/or system actors. While, the class diagram depicts the interconnection among classes in system which explains the classes' behavior (attributes and operations).

4. Results and Discussion

4.1. The Constructed System

The usecase diagram of the constructed system is shown clearly in Figure 2. Here, there are two types of human actor (i.e. expert and local government) and a system actor (i.e. traffic center management information system). There are five usecases here; `Extracting Data`, `Parameterizing`, `FWFA Optimizing`, `Making Decision`, and `Reporting`. Three experts involve in the system to justify the urgency of decision alternatives. Through fuzzy and de-fuzzy process, the decision coefficients are produced. The actor `Local Government` involves in parameterizing, optimizing, decision making, and reporting. Instead, the actor is who has responsibility to implement the decision. Whereas, the system actor `Traffic Center MIS` provides all data and information required by the system. Theoretically, the usecase diagram signifies the components of decision support system [17], and ever used by [18] as well); the usecase `Extracting Data` (with its actor) represents the data base management component, the usecases `Parameterizing` and `FWFA Optimizing` symbolize the model base management component; and the usecases `Making Decision` and `Reporting` characterize the user interface base management.

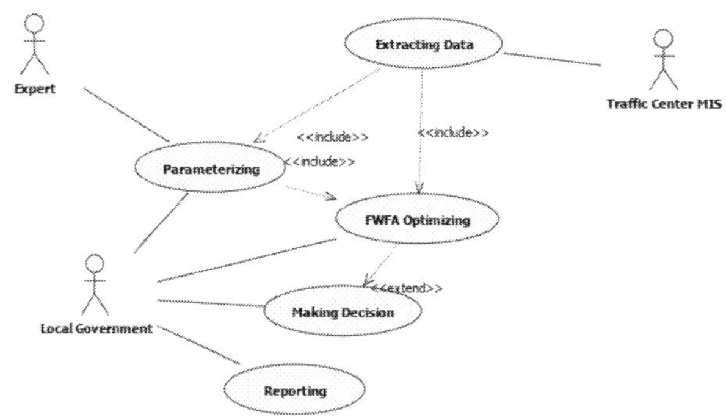

Figure 2. The Usecase Diagram of the Constructed System

The configuration of classes' interconnection of the constructed system is obviously presented in Figure 3. Indeed, it consists of seven main classes; `TrafficRoad`, `UTurn`, `TrafficLight`, `Vehicle`, `Decision`, `ConsideredParameter`, and `FWFAOptimization`. To produce the best decision (class `Decision`), all considered parameters (class `ConsideredParameter`) of traffic road (class `TrafficRoad`) are determined. The value of each parameter is empirically coming from the vehicle (class `Vehicle`) behavior. The best decision itself is generated from process of FWFA optimization (class `FWFAOptimization`). Also, physically, one traffic road can be configured by defining a number of u-turn (class `UTurn`) and traffic light (class `TrafficLight`).

Furthermore, in this study, three types of vehicle are taken into account. They are light vehicle (class `LightVehicle`), heavy vehicle (class `HeavyVehicle`), and motorcycle (class `Motorcycle`). The class `LightVehicle` is representing several types of public (class `PublicTransport`) and private (class `PrivatTransport`) transport vehicle, e.g. mini public transport (angkot in Indonesia), small bus, etc. The class HeavyVehicle symbolizes the heavy vehicles certainly, such as truck, big bus, etc. By using FWFA optimization method, based on a lot of the best decision alternatives (class `BestDecisionAlternative`, the final decision technically is generated. The FWFA optimization method principally utilizes the membership function. A triangular membership function is a membership function type used in this study (class `TriangularMF`) which correlates to limit value (class `LimitValue`).

In addition, there are six direct parameters that are considered in this study. Where each parameter has unique behavior and needs other detailed variables. Classes `RoadVelocity`, `RoadDensity`, `RoadVolume`, `RoadDelay`, `RoadCapacity`, and `RoadSaturationDegree` represent velocity, density, volume, delay, capacity, and saturation degree of the traffic road respectively. The class `RoadDelay` is specifically generated from three types of delay; traffic delay (class `TrafficDelay`), geometry delay (class `GeometryDelay`), and stop point delay (class `StopPointDelay`). Empirically, all parameters are considered and measured based on day (class `Day`) and rush-hour time of each day (class `Time`). The class diagram that describes the detailed classes of class `ConsideredParameter` is illustrated in Figure 4.

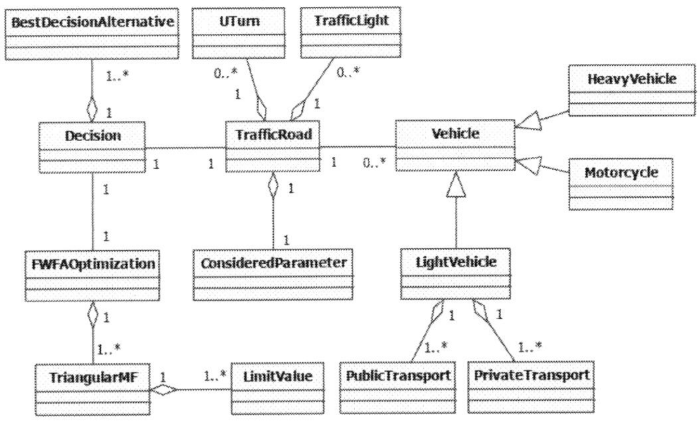

Figure 3. Main Class Diagram of the Constructed System

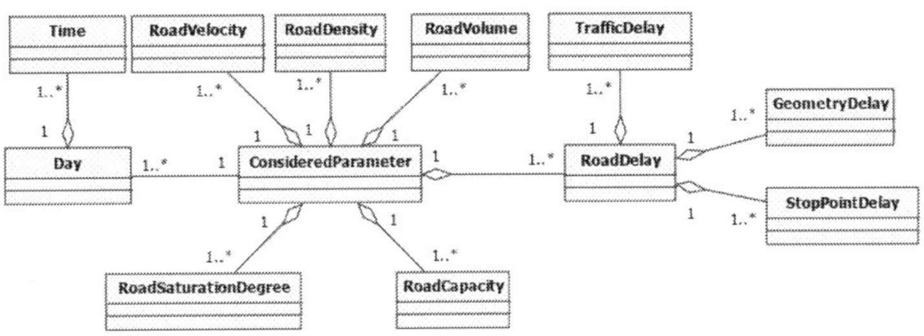

Figure 4. Detailed Classes of Class ConsideredParameter

Nine operations are mechanically determined in class `FWFAOptimization`. The operations `FuzzyProcessing()` and `DefuzzyProcessing()` are used to adjust the coefficient (output crisp value) of the decision alternatives. Where, the input crisp values are coming from the experts' judgment (three expert persons). Principally the mechanism to require the experts' judgment is needed, because it will enrich the final decision objectivity. Technically, the coefficient affects the final relative value of traffic velocity that represents the value of each decision alternative.

By using a triangular membership function (with three fuzzy languages to explain the decision importance; not significant, medium, and significant), the coefficient of six decision alternatives (strategy alternatives of traffic engineering) u-turn elimination, public transport stop point, public transport elimination, motorcycle elimination, traffic light manipulation, and contraflow are 0.43, 0.40, 0.07, 0.05, 0.03, and 0.02 respectively. While, seven other operations (`Splitting`, `Moving`, `Merging`, `Evaporating`, `Precipitating`, `ObtainingBest`, and `Comparing`) and also all mathematical calculations concerning WFA operation are scientifically functioned based on [9] that also ever implemented by [1].

Based on more than 130 decision alternatives that have been checked by using FWFA optimization model (where the optimization process is started from the smallest coefficient) of the constructed system, the highest traffic velocity (in relative value) will be reached (see Figure 5) when the public transport vehicles halt in the right place. It means, the strategy to maintain the public transport to stop in the recognized right place (e.g. bus stop, etc.) is the best decision for road traffic engineering (compared to the other strategies, see Figure 6). It is predicted can improve the traffic velocity three times approximately (from normal velocity 8.64 km/h).

Figure 5. The Alternatives of the Best Decision Produced from FWFA Optimization

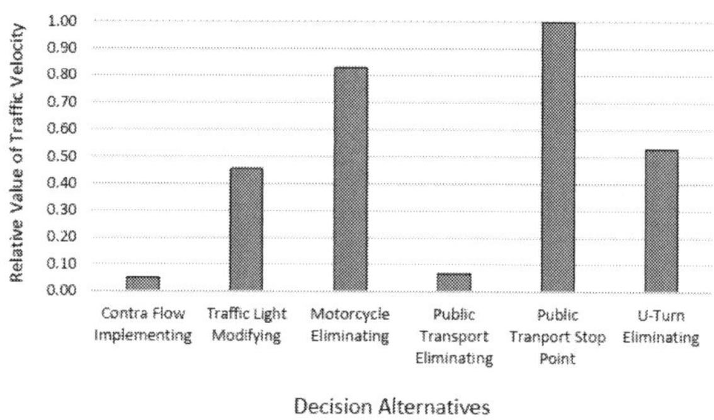

Figure 6. Decision Alternatives Comparison

4.2. *Discussion*

The different decision was proposed here compared to [1]. [1], by using WFA optimization, recommended the u-turn eliminating as the best decision to be taken. It can improved the traffic flow until more than four times. However, when the same data were entered in this proposed FWFA based DSS, where the importance of decision alternatives was allowed to be judged by several experts, the decision alternative to endorse the public transport vehicle to stop in the right place was selected as the best solution. On the other hand, implicitly, the proposed system uses several objective functions of optimization model coming from [1]. As, to select the best decision, the value of objective function of each decision alternative is a value compared. Where practically, the value of objective function characterizes the road traffic velocity.

5. Concussion and Further Works

FWFA is a method that combines the concept of fuzzy-logic and water flow algorithm method. It was used to develop the optimization model that is a part of the constructed DSS. The best decision was proposed as a road traffic engineering to decrease the level of traffic congestion; where it is related to the road traffic velocity. In addition, the concept of fuzzy-logic was practically used to reform the value of expert judgment in determining the urgency value of decision alternatives. It was technically embedded to WFA based optimization model [1] to expand the model. And, the extended model was operated to develop the constructed DSS.

Ecological parameters (eco-parameters), e.g. air pollution degree, CO content in air, fuel consumption level, etc. are attracting parameters to be reasonably taken into account. The consideration will become a new idea of the further study. Besides, other parameters, such as type of city or the parameter of citizens' behavior in driving, can affect the strategy for road traffic engineering.

References

[1] Utama D N, Zaki F A, Munjeri I J and Putri N U 2016 A water flow algorithm based optimization model for road traffic engineering *Proceeding of the 2016 International Conference on Advanced Computer Science and Information Systems* Malang Indonesia
[2] He F, Yan X, Liu Y, and Ma L 2016 A traffic congestion assessment method for urban road networks based on speed performance index *Procedia Engineering* 137 pp 425-433
[3] Xing J and Tanaka J 2016 Effective installation of an auxiliary lane at sag sections to mitigate motorway traffic congestion *Transportation Research Procedia* 15 pp 187-197
[4] Wen W 2008 A dynamic and automatic traffic light control expert system for solving the road congestion problem *Expert Systems with Applications* 34 4 pp 2370-2381
[5] Tan Z and Gao H O 2015 Traffic control for air quality management and congestion mitigation in complex urban vehicular tunnels *Transportation Research Part C: Emerging Technologies* 58 A pp 13-28
[6] Koshizen T 2013 A smartphone-based intelligent vehicle system to minimize the possibility of traffic congestion occurrence *IFAC Proceedings* 46 21 pp 643-644
[7] Davis L C 2012 Mitigation of congestion at a traffic bottleneck with diversion and lane restrictions *Physica A: Statistical Mechanics and its Applications* 391 A pp 1679-1691
[8] Le T P L and Trinh T A 2016 Encouraging public transport use to reduce traffic congestion and air pollutant: a case study of Ho Chi Minh City, Vietnam *Procedia Engineering* 142 pp 236-243
[9] Yang F C and Wang Y P 2007 Water flow-like algorithm for object grouping problems *Journal of Chinese Institute of Industrial Engineers* 24 pp 475-488
[10] Brodić D and Milivojević Z 2010 An approach to modification of water flow algorithm for segmentation and text parameters extraction *Proceedings of Doctoral Conference on Computing, Electrical, and Industrial Systems* pp 324-331
[11] Chang C C and Wu T H 2011 A water flow-like algorithm for celullar manufacturing design and layout *International Conference on Advanced Information Technologies* (AIT)
[12] Srour A, Othman Z A and Hamdan AR 2014 A water flow-like algorithm for travelling salesman problem *Advances in Computer Engineering* pp 1-14
[13] Zainudin S, Kerwad M M and Othman Z A 2015 A water flow-like algorithm for capacitated vehicle routing problem *Journal of Theoretical and Applied Information Technology* 77 125-135
[14] Balakhrishnan S and Sivanandan R 2015 Influence of lane and vehicle subclass on free-flow speeds for urban roads in heterogeneous traffic *Transportation Research Procedia* 10 pp 166-175
[15] Zadeh L A 1996 Fuzzy logic = computing with words *IEEE Transactions of Fuzzy Systems* 4 2 pp 103-111
[16] Mathiassen L, Munk-Madsen A, Nielsen P A and Stage J 2000 *Object-Oriented Analysis and Design* Marko Publishing ApS Aalborg Denmark
[17] Turban E, Aronson J E and Liang T P 2005 *Decision Support Systems and Intelligent Systems, 7th Edition* Pearson
[18] Utama D N, Saputra M D, Wafiroh L N, Putra M A A and Lestari P 2016 F-multicriteria based decision support system for road repair and maintenance (case study: three areas in Tangerang Selatan, Province Banten, Indonesia) *Proceeding of Academics World 39th International Conference* pp 27-31

IOP Conf. Series: Journal of Physics: Conf. Series **801** (2017) 012017 doi:10.1088/1742-6596/801/1/012017

Trajectory Pattern Mining Using Sequential Pattern Mining and K-Means for Predicting Future Location

G Kautsar[1]**, S Akbar**[1]

[1]Informatics, School of Electrical Engineering and Informatics, Institut Teknologi Bandung, Ganesha Street No. 10 Bandung, Indonesia

Email : gifarikautsar@gmail.com, saiful@informatika.org

Abstract. Sequential pattern mining is a method used to find patterns while concerning the sequence of an item set. Sequential pattern mining can be used to find trajectory patterns in moving object data. To implement it in the real life, the spatial attribute of the data needs to be generalized/grouped. In this paper, K-Means is used to group the spatial attribute. In order to group the spatial attribute, the temporal attribute is also considered to see how the patterns are related to time. The resulting trajectory patterns are then used to visualize the habit of the moving object. Therefore, trajectory patterns are used as the reference in this paper to predict the future location of the object. Predicting the future location of the object is performed using the movement history of the object. Result of this research is trajectory pattern which repeat at certain time duration according to its data characteristics.

1. Introduction

Lately, people use cell phone to support almost all of their activities. Therefore, it is normal that the cell phone usage in this era is growing very rapidly. In the year of 2000, Indonesian mobile phone users were about 3.6 million. In 2012, that number increased significantly, reaching a total of 280 million [1]. Cell phone service provider always keep a record every time its user communicates with other users, e.g. when the user make a phone call or send a text message. That data is called Calls Details Record (CDR).

Meanwhile, Location Based Service (LBS) is a service that provides information about a place [2]. LBS can be easily accessed using mobile devices. An example of a system applying LBS is Global Positioning System (GPS). GPS can be used on a vehicle, cell phone, and even animal to find its position at a time. So, using GPS can generate data containing the location of an object. An example of the data is path traversed by a taxi, the migration of animals, and the location of a plane.

Spatio-temporal database is a database managing the spatial attribute that changes over time [3]. A moving object data is an example. CDR and GPS data are another example because the data describes the location of an object at a time. A set of locations in a period of time can show the movement of the object from time to time.

Pattern recognition is a technique to classify or to describe object based on its quantitative characteristics or its main property. Pattern that can be recognized form moving object data includes repetitive pattern, relationship pattern, and frequent trajectory pattern [4]. Repetitive pattern is a pattern formed from an object doing some movement periodically. An example is a bird that every day is flying from its nest to look for food. Relationship pattern is a trajectory pattern resulting from a collection of objects that is moving and each has a relation with each other. Frequent trajectory pattern is a pattern which is the trend of the movement of all objects.

Frequent trajectory pattern mining can be applied in both CDR and GPS data to see the habit of the moving object. This habit can be represented as trajectory pattern. The discovered trajectory patterns are then can be used to predict the future location of an object. In more advanced development, the discovered trajectory patterns can be utilized to know which road segment is often passed by. So, it can be used to give recommendation which road segment needs restoration or to give new routes recommendation. For example, from the trajectory patterns discovered, it is known that movement from location A to location B is quite crowded and the road segment is narrow. From that, the possible recommendation is to improve the road segment from location A to location B.

2. Preliminary Concept

2.1. Sequential pattern mining

Item is the basic value for numerous data mining problems. It can be considered as objects bought by customer, location visited by bus, etc. Item set is a set of items that are grouped by timestamp. Meanwhile, data sequence is a sequence of item set associated to an object [5]. On Table 1, data sequence for object *C2* is "(Camcorder, MiniDV) (DVD Rec, DVD-R) (Video Soft)". It means that the customer bought camcorder and miniDV on the same day, then bought DVD recorder and DVD-R on the next day, and bought video software some days later.

Sequential pattern can be found in a data sequence. For example, "(MiniDV) (Video Soft)" can be found in *C2*, but "(DVD Rec) (Camcorder)" cannot be found in *C2* because it is not in the right order (Camcorder came first). In sequential pattern mining, user has to define the minimum support value. The minimum support value is the minimum number of data sequence that satisfies the sequential patterns so the sequential pattern can be categorized as a frequent pattern. For example, from Table 1, there are four customers in four days. If the value of minimum support is 50% of total customers, then sequence pattern has to be found in minimum two data sequences. A sequential pattern mining performed on the data will find three patterns:

- S1: "(Camcorder, MiniDV) (DVD Rec, DVD-R)"
- S2: "(DVD Rec, DVD-R) (Video Soft)"
- S3: "(Memory Card) (USB Key)"

Table 1. Data sequences of four customers over four days [5].

Cust.	June 04, 2004	June 05, 2004	June 06, 2004	June 07, 2004
C1	Camcorder, MiniDV	Digital Camera	MemCard	USB Key
C2	Camcorder, MiniDV	DVD Rec, DVD-R	-	Video Soft
C3	DVD Rec, DVD-R	MemCard	Video Soft	USB Key
C4	-	Camcorder, MiniDV	Laptop	DVD Rec, DVD-R

Pattern S1 can be found at data sequence C2 and C4, pattern S2 can be found at C2 and C3, and pattern S3 can be found at C1 and C2.

2.2. K-Means

Clustering or cluster analysis is the process of grouping set of objects into clusters. The objects grouped in a cluster will have a high similarity with each other and are different with objects from the other clusters [6]. The higher the similarity and the higher the differences, the better the cluster is.

Prototype-based clustering is a type of clustering. On prototype-based clustering, a cluster is a set of objects with higher similarity to its prototype than to other prototypes [7]. K-Means is an example of prototype-based clustering.

The first step of implementing K-Means is choosing the initial K centroid, with *K* is the number of cluster defined by the user. Each object is assigned to a cluster with the closest centroid. After the assigning process is done, each cluster centroid is then updated based on the objects in the cluster.

These steps are repeated until there are no object moving to other cluster, or in another word, the process will continue until the centroid value doesn't change anymore.

Example of applying K-Means algorithm in clustering object can be seen at Figure 1. On Figure 1 (a), three objects are assigned as the initial centroids. Each object aside from the three before are then assigned to the closest cluster (based on the distance to centroids). On Figure 1, objects are represented in difference shapes, which are: triangle, square, and circle. After the objects are clustered, centroids value is updated and then the objects are clustered again based on the distance to the new centroids. On Figure 1 (b), (c), and (d), it can be seen that the centroids are moving to smaller cluster. K-Means algorithm is stopped (on Figure II.1. (d)) because the objects are no longer moving.

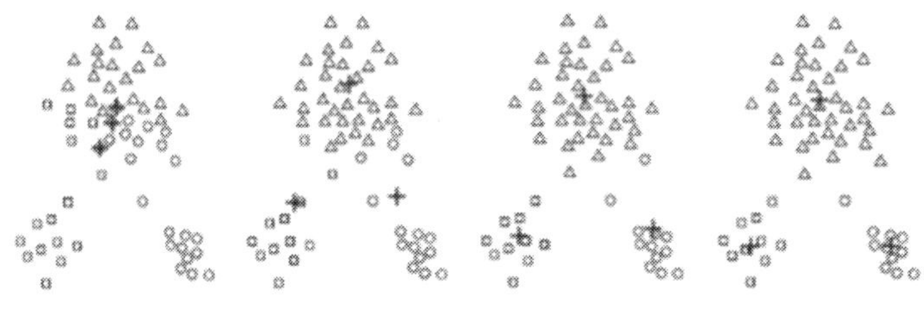

(a) Iteration 1. (b) Iteration 2. (c) Iteration 3. (d) Iteration 4.

Figure 1. An example of clustering objects using K-Means.

3. Related Work

Spatial or spatio-temporal data clustering have been developed by Tork in 2012 using K-Means algorithm. Spatial data clustering method have the same concept like normal clustering. The differences are only on the data type [8]. Data shown in Figure 2 is a sample of simple data (containing no spatial data). Each point is represented in $<x,y>$ form in a 2-D space. It doesn't have any information about the spatial or temporal data of the points. Using this definition, spatial clustering can be simplified into two dimensional vector, like $<x,y>$, but its values are replaced with spatial value, e.g. longitude and latitude.

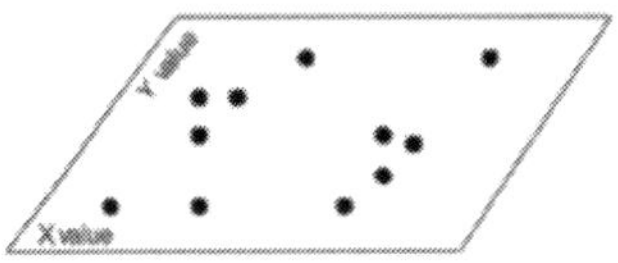

Figure 2. Simplified classical data in 2-D space [8].

Spatio-temporal data clustering is very much the same with with spatial data clustering. The differences between spatio-temporal data clustering and spatial data clustering lies in the distance calculation. For example, take a look at the K-Means distance calculation algorithm. The first step is to change spatio-temporal data into three dimensional vector $<x,y,t>$. Tuple $<x,y>$ is the spatial attribute, and t is the temporal attribute [8]. Using the new three dimensional vector $<x,y,t>$, spatio-temporal distance (using Euclidean distance) can be calculated as:

$$\sqrt{(x_2 - x_1)^2 + (y_2 - y_1)^2 + (t_2 - t_1)^2} \quad (1)$$

Trajectory pattern mining using sequential pattern mining method has been developed by Chung et al. in 2002. The steps are as follow.

1) Make some adjustment to the data representation and arrange the database.
2) Generalize the spatial data.
3) Moving sequence/trajectory extraction. On this step, user have to define the maximum duration between each movement.
4) Trajectory pattern mining from moving sequence/trajectory data. F_k represents pattern-k obtained from data. C_k represents pattern candidate. C_k is self-join from F_{k-1}, ..., $F_{k-1} * F_{k-1}$. The

support value is calculated by checking if the trajectory data satisfies the pattern candidate. If the support value is less than the minimum value, pattern candidate is not assigned as frequent trajectory pattern.

4. Trajectory Pattern Mining

Trajectory pattern mining consists of several steps which are: data pre-processing, clustering the spatial attribute, data adjustment, and pattern mining. On the pattern mining step, a model of the spatial attribute cluster and trajectory pattern are gained. The result is used as reference to predict the future location.

4.1. Data preprocessing

Data preprocessing is a must before the trajectory pattern mining is performed. It consists of two steps. The first step is to cleanse the data and the second step is to generate additional attribute from temporal attribute. The cleansing is done by changing the time attribute into datetime attribute. Datetime is an attribute formed by joining the time attribute and the date attribute. Another adjustment is done to the attribute at each record to fit its definition. The last adjustment is handling the incomplete record or empty record. In the incomplete record case, if the missing attribute is the spatial or the temporal attribute, then the record won't be used in the calculation. This is because trajectory pattern mining can't be done without it.

For clustering step, we need an additional attribute which represents the temporal attribute. The temporal attribute is in datetime type. As it is, temporal attribute is unique and it would be harder to find the pattern. For clustering step, temporal attribute is derived to get a new attribute. A new derived temporal attribute is needed in order to enhance the quality of the trajectory pattern. To simplify the pattern mining, the temporal attribute is changed into two types of hour unit. The first type, called the *daily* type, only uses the hour attribute from the original datetime. The second type, the *weekly* type uses the date and the hour attribute from the datetime. The steps to change the temporal value into *weekly* type are as follow.

1) Change the date into day. For example, for June 14th 2016 is Tuesday.
2) Change the day value into *dayOfWeek* (see Table 2).
3) The new temporal value, the weekly type is calculated with the formula (2) below.

$$t_{weekly} = (dayOfWeek * 24) + hour \tag{2}$$

dayOfWeek is multiplied by 24 because there are 24 hours in a day.

Table 2. Day to dayOfWeek conversion.

Day	*dayOfWeek*
Monday	0
Sunday	1
Tuesday	2
Wednesday	3
Thursday	4
Friday	5
Saturday	6

4.2. Spatial clustering

There are two experiments performed to cluster the data. In the first experiment, data clustering only involves spatial attribute. K-Means algorithm is chosen because it uses the same mechanism to cluster spatial data only and spatial with temporal attribute. In the first experiment, the spatial attribute has latitude and longitude values. Formula (3) is used as distance calculation function in K-Means for spatial data clustering.

$$d = \sqrt{(x_1 - x_2)^2 + (y_1 - y_2)^2} \tag{3}$$

In the second experiment, the data clustered are spatial attribute (latitude and longitude) and temporal attribute (which value had been changed into daily or weekly). Formula (1) is used as distance calculation function for the second K-Means experiment. Spatial attribute (x,y) and temporal attribute (t) each has different unit, but using distance calculation function (see Formula 1), both attributes are considered as equal. Normalization of attribute is done to balance the spatial and temporal attributes. With normalization, there are no specific attribute that would affect the cluster arrangement greatly. The result from data clustering step is a data clustering model. This model is used to define test data cluster before it is used later for predicting future location.

4.3. Generating trajectory data

The trajectory data is generated from clustered data. Object id, datetime, and spatial cluster are the attributes used in this step. Spatial attribute is no longer used because it has been replaced by the spatial cluster. In detail, trajectory data is generated from set of spatial clusters with the same object id in a certain time range. Set of trajectory data can be generated using these following steps.

1) Define the maximum movement duration from one location to the next location.
2) Calculate time differences between location *L* and last location in trajectory *T*.
3) If the time difference is less than or equal with maximum movement duration, then *L* is the next location of trajectory *T*.
4) If the time difference is more than the maximum movement duration, then *L* is the first location of a new trajectory.

4.4. Trajectory pattern mining

Temporal pattern mining, which was developed by Chung et al., 2002, is used for trajectory pattern mining. Temporal pattern mining is sequential pattern mining for moving object data. Temporal pattern mining is chosen because trajectory pattern can be recognized even though there is a missing point in the data. Data used in pattern mining are the data that have been generated into trajectory data. Some examples of generated trajectory data are <A B C D>, <A E C D>, and <B C D A>. The result of pattern mining using temporal pattern mining for the example trajectory data above can be seen at Table 3.

Table 3. Examples of trajectory pattern obtained from trajectory data.

No.	Pattern	Obtained from trajectory...
1.	<A C>	<**A** B **C** D>, <**A** E **C** D>
2.	<A D>	<**A** B C **D**>, <**A** E C **D**>
3.	<C D>	<A B **C D**>, <A E **C D**>, <B **C D** A>
4.	<B C>	<A **B C** D>, <**B C** D A>
5.	<A C D>	<**A** B **C D**>, <**A** E **C D**>
6.	<B C D>	<A **B C D**>. <**B C D** A>

5. Location Prediction

Location prediction is initiated with data pre-processing. Data pre-processing is performed by clustering spatial attribute using clustering model produced in trajectory pattern mining step. After the data is clustered, trajectory data is generated using the same steps used in trajectory pattern mining. Location prediction is done with the following steps.

1) Each trajectory pattern is divided into two parts, the first part is called *route* and the second part is called *nextStep*. The *route* is a set of spatial cluster from the first cluster to n-1 spatial cluster from the trajectory pattern. *nextStep* is the last spatial cluster from trajectory pattern. For example, for trajectory pattern $<S_1\ S_2\ S_3\ S_4 \ldots S_{n-1}\ S_n>$, the value of *route* is $<S_1\ S_2\ S_3\ S_4 \ldots S_{n-1}>$ and the value of *nextStep* is $<S_n>$.
2) Testing is performed to measure the similarity between test data and *route*. If *route* is a sequence pattern from test data, the prediction of future location is the *nextStep* of its *route*.
3) Similarity value of test data and *route* can be measured by two type of pattern quality functions, *value* and *density*. *Value* type is the value of support from trajectory pattern, while

support value is the number of trajectory data which satisfies trajectory pattern [10]. Meanwhile, value of *density* type is obtained using Formula 4.

$$density = support * \frac{\overline{route}}{trajectory\ data\ test} \quad (4)$$

4) If there is more than one route which satisfies the test data, then a route with the highest value of pattern quality will be chosen. Table 4 shows the examples of trajectory patterns.

Table 4. Example of set of trajectory pattern.

Pattern	*support*	*route*	*nextStep*
<A B A>	10	<A B>	<A>
<A D B C D>	6	<A D B C>	<D>
<B A C>	7	<B A>	<C>
<D B D>	3	<D B>	<D>

For example, <C D A D B C> is the test data to be predicted. In Table 5, examples of the similarity calculation between test data and trajectory pattern from Table 4 are shown. The prediction result using *value* type of pattern quality calculation is <A>, while the prediction result using *density*-typed pattern quality is <D>.

Table 5. Examples of similarity calculation between test data and trajectory pattern.

route	*nextStep*	Test Data	*value*	*density*
<A B>	<A>	<C D **A** D **B** C>	**10**	10*2/6 = 3,33
<A D B C>	<D>	<C D **A** D **B C**>	6	**6*4/6 = 4**
<B A>	<C>	-	0	0
<D B>	<D>	<C D A **D B** C> or <C **D** A D **B** C>	3	3*2/6 = 1

6. Experimental Result

In trajectory pattern mining, the value of some variables which will determine the trajectory pattern needs to be determined. The purpose of the experiment is to determine the best combination from the value of variables which would yield the best accuracy. Experiment is done using two datasets. The first dataset is CDR data, and the second dataset is taxi GPS data. Each dataset is divided into two parts, 80% of data as train data, and 20% of data as test data. Variables which values are needed to be determined are:

1) Number of K-Means cluster.
2) Temporal attribute usage in clustering.
3) Pattern quality calculation function.
4) Minimum support value for trajectory pattern.

Experiment is done by trajectory pattern mining on train data using every possible combination of variables value. The evaluation of experiment result is done by calculating accuracy from location prediction of test data.

6.1. Experiment using CDR data

The CDR data used for experiment is CDR data from mobile phone users in Bali and its surrounding areas. The characteristics of CDR data are as follow.

1) Number of records : 1.740.863
2) Number of objects : 20.029
3) Average locations per object : 87
4) Number of distinct locations : 2.416
5) Data duration range : October 6th 2014 - October 12th 2014

Values from variables that are used in experiment using CDR data are as follow.

1) Number of K-Means cluster : 10, 15, 20, 25, and 30.
2) Temporal attribute usage in clustering : *weekly*, *daily*, and *no* (without temporal attribute).
3) Pattern quality calculation function : *value* and *density*.
4) Minimum support value : 15, 20, 25, 30, and 35.

Based on experimental result, the 3 best variables configuration can be seen in Table 6.

Table 6. Best trajectory pattern mining and location prediction configuration for CDR data.

No	*Using Temporal*	*nCluster*	*Min. Support*	*Pattern Quality*	Accuracy *(%)*
1	*weekly*	10	15	*value*	99,43726
2	*weekly*	10	15	*density*	99,43726
3	*weekly*	10	20	*value*	99,3308

Temporal attribute usage in clustering greatly determines accuracy result of location prediction. *Weekly* type temporal attribute usage in clustering resulted in higher accuracy than *daily* type and *no* type. Average accuracy value for *weekly* type is 98,96%, while average accuracy value for *daily* type is 50,44% and for *no* type is 55,60%. Average accuracy value for *weekly* type is higher, that means that trajectory pattern in CDR data is unique for each day in a week (repeated every week).

Average accuracy value of location prediction based on minimum support value is shown in Figure 3. The higher the minimum support value, the smaller the average of accuracy value is. This is because higher minimum support value will cause more pattern candidates did not satisfy the minimum value.

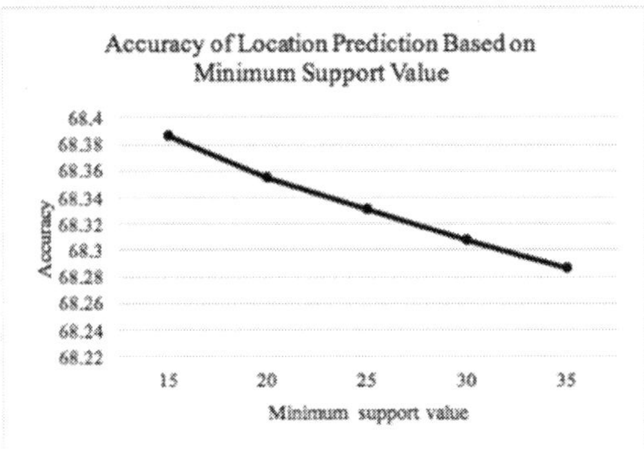

Figure 3. Accuracy of location prediction of CDR data based on minimum supprot value.

6.2. Experiment using taxi GPS data

Characteristic of taxi GPS data used in the experiment are as follows.

1) Number of object : 421
2) Number of record : 34.911
3) Average locations per record : 48
4) Number of distinct location : 1.331.454
5) Data duration range : July 1st 2013 - July 7th 2013

Taxi GPS data type is different with the general moving object data type. Taxi GPS data contains a set of spatial points which are passed in taxi trip. In other words, one record of taxi GPS data represents one trajectory, not only one spatial point. One spatial point in taxi GPS data is recorded every 15 seconds in every taxi trip.

Variables value that are used in the experiment using taxi GPS data are as follow.

1) Number of K-Means cluster : 30, 40, 50, 60, and 70.
2) Temporal attribute usage in clustering : *weekly*, *daily*, and *no* (without temporal attribute).
3) Pattern quality calculation function : *value* and *density*.
4) Minimum support value : 15, 20, 25, 30, and 35.

Based on the experimental result, the 3 best variable configurations can be seen in Table 7.

Table 7. Best trajectory pattern mining and location prediction configuration for taxi GPS data.

No	*Using Temporal*	*nCluster*	*Min. Support*	*Pattern Quality*	Accuracy (%)
1	*daily*	30	15	*value*	99,71202
2	*daily*	30	15	*density*	99,71202
3	*daily*	30	20	*value*	99,71202

Based on experimental result using taxi GPS data, temporal attribute usage in clustering really determines the accuracy of location prediction. *Daily* type temporal attribute usage in clustering has higher average accuracy value than *weekly* type and *no* type. Average accuracy for *daily* type is 99,586%, while the average accuracy value for *weekly* type is 34,296% and the average accuracy value for *no* type is 41,833%. Average accuracy value for daily type is higher, that means the trajectory pattern in taxi GPS data is repeated every day.

The effect of minimum support value in this experiment, is same as the effect of minimum support value in CDR data experiment. The higher minimum support value, the smaller average of accuracy value.

7. Conclusion and Future Works

In this research, trajectory pattern mining and future location prediction has been developed using sequential pattern mining method. Trajectory pattern mining and future location prediction has been tested using CDR data and taxi GPS data. Based on the experiment result, the best accuracy for location prediction using CDR data is 99,437% and the best accuracy using taxi GPS data is 99,712%. K-Means algorithm is used for spatial data clustering. Clustering is done twice, using only the spatial attribute and spatial and temporal attribute. Temporal attribute usage in clustering highly affects the accuracy of trajectory pattern mining. Using temporal attribute, time characteristic of trajectory pattern can be recognized. Based on experiment result of trajectory pattern mining and future location prediction, trajectory pattern of CDR data is repeated every week. Meanwhile, the trajectory pattern of taxi GPS data is repeated every day.

For future research of trajectory pattern mining, spatial clustering can be done using region/area concept. For example, spatial attribute is clustered based on administrative region.

8. References

[1] K Ariansyah 2014 Proyeksi Jumlah Pelanggan Telepon Bergerak Seluler di Indonesia *Buletin Pos Dan Telekomunikasi* **12(2)** 151-166
[2] J Schiller, A Voisard. (Eds.) 2004 *Location-based service* (Amsterdam: Elsevier)
[3] M Nanni 2002 Clustering Methods for Spatio-Temporal Data *PhD thesis, Dipartimento di Informatica, Università di Pisa*
[4] Z Li, J Han, M Ji, L A Tang, Y Yu, B Ding, J G Lee, R Kays 2011 MoveMine: Mining Moving Object Data for Discovery of Animal Movement Patterns *ACM Transactions on Intelligent Systems and Technology (TIST)* **2(4)** 37
[5] F Masseglia, M Teisseire, P Poncelet 2005 Sequential Pattern Mining *Encyclopedia of Data Warehousing and Mining* 1028-1032
[6] H Miller, J Han 2001 Spatial Clustering Methods in Data Mining *Geographic Data Mining and Knowledge Discovery* 188–217
[7] P N Tan 2006 Introduction to data mining (India: Pearson Education India)
[8] H F Tork 2012 Spatio-temporal clustering methods classification *In Doctoral Symposium on Informatics Engineering* 199-209
[9] J Chung, O Paek, J Lee, K Ryu 2002 Temporal Pattern Mining of Moving Objects for Location-Based Service *Database and Expert Systems Applications* 331-340
[10] T H N Vu, K H Ryu, N Park 2009 A method for predicting future location of mobile user for location-based services system *Computers & Industrial Engineering* **57(1)** 91-105

IOP Conf. Series: Journal of Physics: Conf. Series **801** (2017) 012018 doi:10.1088/1742-6596/801/1/012018

Analysis of Human Mobility Based on Cellular Data

F Arifiansyah[1], G A P Saptawati[1]

[1]*School of Electrical Engineering and Informatics Institut Teknologi Bandung Ganesha 10, Bandung, West Java, Indonesia*

Email: fitra@informatika.org, putri@informatika.org

Abstract. Nowadays not only adult but even teenager and children have their own mobile phones. This phenomena indicates that the mobile phone becomes an important part of everyday's life. Based on these indication, the amount of cellular data also increased rapidly. Cellular data defined as the data that records communication among mobile phone users. Cellular data is easy to obtain because the telecommunications company had made a record of the data for the billing system of the company. Billing data keeps a log of the users cellular data usage each time. We can obtained information from the data about communication between users. Through data visualization process, an interesting pattern can be seen in the raw cellular data, so that users can obtain prior knowledge to perform data analysis. Cellular data processing can be done using data mining to find out human mobility patterns and on the existing data. In this paper, we use frequent pattern mining and finding association rules to observe the relation between attributes in cellular data and then visualize them. We used weka tools for finding the rules in stage of data mining. Generally, the utilization of cellular data can provide supporting information for the decision making process and become a data support to provide solutions and information needed by the decision makers.

1. Introduction

Cellular data is one of many data that widely available. Nowadays mobile phone has gradually become a necessity to everyone. First, cellular data is only used as a communication tool. Now, along with the development of technology, Cellular data can also be used as a decission support to create a plan for a smart city [1]. Cellular data is easy to be obtained, as telecommunication companies has been made the record of user's mobile data for company's billing system. Billing data saves the log of cellular data use from the user periodically. That data provides informations about the communication between the users.

Cellular data can be used not only for billing, but also for other domains such as transport, economic, security, the settlement plan, and tourism [1]. In the tourism domain, cellular data can be used to determine which location that becomes a tourist attraction in a city, how long the tourists stayed there, and also the route that usually taken by the tourists. By knowing that points, the government can provide a better transportation to the most favorite tourist destination to get a better profit. Aside from the tourism domain, cellular data can also be used in the economic domain. By knowing the areas where most activities are carried out, the government can plan a place where business or shopping center will be built. Other domains that can utilize cellular data is settlement. A study in Madrid [2] identified five characteristics of the land used by humans based on cellular data using fuzzy c-means algorithm. Cellular data can also be used to estimate the relationship between population density and energy consumption in the domain of settlements and population.

In this paper, we are using data mining technique for analyzing cellular data. After that, the result of data mining visualized to gain prior knowledge for the user about interesting pattern in cellular data. This study uses data records from one of cellular company in Indonesia, Telkomsel in Bali Province for 7 days in October 2014. And then we called this visualization result as a human mobility pattern. Cellular data processing workflow can be seen in Figure 1. The focus of the research are doing the analysis of the results that have been observed which is expected to provide new insights about the utilities of the data on the mobile phone in various domains.

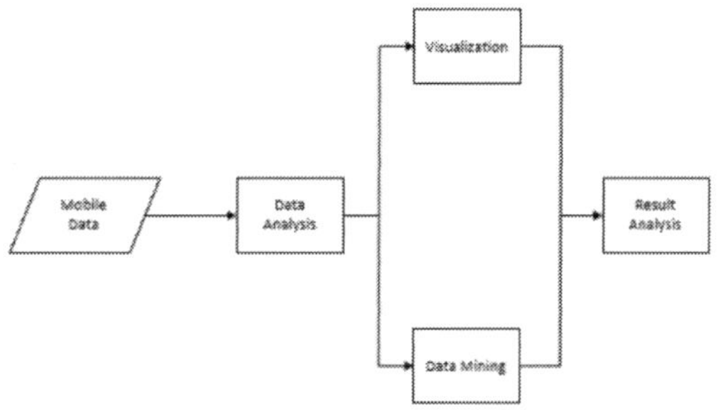

Figure 1. Cellular Data Processing

2. Data Collection

Cellular data records communication among mobile phone users. This study use data that save the location of the user along with time and date where the user are doing the communication. Examples of cellular data used in this study can be seen in Figure 2.

Tanggal	Jam	MSISDN	LAC	CI	LAC CI	Regional	Cellname
2014-06-10	8:00:00	62811100261	8042	16146	8042-16146	BALI NUSRA	MGw614MW1
2014-06-10	8:00:00	62811100261	8042	19572	8042-19572	BALI NUSRA	PELITABANDARABTSMW
2014-06-10	8:00:00	62811100261	8252	35561	8252-35561	BALI NUSRA	TUBANGRAPARIMG1
2014-06-10	9:00:00	62811100261	8065	14754	8065-14754	BALI NUSRA	POLETKUMAR1MX1
2014-06-10	9:00:00	62811100261	8071	10613	8071-10613	BALI NUSRA	PEMOGANMW3
2014-06-10	16:00:00	62811100261	8071	10211	8071-10211	BALI NUSRA	PANJERENONMW1
2014-06-10	16:00:00	62811100261	8065	10421	8065-10421	BALI NUSRA	TUKADAYUNGMW1
2014-06-10	16:00:00	62811100261	8071	39843	8071-39843	BALI NUSRA	TUKADPAKRISANMZ3
2014-06-10	20:00:00	62811100261	8065	17922	8065-17922	BALI NUSRA	DPR792MW1
2014-06-10	20:00:00	62811100261	8065	17925	8065-17925	BALI NUSRA	DPR792MW1
2014-06-10	20:00:00	62811100261	8161	30652	8161-30652	BALI NUSRA	EVCOKTRESNAMG2
2014-06-10	20:00:00	62811100261	8161	30421	8161-30421	BALI NUSRA	EVTUKADAYUNGMG1
2014-06-10	20:00:00	62811100261	8161	30655	8161-30655	BALI NUSRA	EVCOKTRESNAMD1

Long	Lat	Desa	Kec	Kabupaten
115.16866	-8.75691	KEDONGANAN	KUTA	BADUNG
115.18103	-8.74323	TUBAN	KUTA	BADUNG
115.181	-8.74001	TUBAN	KUTA	BADUNG
115.21456	-8.66943	DAUH PURI KLOD	DENPASAR BARAT	KOTA DENPASAR
115.19666	-8.71765	PEMOGAN	DENPASAR SELATAN	KOTA DENPASAR
115.2221805	-8.6898805	SESETAN	DENPASAR SELATAN	KOTA DENPASAR
115.2331	-8.683	PANJER	DENPASAR SELATAN	KOTA DENPASAR
115.22617	-8.68845	PANJER	DENPASAR SELATAN	KOTA DENPASAR
115.236853	-8.65814	SUMERTA KELOD	DENPASAR TIMUR	KOTA DENPASAR
115.236853	-8.65814	SUMERTA KELOD	DENPASAR TIMUR	KOTA DENPASAR
115.2364	-8.6654694	SUMERTA KELOD	DENPASAR TIMUR	KOTA DENPASAR
115.2331	-8.683	PANJER	DENPASAR SELATAN	KOTA DENPASAR
115.2364	-8.6654694	SUMERTA KELOD	DENPASAR TIMUR	KOTA DENPASAR

Figure 2. Examples of Cellular Data

Based on Figure 2, we will do an analysis of the attributes of the data to determine the type of data and its use. There are 13 columns attributes of the cellular data which is 'date' that record the date of the record, 'time' that record the time when user make a call, 'MSISDN' which is the phone number of the user and the primary key because it is unique, 'LAC' noted the location of Base Transceiver Station (BTS) closest of the user position when making call, 'CI' is an ID provided by BTS, 'LAC-CI' is a combination of attributes LAC and CI, 'Regional' is the province of the record, 'Cellname' is the name of the nearest BTS which is used by the user, 'Long' recorded the longitude of the nearest BTS from the user, 'Lat' recorded the latitude of BTS closest from the user, 'Desa' noted the location of the village nearest BTS from the user, 'Kec' noted the location of nearest district from the user, 'Kabupaten' noted the location of nearest kabupaten from the user.

Interpretation of the data in Figure 2 is a user with a MSISDN 62811100491 on October 6, 2014 make a calls at 12 at night in district Denpasar Selatan, then at 2, 4, 6 doing a same call again which the position is still the same in District Denpasar Selatan. User make a move at 7 a.m to District Denpasar Barat. Then back again to its original position in District Denpasar Selatan at 17 in the afternoon.

Based on this information, we can see the movement patterns of users every day. Based on the obtained data, we can assumed that the location of BTS represents the location of the user. This cellular data used to observe the objects over time. By definition of temporal data [3], the observed cellular data included in the temporal data types. Based on the definition of the trajectory [3] which is a series of events in which the movement of objects, spatial location, and time of the visit of the object can be stored for analysis. Then the observed cellular data included in the trajectory data types.

3. Data Mining and Visualization

3.1. Data Mining

Data mining aims to transform information into knowledge [5]. This paper will explain the process to get an information from cellular data. The knowledge obtained from the processing of cellular data is expected to be a tool for decision-making processes in the domains utilization. Knowledge of the transport domain is expected to help the government to make public transport route planning and transport-related infrastructure improvements. Step-by-step application of data mining in general have two step, the first one is preprocessing and the second one is implementation of data mining algorithms.

Data mining process begins with the selection of suitable mining techniques for the cellular data. Pattern mining is the most suitable to be used because we will observed the value or events that occur repeatedly in the dataset to form certain patterns. The chosen approach is a frequent pattern because it doesn't pay attention to the sequence of events, but the frequency of occurrence of an item simultaneously in a transaction dataset. Referring to the characteristics of frequent pattern that doesn't pay attention to the sequence of events but rather focus on the appearance of the same item, the cellular data will be processed using this technique. Because the pattern will be observed in cellular data over to the appeal of patterns in conjunction with the county District A with District B and C by a user, the user here represents a single transaction and District represent the items in a dataset. The dataset used can be seen in Table 1.

Table 1. Dataset of Cellular Data

ID_Trans	Item
628111000225	Denpasar Barat, Denpasar Selatan, Denpasar Utara, Kuta, Kuta Utara
62811100027	Denpasar Barat, Denpasar Selatan, Denpasar Utara, Kediri, Kerambitan, Kuta, Kuta Utara, Kuta Selatan, Marga, Mengwi, Penebel, Tabanan
628111000320	Denpasar Selatan, Kuta, Kuta Selatan
628111000418	Denpasar Selatan, Denpasar Timur, Kuta, Kuta Selatan, Kuta Utara, Sukawati, Ubud
628111000432	Denpasar Barat, Denpasar Selatan, Denpasar Timur, Denpasar Utara, Kuta
628111000521	Kuta
628111000856	Denpasar Barat, Denpasar Selatan, Kuta, Kuta Utara, Mengwi
628111000977	Denpasar Barat, Denpasar Selatan, Denpasar Timur, Kuta, Kuta Selatan

After deciding that the approach chosen for the mining process for cellular data is a frequent pattern, then the algorithm of frequent pattern needs to be selected. Generally, the most common used algorithm is an algorithm Apriori because it is a frequent basis in the search pattern. However, apriori algorithm can't work optimally for relatively small threshold value. The algorithm used in this study is Frequent Pattern (FP)-Growth for finding the frequent pattern on the dataset. This algorithm is more efficient for the process of data mining on a small or large scale with frequent itemset search process is faster than the other algorithms [6].

3.2. Data Visualization

Visualization process is done by using the result of data mining process. Visualization is the process of presenting information to the user called knowledge presentation [4]. Knowledge presentation is done by creating an interactive visualization serving mobile data in map form. The key message in the visualization is the user can view the mapping of cellular data about the route of the user taken in one day and the density of the population in a particular area.

The purpose of cellular data visualization that will be built is doing the mapping of datasets that observed. Then the mapping is done to see the spread of the data used in the province of Bali. One of the method that is suitable for mapping the data is map method. The approach is usually used to present the map dataset with geo-spatial properties. Cellular data have spatial property that can be used to visualize the location of the user. It can be concluded that visualization will be optimal if built through the approach of the map because by using map the distribution of the dataset can be easily observed.

Attributes that used to build a visualization is time and date which used for a marker of time in the map, MSISDN which serves as the user ID in a map, long and lat noted the position of the user in a specific date and time. There are two feature of maps that can be used in the visualization, the first is MSISDN that the user can choose a phone number who wants to be observed on a certain date and we can see the movement of the user. Second, users can see the density in a particular area from the input of the user how to display on view visualization in a certain date. Visualization display can be seen in Figure 3 and Figure 4.

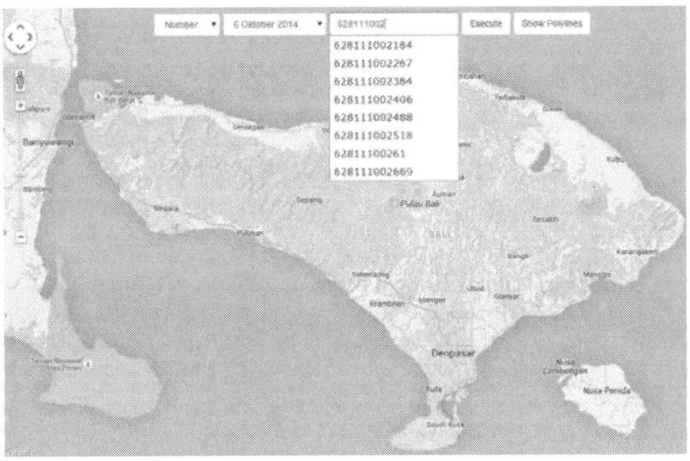

Figure 3. Visualization View

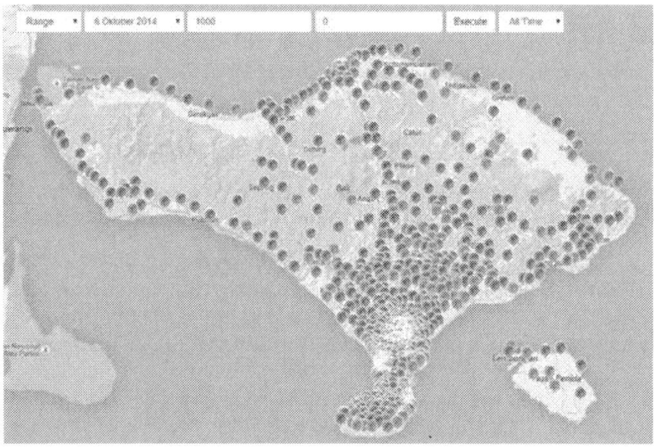

Figure 4. Visualization of the Density

4. Analyzing Mobility Patterns

Based on the results we can observe the density is going on a certain areas in proper time through visualization and data mining. From the results observed then the analysis is made to explain that with visualization and data mining processes, cellular data can be processed properly. In the cellular data processing from the case studies in province bali, can be seen that most of data are spread in southern part of Bali, especially in Kuta area and Denpasar. It is obtained from the results of visualization in Figure 4 and the results of mining in Table 2. However, the cellular data processing for other province or other areas can be done using the same way to produce information.

Table 2. Frequent Item Result

Support		Confidence	
Kuta Selatan ∪ Kuta Utara	195	Kuta	193
Denpasar Barat ∪ Denpasar Selatan ∪ Kuta Selatan	247	Kuta	242
Denpasar Barat ∪ Kuta Selatan	277	Kuta	271
Denpasar Selatan ∪ Kuta Utara	241	Kuta	235
Denpasar Barat ∪ Denpasar Selatan ∪ Kuta Utara	221	Kuta	215
Kuta ∪ Denpasar Timur	206	Denpasar Selatan	199
Denpasar Selatan ∪ Kuta Selatan	326	Kuta	311
Denpasar Barat ∪ Kuta Utara	272	Kuta	252
Denpasar Selatan ∪ Kuta Utara	241	Denpasar Barat	221
Kuta ∪ Denpasar Selatan ∪ Kuta Utara	235	Denpasar Barat	215

The results of data mining shows the human mobility can be view clearly in visualization. It can give the information of an area that have high density. It also suggests alternative routes that can be taken to an area if the normal route is having a congestion. This is very useful for police to control the traffic. Cellular data processing results can also be used as supporting data for the government to make improvements and development of transport infrastructure in the region.

Processing data can show peak hour in an area. This led to the idea to seek alternative routes to avoid congestion. In the case study of Bali province, if the user wants to Denpasar from Kuta vice versa, the user can rotate their route passes near the eastern area of Gianyar. The provision of an alternative route and also the knowledge of the rush-hour at this area that can be used by the police for traffic control.

In addition to these two things, cellular data can also be used as supporting data for the improvement of transport infrastructure if it is supported by the traffic data. From the visualization of the celular data in Bali, it can be seen that the density only occurred in the southern part of Bali. It would raise the question that why the northern part of Bali have less population density and movement. But after seeing the geographical conditions in the northern part of Bali it is understandable because there are many hills and mountains. Based exploration conducted, the road to the northern part of the island prone to landslides. In addition, the vehicle towards the area dominated by large vehicles like trucks because the northern route ends at Gilimanuk. Based on the information obtained, the government can make improvements to the road infrastructure of the north by making a special line for large vehicles and another line for personal vehicles.

Generally, cellular data processing methods by using data mining and visualization can be used for cellular data in other areas. If it is supported by other data, cellular data can provide a good solution for human mobility required by the decision maker which is not limited only to one domain. For example, cellular data can indicate the progress of technology in an area and how that area use of technology. With the right method cellular data can benefit the decision maker.

5. Conclusions

Conclusions obtained during the research is cellular data can be used as supporting data for mapping areas that require improvement of transport infrastructure. Seen from the visualization result and data mining, areas with low density can be assumed that the area is rarely visited. one of the causes is the lack of adequate transport infrastructure in that area. In addition, cellular data can be used to improve the effectiveness of police performance in traffic management such as providing alternate routes when congestion occurs. Mobile data can also provide a complete illustration when supported by other data such as data of economic distribution, demographic data, topographic data, and other data.

For the development of future research, visualization need to be added after data mining process is completed. The visualization is beneficial to display the prediction of user's movement who observed. Data classification needs to be done before the visualization process. In addition, analysis of cellular data utilization is expected to not only focus on one domain but also on other domains connected with the transport domain.

References

[1] Steenbruggen J, Tranos, E Nijkamp P 2014 Data from mobile phone operators: A tool for smarter cities Telecommunications Policy.

[2] Soto V, Frias-Martinez E 2011 Robust land use characterization of urban landscapes using cell phone data 9th International Conference on Pervasive Computing, San Francisco.

[3] Giannoti F, Nanni M, Pedreschi D, and Pinelli F 2007 Trajectory pattern mining ACM SIGKDD Conference on Knowledge Discovery and Data Mining, pp. 330-339.

[4] Kirk and Andy 2012 Data Visualization: a Succesful Design Process.

[5] Han J, Kamber M 2004 Data Mining: Concepts and Technique Morgan Kauffman Publisher.

[6] Han J, Pei J, Yin Y 2000 Mining Frequent Patterns without Candidate Generation Proceeding ACM SIGMOD '00 International Conference on Management of Data vol.29, pp.1-12.

[7] Borgelt, C 2005 An Implementation of FP-Growth Algorithm Proceeding OSDM'05 of the 1st International Workshop on Open Source Data Mining: Freqeunt Pattern Mining Implementations, pp.1-5.

[8] Kang C, et. al. 2007 Analyzing and Geo-visualizing Individual Human Mobility Patterns Using Mobile Call Records.

Corruption Cases Mapping Based on Indonesia's Corruption Perception Index

Noerlina[1], L A Wulandhari[2], Sasmoko[3,4], A M Muqsith[5], M Alamsyah[5]

[1]School of Information System, Bina Nusantara University,Indonesia
[2]School of Computer Science, Bina Nusantara University, Indonesia
[3]Faculty of Humanities, Bina Nusantara University,Indonesia
[4] Research Interest Group in Education Technology, Bina Nusantara University, Indonesia
[5]School of Computer Science, Bina Nusantara University, Indonesia

Abstract. Government plays an important role in nation economic growth. Nevertheless, there are still many occurrences of government officers abusing their offices to do an act of corruption. In this order, the central government should pay attention to every area in the nation to avoid corruption case. Meanwhile, the news media always constantly preach about corruption case, this makes the news media relevant for being one of the sources of measurement of corruption perception index (CPI). It is required to map the corruption case in Indonesia so the central government can pay attention to every region in Indonesia. To develop the mapping system, researchers use Naïve Bayes Classifier to classify which news articles talk about corruption and which news articles are not, before implementing a Naïve Bayes Classifier there are some text processing such as tokenizing, stopping, and stemming.

1. Introduction

Government plays an important role in nation economic growth. Nevertheless, there are still many occurrences of government officers abusing their offices to do an act of corruption. Corruption becomes an undeniable disease and threatens the structure of society (Rahimi & Shakeri, 2016). Generally, corruption limits the governments' institution by disobeying procedure, use of resources, and office occupation without the legal procedure. Corruption also undermines the legitimacy of government and other democratic values (Ghosh & Siddique, 2014). As a national issue, corruption must be dealt with firm and clear with all the potential involvement that exists within the public especially the government and law enforcement officials. The concept of systemic corruption as a collective action has significant implications on policy implementation (Persson, Rothstein, & Teorell, 2013). The media can show the current conditions of the region.

Additionally, the media can also summarize thoughts and ideas of the public (Malik, 2001). Therefore, using the social media as source data to calculate Corruption Perception Index (CPI) is very possible. CPI measures the level of corruption and CPI mapping will help the government to analyze and take measures to prevent corruption in every region (Melgar, Rossi, & Smith, 2010; Rohwer, 2009). CPI may be a reference to map corruption in all regions in Indonesia; CPI mapping will help the government in the analysis and prevention of corruption in each region (Uca, Ince, & Sumen, 2016). Therefore, the researchers propose system development of corruption cases mapping based on Indonesia's Corruption Perception Index. To develop the mapping system, the researchers need to solve the problem of how to classify news articles that talk about corruption case and news articles that do not talk about corruption case.

According to previous research, several algorithms can be used to classify the document including Naïve Bayes Classifier, Decision Trees, and Support Vector Machine, which is Naïve Bayes Classifier is the most accurate algorithm to classify the document (Huang, Lu, & Ling, 2003).

2. News Classification

2.1. Data Collection

The data collected in this research is a news content taken from some news site in Indonesia, there are some considerations made in the selection of news sites for data collection, the consideration include:

2.1.1. Alexa rank.
Alexa rank is a website that calculates the rank of a website based on traffic by counting unique visitors with page views. Alexa rank produces good indicator to determine the credibility of a website (Olteanu, Peshterliev, Liu, & Aberer, 2013)

2.1.2. News content that consistently uses 5W1H format
5W1H is an abbreviation of who, what, where, when, how and why. The information provided by an article will be very informative and clear if it can answer 5W1H (Jang & Woo, 2005)

2.1.3. Site has existed since 2010
To equalize the final result of the calculation system, the site at least has published articles before or at least since 2010.

2.1.4. Have sitemap feature that will be used for web crawling
Sitemap is a single page that displays a collection of structured links that directly connects the pages in a website, sitemap. With the sitemap, web crawling will easily index all the pages in a website (Yang et al., 2009)

2.1.5. Site content can be retrieved using a scraping technique.
There are several websites that prohibit a web crawler to do indexing on the website so that the data of news item could not be retrieved.

There are seven Indonesian news sites considered for data collection. The news sites are Tempo (tempo.co), Detik (detik.com), Kompas (kompas.com), Merdeka (merdeka.com), Liputan6 (liputan6.com), Tribun News (tribunnews.com), and Sindo News (sindonews.com).

There are three techniques for data collection:

- Web Crawling is a software that runs recursively crawl from one page to another page by following the link in each page (Schrenk, 2007), in this research, crawling result will be saved in the database as a news article index for web scraping purpose.
- Web Scraping is a software technique that gathering and extracting information from a web page (Schrenk, 2007), in this research, web scraping is used to get news content based on news article index and saved the content into the database.
- Cron Scheduling is a UNIX and Linux base scheduling to run a command or a shell script periodically, in this research, cron scheduling is used to run web crawling and web scraping periodically.

By using web crawling, web scraping, and cron scheduling, 396565 news articles successfully collected, here is the detail of data collected:

Table 1. *Data Collected.*

News Site	Article Collected
Detik	66826
Kompas	49396
Liputan 6	87950

Merdeka	54475
Sindo News	41016
Tempo	60763
Tribun News	36139
Total	396565

2.2. *News Classification Algorithm*

There are two processes performed in the process of classifying the news content; the processes are text processing and news content classification process.

2.2.1. Text Processing

Text processing is done to represent the text into the same form (Tated & Ghonge, 2015). Text processing has three sub-processes; tokenizing, stopping and stemming:

- Tokenizing is a process that splits the document into a single word; this process split the document by space character and saved into the array of word.
- Stopping is a searching process that searches a stop word in the array of word and removes the found word from the array of word, an Indonesian stop word list is listed from Fadillah Z Tala's research in 2003 (Tala, 2003).
- Stemming is a process that convert each word in the array of word into a root word for example "mencintai" into "cinta" or "memilikinya" into "milik", the process used Nazief and Andriani's algorithm, the algorithm is:
  Step 1, search the word in a root word database, if found then the word is a root word and algorithm is finished.
  Step 2, Inflection suffixes (-lah, -kah, -ku, -mu, or -nya) deletion, if a word is a particle (-lah, -kah, -tah, or -pun), then repeat this step to delete a possessive pronouns (-ku, -mu, or -nya) if exists.
  Step 3, Derivation suffixes (-i,-an,or -kan) deletion, after deletion occurred, search the result word in a root word database, if found, then the algorithm is finished, if not then continue to Step 3.1.
    Step 3.1, if -an was deleted and the last character was -k, then remove the -k, if the result found in root word database, then the algorithm is finished, if not then continue to Step 3.2.
    Step 3.2, restore deleted suffixes (-i, -an, or -kan), and continue into Step 4.
  Step 4, Derivation prefixes (di-, ke-, se-, me-, be-, pe-, or te-) deletion, with the maximum number of iterations is three times:
    Step 4.1, the algorithm finished if:
    - Not allowed prefixes and suffixes combination occurred based on this table (Table 2):

Table 2. *Not allowed prefixes and suffixes combination.*

Prefixes	Not allowed suffixes
be-	-i
di-	-an
ke-	-i, -kan
me-	-an
se-	-i, -kan

te-	-an

- Current detected prefixes equal to last deletion prefixes.
- Three prefixes deletion.

Step 4.2, identify the prefixes type and delete, there are two types of prefixes:

- Standard prefixes (di-, ke-, or se-)
- Complex prefixes (me-, be-, pe-, or te-), this type can create a new word according to the followed root word, so, there are rule that is used to get the right hyphenation (Table 3):

Table 3. Hyphenation rule.

<table>
<tr><th>No</th><th>Word Format</th><th>Hyphenation</th></tr>
<tr><td>1</td><td>berV...</td><td>ber-V... | be-rV...</td></tr>
<tr><td>2</td><td>berCAP...</td><td>ber-CAP... | where C!=r and P!=r</td></tr>
<tr><td>3</td><td>berCAerV...</td><td>ber-CaerV... where C!=r</td></tr>
<tr><td>4</td><td>belajar</td><td>bel-ajar</td></tr>
<tr><td>5</td><td>beC1erC2...</td><td>be-C1erC2... where C1 != {r|l}</td></tr>
<tr><td>6</td><td>terV...</td><td>ter-V... | te-rV...</td></tr>
<tr><td>7</td><td>terCerV...</td><td>ter-CerV ... where C!=r</td></tr>
<tr><td>8</td><td>terCP...</td><td>ter-CP... where C!=r</td></tr>
<tr><td>9</td><td>teC1erC2...</td><td>te-C1erC2... where C1!=r</td></tr>
<tr><td>10</td><td>me{l|r|w|y}V...</td><td>me-{l|r|w|y} V...</td></tr>
<tr><td>11</td><td>mem{b|f|v}...</td><td>mem-{b|f|v}...</td></tr>
<tr><td>12</td><td>mempe{r|l}...</td><td>mem-pe...</td></tr>
<tr><td>13</td><td>mem{rV|V}...</td><td>me-m{rV|V}... | me-p{rV|V}...</td></tr>
<tr><td>14</td><td>men{c|d|j|z}...</td><td>men-{c|d|j|z}</td></tr>
<tr><td>15</td><td>menV...</td><td>me-nV... | me-fV</td></tr>
<tr><td>16</td><td>meng{g|h|q}...</td><td>meng-{g|h|q}...</td></tr>
<tr><td>17</td><td>mengV...</td><td>meng-V... | meng-kV...</td></tr>
<tr><td>18</td><td>menyV...</td><td>meny-sV...</td></tr>
<tr><td>19</td><td>mempV...</td><td>mem-pV... where V!=e</td></tr>
<tr><td>20</td><td>pe{w|y}V...</td><td>pe-{w|y}V...</td></tr>
<tr><td>21</td><td>perV...</td><td>per-V... | pe-rV...</td></tr>
<tr><td>22</td><td>perCAP...</td><td>per-CAP... where C!=r dan P!=er</td></tr>
<tr><td>23</td><td>perCAerV...</td><td>per-CAerV... where C!=r</td></tr>
<tr><td>24</td><td>pem{b|f|V}...</td><td>pem-{b|f|V}...</td></tr>
<tr><td>25</td><td>pem{rV|V}...</td><td>pe-m{rV|V}... | pe-p{rV|V}</td></tr>
<tr><td>26</td><td>pen{c|d|j|z}...</td><td>pen-{c|d|j|z}...</td></tr>
<tr><td>27</td><td>penV...</td><td>pe-nV... | pe-tV...</td></tr>
<tr><td>28</td><td>peng{g|h|q}...</td><td>peng-{g|h|q}</td></tr>
<tr><td>29</td><td>pengV...</td><td>peng-V... | peng-kV...</td></tr>
<tr><td>30</td><td>penyV...</td><td>peny-sV...</td></tr>
<tr><td>31</td><td>pelV...</td><td>pe-lV... exept 'pelajar'</td></tr>
<tr><td>32</td><td>peCerV...</td><td>per-erV... where C!={r|w|y|l|m|n}</td></tr>
<tr><td>33</td><td>peCP...s</td><td>per-CP... where C!={r|w|y|l|m|n} and P!=er</td></tr>
</table>

- Search deleted prefixes result into a root word database, if not found then repeat the Step 4, if found, then the algorithm is finished.

Step 5, if word not found in root word database after Step 4, then repeat the Step 4, in this step, the algorithm is finished and assumed the word is a root word.

2.2.2. News Content Classification

2.2.2.1. Document Training

Documents that are trained in this process is the result of text processing, the document is 30 documents of corruption case, and 30 documents of the non-corruption case, each training result will be saved into a database, which will then be used for Naïve Bayes Classifier algorithm as a reference to an existing document.

2.2.2.2. Document Classification Algorithm

In this classification step, the used algorithm is Naïve Bayes Classifier algorithm, that uses a previous document to calculate the future probability using equation (Murphy, 2006):

$$V_{MAP} = {}^{\arg max}_{V_j e V} \prod_{i=1}^{n} (P(x_i|V_j)P(V_j))$$

$data$	Word from text processing
x_i	Data i $(1,2,3, \ldots, n)$
V_j	Class j $(1,2,3, \ldots, n)$
$P(x_i\|V_j)$	Probability x_i to class V_j, where $P(x_i\|V_j) = \frac{n_k+1}{n+ \|data\|}$
$P(V_j)$	Probability V_j, where $P(V_j) = \frac{\|documents\ j\|}{\|documents\|}$

For $P(V_j)$ and $P(x_i|V_j)$ calculated while data training is being processed, where:

$\|documents\ j\|$	total documents in each class j
$\|documents\|$	total documents in all class
n_k	total frequency of occurrence of each of data
n	total frequency of data appearance of each class

$|data|$ total all data from all class

3. Result and Analysis

To produce good results of Naïve Bayes classifier algorithm, there are 30 news articles that talk about corruption case and 30 news articles that do not talk about corruption case, then do the test of 10 articles that talk about corruption case and 10 articles that do not talk about corruption case, the result of the algorithm is 10 articles that talk about corruption was accurate 100%, and 10 articles that do not talk about corruption is accurate 100%.

Can be concluded, the research method that already described above can produce the accurate results. At this stage, the research can classify news articles that talk about corruption case or talk about non-corruption case.

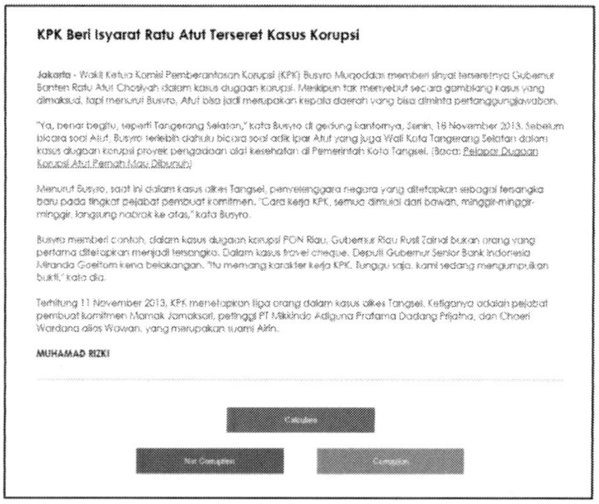

KPK Beri Isyarat Ratu Atut Terseret Kasus Korupsi

Jakarta - Wakil Ketua Komisi Pemberantasan Korupsi (KPK) Busyro Muqoddas memberi sinyal terseretnya Gubernur Banten Ratu Atut Chosiyah dalam kasus dugaan korupsi. Meskipun tak menyebut secara gamblang kasus yang dimaksud, tapi menurut Busyro, Atut bisa jadi merupakan kepala daerah yang bisa diminta pertanggungjawaban.

"Ya, benar begitu, seperti Tangerang Selatan," kata Busyro di gedung kantornya, Senin, 18 November 2013. Sebelum bicara soal Atut, Busyro terlebih dahulu bicara soal adik ipar Atut yang juga Wali Kota Tangerang Selatan dalam kasus dugaan korupsi proyek pengadaan alat kesehatan di Pemerintah Kota Tangsel. (Baca: Pelapor Dugaan Korupsi Atut Pernah Mau Dibunuh)

Menurut Busyro, saat ini dalam kasus alkes Tangsel, penyelenggara negara yang ditetapkan sebagai tersangka baru pada tingkat pejabat pembuat komitmen. "Cara kerja KPK, semua dimulai dari bawah, minggir-minggir-minggir, langsung nabrak ke atas," kata Busyro.

Busyro memberi contoh, dalam kasus dugaan korupsi PON Riau, Gubernur Riau Rusli Zainal bukan orang yang pertama ditetapkan menjadi tersangka. Dalam kasus travel cheque, Deputi Gubernur Senior Bank Indonesia Miranda Goeltom kena belakangan. "Itu memang karakter kerja KPK. Tunggu saja, kami sedang mengumpulkan bukti," kata dia.

Terhitung 11 November 2013, KPK menetapkan tiga orang dalam kasus alkes Tangsel. Ketiganya adalah pejabat pembuat komitmen Mamak Jamaksari, petinggi PT Mikkindo Adiguna Pratama Dadang Prijatna, dan Choeri Wardana alias Wawan, yang merupakan suami Airin.

MUHAMAD RIZKI

Figure 1. Training process of article that talks about corruption.

Ribuan Truk Antre di Pelabuhan Merak

Cilegon - Ribuan truk angkutan barang di Pelabuhan Merak, Kota Cilegon, Banten kembali ramai setelah sepuluh hari dilarang melintas pada H-4 sampai H+4 Lebaran. Meski tidak terjadi penumpukan, kendaraan truk tersebut mendominasi angkutan kapal dibandingkan jenis kendaraan lainnya seperti bus dan roda dua.

Manajer Operasional PT ASDP Indonesia Ferry Cabang Utama Merak, Zailis Anas mengatakan, angkutan truk diperkirakan masih akan terus terjadi hingga akhir pekan ini yaitu 10-11 September 2011. Ini terjadi lantaran banyak perusahaan yang sudah aktif kembali mengirimkan barang baik dari arah Jawa ke Sumatera maupun sebaliknya. "Kami bersyukur arus mudik pada tahun ini berjalan lancar, lebih tertib dibandingkan tahun sebelumnya," kata Zailis, Kamis, 8 September.

Data dari PT Angkutan Sungai Danau dan Penyeberangan (ASDP) Indonesia Ferry Cabang Utama Merak menyebutkan, sejak H+5 truk mulai meningkat dari 302 unit menjadi 904 unit, pada H+6 jumlah truk melonjak menjadi 1.217 unit. Sedangkan jumlah bus menurun dari 532 pada H+5 menjadi 474 pada H+6, begitupun roda dua juga mengalami penurunan dari 1.411 pada H+5 menjadi 1.126 pada H+6. "Truk mulai ramai karena sebelumnya dilarang," kata Zailis Anas.

Kepala Dinas Perhubungan Komunikasi dan Informatika Provinsi Banten Husni Hasan mengatakan, penambahan dermaga di Pelabuhan Merak, Banten dan Bakauheni, Lampung dari lima dermaga menjadi enam dermaga sangat mendesak dilakukan. Hal itu sebagai salah satu solusi untuk mengantisipasi penumpukan kendaraan akibat terus meningkatnya jumlah kendaraan yang melakukan penyeberangan.

"Kami sudah menyampaikannya kepada PT ASDP maupun ke Kementerian Perhubungan saat memantau di Merak. Solusi mengatasi kemacetan atau penumpukan kendaraan di Merak harus menambah dermaga," kata Husni.

Menurutnya, berapa pun jumlah kapal yang disediakan PT ASDP Indonesia Ferry tidak akan mampu mengatasi penumpukan kendaraan. Karena jika dengan lima dermaga itu, kapal yang dioperasikan tetap tidak bisa dimaksimalkan. "Penambahan jumlah kendaraan setiap tahunnya terus bertambah, apa lagi saat arus balik melebihi jumlah yang diprediksikan. Oleh sebab itu, perlu dilakukan langkah-langkah antisipasi dari sekarang dengan penambahan dermaga," katanya.

Pantauan Tempo suasana arus balik di Pelabuhan Merak baik dari arah Sumatera maupun dari arah Jawa masih ramai. Loket penjualan tiket beberapa kali mengalami antrean, meski pun tidak sepanjang beberapa hari lalu. Namun, data dari PT ASDP pada H+6 masih mengangkut 57.562 penumpang dari Bakauheni ke Merak. Sedangkan dari Merak ke Bakauheni 45.605 penumpang. "Biasanya arus balik akan lebih lama, paling tidak lebih dari H+7," kata Kepala Cabang PT ASDP Merak, La Mane.

WASI'UL ULUM

Figure 2. Training process of article that does not talk about corruption.

Figure 3. Testing process of article that talks about corruption.

Figure 4. Testing process of article that does not talk about corruption.

From the process of text mining performed, resulting in the data on the mode of corruption based on the news phenomenon from online media related to corruption, which will be processed further to extract the location referred by the article. Thus, forming the data of the number of articles that have been classified by the scene, which can help the central government to perform certain actions on a particular place based on the data result of the text mining. It is expected to reduce cases of corruption in various regions in Indonesia.

References

[1] Ghosh, R., & Siddique, M. (2014). *Corruption, Good Governance and Economic Development: Contemporary Analysis and Case Studies.*

[2] Huang, J., Lu, J., & Ling, C. (2003). *Comparing naive Bayes, decision trees, and SVM with AUC and accuracy .*
[3] Jang, S., & Woo, W. (2005). 5W1H: Unified User-Centric Context. *Proceedings of the 7th International Conference on Ubiquitous Computing.*
[4] Malik, D. (2001). *Dari Konstruksi ke Dekonstruksi: Refleksi atas Pemberitaan Televisi Kita.*
[5] Melgar, N., Rossi, M., & Smith, T. W. (2010). The Perception of Corruption. *International Journal of Public Opinion Research, 22*(1), 120–131.
[6] Murphy, K. P. (2006). *Naive Bayes Classifiers Generative Classifiers*, 4701**,** 1–8.
[7] Olteanu, A., Peshterliev, S., Liu, X., & Aberer, K. (2013). Web Credibility: Features Exploration and Credibility Prediction. *Lecture Notes in Computer Science (Including Subseries Lecture Notes in Artificial Intelligence and Lecture Notes in Bioinformatics),* 557–568.
[8] Persson, A., Rothstein, B., & Teorell, J. (2013). Why Anti Corruption Reforms Fail — Systemic Corruption as a Collective Action Problem, *Governance*, *26*(3), 449–471.
[9] Rahimi, R., & Shakeri, H. (2016). Reflection on Judicial System's Corruption and Offering Solutions to Promote its Safety. *Journal of Politics and Law, 9*(9), 187.
[10] Rohwer, A. (2009). Measuring Corruption: A Comparison between the Transparency International's Corruption Perception Index and the World Bank's Worldwide Governance Indicators. *DICE*, 42–52.
[11] Schrenk, M. (2007). *Webbots, spiders, and screen scrapers.*
[12] Tala, F. (2003). *A study of stemming effects on information retrieval in Bahasa Indonesia*
[13] Tated, R. R., & Ghonge, M. M. (2015). A survey on text mining - techniques and application. *International Journal of Research in Advent Technology,*(1), 380–385.
[14] Uca, N., Ince, H., & Sumen, H. (2016). The Mediator Effect of Logistics Performance Index on the Relation between Corruption Perception Index and Foreign Trade Volume. *European Scientific Journal, 12*(25), 1857–7881.
[15] Yang, J. M., Cai, R., Wang, C., Huang, H., Zhang, L., & Ma, W. Y. (2009). Incorporating Site-Level Knowledge for Incremental Crawling of Web Forums: A List-Wise Strategy. *Proceedings of the ACM SIGKDD International Conference on Knowledge Discovery and Data Mining,* 1375–1383.

Mining of the social network extraction

M K M Nasution[1*], M Hardi[2], R Syah[3]

[1,2,3]*Informatin Technology, Fasilkom-TI, Universitas Sumatera Utara, Padang Bulan USU Medan 20155, Indonesia.*

*Email: mahyuddin@usu.ac.id

Abstract. The use of Web as social media is steadily gaining ground in the study of social actor behaviour. However, information in Web can be interpreted in accordance with the ability of the method such as superficial methods for extracting social networks. Each method however has features and drawbacks: it cannot reveal the behaviour of social actors, but it has the hidden information about them. Therefore, this paper aims to reveal such information in the social networks mining. Social behaviour could be expressed through a set of words extracted from the list of snippets.

1. Introduction

Based on the superficial method for extracting social network from information sources we have the social network as a resultant [1]. This method is pre-processing for transforming the raw data into the social network whereby Web as one of information sources [2]. Web be a picture of social change dynamically, and Web therefore represent the activities of social actors [3]. In other words, exploring Web as the social media is to allow us acquire the behaviour of either social actor personally or the community of social actors [4]. However, there is no extraction of social networks connected with the exploration of the behaviour of social actors. Thus, this paper aims to disclose formally the mining of the social network extraction: That is to develop the concept formally for extracting social network and propose some problems as basic concept and motivation; to propose an explanation of the problems and prove its importance in an approach systematically; and to conduct the extraction of social network and the related behaviour in an example.

2. Basic Concept and Motivation

Each network naturally we can model it as a graph $G<V,E>$ whereby V is a set of vertices $\{v_i|i=1,\ldots,I\}$ and a set of the name labels $\{a_i|i=1,\ldots,I\}$, and E is a set of edges $\{e_j|j=0,\ldots,J\}$, a set of weights $\{b_j|j=0,\ldots,J\}$ and a set of labels $\{l_k|k=0,\ldots,K\}$, or we can define the template of network as follows [5].

Definition 1. $TN = G<V,E,L_1,L_2,B>$ is a template of network with conditions as follows
T1 $V = v_i|i=1,\ldots,I\}$ as vertices in TN.
T2 $E = \{e_j|j=0,\ldots\}$ as edges in TN.
T3 $L_1 = \{a_i|i=1,\ldots,I\}$ as labels of vertices.
T4 $L_1 = \{l_k|k=0,\ldots\}$ as labels of edges.
T5 $B = \{b_j|j=0,\ldots\}$ as weights of edges.

Each network not only has a set of vertices and a set of edges, but also it has a set of labels for vertices/edges and a set of weights for edges. Therefore, to develop a social network from information sources, we choose the tools that are used to access information. Suppose for a social actor 'Mahyuddin K. M. Nasution' we use search engines, if query q_1 = 'Mahyuddin K. M. Nasution' or q_2 = "Mahyuddin K. M. Nasution" submitted to any search engine, then we get the hit counts like Table 1. So there is a difference between q_1 and q_2. The results of search engine have been contaminated by ambiguity in documents / web pages [6] or the semantic problems: meronymy, holonomy, hyponymy, synonymy, or polysemy [7]. The result of q_1 explained that each social actor socially has many names, aliases or attributes, while the result of q_2 still contains the possibility of same name for different people, although the exact pre-defined name for somebody. Therefore, to gauge out the appropriate information from the stack of documents, we always add an eligible keyword [8].

Table 1. Hit counts

Search engine	q_1	q_2
Google	16,100	1,110
Yahoo	1,980	244
Bing	213,000	2,090
Yandex	57,000	634

In the extraction of social networks, first we declare the social actors with names that are properly defined, to ensure every vertex v in V in template TN represents a social actor: A set of actors $A = \{a_i|i=1,\ldots,I\}$. So there is a function γ_1 that maps one by one actor to vertex, or $\gamma_1(1{:}1)$: A -> V [9]. In this case, if a query q contain a term for a_i in A, then the query classify web pages into one class that is recognizable as a singleton event Ω_{ai} whereby number of occurrences (hit counts) is $|\Omega_{ai}|$ [2].

Lemma 1. If every web page representing the activity of a social actor, then the class of web pages represent the behavior of a social actor.

In addition, for $|\Omega_{ai}| > 0$ the search engine also bring back (return) a list of snippets, every snippet contains ± 50 words, and the label of social actors can be generated from this list [5].

In the classical social networks, the relations between social actors was transformed directly from the data, but in the extraction of a social network, a collection of relationship that may be contained in the document and translated through a search engine in the form of co-occurrence [5]: If the query q contains two terms a_i,a_j in A, then the query classifying web pages into one class that can be recognized as a doubleton event $\Omega_{ai}\cap\Omega_{aj}$ whereby number of co-occurrences (hit counts) is $|\Omega_{ai}\cap\Omega_{aj}|$, with conditions $|\Omega_{ai}\cap\Omega_{aj}| \leq |\Omega_{ai}|$ and $|\Omega_{ai}\cap\Omega_{aj}| \leq |\Omega_{ai}|$ [2].

Lemma 2. If every web page representing the communication of social actor, then the class of web pages represents the behavior of the relation between social actors.

Let $|\Omega_{ai}\cap\Omega_{aj}| > 0$, the search engine returns a list of snippets that also contains information about 50 words in each snippet, and label of a relation can be generated from it through a procedure. All relations that may exist between social actors are accumulated into one and recognizable as weight of the strength relation by using the concept of similarity, for example Jaccard coefficient

$$sim_j = |\Omega_{ai}\cap\Omega_{aj}|/(|\Omega_{ai}|+|\Omega_{aj}|-|\Omega_{ai}\cap\Omega_{aj}|) \quad (1)$$

Definition 2. $SN_e = G<V,E,A,R,\Gamma>$ is an extracted social network with an operation Γ which satisfies the conditions
C1 $\gamma_1(1{:}1) : A \rightarrow V$, γ_1 in Γ.
C2 γ_2: $R \rightarrow E$, γ_2 in Γ.

Gamma function contains a set of processes to gain information from the sources: γ_1 for example is to declare the social actors based on MM method [8], and γ_2 is the steps that must be done for exploring the relations between the social actors like the superficial method [10].

Definition 3. $SN_c = G<V,E,A,R,\Gamma>$ is an labeled social network based on extraction method Γ which satisfies the conditions C1, C2 and/or
C3 γ_3(M:1): $L_1 \rightarrow V$, γ_3 in Γ and $L_1 = \{l_j|j=0,\ldots,J\}$ as labels of vertices, or
C4 γ_4: $B \rightarrow E$, γ_4 in Γ and $B = \{b_k|k=1,\ldots,K\}$ as weights of edges, or
C5 γ_5(M:1): $L_2 \rightarrow E$, γ_5 in Γ, and $L_2 = \{l_t|t=0,\ldots,T\}$ as labels of edges.
Theorem 1. If the behavior of a cluster describes the behavior of a social actor, then the behavior of other actors expressed by the relationship between the clusters.

3. An Approach

In Web as social media, each web page can represent an actor in a way that is the name of actor is a part of text in web page and other words in the same web page as actions, ethics and attitudes of the actor [15]. In literal, the name of actor is a term t_a consists of at least one or a set of words or $t_a = (w_1,\ldots,w_l)$, $l \leq k$, k as a number of words w, l as a number of vocabularies in t_a, $|t_a| = k$ is the size of t_a, and a cluster based on a term we can generate it as follows [12].

Definition 4. Let Ω as a space of the search engine for indexing the web pages. For a search term t_a, where t_a in Σ, that is a set of singleton search term of search engine. There are a dynamic space Ω_a containing the ordered pair of the term t_{ai} $i=1,\ldots,I$ with web pages ω_{aj} $j=1,\ldots,J$, i.e. $(t_{ai},\omega_{aj}) = (t_a,\omega_{aj})_{ij}$ such that $\Omega_a = (t_a,\omega_a)_{ij}$ is a subset of Ω is a singleton search engine event of web pages (briefly we call it as *singleton*) that contains the occurrences of t_a in ω.

A set of web pages as the cluster is the representation of the actor formally (Definition 4), whereby $|\Omega_a|$ is the cardinality of Ω_a (*hit count*), while the expression of behavior we can mine from a list of snippets.

Proposition 1. If $|\Omega_a| > 0$ for a search term t_a, then there is a set of the weighted words W = $\{(w_j,b_j)|j=1,\ldots,J\}$, w_j is a word and b_j is a weight of word.

Proposition 1 as the direct consequence of γ_3 in Definition 3 whereby b_j of words are generated by using term frequency (*tf*) or probability of word $p(w)$ in collection of words from the list of snippets. Therefore, we can rewrite Lemma 1 formally as $|\Omega_a| > 0 \rightarrow \Omega_a \wedge \{(w_j,b_j)\}_a$ and Proposition 1 as the proof of Lemma 1.

Definition 5. [13] Let Ω as a space of the search engine for indexing the web pages. For two search terms t_{ai} and t_{aj}, that is a set of doubleton search term of search engine. There are a dynamic space $\Omega_{ai}\cap\Omega_{aj}$ containing the ordered pair of the terms t_{ai} $i=1,\ldots,I$ and t_{aj} $j=1,\ldots,J$ with web pages ω_{aij}, i.e. $(\{t_{ai},t_{aj}\},\omega_{aij}) = (t_a,\omega_a)ij$ such that $\Omega_{ai}\cap\Omega_{aj} = (t_a,\omega_a)_{ij}$ is a subset Ω is a doubleton search engine event of web pages (briefly we call it as *doubleton*) that contains the co-occurrences of t_{ai},t_{aj} in ω.

Proposition 2. If $|\Omega_{ai}\cap\Omega_{aj}| > 0$ for two search terms t_{ai} and t_{aj}, then there is a set of the weighted words $W = \{(w_k,b_k)|k=1,\ldots,K\}$, w_k is a word and b_k is a weight of word.

Similar to γ_3, Proposition 2 as a consequence of γ_5 in Definition 3 and it prove Lemma 2 if $|\Omega_{ai}\cap\Omega_{aj}| \leq |\Omega_{ai}|$ and $|\Omega_{ai}\cap\Omega_{aj}| \leq |\Omega_{aj}|$ whereby $|\Omega_{ai}\cap\Omega_{aj}|$ is the cardinality of $\Omega_{ai}\cap\Omega_{aj}$, while $\{(w_k,b_k)|k=1,\ldots,K\}$ as the expression of the relation behavior between actors [14].

$|\Omega_{ai}\cap\Omega_{aj}| > 0$ proves the existence of relationship between two social actors, whilst using Eq. (1) gives the strength relation as a weight. A weight of the relation γ_4 be a behavior of same actors based on the relationship between the clusters, whereby $\gamma_4(sim_j)*b_k$ be expression of behavior. It also can proved by the similarity of $W_{ai} = \{(w_k,b_k)|k=1,\ldots,K_i\}$ and $W_{aj} = \{(w_k,b_k)|k=1,\ldots,K_j\}$, $W_{ai}\cap W_{aj} \neq \phi$.

Theorem 1 is proved. Therefore, for exploring the behavior of the social actors and the relation between them as the social network mining [15, 16] we have algorithm as follows.

Algorithm SN_e:

```
Declare a query for each social actor, q←tₐ for a in A,
       A is a set that consist of n social actors.
Submit a query to the search engine, |Ω_ai|←q, a_i in A.
Collect the snippets for t_ai, L_ai←q, L_ai is list of snippets for a_i in A.
Give a weight to every word in L_ai, W_ai←b, W_ai is a set of words for a_i in A.
Submit a query to the search engine, |Ω_aj|←q, a_j in A.
Collect the snippets for t_aj, L_aj←q, L_aj is list of snippets for a_j in A.
Give a weight to every word in L_aj, W_bj←b, W_bj is a set of words for a_j in A.
Submit a query to the search engine, |Ω_ai∩Ω_aj|←q, a_i and a_j in A.
Collect the snippets for t_ai and t_aj, L_ai,aj←q, L_ai,aj is list of snippets for
       a_i and a_j in A.
Give a weight to every word in L_ai,aj, W_ai,aj←b, W_ai,aj is a set of words for
       a_i and a_j in A.
Compute the strength relations by using |Ω_ai|, |Ω_aj|, and |Ω_ai∩Ω_aj|.
Compute the similarity between W_ai and W_bj.
Compute the average for W_ai,aj.
```

4. Extracting social network: A discussion

For the Google search engine, whenever a query containing the search term t_a submitted, we will get hit count $|\Omega_a|$ and a list of snippets. Each snippet consists of the URL address of the web page, the title of the web page, and the abstract of web page. For example, for two social actors obtained a number of snippets, number of words from snippets, and the hit counts like Table 2.

Table 2. The clusters based on search term

Search term	**"Abdullah Mohd Zin"**		**"T Mohd T Sembok"**	
Number of snippets		204		155
Number of words		5,982		4,680
Hit count		5,850		2,740
List of words	w_{ai}	$p(w_{ai})$	w_{aj}	$p(w_{aj})$
	malaysia	0.502	malaysia	0.514
	journal	0.4099	prof	0.4652
	ismail	0.2649	university	0.2264
	nazri	0.2177	ahmad	0.1991
	university	0.1738	abdullah	0.1849
	computer	0.1043	halimah	0.1098
	science	0.0921	journal	0.1071
	prof	0.0733	badioze	0.1005
	network	0.0652	zaman	0.1005
	university	0.0405	ukm	0.0469
	information	0.0276	abu	0.0423

Naturally a social network is built with the social actors. The number of social actors who may be extracted from the Web cannot be predicted, but for resolving this constraint always the extraction of social network based on the community of social actors [4]. In the superficial method, the social network for two social actors formed from two singleton and one doubleton, or formed from three clusters of Web. The clusters have differences and similarities. Each difference causes their behavior differs, for example, if the word "computer" does not exist in the list of words for "T Mohd T Sembok", then it became an expression of behavior differentiator for clusters based on term search "Abdullah Mohd Zin" [5], and this cluster has the weight as follows

$$|\Omega_{\text{"Abdullah Mohd Zin" and "computer"}}| = |\Omega_a|*p(w_i) = 5,850*0.1043 = 610$$

or vice versa

$$|\Omega_{\text{"T Mohd TSembok" and"zaman"}}| = |\Omega_a|*p(w_i) = 2,740*0.1005 = 275$$

Thus, the behavior of any social actor be formed by the expression. It defined by a number of words with its weight that affects the cluster $|\Omega_a|$ [9].

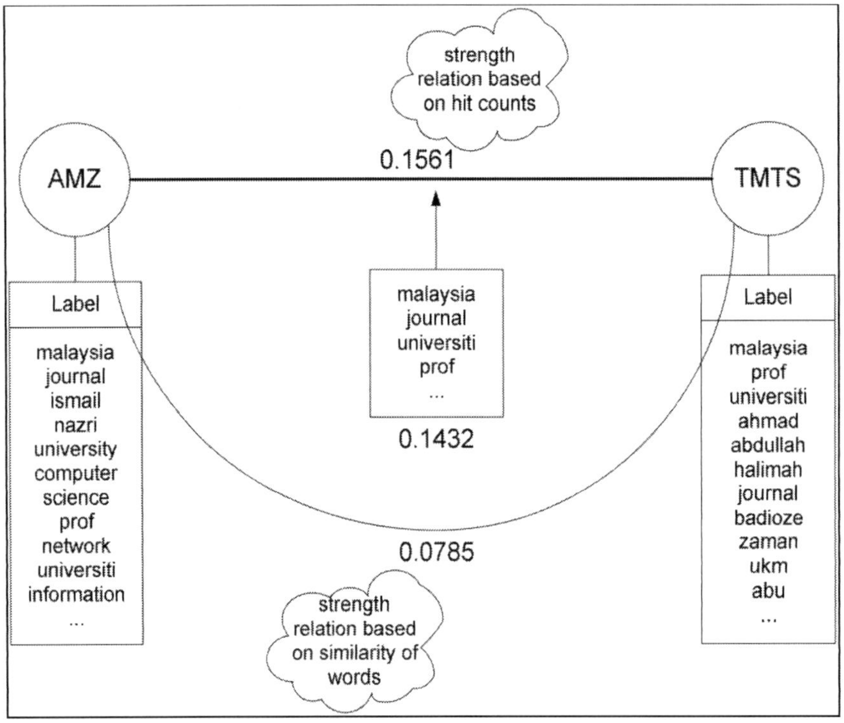

Figure 1. The strength relation between AMZ ("Abdulah Mohd Zin") and TMTS ("T Mohd T Sembok") in the extracted social network

The computing toward three clusters using Eq. (1) yield a strength relation between "Abdullah Mohd Zin" and "T Mohd T Sembok", that is 0.1561 where $|\Omega_{ai} \cap \Omega_{aj}| = 1160$, whereas the relationship behavior affects the behavior of social actor personally or the social actors have the relationship with others. This relation behavior are defined by an expression through a number of words in the same cluster. It is also indicated by the similarity of expression between social actors although from the different clusters, see Figure 1. By using Eq. (1) for two set of words with their weights we obtain the strength relation between "Abdullah Mohd Zin" and "T Mohd T Sembok" based on similarity of words, i.e. 0.0785, while the average of word weights based on list of snippets (of $\Omega_{ai} \cap \Omega_{aj}$) is 0.1432. Therefore, the social network that is extracted through two singletons and a doubleton, can be mined by involving the snippets to give meaning to the strength relation that have been generated.

5. Conclusion

In particular, the behavior of social actors and the relationships between them based on the Web depends on the expression and the behavior of clusters. The superficial method can be developed to express these behaviors and to mine it by following the extraction of social networks. Further study is to reveal the properties associated with the data, information, and analysis of social networks.

References

[1] M K M Nasution, S A M Noah and S. Saad 2011 Social network extraction: Superficial method and information retrieval, *Proceeding of International Conference on Informatics for Development* (ICID'11).

[2] M K M Nasution 2016 Extracted Social Network Mining *Proceeding of International Conference on Information Technology and Engineering Application* (5-th ICIBA), Palembang, February 19-20.

[3] Y Masunaga, K Ito, Y Miyama, N Oyama, C Watanabe and K Tachi 2010 SERPWatcher: A sophisticated mining tool utilizing search engine results pages (SERPs) for social change discovery *IEEE International Conference on Social Computing/IEEE International Conference on Privacy, Security, Risk and Trust.*

[4] M K M Nasution 2016 Social network mining (SNM): A definition of relation between the resources and SNA *International Journal on Advanced Science, Engineering and Information Technology* **6**.

[5] M K M Nasution 2013 Superficial method for extracting academic sosial network from the Web *Ph.D Dissertation*, Universiti Kebangsaan Malaysia.

[6] M K M Nasution and S A Noah 2012 Information retrieval model: a social network extraction perspective *IEEE International Conference on Information Retrieval & Knowledge Management.*

[7] M K M Nasution and S A M Noah 2012 Keyword extraction for identifying social actors arXiv:1212.3023v1 [cs.IR] 13 Dec 2012.

[8] M K M Nasution 2014 New method for extracting keyword for the social actor *Intelligent Information and Database Systems* LNAI **8397**, Heidelberg, Springer.

[9] M K M Nasution and S A Noah 2011 Extraction of academic social network from online database *Proceeding of 2011 International Conference on Semantic Technology and Information Retrieval*, Putrajaya, Malaysia.

[10] M K M Nasution and S A M Noah 2010 Superficial method for extracting social network for academic using web snippets *Rough Set and Knowledge Technology* LNAI **6401**, Heidelberg, Springer.

[11] M K M Nasution, R Sitepu and M Hardi 2016 Using social networks to assess forensic of negative issues, *IEEE CITSM*, Bandung.

[12] M K M Nasution 2012 Simple search engine model: Adaptive properties, arXiv:1212.3906v1 [cs.IR] 17 Dec 2012.

[13] M K M Nasution 2012 Simple search engine model: Adaptive properties for doubleton, arXiv:1212.4702v1 [cs.IR] 19 Dec 2012

[14] M K M Nasution, M Elveny, R Syah, and S A M Noah, 2015, Behaviour of the resources in the growth of social network *IEEE Proceeding of the 5th International Conference on Electrical Engineering and Informatics*, Bali.

[15] M K M Nasution, O S Sitompul, E P Sinulingga, and S A N 2016 An extracted social network mining, *IEEE SAI Computing Conference*, London, UK.

[16] S Vaithyanathan 1999 Introduction: Data mining on the Internet *Artificial Intelligence Review* **13**

Insurance Contract Analysis for Company Decision Support in Acquisition Management

H P Chernovita, D Manongga and A Iriani

Faculty of Information Technology, Satya Wacana Christian University, 52-56 Diponegoro Street, Salatiga, Indonesia

Abstract. One of company activities to retain their business is marketing the products which include in acquisition management to get new customers. Insurance contract analysis using ID3 to produce decision tree and rules to be decision support for the insurance company. The decision tree shows 13 rules that lead to contract termination claim. This could be a guide for the insurance company in acquisition management to prevent contract binding with these contract condition because it has a big chance for the customer to terminate their insurance contract before its expired date. As the result, there are several strong points that could be the determinant of contract termination such as: 1) customer age whether too young or too old, 2) long insurance period (above 10 years), 3) big insurance amount, 4) big amount of premium charges, and 5) payment method.

1. Introduction

One of company activities to retain their business is marketing the products, in the form of goods or service [1]. Maximum marketing activity will increase sales number and impact the increase of company revenue. Customer Acquisition Management [2] is an effort by the company to manage acquisition process or an approach to the customer candidate. This activity includes gaining information about potential customer, measure customer value, and allocating resources to acquire those with greater long-term value [3]. There is tendency that company more invest their resources, like bigger cost to marketing and sales activities [4]. So, the company try hard to gain bigger revenue to return the venture capital.

Marketing aim to attract customer candidates and retain loyal customer who binding contract with the insurance company [5]. Insurance companies will compete each other to gain customers using promising offerings to bring out customer interest. Insurance company often promising protection [6], service [7], and certain bonuses [8] to make customer want to invest their money in the company. Conflict handling according to customer's need includes in customize service [9] that should be done maximally by the company to prevent customer disappointment that could lead the customer to move to competitor company or terminate the contract.

Decision support is vital element in insurance company to help the company to provide proper insurance product offering [8], suggestions, and alternatives for prospective customers in acquisition process. One of many algorithms used is Iterative Dichotomizer Three (ID3) which take form of decision tree with root, branch, and leaf as the parameters [10] which then being calculated and arranged based on the entropy score of each parameter.

Insurance Contract Analysis in this paper will try to create decision support using ID3 algorithm to arrange rules that could be used by insurance company in customer acquisition process. This is necessary for the company to make customer could survive until the end of insurance contract and prevent customer to terminate their contract before its expired date. Parameter used in this research will be from customer side and insurance contract side. Result from ID3 will provide rules to guide insurance agent to offer insurance product and confirmation of contract agreement content that will be undergo by the customer during specified time. Research result could help insurance company to raise customer acquisition and increase customer lifetime. Along with it, the customer trust and satisfaction will go up because the company can offer the proper product according to customer's need, condition, and ability.

2. Related Work

Research by Akinsola [10] used data mining to gather dataset from motor vehicle insurance in Lagos about 17,000 records. Each record contains 57 fields of attributes that through filter process to obtain suitable data and attributes. Insurer risk analysis using ID3 to produce rule to decide insurer risk class: Low Risk, Medium Risk, and High Risk. Result of decision tree will be used by insurance company to be more careful when define the premium policy.

Research by Balaji [11] tried to examine the Decision Tree Induction technique to predict customer preferences over the life insurance policies. Dataset from Life insurance database about 10,000 records from June 2011 to December 2011. The decision tree lead to four kinds of class: Class A, Class B, Class C, and Class D that will be used by the company in order to retain the profitable customer with perform cross-selling and upselling insurance products.

Research by Roberts-Lombard [8] explored two important ways to enhance customer retention: two-way communication and conflict handling. Data gathered using questionnaire to 254 customers. The result shows that close relationship between insurance company and customer will provide a long-term retention, so timeously it will create greater customer loyalty.

Recent study will adopt the ID3 algorithm to analyse insurance contract agreement between customer and the company to create decision support rules in acquisition management. Insurance company could use this rules to offer proper product needed by the customer. Excellent product offerings by insurance company will help the customer to decide what product should they buy, how long the contract could they take, which payment method they will take, and how much the premium they could pay. This effort will make the customer could projecting their financial before binding insurance contract. The main goal of this paper is to increase company service in customer acquisition process that will increase customer satisfaction then lead to customer loyalty to the company.

3. Research Method

This research using data from well-known insurance company in Indonesia, but the dataset came from branch office in Salatiga from January 2016 to May 2016 about 245 records with 37 attributes. The data is insurance claim report submitted by customers to the company which divided in four kinds: completion, termination, retrieval, and school expense. This research using the completion claim about 102 records and termination claim about 27 records to find out characteristic or condition that will lead customer to complete or terminate their insurance contract. The data then cleansed with removing unnecessary attributes from 37 attributes to 7 attributes: product name, customer age when binding insurance contract, insurance period, insurance amount, payment method, premium charge, and final claim. Figure 1 shows table structure of data claim inserted in a spreadsheet and adjusted with the code in Table 1. In the spreadsheet, field name for each attribute will be shortened as: PRODUCT for insurance product name, AGE for customer's age when start the insurance contract, PERIOD for insurance validity period, AMOUNT for sum assured amount, PAYMENT for payment method, PREMIUM for premium charge, and FINAL for final claim.

PRODUCT	AGE	PERIOD	AMOUNT	PAYMENT	PREMIUM	FINAL
PERMATA	5	1	2	5	2	COM
PERMATA	4	1	3	5	2	COM
PERMATA	3	1	4	5	2	COM
PERMATA	4	2	3	5	3	COM
MELATI	4	1	2	2	2	COM
POESAKA	3	1	1	5	1	COM

Figure 1. Table structure of imported data in RapidMiner.

Table 1. Table of attribute code.

Attribute	Class	Code
Customer Age (years old)	17 - 24	1
	25 - 32	2
	33 - 40	3
	41 - 48	4
	49 - 56	5
Insurance Period (years)	4 - 6	1
	7 - 9	2
	10 - 12	3
	13 - 15	4
	16 - 18	5
Sum Assured Amount (IDR)	2,000,000 - 6,600,000	1
	6,600,001 - 11,300,001	2
	11,300,002 – 16,000,002	3
	16,000,003 – 20,700,003	4
	20,700,004 – 25,400,004	5
Premium Charge (IDR)	16,840 – 403,472	1
	403,473 – 790,105	2
	790,106 – 1,176,738	3
	1,176,739 – 1,563,371	4
	1,563,372 – 1,950,004	5
Payment Method	Monthly	1
	Quarterly	2
	Semester	3
	Year	4
	Single	5
Final Claim	Completion	COM
	Termination	TER

Code shown in Table 1 that inserted in the spreadsheet for each attribute represents insurance contract content that have been claimed to the insurance company. If in the insurance contract written that the customer age 30 years old, binding insurance contract for 10 years, with sum assured IDR 20,000,000, with premium charge IDR 500,000, payment quarterly, then the final claim is termination will be inserted as: AGE = 2, PERIOD = 3, AMOUNT = 4, PREMIUM = 2, PAYMENT = 2, and FINAL = TER. The code was made to simplify the contract content to be processed in the next step.

When the spreadsheet is complete with the contract content for each customer, then it imported in RapidMiner Studio 7.3 to be processed with ID3 algorithm with choosing Operators: Modeling → Predictive → Trees → ID3. In the Parameters setting, set the criterion with "accuracy", minimal size of split with 2, minimal leaf size with 1, and minimal gain with 0.1. When the setting is complete and RapidMiner "Run the process", the application will create the decision tree based on the imported data.

4. Result

The result in this research is a decision tree which could being used by insurance agent as decision support when offer insurance product to the prospective customers. Figure 2 shows rules formed based on entropy calculation from each involved attribute. The decision tree has PRODUCT attribute as its root and followed by AGE, PERIOD, AMOUNT, PAYMENT, and PREMIUM attribute as the branches, while FINAL attribute set to be the label or the leaf. The leaf or node have two kinds of colour. Blue leaf represents the completion claim, while red leaf represents the termination claim. There also leaf which have two colours in the same leaf, that means the leaf represents same chance of both completion claim and termination claim in one rule. On the decision tree, there are total 53 rules formed, which consist of 33 rules of completion claim, 13 rules of termination claim, and 7 rules with same chance of completion claim and termination claim.

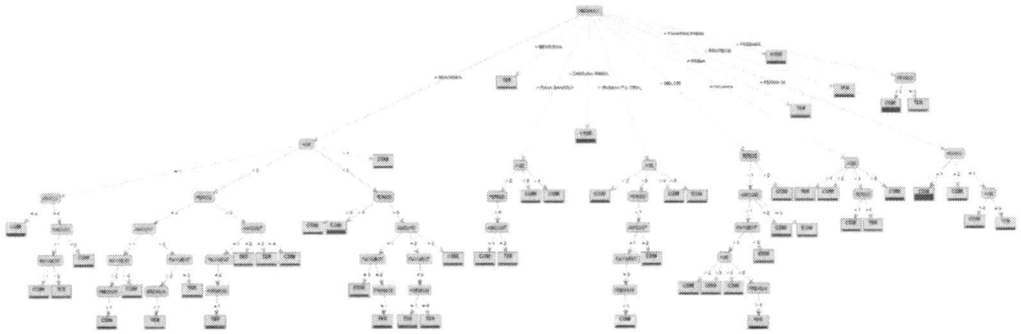

Figure 2. Decision tree based on insurance completion and termination claim.

Decision tree in Figure 2 produce 13 rules of termination claim that will be the main focus of this paper. Below are the rules that show the contract content lead to termination claim:

`PRODUCT = BEASISWA, AGE = 1, PERIOD = 5, AMOUNT = 1, PAYMENT = 2: TER {COM=0, TER=2}` (1)

Contract with product "Beasiswa", customer age between 17-24 years old, insurance period 16-18 years, insurance amount between IDR 2,000,000 - 6,600,000, and payment method quarterly could lead to termination claim. This occurred because of customer age which is too young that predicted have not occupation with settled salary. Besides, the insurance period which too long makes the customer feels burdened to fulfil their contractual obligation. Consequently, this condition could lead the customer to terminate their insurance contract before its expired date.

`PRODUCT = BEASISWA, AGE = 2, PERIOD = 4, AMOUNT = 2, PAYMENT = 3: TER {COM=0, TER=1}` (2)

Contract with product "Beasiswa", customer age between 25-32 years old, insurance period 13-15 years, insurance amount between IDR 6,600,001 - 11,300,001, and payment method semester could lead to termination claim. This occurred because customer in this age group usually on their way to pursue specific career and build household. In this time, there is a lot of changes happened in financial aspect like when the customer has more child, then the customer income automatically will be allocated to sufficient the daily needs. This could lead to contract termination because the cost that originally paid for insurance premium then being used to another necessity. Besides, the long insurance period also become additional issue that will lead to contract termination. Moreover, the payment method also affect the continuity of insurance contract because the longer payment period means the bigger premium charges. This contract has "semester" payment method that should be paid every six month with bigger premium charges that will give burden to the customer to pay such a big amount of money in every payment due date.

`PRODUCT = BEASISWA, AGE = 2, PERIOD = 5, AMOUNT = 2: TER {COM=0, TER=1}` (3)

Contract with product "Beasiswa", customer age between 25-32 years old, insurance period 16-18 years, and insurance amount between IDR 6,600,001 - 11,300,001 could lead to termination claim. This occurred because of the reasons appeared in Rule (2). The difference is at insurance period that longer from previous rule which could lead to termination claim.

`PRODUCT = BEASISWA, AGE = 2, PERIOD = 5, AMOUNT = 3: TER {COM=0, TER=1}` (4)

Contract with product "Beasiswa", customer age between 25-32 years old, insurance period 16-18 years, and insurance amount between IDR 11,300,002 - 16,000,002 could lead to termination claim. This occurred because of the reasons appeared in Rule (3). The difference is at insurance amount that bigger than the previous rule. This means customer have to pay bigger premium charges for a long time. This condition shows that the longer insurance period and bigger premium charges could lead to termination claim.

```
PRODUCT = BEASISWA, AGE = 3, PERIOD = 5, AMOUNT = 2, PAYMENT = 2, PREMIUM =
5: TER {COM=0, TER=1}                                                    (5)
```

Contract with product "Beasiswa", customer age between 33-40 years old, insurance period 16-18 years, insurance amount between IDR 6,600,001 - 11,300,001, payment method quarterly, and premium charges IDR 1,563,372 – 1,950,004 could lead to termination claim. This occurred because the customer age includes in middle-aged category which have settled job and salary but usually have a lot more necessity than the younger customer. Besides, the long insurance period with big premium charge that should be paid quarterly or every three months will load the customer with big responsibility, while the customer should pay other bills in first place like school tuition, electricity, water, daily needs, etc. Those conditions in fact give disadvantages to customer that could lead to termination claim.

```
PRODUCT = BEASISWA, AGE = 3, PERIOD = 5, AMOUNT = 4: TER {COM=0, TER=1} (6)
```

Contract with product "Beasiswa", customer age between 33-40 years old, insurance period 16-18 years, and insurance amount between IDR 16,000,003 - 20,700,003 could lead to termination claim. This occurred because of the reasons appeared in Rule (5). The difference is at insurance amount that bigger from previous rule. The bigger insurance amount with same period will make the customer pay bigger amount of premium charges. This shows that customer in middle-aged category could terminate their insurance contract when the insurance period is too long and the insurance amount is too big.

```
PRODUCT = DANA BAHAGIA, AGE = 2, PERIOD = 4, AMOUNT = 2: TER
{COM=0, TER=1}                                                           (7)
```

Contract with product "Dana Bahagia", customer age between 25-32 years old, insurance period 13-15 years, and insurance amount between IDR 6,600,001 - 11,300,001 could lead to termination claim. This occurred because the customer has long insurance period, while the necessity increase every year, especially when the customer has more child. This also occurred when the customer has unexpected needs that demand a big amount of money, so the customer prefer to terminate the insurance contract.

```
PRODUCT = PELANGI, AGE = 1: TER {COM=0, TER=1}                           (8)
```

Contract with product "Pelangi" and customer age between 17-24 years old could lead to termination claim. This occurred because the customer age considered too young that have not settled job and salary yet. Another supporting reason like the long insurance period, big insurance amount, big premium charges, and unexpected needs could lead the customer to contract termination.

```
PRODUCT = PELANGI, AGE = 4, PERIOD = 4: TER {COM=0, TER=1}               (9)
```

Contract with product "Pelangi", customer age between 41-48 years old, and insurance period 13-15 years could lead to termination claim. This occurred because the customer age which already have settled job and salary but also have more necessity must pay big amount of premium for a long period. The termination claim could be caused by unexpected need like illness, disaster, or non-lethal accident that demand a big amount of money.

```
PRODUCT = PERMATA, PERIOD = 3, AGE = 4: TER {COM=0, TER=2}               (10)
```

Contract with product "Permata", insurance period 10-12 years, customer age between 41-48 years old could lead to termination claim. This occurred because product "Permata" require single payment with big premium charges but after that, customer have to wait until the contract period is completed. The insurance period seems too long and usually there is an unexpected event or needs within a period of 12 years that could lead the customer to terminate their insurance contract. This kind of customer usually consider the insurance as emergency funds that could be claimed at any time.

```
PRODUCT = POESAKA, PERIOD = 3: TER {COM=0, TER=1}                        (11)
```

Contract with product "Poesaka", and insurance period 10-12 years could lead to termination claim. This occurred because the product "Poesaka" more like investment contract that every year passed will add cash value to the insurance earnings. The big premium charges paid in single payment and the customer have to wait until the insurance period is completed. Termination claim in this case is more likely to get the customer investment earnings for several years to be used as emergency fund.

`PRODUCT = PRIMA: TER {COM=0, TER=1}` (12)

Contract with product "Prima" could lead to termination claim. This occurred affected by the other supporting reasons like customer age, insurance period, insurance amount, premium charges, and payment method. But in this case, the customer who binding the contract already in age between 49-56 years old, insurance period 10-12 years, insurance amount IDR 20,700,004 – 25,400,004, payment method yearly, and premium charge IDR 1,563,372 – 1,950,004. Those conditions could be big burdened for the younger customer, but this case include the oldest group of customer which really near to the retirement age. So, from this case, the customer age proven as the strong determinant of contract termination.

`PRODUCT = PROTEKSI: TER {COM=0, TER=3}` (13)

Contract with product "Proteksi" could lead to termination claim. This occurred affected by the other supporting reasons like customer age, insurance period, insurance amount, premium charges, and payment method. In this case, all three customers with this kind of contract choose to terminate their contract because of unexpected event like sudden illness, non-lethal accidents, and dismissal from workplace that require a lot of money. The customer tends to terminate the insurance contract because their money more allocated to fulfil the daily needs than to pay insurance premium for a long time.

5. Conclusion

Contract analysis is important to observe the condition that will lead to the termination claim. The decision tree shows 13 rules that lead to contract termination claim with explanation for each rule. These "termination rules" could be a guide to insurance company as decision support in acquisition management to prevent contract binding with these conditions because it has a big chance for the customer to terminate their insurance contract before its expired date. Besides, there are several strong points that could be the determinant of contract termination such as: 1) customer age whether too young or too old, 2) long insurance period (above 10 years), 3) big insurance amount, 4) big amount of premium charges, and 5) payment method.

6. References

[1] Cespedes F V 1995 *Concurrent Marketing: integrating product, sales, and services* (Boston, Massachusetts: Harvard Business School Press)

[2] Saylor J 2004 The missing link in CRM: customer acquisition management http://www.destinationcrm.com/Articles/Web-Exclusives/Viewpoints/The-Missing-Link-in-CRM-Customer-Acquisition-Management-44024.aspx

[3] Arnold T J, Fang R and Palmatier R W 2011 The effects of customer acquisition and retention orientations on a firm's radical and incremental innovation performance *J. of Ac.of Mark. Sci.* **39** 234-51

[4] Winer R S 2001 A framework for customer relationship management *California Man. Rev.* **43** 4

[5] Ayoubi M 2016 Customer segmentation based on CLV model and neural network *Int. J. of Comp. Sci. Iss.* **13** 31-37

[6] Yadav R K and Mahonia S 2014 Claim settlement process of life insurance services - a case study of ICICI Prudential life insurance company *Int. Let. of Soc. and Hum. Sci.* **35** 26-32

[7] Singhal P K and Gupta M A 2013 Assessment of service quality in insurance sector – a case study of private companies of Haryana state *Int. J. of Res. of IT and Man.* **3** 118-33

[8] Roberts-Lombard M 2011 Customer retention through customer relationship management: the exploration of two-way communication and conflict handling *African J. of Bus.s Man.* **5** 3487-96

[9] Saarijarvi H, Karjaluoto H and Kuusela H 2013 Customer relationship management: the evolution role of customer data *Mark. Intel. & Plan.* **31** 584-600

[10] Akinsola A F, Sokunbi M A, Lawal O N and Okikiola F M 2015 A data mining to insurance risk analysis *Int. J. Of Eng. And Comp. Sci.* **4** 10255-8

[11] Balaji S and Srivatsa S K 2012 Decision tree induction for mining Life Insurance databases *Int. J. of Comp. Sci. and Info. Tech. & Sec.* **2** 699-703

Studies on behaviour of information to extract the meaning behind the behaviour

M K M Nasution[1*], R Syah[2], M Elveny[3]

[1,2,3]Information Technology, Fasilkom-TI, University of Sumatera Utara, Padang Bulan USU Medan 20155, Indonesia.

[*]Email: mahyuddin@usu.ac.id

Abstract. Web as social media can be used as a reference for determining social behaviour. However, the information extraction involves a search engine is not easy to give that picture. There are several properties of the search engine to be formally disclosed to provide assurance that the information is feasible. Although quite a lot of research that has revealed the interest of the Web as social media, but a few of them that have revealed behaviour of information related to social behaviour. In this case, it needs the formal steps to present possibilities related properties. There are 12 properties that are interconnected as behaviour of information and then it reveals several meanings based on the simulation results of any search engine.

1. Introduction

As for making the decisions, any people need a sequence of processes (order): data, information, knowledge, and wisdom [1]. That way, the systems of the world have been organized [2]. There is a part that serves to explain another part: The social media which recorded daily activities about a person or the social, or event of the world into the Internet and it will be a picture of the ongoing social life [3]. However, a concept said that the past determines the present and then both determine the future, and each of time the Web as social media provide it [4]. The research relating to the search engines [5,6], data mining [7,8] and semantic technology [9] have been done by researchers [10], but the little of them that explains the behaviour of search engine results act as the properties (P) to explain the social behaviour. This paper will reveal the meaning consistently to both behaviour so that it can be used as a material consideration in decision making.

2. Basic Concept and Motivation

In a world increasingly smarter, the world is divided into several parts: one part becomes more complex, the other part be simpler and it is not uncommon to get that simpler must go through that more complex, but there are also other parts exist in among them and usually these parts behave as the buffer or the interface between that simpler and more complex [11,12,13]. As the concept, a definition in a system is mathematically to create the demarcation of complexity: A definition of word [14], for example, as a simplification to understand a part of world as follows.

Definition 1. [15] A word w is the basic unit of discrete data and an item from a vocabulary indexed by $i = 1,2,\ldots$ such that $w_i = 1$ if i in I or $w_i = 0$ otherwise, W is a set of words, $W = \{w_i|i=1,2,\ldots,n\}$, and size of W is $|W| = n$.

Each object as entity not only has name in word, but the object has one or more attributes (*at*). The name and attributes can be given literally like the literal text of "Knowledge Technology" as the tangible or intangible knowledge [16]. The name and attributes of the entity refer to the social meaning of the objects. As a system, the World Wide Web (WWW) or Web, is a space of information not only contain the text (words) but can also consist of: pictures (*pi*), voices (*vo*), videos (*vi*) or animations (*an*), but all of them represented by the words, at least the file name of them [17]. By Web as part of our world is increasingly more complex ever since found, because their tags continue developed and enriched. Web allows each person to obtain information and informs something. In this case, the tags are the metadata refer to the identity of attribute (*md*) for create more smart documents in Internet: like foaf, xml, etc. In information system, we define a space as the part of world as below [5].

Definition 2. A dynamic space, we call it as Web and denote it as Ω, is a collection of web pages ω_j, j=1,2,... with $w_k \vee pi \vee vo \vee vi \vee an$, at, and md. or $\Omega = <\{\omega_j{:}j=1,2,\ldots\},(w_k \vee pi \vee vo \vee vi \vee an), at, md>$.

Every day the Web Ω continues to grow and contains many elements that are not known exactly or {...} for recording all activities of social members. So for it we define the other system as the interface as below [5].

Definition 3. A term be $t_k = (w_1)$ contains one word or $t_k = (w_1w_2\ldots w_k)$ the words in pattern, where l is number of tokens (vocabularies) in t_k and k is a number of parameters representing word w. $l \leq k$ for $|t_x| = k$ and $W = \{w_1, w_2, \ldots, w_l\}$.

Definition 4. Ω is a set of web pages indexed by search engine: a set contains the ordered pair of the terms t_k and the web pages ω_k or (t_k,ω_k) such that $\Omega_k = \{(t_k,\omega_k)_{ij}\}$ is a subset of Ω to be the relation table that consists of two columns t_k and ω_k where $i = 1,\ldots,I, j = 1,\ldots,J$.

Based on Definition 4, $|\Omega|$ is the cardinality of Ω or we call it as hit count for t_x, and we can write $\Omega_k = \{(t_k,\omega_k)_{ij}\}$ as $\Omega_k = \{\omega_{k1},\ldots, \omega_{kj}\}$. If t_x has the probability $p(t_k)$ against Ω with uniform mass probability function $P : \Omega \rightarrow [0,1]$, then the properties of search engine to t_x based on $|\Omega_x| \leq |\Omega|$ whereby $|\Omega_x|/|\Omega|$ describes the behavior of property.

3. An Approach

Developing the concept abstraction in mathematics, we use to model the parts of system. A conceptual bridge for building the interface between two different systems in complexity: We declare first a query contains search term t_k, as the bridge of understanding as follows [15].

Definition 5. Let search term t_k as a vector of web pages based on the search engine, i.e. ω_k in Ω in which t_x occurs. q is the query if $\omega \Rightarrow q$ is true or $\omega \Rightarrow t_x$ is true for all ω_x in Ω: $(\omega_x \Rightarrow t_k) = 1$.

Second, for formalizing property we define the singleton and then the doubleton in the modelling and the properties.

3.1. Model

Definition 6. [5,15] Let t_x is a search term, and t_x in S, S is a set of singleton search term for any search engine. A vector space Ω_x is a subset of Ω is singleton search engine event (*singleton*) of web pages that contain an occurrence of t_x in ω_x.

To express the activities of a social member from Web, we can use query with t_x = Mahyuddin K. M. Nasution for example, and then we have a hit count as representation of all activities. Something socially can be ascertained with regard to uncertainty: a name may refer to different persons or a person possesses the literal text of different names.

Suppose t_x and t_y are search terms where $t_x \neq t_y$, based on Definition 6 we obtain as follows,

(i) For $t_x \cap t_y \neq$ empty set and $|t_y| < |t_x|$: If for all w_y in t_y we get w_y in t_x and there is w_x in t_x but w_x is not in t_y, then for all w_y in ω_y we find that w_y in ω_x and there is w_x in ω_x and w_x is not in ω_y such that

$$t_x \cap t_y = t_y \text{ and } t_x \cup t_y = t_x \quad (1)$$

and

$$\omega_x \cap \omega_y = \omega_y \text{ and } \omega_x \cup \omega_y = \omega_x. \quad (2)$$

(ii) For $t_x \cap t_y = \emptyset$: If for all w_y in t_y we find that w_y is not in t_x and for all w_x in t_x we get also w_x not in t_y, then for all w_y in ω_y we find that w_y is not in ω_x and for all w_x in ω_x we find that ω_x is not in ω_y such that

$$t_x \cap t_y = \emptyset \text{ or } t_x \cup t_y = t_y \cup t_x \quad (3)$$

and

$$\omega_x \cap \omega_y = \emptyset \text{ or } \omega_x \cup \omega_y = \omega_y \cup \omega_x \quad (4)$$

(iii) For $t_x \cap t_y = \emptyset$ and $\omega_x \cap \omega_y \neq \emptyset$: For w_x in t_x we find that w_x is not in t_y and for all w_y in t_y we get also w_y is not in t_x then

$$t_x \cap t_y = \emptyset \text{ or } t_x \cup t_y = t_y \cup t_x, \quad (5)$$

but for all w_x in ω_x there is w_x in ω_y and for all ω_y in ω_y there is w_y in ω_x, then

$$\omega_x \cap \omega_y = \omega_x = \omega_y \text{ or } \omega_x \cup \omega_y = \omega_y \cup \omega_x = \omega_y = \omega_x \quad (6)$$

Although ambiguity and bias naturally become the gift of social into social media such as the Web, the social media itself affects social behavior so that information about a person is not only independent but also has a relationship with each other. For example, if the query contains t_x follow the pattern name such as "Mahyuddin K. M. Nasution" (in quotes), the hit count will be obtained which contains no bias, but still have ambiguous. If the bias has already reduced, then at the moment also it did not lose the social natural of Web. Thus, we formulate a property of singleton for reducing ambiguity.

Definition 7. [6,15] Let t_x and t_y are two different search term, $t_x \neq t_y$, t_x,t_y in S as singleton. A doubleton search term is $D = \{\{t_x,t_y\}:t_x,t_y \text{ in } \omega\}$ and its vector space denoted by $\Omega_x \cap \Omega_y$ is a doubleton search engine event (*doubleton*) of web pages that contain a co-occurrence of t_x and t_y such that t_x,t_y in ω_x and t_x,t_y in ω_y, where Ω_x, Ω_y, $\Omega_x \cap \Omega_y$ are subsets of Ω.

3.2. Properties

Each activity carried out at different times and places and by different people, so the social behaviour of the independence of the activity is not the same with each other. In other words, for the different search terms t_x and t_y and in different query, based on set theory clear that if $\Omega_x \neq \Omega_y$ then we obtain the behaviour of the disjoint singletons in first property (P1) as follows

$$|\Omega_x \cap \Omega_y| = 0. \quad (P1)$$

However, socially an activity may be carryout by more than two people, even though their roles may differ, or second property (P2) is

$$|\Omega_x \cup \Omega_y| = |\Omega_x| + |\Omega_y|. \quad (P2)$$

This behaviour also supported by (P1), (5) and (6), and we have Lemma as follows.

Lemma 1. If $t_x \neq t_y$, $t_x \cap t_y \neq \emptyset$, then $|\Omega_x \cap \Omega_y| = 0$ and $|\Omega_x \cup \Omega_y| = |\Omega_x| + |\Omega_y|$.

In other case, although an activity can stand alone, but in social the behavior of the activity does not just apply to personal but involves another person, causing the inclusion of other. Or, based on Eq. (1) and Eq. (2) we have $\Omega_x = \{(t_x,\omega_x)\} = \{(t_x \cup t_y,\omega_x \cup \omega_y)\} = \{(t_x,\omega_x) \cup (t_y,\omega_y)\} = \{(t_x,\omega_x)\} \cup \{(t_y, \omega_y)\} = \Omega_x \cup \Omega_y$. Or, based on Eq. (2) and Eq. (5) we obtain the behavior of inclusion as a property (P3) as follows

$$|\Omega_x| = |\Omega_x| + |\Omega_y|. \quad (P3)$$

For it all, we can prove Lemma as follows.

Lemma 2. If $t_x \neq t_y$, $t_x \cap t_y \neq \emptyset$ and $|t_y| < |t_x|$, then the relation of singletons based on t_x and t_y is $\Omega_x = \Omega_x \cup \Omega_y$ or $|\Omega_x| = |\Omega_x| + |\Omega_y|$ where Ω_x, Ω_y are subsets of Ω.

There the same activities have different behaviors, but also the same behaviors come from the different activities. In other words, based on Eq. (5) and Eq. (6), we obtain $\Omega_x \cap \Omega_y = \{(t_x,\omega_x)\} \cap \{(t_y,\omega_y)\} = \{(t_x,\omega_y)\} \cap \{(t_y,\omega_y)\}$, and because t_y in ω_y the intersection of $t_x \cap t_y$ must be $\{(t_x,\omega_y)\} \cap \{(t_y,\omega_y)\} = \{(t_y,\omega_y)\} \cap \{(t_y, \omega_y)\}$ or $\Omega_x \cap \Omega_y = \Omega_y \cap \Omega_y$ or $\Omega_x \cap \Omega_y = \Omega_y$. Similarly, $\Omega_x \cap \Omega_y = \Omega_x$. Thus, we have the behavior of singleness as a property, i.e.

$$|\Omega_x| = |\Omega_y|. \qquad \text{(P4)}$$

We conclude the Lemma as follows.

Lemma 3. If $t_x \neq t_y$, $t_x \cap t_y = \emptyset$, and $\omega_x \cap \omega_y \neq \emptyset$, then $|\Omega_x| = |\Omega_y|$, Ω_x and Ω_y are subsets of Ω.

Expansion of P3 as the behavior of the singleton into the behavior of a singleton shows the behavior of inclusion socially the persons in the same activities. Or based on Lemma 1 we derive

$$|\Omega_x| = |\Omega_x \cup \Omega_y| = |\Omega_x|+|\Omega_y| = |\Omega_x|+|\Omega_y \cup \ldots| = |\Omega_x|+|\Omega_y|+\ldots = |\Omega_x|+|\Omega_y|+|\ldots \cup \Omega_z|$$

such that the behaviour of P3 in recursive is as follows

$$|\Omega_x| = |\Omega_x|+|\Omega_y|+\ldots+|\Omega_z|. \qquad \text{(P5)}$$

In general, we have the Proposition as follows.

Proposition 1. If some of search terms in $t_z \neq \ldots \neq t_y \neq t_x$ and $t_z < \ldots < t_y < t_x$, then $|\Omega_x| = |\Omega_x|+|\Omega_y|+\ldots+|\Omega_z|$, Ω_x, Ω_y, …, Ω_z are subsets of Ω.

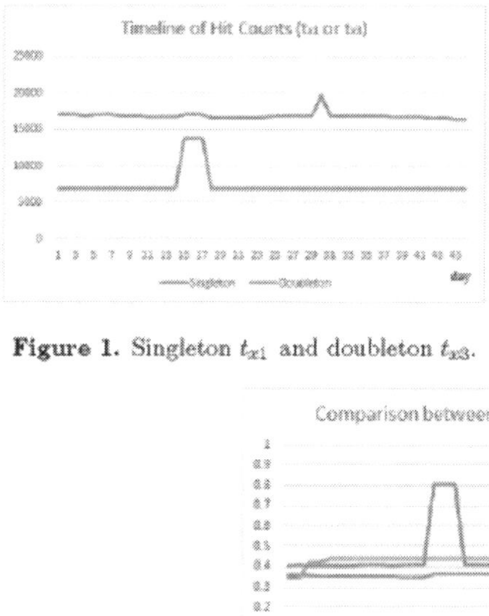

Figure 1. Singleton t_{x1} and doubleton t_{x3}.

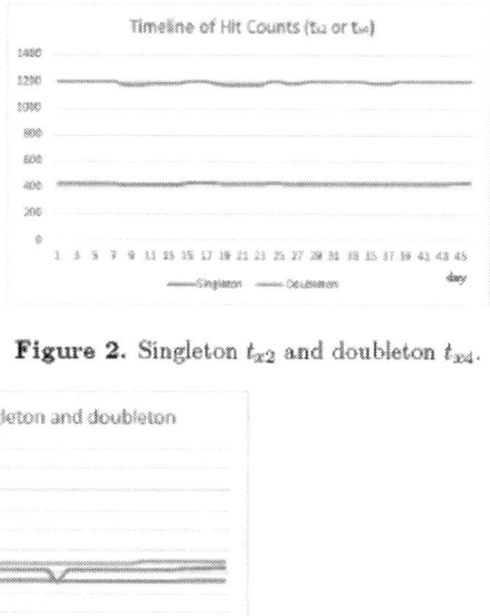

Figure 2. Singleton t_{x2} and doubleton t_{x4}.

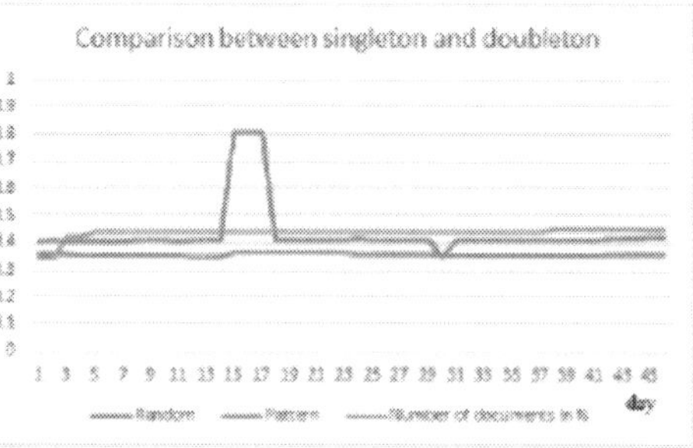

Figure 3. Relation between t_{x1}, t_{x2}, t_{x3} and t_{x4}

As method, we specify some properties associated with singleton and doubleton to understand any system that has the same characters, and then we conduct experiment using Google search engine and

Google Scholar as a part of Google search engine for simulating some properties so that it can express some meaning.

4. Discussion

Based on the concepts, we have disclose some of the properties of the subspaces which are built into the more complex system to disclose any information on the reverse [5,6,18,15]. In other words, in systems that contain information such as the Web, we have been lowered several properties in accordance with the terms prevailing in the system, whereby based on singleton and doubleton properties we reveal characteristics for expressing their meaning.

Information contained by the properties depend on the growth of the documents or web pages on the Web where any search term exists or not. Growth of documents or web pages can be measured by using a search engine for submitting a query that contains term like t_x = 'Mahyuddin K. M. Nasution' some of times regularly in various forms of the following: t_{x1} = Mahyuddin K. M. Nasution (as a singleton); t_{x2} = "Mahyuddin K. M. Nasution" (as a singleton); t_{x3} = Mahyuddin K. M. Nasution,Universitas Sumatera Utara (as a doubleton); and t_{x4} = "Mahyuddin K. M. Nasution","Universitas Sumatera Utara" (as a doubleton). In this discussion, we used the Google search engine as a simulator against singleton and doubleton properties that have been derived. A snapshot about the dynamic of Web can be seen in Figure 1 and Figure 2. We already collect a number of 45 hit counts for representing each the singleton and doubleton per day for t_{x1}, t_{x2}, t_{x3} and t_{x4}. Although the search results of t_{x1} and t_{x3} have showed instability by the presence of fluctuation on the timeline of the hit counts based on day. Matching randomly, however, the singleton of t_{x1} also have showed downward trend insignificantly in some hit counts. This decline due to the reduction for a number of terms not suitable after the presence of the documents that contain the words that only partially match the search terms. It expressed by some properties. Instead, the doubleton of t_{x3} tends to increase insignificantly in a series of hit counts, it has showed that the increasing number of documents relating to 'Mahyuddin K. M. Nasution' which is affiliated to the 'Universitas Sumatera Utara' as a keyword has also increased. On the other hand, on matching based on patterns (the search term in quote), the singleton of t_{x2} and the doubleton of t_{x4} are always on an even keel increasing.

Likewise, a comparison between t_{x1} and t_{x3} already exhibit the same characteristic with each of t_{x1} and t_{x3}. While comparison between t_{x2} and t_{x4} have showed the same thing with the characteristic of t_{x2} or t_{x4}, see Figure 3, and therefore we can say that

$$c_{xy} = |\Omega_x \cap \Omega_y| / |\Omega_x| \quad \text{(P6)}$$

if Ω_y as representation of keyword and $|\Omega_x \cap \Omega_y| \leq |\Omega_x|$.

Table 1. Four items of hit counts for 94 names

Results	z_1	z_2	z_3	z_4
Total	59,207,520	1,338,656	193,656	16,549
s_x^2	1,095,067,218,274,400	203,058,300,676	11,535,788,845	12,718,265
σ_z^2	11,252,918,525,718	1,957,388,639	118,471,902	104,306

Furthermore, we explore characteristic of singleton or doubleton by measuring the internal consistency (reliability). This characteristic based on Cronbach's alpha $\alpha = k(k\text{-}1)(1\text{-}(\sum\sigma_z^2)/s_x^2)$ as the instrument reliability coefficient for k = number of the question items or the amount of matter, while the sum squares is $s_x^2 = \sum_{i=1...n} |\Omega_x|_i^2$, average $\mu = (\sum_{i=1...n} |\Omega_x|_i)/n$, variance is $\sigma_z^2 = (\sum_{i=1...n} |\Omega_{xi}|\text{-}\mu)/(n\text{-}1)$ and total of items variance $\sum\sigma_z^2 = \sigma_{z1}^2 + \sigma_{z2}^2 + \ldots + \sigma_{zi}^2$ with a rule as the marker of: (a) $\alpha \geq 0.9$: The internal consistency is very good, (b) $0.7 \leq \alpha \leq 0.9$: The internal consistency is good, (c) $0.6 \leq \alpha \leq 0.7$: The internal consistency is acceptable, (d) $0.5 \leq \alpha \leq 0.6$: The internal consistency is poor, and (e) $\alpha < 0.5$: The internal consistency is not accepted [19, 20, 21]. In this discussion, we have the hit counts for four items: For example, for name "Urip Harahap" and/or keyword "Universitas Sumatera Utara" by using the Google search engine to the query form of t_{x2} and t_{x4} and we have obtained $|\Omega_x|$ = 16400 and $|\Omega_x \cap \Omega_y|$ = 4270 respectively where the collection of $|\Omega_x|$ as z_1 and the collection of $|\Omega_x \cap \Omega_y|$ as z_2, while by using Google Scholar to the query form of t_{x2} and t_{x4} we have gained $|\Omega_x|$ = 68 and $|\Omega_x \cap \Omega_y|$ = 47 respectively where the collection of $|\Omega_x|$ as z_3 and the collection of $|\Omega_x \cap \Omega_y|$ as z_4 [22]. We collect the hit counts of singleton or doubleton for $n = 94$ names, we have obtained $\alpha = 0.015$ based on Table 1. It reveal characteristic of them is that property of 94 names have no the internal consistency as characteristic. The properties of singleton or doubleton with their characteristics is a behavior of information, while based on property and their characteristic we can try to extract behavior of names (actors).

For extracting meaning behind behavior, we calculate the ratio of the hit count based on Equation (P6) for getting (z_2/z_1) = closeness of the activities with his/her affiliation, z_4/z_3 = closeness of scientific work with their affiliation, and z_3/z_1 = percentage of scientific works toward others activities based on assumption that Google Scholar is part of Google search engine. The result of calculation in Table 2, we have $\alpha = -0.008$.

Table 2. Three information types of hit counts for 94 names

Results (a)	z_2/z_1	z_4/z_3	z_3/z_1	Results (b)	z_2/z_1	z_4/z_3	z_3/z_1
Total ($n = 94$)	190.314	62.754	2.979	Total (n =30)	26.034	21.919	0.848
s_x^2	14,692,353	14,692.353	0.519	s_x^2 ($n = 30$)	37.302	18.487	0.038
σ_z^2	152.203	152.203	0.005	σ_z^2 ($n = 30$)	0.054	0.026	0.0001

Computation 1 (Table 1) and Computation 2 (Table 2 (Results (a))) have showed that the internal consistency of information is not accepted, due to non-fulfillment one of the characteristic of properties through the hit counts, for example a name only consists of one word and be part of another name. In other meaning, we have Computation 3 (Table 2 (Results (b))) with reducing the number of names based on the absence of scientific paper in the Scopus index so that remaining $n = 30$ names only as authors, and based on Table 2(b) we have obtained $\alpha = 0.63$: the internal consistency of information is acceptable.

Based on computations we can say that when hit counts that represent the names from the parts of system be the interrelated things, then the comparison results in accordance with properties whereby semantically properties are the mutual-support meaning, namely the documents have indexed strongly supported by the activities of writing scientific papers, moreover they have the affiliation when published, while other activities only increases traffic information relating to the names associated.

5. Conclusion

We already generate 6 properties. It passed down by the search engine model based on concept of word, terms, and documents/web pages. The properties with their characteristic be the behavior of information. We can be concluded that possibility of internal consistency accepted or not depend on behavior of names: There are relation between the activities of writing scientific paper and the behavior of information. Furthermore, we will be studying the possibilities of data mining involving the academics/researchers.

References

[1] R D Gopal, J R Marsden, and J Vanthienen 2011 Information mining - Reflections on recent advancements and the road ahead in data, text, and media mining *Decision Support Systems* **51**.

[2] W F Nasrallah 2006 When does management matter in a dog-eat-dog world: An "Interaction Value Analysis" model of organizational climate *Comput Math Organiz Theor* **12**.

[3] S Stieglitz, L Dang-Xuan, A Bruns, and C Neuberger 2014 Social media analytics: An interdisciplinary approach and its implications for information systems *Business & Information Systems Engineering* **2**.

[4] S Cable 2013 Beyond the past, present, and future: towards the semantics of 'graded tense' in Gikuyu *Nat Lang Semantics* **21**.

[5] M K M Nasution 2012 Simple search engine model: Adaptive properties, arXiv:1212.3906v1 [cs.IR] 17 Dec 2012.

[6] M K M Nasution 2012 Simple search engine model: Adaptive properties for doubleton, arXiv:1212.4702v1 [cs.IR] 19 Dec 2012.

[7] S Vaithyanathan 1999 Introduction: Data mining on the Internet *Artificial Intelligence Review* **13**.
[8] A A Freitas 2001 Understanding the crucial role of attribute interaction in data mining *Artificial Intelligence Review* **16**.
[9] N Fornara and C Tampitsikas 2013 Semantic technologies for open interaction systems *Artif Intell Rev* **39**
[10] S Heras, K Atkinson, V Botti, F Grasso, V Juli'an, and P McBurney 2013 Research opportunities for argumentation in social networks *Artif Intell Rev* **39**.
[11] M Rigo 2004 Characterizing simpler recognizable sets of integers *Studies Logics* **76**.
[12] C Coen 2009 Simple but not simpler: Introduction CMOT special issue - simple or realistic *Comput Math Organ Theory* **15**.
[13] G M P O'Hare, M J O'Grady, R. Tynan, C. Muldoon, H. R. Kolar, A. G. Ruzzelli, D. Diamond, and E. Sweeney 2007 Embedding intelligent decision making within complex dynamic environments *Artif Intell Rev* **27**.
[14] M K M Nasution, and S A Noah 2012 Information retrieval model: a social network extraction perspective *IEEE International Conference on Information Retrieval & Knowledge Management*.
[15] M K M Nasution 2014 New method for extracting keyword for the social actor *Intelligent Information and Database Systems* LNAI **8397**, Heidelberg, Springer.
[16] M K M Nasution, and S A Noah 2011 Extraction of academic social network from online database *IEEE Proceeding of 2011 International Conference on Semantic Technology and Information Retrieval*, Putrajaya, Malaysia.
[17] M K M Nasution 2013 Superficial method for extracting academic social network from the Web, *Ph.D Dissertation*, Universiti Kebangsaan Malaysia.
[18] M K M Nasution, S A M Noah, and S. Saad 2011 Social network extraction: Superficial method and information retrieval *Proceeding of International Conference on Information for Development*.
[19] M K M Nasution, M Elveny, R Syah, and S A M Noah 2015 Behaviour of the resources in the growth of social network, *IEEE Proceeding of the 5th International Conference on Electrical Engineering and Informatics*, Bali.
[20] M K M Nasution, O S Sitompul, E P Sinulingga, and S A Noah 2016 An extracted social network mining, *IEEE SAI Computing Conference*, London, UK.
[21] M K M Nasution 2016 Social network mining: A definitiion of relation between the resources and SNA, *M-CAIT*, Universiti Kebangsaan Malaysia, Malaca.
[22] M K M Nasution, and S A M Noah 2010 Superficial method for extracting social network for academic using web snippets *Rough Set and Knowledge Technology* LNAI **6401** Heidelberg, Springer.

The impact of e-Procurement practice in Indonesia government: A Preliminary Study (The case of Electronic Procurement Service at Bekasi District)

S Candra[1], F E Gunawan[2]

[1]Bina Nusantara University, School of Business Management, Jakarta, Indonesia
[2]Bina Nusantara University, BINUS Graduate Program, Jakarta, Indonesia

E-Mail: seven@binus.ac.id

Abstract. In digital economy, government of Indonesia uses the internet to deliver services and to communicate with citizens and organizations. One of applications are e-Procurement and LPSE is a unit to hold the service system of e-Procurement. Procurement of goods and services electronically in addition will increase transparency and accountability, improve market access and healthy competition, as well as improving the efficiency of the procurement process. Based on the background and specific objectives to be achieved, then the research will be viewed from e-Marketplace participation, Trust and its impact on the performance of existing procurement within the institution. Methods of data analysis used in this research is by using analysis of Structural Equation Modelling (SEM) by using the program Partial Least Square (PLS) to examine the relationship between variables. The population used in this study is an enterprise as well as other clients using the system LPSE Bekasi District. There are three part in this study population, namely the provision of a company, customer, and supplier of LPSE users amounted to 60 users. From data analysis, there are one hypothesis that rejected. The implication of this study are implementation of e-procurement and e-marketplace participation giving impact to procurement performance.

1. Introduction

The fast growing of information technology has shifting the business process of supply chain management. Supply chain management with utilizing technology has changed how goods and services are purchased in the public sector. E-procurement is an online system that can streamline the procurement process [1]. In digital economy, government of Indonesia uses the internet to deliver services and to communicate with citizens and organizations. The organization that responsible of that is Layanan Pengadaan Secara Elektronik (LPSE). LPSE is a unit formed across Ministries / Agencies / Others institution to hold the service system of procurement of goods and services electronically and facilitate Procurement in carrying out the procurement of goods and services electronically [2].

Procurement of goods and services electronically in addition will increase transparency and accountability, improve market access and healthy competition, as well as improving the efficiency of the procurement process, of course it will also indirectly support the process of monitoring and auditing, and meet the needs of information access real-time order realizing clean and good government in the procurement of government goods and services. LPSE associated with this process, it is necessary to do an evaluation. Where the evaluation carried out to see the impact that occurred on LPSE implementation of the e-Marketplace and Performance. With the hope the results of this study can be a material analysis and future development on this LPSE. So with this evaluation, the increase in service will be better, especially for the field of procurement services related section.

With the e-procurement system, then the performance of existing government institutions, especially in the Bekasi district is expected to increase, because orders are created by selecting items that exist in the system that can reduce the likelihood of errors. All tender selection of goods and services the government done online via the Internet so that the process becomes effective, efficient and transparent. It that were outlined will indirectly affect the performance of operational processes that are in these institutions. Good performance, especially in the areas of procurement will support the running of transparency and fair competition among providers of goods and services as well as for government officials, thus reducing the intensity of direct meetings between suppliers of goods and services procurement committee in support of clean government free from corruption. And trust required will also affect the process of implementing e-procurement application in giving effect to the e-marketplace participation therein.

Based on the background and specific objectives to be achieved, then the research will be viewed from participation e-Marketplace, Trust and its impact on the performance of existing procurement within the institution.

2. Literature Review

E-procurement is a procurement process that refers to the use of the Internet as a means of information and communication [3]. Procurement of goods and services with e-procurement systems utilizing the facility of communication and information technology used to support the public tender process electronically. Procurement of goods / services in Indonesia implemented the guidelines of Presidential Decree 80 of 2003 and amendments. In practice, the process of selecting service providers performed using several methods of election / selection, among others: (1) Auction / common selection is a method for the selection of providers of goods / services is done openly with the announcement widely through the media, (2) Auction / limited selection which is a method of selecting providers of goods / services are limited in terms of the number of providers capable of performing believed to limited, (3) Auction / direct selection is the method for selecting the providers of goods / services to compare at a lack of 3 (three) offerings, (4) is a method of direct appointment provider selection criteria for jobs that meet certain conditions and special circumstances, by direct appointment to one (1) service providers.

From the implementation of e-procurement have gained some benefits as described by Ref [4] divides the advantages of e-procurement into 2 of the direct benefits (improve data accuracy, increase efficiency in operations, process applications faster, reduce costs administration and reduce operating costs) and indirect benefits (e-procurement to make procurement more competitive, improve customer services, and improve relationships with partners). Additionally, Ref [5] also added that e-procurement can reduce the supply cost (an average of 1%), reducing Cost per tender (average 20% cost per tender); e-procurement allows the consolidation of resources so that with the electronic network between organizations is believed to improve coordination between the company and reduce costs in the search for goods and services as appropriate.

Reference [5] states that B2B e-marketplace can be seen as a space where shoppers can find new products and services in which the seller can identify new market opportunities and seek new customers, thereby lowering transaction costs for both. The role of e-marketplace is more to matching buyers and sellers. Some of the factors that affect the participation of e-marketplace are: (1) Efficiency in time and cost savings. E-marketplace can give suppliers access to a higher level of decision-making authority in the buying organization; (2) The legitimacy because the organization must justify their actions and conduct in accordance with social norms and institutional expectations; (3) IT capability is required; companies need a good infrastructure for the benefit of market participation. Building a sustainable competitive advantage and increase the participation of company e-market is not easy or cheap. However, the lack of participation may lead a company to be isolated.

The trust has a moderate effect on processes and behaviors [6]. Reference [7] trust as the belief that a company or institution manufacturer's efforts to meet commitments, to be honest, and do not seek to take unfair advantage of the opportunities. Reference [6] proposed four components trust in B2B relationships: competence, benevolence, integrity and predictability, by adopting the same four dimensions, ability, friendliness, safety and predictability.

Basically, a company or institution would require employees as workers improve product and service quality. Given the employee is considered an important part of the company's assets to contribute to the company to obtain a good performance and be able to compete. Based on above explanation, so the research model for this study as seen on Figure 1 [4], [5], [8], [9]

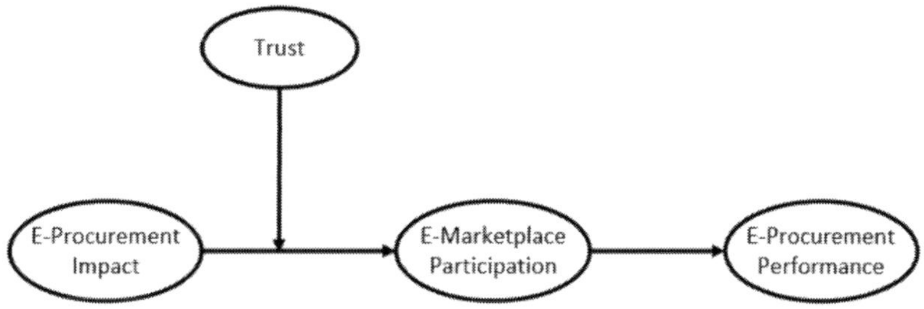

Figure 1. Research Model

3. Research Methods

Researchers decided to use non-probability sampling (non-probability sampling). Techniques used the purposive sampling technique, where the sample selection was based on certain characteristics, which have nothing to do with the relationship characteristics of the population who have previously been known. The population used in this study is an enterprise as well as other clients using the system LPSE Bekasi District. There are three part in this study population, namely the provision of a company, customer, and supplier of LPSE users amounted to 60 users.

Methods of data analysis used in this research is by using analysis of Structural Equation Modeling (SEM) by using the program Partial Least Square (PLS) to examine the relationship between variables. PLS evaluation model based on measurement predictions that have the nature of non-parametric. Then, under certain conditions, PLS working with a relatively small sample size [10].

4. Research Findings

The data is gather by using printed questionnaire in Bahasa. Within one month, all the data can be done and from 60 questionnaires that distributed, only 48 questionnaires that can be using for analysis. Rest of questionnaires are not filled completely by the respondents. In this study, only 80% from total of respondents and this level of response still can be tolerating for analysis the study [11].

As seen Figure 2, most of respondents are coming from suppliers. There are 22 suppliers, 15 customers and the rest are coming from e-procurement officers. Based on Figure 3, there are 36 officers that contribute in this study of e-procurement. And the rest of respondents are management level above, where more than 70% are come from Officer level.

Based on data finding, most of respondents already using e-procurement for at least 5 months. In detail of how long user using e-procurement can be seen on Figure 4. Basically, all of respondents have been using e-procurement for almost 2 months.

First, the test model fit with 3 index test is Average Path Coefficient (APC) with a P-value of <0.05, Average R-Squared (ARS) with a P-value of <0.05, and Average Variance Inflation Factor (AVIF) is greater than 5. and the results show APC at $P <0.001$, ARS with $P <0.001$, and AVIF with a value of 1.807 indicates that all three of these criteria are eligible, which means that the model was fit to the data and models can be used.

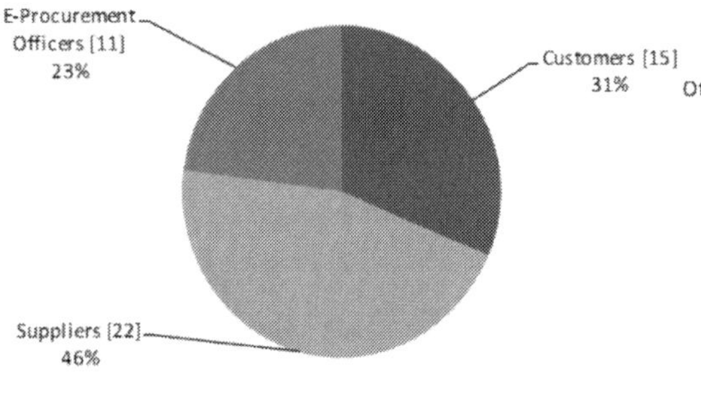

Figure 2. E-Procurement Respondents

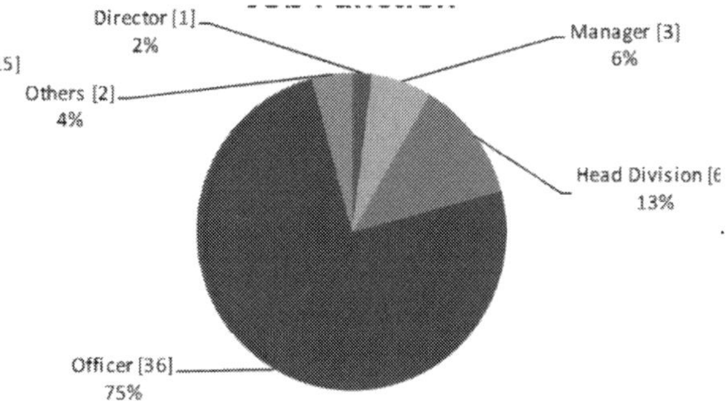

Figure 3. Job Function of Respondents

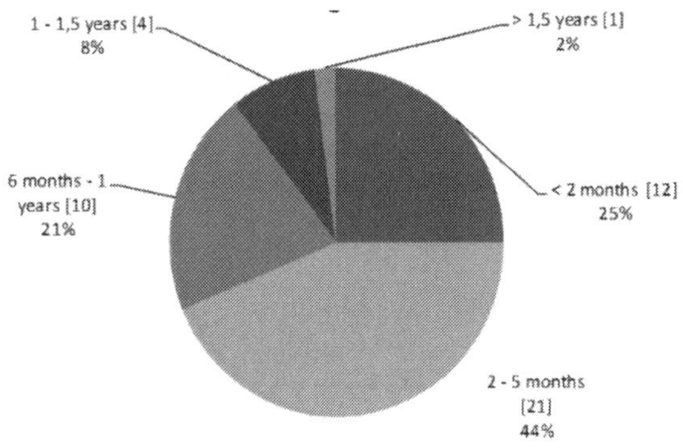

Figure 4. Time using e-Procurement

Further testing the hypothesis on the path coefficients to evaluate the structural relationships between latent variables by comparing the p-value with the alpha (0.05). Figures obtained from the p-value output where the result is significant if less than 0.05. Based on Table 1, result was obtained that result from H1 that has received the results of which means that the variable implementation of e-procurement in local government LPSE Bekasi have a significant impact on the e-marketplace participation variable, with a beta coefficient is positive value of 0.84, which means the better implementation of e-procurement in the Bekasi local government LPSE user participation in the e-marketplace will also increase.

The results of H2 which has rejected the results of which means that the variable trust has no significant effect on the implementation of e-procurement and e-marketplace participation (as a moderator variable), with a beta coefficient is negative at -0.17. Results of H3 which has received significant results e-marketplace participation variables have a significant effect on the performance of the procurement, the value of which is positive beta coefficient of 0.96, which means more and better user participation (participating in the e-marketplace) in the Local Government LPSE Bekasi then the performance will increase procurement. And can be seen with R2 values of 0.918 means that the e-marketplace participation variables affect performance variables LPSE local government procurement in Bekasi by 91.8% and 8.2% were influenced by other variables outside of this study.

5. Conclusions

Based on the research that has been done, it can be concluded that the implementation of e-procurement in LPSE Bekasi has positive and significant impact on the participation of the e-marketplace. It is shown from the beta coefficient is positive, which means that the better the quality of the application or implementation of e-procurement, the greater the participation or the participation of users (users) in the e-marketplace.

The existence of these effects can be caused because indirectly, the use of e-procurement in meeting the needs of users will have an impact on the efficiency of user participation in the e-marketplace. As in previous studies, the results of research conducted by [8] showed that the implementation of e-procurement positive effect on the participation of the e-marketplace, where companies or institutions that implement e-procurement implementation so has the role of contributing both to the participation of e -marketplace. If seen again based on the processing of data, indicators such as time and cost efficiency greatly affect the conformance process of e-procurement itself. Meanwhile, other indicators such as ease in obtaining the information, reduces errors, and makes the data unified, have a higher loading values than others. This suggests that these indicators contribute more in the implementation of e-procurement.

Table 1. Result of Hypothesis

No	the influence between		Beta	R^2	P-*value*	Result
H1	e-Procurement Implementation	e-Marketplace participation	0.84	0.79	<0.001	Accepted
H2	Trust	e-Procurement Implementation and e-Marketplace participation	-0.17	0.109	0.179	Rejected
H3	e-Marketplace participation	Procurement Performance	0.96	0.918	<0.001	Accepted

Then based on the results R^2 is also known that 79% of e-marketplace participation variable is affected by the implementation of e-procurement. So we can conclude that most of the e-marketplace participants obtained based e-procurement (LPSE) is applied to the Bekasi District. So it has been important for the manager of the e-procurement or LPSE to continue to pay attention to and improve the quality of service provision (LPSE), especially for attributes that contribute more to the implementation of e-procurement in order to attract more participants e-marketplace and meet the needs of the user.

Trust as moderator variable does not have a significant effect on the implementation of e-procurement and e-marketplace participation. These results differ from previous studies in which the trust has a fundamental role in affecting the implementation of e-procurement and e-marketplace participation. Based on average scores on the indicators of trust, it is known that the smallest average is an indicator of the e-marketplace is an honest transaction space. This may imply that the user (user) e-procurement have not felt completely trust the e-marketplace as a reliable transaction space.

Reference [6] on research predecessor argued that the trust has a moderate effect on processes and behaviors, whereas Ref [8] in his research found trust is a measure that affects the willingness of a company or institution to participate in the e-marketplace and using e-procurement. In accordance with this theory, it can be concluded that the user (user) e-procurement are participating in the e-marketplace in LPSE District Bekasi not making full use of confidence (trust) as the main reason or reasons of their fundamental when conducting transactions, so there is no effect on the implementation of e-procurement and e-marketplace participation.

e-Marketplace participation in LPSE Bekasi District has a positive and significant impact on the performance of the procurement. It is shown from the beta coefficient is positive, which means that the better the quality of the e-marketplace participation, it will be better the quality of existing performance. The existence of these effects can be caused by indirect, users who participate in the e-marketplace will have an impact on the performance of the procurement part LPSE Bekasi District Government, the more users that participate LPSE then the performance will also be improved.

Performance here mainly include things like customer satisfaction and relationship development with partners and supplier's increases. It is also indicated as such on previous research, the results of research conducted by Ref [8] shows that the participation of e-marketplace positive effect on the performance of the procurement, where the company or institution that participates in the e-marketplace so has the role of contributing both to procurement performance itself. If seen again by the processing of the data, the respondents gave high scores on indicators such as production and transaction costs that can be reduced after participating in the e-marketplace.

Meanwhile, other indicators such as customers who are more satisfied with the performance of the procurement LPSE Bekasi district after participating in the e-marketplace and business relationships with partners are improving after participating in the e-marketplace, has a higher loading than others. This suggests that these indicators contribute to greater participation in the relationship e-marketplace will be the performance of the procurement. Then based on the results R^2 well known that amounted to 91.8% variable procurement performance is influenced by the participation of the e-marketplace. It can be concluded that the majority of user participation in the e-marketplace impact on the performance of the procurement (LPSE) at Bekasi District Government.

So it has been important for providers of e-marketplace to continue to develop the quality of service provision (LPSE), especially for attributes that contribute more to the participation of the e-marketplace in order to continuously improve procurement performance in LPSE Government of Bekasi District and a growing number of companies participating in the e-marketplace as well as the use of electronic procurement services.

References

[1] Rotchanakitumnuai S 2013 The governance evidence of e- government procurement *Transforming Government: People, Process and Policy* **7** no. 3 pp 309–321

[2] LKPP 2012 LKPP - Lembaga Kebijakan Pengadaan Barang Jasa Pemerintah [Online] Available: http://www.lkpp.go.id/v3/#/profil. [Accessed: 12-Mar-2016]

[3] Brandon-Jones A and Silvestro R 2010 Measuring internal service quality: comparing the gap-based and perceptions-only approaches *International Journal of Operations & Production Management* **30** no. 12 pp 1291–1318

[4] Teo T S H, Lin S and Lai K 2009 Adopters and non-adopters of e-procurement in Singapore: An empirical study *Omega* **37** no. 5 pp 972–987

[5] Panayiotou N A, Gayialis S P and Tatsiopoulos I P 2004 An e-procurement system for governmental purchasing *International Journal of Production Economics* **90** no. 1 pp 79–102

[6] McKnight D H and Chervany N L, 2001 What Trust Means in E-Commerce Customer Relationships: An Interdisciplinary Conceptual Typology *Int. J. Electron. Commerce* **6** no. 2, pp. 35–59, Dec. 2001.

[7] Quigley N R, Tesluk P E, Locke E A and Bartol K M 2007 A Multilevel Investigation of the Motivational Mechanisms Underlying Knowledge Sharing and Performance *Organization Science* **18** no. 1 pp 71–88

[8] Chang H H and Wong K H 2010 Adoption of e-procurement and participation of e-marketplace on firm performance: Trust as a moderator *Information & Management* **47** no. 5–6 pp 262–270

[9] Quesada G, González M E, Mueller J and Mueller R 2010 Impact of e- procurement on procurement practices and performance *Benchmarking: An International Journal* **17** no. 4 pp 516–538

[10] Candra S 2014 The impact of enterprise resource planning success in Indonesia company *Advanced Science Letters* **20** no. 1 pp 184–187

[11] Candra S 2014 The Role of Knowledge Capability On Enterprise Resource Planning Success *Unpublished Dissertation* Bina Nusantara University, Jakarta - Indonesia

IOP Conf. Series: Journal of Physics: Conf. Series **801** (2017) 012024 doi:10.1088/1742-6596/801/1/012024

Challenges in developing e-government for good governance in North Sumatra

A Y Siahaan[1]

[1] *University of Sumatera Utara, Medan, Indonesia*

Abstract E-government as one form of public administration reform in Indonesia is increasingly related to the pursuance of good governance. This paper examines the relationship between of e-government and good governance by utilizing the case study design on the implementation of e-procurement in North Sumatra. It reveals centrality of local politics and business culture in understanding resistances of both local government officials and local business which creates 'loopholes' for the practice of 'bad governance' in all phases of e-procurement in North Sumatra province. Data transparency does not equate and guarantee the realization of good governance. Public knowledge and understanding on government decision making processes and accountability (process and policy transparency) are central to achieve good governance through e-procurement. E-procurement system does not automatically change organizational and working culture of the implementers and suppliers. This paper provides insight to the attitude and the perception of private sector engage in e-procurement towards government in implementing e-government. Resistance, digital divide and local politics interrelatedly obstruct the realization of pursuing good governance through e-procurement.

1. Introduction

Globally, the utilization of internet, communication and technology (ICT) in governance and public service delivery is increasingly promoted and acknowledged as effective strategy in pursuing good governance since ICT assumptively reduce the time and distance between service providers and community or customers. The United Nations E-Government Survey 2016 reports that countries which implement e-government in providing on-line public services through one stop-platform continuously increasing. In 2003, only 45 countries had a one-stop-platform, and only 33 countries provided online transactions. According to the 2016 Survey, 90 countries now offer one or more single entry portal on public information or online services, or both and 148 countries provide at least one form of online transactional services. The implementation of e-government is aimed at ensuring that public institutions are more inclusive, effective, accountable and transparent. Across the globe, aiming at enhancing transparency by opening up public information and scrutiny, 128 countries now provide datasets on government spending in machine readable formats. Transparency through e-government is utilized by governments in their effort to open up their data for public information and scrutiny.

E-government has been established as an important part in the bureaucracy reform in Indonesia. In Indonesia the implementation of e-government is strengthened by the issuance of President Instruction No. 3 Year 2003 on E-government National Development Policy and Strategy. This policy is an attempt to develop electronic based government implementation in order to enhance the public quality of service effectively, efficiently, transparently and competitively, fair and accountable. It is supported by the issuance of Law No. 25/2009 on Public Service, especially, chapter 4 on the principles of public services implementation. To date, the implementation of e-government in Indonesia extends to various government roles and functions included the provision of goods and services (e-procurement). Turban et.al (2006) defines e-procurement as referring to the purchase of goods and services for organizations". E-procurement is one of government's activities in implementing public service as described in Law No. 25 Year 2009 chapter 5 on Public Service.

In Indonesia the utilization of e-government has been long known, and to date has been implemented in many national and local government functions. E-government is implemented gradually in stages starting from the central government, province, regency and municipality. It is aimed at achieving efficiency, effectiveness, transparency and economic value of government's service to public. Based on Indonesia's e-government roadmap the implementation of e-government in Indonesia use 4C's approach namely Citizen oriented development, Continuous Improvement, Collaborative processes, Connected with multichannel access [1] However, Huang (2016) study on the implementation of e-government in Indonesia finds out three distinct challenges to digital innovation are evident namely: Competition vs. Collaboration (competition and conflict of institutional interests, both within and across sectors); Integration vs. Disintegration (lack of coordination coordinated digital innovation services); Exclusivity vs. Inclusivity (technocrats and the people). This study examines the relationship between e-government and good governance by using the case study of the implementation of e-procurement in North Sumatra.

2. Literature review

The widespread and urgency of e-government is significantly influenced by public sector reform and globalization which is characterized by the interrelatedness of all sectors of life and multidimensional social, cultural and political dimensions and interdependency of trans-national, global-local, cross cultural forces (Kettani and Bernard Moulin, 2014). World Bank defines e-government as "the use of information and communications technologies to improve the efficiency, effectiveness, transparency and accountability of government". Through the utilization of e-government, services to the public can be improved and thus may compete with private sectors in providing more efficient and effective services. Studies of e-government implementation in developing countries, namely Tanzania, South Africa and China, explain how introducing ICT to government organizations generates concrete improvements: efficiency (cheaper, quicker, more productivity) and effectiveness (better quality and innovations) gains (Kettani and Bernard Moulin, 2014).

E-government does not only aim at efficiency and effectiveness in public service delivery but also serves as the strategy in achieving good governance. UNDP defines good governance as addressing the allocation and management of resources to respond to collective problems, and characterised by participation, transparency, accountability, rule of law, effectiveness and equity (UNDP, 2000). Thus, in good governance the efficiency concerns of public management overlap with the issue of accountability of institutions involved in governance. At this point, e-government is succinctly compatible and conducive for the achievement of good governance. E-governance is also necessarily about the improvement of the performance of civil servants to provide better services to citizens. In principle, it has the potential to enhance decentralisation, accountability and transparency, and can thereby improve the position of developing countries in the global political economy (WSIS 2015). Apart from increasing efficiency and effectiveness in service, the implementation of e-government also may hinder the occurrence of corruption practices since services are conducted transparently so as to extend public participation in the way that public and private sector may relate to government without direct face to face interaction. Assessment on 12 e-government projects in developing countries demonstrates the achievement of e-government in delivering concrete benefits by increasing transparency, reducing corruption, improving service delivery, empowering people and enhancing economic goals of good governance Bhatnagar (Kettani and Bernard Moulin, 2015). Thus, as Heeks (as in UN 2003) argues that e-governance (including e-government) "is the ICT-enabled route to achieving good governance".

The utilization of ICT though increase both efficiency and democratic aspects of governance, yet its implementation also face challenges related to institutional dynamics, regulations, technological difficulties, capacities and resources as well as cultural and developmental dimensions (The United Nations E-Government Survey 2016). The implementation of e-government in OECD (Organization for Economic Cooperation and Development) countries reveals obstructions such as the inflexibility of government framework. Obstruction may also come from the legislative in regulation formulation, finance and digital gap in the utilization of e-government. Challenges also derive from executors in the form of resistance toward the use of new technology such as e-government (Lynn, 2000). Besides issue of e-readiness, other governance-related factors which include a lack of transparency and citizen involvement; resistance by entrenched bureaucracy; corruption; regressive policy and regulatory environments; and, unskilled human resources significantly hinders the achievement of e-government in realizing good governance (Moulin & Kettani, 2015).

3. Results and Discussions

3.1. Resistance

E-government policies which include e-procurement adopted by developing countries are vulnerable to reversal at many subsequent points, depending on whether resistance erupts in the public arena, or remains within the bureaucracy (Larmour, 2000:12). Resistance is "the behavioral expression of a user's opposition to an information system during the implementation" (Klaus and Blanton 2010, p.3). Based on regulations, SKPD should propose the package to be procured to ULP by forward it to the general procurement plan (RUP). This is in line with President Regulation No.70/2012 chapter 112 (2) which obligate government institutions at all level to publish the RUP and information on procurement at their own website and national website through LPSE. The aim of this regulation is that public and ULP can be informed on how many packages will be procured by local government units. Problems occurred when the local government units did not provide technical specification, temporary predicted price and contract design completely. In case of ULP working group did not receive these local government units procurement documents or revision of the documents were yet to be returned by the local government units at determined datelines, e-procurement cannot be conducted. Other problem occurred in research sites was the division of package without clear reason.

Besides limited human resource capacity especially related to information technology, there is high resistance of e-procurement implementers to procurement certification due to the perceived high risk of this occupation cannot be ignored. Although Presidential Regulation No.4/2015 stated that procurement committee will be given law assistance in carrying out their task, yet it is to be realized. Also, e-procurement implementers considered that their salary is insignificant compared to psychological burden they bear in conducting e-procurement activities. Persistent organizational culture which characterized by the rooted corruption, cronyism and nepotism is a crucial problems since it is very difficult for e-procurement implementers to break away from this organizational culture. On the other side, providers also still have old mind-set which perceives that the implementation of e-procurement is similar to previous manual procurement. Providers also still lack of integrity as demonstrated by their irresponsibility to ensure the quality of product or project they won.

3.2. Digital divide

Digital divide is mentioned by most informants from private sector in justifying resistance to the implementation of e-procurement in North Sumatra. Whilst the assumption that transparency is frequently promoted as the achievement of e-procurement, the local private companies argue that e-procurement has created and widened the gap between local companies and those of their competitors from outside Sumatra, especially from Java. E-procurement requires the capability of potential providers to use internet and other electronic technology which is rare especially among small local enterprises in North Sumatra. Thus, e-procurement has restricted rather than increase the access and participation of these small local enterprises to the offered government's goods and services provision. The use of technology is perceived by most small local enterprise as benefitting business with big capital rather than local enterprises which are generally characterized by low skill and knowledge on ICT and the lack of computers and internet access. Furthermore, the implementation of e-procurement which created digital divide was perceived as violating the principle of decentralization and local autonomy which aim at promoting local development and enhancing welfare through increase income of local community of the region.

3.3. Loopholes in e-procurement

Even with the utilization of technology which requires certainty and strict procedures in e-procurement, there are still loopholes within the system of e-procurement in North Sumatra. These loopholes are created by stakeholders involve in e-procurement in gaining their own interest. Despite based on regulation all process of procurement should be conducted through internet and not direct personal (face to face) contact between government (ULP) and suppliers were not allowed, yet, in reality direct interaction still occurs with many/diverse reason. It is interesting to highlight how the non-direct interaction was pointed out by government officials as the most significant advantage of e-procurement, yet, on the other hand, government (ULP) also pointed out that due to some factors especially the un-readiness and limited access of suppliers in understanding e-procurement regulations and process and limited access to internet/ non provision of ICT, electricity blackout, no *wi-fi*, direct personal interaction/meeting between ULP and suppliers are still needed. Although based on regulation and procedures, the system and technology will select and determine the result of e-procurement, yet, there were cases where stakeholders related with e-procurement interact personally with potential providers in bidding and based on agreed arrangement the winner of the bidding is determined not through the system. Operationally, the related officials openly offer the procurement through internet, and when all bidders with their proposals were on the system, then 'preferred potential bidders' will propose their bidding based on the information they got from the e-procurement implementer. The 'preferred potential bidder' was called the 'bride' and the electronic system of procurement was called 'the bedroom'. When all application from other bidders had been in the system, then the 'bride' would be informed of all the proposal bidder at the last minute and when these bidders' proposals had been in the system, the system would be locked. In this case, the 'preferred potential bidders' will win the bidding because they already have information of the proposal of other bidders and forward proposal which is more competitive. Thus, e-procurement was perceived as indifferent with the manual procurement.

3.4. Distrust to government

Trust is an integral component of e-procurement. However, the case of e-procurement in North Sumatra demonstrates the existence of widespread doubt on integrity of e-procurement committee that these officials still conduct corruption and cronyism by favoring providers with close relationship or part of those who connected to high government officials. Organization culture of LPSE, ULP and user was perceived as contain with values of unjust, discrimination and closure. Some suppliers consider that government prefer certain suppliers from whom they got 'benefits' and intentionally and systematically directed the process to win certain suppliers thus discriminate against other suppliers. The decision on who wins the bidding was determined more on closeness or preferences of government officials rather than on principle of efficiency, effectivity and professionalism. This view ignites distrust towards e-procurement committee and system. Some providers did not respond positively to facilities which were provided by LPSE and ULP.

Some providers also convinced that there was tolerance to some providers who were late in registering their interest to participate in bidding process, late in uploading and completing bidding documents. In case that some providers fail to participate in bidding or lost bidding package, then the e-procurement committee was blamed with various speculations.

Based on providers' perspective, this distrust was closely related to the low credibility of e-procurement committee. Nurturing credibility is difficult and requires long time. However, LPSE and ULP need to publish data, process and decision of bidding immediately in their website so that transparency will be realized which then will avoid speculations on bidding and other e-procurement related activities. On the other hand, e-procurement committee argued that providers should not hide behind speculations because it will of their own disadvantage. Presidential Regulation has required that all government goods and services provision should be conducted electronically. Therefore, providers had to change their mindset on participating in procurement. Different from previous manual procurement, e-procurement requires providers to obey all bidding process and stages strictly, thus, they need to continuously gain update information available in LPSE website. Update information will eliminate speculations regarding e-procurement activities and processes. However according to service providers, updating information without addressing problem of trust and perceived of business providers gain on participating in e-procurement will not eliminate speculations. Ballard (2011: 2) argues: "In public procurement the goals of fairness, competition and economic value are paramount. To achieve these goals, effective and efficient procurement process must be established. This includes incorporating adequate controls to promote competition and minimize the risk of fraud, corruption, waste, and mismanagement of public fund. In this context, transparency is considered to be one of the most effective tools to deter corruption and ensure value for money". At this point, trust of government and trust of internet (trust in the reliability of the enabling technology); and sensitivity to culture is crucial in striving for the realization of good governance through e-procurement [2].

4. Conclusions

The utilization of ICT in governance and public service does not automatically result in efficient, transparent, accountable and participative governance which are the pillars of good governance. The introduction of technology in government is not only influenced by institutional capability in building trust but also significantly determined by organizational culture and community values. Trust building is as important as capacity building on technological knowledge and infrastructure. Nurturing credibility is difficult and requires long time. However, LPSE and ULP need to publish data, process and decision of bidding immediately in their website so that transparency will be realized which then will avoid speculations on bidding and other e-procurement related activities.

Shared interpretation and meaning of goals, principles and standards in e-procurement may also be realized through training related to computer based technology information and amendment to regulation on procurement to providers, procurement committee and users. Providers had to change their mindset on participating in procurement. Different from previous manual procurement, e-procurement requires providers to obey all bidding process and stages strictly, thus, they need to continuously gain update information available in LPSE website. Update information will eliminate speculations regarding e-procurement activities and processes. Moreover, the involvement of all related business association as those which give recommendation to providers who intend to participate in training as well as the involvement of head of related procurement agency to evaluate training participants, those who are responsible for the enactment of procurement is crucial. Trainings may enhance integrity of all parties which then contribute to successful implementation of e-procurement.

More attention and emphasis need to be given to the political and cultural aspect of e-government in realizing good governance such as the strengthening of the capacity of community organizations (civil society) that look more at the involvement of the society in e-government rather than merely focusing on enhancing the technocratic capacity of government. For e-governance to be supportive to the achievement of good governance requires enhanced in-depth interactions between the citizens and the government. As Fang (2002) succinctly argues that 'e-governance is not just about website or not merely a digitization of service delivery, It certainly stands on greater definition of engagement and depths of relationship that surrounds both the citizens and the government in hinder corruption and fraud risks while at the same time enhance fair competition and ensuring the value of a price'.

5. References

[1] Anggono, Dwi I 2015 E-Government Indonesia Update 2015 – 2019 Ministry of communication and information technology the republic of Indonesia

[2] Fang Z 2002 e-Government in Digital Era: Concept, Practice, and Development, International Journal of the Computer the Internet and Management 10(2) p 1-22

[3] Driss K, Moulin B 2015 E government for good governance in developing countries Empirical evidence from e-Fez project (Anthem Press: London New York New Delhi IDRC)

[4] Alia S, Sabri O, Al-Shargabi B 2012 A Cultural E-Government Readiness Model Intelligent Information Management

[5] The United Nations E-Government Survey 2016: E-Government in Support of Sustainable Development.

Weighted Components of i-Government Enterprise Architecture

E K Budiardjo[1], G Firmansyah[2] and Z A Hasibuan[3]

[1,3] Faculty of Computer Science, Universitas Indonesia, Depok 16424
[2] Faculty of Computer Science, Universitas Esa Unggul, Jakarta 11510

Email: eko@cs.ui.ac.id, gerry.firmansyah@gmail.com, zhasibua@cs.ui.ac.id

Abstract. Lack of government performance, among others due to the lack of coordination and communication among government agencies. Whilst, Enterprise Architecture (EA) in the government can be use as a strategic planning tool to improve productivity, efficiency, and effectivity. However, the existence components of Government Enterprise Architecture (GEA) do not show level of importance, that cause difficulty in implementing good e-government for good governance. This study is to explore the weight of GEA components using Principal Component Analysis (PCA) in order to discovered an inherent structure of e-government. The results show that IT governance component of GEA play a major role in the GEA. The rest of components that consist of e-government system, e-government regulation, e-government management, and application key operational, contributed more or less the same. Beside that GEA from other countries analyzes using comparative base on comon enterprise architecture component. These weighted components use to construct i-Government enterprise architecture. and show the relative importance of component in order to established priorities in developing e-government.

1. Introduction

According to Haiyan Qian, Director of the Division of Administration and management development, the United Nations Department of Economic and social affairs (UNDESA) states that the Enterprise Architecture (EA) is a strategic planning tool that is effective for the government to facilitate contacts and improve interoperability among government agencies, improve processes internal operations and improve service delivery to the community or residents [1]. EA in the government or the Government EA (GEA) is an important factor for the success of all types, scale and intensity of the e-government program. The main purpose of the GEA to make e-government into citizen-oriented, results-oriented. GEA development usually pass through different stages of evolution. MIT's research center information system identifies four stages of evolution or in GEA: business silos, standardized technology, rationalized the data and applications and business modularity.

According to Anthopoulos [2], the impact of the GEA is not utilized for the development of e-government in Greece, influence the procurement process errors causing financial loss and time. According to Jassen [3] at GEA research in Denmark and the Netherlands noted that the GEA can provide guidance in the development of e-government. GEA implementation will reduce costs as well as the development of e-government can ensure its development as needed.

IOP Conf. Series: Journal of Physics: Conf. Series **801** (2017) 012025 doi:10.1088/1742-6596/801/1/012025

2. Background

Indonesia has a e-government ranking (PEGI), which measures the use of ICT of each ministry and agency in order to obtain a map of the condition of the use of ICT in the national scope. These surveys are regularly conducted since 2007 until now. The survey examined five dimensions of policy, institutional, infrastructure, applications and planning [4]. In the survey allow each dimension has the same weight. With such patterns can be made possible a government agency will get the same rank value with other government agencies even though the amount of the value of different dimensions, though each dimension has a different purpose.

According to Firmansyah [5] in Indonesian studies of e-government component using data PEGI in 2011, found that there are five main components must be considered that the implementation of e-government in Indonesia can be successful. Found those components are: IT governance, e-government system, e-government regulation, e-government management and application key operational. By looking at the results of principal component analysis component of scoring formula it can be seen the weight or priority of each of the components that form each are as follows: IT Governance is 20.026, e-government system is 2.145, e-government regulation is 1.744, e-government management is 1.298, application key operational is 1.189. Each component of such research have different weights, resulting in the implementation of e-government should prioritize the largest weighting in advance so that the chance of success easier.

3. Methodology

This research starts from a quantitative study of the data PEGI in 2011 to seek the inherent structure by using principal component analysis (PCA) which produce the main components of Indonesia e-government. The main component is verified using qualitative methods such as interviews with a number of experts in the field of e-government. The next stage is to identify the components forming qualitatively government enterprise architecture that forms the government generic enterprise architecture. The results of the second stage was the basis for the establishment of enterprise architecture models for the Indonesian government (i-GEA). The stages of this research can be seen in Figure 1.

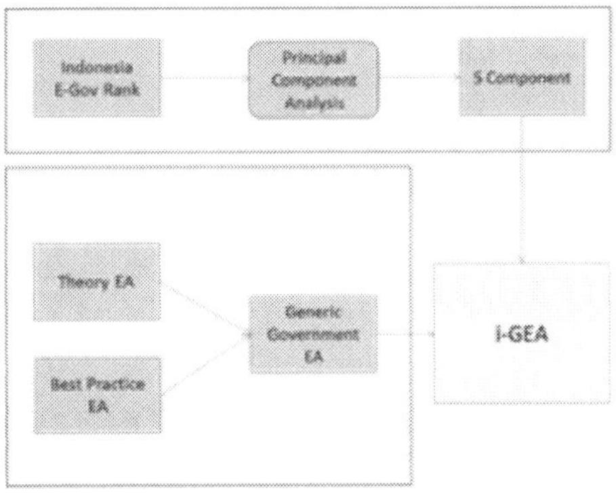

Figure 1. Methodology to create i-GEA

The basic components from enterprise architecture base on theoritical refers to the components of Enterprise architecture Zachman, TOGAF, FEAF, Gartner, as well as parts of other countries who are already using the GEA. Components base on theoritical enterprise architecture consists of : business architecture, application architecture, data architecture, and technical architecture.

Comparison components enterprise architecture in various ways including : comparing based view / perspective, abstraction, SDLC phases [6], or compare by description, component, architectureal view, based on the EA layers.

To compare GEA, some use the comparative approach based policies, actors and structures, governance, architecture frameworks and methodologies, principles and standads architecture, implementations [3]. Each method of comparison is made according to the needs. This research will use EA comparison method by comparing the layer-forming of any enterprise architecture theoritical and best practice. After comparison method it will get the component fundamental building blocks of the GEA.

Developing enterprise architecture using Toaster models. Each top layer supported by the layer below, as well as the outermost layer is supported by the deeper layers, and components depicted vertically supports the components depicted horizontally [7].

4. Results

According to study generic government enterprise architecture (generic GEA), a generic GEA can be built from research on theories EA and the Best practice EA. Theory research examines EA Zachman Framework, TOGAF, FEAF, this research see forming components of the EA as shown in table 1. While EA's Best Practice research, examining the implementation of the EA in the countries with the highest e-government index, Abu Dhabi [8], Australia [9], Canada [10], United Kingdom [11], South Korea [12], Singapore [13] and Egypt [14] as shown in table 2.

Table 1. Component EA base on theories

Framework	Zachman	FEAF	TOGAF
What	Inventory sets (What)	Data Architecture (entities=what)	Data Architecture
How	Process Flows (How)	Application Architecture (activities=how)	Application Architecture
Where	Distribution Networks (Where)	Technology Architecture (Location=where)	Technical Architecture
Who	Responsibility Assignments (Who)	Resources Optimization	IT resources guidance
When	Timing Cycles (When)		
Why	Motivations Intentions (Why)	Business Architecture	Business Architecture

Table 2. Component EA base on GEA others country

Framework	Abudhabi	Australia	Canada	UK	Korea Selatan	Singapura	Mesir
What	Data	Data Architecture	Information Management Architecture	Business Information	Data	Information Architecture	Data
How	Application, Security	Application Architecture	Application Architecture, Security Architecture	Business Process	Application	Application Architecture	Function
Where	Integration, Infrastructure	Technology Architecture	Network Architecture, Platform Architecture	Infrastructure, Security, Integration	Infrastructure	Technical Architecture	Network
Who	Operation		Service	Service Management			People
When	Access & Presentation		Presentation Architecture	Channel			Time
Why	Business	Business Architecture	System Management Architecture	Strategy	Business	Business Architecture	Motivation

IOP Conf. Series: Journal of Physics: Conf. Series **801** (2017) 012025 doi:10.1088/1742-6596/801/1/012025

Based on the results of the study on the components of enterprise architecture, the study of the theory enterprise architecture and government enterprise architecture other countries, the components of an enterprise architecture consists of : responsibility assignment (who), timing cycle (when), business architecture (why), application architecture (how), data architecture (what), infrastructure architecture (where). Proposed form of GEA generic as shown ini figure 2 based on the basic components of an enterprise architecture. From figure 2 described as follows: a group of orange as Responsibility Assignment, red group timing Cycle, a group of purple as business architecture, the deep blue color as Application Architecture, light blue as the data architecture, light blue color as infrastructure architecture, then the color gray as the implementation methodology GEA and black is a stakeholder of the GEA. The preparation of this component refers to the toaster model, where the lower level would support the level above.

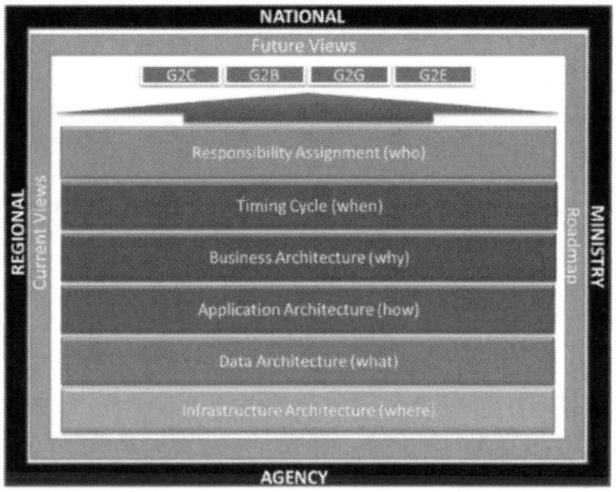

Figure 2. Generic government enterprise architecture

From the results of the PCA, the Indonesian government enterprise architecture enhanced by observing components: IT governance, e-government system, e-government regulation, e-government management, application key operational. To build the GEA with mapping architectural view, generic parts and components PCA results. After mapping component enterprise architecture the GEA be redesigned as show in figure 3.

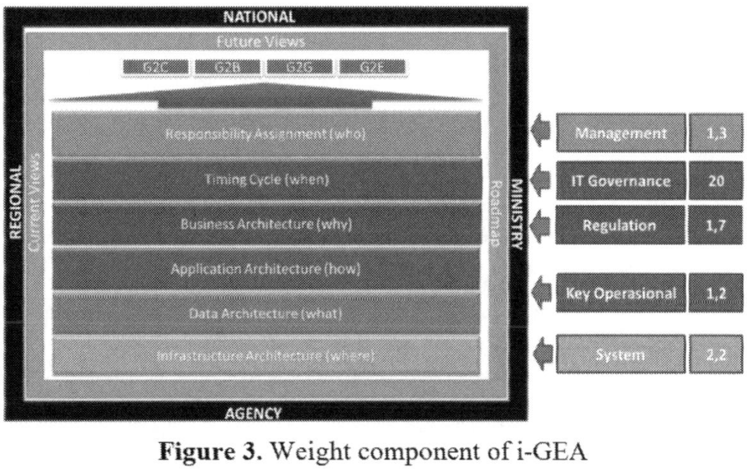

Figure 3. Weight component of i-GEA

5. Conclusion & Future Works

After doing all the research phase which started from the study of literature, the study of best practice EA users in developed countries, the study of e-government ranking outside and within the country, making government EA for Indonesia, performing PCA to the ranking of e-government in Indonesia, setting out the components of e-government as a shaper of EA in Indonesia government, then the researchers made the following conclusions:

1. Enterprise architecture components are theoretically as well as best practice consists of : business architecture, application architecture, data architecture, information architecture.
2. Form government enterprise architecture from a country need to incorporate components of local strong, to acquire inherent structure by PCA.
3. Inherent structure e-government Indonesia consists : IT governance, e-government systems, e-government regulation, e-government management, application key operations.
4. Give the weight / priority of a government enterprise architecture can help focus the implementation of the GEA itself.

The next research is necessary to process the data factor analysis of the e-government ranking of Indonesia as a whole from all the years of the implementation of the research, so we get a strong main components. I-GEA models need to be equipped with an action plan so that the Indonesia Government Enterprise Architecture Framework can be implemented.

References

[1] Saha P 2010 Enterprise architecture As Platform For Connected Government, Government Enterprise architecture Research Project

[2] Anthopoulos, Leonidas G., Vasilis Gerogiannis, and Panos Fitsilis 2010 The impact of enterprise architecture's absence in e-government development: The Greek case. Information Society (i-Society); 2010 International Conference on. IEEE;

[3] Janssen, Marijn, and Kristian Hjort-Madsen 2007 Analyzing enterprise architecture in national governments: The cases of Denmark and the Netherlands System Sciences, 2007, HICSS 2007, 40th Annual Hawaii International Conference on. IEEE

[4] KemKominfo 2011 Pemeringkatan e-government, http://pegi.layanan.go.id/ diakses tanggal 5 November 2013

[5] Firmansyah, G., Zainal A. Hasibuan, Yudho Giri Sucahyo 2014 Indonesia e-government Components: A Principal Component Analysis Approach, ICITSI2014

[6] Urbaczewski, Lise 2006. A Comparison Of *Enterprise architecture* Frameworks, *Issues in Information Systems , Volume VII, No. 2*

[7] Marvin Zelkowitz and Barbara Cuthill, Application of an Information Technology Model to Software Engineering Environments, Computer Systems Laboratog Nat? Institute of Standards and Technology, Gaithersbulg, Maryland, J. SYSTEMS SOFTWARE 1997

[8] Abu Dhabi SIC 2013 https://www.abudhabi.ae/pdfdirectview/en/111274/blaetterkatalog/ blaetterkata log/pdf/complete.pdf

[9] Australia Government. 2013 http://www.finance.gov.au/policy-guides-procurement/ australian-government-archi tecture-aga/

[10] Canada 2013 http://www.tbs-sct.gc.ca/im-gi/ims-sgi/eia-eii-eng.asp

[11] HM Government 2012 https://www.gov.uk/government/uploads/system/uploads/attachment _data/file/60850 /enterprise-architecture.pdf

[12] Song Hee Joon 2006 http://www.opengroup.org/architecture/0310wash/presents/SungBum_ Park_GEAF .pdf

[13] Singapore Gov. 2013 http://www.egov.gov.sg/egov-programmes/programmes-by-gov ernment/singapore-government-enterprise-architecture-sgea

[14] EAWG. 2006 http://www.egypt.gov.eg/english/documents/download/EGEAF_ English _05012010141026 .zip

Probability Model for Designing Environment Condition

Iswar Lubis[1], Mahyuddin K M Nasution[2*]

[1]Pengembangan Wilayah, Universitas Sumatera Utara, Padang Bulan USU Medan 20155, Indonesia.
[2]Information Technology, Fasilkom-TI, Universitas Sumatera Utara, Padang Bulan USU Medan 20155, Indonesia.

*Email: mahyuddin@usu.ac.id

Abstract. Transport equipment has the potential to contribute to environmental pollution. The pollution impact on the welfare of the environment. Thus, the potential of the environment needs to be raised to block the pollution. A design based on probability models of the determining factors in the environment should be a concern. This paper aims to reveal the scenarios as the beginning to express the clues, and based on surveys have been had the ability to choose a design.

1. Introduction

A fun environment is a major part of welfare, whereby residence be dominant basis for generating the travel that caused transportation [1]. One of the reasons why the environment is not pleasant is presence of noise [2]. Noise caused by means of transportation. The noise can be mitigated by involving factors anticipation, and the factors need management to conform to social [3]. Therefore, a probability model is proposed by developing the data collection of factors as data engineering for designing environment condition. A lot of researchers have developed technology to reduce noise, but little study on the management of environmental conditions involve related factors based on probability theory.

This paper aims to address a model for dealing with the data collection to managing the environment conditions. Therefore, after introduction section we introduce some concepts for motivating this research, mainly to review some of researches, and then propose an approach based on scenario. After that, some results discussed by comparing the results.

2. Basic Concept and Motivation

Noise is defined as pollution of voice due to the influx of unwanted noise into an environment that causes environmental degradation, resulting in certain activities such as schools, housing and hospitals [4]. The noise happens to be very disturbing ongoing activities on the environment, even if noise that occurs accepted by humans at a certain time with a certain level will endanger human health itself [5]. Therefore, some areas such as housing allotment endeavoured by imposing noise threshold corresponding to the quality standard that is free from noise pollution. For example, The Decision Minister of Environment No. KEP-48/MENLH/11/1996 on Standards of Noise Level.

Enforcement of a rule is one of attempts that is based on legislation to reduce the noise level to suit the needs of the environment. Another effort is to involve factors is available in the neighborhood itself [6]. Factors that come from the environment is the work and materials. It also related to the budget as external factor. Material factors among them are the fence, the trees, the grass, study room, bedroom, distance, soundproof, wall, window, etc. While work factors like installation, enhancement, rehabilitation, increase, improve, etc. However, how far all these factors favored by social members need to summarize well.

Probability theory has proven itself useful in the study of noise by transportation [7], but the model is to predict flow roads. Other model, but does not involve probability, is used to determine equivalence of annoyance [8]. Research on transportation generally does not rely to cope with noise or relating to environment, but is related to technology and technical transport [9, 10, 11].

3. Proposed Approach

In general, in any event there is a series of interrelated activities [12]. These activities become the factors of success of the event. The success of an event depends on the opportunities and expectations (also called probability) [13]. To determine the probability model among the factors we construct several scenarios as events, i.e:

1. Environmental conditions are offered on scenario-1 (S1) is a reduction in noise levels by 5 dB(A) through improved quality of fence at an additional cost of Rp. 40,000 per month.
2. The scenario-2 (S2) offered an environment where the noise level that exists today remains reduced by 5 dB(A) as well as scenario-1, but it done through efforts to improve the garden at an additional cost of Rp. 50,000 per month.
3. Scenario-3 (S3) offers an environment with reduced noise level of 10 dB(A) through improving the quality of the wall/window by turn to the type of soundproof or another with an additional fee or Rp. 50,000 per month.
4. In scenario-4 (S4), environmental conditions offered experiencing a reduction in noise levels of 10 dB(A) as well as scenario-3, but in the reduction scenario conducted through the rehabilitation of study room / work room with an additional fee of Rp. 25,000 per month.
5. Reduction of the noise level of 20 dB(A) through the rehabilitation of the breakroom/bedroom with an additional fee of Rp. 25,000 per month as offered in scenario-5 (S5).
6. In the scenario-6 (S6), the conditions offered are the reduction of noise by 10 dB(A) through improved quality garden and fence once with an additional fee of Rp. 75,000 per month.
7. An additional fee Rp. 75,000 per month on the reduction of the noise level of 15 dB(A) through improving the quality of fence and wall/window are done at once (Scenario-7 (S7)).
8. In scenario-8 (S8) made bids reduction of the noise level of 15 dB(A) through the rehabilitation of the fence and the study room / work room as well as an additional cost of Rp. 50,000 per month.
9. Scenario-9 (S9), which offered a condition of reduction of the noise level of 25 dB(A) through improving the quality of the fence and breakroom/bedroom with an additional fee of Rp. 50,000 per month.
10. In the scenario-10 (S10), offered environmental conditions is a reduction in the noise level of 15 dB(A) as a result of the efforts made by improving the quality of the garden and the wall/ window at once with the amount of additional cost Rp. 75,000 per month.
11. Environmental conditions are offered on scenario-11 (S11), namely the reduction of the noise level of 15 dB(A) through quality improvement and rehabilitation of the garden and the study room / work room as well at an additional cost of Rp. 60,000 per month.
12. Scenario-12 (S12) offer an environment where there are reduction in the noise level of 15 dB(A) through the quality improvement and rehabilitation of garden and breakroom/bedroom with an additional fee of Rp. 60,000 per month.
13. The last scenario (S13) offered is a reduction in the noise level of 25 dB(A) through improved quality fence, garden and wall/window at once, where the reduction is greatest reduction of scenarios offered at an additional cost of Rp. 75,000 per month.

Of the 13 scenarios we have factors such as Table 1 and Table 2.

Table 1. Variables about noise reduction and additional cost

Id	Noise reduction	Scenario	Id	Additional Cost	Scenario
N1	5dB(A)	1, 2	C1	Rp. 25,000	S4, S5
N2	10 dB(A)	3, 4, 6, 10	C2	Rp. 40,000	S1
N3	15 dB(A)	7, 8, 11, 12	C3	Rp. 50,000	S2, S3, S8, S9,
N4	20 dB(A)	5	C4	Rp. 60,000	S11, S12
N5	25 dB(A)	9, 13	C5	Rp. 75,000	S6, S7, S10, S13

Table 2. Variables about works and material

Id	Works and Materials	Scenario
F1	Improve quality of fence	S1, S6, S7, S9, S13
F2	Improve quality of garden	S2, S6, S10, S11, S12, S13
F3	Improve quality of wall/window	S3, S7, S10, S13
F4	Improve quality of breakroom / bedroom	S9, S12
F5	Improve quality of study room / work room	S11
F6	Rehabilitation of study room / work room	S4, S8, S11
F7	Rehabilitation of breakroom / bedroom	S5, S12
F8	Rehabilitation of fence	S8
F9	Rehabilitation of garden	S11, S12

To build probability model required initial value. Each factor has a probability *p*(factor) in [0,1]. For example, probability of noise reduction for 5dB(A) based on Table 1 is 2/13 or *p*(N1) = 0.15, probability of additional cost for Rp. 75,000 based on Table 1 is 4/13 or *p*(C5) = 0.31, while probability of improve quality of garden in Table 2 is 6/13 or *p*(F2) = 0.46.

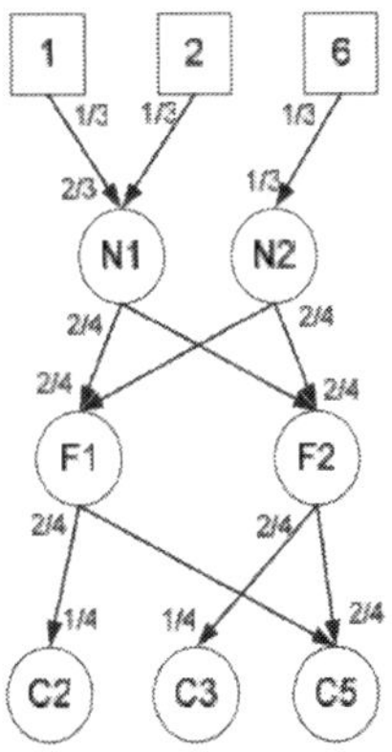

Fig. 1. Network of factors for scenario 1, 2, and 6

Table 3. Probability of Out/In for network in Fig. 1.

Scenario	Out	In	Out	In	Out	In
1	1/3	2/3	2/4	2/4	2/4	¼
2	1/3	2/3	2/4	2/4	2/4	¼
6	1/3	1/3	2/4	2/4	2/4	2/4
6	1/3	1/3	2/4	2/4	2/4	2/4

Suppose taken scenario-1, scenario-2 and scenario-6, we have a network such as Fig. 1, and Table 3. Based on multiplication probability we have probability *p*(scenario-1) = *p*(scenario-2) = 0.0069, while *p*(scenario-6) = 0.0139. Therefore, probability model of a network based on scenario is

$$p(\text{scenario}) = \Sigma_{i=1...row}\Pi_{\text{in-out}}\, p(out/in) \quad (1)$$

where *row* is number of row in the table of out/in. However, the initial value of each scenario is based on answers to the questionnaire by the stated preference methods [14].

4. Results and Discussion

To obtain responses to scenarios conducted a survey of the area. In general respondents on each bid is quite varied. The response options that have been declared in the survey are: (1) definitely disagree, (2) tend to disagree, (3) undecided, (4) tend to agree, or (5) would agree. By involving 38 respondents we obtained responses such as Table 1. Therefore, scenario sure approved has the probability: *p*(scenario-5) = 32/38, *p*(scenario-4) = 29/38, *p*(scenario-13) = 28/38, etc.; scenario tend to agree has the probability: *p*(scenario-8) = *p*(scenario-11) = 30/38, *p*(scenario-12) = 29/38, *p*(scenario-3) = 6/38; while scenario was ignored has the probability: *p*(scenario-10) = 28/38, *p*(scenario-3) = 24/38, *p*(scenario-7) = *p*(scenario-9) = 4/38, etc.; the scenario tend to disagree has the probability: *p*(scenario-1) = 30/38, *p*(scenario-6) = 29/38, *p*(scenario-2) = *p*(scenario-12) = 4/38, etc.; and then scenario definitely disagree has probability: *p*(scenario-2) = 27/38, *p*(scenario-6) = 4/38, *p*(scenario-7) = *p*(scenario-9) = 3/38, etc.

Table 1. Recapitulation of respondents to the scenario offered.

Scenario	Response				
	(1)	(2)	(3)	(4)	(5)
S1	1	30	2	3	2
S2	27	4	2	2	4
S3	2	2	24	6	3
S4	1	3	1	4	29
S5	1	1	3	1	32
S6	4	29	2	2	1
S7	3	1	4	4	26
S8	2	3	2	30	1
S9	3	1	4	4	26
S10	0	3	28	5	2
S11	1	2	1	30	3
S12	1	4	1	29	3
S13	0	3	2	5	28

Suppose the scenario is taken in accordance with the highest ranking is based on the option (5), acquired three top scenarios with the series of probability as follows:

1. S5: (1/3)(1/3)(1/5)(1/5)(1/5)(2/5)
2. S4: (1/3)(1/3)(1/5)(1/5)(1/5)(2/5)
3. S13: (1/3)(1/3)(3/5)(1/5)(1/5)(3/5), etc.

Based on Eq. (1) and network of factors in S5, S4 and S13, we have $p(S5) = p(S4) < p(S13)$. Furthermore, suppose 4 top scenarios based on the option (5), their probabilities as follows:

1. S5: (1/4)(1/4)(1/7)(1/7)(1/6)(2/6)
2. S4: (1/4)(1/4)(1/7)(1/7)(1/6)(2/6)
3. S13: (1/4)(1/4)(3/7)(1/7)(1/6)(4/6), (1/4)(1/4)(3/7)(2/7)(1/6)(4/6), (1/4)(1/4)(3/7)(1/7)(1/6)(4/6)
4. S7: (1/4)(1/4)(2/7)(2/7)(1/6)(4/6), (1/4)(1/4)(2/7)(1/7)(1/6)(4/6)

So the choice remains in the scenario-13 whereby F1 = "Improve quality of fence", F2 = "Improve quality of garden", and F3 = "Improve quality of wall/window" be the activities of social members.

Then, on development and maintenance fences: The results of the survey showed that of the 38 residents of the housing is used as respondents there were as many as 16 respondents were fully aware that the construction of the fence they have done is one of the efforts to block the entry of sound (noise) into the house and from a number of the average respondent to respond that of a number of costs that they incur in the form of development on average Rp. 1.48 million per year; in garden maintenance: The efforts that have been made in form of a garden maintenance, survey results indicate that there are 28 respondents gave answers that have been doing the maintenance of the garden with an average of Rp. 85,000 per month; while for wall/window: Stated preference survey results showed that there were 11 respondents who had made a change/replacement or enhancement of the wall/window with an average cost of Rp. 557,000 per year.

The probability model based on generation of initial value and rank based on the probability of the option provide a selection condition for designing an environment that block the noise.

5. Conclusion

To develop probability model for the object of study can be done by building scenarios. Scenario used to weave related factors, as well as guidance for designing the environmental conditions. So the probability to ensure a successful design. The future work is to develop the probability network for preventing and considering the noise.

References

[1] O M Rouhani, R R Geddes, H O Gao, and G Bel 2016 Social welfare analysis of investment public-private partnership approaches for transportation projects *Transportation Research Part A: Policy and Practice* **88**.

[2] M Foraster, I C Eze, D Vienneau, M Brink, C Cajochen, S Caviezel, H Heritier, E Schaffner, C Schindler, M Wanner, J-M Wunderli, M Roosli, and N Probst-Hensch 2016 Long-term transportation noise annoyance is associated with subsequent lower levels of physical activity *Envronment International* **91**.

[3] R Year, J Castro, and L Martinez 2016 A fuzzy model for managing natural noise in recommender system *Applied Soft Computing* **40**.

[4] L Brocolini, E Parizet, and P. Chevret 2016 Effect of masking noise on cognitive performance and annoyance in open plan offices *Applied Acoustics* **114**.

[5] Z Li, Z Song, T Wang, Y Zheng, and X Ning 2016 Health impacts of construction noise on workers: A quantitative assessment model based on exposure measurement *Journal of Cleaner Production* **135**.

[6] T Wilkening 2016 Information and the persistence of private-order contract enforcement institutions: An experimental analysis *European Economic Review* **89**.

[7] F Li, S S Liao, and M Cai 2016 A new probability statistical model for traffic noise prediction on free flow roads and control flow roads *Transportation Research Part D: Transport and Environment* **49.**

[8] L-A Gille, C Marquis-Favre, and J. Morel 2016 Testing of the European Union exposure-response relationships and annoyance equivalents model for annoyance due to transportation noises: The need of revised exposure-response relationships and annoyance equivalents model *Environment International* **94.**

[9] F Asgari, A Sultan, H Xiong, V Gauthier, and M A El-Yacoubi 2016 CT-Mapper: Mapping sparse multimodal cellular trajectories using a multilayer transportation network *Computer Communications* **95.**

[10] H R Eftekhari and M Ghatee 2016 An inference engine for smartphone to pre-process data and detect stationary and transportation modes *Transportation Research Part C: Emerging Technology* **69.**

[11] D Teodorovic and M Janic 2017 Chapter 11 – Transportation, environment, and society *Transportation Engineering*.

[12] M K M Nasution, and S A M Noah 2011 Extraction of academic social network from online database, *Proceeding of 2011 International Conference on Semantic Technology and Information Retrieval* (STAIRS'11), 64-69, IEEE, Putrajaya, Malaysia (2011).

[13] R A Halim, J H Kwakkel,and L A Tavasszy 2016 A scenario discovery study of the impact of uncertainties in the global container transport system on European ports *Future* **81**.

[14] E P Kroes and R J Sheldon 1998 Stated preference methods *Hague Consulting Group, The Hague, Netherlands.*

E-Learning Personalization Using Triple-Factor Approach in Standard-Based Education

K A Laksitowening[1], H B Santoso[2], Z A Hasibuan[3]
[123] Faculty of Computer Science, University of Indonesia, Depok, Indonesia
[1] School of Computing, Telkom University, Bandung, Indonesia

[1]kusuma.ayu@ui.ac.id, [2]harrybs@cs.ui.ac.id,[3]zhasibua@cs.ui.ac.id
[1]ayu@telkomuniversity.ac.id

Abstract. E-Learning can be a tool in monitoring learning process and progress towards the targeted competency. Process and progress on every learner can be different one to another, since every learner may have different learning type. Learning type itself can be identified by taking into account learning style, motivation, and knowledge ability. This study explores personalization for learning type based on Triple-Factor Approach. Considering that factors in Triple-Factor Approach are dynamic, the personalization system needs to accommodate the changes that may occurs. Originated from the issue, this study proposed personalization that guides learner progression dynamically towards stages of their learning process. The personalization is implemented in the form of interventions that trigger learner to access learning contents and discussion forums more often as well as improve their level of knowledge ability based on their state of learning type.

1. Introduction

Learning is a continuous process performed with the purpose in achieving learning objectives in a specific time period. In standards-based education, there are mechanism in targeting competencies as learning objectives and evaluation to assess the learners' achievement toward the objectives. Learners evolve over time, consequently learning process should also be well prepared to help learners meet the standard.

In the other hand, e-Learning can help learners to learn the way they want without limitation in place and time. Unfortunately, e-Learning commonly does not consider the diversity of learners in term of their ability and characteristics. Usually, the learning paths and contents are given identically for all learners in the same course despite the fact that learning type may differ on one another. To accommodate the variation of learning types, the learning paths and learning contents can be adapted to meet personal needs [1]. Personalization in e-Learning enable learners to have a learning process that is suitable to their learning type. By learning with the appropriate paths and contents, learners are expected to achieve the targeted competency optimally.

Learning types can be identified by taking into account several factors, such as: learning motivation, learning styles, knowledge ability, and others. According to Triple-Factor Approach, there is a close relation between motivation and learning styles with knowledge ability levels. These factors can influence the learning process, which in turn can determine learning outcomes [2].

The factors can be identified through learner's activities in using e-Learning. Learning styles can be identified through the intensity of learning activities on e-Learning [3]. Motivation to learn shown by how often learners interact in discussion forum[4]. Meanwhile, knowledge ability can be known through a series of assessments to figure out learners' the level of knowledge on a particular learning topic[5].

Learning style, motivation, and knowledge ability of learner are very likely to be changed according to their condition at every stage. The intensity of one's learning at the end of the course can be increased or decreased compared to the activities in early topic, as well as learning motivation. In the meantime, the level of knowledge can increase over time and related to learners' knowledge on the discussed topic. Considering that factors in Triple-Factor Approach are dynamic, personalization is better to be flexible with changes that may occurs on motivation, learning styles, and knowledge ability of learners. This study highlights the challenge to provide learning process that helps learners' improvement in achieving expected competency, while also raises the need for personalization that can be presented dynamically according to the situation experienced by learners at every stage of learning.

2. Literature Review

2.1 Personalization

In the context of e-Learning, personalization is a strategy in which the system will be able to provide individual learning for each learner. Personalization in e-Learning enables system to present a scenario adapted to the characteristics of learners[6]. Personalization generally comes from the selection or customization of learning content and learning sequences of activities that are being prepared to meet the needs of individual learners[7].

In recent years, many studies have discussed about personalization in e-Learning through a variety of approaches and techniques. Automation of learning begins with defining the relationship between competence, subject, topic, and learning units. Knowledge domain [8] or knowledge model [9] are the most popular terms to describe this component. Other scholars mentioned it with different terms but essentially described the same meaning. For example, one scholar used the term competency structure [10] while another scholar named this curriculum component as a unified repository [11]. In a study conducted by Gamalel-din [8] ontology is used to represent concepts and relationships concepts. Knowledge domain also represents object of teaching by means of concepts, relations among concepts and teaching preferences connected with concepts[9]. Besides concepts and their relationships, knowledge domain can be used to define competencies and capabilities as the objective of learning process [10]. Knowledge domain is described as relationship between courses, topics, subtopics, and competencies[11].

Most common reference for personalization is learning style, knowledge level, and affective state of learner. Learning style refers to habit or preferred way for learning [3]. There are many learning style models that can be used for personalization, for example Felder-Silverman Learning Style Model (FSLSM) as discussed in [12] and [8], Myers-Briggs Type Indicator (MBTI) as implemented in [13] and Kolb's Model mentioned in [14].

Besides learning styles, personalization can be delivered based on information regarding learners' level of knowledge. Learning Object (LO) can be classified based on its difficulty level, then LO will be presented based on the learner's ability level [15]. Another popular approach is based on affective state of learner such as motivation, effort, and confidence[16].

Personalization can be presented in various forms, among others: personalized interfaces, personalized learning scenarios, as well as personalized learning content. Personalized interface among others proposed by [12] and [17]. Neural networks was applied for learning styles classification in [12], while [17] used support vector machine. Both studied adopted Felder-Silverman learning style model in adjusting the interface of e-Learning system.

Personalization can also be presented as individual learning scenarios [8]. System will analyze the learning style of each learner. The identified learning styles then determine the order of learning unit given to learners individually. For example for global learner, the system will present the general overview at the beginning of each chapter or section and then present the sequence of learning object[8].

Meanwhile a recommender system approach can be utilized for personalizing learning content. System can first analyze the preference of students to this type of learning content. The system will then recommend the learning content that similar to the preference of such learners. We can use the concept of preference score and similarity score to recommend learning content [18]. Preference score is calculated based on the matched object between the learning contents presented by system and learning contents that accessed by user. The similarity score can be derived by the number of same object that accessed between two users. System will recommend learning contents with highest recommendation score based on preference score and similarity score. Learning content recommendation can also be calculated using collaborative filtering. Collaborative filtering is used to predict the preference of one user based on previous ratings by other users to the object[19].

Assessment component also generally implemented in personalized e-Learning. Assessment is used to find out the effectiveness of the learning process. Assessment can be implemented through e-portfolio which include tasks that can be evaluated and provide feedback to learners [20].

In studies that discuss personalization, there are two terms that are often mentioned: adaptive and dynamic. The two terms are often discussed overlapping one to another. The term of adaptive learning denotes to a learning process where the learning contents that are presented "adapts" to individual needs of learner [21]. Thus, adaptive e-Learning system can be defined as system that has ability to deliver learning sequences and contents in accordance with learners' individual characteristics [22]. In the other hand, dynamic refers to something that is not constant. Dynamic approach can be applied, among others, to learner modelling [23] and learning sequence of object [24].

Focusing on personalization techniques, most of the learning scenarios are generated at the beginning of the term and rarely notice to learners' progress in the learning stages. Only few studies re-adjust the given learning scenario throughout the whole learning process, among others utilize Multi-Agent Memetic Algorithm [9]. However, the study only considered cognitive state and preferences of learners in learning presentation (sequences of learning activities) that will be given.

2.2 Triple-Factor Approach

The current study adapts Triple-Factor Approach [3] which uses learning type that consists of learning styles, motivation, and knowledge ability as factors that affecting learning process. This approach was selected since it includes the affective factor (indicated by motivation) and learning styles that reflect the character of learners, in addition to knowledge ability level. Another consideration is that previous studies have shown a close association between factors in the approach [2].

Triple-Factor Approach is based on previous studies which mentioned that there are linkages between factors in the process of e-Learning. Some studies show that learning styles affect learners' learning strategies [25] and ultimately have an influence on learning achievement [25][26]. In addition to learning styles, motivation also has a significant relationship to knowledge ability demonstrated by the results of the assessment [5]. Based on the results of these studies, there had been an analysis of three factors: motivation, learning styles, and the ability of knowledge [2].

The results of the correlation tests showed that motivation to learn affect the level of knowledge ability. Similarly, learning styles have a relationship with the level of knowledge ability. The test results are then led to the learning type in e-Learning process namely Triple-Factor Approach.

Learning style is shown through intensity of learning activity on e-Learning: low, medium, and high intensity. In Triple-Factor Approach, motivation is classified into three levels: low, medium and high. Meanwhile, the learner is based on four levels of knowledge ability, namely: poor, average, good, and excellent[3]. Furthermore, 36 learning types could be obtained from a combination of (3 characteristic learning style: low, medium, and high intensity) * (3 characteristic motivation: low, medium and high) * (4 characteristic knowledge ability: poor, average, good, and excellent).

3. Research Methodology

The present study is conducted through the following steps:

1) Personalized E-Learning System

In this step, the authors analyze the functional requirement of the personalized e-Learning system. The proposed system should covers some components that support personalized learning process from course planning to assessment.

2) Personalization Rule
The next step is designing rules in presenting personalization to user. The personalized learning is based on learning type and intended to improve learner's state in every stage of learning process. The rules are applied topic by topic, to help learner's progress in learning styles, motivation, and knowledge ability.

3) Intervention to improve learning type
The personalization rules are equipped with some additional activities to trigger learner's progress. The intervention cover activities to improve learner's intensity in learning activities and accessing discussion forums, as well as improve their level of knowledge ability.

4. Proposed Design and Discussion

4.1 Proposed Personalized E-Learning

In order to accomplish the objective of enabling learner to learn based on their individual needs and to monitor learner's progress toward targeted outcomes, the authors propose personalized e-Learning system that complies with standards-based education. Personalization provided in the system is driven by learning type based on Triple-Factor Approach that consists of learning style, motivation, and knowledge ability. We can identify learner's learning styles through the intensity of their learning activities on e-Learning. Learning motivation can be seen from the frequency of learner's activity in discussion forum. Knowledge ability can be found through a series of assessments to figure out learners' knowledge on a particular learning topic.

The personalization is presented topic by topic within a subject and completed with an assessment to see how far learner has mastered the targeted competencies. In addition, the assessment is also done to see whether the personalization gives expected results. At the end of each assessment, the system identifies the next state of each learner as the beginning of the new iteration of personalization for the next topic. Thus, personalization operates adaptively to the needs and character of learner as well as dynamically accommodates the state changes of learner in every stage of learning process. Figure 1 describes the proposed personalized e-Learning system.

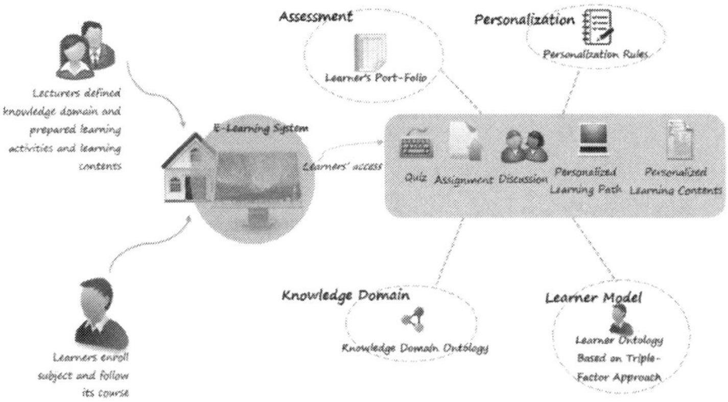

Figure 1. The Proposed System

The proposed system covers several components: knowledge domain, learner model, personalization, and assessment. Targeted competency will drive the knowledge domain as it defines the relationship between competencies, subjects, topics, and learning units. To a accommodate learner's personal needs, systems have to understand the characteristics of learners. Our study adapts the Triple-

Factor in learner modeling. As mentioned earlier, that the standards-based education requires a standard of assessment to measure the learner achievement. Thus, in order to identify the progress of a learner, our study proposes portfolio-based assessment.

4.2 Personalization Rule

The study proposes rules that could be applied in implementing dynamic personalization. The main idea of dynamic personalization is not only the system adaptive to learners' learning type, but also adaptive to learners' change in every stage of the learning process. Considering that learning type are dynamic, the current study proposes that the learning process are evaluated and adjusted at each stage. The next iteration of personalized learning is determined based on the results of learning type identification at the previous stages.

The personalization rules are equipped with some interventions to trigger learner's progress. The intervention cover activities to improve learner's intensity in learning activities, discussion forums, and assessment. In general, the proposed personalization rules shown in Figure 2.

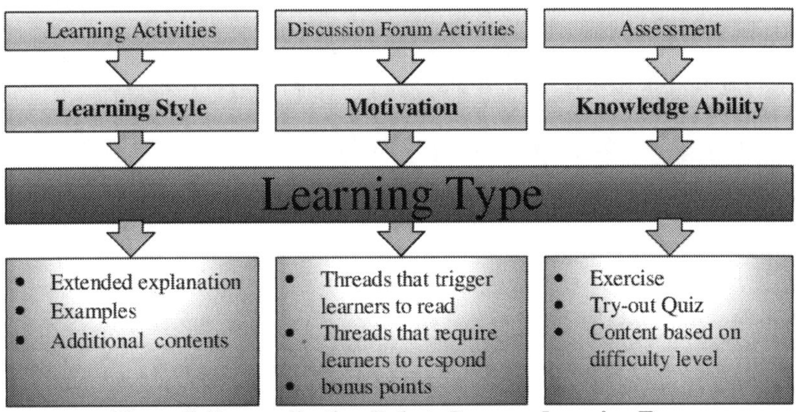

Figure 2. Personalization Rule to Improve Learning Type

The intervention in learning styles can be in form of additional contents such as video, animation, or video conference. The material can be varied by extended explanation and examples. These action is expected to increase learner's learning intensity.

In motivational factor, this study considers some strategies to improve learner's interest in accessing discussion forum, among others: create stimulating threads that trigger learners to read (for example: discussing about the answer of the latest quiz/assignment), posting threads that require learners to respond (for example: discussing about the mechanism and themes of next topic/assignment), and bonus points for every valid action on discussion forum.

To improve knowledge ability, the system can provide exercise activities and try-out just before the quiz period. Exercise and try-out quiz trigger learners to practice more in order to achieve better results in assessment. To excite learner's interest, learners who complete the exercise and try-out quiz can get extra points. Learning contents presented to each learner are adjusted to their level of knowledge ability. Figure 3 illustrates the possible change on the learner's learning type and given personalized learning.

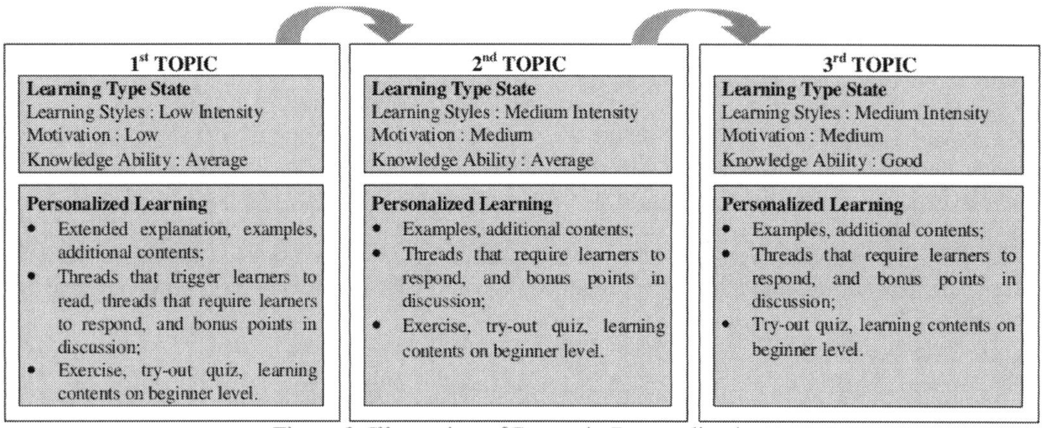

Figure 3. Illustration of Dynamic Personalization

The proposed system offered an approach that could help learners to always improve their selves with the aim to accomplish expected result. The rule is to give learning path and learning contents to improve their state of learning type. As shown in Figure 3, the system offers personalized learning to improve their learning style, motivation, and knowledge ability to a higher level. The personalization is presented in order that the learner would making progress step by step towards the expected learning outcomes.

5. Conclusion

This study proposed personalization that refers to Triple-Factor Approach. The approach identified learners' learning type based on learning style, motivation and knowledge ability. This study considered the possibility that intensity of learners in using e-Learning can fluctuate over time. Moreover the assessment results may differ depending on learners' ability and effort at the time of assessment. Accordingly, this study proposed rules that could be applied in implementing dynamic personalization. The main idea of dynamic personalization is not only a system adaptive to learners' need, but also adaptive to the learners' change that may occurs. The rules are applied topic by topic and implemented in the form of interventions that trigger learner to access learning contents and discussion forums more often, as well as improve their level of knowledge ability.

References

[1] A. Akram, M. Aslam, C. Sc, and E. Uet, "Agent based Intelligent Learning Management System for Heterogeneous Learning Environment," in *IEEE Multitopic Conference (INMIC)*, 2011, pp. 76–81.

[2] Sfenrianto, Z. A. Hasibuan, and H. Suhartanto, "The Influence Factors of Inherent Structure in E-Learning Process," *Int. J. e-Education, e-Business, e-Management e-Learning*, vol. 1, no. 3, pp. 217–222, 2011.

[3] Sfenrianto, "Pendekatan Tipe Belajar Triple-Factor dalam Proses E-Learning Sebagai Basis Rekomendasi dan Personalisasi Pembelajaran (An Approach of Triple-Factor Learning Types in e-Learning Process as a Basis for Learning Recommendation and Personalization.)," Universitas Indonesia, 2014.

[4] F. A. Khan, S. Graf, and E. R. Weippl, "An Approach for Identifying Affective States through Behavioral Patterns in Web-based Learning Management Systems," in *ACM Proceedings of the 11th International Conference on Information Integration and Web-based Applications & Services*, 2009, pp. 431–435.

[5] M. M. Organero and C. D. Kloos, "Using forums and assessments as motivational tools in e-learning courses: A case study," *Proc. - Front. Educ. Conf. FIE*, pp. 0–5, 2007.

[6] S. Ghallabi, F. Essalmi, M. Jemni, and Kinshuk, "Toward the reuse of E-Learning personalization systems," in *IEEE 2013 Fourth International Conference on Information and Communication Technology and Accessibility (ICTA)*, 2013, pp. 3–5.

[7] I. O. Keeffe, A. Staikopoulos, R. Rafter, E. Walsh, B. Yousuf, O. Conlan, and V. Wade, "Personalized Activity Based eLearning," in *ACM Proceedings of the 12th International Conference on Knowledge Management and Knowledge Technologies*, 2012.

[8] S. A. Gamalel-din, "An Intelligent eTutor-Student Adaptive Interaction Framework," in *ACM INTERACCION '12 Proceedings of the 13th International Conference on Interacción Persona-Ordenador*, 2012.

[9] G. Acampora, M. Gaeta, and E. Mu, "An Adaptive Multi-Agent Memetic System for Personalizing e-Learning Experiences," pp. 123–130, 2011.

[10] A. Nitchot, L. Gilbert, and G. B. Wills, "Towards a Competence based System for Recommending Study Materials (CBSR)," in *10th IEEE International Conference on Advanced Learning Technologies*, 2010, pp. 629–631.

[11] I. Rytikova and M. Boicu, "A Methodology for Personalized Competency-based Learning in Undergraduate Courses," in *ACM 15th Annual Conference of Information Technology Education*, 2014, pp. 81–86.

[12] W. Gong and W. Wang, "Application research of support vector machine in E-Learning for personality," *2011 IEEE Int. Conf. Cloud Comput. Intell. Syst.*, pp. 638–642, 2011.

[13] E. El Bachari, E. H. Abelwahed, and M. El Adnani, "E-Learning personalization based on Dynamic learners' preference," *Int. J. Comput. Sci. Inf. Technol.*, vol. 3, no. 3, pp. 200–216, 2011.

[14] D. Samia and A. Abdelkrim, "An adaptive educationnal hypermedia system integrating learning styles : model and experiment," in *IEEE International Conference on Education and e-Learning Innovations An*, 2012.

[15] J. Kim, Y. Jung, Y. Lim, and M. Kim, "An E-Learning Framework Supporting Personalization and Collaboration," in *ACM 3rd International Conference on Ubiquitous Information Management and Communication (ICUIMC)*, 2009, pp. 635–638.

[16] F. A. Khan, S. Graf, E. R. Weippl, and A. M. Tjoa, "An Approach for Identifying Affective States through Behavioral Patterns in Web-based Learning Management Systems," in *ACM Proceedings of the 11th International Conference on Information Integration and Web-based Applications & Services*, 2009, pp. 431–435.

[17] S. V Kolekar, "Learning Style Recognition using Artificial Neural Network for Adaptive User Interface in E-learning," 2010.

[18] L. K. Soo, E. Yeoh, and S. B. Ho, "Ontology Based Personalized Recommendation Model for Learning Objects in a Service Oriented E-learning Environment," in *International Conference on Innovation and Infroamtion Management (ICIIM)*, 2012, vol. 36, no. Iciim, pp. 87–92.

[19] W. Chen, Z. Niu, X. Zhao, and Y. Li, "A hybrid recommendation algorithm adapted in e-learning environments," *Springer - World Wide Web*, vol. 17, pp. 271–284, 2014.

[20] J. M. Vaca, J. E. Agudo, and M. Rico, "Adaptive Competence-Based System in an E-learning Platform: Adapting Support Tasks to Students," in *2012 IEEE 12th International Conference on Advanced Learning Technologies*, 2012, vol. 12, pp. 720–721.

[21] D. S. Chaplot, E. Rhim, and J. Kim, "Personalized adaptive learning using neural networks," *L@S 2016 - Proc. 3rd 2016 ACM Conf. Learn. Scale*, pp. 165–168, 2016.

[22] L. Nguyen, "A User Modeling System for Adaptive Learning," no. December, pp. 864–866, 2014.

[23] S. Baldiris and S. Graf, "Dynamic User Modeling and Adaptation based on Learning Styles for Supporting Semi-Automatic Generation of IMS Learning Design," pp. 218–220, 2011.

[24] G. Nandiboyina and P. Gharpure, "Dynamic Sequencing of Learning Objects Based on Learning Goals," *Technol. Educ. (T4E), 2014 IEEE Sixth Int. Conf.*, pp. 88–89, 2014.

[25] S. Graf, T. C. Liu, and Kinshuk, "Interactions between Students' Learning Styles, Achievement

and Behaviour in Mismatched Courses," *Proc. Int. Conf. Cogn. Explor. Learn. Digit. Age (CELDA 2008)*, no. 1976, pp. 223–230, 2008.

[26] R. S. Shaw, "A study of the relationships among learning styles, participation types, and performance in programming language learning supported by online forums," *Comput. Educ.*, vol. 58, no. 1, pp. 111–120, 2012.

Corruption Early Prevention: Decision Support System for President of the Republic of Indonesia

Sasmoko[1,2], S A Widhoyoko[3], S Ariyanto[4], Y Indrianti[2], Noerlina[5], A M Muqsith[6], and M Alamsyah[6]

[1]Faculty of Humanities, Bina Nusantara University, Indonesia
[2]Research Interest Group in Education Technology, Bina Nusantara University, Indonesia
[3]Forensic Accounting, Podomoro University, Indonesia
[4]Faculty of Economics and Communication, Bina Nusantara University, Indonesia
[5]School of Information System, Bina Nusantara University, Indonesia
[6]School of Computer Science, Bina Nusantara University, Indonesia

Abstract. Corruption is an extraordinary crime, and then the prevention must also be extraordinary, simultaneously (national) in the form of early warning that involves all elements; government, industry, and society. To realize it the system needs to be built which in this study is called the Corruption Early Prevention (CEP) as a Decision Support System for President of the Republic of Indonesia. This study aims to examine 1) how is the construct of the Corruption Early Prevention as a Decision Support System for President of the Republic of Indonesia?, and 2) how is the design form of the system of Corruption Early Prevention as a Decision Support System for President of Republic of Indonesia? The research method is using Neuro-Research which is the collaboration of qualitative and quantitative research methods and the model development of Information Technology (IT). The research found that: 1) the construct of CEP is theoretically feasible, valid and reliable by content to be developed in the context of the prevention of corruption in Indonesia as an early prevention system that diagnoses Indonesia simultaneously and in real time, and 2) the concept of system design and business process of CEP is predicted to be realized in the IT-based program.

1. Introduction

Corruption is essentially a form of the abuse of power as the form of the inefficiency of an institution which is reflected through the weak bureaucratic system, the weak legislative and judicial systems. The increase of 1% corruption level reduces the growth rate by about 0.72% which is affected by political instability, the level of human capital and share of private investment (Mo, 2001). Corruption affects the decrease in productivity because the allocation scheme will not be optimal and the investment quality will also be decreased so that the community needs will be ignored (Andvig et al. 2000 in Ohwer, 2009; Lambsdorff, 2016).

The term corruption has emerged since 1868. Godkin discusses corruption in politics and government and social improvement efforts. In 1961, McMullan exposed the injustice practice, inefficiency, distrustof the government, the waste of public resources, political instability and violations of government regulations as a result of corruption. In 1987, appeared a model that describes opportunities for corruption (Cadot, 1987; Godkin, 1868; Lambsdorff, 2016; McMullan, 1961).

Various attempts were made to prevent and reduce corruption. The Corruption Eradication Commission as an anti-corruption agent works quite effectively (Ernstson, Sörlin, and Elmqvist, 2008). Transparency International (TI) is also an institution that actively contributes to the global movement against corruption through the Corruption Perception Index (CPI) in which the Transparency International Indonesia (TII) is one of the chapters in TI (Transparency International Indonesia, 2016; Simanjuntak, 2008). The country that effectively tackles corruption is Brunei Darussalam, for enforcing legal supervision and non-interference of politicians in the investigation to the government official (Quah, 2016).

In the context of Indonesia, the measurement of corruption has a different paradigm and approach, i.e. the measurement must fit the context of the multi-ethnic Indonesia. The measurement is needed given that the corruption is an extraordinary crime that must be followed up since it can be a major cause of political and power fraud (Akbar &Vukan, 2014). Therefore we need an extraordinary instrument that can monitor in real time the behaviour of corruption and the early detection of corruption. So, it can assist the government in creating a clean community. This idea is in line with Hooker who explained that the difference of the triggers of corruption depends on the damage of internal control system that occurs due to the perception of the offenders towards a system component and the habit of an offender in exploiting the system. Therefore, the quality of internal investigation by an independent body has a significant effect on the reduction of corruption in the government system (Hooker, 1994).

This instrument was named the Corruption Early Prevention (CEP) as a Decision Support System for President of the Republic of Indonesia, i.e. the real-time instrument that can provide early warning for the government (regional and central) on institution or official who is identified with corruption by involving public participation from all the component of good governance (government, community and industry). CEP is designed as an intelligent system that supports the president's decision. The advantages include public perception and assessment of the inspectorate that is independent towards the transparency of the government officials and the identification of corruption risks in the area of regional budget and assessment of government officials personally. This paper focuses on two issues, namely: (1) how is the construct of the Corruption Early Prevention as a Decision Support System for President of the Republic of Indonesia?, and 2) how is the design form of the system of Corruption Early Prevention as a Decision Support System for President of Republic of Indonesia?

2. Research Method

The research method used Neuro-research. It is one of mix methods which is between qualitative research (referred to as exploratory research) and quantitative research (referred to explanatory and confirmatory research). There are three stages in Neuro-research method, which are exploratory, explanatory, and confirmatory stage.

The first stage of exploratory research was conducted by reviewing theoretically of the variables Early Corruption Prevention (CEP) to find a theoretical construct of CEP in the form of CEP conceptual definition regarding Indonesia context, dimensions of CEP formation, and indicators of each CEP dimension as the early warning signs in the corruption prevention in the context of Indonesia society. The society refers to the society that has high disparity in various sectors. After theoretical context CEP has been found, and then the next stage is by organizing Focus Group Discussion (FGD) with Delphi technique.

This technique validates CEP theoretical study by using content validity from six experts, which from Computer Science, Information System, Education, Law, Science, Forensic Accounting, and Psychology. The result is in the form of valid CEP Construct for Indonesia context.

The second stage, based on the result of Delphi technique, is followed by quantitative research which is through explanatory research. In this stage, the instrument is using assessment form with semantic differential scale from 1 to 7, which are Don't Know/No Answer, Never, Seldom, Sometimes, Frequently, Mostly, Always. The explanatory research sample size is 60 respondents.

Third stage, based on analysis result of second stage, is preceded by Confirmatory research. In this stage, several moderator variables from respondent were analysed, such as gender and scientific field.

Algorithms used in this research are:

2.1. Determine constant linear item that consists of 5 (five) categories, Clean, Quite Clean, Quite Corrupted, Corrupted, and Very Corrupted.

2.2. Determine the amount of minimum and maximum data

$$\begin{cases} n : \textit{the amount of maximum data} \\ m : \textit{the amount of minimum data} \end{cases}$$

2.3. Determine the range by subtracting the maximum and the minimum

$$Range = n - m$$

2.4. Determine the class interval from 1-10

$$k = 10$$

2.5. Calculation with Sturges formula

$$f(x) = \left\lceil \frac{Range}{k} \right\rceil$$

3. Result and Analysis

3.1. CEP Construct according to Indonesia Context

Corruption is a social problem that must be eradicated, but the difficulty inherent in this issue is the difficulty of measuring the corruption activity. Becker and Stigler (1974) proposed a way to reduce corruption which is to do the right combination of monitoring and penalties to control corruption (Olken, 2007). Another alternative proposed is to increase the participation of "grass-roots" by the public as the local level observers especially related to the public services (Olken, 2007). Studies from previous research found that the instrument to measure and as a tangible manifestation of anti-corruption has been developed; one of them is the Corruption Perceptions Index (CPI). The main objective of CPI is to provide data about a perception of corruption in various countries.

In 2001, TI also developed The National Integrity System (NIS) to describe the whole law, institution and practice in a country related to the integrity and accountability of the sector (Repucci in Transparency International 2008). Another tool to measure corruption is Transparency International's Global Corruption Barometer 2007, a public opinion survey that seeks to understand how and in what ways corruption affects the lives of ordinary people. (Riano in Transparency International 2008). Global Integrity which is an independent non-profit organization trying to track governance and corruption trends by utilizing grounded research and blind peer review panel (Werve& Heller in the Transparency International 2008).

This study seeks to build construct namely Corruption Early Prevention (CEP) which is designed as a system apps (Android-based). The main philosophy of the CEP is to accommodate public participation as the real-time consideration of the government in formulating policy and set the decision so that it can become an early detection and is expected to mature the capacity of civil society. CEP seeks to consider how to measure the impact of corruption on the legitimacy of the political system as a major element in a democratic political stability. This impact is accurate, considering the elements of society that become the benchmark is the society in the region. So, CEP is the corruption early prevention within Indonesia context. It is categorized by three dimensions which are the external and internal appraisal that create index of corruption prevention, and then the third dimension is text mining dimension as cross-validation for the finding of index of corruption prevention. The result of Explanatory research was analyzed by confidence interval, which can be seen in the below table.

Table 1. The Calculation of Confidence Interval Respond on CEP

Confidence Interval Respond on CEP				
Description			Statistic	Std. Error
CEP	Mean		66.3667	.88649
	95% Confidence Interval for Mean	Lower Bound	64.5928	
		Upper Bound	68.1405	
	5% Trimmed Mean		66.8519	
	Median		67.0000	
	Variance		47.151	
	Std. Deviation		6.86669	
	Minimum		45.00	
	Maximum		75.00	
	Range		30.00	
	Interquartile Range		11.00	
	Skewness		-.828	.309
	Kurtosis		.927	.608

Based on the above table, the lower bound is 64. 5928 and upper bound is 68. 1045 that showed the respondent tends to respond very agree to construct CEP within Indonesia context, significantly at $\alpha < 0.5$.

3.2. CEP Design

One of the most effective methods for the detection of fraud is through a tip given by employees of the relevant organization that is commonly called "whistleblower" (Crosby, Devaney, & Law, 2008). The existence of the whistleblower is very helpful for the work of management inspector in the detection and prevention of fraud (Bastin & Townsend, 2006). By using the flexibility of GRC approach (Governance, Risk Assessment & Compliance), the whistleblower can be used in the risk assessment elements that could include all stakeholders directly or indirectly in the prevention of corruption (Alexander, 2006). In the context of governance at the national scale, the whistleblower is not only located in the area of intra-organization of the government itself because the parties outside the government organization can provide objective information related to the performance of the government apparatus. The consideration of the validity of the evidence of a case is a must. In the realm of law, the verification of case can be done with two types of evidence, namely direct evidence and circumstantial evidence (Crosby, Devaney, & Law, 2008), so that, corruption as the crime behaviour shall be proven directly (confirmatory evidence) (Field, 1926).

Table 2. Whistleblower and available information (internal assessment)

Whistleblower	Type of Information	Role in Measurement (the obtained information)
Industry/Retail	Direct Information (A)	• Conflict of interest • Company tender • Bribes to Government
Civil Society	Supporting Information (B)	• The independent opinions about the development of infrastructure • Perception about the service quality of the government officials. • Bribes
Independent Service Provider	Confirmation Information (C)	1. Opinion of external auditor, namely: • Financial statements (accounts indicated in bribes) • Indications of entertainment cost, gift, etc.

- Individual in Government and the project winner company.

2. Independent survey notice about public dissatisfaction over:
   - The quality of Government service
   - Completion of infrastructure for the region.
   - Additional cost
   - Delays and nominal mismatch of direct cash assistance.

CEP Phase I is called "External Appraisal" which is formed by inserting a perception index of parties that is considered can give an independent view on the performance of government officers.

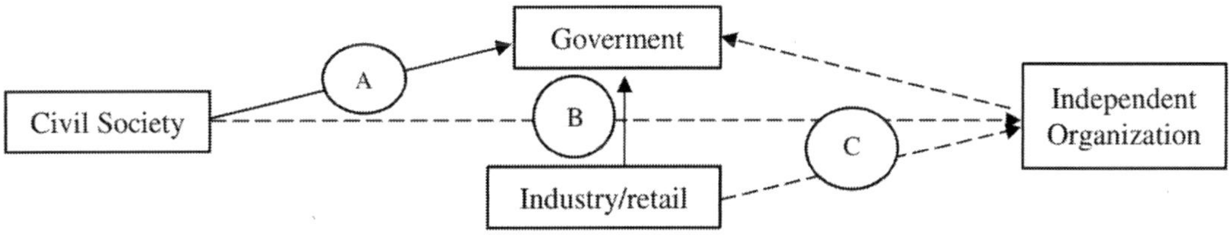

Diagram 1.The process of assessing the performance of government officials externally

Peer review is the process of investigation between departments to ensure the procedure goes according to regulations to be effective in determining the code of professional conduct for accountants in the association (example: ACFE, AICPA and IIA) (Vanasco, 1998). The effectiveness of the process of "peer review" is expected to prevent corruption. In the UK, the peer review process between the health departments has managed to reduce the level of ineffectiveness of the health personnel in the delivery of information and community services (Mangan, Pietroni, & Porter, 2016). With the similar concept and verification of information, the CEP phase II combines parties of governmental departments which are independent and have the authority to make an assessment of government performance called "Internal Appraisal".

Table 3.Whistleblowerand available information (internal appraisal)

Whistleblower	Type of Information	Role in Measurement (the obtained information)
Corruption Eradication Commission	Direct Information (A)	Financial statements Internal control system Conformity of financial report by Government Reports on the results of monitoring follow-up actions The reports of review results of fiscal transparency
Government Employees	Supporting Information (B)	Culture and values application in the government offices Process and the period of community service Internal control over government operational activities Government and private employees relation
Inspectorate	Confirmation Information (C)	Conformity of activity by government officials with the order of the Strategic Plan Conformity of policy towards the vision and mission of government Development activities which are carried out periodically Results of follow-up response after the internal audit activity Local government activities The procurement process of operational support items

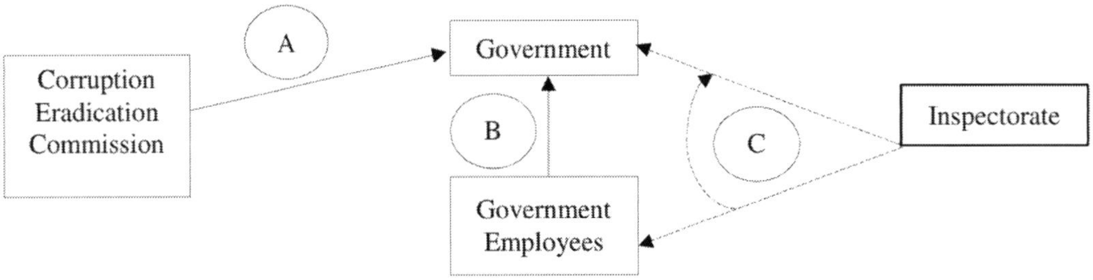

Diagram 1.2 The process of assessing the performance of government officials internally (Internal Appraisal)

CEP phase III sourced on text mining (TM) which is done by collecting a variety of news sources that are relevant to then extracted so that it can be analyzed and classification can be performed so that it can become an assessment source that is objective and accurate.

The next phase in the CEP is to gather input result of assessment partially that can describe the interdependency per dimension and per indicator, per part, per opinion of leaders in the regency/ city, provincial and national. The next phase of this research is by doing the interdependence analysis of EA, interdependence analysis of IA and interdependence analysis of text mining which can be viewed per regency / city, per province and on the national scale. CEP system receives input data from four sources, namely Perception, External, Internal, and Text Mining, each of the sources will be analyzed to generate data and mapped in the form of an interactive map. The requirements to realize the CEP software are the minimum PHP version 7.0, MySQL database and Curl.

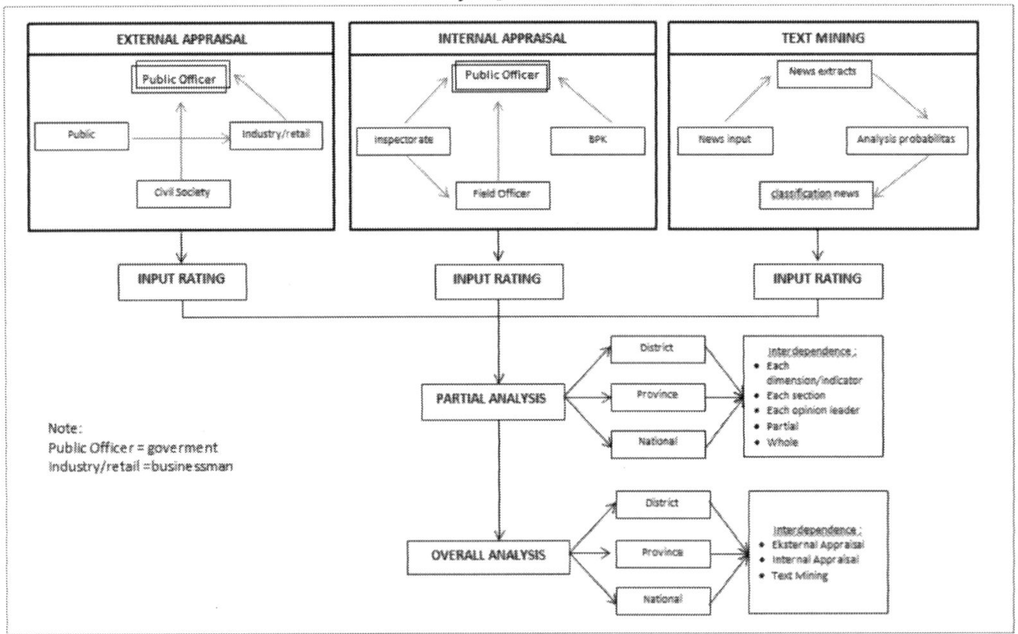

Diagram 1.3.Design of Model of Corruption Early Prevention (CEP)

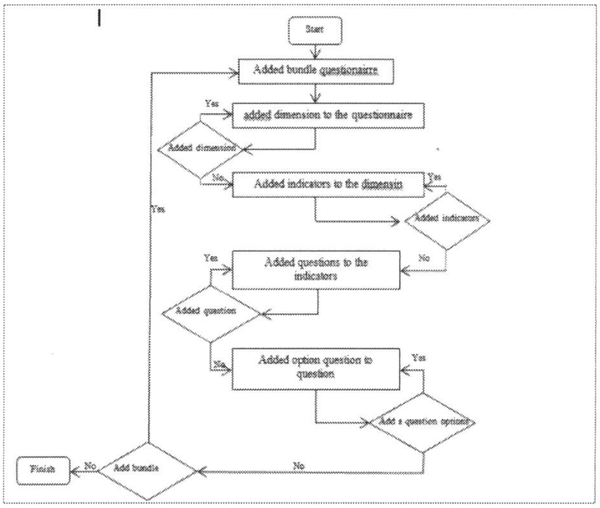

Diagram 1.4 Development of Assessment Sheet of CEP

4. Conclusion and Recommendation

CEP construct is theoretically feasible, valid and reliable by content to be developed in the context of corruption prevention in Indonesia as an early prevention system that diagnoses Indonesia simultaneously and in real time, and the concept of system design and business process of CEP is predicted to be realized in the IT-based program.

The result of exploratory research can be concluded that theoretical construct CEP as content is valid for Indonesia society context. CEP is corruption early prevention that categorized by three dimensions which are external and internal appraisal that create index of corruption prevention, and then the third dimension is text mining dimension as cross-validation for the finding of index of corruption prevention. The results of exploratory research show that the respondent tends to respond very agree to construct CEP within Indonesia context, significantly at α <0.5. While the results of confirmatory research indicate that there is no difference in response to construct CEP within the context of Indonesian society if seen from the difference respondents based the scientific field and gender significantly at α <0.5.

Recommendations of this research is the actualization of Law No. 30 Year 2002 concerning the Commission for Corruption Eradication through the policy of "corruption eradication based on prevention" through an integrated system as an early warning as well as an early detection on the prevention of corruption which leads to the establishment of synergy and integration of education system that can revolutionize human mental in Indonesia for anti-corruption. Based on the findings of CEP construct as a decision support system for the President of the Republic of Indonesia, the design "systems and business process" can be realized and become valid to be implemented by building IT systems in real time for the President of the Republic of Indonesia.

References

[1] Akbar, Y., & Vukan, V. (2014). Explaining corruption: The role of national culture and its implications for international management. *Cross Cultural & Strategic Management*, 191-218.
[2] Alexander, R. (2006). The role of whistleblowers in the fight against economic crime. *Journal of Financial Crime*, 131-138.
[3] Bastin, G., & Townsend, P. (2006). Whistleblowers — A Legitimate Role in Corporate Life?. *Journal of Financial Crime*, 179-182.
[4] Cadot, O. (1987). Corruption as a gamble. *Journal of Public Economics*, 33(2), 223–244.
[5] Crosby, N., Devaney, S., & Law, V. (2008). Benchmarking and valuation issues in measuring depreciation for European office markets. *Property Management & Built Environment*, 7-28.

[6] Ernstson, H., Sörlin, S., & Elmqvist, T. (2008). Social movements and ecosystem services - The role of social network structure in protecting and managing urban green areas in Stockholm. Ecology and Society, 13(2), 38–48.
[7] Field, O. P. (1926). Criminology. Sutherland. *Indiana Law Journal*, 228-243.
[8] Godkin, E. L. (1868). University of Northern Iowa Commercial Immorality and Political Corruption Author (s): E . L . Godkin. *The North American Review*, 107(220), 248-266.
[9] Hooker, J. (1994). Corruption from a cross-cultural perspective, *Cross Cultural & Strategic Management*, 251-267.
[10] Lambsdorff, J. G. (2016). *How Corruption Affects Productivity*, (September).
[11] Mangan, C., Pietroni, M., & Porter, D. (2016). "Being brave": A case study of how an innovative peer review approach led to service improvement. *Journal of Integrated Care*, 201-213.
[12] McMullan, M. (1961). A theory of corruption based on a consideration of corruption in the public services and governments of British colonies and ex-colonies in West Africa. *The Sociological Review*.
[13] Mo, P. H. (2001). Corruption and Economic Growth.1, 79, 66–79.
[14] Ohwer, A. (2009). Measuring corruption: A comparison between the transparency international's corruption perception index and the world bank's worldwide governance indicators, 42–52.
[15] Olken, B. A. (2007). Monitoring Corruption: Evidence from a Field Experiment in Indonesia. *Journal of Political Economy*, 115(2), 200–249.
[16] Quah, J. S. (2016). Combating corruption in six Asian countries: A comparative analysis. *Asian education and development studies*, 244-262.
[17] Sasmoko, & Ying, Y. (2015). Construct validity in neuroresearch. *Advanced Science Letter, 21,* 2438 - 2441
[18] Simanjuntak, F. (2008). *Indeks Persepsi Korupsi Indonesia 2008 dan Indeks Suap*, (Mcc)
[19] Transparency International Indonesia. (2016). *Survei Persepsi Korupsi 2015*. Riset, 1.
[20] Vanasco, R. R. (1998). Fraud auditing. *Managerial Auditing Journal*, 4-71.

IOP Conf. Series: Journal of Physics: Conf. Series **801** (2017) 012029 doi:10.1088/1742-6596/801/1/012029

An ICT Adoption Framework for Education: A Case Study in Public Secondary School of Indonesia

S Nurjanah, H B Santoso and Z A Hasibuan

Faculty of Computer Science, Universitas Indonesia

Email: soficitta@gmail.com, harrybs@cs.ui.ac.id, zhasibua@cs.ui.ac.id

Abstract. This paper presents preliminary research findings on the ICT adoption framework for education. Despite many studies have been conducted on ICT adoption framework in education at various countries, they are lack of analysis on the degree of component contribution to the success to the framework. In this paper a set of components that link to ICT adoption in education is observed based on literatures and explorative analysis. The components are Infrastructure, Application, User Skills, Utilization, Finance, and Policy. The components are used as a basis to develop a questionnaire to capture the current ICT adoption condition in schools. The data from questionnaire are processed using Structured Equation Model (SEM). The results show that each component contributes differently to the ICT adoption framework. Finance provides the strongest affect to Infrastructure readiness, whilst User Skills provides the strongest affect to Utilization. The study concludes that development of ICT adoption framework should consider components contribution weights among the components that can be used to guide the implementation of ICT adoption in education.

1. Background

During the past decade a vast number of reports have been published on global initiatives to improve education quality. Among those initiatives there is an increased government investment to support adopting information and communication technology (ICT) in teaching, learning, and administration at public schools through ICT integration into existing pedagogies. It is expected to transform teaching and learning. Improving education quality in various countries has common long-term objectives including strengthening global competitive, economic growth, social conditions, and human development index. Several statistical analysis have shown that enhancing Education Index (EI) affect significantly ICT Development Index (IDI) which in turns increased Human Development Index (HDI).

ICT usage in daily life activity has become a common practice by almost people in the world. Using technology is conducted to make the work completion more efficient and the result is better. ICT is all technology used to handle (e.g. save, update, manipulate, organize, distribute, forward, etc.) information and communication, such as internet, computer, email, smartphone, including software supported. Internet is the famous growth parameter of ICT. By the end of 2016, seven billion people (95% of the global population) live in an area that is covered by a mobile-cellular network, and 3.9 billion people (53% of the global population) is not using the internet, it means that close to one out of two people (47%) in the world are using the internet, but only one out of seven people in the LDCs [1].

Global ICT development index from ITU released in early 2016 (see Figure 1) show the growth of phone and internet user in last 15 years. Active mobile-broadband user subscription started in 2007 and increase faster than most others. It also shows that a decrease in fixed telephone subscribers (fixed telephone subscription). It happened probably due to the flexibility and ease of buying and installing mobile phone connections than fixed telephone lines.

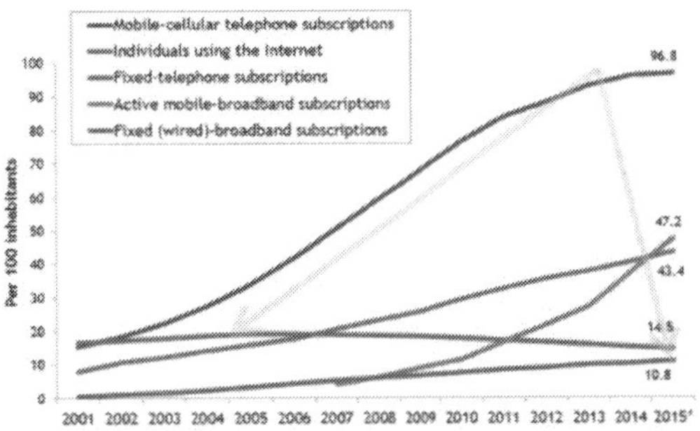

Figure 1. ICT development variable (2001 – 2015)
Source: ICT fact and figures (ITU, 2016)

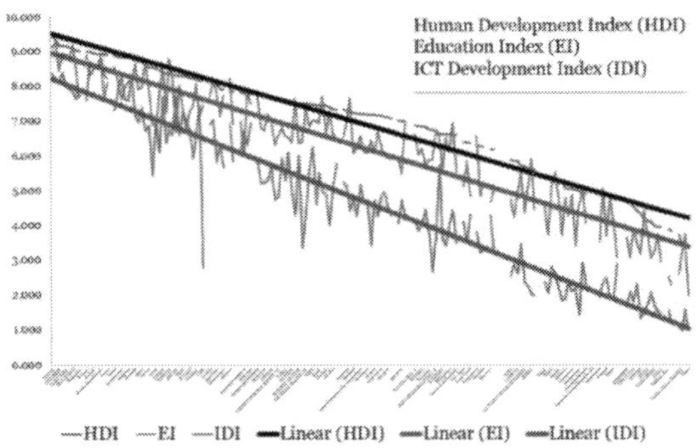

Figure 2. Global ICT development index and human development index (HDI) (2001 – 2015)
Source: Compiled from ITU (2015), UNDP (2015)

ICT usage is adopted in many sectors for specific purpose in many life sectors (health, education, politic, etc.). So, ICT adoption means the use of ICT for particular purpose. ICT adoption influences user in achieving the goal of any purpose. It means ICT adoption in education might impact to the goal of education. At the other view, increasing ICT Development Index (IDI) indicates that Education Index (EI) also increase, where EI is a part of Human Development Index (HDI) as shown as Figure 2.

ICT adoption in education is a common term refers to the use of ICT in teaching, learning, and administration aimed to strengthen the achievement of education goals, priorities, and strategies. There is encouraging evidence that ICT can be an effective tool in supporting teaching and learning. There is a common believe among education policy makers in various countries that the necessary condition for effectively introducing technology into schools are mainly the availability and accessibility of ICT resources. The resources are including: computer hardware, software, communications infrastructure, digital content, technical as well as Finance support for ICT operation and maintenance.

Some researchers found that one of the greatest challenges in ICT use in education is balancing educational goals with economic condition. ICTs in education programs require large capital investments. Some special attention should be paid to areas in developing countries due to the digital divide. To achieve a greater development of ICT-supported education systems will require new approaches to teaching and learning, and new types of technologies to support those new approaches in schools [2]. That is why designing an ICT policy in education is a far more complex task than merely deploying hardware in the schools.

From the best of our knowledge, little has been said clearly about what/how component contribute to the success of ICT adoption framework. Therefore, the objective of this research is two folds: (1) investigating components that affect ICT adoption in education, and (2) proposing a conceptual framework for adopting ICT in education, especially in public secondary schools in Indonesia. Moreover, the contribution is mainly to provide information to education policy maker regarding the issue what components contribute to the success of ICT adoption in secondary schools in Indonesia.

This information is valuable in helping policy maker to ensure that decision making in education will effectively enhance education quality for students.

The rest of the paper is organized as follows. Section 2 describes some related works. Section 3 explains the research methodology. Section 4 explains the results and discussion of the study followed by conclusion and future works in Section 5.

2. Related works

Review of related works cover theories, cases and practices about components of ICT adoption for Education

2.1. Theories related to ICT adoption in education

Regarding to the research question, this research must find the answer about: (1) What components that affect ICT adoption in education including how these components interact within a framework; and like what and how a conceptual framework for adopting ICT in education, especially in public secondary schools in Indonesia. There are six components of ICT adoption in education gathered from literature which mapped as shown in Table 1.

Table 1. Component's map

Source \ Components	Infrastructure	Application	User Skills	Utilization	Policy	Finance
Doshi, 2014. [3]	✓	✓	✓	✓		
HU, 2012. [4]	✓	✓		✓		
Langrova and Poulova, 2013. [5]	✓	✓	✓			
Wang and Zhou, 2013. [2]	✓	✓	✓	✓		✓
Hepp, 2003. [6]	✓	✓	✓	✓	✓	
Poulova and Simonova, 2014. [7]	✓	✓	✓	✓		
Tabira and Otieno, 2014. [8]	✓		✓	✓		
Zain and Murugaiah, 2004. [9]	✓	✓	✓	✓	✓	✓
Goyal and Purohit, 2011. [10]	✓	✓	✓	✓	✓	✓
Sooryanarayan DG and Rekha, 2014. [11]	✓	✓	✓	✓	✓	✓
Ila and Kitapci, 2014. [12]	✓	✓		✓	✓	
Arshad and Noman Saeed, 2014. [13]	✓	✓	✓	✓	✓	✓
Jemni, 2014. [14]		✓			✓	
Bo and Ming-Sheng, 2010. [15]	✓			✓	✓	✓
Haruzuan, Eamas and Forret, 2013. [16]		✓		✓		
Šumak, G. Polančič and Heričko, 2010. [17]		✓		✓		
Tatnall, 2008. [18]	✓	✓		✓		
Zahariev, Ruseva, Hristov and Benchev, 2013. [19]	✓	✓	✓	✓		✓
Panangalage and Pasqual, 2008. [20]	✓		✓	✓	✓	
Rekha and Adimarayanan, 2014. [21]		✓	✓	✓		✓
Sharma, GOH, SUN and HO, 2014. [22]	✓	✓				

There are almost all literatures about ICT adoption in education examine those six components. In each components, there are still factors inside which contribute entirely.

2.2. Cases and practices from various countries

Many studies on ICT adoption in education in various countries have been reported and the summary as Table 2.22. The result of brief review on ICT adoption involving some prominent works from countries is summarized as Table 2.

Table 2. ICT adoption indicators in countries

No	Countries / Factors Mapped into Components	United States	Canada	England	France	Germany	Italy	Japan	Russia	Malaysia	Singapore	Indonesia
I	INFRASTRUCTURE											
1	Computer per schools rate			X			X	X				
2	Computer laboratory per schools rate									X		X
3	Student per computer ratio	X	X	X	X		X	X		X	X	X
4	Schools internet connection							X				
5	Computer connects to internet					X	X					
6	Internet access					X	X	X				
II	APPLICATION											
1	Schools has email											X
2	Schools has website							X				
3	Schools has specific application	X										
III	USER SKILLS											
1	Teacher has ICT skill in teaching student							X				
2	Administration Staff has ICT skill in managing school										X	X
3	Student has basic ICT skill in learning				X	X	X					
IV	UTILIZATION											
1	Student use ICT during studying	X	X	X	X		X	X				X
2	Teacher utilize ICT during teaching	X						X				
V	POLICY											
1	Schools follow Government's program in implementing ICT	X	X	X	X	X	X	X		X	X	X
VI	FINANCE											
2	Schools receive Government's Finance support in implementing ICT	X	X	X	X	X	X	X	X	X	X	X

Source: compiled from cross-national ICT report [23].

2.3. Cases in Indonesia

For additional information, there are few cases ICT adoption in education related to the components specifically in Indonesia:

1. Ministry of Education Indonesia database tells there are 212,671 national schools in Indonesia and shows 157,656 schools have computers, but still 87,213 schools only have 1 or 2 computers. 190,604 schools have email. 180,990 mobile schools have a contact number and 11,370 schools have websites [24].
2. Ministry of Education Indonesia periodically granted to some schools with general specification of ICT supply, such as computers, networks and LCD screens. Schools receive the grant and start using ICT with general guide. There are not yet any comprehensive guidance which can drive schools to adopting ICT in appropriate way.

3. Ministry of Education Indonesia provide Jardiknas as internet backbone to supply internet connection to schools and education centre. Jardiknas also provide network domain name email and clouds service to ministry office [25].
4. Ministry of Education Indonesia provide centralized education database (Dapodik) which support detail historical education information. Almost all ministry's policy use data from Dapodik.
5. To strengthen the learning process, in 2008, the Ministry of Education bought the copyright from the authors of textbooks and waive the use of the book in digital form. The digital textbook, Electronic School Book (BSE). In 2010, the Ministry of Education released a Content Management System (CMS) for a school that is Free Open Source.

3. Research design

The research design is combined explorative and confirmative based on premise that there exists an association between observed variables that construct the investigated latent variables and causal relationship among these latent variables in related to ICT adoption in education [26]. In general, the target population of this research comprises: principals of public secondary school in Indonesia, ICT content developers and education specialist, and policy makers in Ministry of Education Republic of Indonesia.

Sampling technique in this research adopted both purposive and systematic random sampling techniques. Purposive sampling ensured that prominent digital/non-digital educational content developers, and policy makers can be selected in order to collect much information concerning the subject of study. Systematic random sampling, in this preliminary study, was used to select respondents from school principals.

3.1. Research phase and data collection

The research activities can be divided broadly into three phases as follows.

3.1.1. Literature review. In this phase, literatures that have been published on IEEE Explore and ITU.int in the last decade on ICT adoption in education are analysed to gain relevant information.

3.1.2. Explorative and confirmative analysis. This phase comprises of data collection using focus group discussion (FGD) method whose objectives are: (1) exploring main components of ICT adoption framework for secondary education and (2) gathering confirmation from some education stakeholders on the proposed ICT adoption framework alternatives for secondary school. Three groups of education stakeholders who are involved in this activity namely: (a) education content developer and secondary education specialist; (b) policy maker in education; and (c) principals of public secondary schools. The FGD participants (respondents) are selected using purposive sampling technique. At the end of this phase, based on judgment of the education stakeholders as the result of FGDs, the best alternative of the proposed ICT adoption framework for secondary education is selected. Data collecting instruments are video camera, audio recorder, and brief questionnaires.

3.1.3. Testing the proposed ICT framework. This phase comprises several activities including: (1) data collecting from samples of public secondary school principals using online survey technique; (2) data processing and analysis; and (3) developing ICT adoption model. In order to outreach respondents from various provinces, data collection is implemented using online questionnaires as instrument.
The link to the online questionnaire is emailed to the list of school principals in several provinces who are selected using systematic random sampling technique. Back checking process to the returned questionnaires is implemented to ensure reliability and validity of the collected data. Finally, some hypotheses about main variables that contribute to ICT adoption in secondary schools and association among latent variables in ICT adoption in secondary education are tested using statistical methods.

3.2. Data analysis

Data analysis is implemented using qualitative and quantitative data analysis techniques as follows.

3.2.1. Qualitative data analysis. The purpose of qualitative data analysis is to identify: (1) the main observed variables and candidate of latent variables that presumably contribute to the success of ICT adoption in secondary education, and (2) prior model of ICT adoption in secondary schools. Given qualitative data collected from literature analysis and FGD method, a set of observable variables are collected and categorized. Each category of observable variables is then labelled.

3.2.2. Quantitative data analysis. The main purpose of quantitative data analysis is to test hypotheses regarding the effect of each observed variable to the constructed latent variables, and the association among latent variables in the ICT adoption which are represented by prior models.

The results of hypotheses testing on the effect of each variable to its corresponding latent variable are used as basis for variable selection. Similarly, the results of hypothesis testing on the association among latent variables and linear model goodness-of-fitness in the structural equation model are used as basis for building ICT adoption model. Finally, the ICT framework model is derived from the best structural ICT adoption model. Quantitative analysis is conducted using Structural Equation Modelling [27] which combines several statistical analyses such as: factor analysis models, regression models, and complex path models.

After review the theories and cases, it shows that ICT adoption cases in Indonesia are similar from other developing countries. Then, the authors conducted Focused Group Discussion in three respondent categories groups (1) ICT developer who produce education ICT product with five years experiences in implementing to some schools, (2) ICT integrator who design ICT implementation in schools as project manager, and (3) school headmaster who implements ICT-based teaching and learning.

Group 1 conclude that there are general and specific ICT product which used in education. Common ICT product such as computer, internet, projector, etc, and particular ICT product is tools (hardware and software) who produced for teaching and learning purpose only such as LMS and LCMS. The particular one are often difficult to meet the requirement. ICT product that is used in schools should be in accordance with the curriculum applied. It is normal that finally they have their product is not aligned to the school process. So the schools involve teachers in designing the particular product ICT for teaching and learning, until deployment.

Group 2 conclude that they have difficulty in providing ICT support which ideally should be adopted by school effectively. The human resource composition of a school is the teacher whose main job is teach without any ICT skill requirement. This causes the supporting hardware and software is not sufficient, however, ICT skill of teachers should also be improved in order to achieve the expected goals. Enhancing ICT skills of teachers obviously need the support of the principal and sufficient funds. Wisely, teacher must also have motivation to improve themselves.

Group 3 explain that curriculum requires students to have diverse skills. This makes it difficult to achieve the learning process competencies expected from students. ICT reduces the limitations of teachers and students in exploring science. Limitation of learning is about observation learning object characteristics such as too small, too big, too far, too fast, too slow, etc. Learning content is easy to be managed.

Regarding to ITU framework and Components found from many sources, there are view angle like Figure 3 where ICT adoption components are mapped.

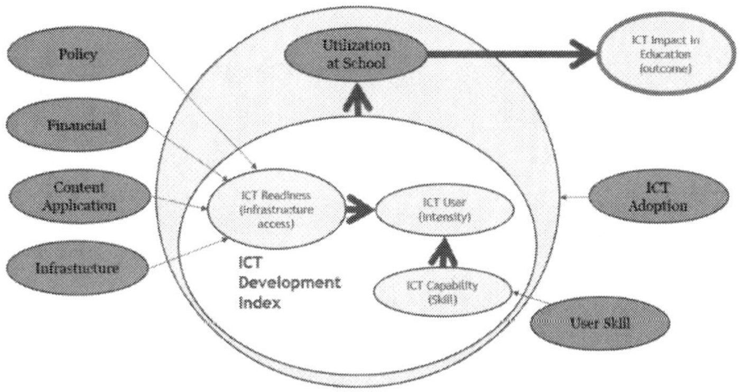

Figure 3. ICT adoption component mapping
Source: compiled from ITU framework and variable from literatures.

4. Results

From literature review, explorative/confirmative analysis, gathered are components and factors of ICT for Education found as shown at Table 1 and the relation shown at Figure 4.

Table 1. Component and factor of ICT for education

Components	Factors	Components	Factors
INFRASTRUCTURE	Availability of: Electricity TV/Radio Telecommunication Computer Internet Bandwidth Printer/Scanner/Fax Smart Device	UTILIZATION	Intensity of: Using infrastructure, application/content in teaching student and manage schools.
APPLICATION	Availability of: Computer System Administration System Learning System Learning Content System Learning Evaluation System	POLICY	Applying: Rule Regulation Term and Condition Guidance
USER SKILLS	Skill of: Basic computer literate Using Application/ Content in teaching and learning	FINANCE	Availability of: Finance source to support infrastructure, application/content and training for teacher.

From testing and data analysis, there are components relationship diagram can be seen in Figure 5, 6 and 7 where each component has weight contribution which calculated by using SmartPLS data analysis tools.

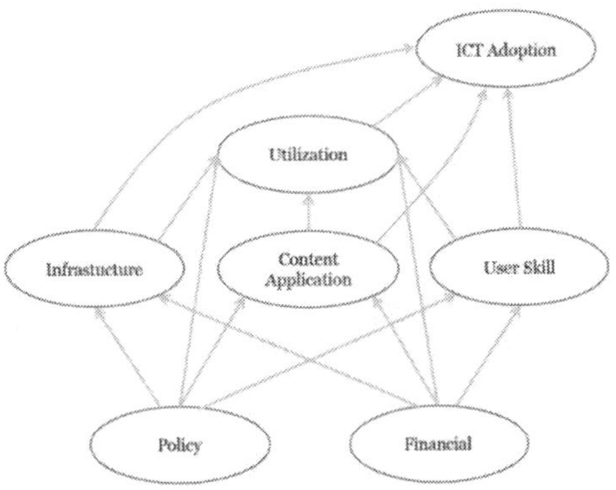

Figure 4. ICT adoption component's linkage from literature

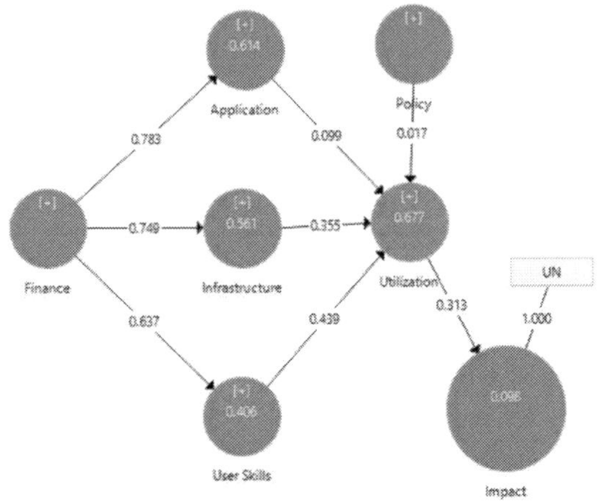

Figure 5. ICT adoption impact to variable of school's national examination score average

Data processed by SEM revealed several components such as: Infrastructure, Finance, Policy, Content, Utilization, and User Skills. After conforming the form and weight of relation, each component's contribution weight can be seen from how much the component affects another components. One of the result shows each component's contribution weight to impact school's national examination score component is described completely in Figure 5.

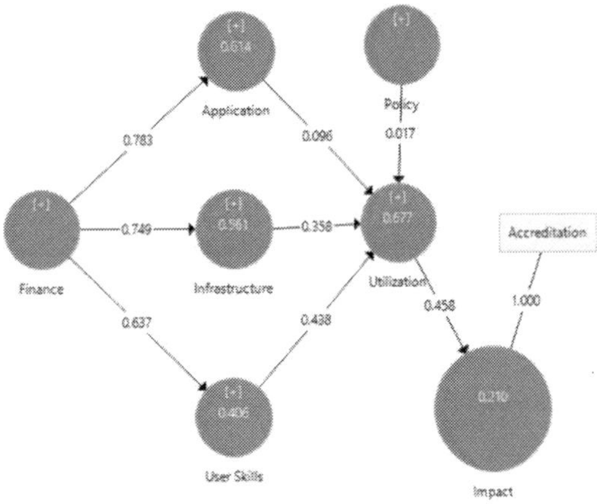

Figure 6. ICT Adoption impact to variable of school's national accreditation

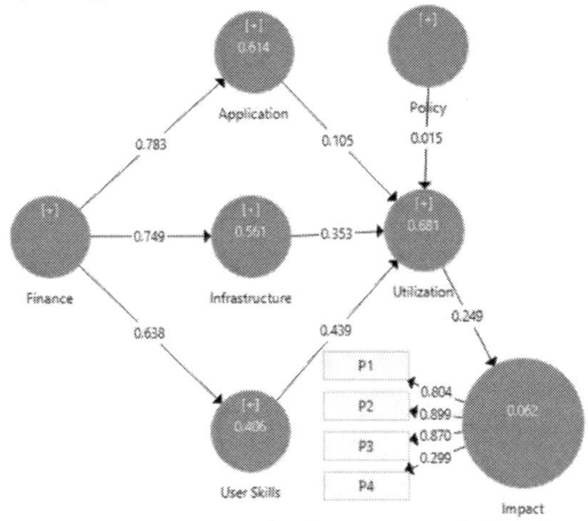

Figure 7. ICT adoption impact to variable of student's competition performance

Finance directly affects Infrastructure (0.749), Application (0.783) and User Skills (0.637). Whilst, Infrastructure (0.355), Application (0.099), User Skills (0.439) and Policy (0.017) directly affect Utilization. Since Utilization is the only one observed component affects to the impact, the result shows specifically that Utilization affects school's national examination impact about 0,313, affect school's national accreditation about 0.458, and affect student's competition performance (stated as P1, P2, P3, and P4) about 0.249. P1 until P4 represents number of student win the competition level sub-district, district, province and national.

Figure 6 and 7 are also conformable. The result indicates that Utilization can be executed after school has Infrastructure and User Skills. Whist Policy contribute Utilization least, and Financial give contribution much, so the proposed ICT adoption framework in education can be created like Figure 8.

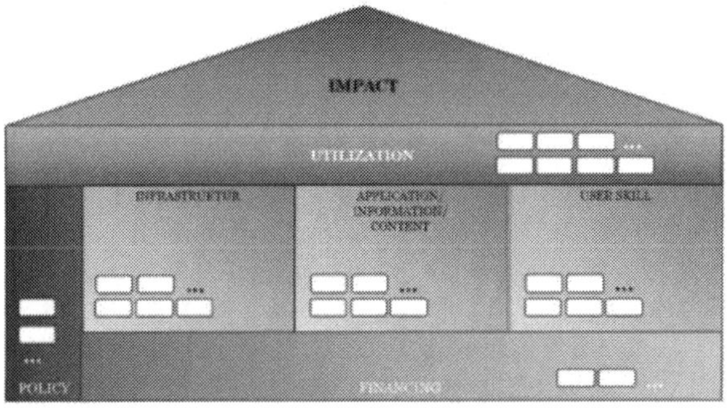

Figure 8. The proposed ICT adoption framework in education

5. Conclusion

Figure 8 shows ICT adoption impact is reached after schools utilized ICT by contribution from infrastructure and application existence, operated by user whose ICT skill is literate. Policy is support component since the weight is not significant, it means that policy is not considered as a primary key in order to utilize ICT in education. The study concludes that development of ICT adoption framework should consider components contribution weights among the components that can be used to guide the implementation of ICT adoption in education.

To get better work, the results need to be refined in the future to get more valid numbers of sample which wider or clustered in order to get the result more accurate.

References

[1] ITU 2016 *ICT Fact and Figure 2016.*

[2] Wang F and Zhou C 2013 *A Theoretical Study on Development of Information and Communication Technology (ICT)-Supported Education Systems* 10th Int. Conf. on Fuzzy Systems and Knowledge Discovery (FSKD).

[3] Doshi J 2014 *ICT Based Solutions for Education in Rural India- A Case Study*, in IEEE 2014 Global Humanitarian Technology Conf., San Jose, CA, USA.

[4] HU J 2012 *Challenges of Utilising a New ICT Approach to Generate a Digital Language Learning Model* The 7th Int. Conf. on Computer Science & Education (ICCSE 2012), Melbourne.

[5] Langrova P and Poulova R 2013 *The IT Impact on Primary Education* 11th IEEE Int. Conf. on Emerging e Learning Technologies and Applications, The High Tatras, Slovakia.

[6] Hepp P 2003 *Critical Factors for an ICT in Education Policy in Developing Countries.*

[7] Poulova P and Simonova I 2014 *The Impact of ICT amongst the Secondary School Graduates towards Higher Education* 8th Malaysian Software Engineering Conf. (MySEC) Langkawi, Malaysia.

[8] Tabira Y and Otieno F 2014 *How to Implement and Diffuse ICT Based Education in Areas with Limited Resources in Developing Countries: Lessons from Rural Kenya* Proc. of PICMET '14: Infrastructure and Service Integration, Japan.

[9] Zain M Z M, Murugaiah P, Atan H and Idrus R M 2004 *Management Practice in Malaysian Smart School: Tasks and Support Analysis of the ICT Implementation* Proc. of the IEEE Int. Conf. on Advanced Learning Technologies (ICALT'04).

[10] Goyal E and Purohit S 2011 *Using Moodle to Enhance Student Satisfaction from ICT* IEEE Int. Conf. on Technology for Education.

[11] Sooryanarayan D G and Rekha S 2014 *Trends in Open Source Software Adoption in Indian Educational Institutions* IEEE Sixth Int. Conf. on Technology for Education.

IOP Conf. Series: Journal of Physics: Conf. Series **801** (2017) 012029 doi:10.1088/1742-6596/801/1/012029

[12] Ila M B and Kitapci H 2014 *Selecting an Effective Information and Communication Technology Architecture for an Education System based on Non-functional Requirements* IEEE 8th Int. Conf. on Application of Information and Communication Technologies (AICT).

[13] Arshad M and Noman Saeed M 2014 *Emerging Technologies For E-Learning and Distance Learning: A Survey* Int. Conf. on Web and Open Access to Learning (ICWOAL), Jazan, Kingdom of Saudi Arabia.

[14] Jemni M 2014 *Promoting the Effective Use of ICT for Enhancing Education in the Arab World* IEEE 14th Int. Conf. on Advanced Learning Technologies.

[15] Bo W and Ming-Sheng H 2010 *Impacts of the Government Action on the Diffusion of ICT Technical Standards.*

[16] Haruzuan M N, Eamas C and Forret M 2013 *Online collaborative learning in tertiary ICT education: Constraints and suggestions for improvement* Int. Conf. on Informatics and Creative Multimedia, Malaysia.

[17] Šumak B, Polančič G and Heričko M 2010 *An Empirical Study Of Virtual Learning Environment Adoption Using UTAUT* Second Int. Conf. on Mobile, Hybrid, and On-line Learning, Netherlands, Antilles.

[18] Tatnall A 2008 *Factors Affecting the Adoption of ICT Curriculum Innovations and Educational Technology* ITI 6th Int. Conf. on Information & Communications (ICICT), Cairo, Egypt.

[19] Zahariev P, Ruseva Y, Hristov G; and Bencheva N 2013 *ICT Convergence Challenges in Education and their Impact on both Instructors and Students* 24th EAEEIE Annual Conf. (EAEEIE), Crete, Greece.

[20] Panangalage R and Pasqual A 2008 *Impact of ICT on Learning and Teaching* IEEE Int. Symposium on Technology and Society.

[21] Rekha S V and Adimarayanan V 2014 *A Free and Open Source Framework for Rural Education.*

[22] Sharma R S, GOH Z H, SUN G and HO T W 2014 *Does ICT Effectively Contribute to the Delivery of Mass Education in Developing Countries?* Proc. of the 2014 IEEE ICMIT, Singapore.

[23] IEA 2009 *Cross-National Infomation and Communication Technology, Policies and Practices in Education* Revised Second Edition ed Plomp T, Anderson R E, Law N and Quale A.

[24] M. o. E. Indonesia, *Data Pokok Pendidikan*, in *http://www.kemdikbud.go.id*, Jakarta.

[25] M. o. E. Indonesia 2015 *Rencana Strategis Kementeriam Pendidikan dan Kebudayaan 2015-2019.*

[26] Curtis E A, Comiskey C and Dempsey O 2014, *Correlational Research: Importance and Use in Nursing and Health Research Authors.*

[27] Kline R B 2011 *Principles and practice of structural equation modeling* © 2011 The Guilford Press, New York.

[28] Wang R, Fu Z and Duan Y 2011 *Understanding ICTs Adoption from an Evolutionary Process Perspective* Int. Conf. on Management and Service Science, Wuhan, China.

University Accreditation using Data Warehouse

A S Sinaga[1], A S Girsang[2]

[1]Department in Informatics Engineering, STMIK Pelita Nusantara, Jl. Iskandar Muda No.1, Medan.
[2] Master in Computer Science , Bina Nusantara University, Jl. Kebon Jeruk Raya No. 27 Jakarta, Email : agirsang@binus.edu

Email : agirsang@binus.edu

Abstract. The accreditation aims assuring the quality the quality of the institution education. The institution needs the comprehensive documents for giving the information accurately before reviewed by assessor. Therefore, academic documents should be stored effectively to ease fulfilling the requirement of accreditation. However, the data are generally derived from various sources, various types, not structured and dispersed. This paper proposes designing a data warehouse to integrate all various data to prepare a good academic document for accreditation in a university. The data warehouse is built using nine steps that was introduced by Kimball. This method is applied to produce a data warehouse based on the accreditation assessment focusing in academic part. The data warehouse shows that it can analyse the data to prepare the accreditation assessment documents.

1. *Introduction*

The value accreditation of a program study is one of important parameter to classify the quality of university in Indonesia especially private universities. The accreditation is measured based on the evidence assessment related to the specific standards and the expert(s)'s judgement or consideration. In Indonesia, the judgment for accreditation higher education, including university is did by accreditation council national, called BAN-PT. The university should prepare the comprehensive data in a template which is provided by BAN-PT. This data consists academic history, profile lecturers and staff, vision mission university, the facility and so forth. One of the important form is called "borang" which consists 7 forms. In STMIK Pelita Nusantara university, the source data is spread in both systems, manual and information system. Some of them are in document file (.docx) , spread sheet file (.xls) and database system. This fact causes gathering the comprehensive data for "borang" to be a serious problem.

Data warehouse is one of tool to gather all the company data in one place in order to get a better view of a business process and improve the performance of the work/organization. Application of data warehouse are implemented to solve some problems [1-7]. Bassil [1] presented a design model for building data warehouse for an university. The proposed model is based on four stages of data migration: data extraction, data cleansing, data transforming, and data indexing - loading [1]. Sappagh et al proposed a model which is built upon the enhancement of the models in the previous models to support some missing mapping features [2]. Data warehouse and business intelligence have been used in the recent years in many areas for analyzing different phenomena. Rabuzin showed how data warehouse has been used in educational field for analyzing university exams[3]. Kurniawan and Halim proposed a model that can be applied in data warehouse and data mining techniques to predict student performance (academic) in schools [4]. The data mining and data warehouse model can help the low achiever students, evaluate the course or module suitability, and tailor the interventions to increase student academic performance in schools [4]. Jindal and Taneja started developing a data warehouse identifying and gathering requirements, designing in the dimensional model followed by testing and maintenance [5]. Luan and Jiang [6] present a system architecture of library books based on data warehouse which uses the accumulated data of college library as data resources, simulate simple data warehouse and use OLAP technology. Namusonge [7] found out the effect of information technology on warehousing management. The researcher used descriptive research design taking Jomo Kenyatta University of Agriculture and Technology as a case for this study. This study recommended continuing investment and training in information technology and adoption of better information sharing tools.

2. *Related Works*

Data warehouse is a collection of large data which is used by the management for making strategic decisions. The data in a data warehouse is gathered from heterogeneous sources and then populated and queried for carrying out the analysis. The data warehouse design must support the queries for which it is being used for [8]. Characteristic of data warehouse can be defined as subject-oriented, integrated, time-varying, and non-volatile collections of data. Nowadays, data warehouse becomes an important strategy to integrate heterogeneous information sources in organizations, and to enable On-Line Analytic Processing (OLAP) [9]. Data warehouse can be built using a number of architectures [10]. It goes beyond the traditional focus on data mining problems to introduce advanced data types such as text, time series, discrete sequences, spatial data, graph data, and social networks [11]. As data warehouse stores huge amount of data, the security of this huge information is crucial for the sustainability and reliability of data warehouse [12]. The quality of correctness and completeness of an information depends on how well the data model is constructed. One way to get a good data model is by utilizing patterns [13].

The integration of data sources is achieved by using ETL (Extract, Transform, and Load) processes. The ETL process is to identify the main characteristics, notation, and activities of company. A mapping study provides a systematic and objective procedure for identifying the nature and extent of the available research by means of research questions [14]. The ETL extraction step is responsible for extracting data from the source systems. The transformation step tends to make some cleaning and conforming on the incoming data to gain accurate data which is correct, complete, consistent, and unambiguous. This process includes data cleaning, transformation, and integration. It defines the granularity of fact tables, the dimension tables, data warehouse schema (stare or snowflake), derived facts, slowly changing dimensions, and fact less/ fact tables. All transformation rules and the resulting schemas are described in the metadata repository. Loading data to the target multidimensional structure is the final ETL step. In this step, extracted and transformed data is written into the dimensional structures actually accessed by the end users and application systems. Loading step includes both loading dimension tables and loading fact tables.

The source data consists event streams that indicate changes in the business process state i.e. progression of the process. The idea is to automate ETL process using scripting technologies [15]. As Business Intelligence evolves from off-line strategic decision making to on-line operational decision making, the design of the back-end Extract-Transform-Load (ETL) processes is becoming even more complex. Many challenges arise in this new context like their optimization and modelling. Wilkinson et al investigated the disconnection between the IT-level view of the enterprise presented by ETL and the business view of the enterprise required by managers and analysts [16].

3. *Proposed Method*

3.1 Preparing Data Source

There are seven forms (standard1, standard2,...,Standard7) of accreditation council national for university (BAN-PT). However, this research focuses only form standard3 and standard4. It is because most of the rest form standards (1,2,5,6,7) are from the data text. For example standard1 is vision mission, standard2 is the leadership. The standard3 is about students and alumni, while standard4 is about human resource. Table 1 shows the standard3 and standard4 of accreditation council (BAN-PT). Not all of items are treated for data warehouse in this study. This research focuses on items a, c, d, e, f, l and m on standard3, and items d, h, k on standard4. To prepare for data staging, three steps of ETL, extraction-transformation-load, are conducted [18] as described on Section 2.

Table 1 Form Standard 3 and 4 of BAN-PT

Standard Form	Value
Standard3 Students and Alumni	a) The ratio candidate students compared to the capacity of the new students. b) The ratio candidate new students compared to the new graduated student. c) The ratio new students transfer compared to regular students. d) Average GPA of students in the last five years. e) Percentage of timely graduation. f) The percentage of students drop out or resign. g) The percentage of students who have access to student services. h) The quality of service to students. i) Efforts tracking and data recording the alumni. j) Use of tracking results for improvement. k) Opinion of industry about quality of the alumni. l) The average waiting time alumni get jobs. m) The percentage of the similarity job with the field of study for alumni. n) Participation alumni participation of academic and non-academic study programs.
Standard4 Human Resource	a) Written catalogue for the selection system. b) Monitoring system. c) Evidence of monitoring and evaluation of lecturers. d) The number of faculty member lecturers who have at least master degree. e) The number of faculty member lecturers who have been associate professor and professor, f) The number of faculty member lecturers who have the certified educators. g) The ratio of the number of students compared to lecturers who are suitable areas of expertise. h) The average of the lectures load (credits) per semester. i) The similarity the lecture's expertise with the course being taught. j) The attendance of the lecturer. k) The percentage of part-time lecturers compared to the entire lecturers. l) The dissimilarity the part time lecture's expertise with course being taught. m) The attendance part-time lecturers in teaching. n) The activities of the expert. o) Improvement by sending the lecturers study which is similar with field program study. p) The activities of faculty member lectures whose expertise appropriate study program. q) The performance in obtaining grants / awards to institutions. r) Reputation and networking lecturers in academia. s) Librarian with qualifications. t) Laboratory assistant, technician, operator and programmer. u) Administration staff.

3.2 Design of Data Warehouse

The design data warehouse is implemented by using Nine-Step Methodology Kimball, consisting of 9 design stages [19]: Choosing the process, Choosing the Grains, Identifying and Conforming the Dimension, Choosing the Fact, Storing Pre-Calculation in the Fact Table, Rounding

Table, The duration of the Election, Slowly Changing Dimension, The determination of the priority of the query model. There is some star schema designed for this case.

a. Registration for new students, can be shown on Figure 1. This star schema can solve the accreditation standard 3 items (a) and (c)
b. Studying, can be shown on Figure 2. This star schema can solve the accreditation standard 3 items (d), (e), (f), (l) and (m)
c. Lecturer, can be shown on Figure 3. This star schema can solve the accreditation standard4 items (d), (h), and (k)

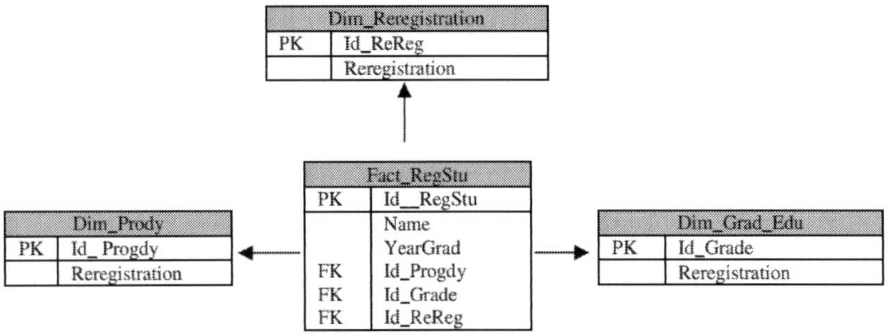

Figure 1. Star schema registration for new students

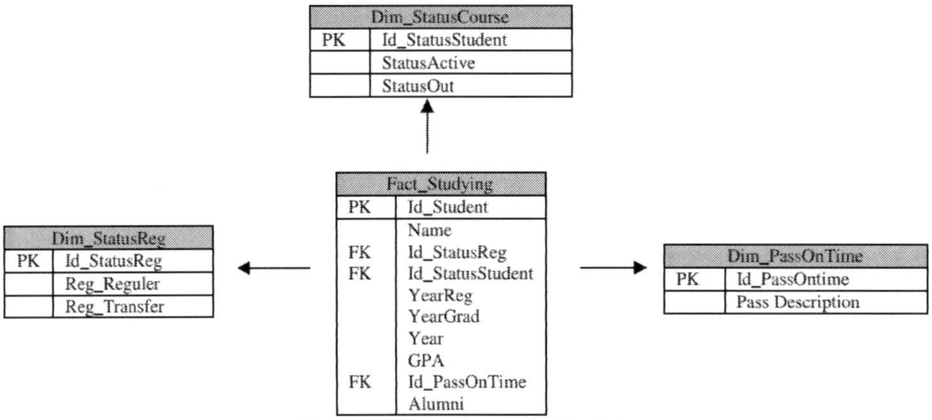

Figure 2 Star schema for Studying

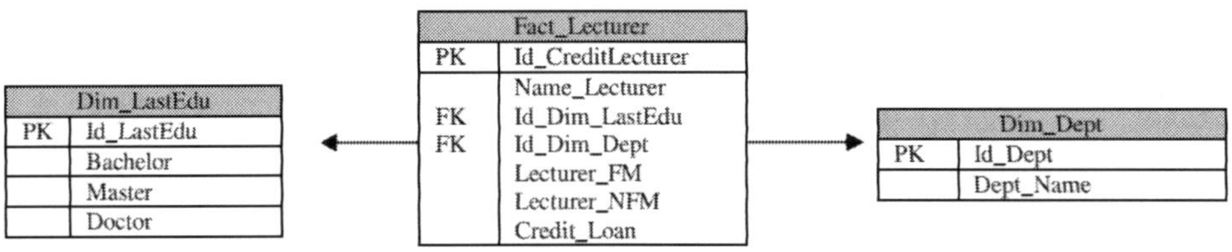

Figure 3 Star Schema Lecturer

To ease building data warehouse, the process ETL of all fact table and its dimension have to be done.

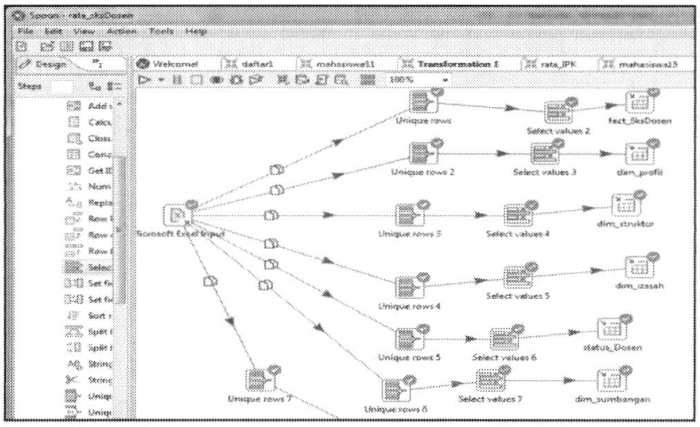

Figure 4 Process of ETL of Fact Registration Student

In this study, the data extraction process is conducted by using a tool called *Pentaho Data Integration / Kettle. Kettle* is a part of *Pentaho* application. The data source are obtained from operational data. Figure 4 shows one of process ETL (Fact Registration Student).

4. *Analysis Result*

This section describes the data which is provided by data warehouse to fulfil the form of BAN-PT. This data is taken 2011-2015. Some of the data are shown as follows. Figure 5 shows Percentage of capacity for new students. This chart can fulfil the requirement standard3 (a). Each academic year conducting new admissions, the capacity of capacity of 2011 = 150 students, the year 2012 = 150 students, 2013 = 300 students, 2014 = 300 students, the year 2011 = 500 students. The ratio of the total number of new students with a capacity that is in 2011 = 50%, in 2012 = 70%, 2013 = 81.0%, 2014 = 97.9%, 2015 = 60.2%.

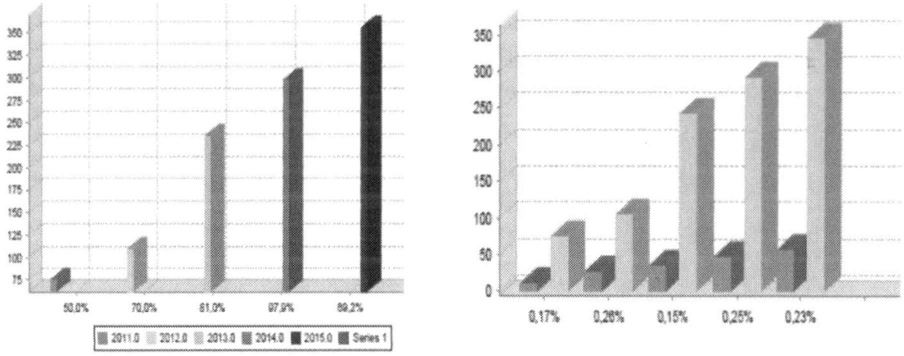

(a) (b)

Figure. 5(a) Capacity percentage candidate students and (b) Ratio percentage transfer students compared to regular students

Each year the university organizes new admissions for some transfer students. Comparison of registered students regularly with transfers students can be seen on Figure 5b. This chart can answer the requirement standard 3(c). Figure 6 (a) shows the graph of the average GPA of students in the last five years (Standard3 (d)). GPA (grade point average) is obtained from the total value of the number divided by the total number of credits of all subjects.

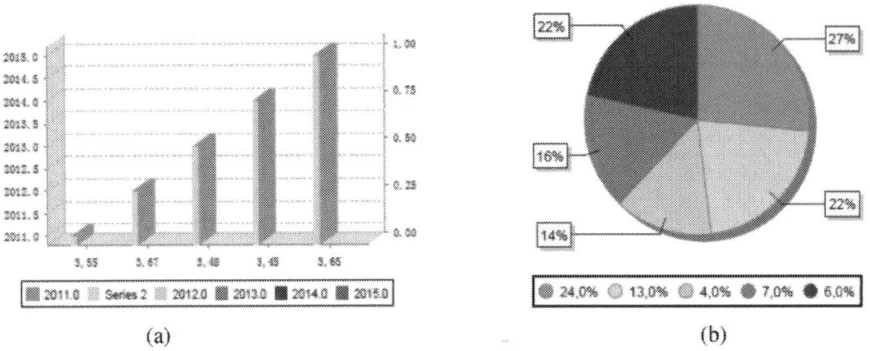

Figure 6 (a) Average GPA, (b) Percentage of drop out students

Figure 6 (b) shows the percentage of students drop out or resign. This chart is useful for Standard3 (f)

5. *Conclusion*

In this paper, we design a data warehouse to produce reports and charts as the representation activity in university to fulfil the requirement of accreditation council-BAN PT. The nine step methodology is used to develop the data warehouse. This data is taken from the data operational of management system on STMIK Pelita Nusantara, Medan. By using the ETL process, the data OLTP can be integrated into data warehouse.

This data warehouse is designed only for fulfil the standard3 and standard4. In future work, the data warehouse can be extended for the other standards. Also, the data which is created by data warehouse can be used for mining research such as prediction, association rule and so forth.

REFERENCES

[1] Bassil Y, "A Data Warehouse Design for A Typical University Information System," J. Comput. Sci. Res., vol. 1, no. 6, pp. 12–17, 2012.
[2] El-Sappagh S H A, Hendawi A M A, and El Bastawissy A H, "A proposed model for data warehouse ETL processes," J. King Saud Univ. - Comput. Inf. Sci., vol. 23, no. 2, pp. 91–104, 2011.
[3] K. Rabuzin K, "The Use of a Data Warehouse For Analyzing Exams at the University," 3rd Int. Conf. Data Min. Intell. Inf. Technol. Appl., pp. 145–148, 2011.
[4] Kurniawan Y, and Halim E, "Use data warehouse and data mining to predict student academic performance in schools: A case study (perspective application and benefits)," in Proceedings of 2013 IEEE International Conference on Teaching, Assessment and Learning for Engineering, TALE 2013, 2013, pp. 98–103.
[5] Jindal R, and Taneja S, "Comparative study of data warehouse design approaches: a survey," Int. J. Database Manag. Syst., vol. 4, no. 1, pp. 33–45, 2012.
[6] Luan X M and Jiang H, "Design of Books Analysis System in University Library Based on Data Warehouse," Appl. Mech. Mater., vol. 513–517, pp. 2121–2124, 2014.
[7] Namusonge K G K, "Role of Information Technology on Warehouse Management in Kenya: A Case Study of Jomo Kenyatta University of Agriculture and Technology," Int. J. Acad. Res. Bus. Soc. Sci., vol. 4, no. 11, pp. 2222–6990, 2014.
[8] Gupta D S L, Pahwa D P, and Mathur M S, "Classification Of Data Warehouse Testing Approaches," International Journal Of Computers & Technology, vol. 3, no. 3a. pp. 381–386, 2012.
[9] Vassiliadis P, "Data Warehouse Modeling and Quality Issues," Natl. Tech., no. June, pp. 1–129, 2000.
[10] Hajmoosaei A, Kashfi M, and Kailasam P, "Comparison plan for data warehouse system architectures," in 3rd International Conference on Data Mining and Intelligent Information Technology Applications (ICMiA), 2011, pp. 290–293.
[11] Aggarwal C C, Data Mining: The Textbook. 2015.
[12] Gosain A and Arora A, "Security issues in data warehouse: A systematic review," in Procedia Computer Science, 2015, vol. 48, no. C, pp. 149–157.
[13] Viqarunnisa P, H Laksmiwati, and Azizah F N, "Generic data model pattern for data warehouse," in

Proceedings of the 2011 International Conference on Electrical Engineering and Informatics, ICEEI, 2011.
[14] Muñoz L, Mazón J N, and J. Trujillo, "ETL Process Modeling Conceptual for Data Warehouses : A Systematic Mapping Study," America (NY)., vol. 9, no. 3, pp. 358–363, 2011.
[15] Radhakrishna V, Sravankiran V, and Ravikiran K, "Automating ETL process with scripting technology," in 3rd Nirma University International Conference on Engineering, NUiCONE 2012, 2012.
[16] Wilkinson K, Simitsis A, Castellanos M, and Dayal U, "for ETL Design," Business, vol. 6412, pp. 15–30, 2010.
[17] D. Warehousing and B. Intelligence, Kimball Group. .
[18] George S, "Inmon vs. Kimball: Which approach is suitable for your data warehouse?," Data Warehous., pp. 1–12, 2012.
[19] Baker P, and Canessa M, "Warehouse design: A structured approach," Eur. J. Oper. Res., vol. 193, no. 2, pp. 425–436, 2009.
[20] S. H. A. El-sappagh, A. Hamed, E. Bastawissy, and A. M. Ahmed, "Original Article A proposed model for data warehouse ETL processes," J. King Saud Univ. - Comput. Inf. Sci., vol. 23, no. 2, pp. 91–104, 2011.
[21] L. Dokumen, "With Pentaho Community Edition."
[22] "Introducing the Pentaho BI Suite 3.5 Community Edition."
[23] Sandoval L J, "Design of business intelligence applications using big data technology," in 2015 IEEE Thirty Fifth Central American and Panama Convention (Concopan XXXV), 2015, no. Concapan Xxxv, pp. 1–6.

IOP Conf. Series: Journal of Physics: Conf. Series **801** (2017) 012031 doi:10.1088/1742-6596/801/1/012031

Prototype Design for Digital Game-Based Learning in Prisoner's Dilemma for International Relations Students of Bina Nusantara University

J V Moniaga; L N R W Kabinawa; S Hambali ; Y Chandra

School of Computer Science and International Relations Department Bina Nusantara University
Jl. Kemanggisan Ilir III No. 45, West Jakarta 11480, Indonesia

Email: jurike@binus.edu ;lratih@binus.edu; sharon001@binus.ac.id; yohanes.chandra@binus.ac.id

Abstract. The traditional method of teaching still lacks some elements that require students to delve and understand the application of subject in various situations. A learning method called Digital Game-Based Learning (DGBL) is used to replace old learning methods with a digital game to learn. This research aims to propose the best method of DGBL design and development to create game prototypes for subject applied by International Relations Department of Bina Nusantara University. This research focuses on developing DGBL for the concept of Prisoner's Dilemma and taking the case study of Trans-Pacific Partnership (TPP) in International Relations Studies. The research uses literature review to back up design theories and survey to determine the basic foundation of prototypes creation. Based on the research surveys, it is concluded that prototypes will apply mobile as the platform, use simulation as the genre, and adopt the style of old American cartoon.

1 Introduction

There are several situations that raise conflicts between individuals and group interest in order to maximize their payoffs; these create a problem of cooperative behavior among individuals that will gradually and eventually affect civilization. The situation is called *Social Dilemma*, and Dawes in [1] specified the first formal definition of Social Dilemma with game-theoretic perspective as a situation where each player has a choice to dominate other players, which will result in a deficient equilibrium.

A study divided Social Dilemma into several types. One type of social dilemma that has been used rapidly on International Relations Studies called Prisoner's Dilemma. Originally framed by Merrill Flood and Melvin Dresher in 1950, Prisoner's Dilemma is defined as a condition where two people are bound to decide in a **non-cooperative way** with consequences that every decision of one person will undoubtedly affect the other person fully. Non-cooperative way mentioned here refers to how each person has to decide unknowingly of the other person's decision. Ohdaira and Terano in [1] explained the possible results of prisoner's dilemma as can be seen in the table below.

Table 1 Payoff Matrix of Prisoner's Dilemma [3]

		Player A	
		Cooperate	**Deflect**
Player B	**Cooperate**	(R,R)	(S,T)
	Deflect	(T,S)	(P,P)

If both players choose to cooperate, they are given the same rewards (R) hence if both players defect; they receive the same punishments (P). It gets a little complicated if player A chooses to cooperate while player B chooses to defect when this is the case player B receives Temptation payoff (T), and player A receives Sucker's payoff (S); this applies vice versa. The basic scoring system is explained as followed: T > R > P > S.

Prisoner's dilemma has always been a tough subject for students of International Relations Department. Thus it requires a medium for students' understanding and subject using technology as alternatives to classroom method. It is the main foundation of this research.

Exploring the advance development of technologies, DGBL thrives over the traditional method of learning. As discussed by Perrotta, *et al* in [3], it offers several ways for leaners to engage fully with topics and also to exercise controls and choices without affecting serious consequences in real world.

This paper will focus on design and development of game prototypes for helping International Relations students focusing in Bina Nusantara University to get a full grip on the concept of prisoner' dilemma, using relevant and happening case study such as TPP.

2 Methods

This research is mainly done in 3 steps: literature reviews, consultation, and needs assessment. After conducting these steps, authors will draw a conclusion, implement the findings in the game design and finally create the prototype of the game based on them.

2.1 Literature Reviews

In this step, authors search other studies related to game-based learning, including serious games, DGBL, and their implementations. Each of the studies is then reviewed to give authors a deeper understanding the topic and how to proceed with the development of the game. After conducting the reviews, two studies were found that can be used in this research. The first one is about DGBL Design Model [6] and the second one is a framework about Educational Game Design [4].

2.1.1 DGBL Design Model

This model incorporates Instructional Design (ID) process to be used explicitly for DGBL development. Instructional Design itself is a process to achieve more efficient, effective and appealing learning experience and environments. One of the most common steps used in ID is the ADDIE (Analysis, Design, Develop, Implementation, and Evaluation) Model [8].Zin & Yue in [6] used the steps of ADDIE to design the DGBL game used in their research. The result can be seen in figure 1 below.

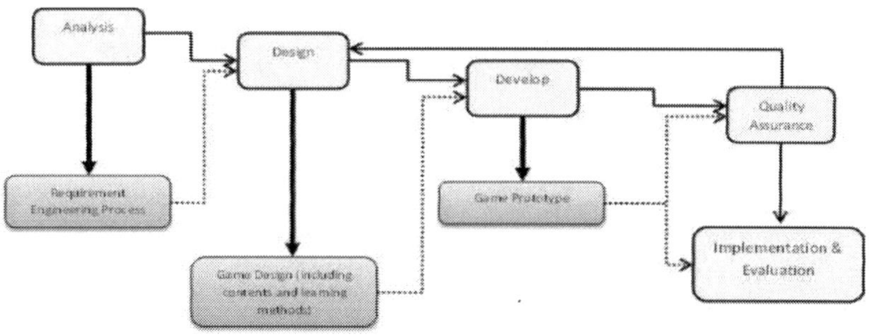

Figure 1 DGBL-ID Model

IOP Conf. Series: Journal of Physics: Conf. Series **801** (2017) 012031 doi:10.1088/1742-6596/801/1/012031

As can be seen in figure 1, DGBL-ID model divides itself into 4 steps. The analysis is where the problems, objectives and other relevant factors of the game including target players and their prior knowledge of the subject are defined. The design is the process of creating the content and the gameplay. Development is integrating the design into the actual program. Quality Assurance is making sure that every factor of the game is up to standard and the game will deliver its educational purposes. Implementation and evaluation are using the game in the actual learning process and measuring its effectiveness as a learning tool.

Zin and Yue used this model in their research [6]. What this research wants to highlight from their research is located on the design step. On the designing phase, Zin & Yue used **DGBL Design model** to design their game. According to it, DGBL software design has four main elements: **interaction**, **knowledge**, **engine** and **level**.

Interaction is how players communicate with the game. It involves the game environment, interface, control, characters and point of view. It is very important to define the game environment clearly, as it refers to the general situation of the game including the atmosphere and the feelings projected while playing the game.

Knowledge is the educational contents of the game. It is what the game wants to deliver to its player. Besides the educational contents, knowledge also includes players' motivation in learning through the game. Instructional models or principles can be used to define this aspect of the game.

DGBL engine is how the knowledge aspect of the game is delivered to the players. It involves what modes the game allows the players to play including the tutorial mode, which gives the player the basic knowledge to manage the game-play. This aspect is related to the next aspect, which is DGBL game level. Game level allows the game to be presented multiple layers to ensure the players' acquisition of the knowledge intended.

2.1.2 Educational Game Design

A framework for designing educational game was created by Aleven, Myers, Easterday and Organ in 2010 [4]. This framework provides many layers at which an educational game operates. That is why this framework is believed to help game designers to create an effective game for educational purposes. The framework is built from 3 essential components: **learning objectives**, **MDA framework,** and **instructional design principles**.

The first component to be defined is the learning objectives of the game. Specifying learning objectives is a crucial first step in designing an educational game as doing so helps designers ensure their game will meet the intended educational goals. Defining learning objectives includes defining the players'' knowledge prior to playing the game and the knowledge gained by playing the game, whether intended or not.

The second component of the framework is called MDA Framework. MDA Framework itself is another framework to help designers and analysts to look at the game in 3 connected layers. These layers are the mechanics, dynamics, and aesthetics. Mechanics are what make up the game system; the rules, the components of actions that can be taken by the players, the conditions for progress and advancement. An example of this is the process of giving points and rewards to players who have made decisions also leaderboards can be used to enhance the gamified system. Mechanics also take care of quests, levels, and etc. Dynamics refer to how the rules develop along with the progress and choices of players, dynamics can be seen by players directly throughout gameplay. Mechanics is simply what the players can do in games when in contact in game mechanics. Aesthetics of games take responsibility of emotional aspects, how the players feel when they interact with the games.

The last component is instructional principles. Instructional principles are research-tested principles that can be used or added into the game. There are many lists of these principles. Designers should research these principles and apply those that align with their objectives.

2.2 Consultation

After collecting relevant studies in the previous step, further discussions about findings are done with fieldwork. Fieldwork consists of consultation activities with the lecturers regarding the concept prisoner' dilemma and case study TPP. This consultation is done to ensure that the design process of the game can indulge all educational aspects and that the game may convey the learning objectives.

2.3 Needs Assessment

Besides consultation with lecturers, a needs assessment is also conducted to gain more data. The assessment was done in the form of a questionnaire given to students who are going to take specific topic on prisoner's dilemma. The total respondents amount to 150 students. This assessment did not aim to be a guideline of the designing process of the game but rather as a mean to find the students' preference to enhance their motivation. Questions asked are mainly about the technicality of the game such as fondness, genre, platform and featured of the game.

3 Result and Discussion

After conducting literature reviews, consultation, and needs assessment, authors decide on a game design to be implemented into an early prototype.

3.1 Game Design

The game will have the players to act as the leader of a country that has not yet become a member of TPP. Authors opt for Indonesia as a designated country. The game will ask the players some questions based on several scenarios and situations regarding Trans-Pacific Partnership and the player will have to choose between cooperating with TPP or not, like in prisoner's dilemma concept. Not cooperating can be "defecting" TPP or maintaining status quo. In order to increase the fun aspect of the game and the players'' engagement, the choosing will be presented as a kind of mini-game.

The design process is based on the educational game design taken from the literature reviews [4].After consultation with lecturers of the field, authors decided to create prototype by considering several components. The first and third components, which are learning objectives and instructional principles, remain unchanged, whereas authors decide to use the elements of DGBL design model as its second component. Details about each component will be written below.

3.1.1 Learning Objectives. Authors have defined the benefit that can be obtained by International Relations students of Binus University from this game-prototype:

- Students are expected to have a better understanding of the concept of Prisoner's Dilemma.
- Students are expected to know the current situation of Trans-Pacific Partnership (TPP).
- Students are expected to understand how to apply Prisoner's Dilemma concept in a case study (TPP) by using visual game.

3.1.2 DGBL Design Elements.

- **Interaction**

Interaction of the game is mainly done with questions. Players will be given questions from scenarios regarding TPP. Players will then analyze the scenario and give the qualified answer. The players will answer the question by playing some kinds of mini-game where their performance will determine their actions to whether cooperate, defect or maintain status quo.

- **Knowledge**

The educational contents of the game will be regarding Indonesia and its relation to the Trans-Pacific Partnership. It will also portray the application of Prisoner's Dilemma concept in a real-life issue like

TPP. Upon playing, players are expected to know more about how TPP will affect Indonesia and Prisoner's Dilemma application in a real life issue.

- **Engine**

The game will be divided into 2 modes. The first one is the presentation. Players will be given the scenario, ending with a question of what they should do. The second mode is the choosing mode. In this mode, players will play some kinds of mini game to determine their action to the previous question,

- **Level**

The questions given will be divided into 4 questions. Each question will examine different aspects of the TPP-Indonesia relations, such as consideration of joining TPP or not through the scenario of investment opportunities, the existence of small and medium enterprises, privatization of state-owned enterprises, and strategic partnership with other countries in the Asia and Pacific region.

3.1.2 *Instructional Design Principles.* Other instructional principles such as engage, explore and evaluate are also added into consideration in the design process. The engage principle is implemented by using the mini-game to answer the question to keep the players interested. Furthermore, the questions will be presented with moving image. This is hoped to keep the players interested. They also have to explore their thinking and understanding of the topic to analyze the question. Lastly, the game will give feedbacks on each choice by the player for them to evaluate their thinking.

3.2 Screenshot of Prototype

Figure 2. Cooperate or Defect Options Screen

The figure above is the screenshot of the game prototype. The screenshot shows the gameplay screen in which the player was presented with one of the case study of Trans Pacific Partnership. As the question text appears, an animation portraying the case will be played. These animations are hoped to enhance players' understanding of the case as well as offering a visual presentation to increase players' enthusiasm in playing the game. After the questions are fully displayed, player will have to choose an option regarding the nation's take on this particular case. Incorporating the concept in prisoner's dilemma, the options presented will be cooperate or defect. Shaking hands means cooperate while the other means defect.

4 Conclusion

As the most commonly applied alternative method to the traditional teaching, this research continues using previous studies as guidelines and comparison. Concept design of the prototype is taken by implementing frameworks found in the literature reviews. The game takes simulation as the genre, where the player acts as the leader of Indonesia and has to choose a course of action by Indonesia in a certain scenario regarding TPP. The scenario is given into four different aspects: foreign investments, strategic partnership with other countries, the existence of state-owned enterprises, and wider market and trade networks with TPP's member countries. Each scenario consists of two options either to cooperate or defect from TPP. These options will result with two different dilemma situations, which direct the students to decide the best options. After the students finish the fourth scenarios, they will be given code number that will be used to record and to analyses their answers. These answers will be examined by the lecturers to assess student's understanding on the application of Prisoner' Dilemma into case studies in International Relations Studies.

References

[1] Dawes R M 1980 *Social Dilemmas*, Annual Review of Psychology ed 31 pp 169-193
[2] Ohdaira T and Terano T 2009 *Cooperation in the Prisoner's Dilemma Game Based on the Second-Best Decision*, Journal of Artificial Societies and Social Simulation, 12(4)
[3] Perrota C, Featherstone G, Aston H and Houghton E 2013 *Game-Based Learning: Latest Evidence and Future Directions.* NFER Research Programme: Innovation in Education
[4] Aleven V, Myers E, Easterday M and Ogan A 2010 *Toward a Framework for the Analysis and Design of Educational Games.* IEEE Conference on Digital Game and Intelligent Toy Enhanced Learning (DIGITEL) pp 69-76
[5] Ruhi U 2015 *Towards a Descriptive Framework for Meaningful Enterprise Gamification.* Technology Innovation Management Review 5(8) pp 5-16
[6] Zin N A and Yue W S 2013 *Design and Evaluation of History Digital Game-Based Learning Software.* Journal of Next Generation Information Technology 4(4)
[7] Sommerville I 2010 *Software Engineering* ed 9, Addison-Wesley
[8] G M Piskurich 2006 *Rapid Instructional Design: Learning ID Fast and Right* ed 2 p 396

IOP Conf. Series: Journal of Physics: Conf. Series **801** (2017) 012032 doi:10.1088/1742-6596/801/1/012032

Assessing Leading ERP-SAP Implementation in Leading Firms in Indonesia

B Syaiful[1], **W Gunawan**[1,2]
[1]Master in Information Systems Management, Bina Nusantara University, Jakarta, Indonesia.
[2]Faculty member, Post Graduate Program.

Abstract. Enterprise resource planning (ERP) enables to bring critical capabilities to an organisation, however, the implementation of such capabilities is often surrounded with problems. The implementing ERP-SAP in Indonesian enterprises are still facing tremendous challenges with the failure rate can reach more than 80% of the cases. The article examines the common problems faced by the consultants whenever they deal with their clients, from the practical perspectives. The article takes the multiple case studies of the leading enterprises in Indonesia, such as: KS (largest steel producer), GEM (large mining producer), and HS (large retailer), with the aim to identify the root of problems of SAP implementation. The outcome of the study is expected to provide the consultants with the guideline to understand the ERP implementation process in their clients and effective solutions to cope with it.

1. Introduction

ERP system has been known as the most popular business management systems, that enables providing real-time benefits and seamless communication for business in large firms (1). An ERP system can integrate information used by accounting, manufacturing, distribution and human resources departments, etc. into a seamless computing system (2). However, due to its success, the ERP implementation still poses major challenges for Indonesian firms, where it has more than 80% failure rates (3).Wong and Tein (4) summarised the critical success factors for ERP projects such as: top management commitment and support; BPR; use project management to manage implementation; change management culture and program; clear goals, focus and scope (business plan and vision); selecting the right team (competence); customization avoidance; project champion; user training and education; effective communication; use of ERP's consultants; vendor package selection; user participation; technical and business knowledge; integration of the system; appropriate management expectation; appropriate business and IT legacy systems; software development, testing and trouble-shooting; vendor partnership; use of vendors' development tools, monitoring and evaluation of performance; management structure, inter-departmental cooperation and communication.

The article takes the multiple case study method to examine the leading ERP software, SAP that are used in the enterprise applications. The article examines the three leading enterprises in Indonesia, as comparison, such as: KS (largest steel producer/national strategic firm), GEM (subsidiary of national conglomerate in energy producer), and HS (subsidiary of largest retailer in Asia region). Those are the leading firms in their fields.

2. Theoretical Foundation

2.1. Factors that affect the ERP implementation

Implementing SAP involves complex process, since it requires very detailed work plans and schedules to ensure a smooth and well executed cross over. It also requires fundamental changes in technology, business process management and job functions, and they will not easily suits with the firm strategy and culture (5). To design an effective ERP system, it needs high-quality data, the notion of data quality should be well understood. Redman (6) states the impact of data quality covers the difficulty to establish and execute strategy, issues of data ownership, harder to align organisation, and distraction for management attention. It also hurts employee morale and increases organisational mistrust. Chung (7) recommends several measures to maximize ERP benefits, such as: (a) the firm should focus more on the quality and scope of the ERP system that matches with the firm's needs. Well defined functions and the right software are mandatory, similar to increasing usefulness of the system; (b) a more realistic schedule and budget should be planned to minimize the negative effects on the quality of the system. This method can satisfy the firm in both progress and quality of the ERP project; (c) choosing qualified consulting partners is required for ERP project success. They can lead the firm in the right direction in term of both progress and quality; (d) internal support is the main determinant of the progress of the ERP project such as: to complete the project on time with initial planned budget, top management support, training and good project planning are required during the ERP implementation. Timbrell (8) in his study found that major issue categories in SAP implementation comprised of: lack of organization-wide knowledge strategy; knowledge required to support and run SAP was not managed effectively; costs are too high compared to the benefits; customization and system integration; SAP system is inadequate or difficult to use; poor management of the implementation project and processes; organizational restructuring affected implementation efforts.

2.2. Data migration Process
Data migration process is the main focus of the ERP implementation. The common SAP implementation phase comprises of (9) (see figure 1): retrieving legacy data; data staging and test environment; loading; and loading in target environment.

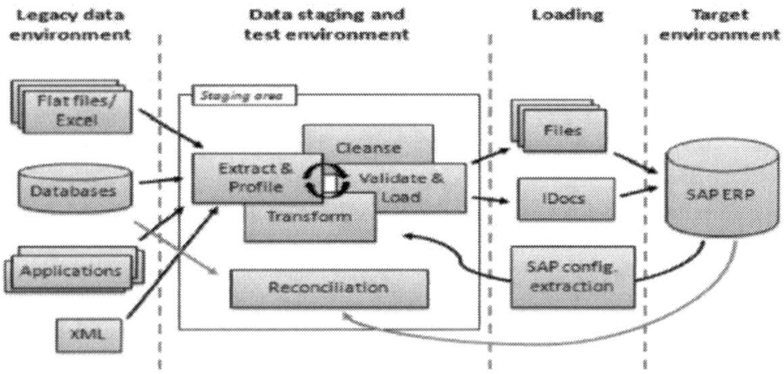

Figure 1. SAP rapid data migration solution architecture.

The cost for data migration projects can summarised into: process, visibility, velocity, and accountability (10). Shatil and Tiwari (11) discovered that more than 80% of data migration projects run over time and/or cover budget, cost overrun average 30%, more than 75% of respondents experienced data management issue late in the project lifecycle. Russom (12) addressed the type of data-movement methods comprised of (13): use ETL-based tools for most data migration solutions; profile source data carefully; expect data migration like all variants of integration to expose data quality issues; use an interactive process to develop a data migration solution; allocate considerable time to modelling the target and mapping data to it; do not overlook stored procedures and other in-database procedural logic; decide what size and content of source data to test with; keep the project's momentum rolling after deployment; expect that redundant instances of platforms will be required; and do not be in a hurry to unplug the legacy platform. The article examines the data migration process based on their capabilities (14): data source, data type, data reliability, interaction with other systems, and user access.

2.3. Knowledge Transfer
ERP systems can take benefit from knowledge management system (KMS) utilisation, including supporting the knowledge transfer in the implementation difficulties and managing system changes (15). Developing ERP systems requires that multiple users and developers to collaborate their implementation efforts, such as keeping a record of those problems, and to make sure the problems are addressed properly and solutions are found. Ko (16) addressed the important factors in knowledge transfer from consultants to clients such as: (a) shared understanding, allows them to effectively transfer knowledge. absorptive capacity and non-arduous relationship to minimize barriers in transfer knowledge. Identifying the individuals with similar work values, norms, and problem-solving approaches to work

with each other, firms can initiate activities.; (b) motivational factor, especially intrinsic values strongly influences knowledge transfer in complex IS implementations; (c) communication factors influence the transfer of ERP knowledge between consultants and clients both directly and indirectly. The ability to express idea clearly and be easily understood by the client determines the successful transfer of ERP-related knowledge; (d) Good business knowledge. consultant with a good knowledge of business and understanding the thinking of business people, as well as general interpersonal skills.

2.3. ERP Testing methods
Gerrard (17) recommends the integrated set of test methods and tools is required to support the ERP implementation as follows: (a) results-based management, modelling of results-chains to provide input to a benefits-based test approach; (b) experience-based risk assessments that can provide input to a risk-based test approach; (c) modelling of integrated user, automated and manual processes and interactions in a format useful to testers; (d) derivation of objective, measurable and manageable test objectives from these three models; (e) to integrate these three test objectives into a non-proprietary life-cycle test methodology that can be managed systematically. Timbrell (8) emphasizes that SAP testing system is one of the major issues that need to be addressed in SAP implementation, since it reflects the availability and reliability of the entire ERP system.

The article takes the common crucial testing methods faced by the SAP consultants in the selected firms such as (18).: functional (check functionality of each desired module), integration (combining multiple systems together to test output of integrated system), regression (testing full application/system for modification), authorisation (identity test) and performance testing (check whether the system meets the performance requirements).

2.4. Interface Integration with Legacy applications
Themistocleous et.al. (19) propose that enterprise architecture integration is summarised at 3 integration layers such as: transportation layer, transfers the information from source application to integration infrastructure and latter to the target application; transformation layer, translates the information from source application format to target system structure; process automation layer, integrates the business processes and controls the integration mechanism (see figure 2)

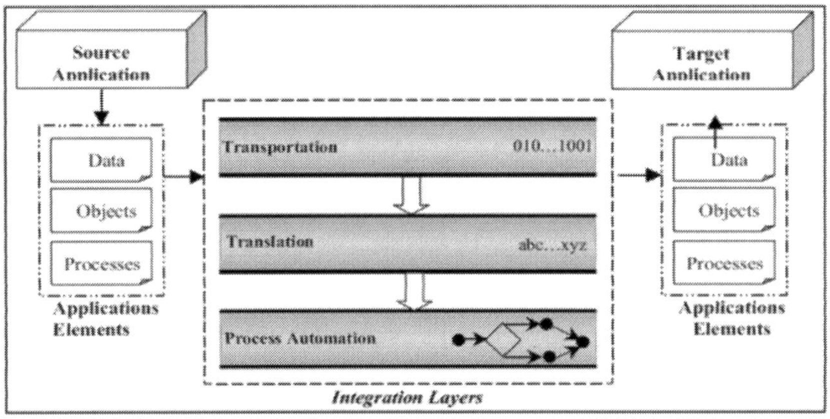

Figure 2. Integration layers in ERP environment.

The article examines the interface integration from the following factors (20): availability technology and support, information system characteristics, system flexibility, ease of maintenance and control access and data.

3. Methodology
3.1. SAP Accelerated Methodology

The three firms in the case study applies the SAP accelerated methodology comprised of several phases such as (21): project preparation; business blueprint (project preparation & refine goals); realization (implement system, test and configure); final preparation (system testing, end user training and fine tuning); Go-Live and support (move from pre-live to a live-operation, including end-user support).

3.2. Research method
The article applies the multiple-case studies for SAP-ERP implementation in the leading enterprises, that enables to examine the differences within and between cases (22) (23). The case study method involves using interviews and focus group meeting with the selected respondents. The objective is to replicate findings across cases with the purpose to forecast the similar results amongst SAP-based enterprises in Indonesia. The article applies the propositions as follows: (a) Proposition 1. Data migration process improves the ERP performance; (b) Proposition 2. Knowledge transfer mechanism improves the ERP performance; (c) Proposition 3. ERP testing methods improves the ERP performance; (d) Proposition 4. Interface integration with legacy methods improves the ERP performance.

3.3. *Data gathering method*
The questionnaires were designed with the Likert scale 1-5, and distributed to 118 respondents in those three enterprises, comprised of: 92 males (81.36%) and 22 Females (18.64%). The interviews and focus group also had been conducted to the selected respondents to explore their thoughts and data findings. The majority of the respondents age were below 30 years, with minimum education background bachelor or above. The distribution of respondents is shown in Table 1 below.

Table 1. Distribution the respondents

No.	Respondents	KS	GEM	HS	Total	percen-tage (%)
1	**Age:**					
	< 30 years	48	19	5	72	61.02%
	31 – 40 yrs.	9	15	5	29	24.57%
	41 – 50 yrs.	8	1	6	15	12.71%
	> 50 yrs.	2	0	0	2	1.70%
2	**Education:**					
	High School	4	0	1	5	4.24%
	Vocational	6	1	4	11	9.32%
	Bachelor	52	28	8	88	74.58%
	Post Graduate	5	6	3	14	11.86%

No.	Respondents	KS	GEM	HS	Total	percen-tage (%)
3	**Working Exp:**					
	< 3 yrs.	32	20	6	58	49.15%
	3 – 5 yrs.	21	12	4	37	31.36%
	6 – 10 yrs.	6	3	4	13	11.02%
	> 10 yrs.	8	0	2	10	8.47%
4	**Title:**					
	Staff (*Analyst*)	37	33	15	85	72.03%
	Consultant	26	0	0	26	22.03%
	ERP PM	2	1	1	4	3.39%
	Head of Department	2	1	0	3	2.54%

4. Case Study
4.1. Overview Enterprise
KS is one of the largest steel producer in Indonesia, and national strategic enterprises. KS has shown a significant progress with the addition of a number of production facilities such as: sponge iron plant, steel billet plant, wire rod plant, as well as the supporting infrastructure including power plant generator, water treatment plant, ports and telecommunication system. KS serves the steel demands both domestically and international markets. In domestic market, KS also serves the defence industry. Since 1995, KS has implemented the SAP R/3 modules such as: FICO (Financial Controlling), SD (Sales and Distribution), HR (Human Resources), PSIM (Project System & Investment Management), MM (Material Management), QM (Quality Management), PM (Plant Maintenance), PP (Production Planning), and XI (Xchange Interface).

GEM is engaged in the business of trading sector of mining products and mining services. GEM was awarded as one of the 50 firms with the best Good Corporate Governance by Indonesian Institute for Corporate Directorship (IICD) for the last 3 years in 2015. Since 2010, GEM has implemented SAP modules such as: FICO (Financial Controlling), MM (Material Management), SD (Sales and Distribution), and PP (Production Planning). GEM and its subsidiaries operate large mining fields in Jambi and Borneo (Kalimantan).

IOP Conf. Series: Journal of Physics: Conf. Series **801** (2017) 012032 doi:10.1088/1742-6596/801/1/012032

HS is a subsidiary of largest retailer in Asia region with more than 700 stores in Indonesia. HS has subsidiaries such as supermarkets, health and beauty stores, convenience stores, and mini markets. HS group receives the 7th IICD Corporate Governance Conference and Award (Top 50 Public Listed Companies), community care firm of the year 2015, Asian version Corporate Excellence and Sustainability award (ACES). Since 2010, HS has implemented SAP modules such as: FI (Financial), ME (Merchandise), ME (Merchandising) and RL (retail).

4.2. Findings and Analysis: Technical Perspectives

After the implementation, the questionnaires were distributed to all respondents with the following results:

Table 2. Results of SAP Implementation

	KS		GEM		HS	
Parameters	**Mean**	**SD**	**Mean**	**SD**	**Mean**	**SD**
1. Data Migration	4.30	5.27	4.15	2.56	4.26	1.98
2. Knowledge transfer	4.29	5.34	4.07	3.48	4.36	2.51
3. ERP testing methods	4.39	5.53	4.22	2.88	4.49	2.14
4. Interface integration	4.25	5.45	4.14	2.66	4.51	1.85

The results above show that all respondents are quite happy with the implementation. The interviews and focus groups were conducted to explore these results, with the representative from staffs, consultants, PM, and head of departments with the findings are shown as follows (see Table 3):

- Data migration process, dominated with data management (24) and technical expertise of project manager (PM) and team. The data management problem (25) was inherited within the organisational immaturity related to ERP implementation. Consultants and PM don't realise the organisation immaturity is the root cause of the communication struggle that affect to creating information quality and technical expertise associated with it (26). The business and IT managers need to understand there is urgency to revising the cultural issues in order to support data migration process.
- Knowledge Transfer, dominated with crucial share understanding between users and consultants, in developing appropriate KMS that supports the application of BPR (27), the need of KMS to identify critical knowledge, manage collection, storage and organisation those knowledge, solution manual facilitates communication, and developing good business/technical knowledge. ERP implementation should be understood from the perspective of the entire organization with its environment, and not just a software installation ERP (28).
- Testing methods, dominated with the need of early introduction BPR change initiative (29) and Test plan. Data findings show that both BPR initiative and Test plan are highly related each other and affects the entire ERP implementation process. Organisational strategies are needed in testing methods and quality assurance to assist the synchronisation organizational needs in ERP systems implementation and also increasing user acceptance (30).
- Interface integration, dominated with BPR efforts, supports of top management (28), and having good of technical knowledge (related to cost effective solutions).

Table 3. Summary Issues related to the Interviews and Focus Group.

No	Critical points in	KS	GEM	HS	Key Factors to highlight
1	Data Migration Process	Different type and data format, has delayed the data mapping (from legacy to new system)	Hierarchy of data approval takes times (bureaucracy of approval)	Large volume of data in legacy system, and it requires separate team to manage it.	Data management is often under-estimated and not understood properly.
		Inadequate current software tools, so it requires additional tools to conduct data transfer.	Frequent data revision supplied by users, and data needs frequent validation.	Data validation needs frequent confirmation from other departments/ business groups.	Expertise of PM is required to ensure the end-to-end data management runs smooth (31)
2	Know-ledge Transfer	KS has defined the ERP-integrated architecture vision, however, there is a gap in developing shared-understanding between senior	Consultant did not develop adequate shared under-standing with the users, and It affects the entire modules implementations.	Consultant and HQ technicians have lack of shared understanding with the local users in developing applications. Any urgent change problem cannot taken care immediately.	Share understanding plays important roles in transferring important knowledge from consultants to users and vice-versa (32).
		Lack of encouragement to use KMS	No KMS is available	KMS is only used for storage purpose, such as: SAP documentation related to change program.	KMS enables to facilitate BPR, along with SAP implementation (31).
		Solution manual is not applied effectively; it is used more as directory of documentation.	Solution manual has not yet been implemented.	Solution manual is used in a workflow of transport change request.	Solution manual enables to facilitate communication between users and consultant.
		There are always unexpected things required by the users or top management (beyond initial plan), i.e. additional dual currency issues (FICO).	Multiple roles of PM. GEM require the task of PM should be ready for all projects associated with SAP applications.	PM should develop adequate knowledge related to retail business. Current PM has high dependency on the knowledge of key user in each department.	Having good business knowledge related to business process and nature of the firm will help the PM completes his/her tasks effectively (31).
		Slow hierarchical approval for fixing the defect applications that prolong the testing phase. It delivers impacts to the next business process.	Unreliable testing facilities such as availability of robust servers delay testing activities, ie. Development server experiences frequent downtime.	Testing server for QA has different data format compared to the ones stored in the production server, so it delays the testing process.	BPR and Test plan should be introduced far earlier. Lack of these preparation delays the entire testing process and creates hidden problems in the future.
		Lack of BPR efforts has been the major challenges in interface integration, i.e. it delayed the MES (legacy apps) and SAP system integration to several years ahead. Initiating the BPR efforts requires relentless efforts.	Firm has complete licenses for all SAP modules, however, the integration with legacy applications still runs slow due to technical expertises. Major legacy applications are left stand-alone.	Cost saving and lack of focus are the major issues in performing interface integration. Staffs need to figure out the most cost effective way to integrate the legacy apps in the SAP environments, and they are required be able to	BPR efforts, supports of top management and Having good technical knowledge (middle-ware technologies) are crucial for interface integration (33).

Table 3. Summary Issues related to the Interviews and Focus Group (continued).

No	Critical points in	KS	GEM	HS	Key Factors to highlight
				run both legacy and SAP systems smoothly.	
3	ERP Testing methods	Slow hierarchical approval for fixing the defect applications prolong the testing phase and delivers impacts to the next phase.	Unreliable testing facilities delays testing activities, ie. Development server experiences frequent downtime.	Testing server (QA) has different data format compared to the ones stored in the production server, so it delays the testing process.	BPR and Test plan should be introduced far earlier. Lack of these preparation creates hidden domino effects (problems) in the implementation (34).
4	Interface Integration	Lack of BPR efforts has been the major challenges in interface integration, it delayed integration efforts (MES-SAP) to several years ahead. Initiating the BPR efforts requires relentless efforts.	Firm has complete licenses for all SAP modules, however, the integration with legacy applications still runs slow due to lack of technical expertise. Major legacy applications are left stand-alone.	Cost saving and lack of focus are the major issues in performing interface integration. Staffs need to figure out the most cost effective way to integrate and manage the legacy apps in the SAP environments.	BPR efforts, supports of top management and having good technical knowledge (middle-ware technologies) (13) are crucial for interface integration (34) (33).

5. Lessons Learned

The successful ERP project depends on clear and change managed objectives (such as BPR initiative and cultural issues), an attention to risk management and the monitoring of data quality (data migration process), schedule (knowledge transfer) and cost-effective solution (interface integration). Successful PM should develop the necessary knowledge that enables to manage the (technical integration) achievements (31), not activities. It infers that project deliverables should be monitored, rather than the completion status of activities. It includes organization preparation (includes top management supports, BPR, developing appropriate KMS tools), user issues (knowledge transfer), change management (data migration and interface integration), as well as testing plan and QA (17).

6. Summary

The multiple case studies have been presented to examine the common problems faced by the consultants related to: data migration process, knowledge transfer, testing methods, and interface integration. The foundation problems beneath these factors such as: introducing the change management, BPR initiative and test plan as early as possible, developing appropriate KMS that enables knowledge transfer, top management supports and developing good business and technical knowledge.

References

[1]. Seo G. Challenges in implementing enterprise resource planning (ERP) system in large organizations: similarities and differences between corporate and university. 2013;(May). Available from: http://dspace.mit.edu/handle/1721.1/80683
[2]. Motwani J, Subramanian R, Gopalakrishna P. Critical factors for successful ERP implementation: Exploratory findings from four case studies. Comput Ind. 2005;56(6):529–44.
[3]. Dantes GR, Hasibuan ZA. Comparison of ERP Implementation Process of Two Organizations: An Exploratory Research. Int Conf Soc Sci Humanit [Internet]. 2011;5:295–9. Available from: http://www.ipedr.com/vol5/no1/63-H00145.pdf
[4]. Wong B (University of TS, Tein D (Prisocroncipal ESM. Critical Success Factors for ERP Implementation. SSRN Electron J. 2013;(June 1999):30–6.
[5]. Warriss G. Data Migration from various Legacy Systems to SAP R/3. 2000.

[6]. Redman TC. The Impact of Poor Data Quality on the Typical Enterprise. Commun ACM. 1998;41(2):79–82.
[7]. Chung BY. An Analysis of Success and Failure Factors for ERP Systems in Engineering and Construction Firms. ProQuest. 2007;
[8]. Timbrell G. A Meta-Study of SAP Financials in the Queensland Government by Doctor of Philosophy. 2006;
[9]. Nyrhilä P. PETTERI NYRHILÄ IMPROVING MASTER DATA QUALITY IN DATA MIGRATION OF ERP IMPLEMENTATION PROJECT. Tampere University of Technology; 2014.
[10]. Geberit I. I @ W Geberit Migration to SAP HANA. In 2015.
[11]. Shatil O, Tiwari S. Legacy Data Migration to SAP. In 2011.
[12]. Russom BP. Best Practices in Data Migration Best Practices in Data Migration Table of Contents. 2006;1–12.
[13]. Alizai F. A model for the implementation of ERP systems in midsize businesses. 2014;(November).
[14]. Isik O. Business intelligence success: An empirical evaluation of the role of bi capabilities and the decision environment. ProQuest Diss Theses. 2010;181.
[15]. O'Leary D. Knowledge management across the enterprise resource planning systems life cycle. Int J Account Inf [Internet]. 2002;3:99–110. Available from: http://www.sciencedirect.com/science/article/pii/S1467089502000386
[16]. Ko D, Kirsch LJ, King WR. Antecedents of Knowledge Transfer From Consultants To Clients in Enterprise System Implementations. MIS Q. 2005;29(1):59–85.
[17]. Gerrard P. Test methods and tools for ERP implementations. Proc - Test Acad Ind Conf Pract Res Tech TAIC PART-Mutation 2007. 2007;40–3.
[18]. Point T. About the Tutorial Disclaimer & Copyright. In 2016.
[19]. Themistocleous M, Irani Z, Love PED. Enterprise Application Integration: An Emerging Technology For Integrating ERP And Supply Chains. Ecis. 2002;(2002):1087–96.
[20]. Vatcharaporn E, Piyanan N. CONNECTIVITY. 2004;165–74.
[21]. Yusuf Y, Gunasekaran A, Abthorpe MS. Enterprise information systems project implementation: A case study of ERP in Rolls-Royce. Int J Prod Econ. 2004;87(3):251–66.
[22]. Baxter P, Jack S. Qualitative Case Study Methodology : Study Design and Implementation for Novice Researchers Qualitative Case Study Methodology : Study Design and Implementation. 2008;13(4):544–59.
[23]. Yin RK. Case study research: Design and methods. Sage publications; 2013.
[24]. Columbus. ERP IN 2014 and BEYOND Special Report Series. 2014.
[25]. Keller P. SAP Test Data Migration Server: Creating High-Quality Test Systems with a Reduced Set of Data. 2009.
[26]. Dravis F. Generations of information quality A practice paper Your Speaker ... Generations of Information Quality. Proc Sixth Int Conf Inf Qual. 2003;12–9.
[27]. Al-fawaz KMS. Investigating Enterprise Resource Planning Adoption and Implementation in Service Sector Organisations by. 2012;
[28]. Hwang W. The Drivers of ERP Implementation and Its Impact on Organizational Capabilities and Performance and Customer Value. 2011;(June):1–301.
[29]. Graham JF. Enterprise Resource Planning Implementation in Higher Education. University of Missouri-Columbia; 2009.
[30]. Kumar V, Maheshwari B, Kumar U. An investigation of critical management issues in ERP implementation: Emperical evidence from Canadian organizations. Technovation. 2003;23(10):793–807.
[31]. Bishnoi V. Thesis on IMPLEMENTATION OF ENTERPRISE RESOURCE PLANNING (ERP) SOFTWARE AND ECONOMIC ANALYSIS OF ITS IMPLEMENTATION IN INDIAN MULTINATIONAL ORGANISATIONS " Submitted for the award of DOCTOR OF PHILOSOPHY Degree in Business Management By Vishal Bishnoi R. Vol. 250110. Shobit Institute of Engineering and Technology; 2011.

[32]. Deng XN. Understanding ERP Requirements Analysis and Design : User Participation , Knowledge Transfer and Shared Mental Models. 2006;
[33]. Almgren K, Bach C. ERP Systems and their Effects on Organizations : A Proposed Scheme for ERP Success. 2014;
[34]. Zach O. ERP System Implementation in Small and Medium- Sized Enterprises [Internet]. 2012. 82 p. Available from: http://brage.bibsys.no/xmlui/handle/11250/136206

IOP Conf. Series: Journal of Physics: Conf. Series **801** (2017) 012033 doi:10.1088/1742-6596/801/1/012033

Scheduling Diet for Diabetes Mellitus Patients using Genetic Algorithm

M F Syahputra[1], V Felicia[1], R F Rahmat[1], and R Budiarto[2]

[1]Department of Information Technology, Faculty of Computer Science and Information Technology, University of Sumatera Utara, Medan, Indonesia

[2]Department of Information System, College of Computer Science and Information Technology, Albaha University, Saudi Arabia

Email: nca.fadly@usu.ac.id, vanesa_felicia92@students.usu.ac.id, romi.fadillah@usu.ac.id, rahmat@bu.edu.sa

Abstract. Diabetes Melitus (DM) is one of metabolic diseases which affects on productivity and lowers the human resources quality. This disease can be controlled by maintaining and regulating balanced and healthy lifestyle especially for daily diet. However, nowadays, there is no system able to help DM patient to get any information of proper diet. Therefore, an approach is required to provide scheduling diet every day in a week with appropriate nutrition for DM patients to help them regulate their daily diet for healing this disease. In this research, we calculate the number of caloric needs using Harris-Benedict equation and propose genetic algorithm for scheduling diet for DM patient. The results show that the greater the number of individuals, the greater the more the possibility of changes in fitness score approaches the best fitness score. Moreover, the greater the created generation, the more the opportunites to obtain best individual with fitness score approaching 0 or equal to 0.

1. Introduction

Scheduling is resources allocation needed to do tasks in certain time to generate operations [1]. It is used to minimalize required time so it can improve productivity. For every Diabetes Mellitus (DM) patient, scheduling diet, like regulating balanced and healthy diet, is required to heal the disease. Diabetes Mellitus is the symptoms caused by elevated levels of blood glucose due to a decrease in progressive insulin secretion motivated by the insulin resistance [2]. Based on Basic Health Research [3], around 1.5-2.3% of population aged > 15 years old in Indonesia who suffers Diabetes Mellitus II. According to World Health Orgnization (WHO), the number of DM patients will increase to 366 million people in 2025. It is caused by lifestyle changes, like unhealthy or irregular dietary habit, and lack of information in consuming food. Scheduling diet and consuming healthy food with appropriate nutrition certainly can prevent this disease from human body. Therefore, a scheduling diet system is required to help DM patients get much information about a proper diet in accordance with the caloric needs of patients in a day based on several parameters, such as age, sex, weight, height, level of activity and allergy.

Many researchers have done research about diet for diabetics, but they did not include complete required parameters for calculation of calorie needs in a day. They classified allowed food for diabetics using various method. In 2010, Phanich, et al. use Self-Organizing Map (SOM) and K-meand Clustering to classify allowed foor for diabetics [4].

The next research in 2011, scheduling diet for cancer patient had been conducted by Husain, et al. using Genetic Algorithm (GA) [5]. They succeded to schedule diet with high accuracy, but they limited the data of food and generated it only per day.

Previously, GA was often used for recommendation research in various kind of problem. Another related research using GA was done by Tias, et al. for determination of healthy food menu. From their testing result, they obtained healthy food menu solutions in generation 100 with the fitness score of 0449 [6].

In 2010, GA was used for solving the optimization of function problem with encoding integer [7]. They concluded that the high number of population is not necessarily going to generate better fitness score. It is caused by the weakness of GA, probabilistic, always done randomly. Scheduling courses for students and lecturer in Department of Information System ITS was also done by Puspaningrum in 2014 using GA [8]. They reached 10 of best fitness score and 275 of generation number.

Another related research about DM and GA was done by Kushardiana in 2013, discussed about the composistion of diet for DM patients. They succeeded to determine the diet, however they only used age parameter for determining the calorie needs in a day.

In this research, we propose genetic algorithm for scheduling diet for Diabetes Mellitus patients. The rest of the paper will be organized as follows: Section II describes the proposed method. Results and discussions are presented in Section III. Section IV provides summary and suggestions for future research.

2. Research methodology

In our proposed method, there are several steps for scheduling diet for diabetics, namely collecting data of food nutrition and patient data; calculation of calorie needs using Harris-Benedict equation; determination of diet type; implementation of genetic algorithm, including initializing population, calculating fitness score, selecting the best individuals, performing crossover, performing mutation to change the missing genes and generate new chromosome. The obtained output of this system is a schedule menu in a week for DM patients. Each of steps will be explained in the next subsection. Fig. 1 shows the general architecture of our proposed method.

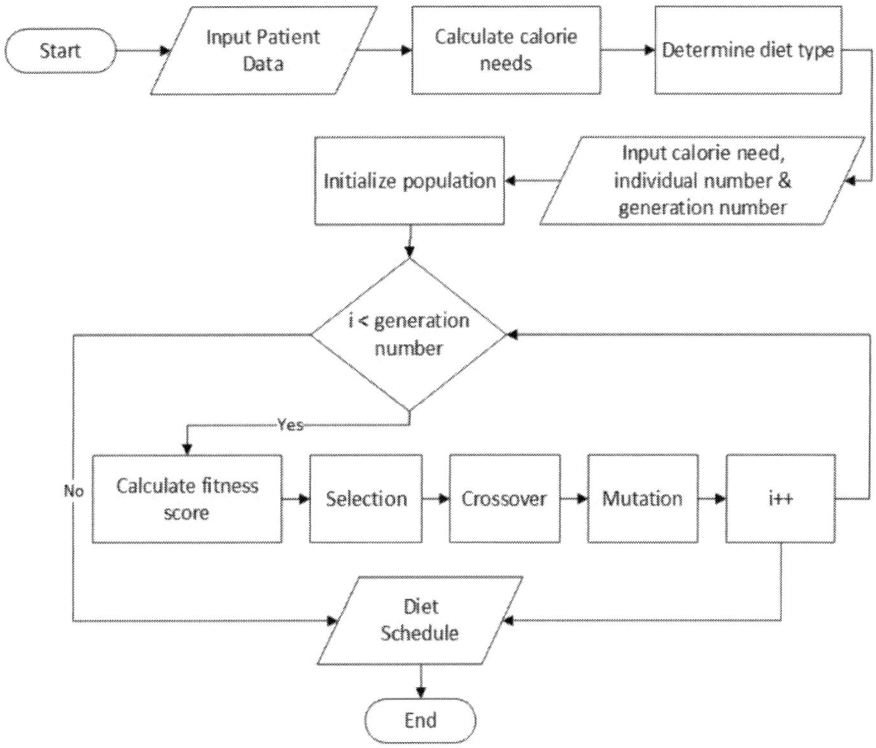

Figure 1. General Architecture

2.1. Data

The used data in this research consists of two data types, namely DM patient data and food nutrition data. The data of patients is used to calculate calorie needs then used to determine the type of diet. The used parameters in this research are age, sex, weight, height, level of activity and allergy. The data of food nutrition is ingredients data in 100 grams obtained from Nutrisurvey Indonesia in 2007. This data is used to generate a menu in accordance with the calorie needs. The used parameters are food name, type of food (staple food, side dishes, vegetables, fruits, and complement), energy (kcal), protein (gr), fat (gr) and carbohydrate (gr).

2.2. Calorie Needs Calculation

Calorie needs can be calculated using Harris-Benedict equation which estimates resting energy expenditure (REE). The calculation of calorie needs is obtained from weight, height, sex, age and level of activity. These following steps are used to calculate the calorie needs.

Step 1 : Calculate Ideal Body Weight (kg) based on rule in table 1.

Table 1. Rule of Ideal Body Weight calculation.

Sex	Height	Formula
Male	< 160 cm	$(Height - 100)\ x\ 1\ kg$
Female	< 150 cm	
Male	≥ 160 cm	$((Height - 100)\ x\ 1\ kg)\ x\ 90\%$
Female	≥ 150 cm	

Step 2 : Calculate Basal Calorie (kcal) based on rule in table 2.

Table 2. Rule of Basal Calorie calculation.

Sex	Formula
Male	$Ideal\ Body\ Weight\ x\ 30\ kcal$
Female	$Ideal\ Body\ Weight\ x\ 25\ kcal$

Step 3 : Calculate the subtraction for activity factor (kcal) based on rule in table 3.

Table 3. Rule of subtraction for the activity calculation.

Category	Activity	Formula
Light	teaching, reading, walking, working in the office, fishing, etc.	$Basal\ Calorie\ x\ 10\%$
Medium	cycling, bowling, gardening, house cleaning, etc.	$Basal\ Calorie\ x\ 30\%$
Hard	aerobic, jogging, dancing, etc.	$Basal\ Calorie\ x\ 40\%$

Step 4 : Calculate the subtraction for age factor (kcal) based on rule in table 4.

Table 4. Rule of subtraction for the age calculation.

Age	Formula
40 - 59	$Basal\ Calorie\ x\ 5\%$
60 - 69	$Basal\ Calorie\ x\ 10\%$
> 70	$Basal\ Calorie\ x\ 20\%$

Step 5 : Calculate calorie needs (kcal) using equation (1)

$$Calorie = Basal\ Calorie - (Activity + Age) \tag{1}$$

Step 6 : Calculate Body Mass Index using equation (2)

$$BMI = \frac{Weight}{Height^2} x\ 100\% \tag{2}$$

Step 7 : Calculate total calorie needs based on Body Mass Index based on rule in table 5.

Table 5. Rule of total calorie needs calculation.

Body Weight Status	Body Mass Index	Formula
Obesity	BMI > 28	$Calorie - 700\ kcal$
Fat	BMI > 25	$Calorie - 500\ kcal$
Normal	18 < BMI < 25	$Calorie$
Thin	BMI < 18	$Calorie + 500\ kcal$

2.3. Diet Type Determination for DM Patient

Diet for DM patients is classified into eight types. Table 6 shows the nutrition ingredients, such as energy, protein, fat, carbohydrate, cholesterol, fiber, sodium and sucrose, for each type [10].

Table 6. Diet type classification.

Diet Type	Energy (kcal)	Protein (g)	Fat (g)	Carbohydrate (g)	Cholesterol (mg)	Fiber (g)	Sodium (mg)	Sucrose (mg)
I	1100	43	30	172	303	37	< 2300	< 5
II	1300	45	35	192	303	37	< 2300	< 5
III	1500	51.5	36.5	235	303	37	< 2300	< 5
IV	1700	55.5	36.5	275	303	37	< 2300	< 5
V	1900	60	48	299	303	37	< 2300	< 5
VI	2100	62	53	319	303	37	< 2300	< 5
VII	2300	73	59	369	303	37	< 2300	< 5
VIII	2500	80	62	396	303	37	< 2300	< 5

2.4. Genetic Algorithm Model

2.4.1. Genetic Algorithm Structure. Genetic algorithm is based on an analogy with the genetic structure, which consists of three main component, namely individuals, chromosome and genes. In this research, there are three chromosomes in each of individual, namely breakfast, lunch and dinner. Each of chromosome has two values, namely fitness score and the total of calorie. One individual consists of 15 genes which contains the total of calorie for each kind of food, namely staple food, side dishes, vegetables, fruits, and complement. The value in each of gene is called by allele. Figure 2, figure 3 and figure 4 sequentially shows the structure of individual, chromosome and genes.

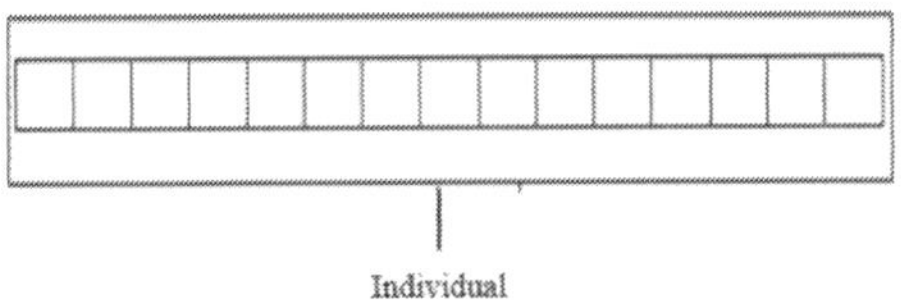

Figure 2. Individual Structure

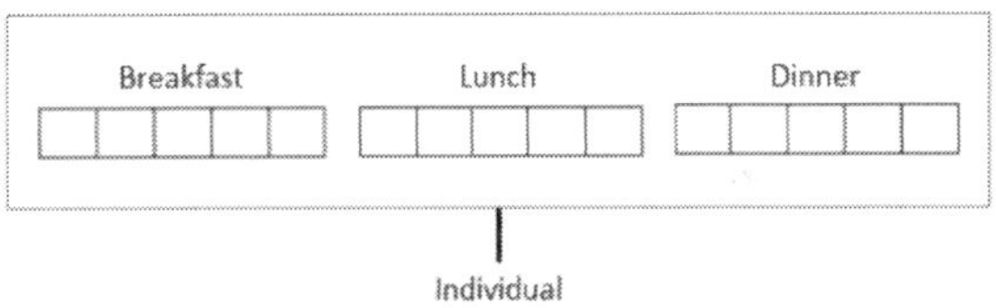

Figure 3. Chromosome Structure

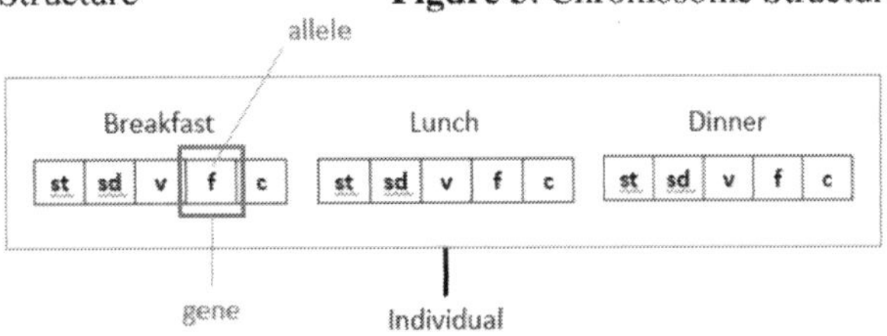

Figure 4. Gen Structure

2.4.2. Initialization. Genetic algorithm always starts from randomly generating the initial population or individuals. The population size depends on the nature problem and types of implemented genetic operators. In this research, the individual is food menu from food database. The greater the size of individual, the more the generated diet variation. Several steps below are done to initialize the individual.

Step 1 : Normalize food menu which aims to adjust the generated diet.

a. Divide the total of calorie needs

Total of calorie needs is divided into 3 chromosomes, namely breakfast, lunch and dinner. It aims to make the diet schedule consider the balance of total calorie for breakfast, lunch and dinner. It is done as formulated in equation (3).

$$Total\ calorie\ per\ chrom = \frac{Total\ of\ calorie\ needs}{3} \quad (3)$$

b. Calculate total calorie for every kind of food
Each of chromosome consists of 5 genes of food type, namely staple food, side dishes, vegetables, fruits, and complement, which respectively has a weight of 45%, 20%, 15%, 15%, and 20%. It can be calculated using equation (4) until equation (8).

$$staple\ food = 45\%\ x\ total\ calorie\ per\ chrom \quad (4)$$
$$side\ dishes = 20\%\ x\ total\ calorie\ per\ chrom \quad (5)$$
$$vegetables = 15\%\ x\ total\ calorie\ per\ chrom \quad (6)$$
$$fruits = 15\%\ x\ total\ calorie\ per\ chrom \quad (7)$$
$$complement = 20\%\ x\ total\ calorie\ per\ chrom \quad (8)$$

c. Calculate each food weight in 100 gram
Each of food in database is expressed in units of 100 grams. In the database, there are a few of menu that has calorie more than needed calorie for the food. To generate the number of calories and weight that corresponds to the total calorie needs of DM patient, we do the calculation as formulated in equation (9).

$$Food\ weight\ (gr) = \frac{Calorie\ needs}{Food\ calorie} x100 \quad (9)$$

Step 2 : Random and generate 15 normalized food menu as an individual.
Step 3 : Input the calorie of food into chromosome. The generated food menu from previous step will be divided into 3 chromosomes, namely breakfast, lunch and dinner, so each of chromosome has 5 genes which contain the calorie of staple food, side dishes, vegetables, fruits, and complement.

2.4.3. Fitness Function Value Evaluation. Fitness score represents the ability of an individual assigned to generated solution. It also can be used as reference in achieving optimal value in genetic algorithm. In this research, GA is used to find an individual with the most minimum fitness score. Fitness score can be calculated using equation (10), while the average of fitness score for each generation can be calculated using equation (11).

$$Fitness\ score = |total\ calorie\ needs - total\ food\ calorie\ in\ a\ day| \quad (10)$$

$$fitness\ average = \frac{\sum_{k=1}^{7}(a_k)}{7} \quad (11)$$

where :
a_k = fitness score for each individual
k = individual in a week (1-7)

2.4.4. Selection. Selection aims to give preference to better individual, allowing them to pass on their genes to the next generation. The goodness of each individual depends on its fitness score. The higher the fitness score of individual, the higher the possibility of selecting the individual. In this research, we use Rank Based Fitness Assignment method to do selection. It sorts the individual based on the fitness score ranged from lowest to highest. It serves to make each individual have the possibility to be selected. The illustration of Rank Based Fitness Assignment method is shown as figure 5.

Individual	Fitness Score
1	90
2	100
3	65
4	40
5	110
6	75

→

Individual	Fitness Score
4	40
3	65
6	75
1	90
2	100
5	110

Figure 5. Rank Based Fitness Assignment Illustration

2.4.5. Crossover. After selecting the individuals, then it will be used to do crossover. We set 0.1 as the percentage of crossover to determine how many parents are selected to do crossover. After that, determination of gen cut point is done randomly using two point crossover method. It almost same as one point crossover method, the difference is that there are two cut points randomly generated in this method [11]. These following steps will describe the two point crossover method.

Step 1 : Input the parent or selected individuals to do crossover
Step 2 : Determine two random number of cut point along the bit number of chromosome in each individual
Step 3 : For each gen in parents chromosome, do :

a. For the initial index is smaller than the final index, do :
   - For genes located before initial index and after the final index, copy gen value from first parent to be given to the next generation (offspring).
   - For genes located between the initial and the final index, genes in offspring will be genes from the second parent.
b. For the initial index is higher than the final index, do :
   - Crossover starts from the initial index to the final index
   - For genes located before initial index and after the final index, copy gen value from first parent to be given to the next generation (offspring).
   - For genes located between the initial and the final index, genes in offspring will be genes from the second parent

2.4.6. Mutation

Mutation, done after doing crossover, aims to change the new offspring by exchange 0 to 1 and 1 to 0. It purposes to maintain diversity within the population and prevent the local optima in search space. In this research, we use random method to implement the mutation. Before doing mutation, the number of genes, which will do mutation, can be calculated based on Probability of Mutation (Pm) and number of gen as formulated in equation (12).

$$Pm\ x\ total\ population\ x\ total\ gen \tag{12}$$

Equation (12) will generate the number of gen which will do mutation, and then its position will be randomly determined. The selected gen will perform mutation with the gen which has the same type in the next chromosome. These following steps will describe the process of mutation.

Step 1 : Determine the number of gen to do mutation
Step 2 : Random a number as gen position from 0 until the number of chromosome.
Step 3 : Perform mutation by replacing the gen value into the value of gen in selected index

a. If the selected index is on chromosome 1 (one) or 2 (two), then the value is replaced with the value of gen that has the same type in the next chromosome.
b. If the selected index is on chromosome 3 (three), then the value is replaced with the value of gen that has the same type in the previous chromosome.

GA will be repeated in accordance with the number of individual and generation. The result from this method is schedule of diet from 7 (seven) better individuals with lower fitness score. The schedule will be displayed in a week (7 days) for breakfast, lunch and dinner menu. Each of schedule consists of five types of food, namely staple food, side dishes, vegetables, fruits, and complement.

3. Result and discussion

The result of this research is schedule of diet for DM patient. The testing is conducted twelve times with different number of individual and the number of generation. The number of individual, generation and fitness score is generated from the patient data, then calculated the calorie needs using Harris-Benedict equations. Table 2 shows the testing result of the proposed method and the correlation among the individual, generation and fitness score is shown as figure 11.

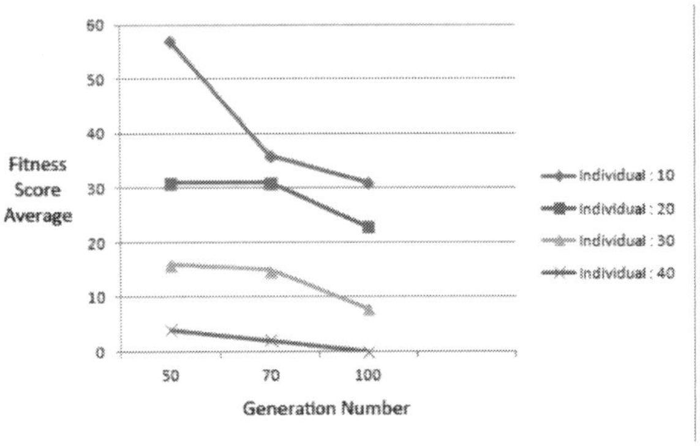

Fig 11. Correlation among individual, generation and fitness score

Table 2. F-Score value after evaluation

Total Individu	Total Generation	Best avg fitness value
	50	57
10	70	36
	100	31
	50	31
20	70	31
	100	23
	50	16
30	70	15
	100	8
	50	4
40	70	2
	100	0

Based on Fig. 11, it can be informed that the number of individual and generation affect on fitness score. The greater the number of individual, the greater the generated individuals, which affects on fitness score. The greater the generated individual, the more the possibility of fitness score changes into approaching best fitness score. It also can be informed that the greater the created generation, the more the possibility to get best individual with fitness score approaching 0 or equal to 0. Best individual has the smallest fitness score.

4. Conclusion

Based on all testing results, we conclude that genetic algorithm can be implemented to find optimal solution for scheduling diet for DM patients and works well. The weakness of genetic algorithm is probabilistic, caused by randomly initializing the individual. The greater the number of individual, the more the possibility of fitness score changes into approaching best fitness score. The greater the created generation, the more the possibility to get best individual with fitness score approaching 0 or equal to 0.

Future work can be done by adding generated food data, so it can be more varied and appropriate. Another research possibly done is scheduling diet for DM patient who has another complication of disease. Adding another parameter to calculate the calorie needs, such as metabolic stress level, is also possibly done.

5. References

[1] Baker K R 1974 Introduction to Sequencing and Scheduling *John Wiley & Sons*

[2] Soegondo S, Soewondo P and Imam S 2009 *Penatalaksanaan Diabetes Melitus Terpadu, Edisi Kedua* (Jakarta : Balai Penerbit FKUI)

[3] Badan Penelitian dan Pengembangan Kesehatan 2007 *Riset Kesehatan Dasar (RISKESDAS 2007)* Kementrian Kesehatan Republik Indonesia Jakarta

[4] Phanich M, Pholkul P and Phimoltares S 2010 Food recommendation system using clustering analysis for diabetic patients *Int. Conf. on Information Science and Applications (ICISA)* pp 1–8

[5] Husain W, Wei, L J, Cheng S L and Zakaria N 2011 Application of data mining techniques in a personalized diet recommendation system for cancer patients *IEEE Colloquium on Humanities, Science and Engineering Research (CHUSER 2011)* pp 239-244

[6] Tias R A, Rahman M A and Dewi C Implementasi algoritma genetika untuk optimasi 0/1 multi-dimensional knapsack problem dalam penentuan menu makanan sehat *Jurnal Mahasiswa PTIIK UB* **1**

[7] Indrianingsih Y 2010 *Algoritma genetika untuk menyelesaikan masalah optimasi fungsi berkendala dengan pengkodean bilangan bulat* Thesis Sekolah Tinggi Teknologi Adisutjipto

[8] Puspaningrum W A, Djunaidy A and Vinarti R A 2014 Penjadwalan mata kuliah menggunakan algoritma genetika di jurusan sistem infiormasi ITS *Jurnal Teknik POMITS* **2** 127-133

[9] Kushardiana R 2013 *Penetuan Komposisi Menu Makanan untuk Penderita Diabetes Mellitus Menggunakan Algoritma Genetika* Thesis Universitas Brawijaya

[10] Almatsier S 2004 *Penuntun Diet* (PT. Gramedia Pustaka Utama: Jakarta)

[11] Kusumadewi S and Purnomo H 2005 Penyelesaian Masalah Optimasi dengan Teknik-Teknik Heuristik (Yogyakarta : Graha Ilmu)

Design of Knowledge Management System for Diabetic Complication Diseases

Cut Fiarni

Department of Information System, Institut Teknologi Harapan Bangsa (ITHB)
Jl Dipati Ukur 80-84 Bandung, 40132. Indonesia

Email : cutfiarni@ithb.ac.id

Abstract. This paper examines how to develop a Model for Knowledge Management System (KMS) for diabetes complication diseases. People with diabetes have a higher risk of developing a series of serious health problems. Each patient has different condition that could lead to different disease and health problem. But, with the right information, patient could have early detection so the health risk could be minimized and avoided. Hence, the objective of this research is to propose a conceptual framework that integrates social network model, Knowledge Management activities, and content based reasoning (CBR) for designing such a diabetes health and complication disease KMS. The framework indicates that the critical knowledge management activities are in the process to find similar case and the index table for algorithm to fit the framework for the social media. With this framework, KMS developers can work with healthcare provider to easily identify the suitable IT associated with the CBR process when developing a diabetes KMS.

1. Introduction

Diabetes is a complex, chronic illness that requires continuous medical care with multifactors risk reduction strategies to control the blood sugar level. Diabetes could increase the risk of developing a series of serious health problems. This happens because the high level of glucose could lead to serious diseases affecting the heart and blood vessels, eyes, kidneys, nerves and teeth. In addition, people with diabetes also have a higher risk of developing infections. People with diabetes need self-management education and they also need support for preventing and reducing the risk of other complications diseases. People with diabetes should receive medical care from a collaborative, integrated team with expertise in diabetes and also a support group from family, friends and other people that diagnose with diabetes. On the other hand, it is well known that every medical treatment has risks as well as benefits. The medical ethical principle state that a patient has the right to decide what's appropriate treatments for them, taking into account their personal circumstances, lifestyle, beliefs, and priorities. Moreover, to choose the right medical treatment is a long and confusing process.Especially if one of the options is whether to take surgery or other high risk and expensive treatments. Informed decision-making is part of process to give a relevant information regarding to the medical treatments, by making collaborative communication between patient, family and one or more medical practitioners. But sometime, this information still not enough, patient also has the need for finding and get information from another patient with similar case to make the most suitable decision. The better information and knowledge to other people with a similar case, lifestyle and treatments history, personal circumstances and lifestyle are crucial in making the decision [2].

In this digital era that we live in, its natural to gain information and knowledge regarding the disease and its medical treatment through the web and search engine. But the problem is because of the vast and complex knowledge and the overload information on the web. Especially, the information regarding diabetes risk health, complication disease and healthy lifestyle, not yet well exist, manage, share and informed. Moreover, the information regarding diabetes complication disease and treatments, is scattered with unstructured format, it makes the decision process becomes more complex and confusing.

We address this issue in this paper by presenting an integrated framework for the diabetes knowledge management system with social networks. In this paper the analysis focused on designing a model of knowledge and the system for mapping, gathering, acquisitions, filtering information and share the knowledge of diabetes complication and treatments from social media.

2. Related Work

Knowledge management systems refer to a class of information systems applied to manage and facilitate knowledge sharing among the system's user effectively and efficiently. There are seven general techniques for knowledge management: knowledge management framework, knowledge based systems, data mining, information and communication technology, artificial intelligence, database technology and modeling, along with their implementation for different problem domains [3]. The problem is, medical and health knowledge highly related to experience and each patient has its unique condition that lead to unique treatment. There are already several research on medical and health knowledge management system. Koutkias et al designed a data acquisition mechanism for multi agent chronic disease management system to enhance homecare health service [4]. Pham et al presented research on how medical knowledge management includes patients' information gathering, medical expert note and experiences, and tests result could be a used as critical component is diagnosis and treatment planning [5]. There also research to design the framework a knowledge management system to establish key relationships between Western Medicine and Complementary and Alternative Medicine in order to find an effective exploration of ideas and possible solutions of medical diagnosis [6]. Chen et al using machine learning techniques and propose an "Eat, Trust, and Correct" (ETC) framework to find pattern of the behavior of the diabetes patients who use insulin pumps [7]. Klenk et al presented a method to calculate similarities of patient profiles for recommending people to other members in social networks in [8]. But there is still no similar research regarding a knowledge management system of complication diseases and treatments related to diabetes condition.

Therefore, the objective of this research is to propose a conceptual framework for designing a knowledge management system of diabetes complication diseases and treatments. This proposed system will integrate the relevant enabling information and communication technology into the knowledge management process that would support informed medical decisions for diabetes patient. Parallel with the framework for a knowledge management system to organize, store, and manage the abundant data available for diabetes management, one of important output of the proposed system is the potential to give recommendations of the most suitable treatment for the patient based on their living condition and lifestyle. This knowledge gain by manage and find other patients with similar case by using a profile matching technique on the proposed system.

3. Research Approach

Given the growing prevalence of diabetic patient, especially in Indonesia, which is the seventh largest number of diabetic patients in the world. It is including both diabetes type 1 and 2 in individuals aged between 20–79 years [1]. The establish knowledge regarding comprehensive diabetes risk factors and treatment management could significantly reduce diabetes complication diseases and could also improve patient quality of life. We developed a framework of a knowledge acquisition model from social media, regarding diabetes complication diseases and treatments. This knowledge management system (KMS) will build based on diabetes management plan that would also consider the patient profile, symptoms, diagnosis and recommended treatment option.

This research will focus on how to design a model of KMS that could give an informed medical option for diabetic patients. The idea is to help the diabetic patient to get, organize, and share knowledge related to factors that could worsen their condition based on each patient profile. In general, the process to decide the most suitable and correct medical decision is not easy. It is because each patient has different life condition and lifestyle that could influence on how they make decisions and the outcome of treatment they took [2]. The proposed system also designed to give a better communication and sharing information among diabetes patients and between their medical experts, and also to help patients to get recommended medical care based on their preference and similar condition as the form of collaborative systems. In this research, we adopt a data mining technique to find a similar pattern of the diabetes patient condition and treatment, to be implemented as algorithm of the proposed system.Our proposed design of the knowledge acquisition model in diabetes medical treatments is to assist the patient in acquiring needed information and knowledge to facilitate the process of decision making. To achieve those goals, the proposed system should have these following characteristics:

- Build a diabetes social network to provide collaborative ecosystem for people with diabetes and the medical expert to exchange thoughts, experiences and share others related medical information
- Store patient profile, complication disease and treatment they have taken, and other trivial information regarding the outcome of treatment and extract data mart from those data
- Have privacy filter, so the patient could hide their identity on other private matter.
- Have a knowledge management system, by combining data from user profile and diabetes information and knowledge.
- Allow patient to find solution to their problem by obtaining another patient(s) with similar case and conditions and weighing of personal value and similarity measurement.
- Inform patient of KMS limitation and mission as educational tools, clearly disclaiming the proposed system as substitutes for medical specialist diagnosis and care .

All sixth characters of the proposed system as stated above are important to develop a diabetes complication diseases and treatment. The important characteristic of this proposed system is their ability to proposed solution on user problems, based on similar case. For instance, the fact that two diabetes patients have a different medical treatment recommendation, because they have different age, lifestyles and other medical complexity problem. This leads to a profile matching and provides better recommendations on who might have had similar experiences or who might have knowledge other patients can get benefit from. Funding relationship and extracted the knowledge is the very basis of this proposed system.

For that proposed, we adopt the Case Based Reasoning (CBR) method in this research. The CBR method is inspired by human reasoning, i.e., solving a new problem by applying previous experiences adapted to the current situation. A case (an episodic experience) normally contains a problem, a solution, and its result [9]. Today, CBR is a recognized and well-established method for research in health sciences. Aamodt and Plaza have outlined a life cycle of CBR with four main steps (retrieve, reuse, revise, and retain). There are four steps in the CBR step. first is retrieved, in this step a new problem is matched against the previous cases in the case library. The next step has reused previous case by determining how similar it is to the current case, the important process in this step is the similarity measurement that could estimate of how suitable the previous solution is for the current problem. The most relevant solutions are proposed to solve the current problem (after some adaptations if necessary). The selected solution is revised before it is reused. Then, the new problem and its solution are retained in the case library for any future use.

4. Architecture of The Proposed System

In this section we would discussed the architecture of the proposed system. Fig 2 illustrated the schematic model of proposed system. As illustrated on Fig.2, there are three main phase of this proposed system: data acquisition from social network modeling, data organization and knowledge creation.

4.1. Data acquisition and modelling.

In this phase, the system will collect two type information which are user profile and complication diseases. User profile consists data of demographic, medical condition, diagnosis and lifestyle, medical treatments (drugs, therapies and surgical options), healthcare provider and outcome of the treatment. Diabetes complication diseases consist data and information from explicit resources regarding diabetes management and list of alternative medical treatments, a list of the treatment outcome with its degree of success, a list of health care provider histories. Information regarding diabetes complication such as retinopathy, cardiovascular, gastrointestinal, foot care management, etc. and their risk factors also recorded in this phase. The idea is to give relevant information and knowledge to diabetes patient and their health care proxy in order to reduce risk factors and increase patient quality of life.

4.2. Data organization, storage and transfer and information distribution.

This phase divided into three stages, which are feature extraction, similarity measurement, profile matching for the diabetes patient. In the first stage, because of the various and important attribute from the user profile, we need to reduce the dimensionality attribute in order to group the similar diabetes complication and treatment case.

Dimensionality reduction is one of the most crucial stages of modeling data and it will also remove irrelevant and unrelated attribute. Both Feature extraction and feature selection are capable of improving learning performance, lowering computational complexity, building better models, and built an efficient storage. In this research, we used feature extraction and selection to select demographic from a user profile dataset of and then classifying them to the respective class of similar medical condition and lifestyle. So dimensionality reduction is done by selecting features that are capable of discriminating attributes (token) that belong to different classes. And then the next stage is similarity measurement. The proximity measures of these objects based on it weight of different proportion of data to find similarity are calculated. The Similarity Measurement (SM) coefficient is useful because attributes from dataset on has both positive and negative values with equal degree of information. For example, age and treatment outcome are a symmetrical attributes because the number of males and female, or satisfied and dissatisfied provides an equal degree of information. Then in the last stage is profile matching of the diabetes patient in order to gain and sharing knowledge on the personalize level. Based on the previous step, convergence of patient profile, their treatments and diabetes information through profile matching to classify the similar case. These classified diabetes treatment alternative option added in extended metadata in the metadata repository. d) Proposed alternative solution. This part, determines the need for diabetes information and provides the most suitable alternative solution of the current case by finding it similarity from the profile matching algorithm. Then we build the integration and combination of alternative solutions that returned by queries on the data storage, forming set of list that gives the user a ranked list of alternative treatments.

4.3. Learning and knowledge creation.

In this phase, we built model for diabetes knowledge creation from previous organizing and storing dataset. The idea is to adopt CBR concept, so it could shown relationship between dataset properly. By adopting CBR, the initial data can be processed into meaningful information for learning and knowledge creation. This following section will explained in detail how data CBR and profile matching integrated to gather the informed medical option with its degree of satisfied outcome. The first function of this proposed knowledge management system is to build a social network of diabetes patient, with two general approaches:

- Content based KMS which uses the information and knowledge the user input into the proposed social network application.
- Relationship between information traces to find another diabetes patient with similar cases and conditions. The proximity measures of these objects based on its weight of different proportion of data to find similarity are calculated. To find the similarity we will used profile matching coefficient as state on table 1. Then, to compute different matching we used allocation scheme we used this equation [10].

$$\bullet\ \mathbf{SM} = \frac{a+d}{a+b+c+d} \qquad (1)$$

The proposed system is characterized by working with cases defined by the characteristics of the diabetes complication diseases and treatment outcome. Each case is divided into two main components, which are a set of attributes referred to as a target, which contains the definition of the problem and a set of attributes associated to the previous user interactions. Once the definition of the problem is formed in terms of attributes, the objective of CBR is to generate the ranking of these treatment outcome in response to user characteristics that are reflected in the characteristics of user demographic available.

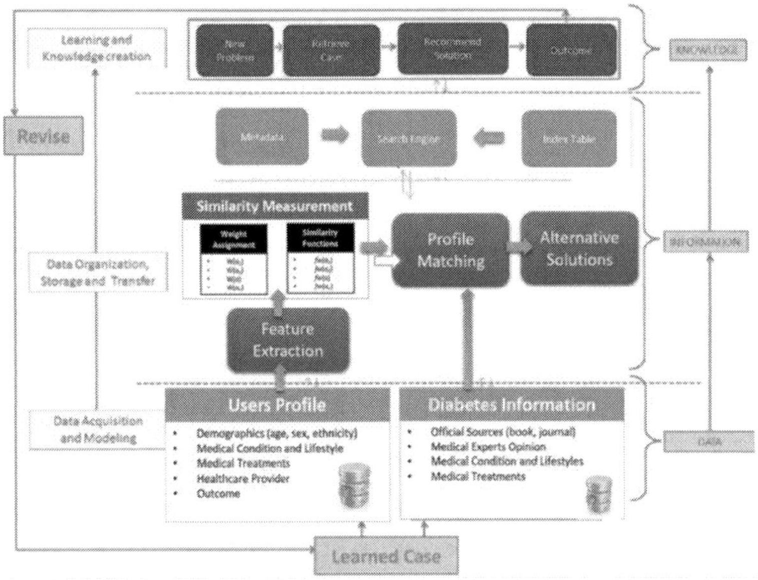

Fig. 1. Schematic Model of Diabetes Complication Diseases and Treatments Knowledge Management System.

The proposed KMS is initiated by a new request made by the user who is searching for medical advised related to diabetes health condition. At that moment, the CBR system is executed. The information contained in the new case at the beginning of the execution cycle of the CBR system is defined by the following tuple:

$$c = \{ \mathrm{T}, u_i, x_i \} \qquad (2)$$

Where T refers to the set of attributes defined in the target extracted mainly from the information in the markup language in accordance with standard tagged used i.e. T = { age, sex, blood sugar level, years since diagnosed with diabetes,... }. The user identifier is ui and xi is the value associated with the proposed alterative solution. Using the information defined by equation 2, the reasoning cycle for the CBR system is initiated. This process illustrated on the top area of Figure 1. During the retrieve phase the metadata for the similar case are downloaded from data storage. Profile matching algorithm applied during the reuse phase in order to predict the degree of similarity of the recovered medical case. Finally, during the revise and learning phase, information related to the user's final decision on alternative treatment and its outcome is stored.

Table 1. Allocation Scheme for Matching Coefficients [2]

		Object 1	
		Number of variables with categories treatment option 1	Number of variables with categories treatment option n
Object 2	Number of variables with categories treatment option 1	a	c
	Number of variables with categories treatment option n	b	d

5. Conclusion

Before taking any medical treatments, patient and their medical proxy facing the biggest task, which is to decide the best treatment they would take. It not only to facilitate their success rate and risks, but also their medical and lifestyle condition. Unfortunately, they often make the decisions without completely understanding their options that could lead to choose the wrong treatments. to make decision, patient also need advice not just from medical expert but also from other people who already facing the same dilemma. There are numerous medical and health information available on the internet. But the information is scattered, overload and most of the time not related to the medical advice that needed, so instead of getting information, patient get confused and take more time in deciding the best treatment that could lead to health condition worsening.

We have presented a model of diabetes knowledge management system that could calculate similarity of patient profiles for recommending the most suitable treatment by using user data. The real power of this proposed system is to provide high-quality information to ease sharing information between patient and physician in order to make a better informed medical decision through a web based system. In order to adopt integrating profile matching and CBR cycle on the proposed KMS, the acquisition and organizing data process use to solve the problem of binary class (satisfied or unsatisfied outcome of treatment). . The result of our work offer many perspectives or further research at both the theoretical and practical level by developing models for KMS that still need detailed algorithm and social network model for medical information sharing. This study will help diabetes patients to have a better understanding of their possibility of complication diseases, its treatments option and outcome, that could lead to improvements in their medical treatment decision making. It also to identify those trigger factors and taking appropriate medical action at the right time. The future work of this research is to build the prototype system and generated data for the knowledge acquisition, storage, manage and sharing.

References

[1] Soewondo, Pradana, Ferrario, Alessandra and Tahapary, Dicky Levenus (2013) Challenges in diabetes management in Indonesia: a literature review.Globalization and health, 9 (1). p.63.

[2] Fiarni C., "Design of Knowledge Acquisition Model in Glaucoma Medical Treatments Recommender System," International Seminar on Industrial Engineering and Management, 2015.

[3] Liao. S. "Review: Knowledge Management Technologies and

[4] Application-Literature Review from 1998 to 2002," Expert System with Application Vol 25, pp 155-164. 2003

[5] Koutkias V., Chouvarda I., Maglaveras. N. "A Multiagent System Enhancing Homecare Health Services for Chronic Disease Management," IEEE Transactions on Information Technology in Biomedicine. Vol 9, No 4. 2005.

[6] Pham D.L., Xu C., Prince. J.L. "Current Methods in Medical Image Segmentation". Annual Rev. Biomedica Engineering. Vol 2 pp 315-337. 2000.

[7] Herrera-Hernandez. M.C. "Engineering of a Knowledge Management Systemfor Relational Medical Diagnosis," Theses. University of South Florida, 2012.

[8] Chen S., Feng L., Rickels M.R., Peleckis A., Sokolsky. O. "A Datadriven behavior modeling and analysis framework for diabetic patients on insulin pumps". Departmental paper (CIS). University of Pennsylvania ScholarryCommons. 10-2015

[9] Klenk S., Fritz P. and Ertl. T. "Web data and data driven medical decision support" . Intelligent Systems Group, Stuttgart university, Germany.

[10] Begum S., Ahmed M.U., Funk P., Xiong N. N, Folke. M. "Case-based reasoning system in health sciences : a survey of recent trends and developments," IEEE Transaction on system, man and cybernetics-part C: Application and Reviews, Vol 41, No 4, July 2011

Understanding social context on TB cases

Y Ariyanto and D M Wati

Department of Epidemiology, Biostatistics, and Population Study, Faculty of Public Health, University of Jember, Kalimantan Rd. 37 Jember, East Java, Indonesia 68121
yunus.ariyanto.fkm@unej.ac.id

Abstract. Tuberculosis (TB) nowadays still becomes one of the world's deadliest communicable disease. More than half were in South-East Asia and Western Pacific Regions, including Indonesia. As developing country, Indonesia remains classic problems in overcoming TB, that is discontinuation on treatment. Most of discontinuation on treatment among TB patients are affected by diagnostic delay that caused by patient delay. These phenomena occur in many areas, rural to suburb, coastal to plantation, and so on, and they are related with social context among community that could be social capital for each community to deal with TB. Jember as one of county in East Java is known as plantation area. It also has a high prevalence of TB. This study focused on understanding about social context among community, especially on plantation area. This cross-sectional study involved in three districts of Jember, those are Tanggul, Pakusari, and Kalisat. The data were obtained directly from the TB patients, local community, and Primary Health Care (PHC) where the patients recorded. Spatial analysis and social network analysis (SNA) were applied to obtain health seeking behavior pattern among the TB patients coincide the community. Most of TB patients had already chosen health professionals to lead the treatment, although some of them remained to choose self-medication. Meanwhile, SNA showed that religious leader was considered as main part of countermeasures of TB. But they didn't ever become central figures. So it can be concluded that there are other parts among community who can contribute due to combatting on TB.

Keywords: social context, social network analysis, spatial.

1. INTRODUCTION

Tuberculosis (TB) is an infectious disease. Early in Europe TB is known as "consumption" and is considered a hereditary disease. Then once known etiological agent, the agent of tuberculosis known by a variety of new term, Koch bacillus (due in 1882 discovered by bacteriologists Germany, Robert Koch), *bacillus tuberculosis* (*tubercle Bacille*), and the last *Mycobacterium tuberculosis*. These bacteria easily mutate, spread from person to person, most often through inhalation of aerosols and droplets of sputum that is emitted from a cough. In the majority of cases, infection causes lesions in the respiratory tract, although it does not show any symptoms of these evidences by immune reactions (eg skin test / Mantoux)(1).

Indonesia also appears to be a serious problem. World TB report by the World Health Organization (WHO) in 2006, placing Indonesia as the third-largest contributor of TB in the world after India and China with approximately 539,000 new cases and approximately 101,000 deaths per year. This position is still persisting in 2007 so as to make the region as a region with the largest TB burden (55%). Data derived from the Basic Health Research in 2007 for pattern cause of death in all age puts TB as a second cause of death by 7.5% after a stroke (15.4%). And became the leading cause of death (27.8%) in the group of death from infectious diseases [1] [2].

TB is always linked to the social context. It cannot be resolved on the side of health care, but also requires empowering people to participate in combatting TB. TB patients have a tendency to consider the spatial aspects in the search for treatment [3]. The tendency of late patients found to be caused away from health services. Reviews Recent medical literature has begun to note the importance of social networks on health outcomes [4].

Jember is one of district in East Java that has 329,333.94 Ha wide area and more than 11% area are plantation area. There are so many inhabitants that drape their livelihood to this field. These areas are plateau with a high rainfall and humidity, which is not only suitable for plantation, but also for spreading of TB. TB is also known as one of infectious diseases that many inhabitants suffered from it [5].

The classic problem of combatting on TB, including on Jember, is the highest drop out of TB treatment and also the highest of diagnostic delay which is caused by patient delay. The highest of diagnostic delay could lead to increasing of TB transmission among inhabitants. The endemics of TB that followed by the increasing of MDR-TB cases in Jember, had made it become urgent to be researched in order to identify in combatting TB.

As stated before, combatting TB doesn't only recline on medical treatment that provided by health care provider, but also community to get involved in it. If combatting TB efforts didn't do comprehensively by involving the community, the transmission of TB would become uncontrollable, which in turn could reduce the quality of life of TB patients. This could result in poverty. Therefore, this research aimed to identify the social capital among community, especially on plantation area, in combatting TB.

2. METHOD

A cross-sectional study was conducted at the three sub districts of Jember, i.e. Tanggul, Pakusari, and Kalisat, that known as areas with high prevalence of TB. There were 38 TB patients that invited to participate in this research on April 2015. They were chosen randomly based on the PHC's records.

We defined TB patients as persons who had been diagnosed with TB by PHC's TB officer. They were asked about preferences of information resources, preferences of supporting treatment resources, as well as role model that they embrace. The questionnaire also covered the position, geographical position, of each TB patients and also the supporting treatment resources. All of the geographical positions were analyzed spatially using Social Network Analysis (SNA) to get seeking treatment behavior pattern of TB patients and supporting treatment pattern among neighborhood.

3. RESULTS

A total of 38 TB patients were invited to participate in this research. All of them are completing in answering the questionnaire. Demographic and socio-economic characteristics are summarized in Table 1 and 2.

Table 1 shows that TB is more on male than female. It may be possible because male have more activities outside the home than female. Therefore, male is more at risk on TB. Furthermore, most of TB patients have less education. Less than 25% of them have a higher educational level. It means that the capability of them in adopting the information, including information of TB, are very low. They couldn't adopt about what TB is, how TB transmitted, and how TB could be prevented. With less information they have, they even couldn't have any capability in managing their environment as one of efforts in preventing TB.

Table 1. Socio-demographic characteristics of TB patients

Characteristics	n	%
Sex		
Male	26	68.0
Female	12	32.0
Formal education level		
No education	14	36.8
Elementary	12	31.6
Junior high school	4	10.5
Senior high school	6	15.8
Under graduate	2	5.3
Occupation		
No occupation	4	10.5
Housewife	2	5.3
Scavengers	2	5.3
Farm workers	4	10.5
Farmer	11	28.9
Self-employment	9	23.9
Pensionary	3	7.9
State officer	1	2.6
Others	2	5.3

Less education attainment on TB patients can cause they have no choices in taking the desired job. Most of them worked informally, such as farm workers, scavengers, etc. Therefore, they have less income with more expenditure. That's why they have more debt insured while have no savings at the same time. By looking at the distribution of income, expenditure, savings, and debt insured among TB patients, it is known that its composition isn't equal. Most of them have debt insured so it can diminish the expenditure of TB treatment. Although the cost of TB treatment on PHC is free, but accessing PHC isn't free for them. This condition is compounded by the absence of any savings and the absence of any health insurance on more of them.

Table 2. Economic characteristics of TB patients

Characteristics	Minimum	Maximum	Median	Mean	Standard deviation
Income	200,000.00	6,000,000.00	950,000.00	1,462,500.00	1,29,.436.34
Expenditure	300,000.00	6,000,000.00	900,000.00	1,284,210.53	1,136,025.87
Savings	0	2,000,000.00	88,947.37	0	340,736.80
Debt insured	0	113,000,000.00	600,000.00	4,872,368.42	18,296,786.37

Furthermore, it is given information about seeking treatment behavior pattern among TB patients until fourth phase. The distribution of educational level on TB patients according to seeking treatment behavior pattern are given on Table 3a until 3d respectively. Each table shows the frequency of TB patients by educational level on each phase of seeking TB treatment.

From the four tables below, it is known that PHC and hospital are chosen by most of TB patients on first phase. But on first phase it also knows that there are 15 TB patients that have already chosen another way for treating TB, i.e. self-medication by consuming the drugs that bought at nearest drug store or others. On second phase, the number of TB patients that choose self-medication and other way for treating TB is declined. This continues until the next phase.

On third and fourth phase, it is known that there are no one who choose self-medication and other way for treating TB. All of TB patients here come to PHC or hospital to get more treatment. In this phase, they become more realize that TB should be cured by health professionals. It may be possible because of the pain that they felt.

Table 3a. Characteristics of TB patients according to seeking treatment behavior pattern on first phase

	Self-medication	PHC	Hospital	Others
No attainment	2	9	1	2
Elementary*	2	6	1	3
Junior high school*	0	0	2	2
Senior high school*	0	1	1	4
Under graduate	0	1	1	0
Total	4	17	6	11

*of the same level

Table 3b. Characteristics of TB patients according to seeking treatment behavior pattern on second phase

	Self-medication	PHC	Hospital	Others
No attainment	0	4	3	3
Elementary*	1	2	4	2
Junior high school*	0	1	3	0
Senior high school*	0	3	2	1
Under graduate	0	0	2	0
Total	1	10	14	6

* of the same level

Table 3c. Characteristics of TB patients according to seeking treatment behavior pattern on third phase

	PHC	Hospital	Others
No attainment	3	3	1
Elementary*	2	2	0
Junior high school*	2	0	1
Senior high school*	0	1	0
Under graduate	1	1	0
Total	8	7	2

* of the same level

Table 3d. Characteristics of TB patients according to seeking treatment behavior pattern on fourth phase

	PHC	Hospital	Others
No attainment	2	0	1
Elementary*	1	1	0
Under graduate	1	0	0
Total	4	1	1

* of the same level

Next session of the results is given explanation about the role model that being referred by TB patient in seeking TB treatment. All of TB patients had mentioned for about four types of role model they referred. They are explained bravely in table 4. From the table, it is known that more of TB patients had mentioned the religious leaders as their role model for seeking TB treatment than any other role model. Most of them that mentioned the religious leaders came from Kalisat than any other sub district.

Table 4. The typical of role models that referred

Role Model	Kalisat	Pakusari	Tanggul	Total

	n	%	n	%	n	%	n	%
Government figures	4	14,3	6	23,1	4	21,1	14	19,2
Religious leaders	22	78,6	14	53,8	13	68,4	49	67,1
Neighbors	2	7,1	4	15,4	2	10,5	8	10,9
Others	0	0	2	7,7	0	0	2	2,8
Total	28	100	26	100	19	100	73	100

Results of SNA provides an overview for researchers, who's a role model are should be involved in TB control at the district level. Degree Prestige Index analysis (DPI) used as SNA approach for this research. It is used for making nodes that represent both of TB patients and the role models. Then we made hyphens between nodes that represent preference of the role model they referred. This analysis is conducted using Socnetv. The results then could describe how central figure who serve as role models there.

The mean nodal indegree on Pakusari is 0.648 and the DP classes is 4, which is the highest DP class among the others. The GDP on Pakusari is about 0.0644, which close to zero, either the others. They mean that there is no role model among the three sub districts that are really as central figures. Each of the role model has DP of the same level. More comprehensive explanation about these findings can be seen on graph 1a –c below.

(a)

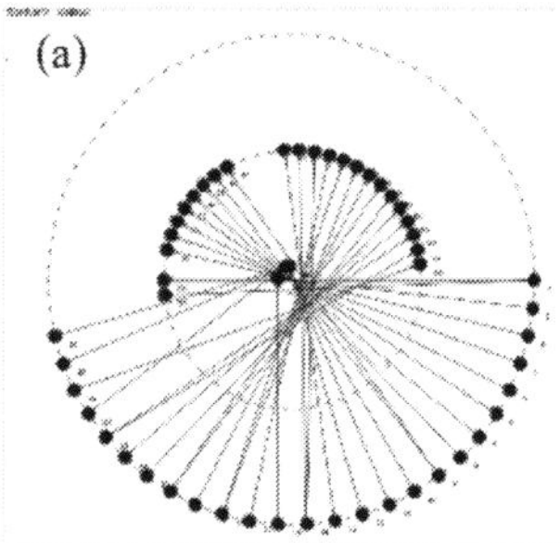

Mean Nodal InDegree = 0.593
InDegree Variance = 0.39
Max DP' = 0.0377
Min DP' = 0
DP classes = 3
GDP = 0.0271

(b)

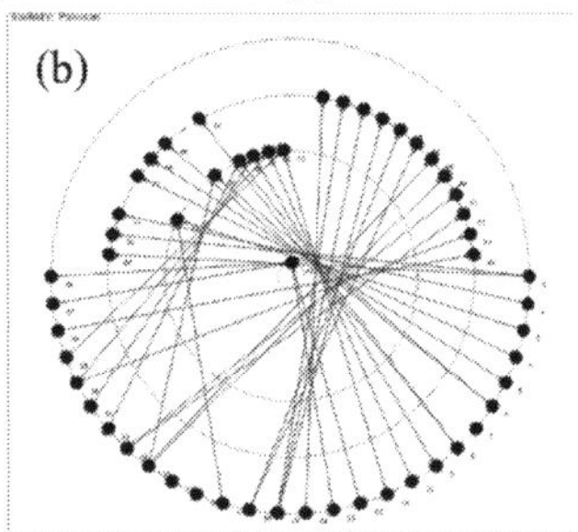

Mean Nodal InDegree = 0.648
InDegree Variance = 0.672
Max DP' = 0.0755
Min DP' = 0
DP classes = 4
GDP = 0.0644

(c)

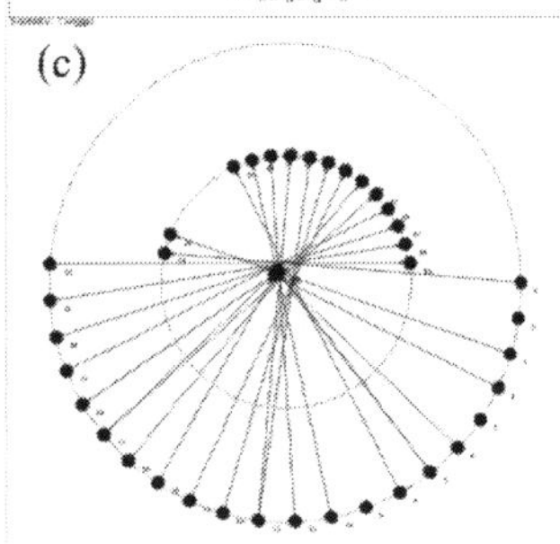

Mean Nodal InDegree = 0.561
InDegree Variance = 0.441
Max DP' = 0.05
Min DP' = 0
DP classes = 3
GDP = 0.0369

Graph 1. Overview of connected points and DPI results between TB cases and their Role models each sub-district: (a) Kalisat; (b) Pakusari; (c) Tanggul

From the DPI analysis above, we then want to describe how spatial network analysis describing whether line characterizations among the three sub districts are spatial centralized or separated. All of node here are plotted, including the node of their models that referred. Then we analyzed these plotted nodes using Quantum GIS. The results show that the reach of the role model on Pakusari is further than other area. It means that the role model there had influenced as many as TB patients to get TB treatment from health care provider, i.e. PHC or hospital.

(a)
(b)
(c)

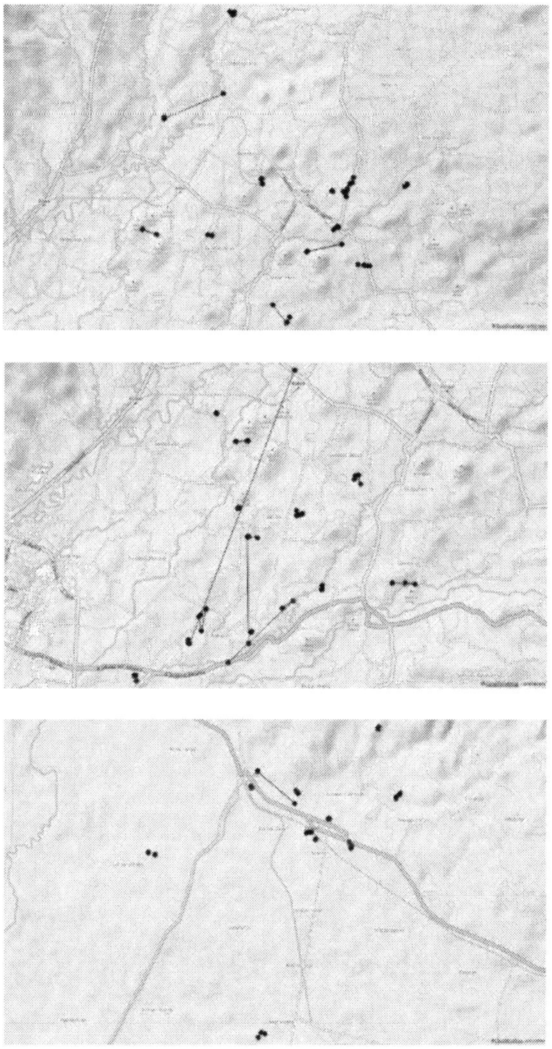

Graph 2. Overview of Spatial network analysis between TB cases and their Role models each sub-district: (a) Kalisat; (b) Pakusari; (c) Tanggul

Then we complete those graphics with several calculations of their properties, such as minimum and maximum reach, mean and standard deviation of reach, and also their median. The complete results are presented on table 5 below.

Table 5. Properties of spatial network analysis

Properties	Kalisat	Pakusari	Tanggul
	In kilometers		

Mean	0.259743	0.492603	0.134851
Standard deviation	0.382777	1.253340	0.279709
Median	0.127110	0.148520	0.084841
Minimum	0.010030	0	0
Maximum	1.862790	7.098130	1.417830

Based on the results of SNA and spatial analysis above, we describe that there is no central figure at the district level, but more scattered in each village. Therefore, TB control interventions and community mobilization program are not quite done at the sub-district level, but it must be entered each village to be effective.

4. CONCLUSIONS

We have shown that there still remains problem in combatting TB among community. The problem here is related to diagnostic delay which is caused by patient delay. We also have known that these such of problem can cause increasing of TB transmission among community. In order to minimize the diagnostic delay, we need more media or such instruments to intensify TB case findings among community. The community, especially in plantation area, has many components who are movable to intensify TB case findings. There are religious leaders, government figures, neighbors, and others who considered as role models for TB patients in seeking TB treatment. Eventually this kind of seeking TB treatment is adopted by TB patients and being patterned. However, these role models have not played role as central figures in Sub District level. There is a tendency for people in Jember personifying religious leaders as role models. On the other side we also need to identify other components on community that can be mobilized and empowered for contributing in combatting TB.

5. ACKNOWLEDGEMENT

We thank to the DP2M DIKTI Ministry of Research Technology and Higher Education and the Research Center of University of Jember for funding this research using HIBER scheme on 2015.

6. REFERENCES

[1] M. o. Health, Pedoman nasional penanggulangan tuberkulosis, Jakarta: Ministry of Health, 2007.

[2] WHO, WHO Report 2009, Global Tuberculosis Control: Epidemiology, Strategy, Financing, Geneva: WHO, 2009.

[3] Y. Ariyanto, "Pengaruh Faktor Sosio Ekonomi terhadap Keterlambatan Penemuan Kaus Tb di Wilayah Kabupaten Jember," *IKESMA: Jurnal Ilmu Kesehatan Masyarakat,* vol. 7, no. 1, 2011.

[4] K. C. N. Smith, "Social networks and health," *Annual Review of Sociology,* vol. 34, pp. 405-429, 2008.

[5] R. Karisma, "Analisis Determinan Terjadinya Tuberculosis pada Pekerja PT. Perkebunan Nusantara XII (Persero) Kabupaten Jember. Script. Not published.," Jember, 2011.

[6] D. Barnes, The making of a social disease: tuberculosis in nineteenth-century France, Los Angeles: University of California Press, 1995.

E-Service Quality Evaluation on E-Government Website: Case Study BPJS Kesehatan Indonesia

A Rasyid and I Alfina

Faculty of Computer Science, Universitas Indonesia, Depok, Indonesia

Email: aulia.rasyid44@gmail.com, ika.alfina@cs.ui.ac.id

Abstract. This research intends to develop a model to evaluate the quality of e-services on e-government. The proposed model consists of seven dimensions: web design, reliability, responsiveness, privacy and security, personalization, information, and ease of use. The model is used to measure the quality of the e-registration of BPJS Kesehatan, an Indonesian government health insurance program. The validation and reliability testing show that of the seven dimensions proposed, only four that suitable for the case study. The result shows that the BPJS Kesehatan e-registration service is good in reliability and responsiveness dimensions, while from web design and ease of use dimensions the e-service still needs to be optimized.

1. Introduction

Indonesia is one of the developing countries that already used e-government to improve the quality of its existing services. Implementation of e-government is divided into five stages: the existence of web, interaction, transaction, transformation, and e-democracy [1]. Nowadays, Indonesia e-government implementation has reached the third stage [2]. The third stage indicates that the Indonesian government has been using online transactions to improve its services [1]. Such online transaction, also known as e-service, has been widely adopted by various government sectors in Indonesia.

The implementation of e-services is expected to change the traditional transaction processes that take time and resources, to become more modern and practical. The government should not only maximize the benefits offered through services owned, but also should reduce even avoid the loss generated if the government cannot adapt to the rapid technological change [3]. With the increasing number of e-services that are starting to be used by various government sectors, it is important for developers to pay attention to the quality of e-services they developed.

Assessing the quality of e-services in e-government has attracted the attention of many researchers in the past few years [4, 5, 6]. In [4], e-GovQual, a multiple-item scale for assessing e-government service quality was proposed, while [5] modified an already existed model (SERVQUAL) to assess the quality of e-services in e-government. Another work by [6] conducted a study to determine what factors affect the quality of e-services in e-government sites in Oman. In Indonesia, there are a few studies that assess e-service on e-government, such as the research conducted to assess the effectiveness of e-service on local government of Serang [7]. However, the literature related to the quality of e-services especially in Indonesia's e-government is still rare.

One of the Indonesian government sectors that have already implemented e-government is BPJS Kesehatan (Badan Penyelenggara Jaminan Sosial Kesehatan), a government health insurance program. The law states that every citizen of Indonesia is obliged to become the member of BPJS Kesehatan. It is reported that in 2010, Indonesian population has reached 237,641,326[1] and this huge number of people are expected to join the program. The program provides two ways for citizen to register, by coming directly to one of the branch offices of BPJS Kesehatan or by using the e-registration service[2]. Due to the huge number of citizen that needs to register, the role of the e-registration service is very important in helping people that have no time to come directly to the respected BPJS offices. This e-registration service should have good quality so that people from different range of social and academic background can use it easily.

The objective of our research is to propose a model to evaluate e-service quality and then implement the resulting model to measure the quality of e-registration of BPJS Kesehatan.

This paper is organized as follows. Section 2 describes the previous researches in assessing the quality of e-service. Section 3 discussed the research methodology. Our proposed model is presented in Section 4. Section 5 describes the results and analysis. Finally, we discuss about the conclusion in Section 6.

2. Related Works

This research was conducted based on three previous studies on e-government service. The first study [5] modified an already existed model (SERVQUAL) that evaluates the quality of e-service in general. The research found that there were five dimensions on the model that needs to be transformed. The five dimensions are *website* design, reliability, responsiveness, security and privacy, personalization, information and easy to use. There were also new dimensions that need to be added to make the instrument able to evaluate e-government service quality: information and easy to use.

The second work [4] developed and validated an instrument to evaluate which factors that affect citizen's perception upon e-government service. The result of this research was a multi-item instrument named e-GovQual. The result of this research shows that efficiency, trust, reliability, and citizen support are important when evaluating the quality of e-government service.

The third work [6] determined which factors (or dimensions) are critical when evaluating e-government service. The number of this research respondent was 324 participants who are familiar with e-government site in Oman. The study shows that efficiency is the most significant factor, followed by security, reliability, and responsiveness, when assessing the quality of e-government service.

3. Methodology

This research used five steps: Research Design, Instrument Evaluation, Hypothesis Definition, Data Collection, and Data Processing.

3.1. Research Design

In this step, all the previous models are analysed in order to define a new model. The new model was defined by combining all dimensions and attributes from previous researches. The attributes that have similar meanings but have different name in previous studies are represented by only one of those attributes. On the other hand, the attributes that have different meanings but still in the same context, was added to the proposed model. Meanwhile, the attributes with a different meaning and different context were removed.

Based on the resulting proposed model, a questionnaire was created and a scenario in using e-registration service was made. Our respondents are asked to do the scenario before filling the questionnaire.

3.2. Instrument Evaluation

This phase tested the proposed model that was made at research design phase. First, we conducted a readability test to evaluate whether the questionnaire is easy to read and understand. Second, we performed validity and reliability test. The evaluation produced a final questionnaire that was valid

[1] https://www.bps.go.id/linkTabelStatis/view/id/1267

[2] https://daftar.bpjs-kesehatan.go.id/bpjs-online/

and reliable for later use in data collection. The validity test used Corrected Item-Total Correlation to see the validity of each attribute, while the reliability test was used to test the reliability of each dimension using Cronbach's Alpha (CA).

3.3. Hypothesis Definition
We conducted two-tailed hypothesis testing for one-sample case for each attribute on every dimension. For each hypothesis testing, we defined null hypothesis and alternative hypothesis. Below is an example of hypotheses for a variable.

> H_0: *The mean value is equal to the value 3 in the web design dimension on the BPJS Kesehatan e-service from the visual attribute side when doing registration.*

> H_1: *The mean value is not equal to the value 3 in the web design dimension on the BPJS Kesehatan e-service from the visual attribute side when doing registration.*

3.4. Data Collection
We used purposive sampling technique to choose 60 respondents. They were students of Universitas Indonesia who have never used e-service on BPJS Kesehatan website. The respondents were asked to follow premade scenarios and preceded to fill out the questionnaire. The survey was conducted face to face, thereby reducing fraud and bias in the collection of data.

3.5. Data Processing
On this phase, the collected data was processed based on the attribute normality. If an attribute was found normal, then its hypothesis would be tested with t-test for one sample case. If an attribute was found not normal, then its hypothesis was tested using the Wilcoxon Sign Rank Test for one sample case. The tool that used for data processing was IBM SPSS Statistics 24.0 software. We have defined the interpretation of mean value into the respective quality label. The details of the interpretation is shown in Table 1.

Table 1. Interpretation of mean value.

Mean value	Quality Interpretation
Less than 3.00	Poor
3.00	Average
More than 3.00	Good

4. Proposed Model
This model was the result of the research design phase. Our study used a research model proposed by [5] as the main model for the instrument. They proposed that in assessing the quality of e-services in e-government, there are seven dimensions that need to be reviewed, namely website design (site design), reliability (reliability), responsiveness (responsiveness), security (privacy), personalization (empathy), information (addition), and ease of use (addition). Every dimension and its attributes on the main model will be combined with two other studies: [4] and [6]. Our proposed model consists of seven dimensions and 35 attributes. The details of proposed model can be seen in Table 2.

5. Result and Analysis
In this section, we will see the result of each analysis that was done on this research.

5.1. Validity and Reliability Testing
An instrument was considered valid when the value of Corrected Item-Total Correlation is greater than r-table testing [8]. For our data, the threshold is 0.361, based on the number of respondents involved. The level of significance used was 0.05 (5%).

Table 2. List of dimensions and attributes of the proposed model

Dimension	Attribute	Reference	Valid & Reliable
Web Design	Visual appeal	[5]	Yes
	Layout	[5][4][6]	Yes
	Site design	[5]	Yes
	Accessibility	[5]	No
	Availability	[5]	Yes
	Crash rate	[5]	No
	Input Response	[5]	No
Reliability	Transaction Confirmation	[5]	No
	Service Accuracy	[5]	Yes
	Transaction tariff	[4] [6]	No
	Download speed	[4] [6]	Yes
	Consistency	[4]	Yes
	Browser compatibility	[4]	Yes
Responsiveness	Response time	[5]	Yes
	Response Availability	[5]	Yes
	Information	[6]	Yes
	Help offer	[6]	No
	User's expectation	[5]	Yes
Privacy and Security	Security assurance	[5][6]	No
	Information keeping	[5]	No
	User's trust	[5]	No
	Transaction status	[5]	No
	Transaction cancellation	[6]	No
	Authentication	[4]	No
	Information management	[4]	No
Personalization	Personalization choices	[5]	No
	e-government link	[5]	No
	e-government service option	[5]	No
	Product delivery	[5]	No
Information	Information accuracy	[5] [4]	No
	Updated information	[5] [6]	No
Ease of Use	Ease of use	[5] [4]	Yes
	Site Structure	[5] [6]	Yes
	Information shape	[5] [4]	Yes
	Information to complete transaction	[5] [4]	No

As for the reliability test, our research used ranged value to qualify each dimension. The conditions were: CA under 0.6 was rejected, above the 0.7 figure was accepted and classified as reliable, and dimension with CA above 0.8 was also accepted and classified as very good.

From the seven dimensions, three dimensions (web design, reliability, ease of use) were found reliable and acceptable because the CA value was inside the range of 0.7 to 0.8. The responsiveness dimension was found reliable and questionable because the CA value was inside the range of 0.6 to 0.7. However, one studies found that a CA value depends on the number of items that have a dimension [9], so that the dimension of responsiveness remains into the final questionnaire. Other three dimensions were not reliable: privacy and security, personalization, and information. The reason was because the amount of valid attributes in each dimension is not sufficient for its dimensions to be calculated of its reliability. The explanation why these three dimensions were not suitable for e-registration of BPJS Kesehatan case study was because the research is limited to a specific e-service in e-government that is not so related with those three dimensions that made the respondents confused in answering the related questions.

In the end, from the seven dimensions and 35 attributes of the proposed model, only four dimensions and 15 attributes are valid and reliable to be used in assessing the e-registration quality of BPJS Kesehatan. The detailed result is shown in Table 2.

5.2. Hypothesis Testing Result and Analysis

Details of the hypothesis testing results can be found in Table 3. We can see that there are only three variables that its respective null hypothesis is accepted. In other words, we accepted the hypothesis that the mean for those variable are 3 that we labeled as "average" quality. For the rest 12 variables, null hypothesis is rejected, so we conclude that the mean is not equal 3. Based on its mean value, then we decide the quality of each variable. There are two variables that categorized as "poor" and ten variables are considered "good".

Table 3. Hypothesis Testing Result

Dimension	Attribute	*Decision on H_0*	Mean	*Quality Interpretation*
Web Design	Visual Appeal	Accepted	2.9	Average
	Site design	Rejected	2.7	Poor
	Layout	Rejected	2.7	Poor
	Availability	Rejected	3.5	Good
Reliability	Service accuracy	Rejected	3.35	Good
	Download speed	Rejected	3.56	Good
	Consistency	Rejected	3.48	Good
	Browser Compatibility	Rejected	3.76	Good
Responsiveness	Response time	Rejected	3.51	Good
	Response availability	Rejected	3.35	Good
	Information	Rejected	3.3	Good
	User's expectation	Rejected	3.41	Good
Ease of Use	Ease of use	Accepted	3	Average
	Site structure	Accepted	3	Average
	Information shape	Rejected	3.25	Good

If we observe into each dimension, web design dimension of BPJS Kesehatan has only one of four attributes that considered as "good". Since the majority of attributes are not good, we conclude that the evaluation result for this dimension is not good. For reliability and responsiveness dimensions, since all attribute qualities value are good, we conclude that the evaluation of these two dimensions are good. For the ease of use dimension, since only one of three attributes is good, we conclude that the evaluation of this dimension is not good.

6. Conclusion

This research has proposed a model to evaluate the quality of an e-service on e-government. The model was defined by analyzing three related researches on e-government service. The model consists of seven dimensions: web design, reliability, responsiveness, privacy and security, personalization, information, and ease of use. Under those seven dimensions, there are 35 different attributes. The proposed model had been used to evaluate the quality of e-service of BPJS Kesehatan, especially its e-registration. The validation and reliability testing show that of the seven dimensions and 35 attributes proposed, only four dimensions and 15 attributes that suitable for the case study. The result shows that the BPJS Kesehatan e-registration service is good in reliability and responsiveness dimensions, while from web design and ease of use dimensions the e-service still needs to be optimized.

The government can use heuristic evaluation to improve those aspects. The e-government service still needs much to be explored, especially in Indonesia. This is important because the government service is important in making the citizen satisfied. We hope by doing this research we can inspire other research to look more on e-service and its quality.

References

[1] Siau K and Long Y 2005 Synthesizing e-government stage models - a meta-synthesis based on meta- ethnography approach *Industrial Management & Data Systems* 105 443-456

[2] Hermana B, Tarigan A, Widyawati H and Silfianti W 2012 E-Government Implementation in Indonesia: Financial Transparency on the Web *3rd Int.l Conf. on e-Education, e-Business, e-Management and e-Learning* (2012: IACSIT Press) pp 194-9

[3] Asgarkhani M 2005 The Effectiveness of e-Service in Local Government: A Case Study *Electronic Journal of e-Government* 3 157-66

[4] Papadomichelaki X and Mentzas G 2012 e-GovQual: A multiple-item scale for assessing e-government service quality *Government Information Quarterly* 29 98-109

[5] Alanezi M, Kamil A and Basri S 2010 A proposed instrument dimensions for measuring e-government service quality *International Journal of u- and e- Service, Science and Technology* 3

[6] Sharma S, Al-Shihi H and Govindaluri S 2013 Exploring quality of e-Government services in Oman *Education, Business and Society: Contemporary Middle Eastern Issues* 6 87-100

[7] Rahmawati 2011 Efektifitas E-Service di Pemerintah Daerah (Kasus di Kabupaten Serang) *Proc. Simposium Nasional Otonomi Daerah* (Banten: Lab-ANE FISIP Untirta) pp 190-6

[8] Priyatno D 2012 *Belajar cepat olah data statistik dengan SPSS* (Yogyakarta: Andi Publisher)

[9] Gliem J and Gliem R 2003 Calculating, Interpreting, and Reporting Cronbach's Alpha Reliability Coefficient for Likert-Type Scales *the Midwest Research-to-Practice Conf. in Adult, Continuing, and Community Education* (Columbus: The Ohio State University)

IOP Conf. Series: Journal of Physics: Conf. Series **801** (2017) 012037 doi:10.1088/1742-6596/801/1/012037

A Conceptual Data Model for Flood Based on Cellular Automata Using Moving Object Data Model

R S Rachmatullah[1], F N Azizah[2]

School of Electrical Engineering and Informatics, Institut Teknologi Bandung, Jln. Ganesha no. 10 Bandung, Indonesia

[1] 13512036@std.stei.itb.ac.id, [2] fazat@stei.itb.ac.id

Abstract. Flood is considered as the costliest natural disaster in Indonesia due to its frequent occurrences as well as the extensive damage that it causes. Several studies provide different flood prediction models based on various hydrological factors. A lot of these models use grid-to-grid approach, making them suitable to be modelled as cellular automata. This paper presents a conceptual data model for flood based on cellular automata model using spatio-temporal data model, especially the moving object data model, as the modelling approach. The conceptual data model serves as the model of data structures within an environment for flood prediction simulation. We describe two conceptual data models as the alternatives to model the data structures of flood model. We create the data model based on the study to the factors that constitute the flood models. The first conceptual data model alternative focuses on the cell/grid as the main entity type. The changes of the states of the cells are stored as moving integer. The second alternative emphasizes on flood as the main entity type. The changes of the flood area are stored as moving region. Both alternatives introduce some advantages and disadvantages and the choice rely on the purpose of the use of the data model. We present a proposal of the architecture of a flood prediction system using cellular automata as the modelling approach. As the continuation of this work, further design and implementation details must be provided.

1. Introduction

Flood is one of natural disasters that frequently occur in Indonesia. According to the data of Badan Nasional Penanggulangan Bencana Indonesia (Indonesia's National Agency for Disaster Management), up to the month October, there have been 639 occurrences of flood throughout Indonesia in 2016 in which 139 people died or missing, more than 2 million people were displaced temporally or permanently from their houses, as well as hundreds of millions of rupiahs of material loss [1]. Due to the frequent occurrences and extensive damage that it causes, flood is considered as the costliest natural disaster. In the attempt to reduce the loss, several initiatives have been taken by the Indonesian government, such as constructions of reservoirs, river dredging, and the use of flood prediction systems to determine the location and time of the incidents.

Several studies on flood prediction model use various hydrological cycle factors depending on the focus of studies. A lot of these models use grid-to-grid (G2G) approach to determine the spread of flood. A flood is modeled as the dispersal of water from one grid to its neighbors per unit of time [2]. This characteristic makes cellular automata a suitable approach to model it.

In this paper, we present a study on several existing flood prediction models to determine the important factors that make up a flood, as well as the formulas used to calculate the spreading of flood.

We especially examine how the cellular automata concept is used to model the flood. We incorporate the result of the study to create a conceptual data model using spatio-temporal approach, especially the moving object data model [3]. Moving object data model is chosen for its maturity as a spatio-temporal data modelling concept.

The conceptual data model serves as the model of the data structures within an environment for flood prediction simulation using cellular automata and spatio-temporal data modelling as the main approach. Eventually, our final goal is not only for simulating flood, but also providing a spatio-temporal model based environment for simulating problems that can be modeled as cellular automata, such as: fire (e.g. [4]). Flood, in this case, is an interesting case because it has a complex nature in term of different factors that we must deal with and complex formula for calculating the spreading of flood.

The paper is organized as the following: in the next chapter, we present some theoretical background regarding flood prediction models, cellular automata, as well as moving object data model. Subsequently, we discuss two alternatives of the conceptual data models based on the result of the study on the factors involved in the formation and spreading of flood. We present the impact of each alternative on the algorithms used for simulating the cellular automata as well as its subsequent implementation in the simulation environment.

2. Theoretical Background

2.1. Basic Equations for Flood Models

Flood is rainfall excess (surface runoff) that cannot be covered neither by infiltration nor by drainage system [2]. The flood models are usually created based on the hydrological cycle models that are used to model fluid and water circulation. An important key component of the model is the water balance equation which is calculated based on the Land Surface Model (LSM). The LSM is used to calculate the water balance in soil by using precipitation, infiltration, and evapotranspiration factors [5].

The water balance equation calculates the amount of water in soil so that the input and the output is balanced [2][5]. The input is precipitation, i.e. the total of the volume of water or rainfall per unit of time. The output is divided into three parts: infiltration, evapotranspiration, and runoff. Infiltration components are calculated based on the soil types and land use. Meanwhile, the evapotranspiration factor is divided into two components, namely the evaporation and the transpiration. Evapotranspiration can be calculated using Penman-Monteith equation stated in [6].

2.2. Cellular Automata Approach for Flood Spreading

The result of the calculation on the surface runoff from water balance equation is used by the inundation (flood) model to determine the flow of water and to create puddles [2]. The flow of water is determined using flat-water concept. The flat-water concept assumes that the nature of water is always flat without considering the friction caused by gravitation or inertia and influx or wave. Because of flat water concept, the spreading of flood can be determined just by using topographic data from the Digital Elevation Model (DEM) [2][7]. The cellular automata model approach is used to determine the spread of flood by utilizing the properties of the water that flows from the high place to a low place.

In representing the flood model using the cellular automata model, grids or cells are divided into two types: the wet cells for flooded soil and water objects such as river; and the dry cells for soil that has not yet been flooded [7]. The starting point of calculation using the grid-to-grid approach is determined from the cell that has the highest height of cell. Height of a cell is composed by the height of water and the topography. That cell will be the central cell. The neighboring cells have to be ready to accommodate any excess water from the central cell.

2.3. Moving Object Data Model

Spatio-temporal data poses a challenge in database management system since handling spatio-temporal data is not a straightforward task due to the complexity of the data structures. It requires careful analysis in dimensions together with the representation and manipulation of the data involved. Moving object data model [8] assumes the aspect of time as an integral part of the spatial aspects of the object. This modelling approach captures the change of geometry and the movement of the object. Therefore, it is suitable for all queries required in spatio-temporal problems.

The moving object data model introduces several data types. For each data type, operations are defined. They can be divided into two groups [9]: non-temporal data types and temporal data types. The temporal data types provide the combination of the base or spatial types with the temporal aspect of the data. The new types are called the moving data types, for example: moving real (*mreal*), moving int (*mint*), moving point (*mpoint*), moving line (*mline*) and moving region (*mregion*). Another temporal data type is the *intime* data type which is used to define a single pair of (*instant*, value).

3. The Modelling Methodology

This work is aimed at providing the conceptual data model of flood models which considers cellular automata as the main flood modelling approach and moving object data model as the spatio-temporal modelling approach. As can be observed in the previous section, various factors must be considered in the modelling of flood and therefore, they must examine further in order to provide an appropriate approach on how to incorporate them in the conceptual data model. The cellular automata model determines how we store the flood event in which cell states change over time depending upon the calculation of the surface runoff on each cell.

To present our conceptual data model, we use Entity-Relationship (ER) model which uses ER diagram as the main tool to visualize the data model. We incorporate the use of moving object data types in the data model as the types for the attributes of entity types. Figure 1 shows the methodology that we use in this paper to define the conceptual data model of

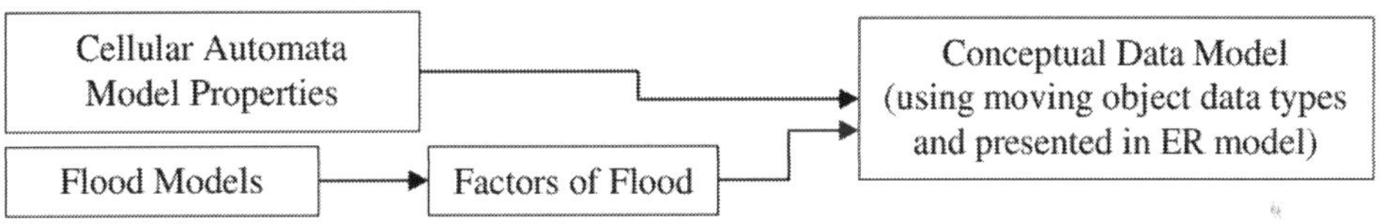

Figure 1. The Modelling methodology.

4. The Alternatives of Conceptual Data Model for Flood

4.1. Study on the Factors that Constitute Flood

Several studies on flood models use different formulas in the process and require different data. The more complex and complete the data, the more accurate the prediction results will be. As observed in water balance equation, rainfall or precipitation is the foremost data that must be provided. Evapotranspiration can be estimated using Penman-Monteith equation [6]. By contrast, Chen et al. [2] stated that evapotranspiration factor in urban areas can be ignored because it is not considered significant.

The calculation of infiltration can be done by using the VIC model of water balance in every layer ground [5]. The experiment conducted by Chen et al. [2] adds drainage capacity as a factor in infiltration capacity especially in urban areas. The experiments conducted by Prasetya et al. [7] uses the Horton equation to determine the capacity of water absorption per unit time. Standard absorption rate is determined as a constant, based on MODIS land cover classes. To determine the surface runoff and infiltration, additionally, a flood model requires the results of the morphology of satellite images or GIS (Geographic Information System) data to determine the soil type and MODIS land cover class or Land Use. Furthermore, the models require DEM data for ground topology. Subsequently, VIC model can be applied to calculate the distribution of water runoff only when all data are present. Other flood models may not require such requirement.

After examining the theories on the flood models, we conclude that some of the factors that determine the flood models are mandatory (it means that the data must be provided), while some others are optional (it means that they can be omitted when they are not present). The factors can also be viewed from the global or local perspectives. Global factors are shared and influence all process and cells in the cellular automata model. Local factors, on the contrary, influence and are used exclusively within a cell. Several factors can give a broad impact to the area of prediction, for example: daily mean temperature and wind speed. Therefore, they are designated as global factors.

4.2. The Proposed Conceptual Data Models

To model the factors that constitute a flood, we define several entity types, i.e. the Soil Reference, Soil Type, Land Use, Drainage, Topography and District. The entity Land Use defines the usage of a certain piece of land. It contains the attributes to define the type of land use (e.g.: forest, farm, field, residence) as well as the area (spatial type: region) covered by the land. The entity Soil Type defines a certain area of soil and what kind of soil type it belongs to. The attributes define for the Soil Type are the type name as well as the area covered (spatial type: region). The entity Soil Reference refers to the soil condition which can be imposed either to a Land Use or a Soil Type.

The existence of drainage systems is captured via the entity Drainage which contains the attributes the volume of the drainage, type, as well as the area covered by the drainage. The entity Topography is used to store the height of land area. The entity District covers a certain area, usually in the form of governmental district, in which we record the rain precipitation, the radiation flux, humidity, wind speed, aerodynamic resistance, mean saturated vapor pressure and daily mean temperature, all of which are of the type moving real as well as the air density. The attributes are modeled as moving real since the values changes over time. The ER diagram that covers the factors that constitute the flood models can be seen on Figure 2.

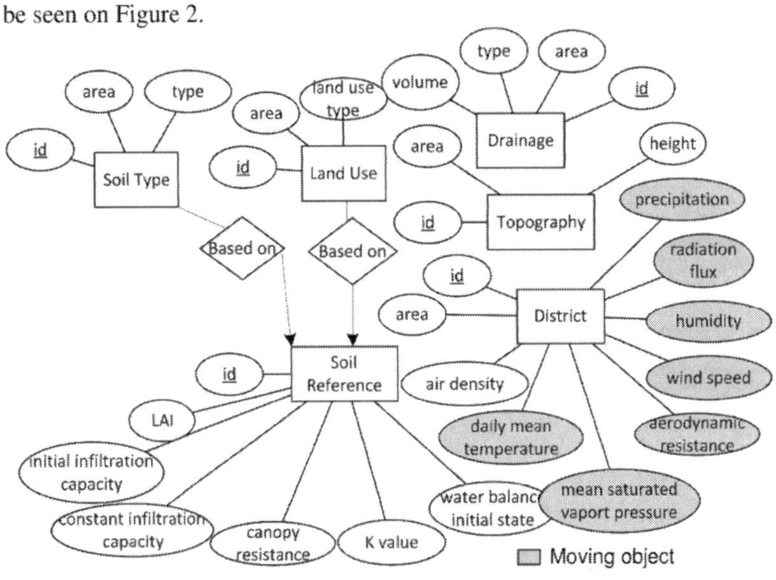

Figure 2. The ER diagram for the factors of flood.

To model the area affected by the flood, we define two alternatives of modelling approach:
1. focusing on the nature of cellular automata model, we store the cells and the changes of their states (as well as other attributes) over time;
2. focusing on the flood as an entity, we store the flood area and its changes over time.

Using the first approach, we store each cell as an entity. Thus, we have the Cell entity type. The state of a cell may be: (1) wet; (2) dry (not wet). A cell is in the state of wet when it is submerged by water. Since the state of each cell can change over time, it is stored as a moving object attribute (type:

moving integer). Other contributing factors are also measured: the water height and the water balance, both also changes over time. Thus, they are stored as moving object attributes (type: moving real). See Figure 3 (a) for the ER diagram of the model.

The second approach considers flood as an entity whose area changes over time. Thus, we define the Flood entity type. Therefore, we store the flood area as a moving object attribute (type: moving region). Other attributes to be stored for a flood are: the average of water height and water volume for the whole flood area per time. Both are also defined as moving object attributes (type: moving real). See Figure 3 (b) for the ER diagram of the model.

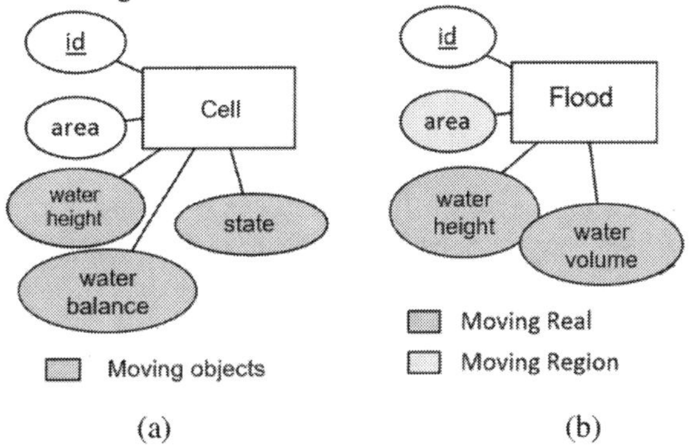

Figure 3. Two modelling approaches of flood as a cellular automata model: (a) 1st approach: focusing on the nature of cellular automata, cells are stored as a moving object entity; (b) 2nd approach: focusing on the flood as an entity, flood is stored as a moving object entity

How flood is stored influences the algorithms for simulating the flood using cellular automata approach. The first approach is obviously a natural way of storing the results of the cellular automata calculation. To calculate the state of each cell, we need to intersect the area of the cell with the areas of the other factors of flood: the soil type, the land use, the drainage, the topography and the region. Thus, we can get the value of each contributing factor and use the formulas to get the state of the cell.

The basic algorithm to do the flood simulation is as the following:

1. Intersect the area of each cell with the area of the entity Soil Type, Topography, Drainage and Land Use (the Soil Reference entity involves accordingly).
2. For each unit of time: do the calculation of the equation 1 (do first the calculation to get the evapotranspiration and infiltration, involving also the data of the entity District) and check if there is a runoff in each cell.
3. If there is a runoff, change the status of the cell to 'wet' and then give the water runoff to neighbor cells based on Prasetya et al. inundation algorithm [7] or Chen et al. algorithm [2].
4. Once the water runoff has been calculated and spread properly, keep the state of each cell for the unit of time in the attribute of state of a Cell, and then, re-do step 2.
5. Process 2-4 is repeated until the rainfall data are all processed or the simulation is terminated.

The second approach stores the changes of flood area over time. In each unit of time, after the state of each cell is calculated, all adjacent wet cells are combined to form the spatial object of flood. Therefore, we still require the steps to calculate the state of each cell for each unit of time, except that we are not required to store the state of each cell. Therefore, we need to adapt the 4th step of the algorithm into:

4. Once the water runoff has been calculated and spread properly, perform an aggregation operation on the area of wet cells to form an area of the flood per time and then store it as a part of the moving region of Flood. Subsequently, re-do step 2.

An important note to take is that calculating the state of cells per unit of time require a great amount of work. Suppose we have a *n x n* grid on the cellular automata model, then we must do *n x n* cell processing or in term of algorithm complexity, we will have a quadratic time complexity of $O(n^2)$. In this paper, we will not discuss how to deal with the challenge to create an efficient algorithm to reduce the time complexity required to calculate the cell states. Nevertheless, this challenge must be kept in mind when discussing the approach to store the data.

The first data modelling approach (model shown in Figure 3 (a)) provides the direct correspondence between the cells in the cellular automata and the cells in the data store. No further processing is required once the state of a cell per unit of time is determined: the resulting calculations of state, water height and water balance of each cell are directly stored in the appropriate attributes. As the calculation of the state of each cell requires great effort, directly store them will reduce the processing time. Suppose we have a *n x n* grids, then we will need a space to store n^2 cell entities, or the space complexity is $O(n^2)$.

Using the second data modeling approach (model shown in Figure 3 (b)), however, after we calculate the state of each cell, we need extra time and other resources to aggregate the wet cell area to form the area of a flood per unit of time. If we have n wet cells, the time complexity to aggregate the area will be $O(n)$. As mentioned earlier, calculating the state of the cells per unit of time can already be considered as a horrendous task. Adding the task to aggregate them into a flood area will not give a better picture. Nevertheless, in term of space requirement, for each flood event, we need only to store 1 (one) entity, i.e. the flood itself. Therefore, we can say that the space complexity is constant. The changes of the flood area are stored as a moving area attribute, i.e. for each flood, it is the aggregation of the wet cells. This will significantly reduce the amount of space required to store the data in comparison to storing the changes of all cells.

The second approach may also be more advantageous for further exploration and research. If we want to do further analysis on the flood data or do some statistics, especially the ones involving the area of the flood, the second approach may be more suitable than the first approach. Suppose we want to get the information about the coverage of a flood (at all times or at a certain time slice). Using the first approach, we will need to aggregate all the wet cells into the flood area. Using the second approach, the flood area has been provided in the storage.

The first approach is strictly tight to the grid-to-grid/cellular automata model. Thus, if we want to do simulation using different flood models that do not involve the grid-to-grid approach, we will find difficulties to adapt to the data structure. As for the second approach, since we store directly the flood, depending on how the flood model works, the problem to adapt to the structure will not be worse than the problem it has with the cellular automata approach. It will work even better when the flood model gives the flood area as its direct result.

Table 1 shows the summary of the comparison between the two approaches.

Which one of the two approaches is better? The answer, unfortunately, relies heavily on the purpose of the use of the conceptual data model. If the data structure is required in an environment in which simulation is purposefully done using the cellular automata model, then the first approach is favorable. The second approach, as discussed, actually provides a more general approach for more flood modelling and analysis.

5. The Proposed Architecture of The Flood Prediction System

As earlier described, the development of the conceptual data model of flood is aimed at providing the structure for the data associated with flood models and providing an environment on which experts can simulate flood events using cellular automata as the flood modelling approach. Figure 4 shows a proposal of the architecture for the simulation environment. The conceptual data model serves the blue print of the data stored in a moving object database management system (MO-DBMS). We reserve that no mature moving object database management system is ready to be used. Therefore, we propose the use of current existing database technology, equipped with the moving object data types and operations. On the application level, we have the flood prediction tool which consists of two important modules: the Cellular Automata Module and the Visualization Module. During a simulation, the Cellular Automata Modules calculates the model based on the stored data and store the results back to the database. The Visualization Module visualizes the result of the simulation process.

Table 1. Comparisons between the conceptual data model alternatives.

	First Approach	**Second Approach**
Description	Focusing on the storage of cells, resulting in Cell as a moving object entity	Focusing on flood event, resulting in Flood as a moving object entity
Time complexity for cell calculation per unit of time	Cell processing: $O(n^2)$ n x n is the number of cells	Cell processing: $O(n^2)$ Aggregation of cells: $O(m)$ n x n is the number of cells m is the number of wet cells
Space complexity for one flood event simulation	$O(n^2)$ n x n is the number of cells	$O(1)$
Further information or statistics involving the flood area	Difficult, always need to aggregate the area of the wet cells	Easier, take the information from the moving area attribute of flood
Possibility to use other approaches for flood modelling	Difficult, since the model is tightly related to the cellular automata model	Easier to adapt since the model stores directly the flood data

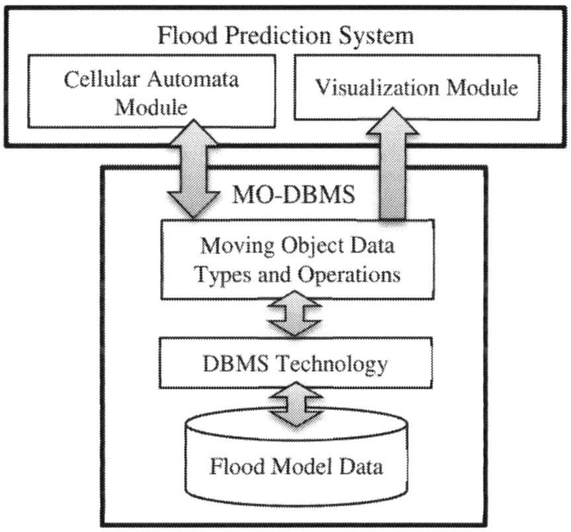

Figure 4. The proposed architecture of the flood prediction system

Since the system focuses on the use of cellular automata as the modeling approach, it is natural to use the first conceptual data model alternative (see Figure 3 (a)) as the appropriate data modelling approach. We will examine the modelling choices and the simulation architecture environment further as the continuation of this work.

6. Conclusions and Future Works

Flood simulation system is important in supporting the efforts of reducing the effect of flood as the costliest natural disaster in Indonesia. In this paper, we describe two conceptual data models as the alternatives to model the data structures of flood model using cellular automata as the main approach. We utilize the moving object data model approach to model the changes in the simulation environment. We create the data model based on the study to the factors that constitute the flood models. Several mandatory and optional factors of flood models are defined as the result of the study and based on that, a conceptual data model for the factors is defined. This data model is a common part for both conceptual data model alternatives.

The first conceptual data model alternative focuses on the cell/grid as the main entity type. In particular, it stores the changes of the states of the cells as moving integer. This approach produces a one-on-one correspondence between the cells in the cellular automata approach and the cells in the data storage, making it the best option for simulating the flood in a cellular automata environment.

The second alternative emphasizes on flood as the main entity type. It stores the changes of the flood area as moving region. In a cellular automata based simulation environment, this approach introduces extra effort to aggregate the flood area based on the wet cell area which must be done after the calculation of the states of each cell. On the other hand, this approach is probably a better choice for a more general purpose of flood modelling and analysis.

The conceptual data model serves as a basis for the development flood prediction system which involves simulating the flood models. We present a proposal of the architecture of a flood prediction system in section 5. Our next plan is to provide details in the implementation of the flood prediction system. We still need to provide the logical and physical design of the data as well as choosing and developing the appropriate database technology for storing the moving object data. We need to implement the cellular automata algorithm based on the chosen data design as well as providing a mechanism to visualize the result. Furthermore, we are still required to test the system to evaluate the accuracy of the model and its implementation.

References

[1] Badan Nasional Penanggulangan Bencana 2016 *Data dan Informasi Bencana Indonesia* http://dibi.bnpb.go.id

[2] Chen J, Hill A A and Urbano L D 2009 A GIS-based Model for Urban Flood Inundation *J. Hydrology* **373** (1–2) p 184–192

[3] Erwig M, Güting R H, Schneider M and Vazirgiannis M 1999 Spatio-Temporal Data Types: An Approach to Modeling and Querying Moving Objects in Databases *GeoInformatica* **3** (3) p 269–296

[4] Ohgai A, Gohnai Y and Watanabe K 2007 Cellular Automata Modeling of Fire Spread in Built-Up Areas—A Tool to Aid Community-Based Planning for Disaster Mitigation *Computers Environment and Urban Systems* **31** (4) p 441-460

[5] Liang X, Lettenmaier D P, Wood E F and Burges S J 1994 A Simple Hydrologically Based Model of Land Surface Water and Energy Fluxes for General Circulation Models *American Geophysical Union* **99** p 14415–14428

[6] Zotarelli L, Dukes M D, Romero C C, Migliaccio K W and Morgan K T 2010 *Step by Step Calculation of the Penman-Monteith Evapotranspiration (FAO-56 Method)* (Florida: The The Institute of Food and Agricultural Sciences)

[7] Prasetya R B, Fariza A, Hasim J A N and Basuki 2015 A River Flood Spreading Prediction System Using Cellular Automata (Case Study Bengawan Solo River) *Proc. Int. Conf. on Data and Software Engineering (Yogyakarta)*

[8] Pelekis N, Theodoulidis B, Kopanakis I and Theodoridis Y 2004 Literature Review of Spatio-Temporal Database Models *Knowledge Engineering Review* **19** (3) p 235–274

[9] Güting R H and M. Schneider M 2005 *Moving Object Databases* (San Fransisco: Morgan Kaufmann)

Food Delivery System with the Utilization of Vehicle Using Geographical Information System (GIS) and A Star Algorithm

B Siregar[1], D Gunawan [1], U Andayani[1], Elita Sari Lubis[1] and F Fahmi[2]

[1]Dept. of Information and Technology, Faculty of Computer Science and Information Technology, University of Sumatera Utara Jl. Dr. Mansur 9 Medan Indonesia
[2]Dept. of Electrical Engg, Faculty of Engineering, University of Sumatera Utara Jl. Dr. Mansur 9 Medan Indonesia

Email: fahmimn@usu.ac.id

Abstract—Food delivery system is one kind of geographical information systems (GIS) that can be applied through digitation process. The main case in food delivery system is the way to determine the shortest path and food delivery vehicle movement tracking. Therefore, to make sure that the digitation process of food delivery system can be applied efficiently, it is needed to add shortest path determination facility and food delivery vehicle tracking. This research uses A Star (A*) algorithm for determining shortest path and location-based system (LBS) programming for moving food delivery vehicle object tracking. According to this research, it is generated the integrated system that can be used by food delivery driver, customer, and administrator in terms of simplifying the food delivery system. Through the application of shortest path and the tracking of moving vehicle, thus the application of food delivery system in the scope of geographical information system (GIS) can be executed.

1. Introduction

Geographical Information System (GIS) is used in terms of supporting the decision making for planning and managing the land usage [1]. Many companies use GIS to execute market analysis and to find the optimal delivery route [2]. The problem in finding the optimal delivery route has function to minimalize the amount of transportation, supply, and resource costs [3]. Many problems occur when the food delivery process is held to each customer. One of them is the affair when customer who has already ordered foods cannot track food delivery vehicle. This causes deficits for the food delivery courier that has to pay punishment risk from customer, and the most serious case is losing customer [3]. Thus, the ability of customer to find out information and location of food delivery vehicle significantly impacts the delivery workload and the benefit from the restaurant side.

Furthermore, it is needed to develop the method to generate optimal vehicle route model [3]. The problem of shortest path is needed to find the least travel value of one or more initial nodes into one or more goal nodes inside the connected network. This problem is very crucial because of the broad scope in its application seen from transportation side [4]. Using the Global Positioning System (GPS) technology and the advanced cellular network, the real-time vehicle tracking for the better transportation management has been possible [5].

A star algorithm was firstly initiated by Peter Hart, Nils Nilsson, and Bertram Raphael who proposed this algorithm in the year of 1968 [6]. The aim of this algorithm is to find the optimal path inside of graph by combining cost searching and heuristic feature. According to research done by Computer Science researcher from Stanford University, it is concluded that A star algorithm is very applicable in terms of combining the advance of heuristic approach owned by Greedy Best-First-Search and the advance of general algorithms, such as Dijkstra algorithm [7].

In this research, writer proposes A star algorithm to find the shortest path in food delivery system. In part II, it will be explained the problem identification of this research. In part III, it is descripted some previous researches that have been conducted for shortest path and LBS. Section III describes the proposed methods. Result from the research is depicted in Section V. Section VI consists of the research conclusion followed by advices for upcoming researches.

2. Material and methods

Method that is being proposed in this research consists of four phases, that are data input, data saving, shortest path finding, and real-time JSON parsing of vehicle coordinate.

Several assumptions have been made to limit the scope of workload in this research, which are: mapping route only covers some districts in the city of Medan, only simulates 12 restaurants, road calculation and usage are limited into main road merely and calculation in shortest path finding uses the vehicle with the equal and constant pace assumption. Moreover, the architecture of the system can be seen in Figure 1.

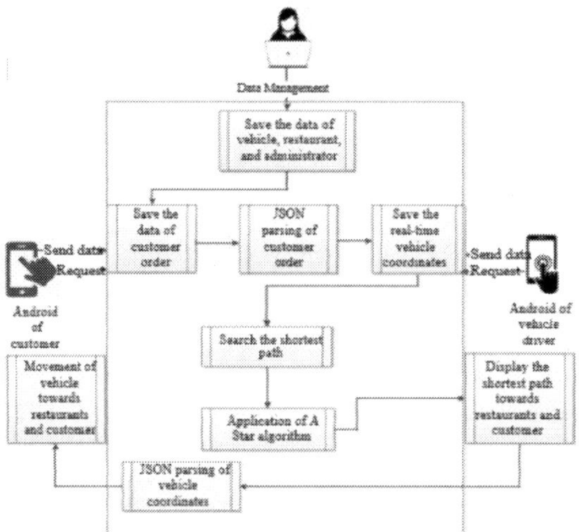

Figure 1. General Architecture

2.1 Data Input

In the phase of data input, there are several inputs from three elements of users, as described below: First input is conducted by admin. This aforementioned admin inputs the simulated 12 restaurant objects along with their attributes into server. Their attributes include the coordinate (latitude, longitude), restaurant address data, and others. Second input is entered by courier. This courier inputs the data and sends real-time data into server from his/her smart phone which has the built-in GPS. Third input is filled out by customers through their smart phone that has built-in GPS. This input consists of the customer log-in data, customer location, and orders.

2.2 Data Storage

The second phase is data storage. The process of data storage done as follows: First process covers the attribute storage of simulated 12 restaurants which have data and coordinates. These restaurant coordinates have function to simplify the ways of customer to notify the restaurant location of their choosing. These 12 restaurants eventually will have legend or restaurant badges, thus the different 12 restaurants can be easily notified by the customer. Second process is the real-time data storage and courier coordinates. Third process is data and customer coordinate storage.

2.3 Shortest Path Searching

The third phase is the shortest path searching by applying A star algorithm. This A star algorithm process that eventually generates shortest path as follows: Initialization of initial node and goal node. There are two paths that can be taken and marked using node. Each node will have each function f(x), g(x), and h(x).

It will be shown in each node and indicated the next node that will be visited according to cost of shortest path using formula f(x) = g(x) + h(x), where:

- G(x) is the total distance from the initial position into current location.
- H(x) is the heuristic function used to estimate the distance from current location into the destination location.

A star algorithm uses the shortest path into destined node that makes it the best first search using the formula f(x) = g(x) + h(x). As mentioned beforehand, the A star will choose the lowest f(x). When the destined node is popped up in priority list, then the search is terminated.

2.3.1 Pseudo-Code of A starAlgorithm. Pseudo code of A star algorithm used to solve this system can be seen in Figure 2.

```
Function A*(start,goal)
    closed = empty set
    q = makequeue(path(start))
    while q is not empty do
        p = removefirst(q)
        x = lastnode(p)
        if x in closed then
        end if
        if x = goal then
            return p
        end if
        add x to closed
        for y successor(x) do
            enqueue(q, p, y)
        end for
        return Failure
    end while
end function
```

Figure 2. Pseudo-Code of A Star Algorithm

2.3.2 Implementation of A Star Algorithm on System. The picture below is one of graph samples for the shortest path searching using A star algorithm.

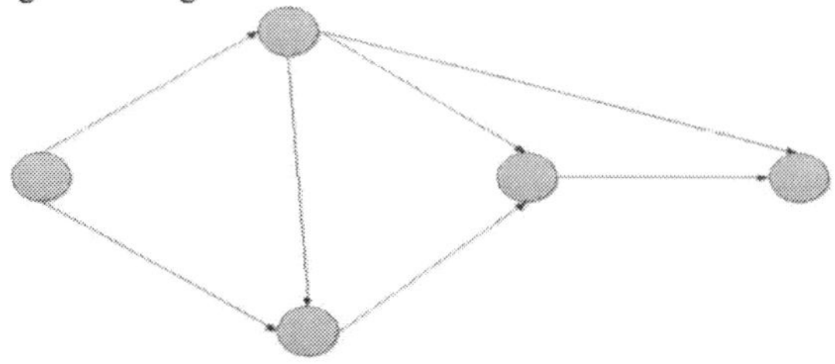

Figure 3. Sample of Graph for Shortest Path Searching

The explanation of the picture above can be seen in Table 1.

Table 1. Explanation of Shortest Path Element

No.	Node	Coordinate	Name
1	S	1,1	Pempek Palembang Setiabudi
2	A	3,1	Intersection of Dr. Sumarsono Street
3	B	2,6	Intersection of Politeknik/ Tri Dharma Street
4	C	6,1	Intersection of Gate3, University of North Sumatera (USU)
5	G	6,5	Faculy of Medicine, University of North Sumatera (USU)

The calculation of shortest path for this case is started with the initial node of node S or Pempek Palembang Setiabudi to the goal node of node G or Faculty of Medicine, North Sumatera University (USU). Relation and distance among nodes on database can be seen on Table 2.

Table 2 Relation and Distance among Nodes

No.	Node Relation	Distance in Kilometers (km)
1	S-A	1
2	S-B	4
3	A-B	2
4	A-C	5
5	A-G	12
6	B-C	2
7	C-G	3

2.3.2.1 Heuristic Calculation

Process of heuristic calculation using A star algorithm is done with the equation below:

$$d(x,y) = \left|\sqrt{(x_1-y_1)^2 + (x_2-y_2)^2 + \cdots + (x_n-y_n)^2}\right| \quad (1)$$

By applying the aforementioned equation, the heuristic result generated as follows:

A. Heuristic S(1,1)□ G(6,5) d(x,y) =

$d(S,G) = \sqrt{(x_1-y_1)^2 + (x_2-y_2)^2}$

$= \sqrt{(1-1)^2 + (6-5)^2}$

$= \sqrt{1.0)^2 + (1)^2}$

$= 1$

B. Heuristic A(3,1)□ G(6,5) d(x,y) =

$d(A,G) = \sqrt{(x_1-y_1)^2 + (x_2-y_2)^2} = 2{,}24$

$\sqrt{(3-1)^2 + (6-5)^2}$

$\sqrt{(2)^2 + (1)^2}$

$\sqrt{5}$

C. Heuristic B(2,6)□ G(6,5) d(x,y) =

$$d(B,G) = \sqrt{(x_1 - y_1)^2 + (x_2 - y_2)^2}$$
$$= \sqrt{(2-6)^2 + (6-5)^2}$$
$$= \sqrt{(-4)^2 + (1)^2}$$
$$= \sqrt{17}$$
$$= 4{,}12$$

D. Heuristic C(6,1)□ G(6,5) d(x,y) =

$$d(C,G) = \sqrt{(x_1 - y_1)^2 + (x_2 - y_2)^2}$$
$$= \sqrt{(6-1)^2 + (6-5)^2}$$
$$= \sqrt{(5)^2 + (1)^2}$$
$$= \sqrt{26}$$
$$= 5{,}1$$

E. Heuristic G(6,1)□ G(6,5)
d(x,y) = 0

Table 3. The Heuristic List of Each Node

No.	Node	Heuristic
1	S	1
2	A	2,24
3	B	4,12
4	C	5,1
5	G	0

2.3.2.2 The Calculation of Shortest Path Using A Star

After generating the heuristic through process above, The next phase is the calculation of shortest path using A star algorithm executed using the equation: f = g + h, where:g = distance and h = heuristic

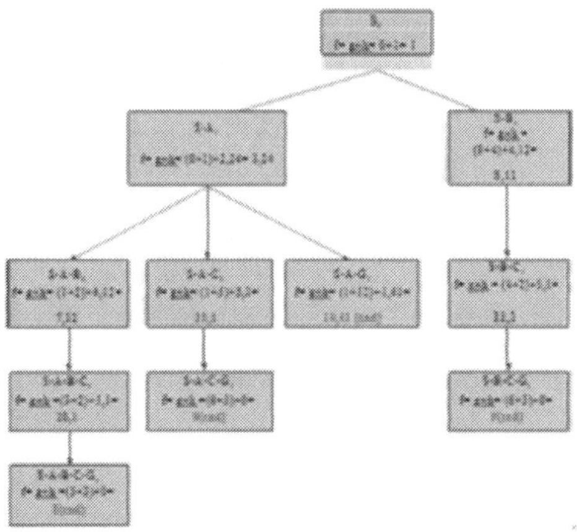

Figure 4. The Process of Shortest Path Calculation

From the searching above, we can conclude that the shortest path from node S into node G using A star algorithm is through S-A-B-C-G with the distance weight of 8 kilometers.

2.3.2.3. Real-Time JSON Parsing of Vehicle Coordinate

The last phase is the real-time JSON parsing of vehicle coordinate that is handful to track the motion of vehicle. By using this feature, thus the food delivery customer can track the movement of vehicle, the vehicle arrival time, and the distance from vehicle towards the customer location.

3. Results

The implementation is done using Android smart phone with Samsung Core Duos type. Implementation is done according to the interface architecture designed in both mobile and web-based application.

3.1 Display of Shortest Path

After user who is the vehicle driver pushes the button Find Shortest Path on layout in Figure 4.2, next thing popped up is the automatic form contains customer and destined restaurant location, as follows:

Figure 5. Display of Customer Location and Restaurant Name

The interface for destination with maximum number of 3 restaurants is below :

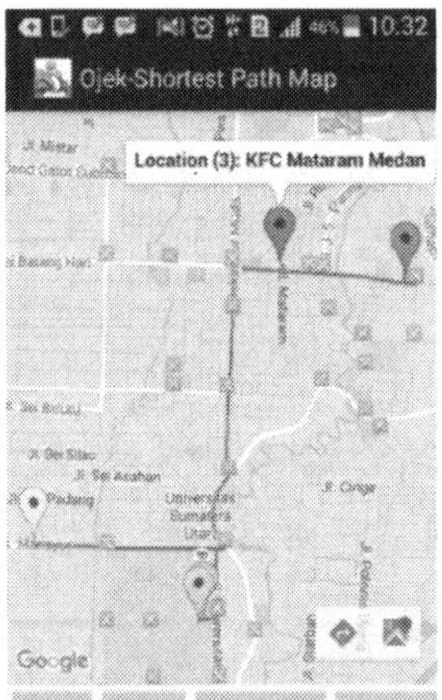

Figure 6. Result of Shortest Path with Destination 3 Restaurants

From the aforementioned system trial result, it can be seen that the shortest path calculation and time to display the movement of food delivery vehicle tracking really depends on the internet data transfer speed used by user. The faster the internet connection, then the faster process and the display of shortest path and the vehicle movement displayed in smart phone screen.

4. Discussions

After doing all the implementation phase and trial towards the system, thus there are some conclusions generated in this research, which are: by applying the A strar algorithm application to solve the shortest path case, Ojek Food Delivery system operated by user that is vehicle driver can be executed, calculation of shortest path is done with the purpose of finding the shortest path from vehicle location into restaurant and customer location in the area of North Sumatera University (USU) listed as Ojek Food Delivery customer. The second purpose is to simplify the customer way in order to choose wanted restaurant, to track the location, distance, and arrival time of food delivery vehicle, by applying the tracking application with the Java Android, PHP, and MySQL functions to implement tracking, thus system of Ojek Food Delivery can operate.

5. Conclusion

Considering traffic jam element and other traffic elements in terms of determining the shortest path, system that has been designed has the purpose to simplify and to help the digitation process of food delivery system with the assistance of freelance vehicle firstly.

References

[1] Batubara EA 2015 Implementasi algoritma learning vector quantization pada prediksi produksi Palm oil di PT. Perkebunan Nusantara I Pulau Tiga *under publication*

[2] Kacaribu RA 2013 Aplikasi Peramalan Produksi Palm oil dengan Metode Regresi Ganda dan *Exponential Smoothing*. *under publication*

[3] Triantoro T, Batubara FR and Fahmi F 2014 Image based water gauge reading developed with ANN Kohonen Proceedings of 2014 *International Conference on Electrical Engineering and Computer Science, ICEECS*

[4] Siswanto A, Tarigan P and Fahmi F 2013 Design of contactless hand biometric system with relative geometric parameters *Proc. of 2013 3rd Int. Conf. on Instrumentation, Communications, Information Technol., and Biomedical Engineering: Science and Technol. for Improvement of Health, Safety, and Environ., ICICI-BME*

[5] Hermantoro and Purnawan RA 2009 Prediksi produksi Palm oil berdasarkan kualitas lahan menggunakan model Artificial Neural Network (ANN) *Agroteknose* **4(2)**: 1-6.

[6] Jannati R 2015 Prediksi produksi panen Palm oil menggunakan jaringan saraf *radial basis function* (RBF) *under publication*

[7] Salamena GG 2011 Pengujian model peramalan deret waktu *sea surface temperature* (SST) teluk ambon luar dengan metode exponential smoothing *Oseanologi dan Limnologi di Indonesia* **37(1)**: 43-55.

[8] Maranata L 2013 Peramalan jumlah produksi kakao di sumatera utara dan konsumsi kakao di Indonesia dengan pemulusan eksponensial ganda metode linier satu parameter dari brown. *under publication*

[9] Siang JJ 2009 Jaringan saraf tiruan & pemrogramannya menggunakan MATLAB. *Penerbit ANDI*: Yogyakarta.

[10] Makridakis S, Wheelwright SC and McGee VE 1999 Metode dan Aplikasi Peramalan *Erlangga*: Jakarta.

IOP Conf. Series: Journal of Physics: Conf. Series **801** (2017) 012039 doi:10.1088/1742-6596/801/1/012039

Detection of Hypertension Retinopathy Using Deep Learning and Boltzmann Machines

B K Triwijoyo[1], Y D Pradipto[2]

[1]Doctor of Computer Science, BINUS University Jakarta, Indonesia
[2]Psychology Department, Faculty of Humanities, BINUS University Jakarta

[1]bambang.triwijoyo@binus.ac.id, [2]ypradipto@binus.edu

Abstract. hypertensive retinopathy (HR) in the retina of the eye is disturbance caused by high blood pressure disease, where there is a systemic change of arterial in the blood vessels of the retina. Most heart attacks occur in patients caused by high blood pressure symptoms of undiagnosed. Hypertensive retinopathy Symptoms such as arteriolar narrowing, retinal haemorrhage and cotton wool spots. Based on this reasons, the early diagnosis of the symptoms of hypertensive retinopathy is very urgent to aim the prevention and treatment more accurate. This research aims to develop a system for early detection of hypertension retinopathy stage. The proposed method is to determine the combined features artery and vein diameter ratio (AVR) as well as changes position with Optic Disk (OD) in retinal images to review the classification of hypertensive retinopathy using Deep Neural Networks (DNN) and Boltzmann Machines approach. We choose this approach of because based on previous research DNN models were more accurate in the image pattern recognition, whereas Boltzmann machines selected because It requires speedy iteration in the process of learning neural network. The expected results from this research are designed a prototype system early detection of hypertensive retinopathy stage and analysed the effectiveness and accuracy of the proposed methods.

1. Introduction

High blood pressure can cause a lot of disturbance on the retina of the eye, one of which is hypertensive retinopathy (HR) where systemic changes produced by arterial hypertension is reflected in the blood vessels of the retina. Most heart attacks occur in patients with undiagnosed high blood pressure symptoms. Hypertensive retinopathy is a hallmark of high blood pressure. Symptoms of hypertensive retinopathy form arteriolar narrowing, while the other major signs are the presence of retinal hemorrhage and Cotton wool spots [1].

Then the diagnosis of early symptoms of hypertensive retinopathy is indispensable for the prevention and accurate treatment. The traditional way to do is through a fundus image analysis to evaluate the presence of hypertensive retinopathy disease and establish a phase of evolution, because the retina is the only place that allows the doctor to examine the blood vessels of the retina directly and by means of non-invasive and the ethics of science this is related to patient privacy [2].

The diameter of retinal blood vessels is one of the most important characteristics for the diagnosis of hypertension, health professionals have been using the ratio of arterial blood vessels of the retina (AVR) to establish the presence of high blood pressure. The diameter of retinal blood vessels is one of the most important characteristics for the diagnosis of hypertension, health professionals have been using the ratio of arterial blood vessels of the retina (AVR) to establish the presence of high blood pressure. This paper proposed the framework to implement Deep Neural Networks and Boltzmann Machines approach to the classification of hypertensive retinopathy stages.

2. Literature Review

A computerized system to facilitate the detection of retinal disease ophthalmologists to review diagnosis and treatment of diseases accurate planning. Hypertension or high blood pressure causes Severe damage to the human eye and vision as hypertensive retinopathy, optic neuropathy and choroidopathy [3]. Retinal disorders (Network Behind the eye) caused by high blood pressure is called hypertensive retinopathy (HR) [4]. Fundus digital image analysis is used to diagnose HR review with track changes picture. The changes including tortuous arterioles, bleeding, hard exudates, cotton wool spots and papilledema.

Important Symptoms of HR abnormal vein width that leads into low ratio of the average diameter of arterial venous (AVR). Retinal diseases can be tracked by changes in blood vessels of the retina hearts. Figure 1 shows the normal fundus image and the image has symptoms of retinal from HR. Research on the identification of hypertensive retinopathy through retinal images have been done before, there are many techniques presented literature to review the automatic detection of HR. [5] proposed a support system for the review in which the HR detection blood vessels are segmented using the Radon transform and the optical disk is detected using the Hough transform and finally counting AVR.

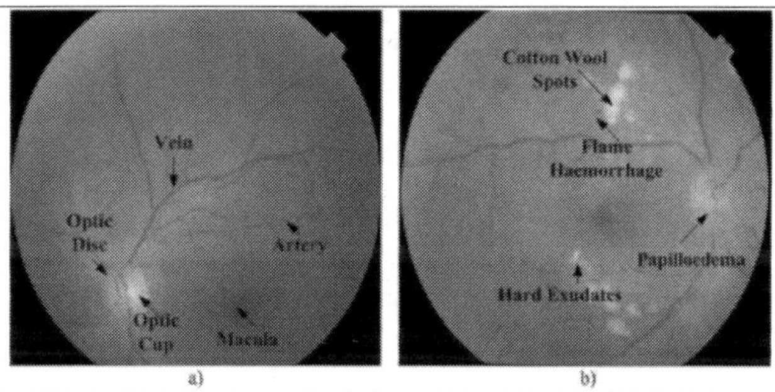

Figure 1. Effect of hypertensive retinopathy on digital fundus image: a) Normal fundus image, b) Fundus image showing symptoms of hypertensive retinopathy

The proposed algorithm was tested on a database DRIVE and showed 92% accuracy. [5]. HR other systems for diagnosis developed by [6]. [7] proposed a system for early detection of HR. Ship segmented using multiscale filtering and identification by region. The width of the vessel and AVR are calculated to find HR. This system was tested in DRIVE and STARE database had an accuracy of 93.7% and 93.1% respectively. The author [8] presents a method for the detection of HR where vessels are segmented using a torque based and features a gray level and support vector machine (SVM) for classification.

The intensity and color information used to classify the vessel as arteries and veins were then used to calculate the ratio of the width of the ship and AV. The proposed system showed 93% accuracy on VICAVR database. [9] proposed a classification method for the measurement of AVR auto ship. Ships segmented and some features based on the intensity of color is extracted for classification of vessels such as arteries and veins by using SVM and ship discriminant analysis (LDA). DRIVE database used to test algorithms and the system showed 96% accuracy. Another HR automated detection method proposed by [10]. The blood vessels of the retina that is detected by the cap and filter transformation of the double ring. Temporal region of the optical disk selected as a destination to calculate the ratio of AV. Features taken from the vessel in the ROI to classify them as arteries and veins using LDA.

2.1. Deep Learning Neural Network

Although many previous studies that utilize ANN-MLP for medical image classification, but the few studies that use Deep Learning for classification of hypertensive retinopathy. Convolutional Neural Network (CNN) is development of Multilayer Perceptron (MLP) that is designed to process two-dimensional data. CNN included in this type of Deep Neural Network for high tissue depth and widely applied to image data.

In the case of image classification, the MLP is less suitable for use because it does not store the spatial information of the image data and consider each pixel is an independent feature that produces poor results as well. CNN first developed under the name NeoCognitron by [11]. The concept is then finalized by [12]. Model CNN named Lenet successfully applied by [11] on his research on the numbers and handwriting recognition. How to work on the MLP CNN have in common, however on CNN every neuron is presented in the form of two-dimensional, unlike the MLP that each neuron is only measuring one dimension of a MLP as in Fig. 3. have the i layer (Red and blue boxes) with each layer containing neurons ji (white circles). MLP received a one-dimensional data input and propagates that data on the network to produce output. [13].

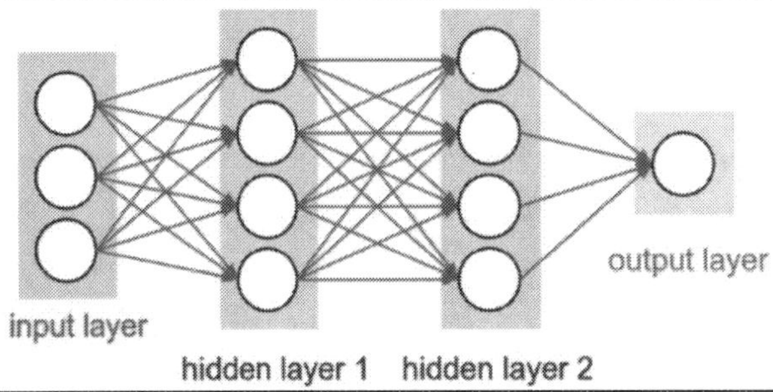

Figure 2. Simple MLP architecture.

Each neuron in the relationship between two adjacent layers have a weight of one-dimensional parameter that determines the quality mode. In each of the input data on a layer linear operation performed by the weight value is, and then computing the results will be transformed using a non linear operation which is called the activation function. On CNN, the data is propagated on the network is a two-dimensional data, so that the linear operation and parameters of different weights on CNN. On CNN linear operation using convolution operation, while the weights are no longer one-dimensional, but the form is a collection of four-dimensional convolution kernel as in figure 3 Dimensional weight on CNN is: neuron input x output x height x width Due to the nature of convolution process, then it can only CNN used on data that have two-dimensional structure such as images and sounds.

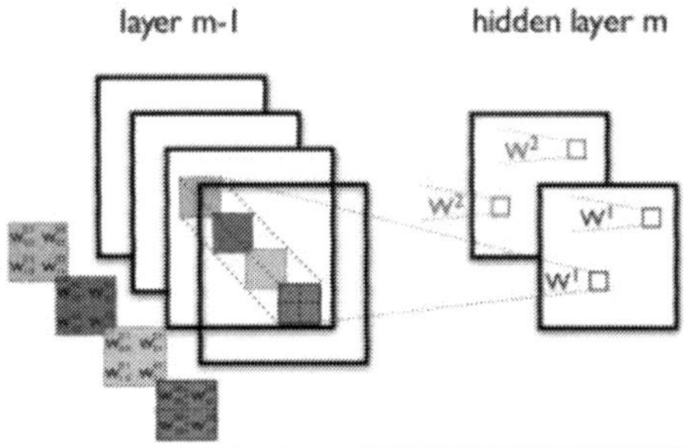

Figure 3. Convolution process on CNN.

2.2. Boltzmann machine

The Boltzmann machine is a particular type of energy-based model with hidden variables, and RBMs are special forms of Boltzmann machines in which P(h|x) and P(x|h) are both tractable because they factorize. In a Boltzmann machine, the energy function is a general second-order polynomial:

$$\text{Energy}(x,h) = -b'x - c'h - h'Wx - x'Ux - h'V\,h \tag{1}$$

There are two types of parameters, which we collectively denote by θ: the offsets b_i and c_i (each associated with a single element of the vector x or of the vector h), and the weights W_{ij} , U_{ij} and V_{ij} (each associated with a pair of units). Matrices U and V are assumed to be symmetric 1, and in most models with zeros in the diagonal. Non-zeros in the diagonal can be used to obtain other variants, e.g., with Gaussian instead of binomial units. The Boltzmann machine energy function can be rewritten by putting all the parameters in a vector d and a symmetric matrix A. [14].

$$\text{Energy}(s) = -d's - s'As \tag{2}$$

Let d−i denote the vector d without the element di, A−i the matrix A without the ith row and column, and a−i the vector that is the ith row (or column) of A, without the ith element. Using this notation, we obtain that P(si|s−i) can be computed and sampled from easily in a Boltzmann machine.

One study using Neural Network for classification hypertensive retinopathy is [15], which utilizes Neural Network and Decision Tree (DT) to determine the area of the arteries or veins using a Naive Bayes classifier and Support Vector Machine (SVM). From the literature review conducted in this study are gaps (gap) that can be used as a reference to subsequent research, namely:

- Required to standardize image acquisition protocols to guarantee the same conditions, retinal images to be taken by trained personnel only.
- It takes a more objective system with the help of automated tools that can monitor disease hypertensive retinopathy.
- It takes a more accurate measurement of AVR for the classification of the arteries and veins to detect retinal disorders.
- The detection of Cotton Wool Spots CWS on the retinal image more accurate eye essential for building a computer system to analyse the symptoms of disease diagnosis at an early stage and at an advanced stage.
- Need to repair blood vessel segmentation algorithm on the image of the eye retina to obtain accurate measurement results ore.
- Keep registration fundus copy image of the same patient on a regular basis (i.e. 6 to 12 months). For comparison automatically changes as the vessel monitoring system that is more efficient and accurate in the diagnosis of hypertensive retinopathy.

The contribution of this research is a computer-based automatic aid system that can monitor disease states hypertensive retinopathy more accurately through the image of the retina.

3. Methodology

Based on the objectives of the research that has been described previously, the following is a proposed framework or approach that will be proposed as shown in figure 4 are as follows:

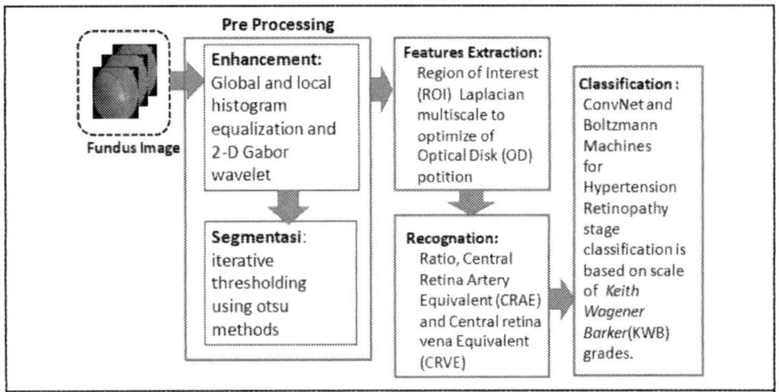

Figure 4. Flowchart of Proposed Methodology.

3.1. Features.
Features used is a combination of the average ratio of the width of arteriovenous (AV) and Optic Disk (OD).

3.2. Data Sets
At the initial staged will use 40 the retinal image size of 768 x 584 pixels in JPEG format on the database DRIVE well as 20 images of the fundus camera the size of 700 x 605 pixels in format PPM STARE database. While at an advanced stage will use 100 fundus-copy images of Eye Hospital, 20 patients were taken at regular intervals (6 to 12 months). to test prototype monitoring system for the diagnosis of stage symptoms of hypertensive retinopathy.

3.3. Stages of the process:
- Enhancement:
  Using Global and local histogram equalization as well as 2-D Gabor wavelet to improve the visibility of the arteries and veins.
- Segmentation:
  Using iterative Otsu thresholding method for determining the threshold value.
- Features Extraction:
  Using the Set Region of Interest (ROI) as a measurement area of arteries and veins of the retina. Optical Disk (OD) of the retina is used to find the ROI where ROI is set three times larger radius OD.
- Recognition:
  Using Multiscale Laplacian to estimate the radius Optic Disk (OD). Calculate the ratio, Central Retinal Artery Equivalent (CRAE) and Central retinal vein Equivalent (CRVE), which is determined by the measurement of the arteries and veins.
- Classification:
  Using Convolutional Neural Network and Boltzmann machines for Retinopathy Hypertension stage classification is based on a scale of Keith Wagener Barker (KWB) grades.

4. Conclusion
The proposed of computer-based automatic aid framework that can monitor disease states hypertensive retinopathy has been presented. the diagnosis of early symptoms of hypertensive retinopathy is indispensable for the prevention and accurate treatment. The diameter of retinal blood vessels is one of the most important characteristics for the diagnosis of hypertension.

This paper proposed the framework to implement Deep Neural Networks and Boltzmann Machines approach to the classification of hypertensive retinopathy stages. The future work of this this research are to design a prototype system early detection of hypertensive retinopathy stage and analyze the effectiveness and accuracy of features combined ratio of diameter arteries and veins (AVR) with a change in the position of Optic Disk (OD) in the retinal image for the classification of hypertensive retinopathy using Deep Neural Network and Boltzmann machines.

References

[1]. Irshad, Samra., Salman, Muhammad.,Akram, M. Usman.,Yasin, Ubaidullah. (2015). Automated detection of Cotton Wool Spots for the diagnosis of Hypertensive Retinopathy. *Proceedings of the 7th Cairo International Biomedical Engineering Conference, CIBEC*. p.121-124.

[2]. Moor, H.J. 1985. What Is Computer Ethics? Metaphilosophy, **16** (4): 266–275.

[3]. R. Katakam, K. Brukamp, R. R. Townsend. (2008). What is the proper workup of a patient with hypertention ?. *Cleve Clin J Med*, Vol. **75**, pp. 663-72.

[4]. Garner, N. Ashton. (1979). Pathogenesis of hypertensive retinopathy: A review, *J R Soc Med*, Vol. **72**, pp. 362-365.

[5]. K. Noronha, K.T. Navya, K.P. Nayak. (2012). Support System for the Automated Detection of Hypertensive Retinopathy using Fundus Images, *International Conference on Electronic Design and Signal Processing (ICEDSP)*, pp.7-11.

[6]. Ortíz, Daniel., Cubides, Mauricio., Suárez, Andrés., Zequera, Martha., Quiroga, Julián., Gómez, Jorge., Arroyo, Nubia. (2010). Support system for the preventive diagnosis of hypertensive retinopathy. *Annual International Conference of the IEEE Engineering in Medicine and Biology Society*, EMBC'10. p.5649-5652.

[7]. Manikis, Georgios C., Sakkalis, Vangelis., Zabulis, Xenophon., Karamaounas, Polykarpos. ,Triantafyllou, Areti., Douma, Stella. (2011). An Image Analysis Framework for the Early Assessment of Hypertensive Retinopathy Signs. *Proceedings of the 3rd International Conference on E-Health and Bioengineering - EHB*, 24th-26th November, 2011, Iaşi, Romania.

[8]. K. Narasimhan, V.C. Neha, K. Vijayarekha (2012). Hypertensive Retinopathy Diagnosis from fundus images by by estimation of AVR, *International Conference on Modeling Optimization and Computing (IC- MOC), Procedia Engineering*, Vol. **38**, pp. 980-993.

[9]. Q. Mirsharif, F. Tajeripour, H. Pourreza. (2013). Automated characterization of blood vessels as arteries and veins inretinal images, *Computerized Medical Imaging and Graphics*, Vol. **37**, pp. 607- 617.

[10]. Muramatsu, Y. Hatanaka, T. Iwase, T. Hara, H. Fujita. (2010). Automated detection and classification of major retinal vessels for determination of diameter ratio of arteries and veins, *Proc. of SPIE* Vol. **7624**, Medical Imaging.

[11]. Fukushima, K. (1988). Neocognitron: A hierarchical neural network capable of visual pattern recognition. *Neural Networks*, **1**(2), 119–130.

[12]. LeCun, Y., Jackel, L. D., Bottou, L., Cortes, C., Denker, J. S., Drucker, H.,Vapnik, V. (1995). Learning algorithms for classification: A comparison on handwritten digit recognition. *Neural Networks: The Statistical Mechanics Perspective*, 261–276.

[13]. Shan, Juan, Li, Lin. (2016). Review: A Deep Learning Method for Microaneurysm Detection in Fundus Images. *IEEE First Conference on Connected Health: Applications, Systems and Engineering Technologies*, 978-1-5090-0943-5/16, pp 357-358.

[14]. Nair, V., & Hinton, G. E. (2010). Rectified Linear Units Improve Restricted Boltzmann Machines. *Proceedings of the 27th International Conference on Machine Learning*, 807–814. https://doi.org/10.1.1.165.641.

[15]. Abbasi, Uzma Gulzar, Akram, Usman M. (2014). Classification of blood vessels as arteries and veins for diagnosis of hypertensive retinopathy. *International Computer Engineering Conference*. 5-9.

IOP Conf. Series: Journal of Physics: Conf. Series **801** (2017) 012040 doi:10.1088/1742-6596/801/1/012040

Russian Character Recognition using Self-Organizing Map

D Gunawan[1*], D Arisandi[1], F M Ginting[1], R F Rahmat[1] and A Amalia[2]

[1]Department of Information Technology, Faculty of Computer Science and Information Technology University of Sumatera Utara, Jl. dr. Mansur No. 9 Kampus USU Medan 20155

[2]Department of Computer Science, Faculty of Computer Science and Information Technology University of Sumatera Utara, Jl. dr. Mansur No. 9 Kampus USU Medan 20155

[*]Email: danigunawan@usu.ac.id

Abstract. The World Tourism Organization (UNWTO) in 2014 released that there are 28 million visitors who visit Russia. Most of the visitors might have problem in typing Russian word when using digital dictionary. This is caused by the letters, called Cyrillic that used by the Russian and the countries around it, have different shape than Latin letters. The visitors might not familiar with Cyrillic. This research proposes an alternative way to input the Cyrillic words. Instead of typing the Cyrillic words directly, camera can be used to capture image of the words as input. The captured image is cropped, then several pre-processing steps are applied such as noise filtering, binary image processing, segmentation and thinning. Next, the feature extraction process is applied to the image. Cyrillic letters recognition in the image is done by utilizing Self-Organizing Map (SOM) algorithm. SOM successfully recognizes 89.09% Cyrillic letters from the computer-generated images. On the other hand, SOM successfully recognizes 88.89% Cyrillic letters from the image captured by the smartphone's camera. For the word recognition, SOM successfully recognized 292 words and partially recognized 58 words from the image captured by the smartphone's camera. Therefore, the accuracy of the word recognition using SOM is 83.42%

1. Introduction

Typing Cyrillic letters to search for Russian word translation in smartphone's digital dictionary might be as difficult as searching by using a dictionary book. As well as its other neighbor countries, Russia uses letters called Cyrillic instead of Latin. Most of the people who visit Russia are not familiar with the Cyrillic letters. This problem will affect the ability of the visitors to translate the Russian word.

From the statement mentioned earlier, the problem in translating Russian word is how to provide the input word to the digital dictionary. It is difficult to search the Russian word in dictionary book or type it in digital dictionary either. According to those problems, how if we change the way users input the word to the dictionary? Image that is captured by smartphone's camera can be the alternative way to input the word to the digital dictionary. With this approach, we expect the application will help 28 million visitors who visit Russia [1] to translate the Russian word easier.

However, using the image from the smartphone's camera emerges a new problem. As the computer cannot read the text directly from the image, it needs an image recognition technique to read the text in the image. Quraishi provided a new approach for image recognition using Artificial Neural Networks [2]. It calculated total average error on generated data matrix networks on test data matrix to check whether proper identification can be made or not. The observation result that the value of average error is less than that of the test image without application of artificial neural network. In image character recognizing, Bai proposed a Shared-Hidden-Layer Deep Convolutional Neural Network (SHL-CNN) [3]. SHL-CNN can reduce recognition error by 16-30% relatively compared with conventional CNN. This result verified the effectiveness of the learned SHL-CNN on both English and Chinese image character. Another research that utilizes Self-Organizing Map (SOM) to recognize speech has been conducted [4]. It obtains maximum feature Intensity as 98.17% for Mean-SOM performance and 98.54% for Median-SOM performance.

This paper is organized as follows. In section 2, we describe the research methodology to recognize the Russian character. In section 3, we discuss the result of the research. Finally, we draw some conclusions section 4.

2. Research Methodology

Generally, as shown in figure 1, there are several steps to recognize Russian character such as image input, cropping preferred word, image pre-processing, feature extraction and image recognition using Self-Organizing Map algorithm [5]. The client provides image input either from smartphone's gallery or directly from camera. As the image input may contain more than one word or other objects, it should be cropped to separate it with unnecessary objects in the image. As the result, the image input will contain only the preferred word.

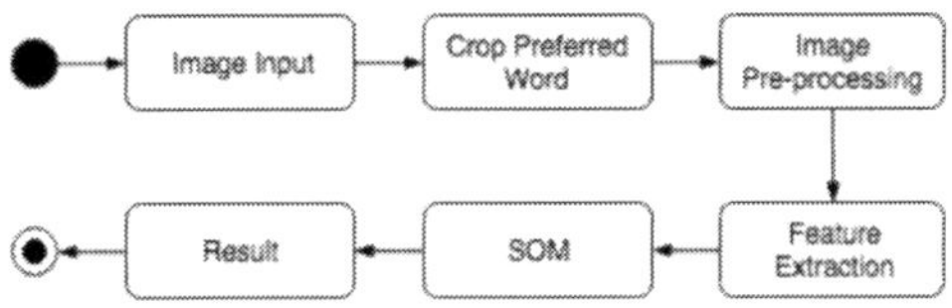

Figure 1. Russian character recognition method

To enhance the image input, some image pre-processing such as noise filtering, binary image processing, segmentation, scaling and thinning are applied. The first step of image pre-processing is noise filtering. Usually images taken by the digital camera will contain noise. In this research, application utilizes median filter which is much better at preserving straight edge structure than Gaussian smoothing [6] to reduce noise. Application applies median filter with neighbor size 3x3. The process will be started from the first pixel in coordinate 0,0 to the last pixel. Neighbor size 3x3 means the process picks the neighbor coordinate of the current coordinate starts from x-1, y-1 to x+1, y+1. Notation x and y represent current coordinate (x, y). The result will be assembled to a matrix. From the matrix, the color values are sorted and then the median value of red, green and blue (RGB) are calculated.

The next step after noise filtering is binary image process. Binary image means the image only consists of two colors, black and white. In this step, binary image is yielded by implementing Otsu algorithm [7]. This algorithm only accepts grayscale image as input. Therefore, grayscale process is applied to the image before implementing Otsu algorithm. This algorithm needs a threshold value. Pixel color which exceeds threshold value will be converted to white, otherwise it will be converted to black. The expected result of this step is the characters in the image will be black and the background will be white.

After binary image processing, the next step is segmentation. In this research, Self-Organizing Map is designed to recognize a character. The word should be separated into characters which is represented in black pixel. The process involved filtering out the white pixel and keeping the black pixel. The segmentation process will consider two stick objects in the image as one object. Therefore, this system will not recognize handwriting font. As the result, segmentation process yields images which contain each character from the word.

The images from the segmentation process will have different size. The size is scaled to 20x25 pixels to uniform the image size for thinning process. The thinning process is done to obtain character's skeleton. The character is thinned down to a skeleton of unitary thickness, one pixel. This research utilizes Zhang-Suen algorithm to obtain the character's skeleton [8]. This process should be done because the thickness of the same character of different fonts are not equal. Usually the same character of different fonts will share the similar character's skeleton.

At the end of image pre-processing, the result is array of images that contain a character for each image. Every character in the image will be extracted with pixel mapping method. Value 1 will be given to the black pixel, and 0 will be given to the white pixel. The extracted value will be saved in a string. Because the size of each image is 20x25 pixels, then the size of the string is 20x25 = 500 characters. Value extracted from pixel in coordinate (0,0) and coordinate (0,1) will be put in string index 0 and 1 respectively, and so forth. String from the feature extraction process will be sent to server for character recognition process.

Character recognition is done by utilizing Self-Organizing Map algorithm, one of the neural network branch. Training is required to be done to the system in order to utilize this algorithm for character recognition. Training uses unsupervised method with learning rate 0.1. The system is trained with 11 random fonts which support Cyrillic letters. Every font consists of 33 lowercase and 33 uppercase letters.

Three methods are used to test the accuracy of Self-Organizing Map implementation. The first method generates images to represent each character of 15 random fonts that support Cyrillic character. Computer will generate 33 images of lowercase letters and 33 images of uppercase letters for each font. These images will be tested against the earlier training dataset. The second method, three fonts are randomly chosen. All the lowercase and uppercase letters of the chosen fonts are printed in white background paper. Those characters will be captured by using smartphone's camera in a room. The camera resolution is 8 Megapixel. Third method, the test is done by capturing the Cyrillic word from posters, leaflets, and newspapers using the smartphone's camera. The accuracy of testing dataset will be calculated with the formula in (1). All photos are taken twice, and the best picture will be taken to be the dataset input.

$$Accuracy\ (\%) = \frac{Recognized\ Pattern}{Sum\ of\ Testing\ Pattern} x100\% \tag{1}$$

3. Result

The process of character recognition is divided into two separate parts, user's smartphone and server. User's smartphone is used as image input and image enhancement. Server is used to do the character recognition process.

Some image pre-processing such as noise filtering, binary, segmentation and scaling are done before applying Self-Organizing Map to a cropped image. For example, we will use the cropped image as shown in figure 2.

Figure 2. Cropped image as input

The first image pre-processing step is noise filtering. As mentioned above, in this research median filter is used to reduce noise in an image. The result of the noise filtering is shown in figure 3.

Figure 3. Image after noise filter being applied

The next step is converting the image to black and white image. This step is called binary process. The color which exceed the threshold will be treated as white pixel, otherwise will be treated as black pixel. The result after binary process is shown in figure 4.

Figure 4. Image after binary image process

Self-Organizing Map will not recognize the whole word. It will recognize a character only. To support that requirement, character separation from the word is applied to the image. This action is called segmentation. As the result of segmentation process, there are new images for each character as shown in figure 5. In this process, failure to separate character might happen. As shown in figure 6, this failure is caused by more than one character are too close or stick with the character next to it.

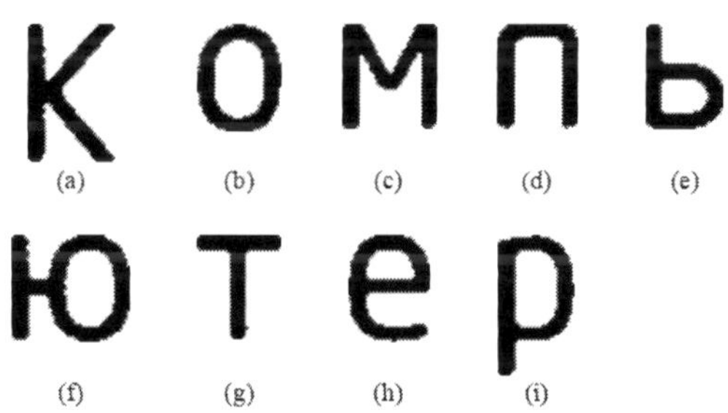

Figure 5. New images (a-i) after segmentation

Figure 6. Segmentation failure

Each character has its own image after segmentation process. The next process is scaling. The image is scaled to 20x25 pixels. The result is shown in figure 7.

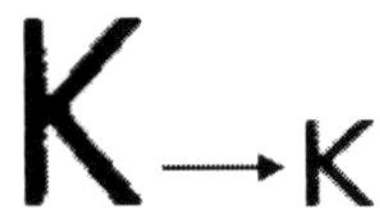

Figure 7. Image after scaled down to 20x25 pixels

After being scaled, all new images are thinned down to yield character's skeleton image. The result of the thinning process is shown in figure 8.

Figure 8. Image after thinning process

After thinning process, the image is ready to be extracted. Extracted feature string size is 500 characters long, based on image width and height.

For training dataset, the system has been trained with 11 random fonts that support Cyrillic. They are, Arial, Arial Black, Batang, Calibri, Cambria, Comic Sans, Courier New, Lucida Sans, Tahoma, Times New Roman, and Verdana.

To test the character and word recognition, three testing methods has been conducted. The first method, computer generates images of each character from 15 random fonts. The character is represented by the black pixel, and the background is white. As shown in table 1, Copperplate Gothic and Corbel font are fully recognized by the system. After being tested against training data, the accuracy of the data testing is 89.09%.

Table 1. Data Testing for Method 1

Font type	Recognized	Not Recognized
Arial Narrow	65	1
Book Antiqua	54	12
Lucida Console	63	3
Calisto MT	57	9
Candara	65	1
Century Gothic	55	11
Consolas	63	3
Constantia	53	13
Copperplate Gothic	66	0
Corbel	66	0
Georgia	55	11
Impact	46	20
MS Sans Serif	64	2
Palatino Linotype	46	20
Segoe UI	64	2

Second testing method is similar with the first method, unless all of the letters are printed and then they are captured by the smartphone's camera. Also, in this method, only three fonts are used. They are

Arial Narrow, Book Antiqua and Lucida Console as shown in Table 2. If we compare to the first method, the character recognition is slightly drop to 88.89% when images were taken by the smartphone's camera.

Table 2. Data Testing for Method 2

Font type	Method 1		Method 2	
	Recognize	Fail	Recognize	Fail
Arial Narrow	65	1	64	2
Book Antiqua	54	12	64	2
Lucida Console	63	3	48	18

The third method, testing is done to recognize a full word. The word image is taken by using smartphone's camera. Source of the image is posters, leaflets and newspapers. They are taken directly in outdoor at noon. From 350 photos of different words taken, the system is successfully recognized 292 words, while 58 other words were partially recognized. Some letters were not successfully recognized. The accuracy for the word recognition is 83.42%.

4. Conclusion

Character recognition by Self-Organizing Map (SOM) from images that generated by computer or captured by smartphone's camera has slightly different accuracy. This means the image source does not have much influence to the character recognition accuracy.

Partial word recognition is caused by the failure of recognizing some letters, not the whole letters in a word. For future works, there are some approaches that can be taken to increase word recognition accuracy. First, enhance the segmentation process in order to obtain clearly separate images for closed characters in a source image. Second, apply a prediction algorithm to predict the word based on the partially recognized character.

References

[1] UNWTO 2014 UNWTO Tourism Highlight, 2014 (World Tourism Organization)

[2] Quraishi M I, Choudhury J P and De M 2012 Image recognition and processing using Artificial Neural Network *Recent Advances in Information Technology (RAIT), 2012 1st Intl. Conf. on* 95-100 (IEEE)

[3] Bai J, Chen Z, Feng B and Xu B 2014 Image character recognition using deep convolutional neural network learned from different languages *Image Processing (ICIP), 2014 IEEE Intl. Conf. on* 2560-2564 (IEEE)

[4] Venkateswarlu R L K and Kumari R V 2011 Novel approach for speech recognition by using self — Organized maps *Emerging Trends in Networks and Computer Communications (ETNCC), 2011 Intl. Conf. on* 215-222 (IEEE)

[5] Kohonen T 1990 The self-organizing map *Proc. of the IEEE* 1464-1480 (IEEE)

[6] Smith S M and Brady J M 1997 SUSAN—A New Approach to Low Level Image Processing *International Journal of Computer Vision* **23** 45-78

[7] Otsu N 1979 A Threshold Selection Method from Gray-Level Histograms *Systems, Man and Cybernetics, IEEE Transactions on* **9** 62-66 (IEEE)

[8] Zhang T Y and Suen C Y 1984 A fast parallel algorithm for thinning digital patterns *Communications of the ACM* **27** 236-239 (ACM)

ANOMALY DETECTION OF FLIGHT ROUTES THROUGH OPTIMAL WAYPOINT

M Y Pusadan[1,2], J L Buliali[1] and R V H Ginardi[1]

[1]Informatics Department, Institute Technology of Sepuluh Nopember
[2]Informatics Department, University of Tadulako

Abstract. Deciding factor of flight, one of them is the flight route. Flight route determined by coordinate (latitude and longitude). flight routed is determined by its coordinates (latitude and longitude) as defined is waypoint. anomaly occurs, if the aircraft is flying outside the specified waypoint area. In the case of flight data, anomalies occur by identifying problems of the flight route based on data ADS-B. This study has an aim of to determine the optimal waypoints of the flight route. The proposed methods: i) Agglomerative Hierarchical Clustering (AHC) in several segments based on range area coordinates (latitude and longitude) in every waypoint; ii) The coefficient cophenetics correlation (c) to determine the correlation between the members in each cluster; iii) cubic spline interpolation as a graphic representation of the has connected between the coordinates on every waypoint; and iv). Euclidean distance to measure distances between waypoints with 2 centroid result of clustering AHC. The experiment results are value of coefficient cophenetics correlation (c): $0{,}691 \leq c \leq 0974$, five segments the generated of the range area waypoint coordinates, and the shortest and longest distance between the centroid with waypoint are 0.46 and 2.18. Thus, concluded that the shortest distance is used as the reference coordinates of optimal waypoint, and farthest distance can be indicated potentially detected anomaly.

Keyword: Waypoint; Agglomerative Hierarchical Clustering(AHC); ADS-B; Anomaly Detection

1. Introduction

Flight is an integrated system consisting of the use of airspace, aircraft, airports, air transport, air navigation, safety and security, environment, and support facilities and other public facilities. The flight route is part of the data sources monitoring and surveillance of the flights. The flight route contains at least five parameters, including: a) the name of the flight path; b) the reference point and the coordinates (longitude and latitude); c) direction (track) leading to or from a point of reference; d) the distance between the reference point; and e) limits the lowest safe altitude (Minimum Safe Altitude / MSA) [1]. Waypoint on the flight is the reference point / set of coordinates (latitude and longitude) used for the purposed of air navigation. According to the Regulation of the Minister of Transportation No. 14 of 2009 on the Rules of Civil Aviation Safety (Civil Aviation Safety Regulations) sub part 170 051, that: observation data (surveillance) is data derived from radar equipment primary and secondary, or other systems (ADS-B, ADS-C) is used as air traffic services, should be recorded (record) automatically for use in accident and incident investigations, search and rescue, as well as the evaluation of air traffic control and system monitoring and surveillance [2].

Anomaly detection is part of data mining to determine a data (objects) that have different characteristics with the other data sets, called is exception mining. In the case of flight, one of the anomaly detection approach based on the anomalies that occur on the flight route. Anomaly detection of flight route can be obtained by determining the optimal waypoint, therefore the flight route that is far from optimal waypoint can be indicated potential anomalies.

2. Literatures Review

Anomaly detection/outlier detection can be seen several approaches[3]. The detection by using of cluster analysis algorithm "agglomerative hierarchical clustering". These methods: 1). Begins by sampling data from the data source; 2) on the data samples, obtained by clustering the data based algorithms "agglomerative hierarchical clustering"; 3). The results of the clustering process as an accurate sample[4],

This process if found gaffe in the build of noise that value is less than δ, then released; 4) the mathematical approach μ and standard deviation σ obtained until accurate. For the determination of outliers:

$$\mu = E(X)\ dan\ \sigma = \sqrt{D(X)} \quad (1)$$

The Outlier detection method with the approach distance[5] which is an outlier detection method of calculating the distance between data points. In the DB (Distance Based)-Outlier detection, which is the centre of Nearest Neighbourhood. At the time it was founded more than data points that are in the M-Neighbourhood D object, then the object is not an outlier. The dataset used in high-dimensional data characteristics.

In the case of aviation, conducted by the clustering[6][7]. The foremost step is outlier/anomaly analysis, by dividing of flight position (specific event) with two phases, namely the takeoff phase with 68 parameters and the approach phase with 69 parameters, then did the dimension reduction (using PCA), and to examine the cluster be used "DBSCAN (Density-based Spatial Clustering Algorithm with Noise) as the cluster method with density approach.

Some other approach is statistical[8]. The method used is ***two-side median filtering*** by calculating the median m_t, of the data point's local (signal altitude) y_t, then compared the specific threshold value.

$$if\ |y_t - m_t^k| < \tau, \text{keeps } y_t; \quad (2)$$

else $|y_t - m_t^k| \geq \tau$, y_t as **outlier** and replaced y_t with m_t^k

Along the flight routes, must be determined based on traffic of the aircraft. It intended that one aircraft to another aircraft that are in the same traffic determined by separation indicator value. So that anomalies (conflict/collision)[9] data on the detection based on the separation ($R_{CAZ} = 528ft$ and $H_{CAZ} = 200ft$, R = Radius, H = Horizontal, CAZ = Aircraft Collision Zone). Therefore, the flight position or flight level consists of takeoff, cruise, approach, and landing, which allows for the accurate detection of anomalies of the aircraft cruising level or in cruise phase[10].

The most important thing is also considered in the scope of anomaly detection of flight is the determination of flight waypoints[11] which is the specific location of the flights that the accurately to flying past the points of reference before reaching its destination.

The following point of finding the optimal waypoint based on the reference dot on the map of flight routes which occurs in one flight routes (call sign of flight). Established along the point of reference, the cubic spline interpolation process is carried out[12], so get waypoints.

From the waypoints, segmentation and clustering processed[13]. so that, the cluster can be analysed with to calculating the cophenetic correlation coefficient (C)[14][15] for each segment, so that obtained *centroids* that become the new points of reference in waypoints.

3. Proposed Method

The research framework shown in Figure 1, as the sequence of the process of determining the optimal waypoint:

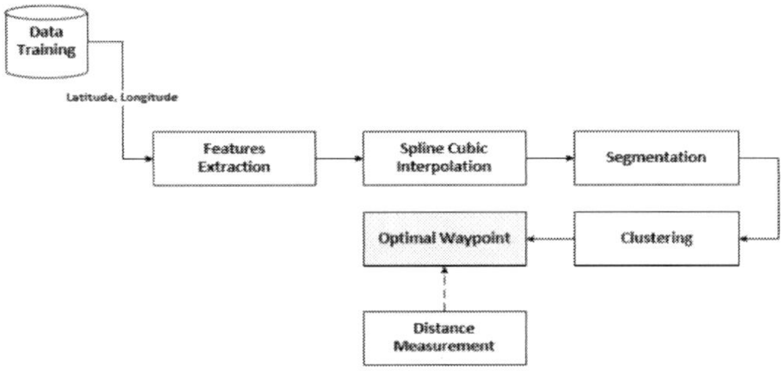

Figure 1. Block Diagram Process

3.1. The Algorithm Determination of the Optimal Waypoint

The algorithm that determination of the optimal waypoint is shown as follows:

```
pointsRf={(LatRf1,LongRf1),(LatRf2,LongRf2),…,(LatRfn,LongRfn)}
waypointRf←pointsRf
get dataTrain

datastart←waypointRf1
dataend←waypointRfk
i←1
while (dataTrain ≥ datastart and dataTrain ≤ dataend) do
  segment←dtTrain
  i←i+1
endwhile

[c,centroid]←clusterAHC(segment,k←2) //prosess clustering AHC
Result: centroid1 and centroid
//mengukur distance menggunakan eucliean distance
dist1←dist(waypointRf, centroid1)
dist2←dist(waypointRf, centroid2)
if dist1<dist2 then
      waypointoptimal ← dist1
  else
      waypointoptimal ← dist2
end
```

3.2. Data Source (ADS-B Data)

ADS-B data used in the research originated from flightradar24[16] with the data format is Comma Separated Values (CSV), with attributes as follows: timestamp, UTC, call sign, position, altitude, speed, and direction. The following on table 3.1 shows the attributes of data ADS-B:

Table 1. Data Source (ADS-B Data)

TIME-STAMP	UTC	CALL-SIGN	POSITION	ALTI-TUDE	SPEED	DIREC-TION
1472721080	2016-09-01T09:11:20Z	LNI860	-0.88474,119.856743	3650	173	161
1472721070	2016-09-01T09:11:10Z	LNI860	-0.87789,119.854507	3850	173	161
...	...	...	...	...	...	...

3.3. Features Extraction

Data sources (ADS-B data), converted to the DBMS (MySQL) tables, to the computational. The following on table 3.2 below, shows the latitude and longitude attributes, parsing from attribute "position" that's shown on the coordinates x, y.

Table 2. Table (DBMS MySQL) with attributes: Latitude and Longitude are expressed individually

TIME-STAMP	DATE	CALL-SIGN	ALTI-TUDE	SPEED	LATI-TUDE	LONGI-TUDE
1472721080	01-09-16	LNI860	3650	173	-0.88467	119.8567
1472721070	01-09-16	LNI860	3850	173	-0.8779	119.8545
...	...	...	...	...	...	...

The following shows the flight routes to the destination of Surabaya to Palu (call sign: LNI860) and the distribution of data for 30 days (01st Nov 2016 until 30th Nov 2016).

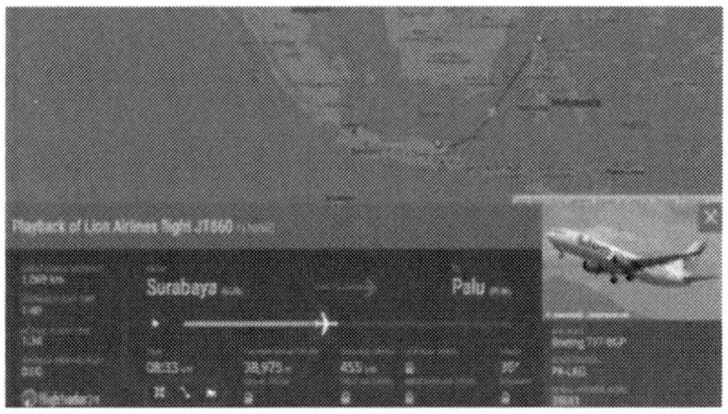

Figure 2. Flight route (call sign: LNI860/SUB-PLW)

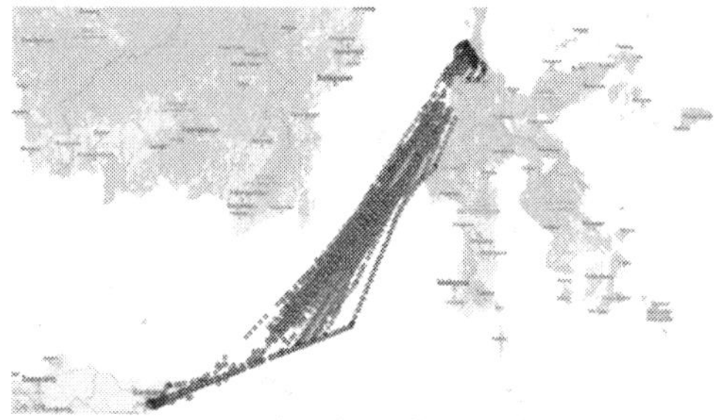

Figure 3. Mapping data distribution of flight with based on call sign aircraft: LNI860 (SUB-PLW)

3.4. Spline Cubic Interpolation

Assumed that the points of reference interpolation are: W_1 = (-2.4687,118.84); W_2 = (-4.172, 117.485001); W_3 = (-5.8717, 117.221703); W_4 = (-6.5922, 115.0785); and W_5 = (-7.0568, 113.8238). Established along the method of cubic spline interpolation equation:

$$S = \begin{cases} S_1(x), \ if \ x_1 \le x \le x_2 \\ S_2(x), \ if \ x_2 \le x \le x_3 \\ \vdots \\ S_{n-1}, \ if \ x_{n-1} \le x \le x_n \end{cases} \tag{3}$$

$$S_i(x) = a_i(x - x_i)^3 + b_i(x - x_i)^2 + c_i(x - x_i) + d_i \tag{4}$$

$$S_i'(x) = 3a_i(x - x_i)^2 + 2b_i(x - x_i) + c_i \tag{5}$$

$$S_i''(x) = 6a_i(x - x_i)^2 + 2b_i \tag{6}$$

Obtainable, equation to interpolation of spline cubic:
$S_1(x) = -1.25x^3 + 2.3x^2 + 1.9x + 113.82$; $S_2(x) = -1.25x^3 - 0.56x^2 + 3.23x + 115.078$; $S_3(x) = 0.6x^3 - 2.15x^2 + 2.075x + 117.22$; $S_4(x) = -0.25x^3 + 0.1x^2 - 0.032x + 117.48$; and $S_5(x) = -0.25x3 - 0.36x2 + 0.1x + 118.84$.

Thus, graph of equation to interpolation of spline cubic is generated:

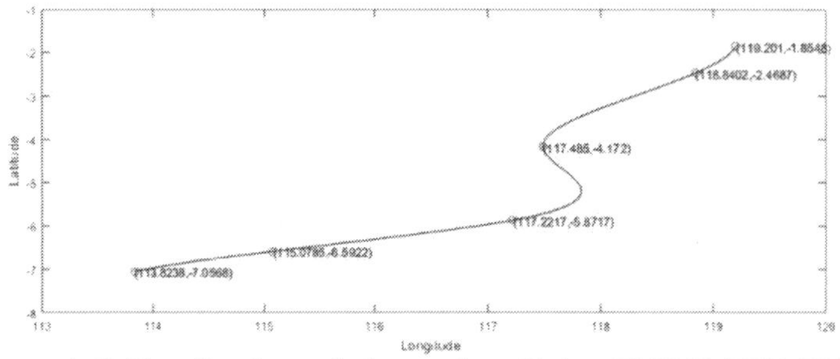

Figure 4. Cubic spline interpolation on the call sign: LNI860 (SUB-PLW)

3.5. Segmentation

Segmentation is the area located between the one waypoint to another waypoint in the same flight routes. Its formulation, as follows: $W_1 \leq SEGMENT_1 \leq W_2$, $W_2 \leq SEGMENT_2 \leq W_3,\ldots,W_n \leq SEGMENT_n \leq W_{n+1}$ (n=1,2, … , k). Hence, range area was obtained:

Table 3. Segmentation based on area range waypoints

RANGE AREA (Latitude, Longitude	SEGMENTASI	WAYPOINTS
(-2.4687, 118.84) – (-1.8548, 119.20)	$SEGMENT_1$	$W_1 - W_2$
(-4.1720, 117,48) – (-2.4687, 118.84)	$SEGMENT_2$	$W_2 - W_3$
(-5.8717, 117.22) – (-4.1720, 117.48)	$SEGMENT_3$	$W_3 - W_4$
(-6.5922, 115.08) – (-5.8717, 117.22)	$SEGMENT_4$	$W_4 - W_5$
(-7.0560, 113,82) – (-6.5922, 115.07)	$SEGMENT_6$	$W_5 - W_6$

3.6. Clustering

Cluster method used is Agglomerative Hierarchical Clustering (AHC), because: the data to be grouped in a cluster is determined by the distance (Euclidean distance), so that the proximity between data become determined to forming a cluster. In this experiment, the "complete linkage" in the method of agglomerative hierarchical clustering (AHC) is used. The AHC formula with complete linkage clustering:

$$d(A,B) = \frac{max}{x \in A, y \in B} \{S_{xy}\} \quad 7$$

S_{xy} : the distance between the two data (x and y) are each derived from cluster A and B.
then, analyze the cluster based on the value of cophenetic correlation coefficient (c). And then, that the value: $0 \leq c \leq 1$, formula c:

$$c = \frac{\sum_{i<j}(x(i,j)-x)(t(i,j)-t)}{\sqrt{[\sum_{i<j}(x(i,j)-x)^2][\sum_{i<j}(t(i,j)-t)^2]}} \quad 8$$

x(i,j) : distance between object i and object j in x;
t(i,j): cophenetic distance between object i and object j of t(:,3);

3.7. Clustering Segmentation

Some of the steps are performed on clustering segmentation, such as: 1). Data training from the first day until thirtieth day, each the segment divided based on the range area of waypoints; 2). Clustering (AHC Clustering) in "data training" each the segment for one day of flight. Hence, obtained centroid of each cluster, as follows:

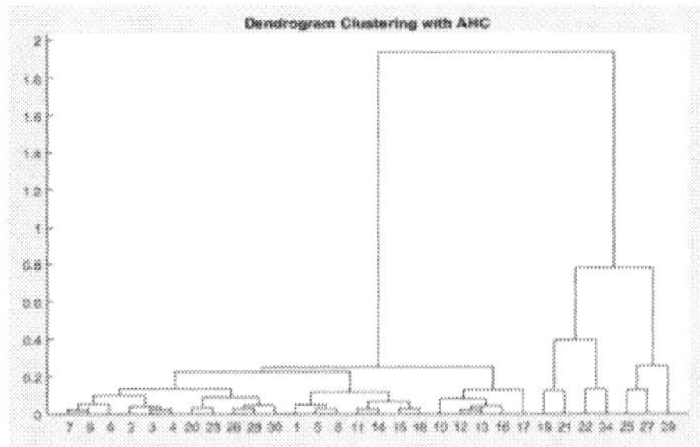

Figure 5. Dedrogram clustering AHC

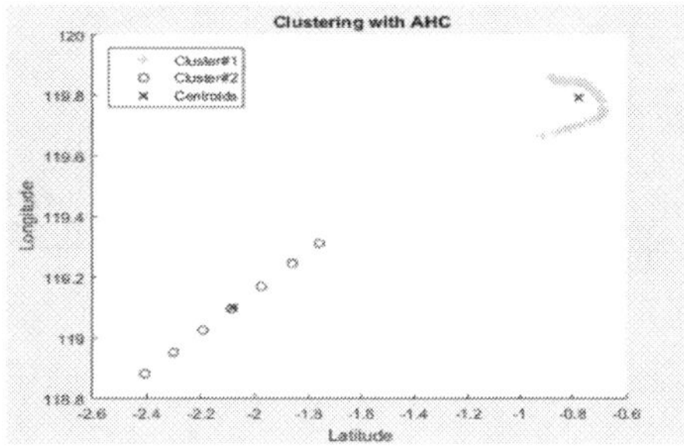

Figure 6. Cluster-1 and Cluster-2

(Data processing in segment-1)

4. Experiment Results

Based on the results of clustering AHC of data in segmentation (segment 1 to segment 5), it obtains the results shown in the following table:

Table 4. Centroid, Segmentation, and Cophenetic Correlation Coefficient

SEGMENTATION	Clustering AHC		
	Centroid-1	Centroid-2	Cophenetic Corr. Coeff.
$SEGMENT_1$	-0.784, 119.793	-2.083, 119.098	0.97424
$SEGMENT_2$	-2.764, 118.610	-3.626, 117.821	0.69170
$SEGMENT_3$	-5.544, 116.041	-4.730, 116.798	0.70207
$SEGMENT_4$	-5.999, 115.616	-6.450, 115.194	0.89212
$SEGMENT_5$	-7.3481, 113.012	6.8742, 114.340	0.88480

4.1. Distance Waypoints and Centroids Segment

To generate optimal waypoints, be determined by measuring the distance between the waypoint and the centroid. Smaller distance as points of reference for the new one. For the measure distance using the Euclidean Distance equation:

$$d(x,y) = \sqrt{(x_1 - y_1)^2 + (x_2 - y_2)^2 + \cdots + (x_n - y_n)^2} \quad 9$$

The following table shows the value of distance between waypoints and centroids.

Table 5. Distance waypoints and centroids

WAYPOINTS		CENTROIDS SEGMENTATION		DISTANCE
RANGE WAYPOINTS	POINTS	SEGMEN-TATION	CENTROID	
$W_1 - W_2$	(-1.8548, 119.201) –	$SEGMENT_1$	C_{11} = (-0.77843, 119.7926)	1.2282
	(-2.4868, 118.840)		C_{12} = (-2.08283, 119.0977)	**0.4638**
$W_2 - W_3$	(-2.4687, 118.840) –	$SEGMENT_2$	C_{21} = (-3.77079, 117.6875)	1.7378
	(-4.1720, 117.4850)		C_{22} = (-2.90686, 118.4808)	**1.6089**
$W_3 - W_4$	(-4.1720,117.4850) –	$SEGMENT_3$	C_{31} = (-5.40080, 116.1750)	1.7927
	(-5.8717, 117,2217)		C_{32} = (-4.58718, 116.9320)	**1.3166**
$W_4 - W_5$	(-5.8717,117,2217) –	$SEGMENT_4$	C_{41} = (-5.99902, 115.6165)	2.0971
	(-6.5922, 115.0785)		C_{42} = (-6.45084, 115.1944)	**0.7987**
$W_5 - W_6$	(-6.5922,115.0785) –	$SEGMENT_5$	C_{51} = (-7.34812, 113.0122)	2.1859
	(-7.056, 113.82380)		C_{52} = (-6.87422, 114.3403)	**0.5443**

4.2. Spline Cubic Interpolation pada Centroid

There are two groups of centroid is be determined based on the distance ($dist_{centroid}$), namely: $dist1_{centroid}$ and $dist2_{centroid}$, obtainable is: $dist1_{centroid} \geq dist2_{centroid}$.

dist1$_{centroid}$ = {C_{11} = (-0.77843, 119.7926), C_{21} = (-3.77079, 117.6875), C_{31} = (-5.40080, 116.1750), C_{41} = (-5.99902, 115.6165), C_{51} = (-7.34812, 113.0122)}

Equation Cubic Spline Interpolation in **dist1**$_{centroid}$:
$S_{11} = 0.27x^3 - 1.40x^2 + 3.33x + 113.01$; $S_{12} = 0.27x^3 - 0.31x^2 + 1.02x + 115.62$; $S_{13} = -0.02x^3 + 0.05x^2 + 0.90x + 116.04$; $S_{14} = -0.02x^3 - 0.08x^2 + 0.82x + 118.61$

dist2$_{centroid}$ = {C_{12} = (-2.08283, 119.0977), C_{22} = (-2.90686, 118.4808), C_{32} = (-4.58718, 116.9320), C_{42} = (-6.45084, 115.1944), C_{52} = (-6.87422, 114.3403)}

Equation Cubic Spline Interpolation in **dist2**$_{centroid}$:
$S_{21} = 0.19x^3 - 0.99x^2 + 2.40x + 114.34$; $S_{22} = 0.19x^3 - 0.75x^2 + 1.66x + 115.19$; $S_{32} = -0.07x^3 + 0.23x^2 + 0.76x + 116.80$; $S_{42} = -0.07x^3 - 0.01x^2 + x + 117.82$

The graph of Cubic Spline Interpolation:

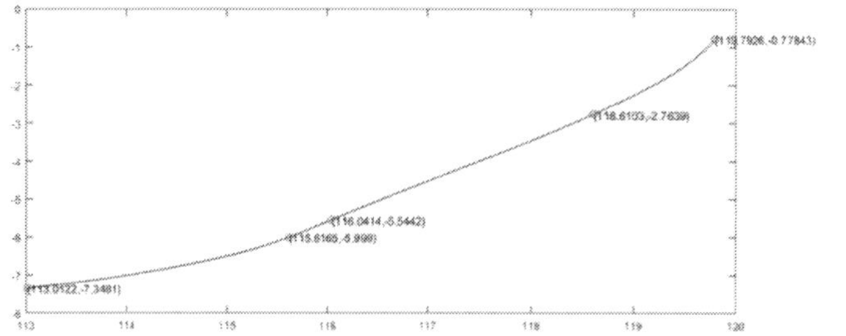

Figure 7. Graphic Spline Cubic Interpolation (**dist1**$_{centroid}$)

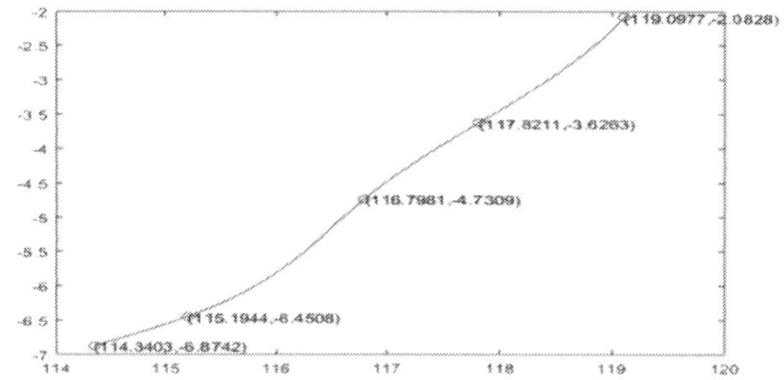

Figure 8. Graphic Spline Cubic Interpolation (**dist2**$_{centroid}$)

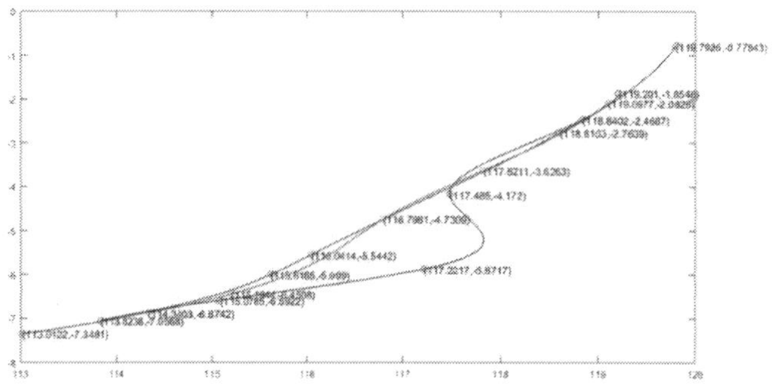

Figure 9. Graphic Spline Cubic Interpolation (waypoint, **dist1**$_{centroid}$, **dist2**$_{centroid}$)

5. Conclusions

The results achieved in this study are: optimal waypoint that obtained from the centroid with the smallest distance from the waypoint ($centroid_2$) in every cluster of the segmentation. The resulting coordinates are C_{12} = (-2.08283, 119.0977), C_{22} = (-2.90686, 118.4808), C_{32} = (-4.58718, 116.9320), C_{42} = (-6.45084, 115.1944), C_{52} = (-6.87422, 114.3403). The coefficient cophenetics correlation (c): $0{,}691 \leq c \leq 0974$.

6. Reference

[1] K. Perhubungan, "Undang Undang No. 1 tahun 2009 tentang Perhubungan Udara." [Online]. Available: http://hubud.dephub.go.id/?en/uu/download/5.
[2] K. Perhubungan, "Peraturan Menteri Pehubungan No. 14 tahun 2009 tentang Peraturan Keselamatan Penerbangan Sipil." [Online]. Available: http://jdih.dephub.go.id/assets/uudocs/permen/2009/KM. 14 TAHUN 2009.pdf.
[3] J. Xi, "Outlier Detection Algorithms in Data Mining," *2008 Second Int. Symp. Intell. Inf. Technol. Appl.*, vol. 1, no. 4, pp. 94–97, 2008.
[4] Y. Zhang, J. Liu, and H. Li, "An Outlier Detection Algorithm based on Clustering Analysis," pp. 1127–1129, 2010.
[5] Y. Li and H. Kitagawa, "DB-Outlier Detection by Example in High Dimensional Datasets," pp. 73–78, 2007.
[6] L. Li, M. Gariel, R. J. Hansman, and R. Palacios, "Anomaly detection in onboard-recorded flight data using cluster analysis," *2011 IEEE/AIAA 30th Digit. Avion. Syst. Conf.*, p. 4A4-1-4A4-11, 2011.
[7] L. Li, S. Das, R. John Hansman, R. Palacios, and A. N. Srivastava, "Analysis of Flight Data Using Clustering Techniques for Detecting Abnormal Operations," *J. Aerosp. Inf. Syst.*, vol. 12, no. 9, pp. 1–12, 2015.
[8] X. Jiao, "One Effective Method of Outlier Detection in Flight Data If |," pp. 797–801, 2009.
[9] M. Gariel, F. Kunzi, and R. J. Hansman, "An algorithm for conflict detection in dense traffic using ADS-B," *AIAA/IEEE Digit. Avion. Syst. Conf. - Proc.*, pp. 1–12, 2011.
[10] E. Chu, D. Gorinevsky, and S. Boyd, "Detecting aircraft performance anomalies from cruise flight data," *AIAA Infotech Aerosp. Conf.*, no. April, p. 3307, 2010.
[11] Z. Liang, Q. Li, and Z. Ren, "Waypoint constrained guidance for entry vehicles," *Aerosp. Sci. Technol.*, vol. 52, pp. 52–61, 2016.
[12] Sky McKinley and Megan Levine, "Cubic Spline Interpolation." [Online]. Available: http://pages.intnet.mu/cueboy/education/notes/numerical/cubicsplineinterpol.pdf. [Accessed: 19-Nov-2016].
[13] J. F. Cui and H. S. Chae, "Applying agglomerative hierarchical clustering algorithms to component identification for legacy systems," *Inf. Softw. Technol.*, vol. 53, no. 6, pp. 601–614, 2011.
[14] O. Access, "Comparison of hierarchical cluster analysis methods by cophenetic correlation," pp. 1–8, 2013.
[15] MathWorks, "Cophenetic correlation coefficient." .
[16] Flight Radar 24, "ADS-B Data Source." .

Who Recognized Fake Image Better: A Comparison Study between Human and Computer

R Bahana[1], A Nurhudatiana[2], M Hidajat[1] and T A Budiono[1]

[1]Computer Science Program, Bina Nusantara University, Jakarta, Indonesia
[2]Information Systems Program, Bina Nusantara University, Jakarta, Indonesia

E-mail: rbahana@binus.edu

Abstract. Along with the rapid development of multimedia and information technology in the recent years, a large number of photos and videos can be easily created and distributed by each and every member of our society. Some of these images are genuine, while some others have gone through modifications. Although the creation and posting of edited photographs in media are basically not prohibited, it becomes a problem when they are used improperly. Until now, the common practice in Indonesia when assessing whether an image has been modified or is original is still heavily dependent on human expert assessment that could be subjective. This research offers an alternative solution that is objective by developing a systematic image analysis method for distinguishing between original and modified image. The method utilizes high-level feature, namely the metadata of the image file, to differentiate between original and modified image. A prototype system was implemented in Visual Basic as a desktop application. The application used MS Windows GDI+. This prototype application can be used as a reliable and faster means to assist in fake image detection. However, as metadata-based analysis only detects whether a photograph has possibly been edited and it does not show which part of the photograph has been manipulated, future work includes development of pixel-based analysis approach.

1. Introduction

Multimedia technology nowadays has made it very easy for anyone to create and share edited images publicly on the Internet. While many of these images are created with no ill intention (e.g., humor), there are fake images intentionally produced to create rumors, sensations, or even defamation of public figures. Fig. 1 (a) and (b) show examples of easily recognized fake images with humor intention (e.g., celebrity meme) [1] while Fig. 1(c) shows another fake image widely circulated on the Internet known to be a form of character attack to Abraham Samad, the previous Chairman of the Corruption Eradication Commission [2]. Unlike the typical meme images in Fig. 1(a) and (b) which seem unrealistic, Fig. 1(c) looks more natural and can be deceiving to people who see it, especially those without any background in photography or image/video analysis.

When controversial photographs become viral in public, it is often not easy to recognize whether the image is actually real or hoax [3]. The common practice in Indonesia to handle such issue is to ask human experts to give their analyses on the photographs. The problem with this approach is that this kind of subjective assessment may lead to varying conclusions depending on each expert's background and experience. Furthermore, with the massive growth of manipulated images on the Internet, the speed of visual evaluation by human eyes will not catch up with the speed of the production of the images themselves. This research aims to provide a preliminary study on how a simple fake image detection system can be used to assist in the recognition of fake images in the public media. In this research, we hypothesized that the automated system would at least give a comparable performance to human assessment in terms of accuracy with significantly faster performance.

(a) (b) (c)

Figure 1. Examples of manipulated images obtained from Google Search (Pinterest and blogs)

2. A Metadata-Based Fake Image Detection System

2.1. Metadata-based approach

Many image forgery detection algorithms have been proposed in literature [4-11]. In general, these techniques can be categorized into high-level feature analysis, where analysis is performed on the whole image as a file (e.g., file metadata), and low-level feature analysis, where analysis is based on parts of the file (e.g., image pixels). Some of the features proposed for use in pixel-based techniques include the use of motion blur metric [9], color filter array analysis [10], and direction of sunlight [11].

Metadata contains information on the structure and properties of a file. The information contained within the EXIF (Exchangeable Image File Format) metadata of a photograph typically contains information as follows: camera models, the manufacturer, and the settings at the time the picture is taken, such as aperture, shutter speed, focal length, metering mode, and ISO speed, the characteristics of the image, the date and time the picture taken, and a thumbnail or a mini version of the image [12]. An example of publicly available image forensic service which uses metadata to check input image is Ghiro [13]. Metadata also contains information about the file compression rate. Since a lot of useful information is available in metadata, it can be used to identify the source file and processing history that ever happened to the file. According to the study by Kee et al. [14], JPEG Headers can provide useful information for metadata-based analysis. JPEG is a compressed format and it is the most common format for storing image files. Every camera model basically produces JPEG Headers information differently from one another [14].

2.2. Prototype of the system

The system was implemented as a desktop application by using Visual Basic programming language. The application used MS Windows GDI+, which is a class-based API. The Microsoft Windows graphics device interface (GDI) enables applications to use graphics and formatted text on both the video display and the printer. Applications based on the Microsoft Win32 API do not access graphics hardware directly. Instead, GDI+ interacts with device drivers on behalf of applications. GDI+ is also supported by Microsoft Win64. GDI+ stores an individual piece of metadata in a PropertyItems object. The PropertyItems is used to retrieve metadata information from a file in the format of array. EXIF information is one of the information contained in PropertyItem Objects. A flowchart to illustrate the method is given in Fig. 2.

For each image, the system takes approximately 1 second to produce a binary output: 0 (original) or 1 (possibly manipulated). Through the Graphical User Interface (GUI), user only needs to open an image or a photo by clicking the button "Open Image" and choose an image or photo he/she wants to check. When the image is original, it will show output like in Fig. 3(a). If the image no longer contains EXIF information, which means that the EXIF metadata has been lost, the image might have been altered or engineered, and thus has possibly been modified (see Fig. 3(b)).

3. Results and Analysis

For this study, Computer Science students were recruited as our human experts. These students were 3rd year students who had taken Multimedia course, so they already had basic knowledge of image editing. Two stages of manual detections were conducted. In the first stage, 15 pictures (i.e., 6 modified and 9 original) were given to 21 panelists. The panelists were asked to determine: 1) whether the image was original or manipulated, and 2) if they thought that the image had been manipulated, which part of the image seemed to have been modified. The difficulty level across the dataset was varied.

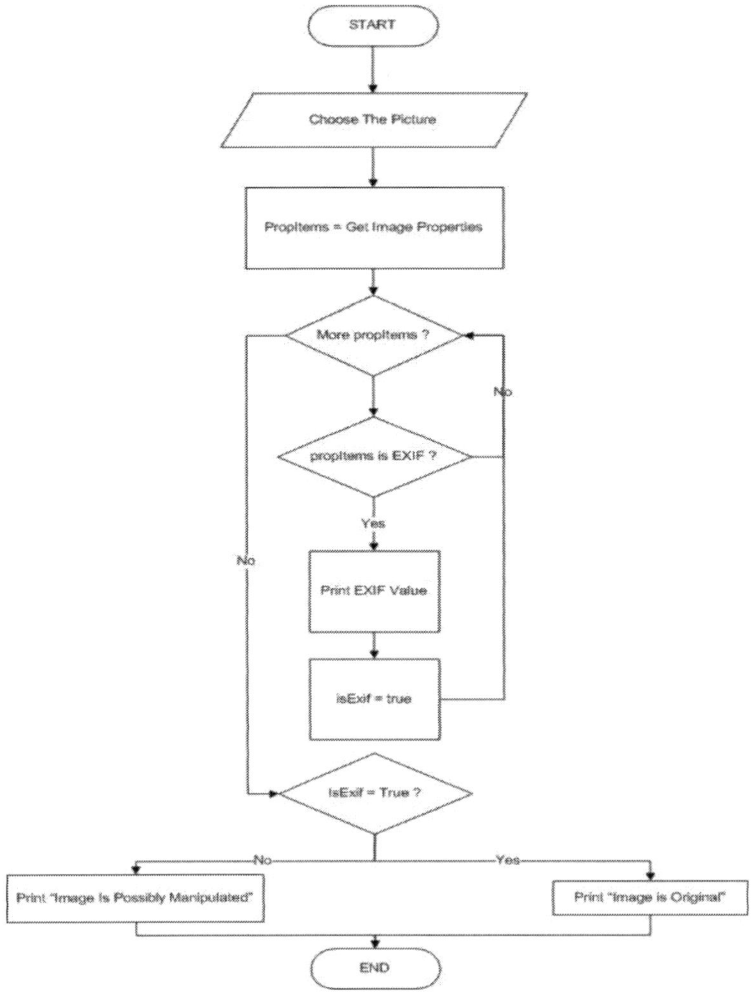

Figure 2. Flowchart of the modified image detection method

Figure 3. (a) Application shows image is original; (b) application shows image has possibly been altered

In the second stage of the experiment, 10 pairs of pictures were shown side by side and 6 panelists were asked to determine which side the engineered image was and to indicate the element of the image which seemed to have been modified. The image datasets used for this research are available in the following two URLs [15, 16].

Table 1 shows the percentage of correct and incorrect answers made by the panelists for each image in Dataset 1. From the 15 images displayed to the panelists, most of them (71% of them) incorrectly guessed photo number 6 as original (see Fig. 4). The license plate of the motorcycle in photo number 6 had actually been changed by duplicating the first character from "B" to "BB". For the image with the most correct answer, 95% of the panelists answered correctly for image number 14 because the trace of copy-move was very visible. In average, only 61% of test participants answered correctly for Dataset 1. This shows that it can be very difficult to recognize manipulated images.

Table 2 shows the percentage of correct and incorrect answers made by the panellists for each image in Dataset 2. From the 10 pairs of images displayed as test materials, most of the test participants (87%) answered correctly for image number 4 (see Fig. 5(a) for the original image and Fig. 5(b) for the modified image) and more (60%) gave incorrect answers for image number 7 (Fig. 6(a) for the original image and Fig. 6(b) for the modified image). In average, 67% of test participants answered correctly, slightly higher compared to the first stage of experiment. A possible reason is because in the second stage, the panellist viewed the two photos altogether, so they could compare both photos and they already knew that one was original and another one was manipulated. The result shows that even by showing the original photo as reference, humans could still find it difficult to recognize edited images.

Table 1. Percentage of correct and incorrect answers – Test I

Photo	% correct answer	% wrong answer
1	71%	29%
2	43%	57%
3	33%	67%
4	62%	38%
5	67%	33%
6	29%	71%
7	71%	29%
8	57%	43%
9	52%	48%
10	38%	62%
11	90%	10%
12	71%	29%
13	57%	43%
14	95%	5%
15	71%	29%
Average	61%	39%

Table 2. Percentage of correct and incorrect answers – Test II

Photo	% correct answer	% wrong answer
1	67%	33%
2	73%	27%
3	67%	33%
4	87%	13%
5	60%	40%
6	73%	27%
7	40%	60%
8	47%	53%
9	80%	20%
10	73%	27%
Average	67%	33%

Figure 4. Photo number 6 in Dataset 1 is the most difficult to recognize

Figure 5. (a) original photo of image number 4 in Dataset 2; (b) a green tile was copied and moved to the blue tiles

Figure 6. (a) original photo of image number 7 in Dataset 2; (b) blue light was duplicated into 3 lines from originally 1 line

Based on the correct answer from Table 2, when the panelists were asked to point which part in the image had been edited, the results are shown in Table 3. It can be seen that only 59% could point out the edited location in the image correctly. There is a possibility that the panelists only guessed the answers during second stage as they could not point out the right location where the image had been edited. With the sophistication of image-editing applications which can manipulate photos easily and are accessible by many people, this could be a worrying thing, especially when the images are intended for a crime or spreading wrong perceptions in society. Educating the general public would not be easy without the help of an application that can detect the originality of an image.

Table 3. Percentage of correct and wrong place – Test II

Photo	% correct place	% wrong place
1	87%	13%
2	73%	27%
3	80%	20%
4	87%	13%
5	67%	33%
6	27%	73%
7	7%	93%
8	27%	73%
9	67%	33%
10	73%	27%
Average	59%	41%

3.1. Comparison with the prototype
All photos in the first and second stages of experiment had already been checked using the prototype application. The prototype application could detect the EXIF information from all original photos and did not display any EXIF metadata information when edited photos were inputted.

4. Conclusion and Recommendation
In this research, we have developed an alternative solution to the current practice of manual recognition of manipulated images by developing a simple Visual Basic system which can be used as a reliable and faster means to assist in fake image detection. One flaw of this method is that metadata is so easy to access, making it easy to remove or replace. In addition, metadata-based analysis can only suspect whether a photograph has possibly been edited and it does not show which part of the photograph has been manipulated. Future direction of this research would be to implement content-based approach to give more fine-grained analysis of the photograph.

Acknowledgment
This research is supported by DIKTI Hibah Bersaing research funding 2016.

References
[1] Kaskus 2014 Super-ngakak-meme-foto-syahrini-tiduran-di-padang-rumput-di-itali-yang-unyu-unyu Retrieved from http://www.kaskus.co.id/thread/53ce6cf45a5163b46f8b46f8/super-ngakak-meme-foto-syahrini-tiduran-di-padang-rumput-di-itali-yang-unyu-unyu/82
[2] Global Indonesian Voices 2015 Retrieved from http://www.globalindonesianvoices.com/18576/foto-mesra-ketua-kpk-abraham-samad-says-intimate-photos-with-elvira-devinamira-are-fake/
[3] Forester T and Morrison P 1994 Computer ethics: Cautionary tales and ethical dilemmas in computing, 2nd ed Cambridge MA: MIT Press
[4] Scientific Working Group Imaging Technology 2007 Section 14: Best practices for image authentication Retrieved from http://www.crime-scene-investigator.net/swgit-section14.pdf
[5] Farid H 2009 A survey of image forgery detection IEEE Signal Processing Magazine, 2(26), 16-25
[6] Sencar H and Memon N 2009 Overview of state-of-the-art in digital image forensics Algorithms, Architectures and Information Systems Security, p 325
[7] Mahdian B and Saic S 2010 A bibliography on blind methods for identifying image forgery. Image Communication, 25(6), 389-399
[8] Mahdian B, Saic S and Nedbal R 2010 JPEG quantization table forensics: A statistical approach. Proceedings of International Workshop on Computational Forensics, 150-159

[9] Kakar P, Sudha N, and Ser W 2011 Exposing digital image forgeries by detecting discrepancies in motion blur IEEE Transactions on Multimedia, 13(3), 443-452

[10] Popescu A C and Farid H 2005 Exposing digital forgeries in color filter array interpolated images IEEE Transactions on Image Processing, 53(10), 3948-3959

[11] Kakar P and Sudha N 2012 Authenticating image metadata elements using geolocation information and sun direction estimation. Proceedings of International Conference on Multimedia and Expo (ICME), 236-241

[12] Metadata Working Group 2010 Guidelines for handling image metadata v2.0. Retrieved from http://www.metadataworkinggroup.org/specs/

[13] Ghiro [Software]. Available from http://imageforensic.org/

[14] Kee E, Johnson M K and Farid H 2011 Digital image authentication from JPEG headers IEEE Transactions on Information Forensics and Security, 6(3), 1066-1075

[15] Dataset 1. Available: http://raymondbahana.blog.binusian.org/rekayasa-photo-i

[16] Dataset 2. Available: http://raymondbahana.blog.binusian.org/rekayasa-photo-ii

Image Restoration on Copper Inscription Using Nonlinear Filtering and Adaptive Threshold

A Chairy[1], Y K Suprapto[2], E M Yuniarno[3]
[1,2,3]*Electrical Engineering Department, Industrial Technology Faculty, Institute Technology Sepuluh Nopember (ITS), Surabaya, Indonesia,*
[1]*amalia.chairy15@mhs.ee.its.ac.is*

Abstract. Inscription is an important document inherited by history of kingdom. Inscription made on hard stuff such as stone and copper. Therefore it is necessary digitizing documents, to keep the authenticity of the document. But the document of the historical heritage have disruption on inscription plate which be called noise. So that, it is necessary to reduce the noise in the image of the inscription, to ease the documentation of historical digital. Then, separation between the background and the writing object carved on inscription is conducted so easy to read. This research is using nonlinear filtering method to reduce the noise and adaptive threshold to separate between the background and letter inscription. Nonlinear filtering method used is median filter, harmonic mean filter and contra harmonic mean filter, whereas in the adaptive threshold using adaptive mean and adaptive median threshold. The results of this research is using measurement methods MSE (Mean Square Error), PSNR (Peak Signal to Noise Ratio) and SNR (Signal to Noise Ratio).

Keyword. Inscription, Digital Documentation, Nonlinear Filtering, Adaptive Threshold, MSE, PSNR and SNR.

1.Introduction

Inscription is an important document inherited by history of kingdom. Inscription made on hard stuff such as stone and copper. The script on inscription usually uses sculpture. Inscription, mostly found in Asian countries, such as Indonesia, India, Bangladesh, and Thailand. For maintaining the ancient heritage, especially inscription, It is necessary to document for research purposes, as well as publications on the content of the inscription. Documentation in the form of digital image, is more appropriate effort because it is easier to be saved and published. The previous research had done digital documentation on an old china calligraphy object, which was carved on stone. Then the improvement using digital technology is conducted, which is the script on the calligraphy stone scanned and save in the computer. The purpose is to compare paper rubbings and digital rubbings to be used to learn to people[1].

In this research, using a type of copper inscriptions. Type of inscription used is Adan-adan inscription. This inscription came from Indonesia and found in Mayangrejo village, Kalitidu District, Bojonegoro on March 2, 1992. It was found buried in the ground in depth about 50cm from ground surface. It was 17 pieces made from copper with length 37.5cm, wide 12 cm and thickness 4mm. This inscription has two types emerging patina, such as dark brown patina and green patina. As you can see on figure 1 are copper inscriptions which have dark brown patina and green patina.

(a) (b)

Figure 1. Copper inscription (a) Dark brown patina and (b) Green patina

On Figure 1. has a defect where the dark brown patina and green patina cover the inscriptions sculpture. so that make the alphabets on inscription become incomplete and difficult to read. Physically, there are various attempts to fix the copper inscription. But many concerns, one of which increased more damage. To repair the image interpretation, some image restoration techniques on previous research has been conducted [2][3]. The previous research had done edge preserving and unseen able details on stone and decreasing noises. Conducted comparison to some denoising image method with relying anisotropic diffusion filter, wiener filter, TV (total variation), NLM(non-local means, NLM), bilateral filtering. A quantity comparison uses PSNR, MSE, SNR, UQI and SSIM [1]. Then, doing improvement of script textures on inscription and ancient script. One of the method is repaired and segmented automatically or manually[2]. On this research, we will do image improvement, to reduce noise on inscription image then doing separation between background and object on inscription image. The method which is used for reducing noise on image is non-linear filtering and adaptive threshold to separate between background and object on inscription image.

2. Related Work

Noise that has been added into image generally has a higher frequency spectrum than image components'. Generally, noise reduction can be done with linear or non-linear approach. Linearly noise reduction can be done either in spatial or frequency. Whereas, Nonlinear approach is on spatial such as Mean filter, Median filter, Harmonic mean filter, contra harmonic mean filter[6][7]. There are many types of threshold method, such as global threshold[8][9][10[11]. Some research utilized colour intensity value from foreground and background by using adaptive local threshold method[12]-[13]. The inscription made from being carved has a less different colour than foreground and background which needs different separate technique[15].

3. Method

On this research, we will compare the proper method to reduce noise and separate between background and object on inscription image. Part of reducing noise on inscription, researcher use nonlinear filtering method. Researcher used a variety of methods from nonlinear filtering such as median filter, harmonic mean filter, and contra harmonic mean filter. While in the part of separation background and object on inscription image, researcher used adaptive mean threshold and adaptive median threshold.

3.1 Median Filter

Median filter is a window which contains amount of odd pixel filter shifted per-pixel point on a whole image. The values on window are sorted by ascending, then it is count the median. the median value will replace the value on centre of window. Where the formula for median is, as follows:

$$f(x,y) = Median\{g(x-i,y-j),(i,j) \in w\} \tag{1}$$

Where $f(x,y)$ is image which produced from image $g(x,y)$ with w as window which is placed on image area and (i,j) element from window.

3.2 Harmonic Mean Filter

Harmonic mean filter is one of nonlinear filter. As you can see in the equation of harmonic mean filter, as follows:

$$\hat{f}(x,y) = \frac{mn}{\sum_{(s,t)\in S_{x,y}} \frac{1}{g(s,t)}} \tag{2}$$

Where *m x n* is *m*-rows and *n*-columns of kernel harmonic mean filter, *G(s,t)* is pixel row and column that will be processed.

3.3 Contraharmonic Mean Filter

Contraharmonic mean filter is one of nonlinear filter. As you can see the equation from contraharmonic mean filter, as follows:

$$\hat{f}(x,y) = \frac{\sum_{(s,t)\in S_{x,y}} g(s,t)^{Q+1}}{\sum_{(s,t)\in S_{x,y}} g(s,t)^{Q}} \tag{3}$$

Where *G(s,t)* is pixel row and column pixel which will be processed. *Q* is order from filter.

3.4 Adaptive Threshold

Adaptive threshold is threshold which used a local threshold value, counted adaptively based statistic neighbour pixel. This is based on reality that small part of image has same illumination, so that it is proper if threshold value counted based on small part of image and not based on whole pixel image.

One of adaptive threshold approaches mean of local intensity on image. Where The constant value involved. As you can see the equation, as follows:

$$T = \frac{\sum_{(y,x)\in W} f(i,j)}{Nw} - C \tag{4}$$

Where *w* states as window on image, N_w states amount of pixels on window and *C* is constant. The second adaptive threshold approach is median of local intensity on image. as you can see the equation, as follows:

$$T = \text{median}(f(x,y),(y,x) \in w) - C \tag{5}$$

Variable *w* states as window on image and variable *C* is constant.

3.5 Paramater of Measurement

3.5.1 *MSE (Mean Square Error)*

Mse is mean of squared fault value between original image with image result, it can be formulated as follows:

$$MSE = \frac{1}{MN}\sum_{x=0}^{M-1}\sum_{y=0}^{N-1}\left|(f(x,y)-g(x,y))^2\right| \tag{6}$$

Where *f(x,y)* is previous image before noise and *g(x,y)* is image which has been restored. The less *MSE* value, the more image restoration performance.

3.5.2 *PSNR (Peak Signal to Noise Ratio)*

PSNR is comparison value between maximum value of color on image result and disruption quantity which is stated in decibel (dB) unit. *PSNR* value can be formulated as follows:

$$PSNR = 10\log 10\left(\frac{(255)^2}{\sum_{x=0}^{M-1}\sum_{y=0}^{N-1}\left|(f(x,y)-g(x,y))^2\right|}\right) \quad (7)$$

3.5.3 *SNR(Signal To Noise Ratio)*

SNR(signal to noise ratio) is standard used to evaluate impact of noise from outside. Which compares the value of image result and noise power which are found on image. Mathematically, *SNR* is stated in decibel(dB) unit by using the formula as follows:

$$SNR = 20\log 10\left(\frac{a\max - a\min}{Sn}\right) \quad (8)$$

Where is variable *amax* is maximum value of pixel from intensity, otherwise *amin* is minimum value of pixel from intensity on image. Variable *Sn* is standard deviation measurement to measure noise value on image.

4. Result And Discussion

The experiment on this research used inscription image which has dark brown patina. The data conducted are five data inscription images of dark brown patina. The first experiment conducted noise on inscription image by using window 3x3 and using median filter, harmonic mean filter, contra harmonic mean filter method. Can be noticed the result of several experiments noise reduction, as follows:

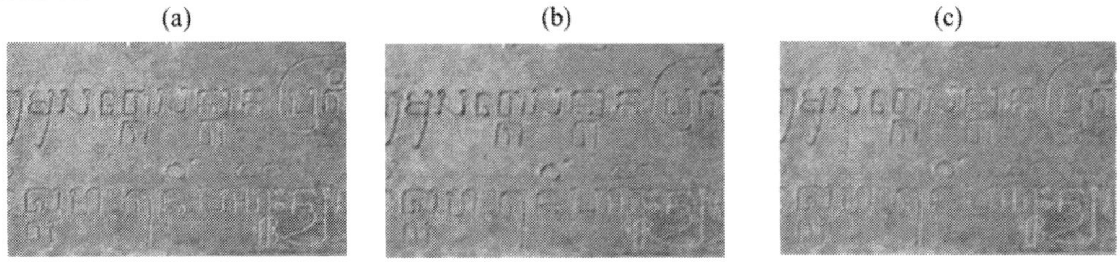

Figure 2. The result of noise decreasing on image inscription using method (a) median filter, (b) Harmonic mean filter and (c) Contra harmonic mean filter

Second experiment conducted is separated between background and object on inscription image using adaptive threshold. On Figure 3. Using image result from noise reduction, median filter and adaptive mean threshold by using value of w = 5 c = 0, w =5 c=5, w=5 c=15, and w=10 c= 20. The separation of background and image objects that use the value w = 5 c = 5 successfully separated. However, for the value of w = 5 c = 0, w = 5 c = 15 and w = 10 c = 20 has not been succeeded, yet. *c* and *w* variable values influence the adaptive threshold on the separation of background and image objects. In the image that has done much noise reduction, the results were shown imperfect. This is due to the colour intensity values to the same object with the background to make the separation less than excellent.

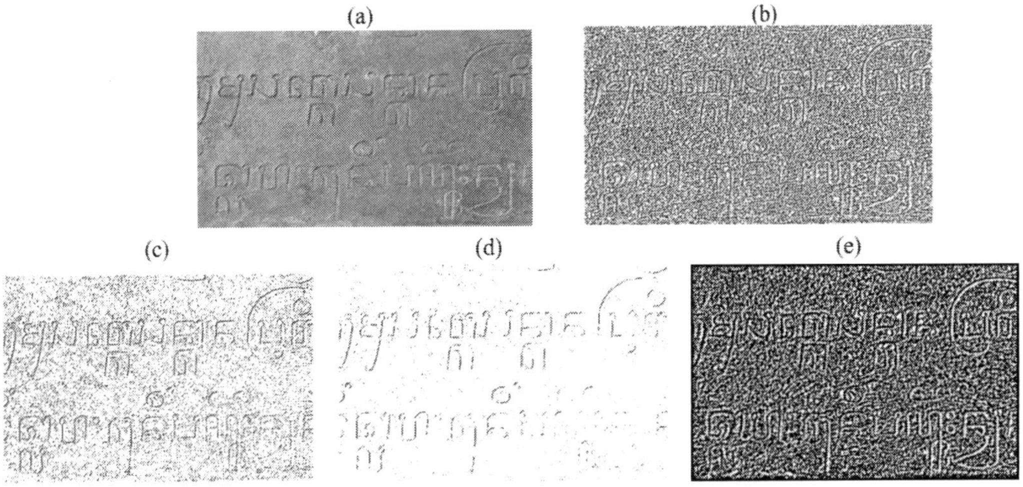

Figure 3. The result of training by using adaptive threshold where (a) original image, (b) using value w = 5 c = 0, (c) using value w =5 c=5, (d) using value w=5 c=15, and (e) using value w=10 c= 20.

We had been doing noise reduction on the image object using the Median filter, harmonic mean filter, and contra harmonic mean filter. Measurement results using MSE, PSNR and SNR Parameter.

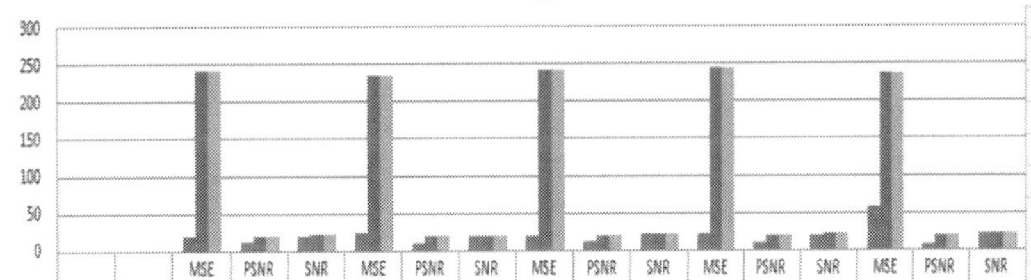

	pic1			pic2			pic3			pic4			pic5		
	MSE	PSNR	SNR	MSE	PSNR	SNR	MSE	PSNR	SNR	MSE	PSNR	SNR	MSE	PSNR	SNR
Median Filter	18.9072	11.2991	20.1956	23.0719	10.4346	19.9305	19.471	11.1715	20.6195	21.8944	10.6621	19.9506	56.2875	6.56128	20.7407
Harmonic Mean Filter	243.31	19.7962	20.6529	235.04	19.646	19.5185	243.494	19.7995	20.8055	246.255	19.8484	20.4433	238.007	19.7005	21.5064
Contraharmonic Mean Filter	243.291	19.7959	20.449	235.017	19.6456	19.7878	243.474	19.7991	21.2452	246.234	19.8481	20.4667	237.986	19.7001	21.0338

Figure 4. The accuration of graphic on *Mean Square Error (MSE)*, PSNR (Peak Signal to Noise Ratio),and SNR (Signal to Noise Ratio)

The result of noise reduction experiment on inscription image is conducted by measuring accurately to find a good and proper method to decrease noise on image. As you can see on Figure 4. showed that *MSE (Mean Square Error)* value on median filter method is smaller than other *MSE* values. It shows that median filter method worked well for reducing noise.

5. Conclusion

Based on the result of experiment that had been done by researcher, could be concluded that median filter method is the best method for noise decreasing. It could be seen that the accurate value of *MSE* 27.9264, *PSNR* 10.0257 and *SNR* 20.2873. Homogeneous colour on the inscription causes the failure of the direct use of threshold method. However, to separate between the background and the script object on inscription still couldn't be tackled by using adaptive threshold method. It causes some noise on inscription becomes too hard to be deprived. On the future, the researcher propose other methods to separate between the background and the object on image. The methods proposed such as the method of Otsu, sauvola, nickblack, NGF-ICA and other methods. This research could also be developed for classifying several features using the method of classifying such as SVM (Support vector machine), NN (Neural Network) and other classifying methods.

6. Reference

[1] Z.-K. Huang, Z.-H. Li, H. Huang, Z.-B. Li, and L.-Y. Hou, "Comparison of different image denoising algorithms for Chinese calligraphy images," *Neurocomputing*, vol. 188, pp. 102–112, May 2016.

[2] I. Sreedevi, R. Pandey, N. Jayanthi, G. Bhola, and S. Chaudhury, "Enhancement of Inscription Images," *ISRN Signal Process.*, vol. 2013, 2013.

[3] O. Hibet-allah, "Hessian-Based Multiscale Filtering and Local Entropy Thresholding," pp. 325–329, 2016.

[4] R. C. Gonzalez, R. E. Woods, A. H. Scott, and P. P. Hall, *Digital Image Processing*. United State: Pearson Education, Inc., 2008.

[5] M. Johansen, "Nonlinear filtering in digtal image processing," no. August, 2011.

[6] J. Chen, X. Kang, Y. Liu, and Z. J. Wang, "Median Filtering Forensics Based on Convolutional Neural Networks," vol. 22, no. 11, pp. 1849–1853, 2015.

[7] H. Liu, R. Xiong, J. Zhang, and W. Gao, "Image Denoising via Adaptive Soft-Thresholding Based on Non-Local Samples," no. 4132039, pp. 484–492, 2015.

[8] P. Smith, D. B. Reid, C. Environment, L. Palo, P. Alto, and P. L. Smith, "A Threshold Selection Method from Gray-Level Histograms," vol. C, no. 1, pp. 62–66, 1979.

[9] C. Wang, H. Zhang, and C. Yu, "Research on color recognition of urine test paper based on learning vector quantization (LVQ)," *Proc. 2012 2nd Int. Conf. Instrum. Meas. Comput. Commun. Control. IMCCC 2012*, pp. 850–853, 2012.

[10] A. Mirkazemi and S. E. Alavi, "Fast Image Segmentation Based on Adaptive Histogram Thresholding."

[11] P. D. R. Raju and G. Neelima, "Image Segmentation by using Histogram Thresholding," vol. 2, no. 1, pp. 776–779, 2012.

[12] C. Kim and Y. Lee, "Medical Image Segmentation by Improved 3D Adaptive Thresholding," pp. 263–265, 2015.

[13] E. Alegre, M. Biehl, N. Petkov, and L. Sánchez, "Automatic classification of the acrosome status of boar spermatozoa using digital image processing and LVQ," *Comput. Biol. Med.*, vol. 38, no. 4, pp. 461–468, 2008.

[14] J. Li, "The pattern recognition of inscriptions on oracle bones by esthetics analysis in computer image," 2010.

[15] S. T. Rasmana, Y. K. Suprapto, and K. E. Purnama, "Color Clustering in the Metal Inscription Images Using ANFIS Filter," vol. 11, no. 3, pp. 529–536, 2013.

Classification of *Acute Myelogenous Leukemia (AML M2 and AML M3)* using Momentum Back Propagation from Watershed Distance Transform Segmented Images

Esti Suryani[1], Wiharto[2], Sarngadi Palgunadi[3], and Nurcahya Pradana TP[4]

Ilmu Rekayasa dan Komputasi Research Group, Sebelas Maret University Jl. Ir. Sutami No 36 Kentingan Surakarta
[1]estisuryani@staff.uns.ac.id, [2]wi_harto@staff.uns.ac.id , [3] palgunadi@staff.uns.ac.id ,
[4]nurcahya.pradana@yahoo.com

Abstract. This study uses image processing to analyze white blood cell with leukemia indicated that includes the identification, analysis of shapes and sizes, as well as white blood cell count indicated the symptoms of leukemia. A case study in this research was blood cells, from the type of leukemia Acute Myelogenous Leukemia (AML), M2 and M3 in particular. Image processing operations used for segmentation by utilizing the color conversion from RGB (Red, Green dab Blue) to obtain white blood cell candidates. Furthermore, the white blood cells candidates are separated by other cells with active contour without edge. WBC (White Blood Cell) results still have intersected or overlap condition. Watershed distance transform method can separate overlap of WBC. Furthermore, the separation of the nucleus from the cytoplasm using the HSI (Hue Saturation Intensity). The further characteristic extraction process is done by calculating the area WBC, WBC edge, roundness, the ratio of the nucleus, the mean and standard deviation of pixel intensities. The feature extraction results are used for training and testing in determining the classification of AML: M2 and M3 by using the momentum backpropagation algorithm. The classification process is done by testing the numeric data input from the feature extraction results that have been entered in the database. K-Fold validation is used to divide the amount of training data and to test the classification of AML M2 and M3. The experiment results of eight images trials, the result, was 94.285% per cell accuracy and 75% per image accuracy

1. Introduction

Acute Myelogenous Leukemia (AML) is a heterogeneous group of hematologic malignancies that are classified according to the predominant type of malignant cells [1]. Myeloid cells are clearly derived from myeloblasts include basophils, neutrophils, eosinophils, monocytes, megakaryocytes and erythrocytes [1]. In 1976, FAB (French American British) Cooperative Group Developed a classification system, classifying AML based on Marrow (marrow), blood (blood morphology) and Cytochemical staining. According to the FAB classification of AML based on morphologic, there are eight, namely M1, M2, M3, M4, M4 E0, M5, M6 and M7 [1]. Particularly in AML M2 characteristic is myeloblastic leukemia with maturation, and there is little Auer rod, round or oval shape of the nucleus and smooth cytoplasm [2], [1]. While the AML M3 is acute promyelocytic leukemia (APL), it has more auer rod, round and lobulated nucleus and smooth cytoplasm [2], [1].

A way to identify leukemia using the traditional procedure to calculate blood cells under a microscope manually, which is time-consuming, laborious, and one of the most expensive routine tests in clinical hematology laboratory [3]. Mechanical counting blood cells that WHO recommended is the technique of using Immunophenotyping. Immunophenotyping is a technique using samples of blood that flowed through the detector (flowcytometer) and then discarded. It is still caused waste, whereas in medical diagnostics is very necessary to record the accuracy of diagnosis [4]. Problems regarding the length of time the calculation and tracking of the diagnosis can be overcome by using image processing techniques to identify blood cells from images taken from digital microscope [4].

Many types of research have been done on the identification of blood cells by image processing techniques, among others, the segment blood particles contained noise by separating the red blood cells and leukocytes [5]. The segmentation process begins with an image denoising walvet bivariate method and cell edge obtained by the method Kuwahara, binarize images process using a combination of Otsu and Niblick methods, after the process of calculating the blood particles used Immersion Watershed algorithm [5]. The last research carried by authors are identifying leukemia types ALL, based on the morphological image of the white blood cells with Fuzzy Rule Based System, with resulting accuracy of 73.68% [6]. Followed by identifying leukemia types ALL and AML M3 based on the morphological image of white blood cells by the Fuzzy Rule based, the identification process is done by the WBC area ratio and the ratio of nucleus granules with 83.65% system accuracy results [7]. Segmentation is done by using Canny Detection and Ellipse Detection. The study was developed further in 2014, with segmentation using a bounding box for case studies of AML detection that kind of M2 and M4 [7]. The weakness of the research is on segmentation by using Ellipse Detection, and Bounding Box can not separate the good cells that overlap with [6] [7]. This study was done to resolve the problems on segmentation cells overlap, i.e., the method of Active Contour Without Edge (ACWE) and Watershed Distance Transform with case studies AML M2 and M3. ACWE method considered to be suitable for segmentation because it can be used to segment images like pictures with low contrast, varied objects, objects with different intensity, smooth contours and high noise [8]. Watershed Distance Transform is suitable for separates the blood cells that overlap by utilizing the watershed transformation distances combined with roundness cells information [9].

2. Proposed Method

Figure 1 Shows the steps of this research:

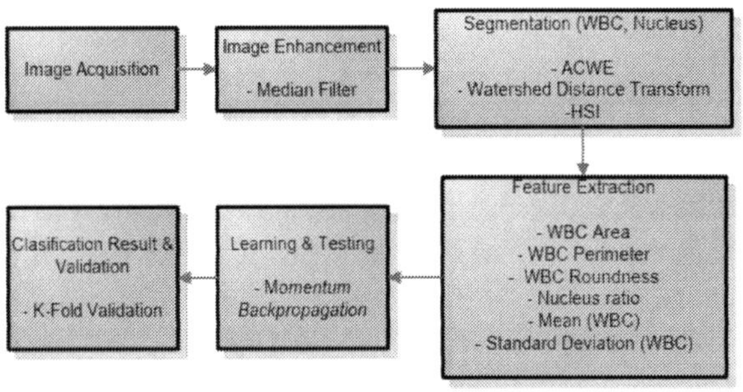

Figure 1. Research Steps

2.1. Images acquisition

Image acquisition is made by collecting images of AML M2 and M3 blood obtained from the image microscopic blood smear of bone marrow. The image is reserved in Installation of Clinical Pathology, Dr. Moewardi Hospital. The process of collecting data was done by a specialist doctor from the Clinical Pathology Clinical in Dr. Moewardi Hospital. Digital images are obtained by using a digital microscope with a magnification of 1000 times of bone marrow blood smear preparations with Giemsa staining.

2.2. Image Enhancement and WBC Segmentation

Figure 2a is an example of the image which identified as AML M2 that will be used as the input image. WBC cells contained in the figure 2a.

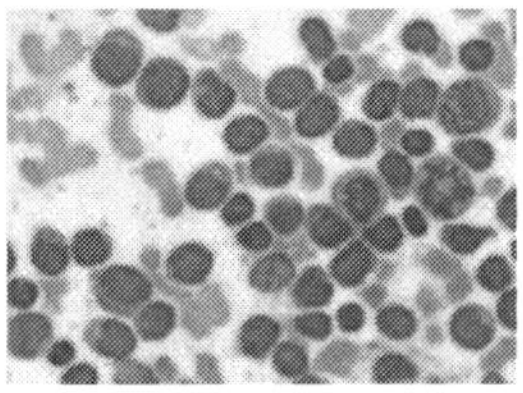

(a)

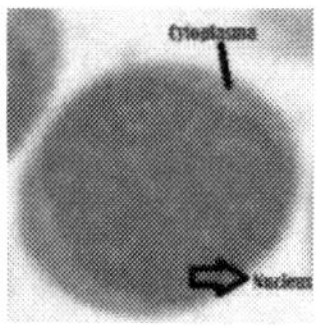

(b)

Figure 2. a. AML M2 Image; b.WBC Cell with nucleus and cytoplasm

Image data which will be processed (see Figure 2a), must first be repaired with the image enhancement process using a median filter to remove noise image. After enhancement, the clear image then will be segmented to get the WBC cell. Figure 3 shows the steps of WBC segmentation.

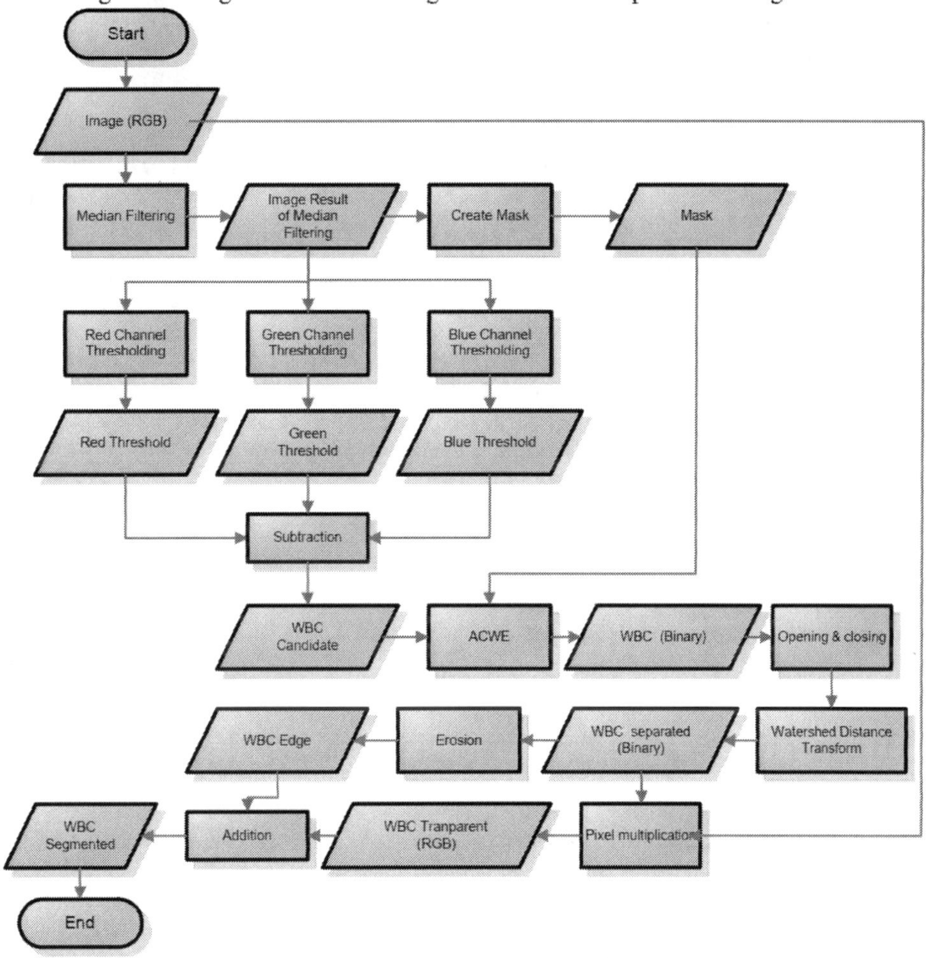

Figure 3. WBC Segmentation Steps

Input for the steps shown in Figure 3 is Figure 2a (AML M2). Median Filtering is done to eliminate noise or small spots contained in images that may interfere in the subsequent segmentation process Furthermore; the following process is creating a mask to get the WBC candidates. Candidates of white blood cells obtained from RGB threshold for each channel red, green, and blue, after a median filter process. The next step is to calculate the size of the resolution of the original image then shaped curve or a mask with a margin from the edge of the image is 100 pixels. Having obtained the candidate of white blood cells and mask, then both combined to do the segmentation process ACWE by calculating

IOP Conf. Series: Journal of Physics: Conf. Series **801** (2017) 012044 doi:10.1088/1742-6596/801/1/012044

the comparable levels of energy outer and inner energy. The results of ACWE process were a binary image of white blood cells that still contained some cells overlap, and therefore needs to be done by using a cell separation distance transform watershed. Catchment basin needs to be sought by the flooding to some extent to get the position of the cell that can be separated by watershed. The flooding process is done with due regard to the existence of local minima value on each object. From the catchment basin, the centroid point can be obtained which later became the benchmark watershed in conducting separation. The results of the WBC watershed separation process is a WBC binary image (see Figure 4a). The binary image is then multiplied by the pixels of the original image and then counted the number of WBC cells (54 cells), (See Figure 4b).

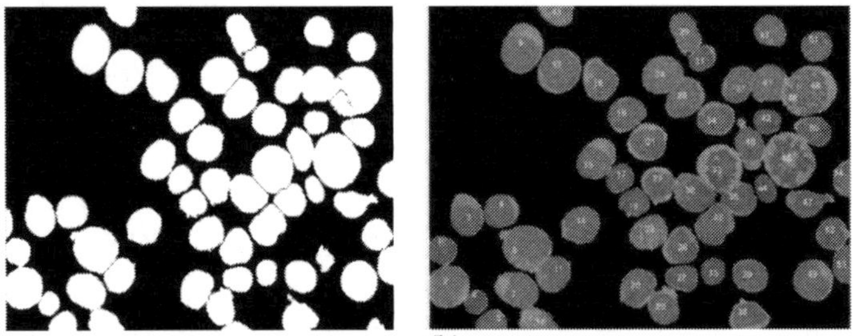

(a) (b)

Figure 4. a. WBC (binary) ; b. WBC (segmentation result)

In some cases where white blood cell objects are overlapping each other, there are some parts that can not be separated well. Figure 5a shows that there are three cells that overlap each other and form a series of fused cells. The catchment basin can be obtained by flooding the object (see Figure 5b). After the watershed transform based on the distance the catchment basin, it acquired four cells (see Figure 5c). A catchment basin is calculated by finding the value of local minima on the object. Therefore, error separation on the object instance of white blood cells (Figure 5a) is because there are four minima (Figure 5b) so that the cell is separated into four parts, namely object 37, object 41, object 46 and object 48 (see Figure 5c).

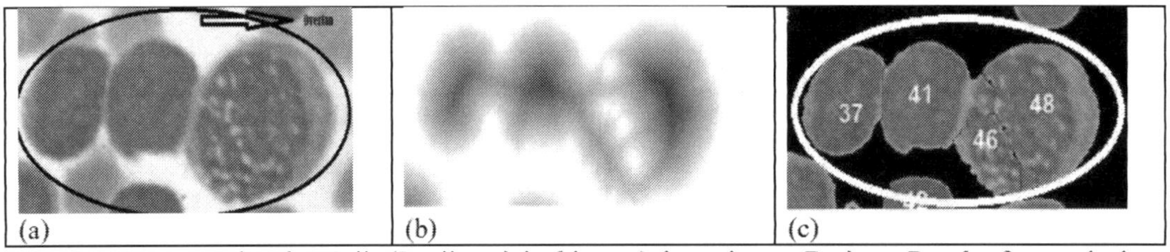

(a) (b) (c)

Figure 5. a. Overlapping Cells (3 cells, original image); b.catchment Basin; c. Result of watershed segmentation (4 cells identified)

2.3 Nucleus Segmentation

The next step is to separate the WBC nucleus. In the process of segmentation of the nucleus, it should be conducted in stages, shown in Figure 6.

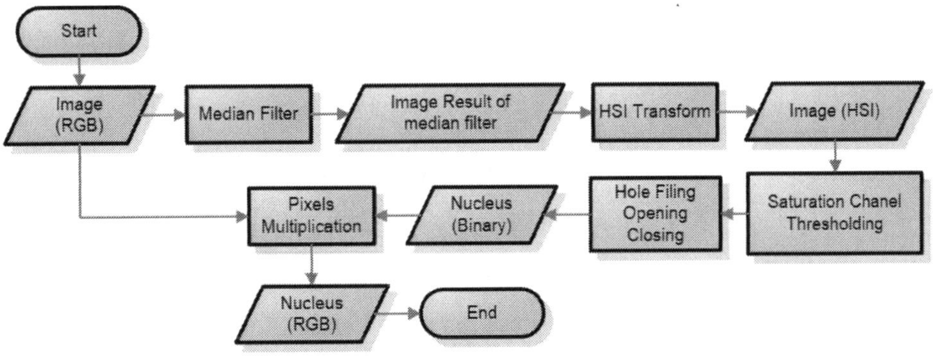

Figure 6. Nucleus Segmentation Steps

Nucleus candidates are obtained by transforming the RGB-based image into HSI. (see Figure 7a). The nucleus has higher saturation than other objects in the image, so it is just enough to perform thresholding on the saturation channel can produce binary nucleus. Then do the filling hole morphological operations, opening and closing on the binary nucleus, to obtain binary object cleaner, which is free from small spots of objects that are not the nucleus. Multiplication of binary object with the original image WBC nucleus will generate objects in RGB color (see Figure 7b).

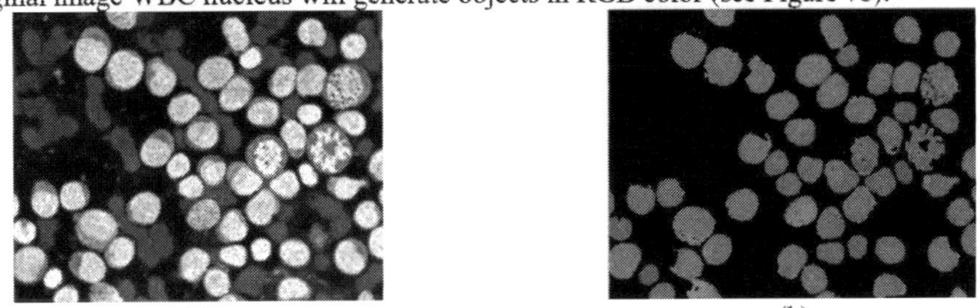

(a) (b)

Figure 7. a. Nucleus Candidate; b. Nucleus Segmentation result

2.4 Feature Extraction

The features used are namely: WBC area, perimeter WBC, WBC roundness, nucleus ratio, mean (WBC), and the standard deviation (WBC).[10][11]. For example, the object number 15 in the image M2 (33) .jpg (see Figure 8), has a cytoplasm and nucleus. Values obtained from the characteristics of the object number 15 in a row are: area of 17 466, 481 perimeters, 0.93639 roundness, 0.74229 ratios, 160.1461 mean and 21.1403 standard deviation

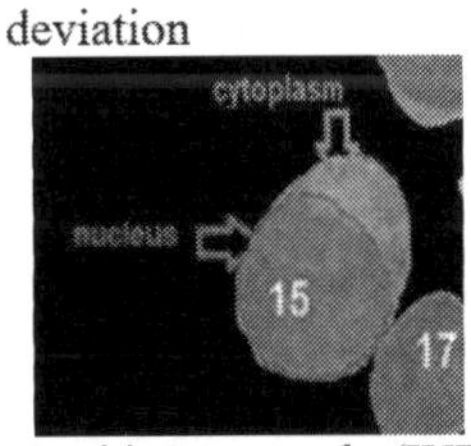

Figure 8. object example (WBC Cell)

Feature extraction performed on eight pieces of the image and obtained 315 objects of white blood cells which will then be used as input in the process of learning and testing.

3. Experimental Results

The input parameters consist of six neurons, namely WBC area, perimeter WBC, WBC roundness, nucleus ratio, mean (WBC), and the standard deviation (WBC), the result consists of two neurons that M2 and M3 which is a subtype of AML. The testing process on the classification would choose a neuron with highest accuracy value as a result. The training process uses a binary sigmoid as activation function and the momentum of its propagation as a classifier. Learning rate (α) : 0.1, 0.5, and 0.9, hidden layer neuron : 6, 8 and 10. Momentum : 0.1, 0.5, and 0.9. Error tolerance limits which are equal to 0.000001. Determination of these parameters aimed at limiting the number of combinations of parameters used. Training using the k-fold cross validation to avoid overfitting. Fold amount used in this training are three-fold with a random distribution. The results of 27 times of training showed that the best accuracy is obtained from a combination of learning rate 0.1, hidden layers 8 and 0.9 momentum which is equal to 94.285% (see Table 1).

Table 1. Training using K-Fold Cross Validation

Learning rate	Hidden layer	Momentum	Time	Accuracy
0.1	6	0.1	39,472	89,673
0.1	6	0.5	30,89	91,184
0.1	6	0.9	25,134	90,176
0.1	8	0.1	36,855	94,285
0.1	8	0.5	28,644	91,184
0.1	8	0.9	23,894	89,421
0.1	10	0.1	34,119	91,436
0.1	10	0.5	27,526	90,932
0.1	10	0.9	36,004	89,421
0.5	6	0.1	22,831	89,421
0.5	6	0.5	20,222	88,917
0.5	6	0.9	20,551	88,917
0.5	8	0.1	22,928	89,421
0.5	8	0.5	22,435	89,169
0.5	8	0.9	21,485	89,421
0.5	10	0.1	22,481	89,421
0.5	10	0.5	23,395	89,673
0.5	10	0.9	22,592	88,161
0.9	6	0.1	22,024	89,421
0.9	6	0.5	20,699	88,413
0.9	6	0.9	20,424	88,413
0.9	8	0.1	23,315	88,917
0.9	8	0.5	22,678	88,917
0.9	8	0.9	20,403	88,413
0.9	10	0.1	21,267	88,917
0.9	10	0.5	21,162	88,917
0.9	10	0.9	19,854	88,161

Testing the whole cells carried on a combination of parameters with the best accuracy. The results obtained are based on the amount of the corresponding object classification compared to the number of objects that do not fit the classification. Total comparison both define a class type of an image. Table 2 shows the testing based on the number of cells in each image.

Table 2. Classification Result

Num	Image	Correct cells	Wrong cells	Type
1	M2 (10).jpg	89	1	M2
2	M2 (14).jpg	76	0	M2
3	M2 (15).jpg	63	0	M2
4	M2 (33).jpg	53	2	M2
5	M3 (2).jpg	0	5	M2
6	M3 (3).jpg	3	6	M2
7	M3 (6).jpg	10	1	M3
8	M3 (7).jpg	3	3	M3

Table 2 shows that from eight pieces of the images used for the experiment, there are two pieces of the image that are misclassified. Misclassified occurred because the number of cell object belongs to which class M3 relatively less than classes M2 so that the patterns formed better recognize the objects in the class M2. The value of accuracy from the classification per cell is:

$$Cell\ Accuracy = \frac{correct\_cell}{total\_count\_cell} * 100\% = \frac{297}{315} * 100\% = 94{,}285\%$$

While the value of the accuracy of the classification per image is:

$$Image\ Accuracy = \frac{correct\_image}{total\_count\_image} * 100\% = \frac{6}{8} * 100\% = 75\%$$

Thus, the test using momentum backpropagation classifier was able to classify subtypes of AML M2 and M3 where per cell classification results show an accuracy of 94.285% and per image classification results have an accuracy of 75%.

4. Conclusion

Watershed Segmentation Distance Transform can separate the white blood cells that overlap. However there is still an error occurred object separation of white blood cells that overlap, due to the minimal local number found during the process of calculating catchment basin on watershed segmentation. Based on experiments conducted on eight pieces of images, obtained 315 pieces of white blood cells which will be used as input in the process of learning and testing. The test using momentum backpropagation classifier was able to classify subtypes of AML M2 and AML M3 where per cell classification results show an accuracy of 94.285% and per image classification results have an accuracy of 75%.

5. Acknowledgment

This research was supported by Grant PNBP UNS 2016 and the image's data AML M2 and AML M3 obtained from Dr Moeward Hospital, Surakarta, Central Java.

6. References

[1].Bell, A &Sallah S 2005 *The Morphology of Human Blood Cells*, Seventh Edition, Abbott, A Promise for Live.

[2].Hamid AG 2011 *Classification of Acute Leukemia, Acute Leukemia - The Scientist's Perspective and Challenge*. Prof. Mariastefania Antica (Ed.), ISBN: 978-953-307-553-2, InTech.

[3] Houwen B 2001 *The Differential Cell Count* Loma Linda, California: Carden Jennings Publishing Co., Ltd.

[4] Supriyadi E dkk 2011 *Immunophenotypic Patterns of Childhood Acute Leukemias in Indonesia* AsianPacific Journal of Cancer Prevention, Vol 12.

[5] Habibzadeh M et al. 2011 *Counting of RBCs and WBCs in noisy normal blood smear* Montréal, Canada; Tehran, Iran: Dept. of Computer Science and Software Engineering, Concordia University, Dept. of Electrical Engineering, Sharif University.

[6] Suryani E, Wiharto, and Polvonov N 2013 *Identifikasi Penyakit Acute Lymphoblastic Leukemia (ALL) Menggunakan Fuzzy Rule Based System Berdasarkan Morfologi Sel Darah Putih* Prosiding Seminar Nasional Teknologi Informasi dan Komunikasi Terapan (SEMANTIK), Udinus, Semarang, ISBN:979-26-0266-6.

[7] Suryani E, Salamah U, Wiharto, and Polvonov N 2014 *Identification and Counting White Blood Study of Leukemia.* International Journal of Computer Science & Network Solutions. Juni 2014. Vol. 2 No. 6. ISSN: 2345-3397

[8] Chan, T. F., dan Vese, L. A., 2001. *Active Contour Without Edges.* IEEE Transactions on Image Processing, *10*(2), 266-277.

[9] Putzu, L., Ruberto C.D., 2013. *White Blood Cells Identification and Counting from Microscopic Blood Image. World Academy of Science, Engineering and Technology.* International Journal of Medical, Health, Biomedical, Bioengineering and Pharmaceutical Engineering Vol:7, No:1, 2013.

[10]. Pradana T P N, Suryani E, and Wiharto 2013 *Pemanfaatan Seed Region Growing Segmentation dan Momentum Backpropagation Neural Network Untuk Klasifikasi Jenis Sel Darah Putih* Seminar Nasional Teknoin 2013 FTI UII.

[11] Nasir A A_1, Mashor MY_1, and Hassan R_2 2013 *Classification of Acute Leukaemia Cells using Multilayer Perceptron and Simplified Fuzzy ARTMAP Neural Networks* $_1$Electronic and Biomedical Intelligent Systems Research Group, Universiti Malaysia Perlis, Malaysia. $_2$Department of Haematology, Universiti Sains Malaysia, MalaysiaPatil T.G., Prof. Mr. V. B. Raskar, 2015. Automated Leukemia Detection By Using Contour Signature Method. E & TC Department JSPM's, Imperial College of Engineering & Research Wagholi, Pune. The International Arab Journal of Information Technology, Vol. 10, No. 4, July 2013.

Recent development of feature extraction and classification multispectral/hyperspectral images: a systematic literature review

A Setiyoko[1], I G W S Dharma[2], and T Haryanto[3]

[123]Faculty of Computer Science, University of Indonesia Depok, Indonesia, andie.setiyoko@ ui.ac.id, i.gede614@ui.ac.id, toto.haryanto@ui.ac.id

[1]Remote Sensing Technology and Data Center-LAPAN, Jl Lapan 70, Pekayon, Pasar Rebo, Jakarta Timur, Indonesia 13710, andie.setiyoko@lapan.go.id

Abstract. Multispectral data and hyperspectral data acquired from satellite sensor have the ability in detecting various objects on the earth ranging from low scale to high scale modeling. These data are increasingly being used to produce geospatial information for rapid analysis by running feature extraction or classification process. Applying the most suited model for this data mining is still challenging because there are issues regarding accuracy and computational cost. This research aim is to develop a better understanding regarding object feature extraction and classification applied for satellite image by systematically reviewing related recent research projects. A method used in this research is based on PRISMA statement. After deriving important points from trusted sources, pixel based and texture-based feature extraction techniques are promising technique to be analyzed more in recent development of feature extraction and classification.

1. Introduction

A multispectral band is a representation of the spectral resolution and a panchromatic band is representativeness of spatial resolution or pixel size of a particular image. Spectral resolution with regard to how detailed the division of the wavelength on the satellite sensor. For example Landsat 8 with OLI (Operational Land Imager) multispectral bands that divided into 8 bands (1-7,9) or WorldView 3 that contained 8 bands. Development of spectral resolution leads to hyperspectral data, which divide wavelength in very narrow range. Hyperspectral data contained a large number of bands, AVIRIS with 224 bands and HYDICE with 210 bands [1]. Multispectral and hyperspectral data have the ability to do classification and feature extraction process because each band has a character to display certain objects. For example, Landsat 8 band 1 representing interactions at wavelengths can detect water quality [2]. Multispectral and hyperspectral data has been used for various applications such as land cover classification for agriculture, for example, there was a research project conducted crop type mapping in a highly fragmented and heterogeneous agricultural landscape by using multitemporal Landsat 8 [3]. There was also project research to do an analysis of land cover change of greenness trends using 3 generations of Landsat images [4]. Some examples of cases also used on specific interests such as identifying burned area by using Landsat TM/ETM+ [5], drought monitoring using Landsat 8 OLI imagery [6].

Multispectral images with various band included in it, have information that can be captured. Feature extraction has an important role to capture meaningful information from the data so can give more precision on supervised or unsupervised technique.

Principal Component Analysis (PCA) besides as a feature extraction, it is a technique for reducing high dimensional of data. PCA has been applied successfully on Landsat-7 ETM imaginary for a high dimensionality reduction method and also image enhancement[7]. Combination PCA with classification such Support Vector Machine (SVM) has been succeeded applied for handling image fusion with filtering called image fusion and recursive filtering (IFRF)[8]. The salt field from Fused Landsat TM data have been mapped based on reflectance-spectra. The inverse PCA in this research applied to transform back to the seven band [9].

Since texture analysis proposed by Haralick as feature extraction on image processing [10], this technique has been applied on Remote Sensing research field. One of the texture analysis using Grey Level Co-occurrence Matrix (GLCM). GLCM has been conducted for detection ice detection on Bohai sea in Tiongkok[11]. The result has been validated using co-temporal HJ1B-CCD on 30 m imaginary and confusion matrix for accuracy. GLCM also has been implemented to analyze texture from Synthetic Aperture Radar (SAR) data for mapping global urban area at 30 m resolution. Another implementation of GLCM is for analyzing multiband texture on multispectral images [12].

A hyperspectral image is containing a lot of bands that have a specific purpose to be classified. Those regions pixel can be assigned as a unique label that able to specify belonging classes [13]. However a hyperspectral image relatively containing low spatial resolution and unavoidable noise which obtained from transmission and reception limit [14]. To classify the region from hyperspectral image, spectral-spatial classification network proposed by Pan, with input from PCANet that label region pixels already determined [15], while Li proposed Framework of Spectral-Mixture-Analysis-Basis HIS SR as a framework to reconstructed low resolution hyperspectral. Before coming into a framework, PCA is applied to a hyperspectral image to de-correlate the atom in the spectral dictionary [14]. In contrast, Lu proposed two main steps in set-to-set distance-based spectral-spatial classification. The first step, they generate both of training and testing sets and the second step is set-to-set distance-based classification. In step two, set-to-set distance-based classification is divided into two subtasks, representation of sets and label assignment. The Affine Hull (AH) model is used to adaptively represent the characteristic of each set [13].

The objective of this systematic review is to develop a better understanding regarding object feature extraction and classification based on an image, particularly satellite image. The research will compare a number of techniques to extract feature information and to classify the specific object from satellite image in the recent time.

2. Method

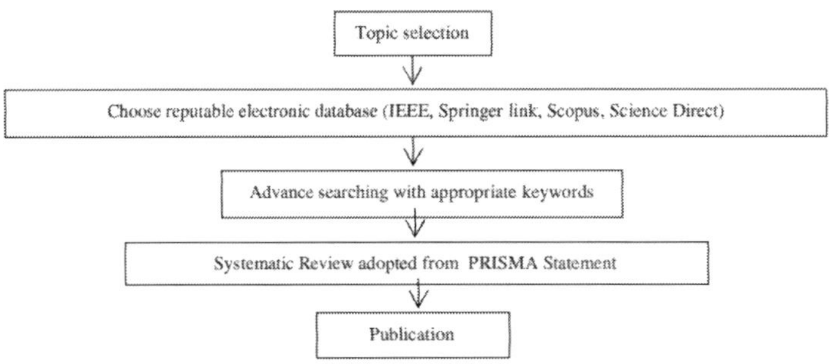

Figure 1. Flowchart of the systematic literature review

A method used in this paper is based on PRISMA Statement developed by Moher et al [16]. Literature searching were involved comprehensive and scooping steps has been done by Catenanu et al [17]. Scientific papers regarding feature extraction and classification on Satellite data are selected from reputable four electronic included IEEE, Springer link, Scopus, Science Direct to give comprehensive result and discussion as representative of updated technique in feature extraction and classification.

To ensure that relevant papers obtained, appropriate keywords are selected. For instance, the keyword ("PCA") and ("multispectal"), ("GLCM") and ("Multispectral"), ("texture feature") and ("hyperspectral"), ("hyperspectral") and ("classification") conducted for searching the journal. Briefly, there are five steps review the paper started from Topic selection finished with paper publication. Flowchart of systematic review shown on the figure 1.

3. Result

There are about 20 paper journals selected for this systematic review. Most of them focus on feature extraction and classification satellite images. PCA and texture analysis like GLCM widely applied on various images. We only include the journal with scope related to the topic. One of the specific journals for this topic is ISPRS Journal of Photogrammetry and Remote Sensing published by Elsevier (1,7,11,12,19,20,21). The others paper journal selected are from Journal Remote Sensing of Environment on Science Direct (2, 4 and 5). The review process considering parameters from PRISMA statement such as 1) study selection study characteristics, description study, data quality and discussion.

3.1 Study Selection

Overall, 17 potentially papers relevant to updated techniques of feature extraction and classification were identified and selected to be reviewed. All papers derived from trusted online journal.

3.2 Study Characteristic

There are various techniques nowadays to be used for feature extraction and classification data, particularly for multispectral data or hyperspectral data. 4 papers regarding Eigen value has been chosen to be reviewed in this project. Murugan *et al* have developed experiment research by conducting PCA (principal component analysis) to extract information feature from remote sensing image [18]. A similar technique has been used by Lin *et al* [19], by analyzing Eigen feature to classify LIDAR cloud point. There were also newly developed techniques as modified PCA to do feature extraction or classification. Zhong et al developed the new technique called blind spectral inmixing based on sparse component analysis (SCA) for hyperspectral data[1]. While Zalbaba *et al* also developed novel-folded PCA to improve feature extraction and data reduction applied to hyperspectral image and SAR (synthetic aperture radar) in remote sensing [20].

Furthermore, Lu et al [7], developed the concept of spectral-spatial classification method where each test and training sets are adaptively represented by AH model. This model utilizes both similarity and variance of pixels to adaptively characterize the sets. In addition, AH-model-based able to represent each set with a similarity between testing and training sets is transformed into a geometric distance. Finally, labeling test sets are coming from a minimum distance between test and training sets.

A proposed method classification accuracy improved greatly from combining both spatial and spectral information of the HSI in a set-by-set manner. In contrast to Lu et al, both Pan et al [9] and Li et al [8] are using PCA as method to reconstructed the hyperspectral images. Li et al used PCA as a method to reconstruct the hyperspectral images. Li et al used PCA as a method to decorrelate the atoms in the spectral dictionary then followed by eigenvectors due to their powerful expressive ability. Moreover, after spectral dictionary describing the spectral characteristics, singular value decomposition (SVD) is used in group-based sparse representation to obtain the adaptive dictionary for each group to effectively analyze spatial similarity and spectral correlation from two different dimensions of a group. Slightly similar to proposed method that Li et al proposed, Pan et al [9] proposed classification method by using modified PCA named PCANet and their classification method named spectral-spatial classification network.

Table 1. General Description of Studies

Study	Objective	Method	Result	Further Research Proposed
Murugan *et. al* (2011), India	Reduction in the dimension of RSI by data compression technique using Eigen matrix	PCA	A prominent hill structure at the middle of the RSI is shown very clearly. The linearity of the structural hill is lucidly displayed sharply	Applying various algorithms such as filtering, edge detection and classification
Zhong *et .al* (2016), China	To consider the properties of sparsity for hyperspectral remote sensing imagery	Sparse component analysis (BSUSCA)	The results show that BSUSCA performs better than the other unmixing algorithms, in an overall view	SCA-based endmember number estimation algorithm will be added to the model
Zalbaba *et. al* (2014), UK	To address challenges regarding memory management	Novel Folded-PCA (F-PCA)	Overall computational cost reduced to roughly 10%, Folded-PCA requires less than 1% of memory as those in conventional PCA	To apply this approach in other analysis of large dimensional dataset is required
Lin *et. al* (2014), Taiwan	The introduction of a method to extract Eigen-features from a local point set by using the weighted PCA	SVM classification and Eigen feature analysis	The accuracy of our classification improves compared to the classification using standard Eigen-features	To investigate the application of the method on the point clouds captured from laser terrestrial mobile mapping systems
Kautsias *et.al* (2009), Greece	Enhancing signal of burnt surface	Forward/ backward PCA	Forward / backward PCA improved spectral discrimination	Not mention specifically
Kang *et.al* (2013), China	To propose Image Fusion Recursive Filter (IFRF) for feature extraction of hyperspectral image	IFRF Support Vector Machine (SVM)	IFRF efficiently can reduce the dimension of the hyperspectral image.	Adoption of EPFs method to process fused band images and investigation of the considering the correlation degree of adjacent band
Zhang *et. al* (2011), China	The objective of research is Mapping the salt filed from Fused Landsat TM data	PCA	Salt farm mapped 91.95 % correctly and rose up to 94.6 % for panchromatic band (15 m). The Highest accuracy with PCA-fused data is 98.8%.	Not mention specifically
Su, *et. al* (2013), China	To detect Sea Ice base on texture information	GLCM	GLCM texture analysis is the best method compared to other methods based on confusion matrix. The result are validated by co-temporal HJ1B-CCD 30m imagery by visual interpretation	Texture analysis will be carried out for different types of Sea Ice using remote sensing technique.
Ban *et al* (2014), Sweden	To evaluate spaceborne SAR data for improved global urban mapping	KTH-Pavia urban extractor, GLCM	KTH- Urban extractor is effective technique for extracting urban from SAR image data	KTH-Pavia can be applied for monitoring urbanization if we have historical images.
Lu *et.al* (2016), China	To classify hyperspectral images (HSI)	Distance Based Classification	Distance based classification have the highest accuracy and faster running time among others methods such as SVM, SVM -CK, LBP etc	Investigating about ability of set-based method for semi-supervised or semi-unsupervised framework
Li *et. al* (2016) , China	The Development the framework for utilizing spectral mixture analysis spatio spectral group sparsity	Sparse representation matrix	Sparse matrix representation not only maintain spectral consistency but also produce image more detail	Computation times within an acceptable range and the acceleration of the algorithm will be focused
Pan et al (2016), China	To construct simple nonlinear spectral-spatial network	Deep Learning Neural Network	Nonlinear Spectral-Spatial Network (NSSNet) is simplified Deep Learning based performs some deep learning method	The future work will be devoted to reducing training data and improving the accuracies

The literature has been reviewed were quantitative experimental research. It can be understood from research design or research methodology conducted and how the result validated.
Commonly the multispectral/hyperspectral data analysis used to be recognized whether supervised or unsupervised technique. From 17 paper selected, all paper regarding the supervised technique. Although [7][8][9][11][12] [21] focus on the feature extraction technique, a classification method is a way to recognize the images. Meanwhile, [13]–[15] specifically focus on the classification for multispectral/hyperspectral images.

3.3 Description of the Study
Research selected related to Eigen value came from multi countries such as India, United Kingdom, China, Taiwan, Greece, and Sweden. According to data used, Murugan et al were using multispectral data in their research [18]. Zhong et al and Zalbaba et al were using hyperspectral data [1] [20], but Zalbaba et al were also added SAR image in their experiment [20]. Lin et al applied their project to LIDAR cloud point [19]. Table 2 shows specification of data used by each research.

Table 2. Literature Data Used

`Study	Type	Data	# of bands
Murugan *et al* (2011), India	Multispectral	IRS-1C LISS III image	4
Zhong *et al* (2016), China	Hyperspectral	AVIRIS, HYDICE	224, 210
Zalbaba *et al* (2014), UK	Hyperspectral and SAR image	AVIRIS, micro-Doppler	224, -
Lin *et al* (2014), Taiwan	LIDAR cloud point	LIDAR	1
Kautsias *et.al* (2009), Greece	Multispectral	Landsat 7 ETM+	7
Kang *et.al* (2013), China	Hyperspectral	Indiana Pine, Pavia Salinas	220
Zhang *et. al* (2011), China	Multispectral	Landsat TM Plus (ETM+)	7
Su, *et. al* (2013), China	Hyperspectral	MODIS	4 (r,g,b, nir)
Ban *et al* (2014), Sweden	SAR-CVV	ENVISAT ASAR	1
Lu *et.al* (2016), China	Hyperspectral	AVIRIS, ROSIS-03	200, 204. 103
Li *et. al* (2016) , China	Hyperspectral & Multispectral	ROSIS-03, HIS, PHI Xiaqiao	103, 191, 80
Pan *et al* (2016), China	Hyperspectral	AVIRIS, ROSIS-03	200, 103

3.4 Discussion

Regarding PCA method and its modification used in many research papers mentioned previously, there are advantages could be taken from these methods. PCA derives Eigenvalue of a bundle of images whether it is multispectral or hyperspectral to get the best value of image in the form of Eigen matrix image. There were various improvements after applying the method. Murugan et al and Zhong et al were proved that objects in the image were more clearly seen visually [18] [1]. Murugan et al focused on prominent hill structure, linearity of hill structure, water bodies, river coarse that were clearly identified after PCA was applied [18]. Zhong *et al* were interested in interpreting many complicated mineral types, including alunite, kaolinite, chalcedony, muscovite, montmorillonite, andradite, buddingtonite [1]. The method developed, BSUSCA can clearly identified the similar but different mineral species. Zhong et al have stated also that road and roof can be classified sharply [1]. Either than that, Lin et al applied modified Eigen-feature technique to classify LIDAR cloud point [19]. The result of the process was 2 classes, building and no-building. This result is similar with 2 previous papers, that the method was strong to detect the edge in the image in building the contrast between two objects.

PCA and its variance still become an interesting technique for capturing satellite images. Some variance of PCA such as Independent Component Analysis (ICA), forward backward PCA, inverse PCA are adopted for solving the issues on multispectral analysis. Another issue regarding PCA and its modification method is efficiency. By applying novel Folded-PCA (F-PCA), Zalbaba et al have been proved that overall computational cost reduced to roughly 10% and Folded-PCA requires less than 1% of memory as those in conventional PCA [20]. This research can be applied to process big data, such as multispectral data or even hyperspectral data that contained a large number of bands. Texture based analysis also applied by researchers for multispectral images. GLCM technique can be conducted to enrich pixel by pixel information of the data. GLCM give more comprehensive analysis and can improve global classification rate up to 13.5% [21]. On the other hand, the development of machine learning technique has an important role in this research. Even [15] claimed that their method, (NSSNet) outperform from some deep learning technique.

4. Conclusion

The review has shown that use multispectral images with its behavior have an important role in any research field. The technique of feature extraction and classification are still to be developed by researchers for better accuracy. Pixel-based and texture-based feature combined with robust supervised algorithm ensure that the research is promising to be continued for the future work.

5. Reference

[1] Y. Zhong, X. Wang, L. Zhao, R. Feng, L. Zhang, and Y. Xu, "Blind spectral unmixing based on sparse component analysis for hyperspectral remote sensing imagery," *ISPRS J. Photogramm. Remote Sens.*, vol. 119, pp. 49–63, 2016.

[2] J. A. Concha and J. R. Schott, "Retrieval of color producing agents in Case 2 waters using Landsat 8," *Remote Sens. Environ.*, vol. 185, pp. 95–107, 2015.

[3] A. Asgarian, A. Soffianian, and S. Pourmanafi, "Crop type mapping in a highly fragmented and heterogeneous agricultural landscape : A case of central Iran using multi-temporal Landsat 8 imagery," *Comput. Electron. Agric.*, vol. 127, pp. 531–540, 2016.

[4] Z. Zhu *et al.*, "Including land cover change in analysis of greenness trends using all available Landsat 5, 7, and 8 images: A case study from Guangzhou, China (2000-2014)," *Remote Sens. Environ.*, vol. 185, pp. 243–257, 2015.

[5] A. Bastarrika, E. Chuvieco, and M. P. Martín, "Mapping burned areas from landsat TM/ETM+ data with a two-phase algorithm: Balancing omission and commission errors," *Remote Sens. Environ.*, vol. 115, no. 4, pp. 1003–1012, 2011.

[6] E. Ozelkan, G. Chen, and B. B. Ustundag, "Multiscale object-based drought monitoring and comparison in rainfed and irrigated agriculture from Landsat 8 OLI imagery," *Int. J. Appl. Earth Obs. Geoinf.*, vol. 44, pp. 159–170, 2016.

[7] N. Koutsias, G. Mallinis, and M. Karteris, "A forward/backward principal component analysis of Landsat-7 ETM+data to enhance the spectral signal of burnt surfaces," *ISPRS J. Photogramm. Remote Sens.*, vol. 64, no. 1, pp. 37–46, 2009.

[8] X. Kang, S. Li, and J. A. Benediktsson, "Feature extraction of hyperspectral images with image fusion and recursive filtering," *IEEE Trans. Geosci. Remote Sens.*, vol. 52, no. 6, pp. 3742–3752, 2014.

[9] Y. Zhang, P. X. Hu, and J. Gao, "A reflectance spectra-based approach to mapping salt fields using PCA-fused Landsat TM data," *Adv. Sp. Res.*, vol. 47, no. 9, pp. 1490–1496, 2011.

[10] R. M. Haralick, K. Shanmugam, and I. Dinstein, "Textural Features for Image Classification," *Syst. Man Cybern. IEEE Trans.*, no. 6, pp. 610–621, 1973.

[11] H. Su, Y. Wang, J. Xiao, and L. Li, "Improving MODIS sea ice detectability using gray level co-occurrence matrix texture analysis method: A case study in the Bohai Sea," *ISPRS J. Photogramm. Remote Sens.*, vol. 85, pp. 13–20, 2013.

[12] Y. Ban, A. Jacob, and P. Gamba, "Spaceborne SAR data for global urban mapping at 30m resolution using a robust urban extractor," *ISPRS J. Photogramm. Remote Sens.*, vol. 103, pp. 28–37, 2014.

[13] T. Lu *et al.*, "Set-to-Set Distance-Based Spectral – Spatial Classification of Hyperspectral Images," vol. 54, no. 12, pp. 7122–7134, 2016.

[14] J. Li, Q. Yuan, H. Shen, and S. Member, "Hyperspectral Image Super-Resolution by Spectral Mixture Analysis and Spatial – Spectral Group Sparsity," vol. 13, no. 9, pp. 1250–1254, 2016.

[15] B. Pan, Z. Shi, N. Zhang, and S. Xie, "Hyperspectral Image Classification Based on Nonlinear Spectral–Spatial Network," *IEEE GEOSCIENCE AND REMOTE SENSING LETTERS*, pp. 3286–3289, 2016.

[16] D. Moher *et al.*, "Preferred reporting items for systematic review and meta-analysis protocols (PRISMA-P) 2015 statement," *Syst. Rev.*, vol. 4, no. 1, p. 1, 2015.

[17] A. Cetateanu and A. Jones, "How can GPS technology help us better understand exposure to the food environment? A systematic review," *SSM - Popul. Heal.*, vol. 2, pp. 196–205, 2016.

[18] S. Murugan, "Significance of Eigen Matrix in Spectral Domain of Remote Sensing Images (RSI)," *International Journal of Computer Applications (0975 – 8887)*, vol. 35, no. 10, pp. 1–6, 2011.

[19] C. H. Lin, J. Y. Chen, P. L. Su, and C. H. Chen, "Eigen-feature analysis of weighted covariance matrices for LiDAR point cloud classification," *ISPRS J. Photogramm. Remote Sens.*, vol. 94, pp. 70–79, 2014.

[20] J. Zabalza *et al.*, "Novel Folded-PCA for improved feature extraction and data reduction with hyperspectral imaging and SAR in remote sensing," *ISPRS J. Photogramm. Remote Sens.*, vol. 93, pp. 112–122, 2014.

[21] A. Safia and D. C. He, "Multiband compact texture unit descriptor for intra-band and inter-band texture analysis," *ISPRS J. Photogramm. Remote Sens.*, vol. 105, pp. 169–185, 2015.

IOP Conf. Series: Journal of Physics: Conf. Series **801** (2017) 012046 doi:10.1088/1742-6596/801/1/012046

Contiguous Uniform Deviation for Multiple Linear Regression in Pattern Recognition

D Andriana[*#], A S Prihatmanto[*], E M I Hidayat[*], I Supriana[*], C Machbub[*]

[#]Indonesian Institute of Sciences, Research Centre for Informatics
[*]Institute of Technology Bandung, School of Electrical Engineering and Informatics

Email: dian.andriana@gmail.com, asetijadi@lskk.ee.itb.ac.id, egi.hidayat@lskk.ee.itb.ac.id, iping@informatika.org, carmadi@lskk.ee.itb.ac.id

Abstract. Understanding images by recognizing its objects is still a challenging task. Face elements detection has been developed by researchers but not yet shows enough information (low resolution in information) needed for recognizing objects. Available face recognition methods still have error in classification and need a huge amount of examples which may still be incomplete. Another approach which is still rare in understanding images uses pattern structures or syntactic grammars describing shape detail features. Image pixel values are also processed as signal patterns which are approximated by mathematical function curve fitting. This paper attempts to add contiguous uniform deviation method to curve fitting algorithm to increase applicability in image recognition system related to object movement. The combination of multiple linear regression and contiguous uniform deviation method are applied to the function of image pixel values, and show results in higher resolution (more information) of visual object detail description in object movement.

1. Introduction

Face detection as an example of object detection algorithm has been developed and shows success in featuring facial detail landmarks such as corners of the eyes, the tip of the nose, the mouth, the eyebrows, and the face boundaries using regressing trees [1]. However, the facial landmarks method is still not able to describe enough information (lower resolution of information) of facial features compared to true human perception capabilities of face recognition. Many statistical and machine learning methods need huge amount or training data taken from human perception or knowledge about objects description [1], but it is still incomplete collected training data [2].

Besides statistical methods, syntactical methods include hierarchical, relational, structural, and morphological methods have been developed for face recognition [3][4][5]. Structural hierarchical and relational methods have been done for high level abstraction of image objects modelling and can provide more information about image objects, but they still have problems for low level implementation [3][4].

Image data understanding is also a pattern recognition problem. Image objects movements can be considered as pixel value patterns. Research has been done [5] by approximating a structure pattern of data signal in time-series. The work [5] uses a hybrid method which combines statistical methods for discriminating objects, and structural methods for describing morphological (shape-based) features. Usually, shapes of objects in morphological feature extraction are described with edges, curves, and regions. Feature extraction techniques include chain code, piece-wise linear regression, and curve fitting. Multiple linear regression in Choquet integral is also used in face recognition improvement to enhance learning result from neural networks [6][7]. In ref. [6], the Choquet integral was used as a mechanism to integrate information resulted from neural networks to learn Sobel edges and morphological gradients, but still needs more testing and training samples. Facial landmarks in [1] are also learned using SVM (Support Vector Machine) with combination of huge training samples. Neither ref. [1] nor [6] used mathematical function to describe specifically the face objects details. In ref. [8], gray-level morphological operators such as dilation and erosion are used, but still do not show further description of shape characteristics and its transformation functions in objects movement.

2. Features Modelling Using Choquet Integral

Structural (shape based) detail features description of images can use edges and curves in curve fitting [5]. Choquet Integral was used to approach curve fitting in ref. [9]. Choquet Integral was used to improve face recognition together with Artificial Neural Networks [6][7]. Choquet Integral formulation [9] is described as follows:

$$C_\mu(f) = \sum_{i=1}^{n} w_i \cdot f(c_i) \quad (1)$$

$$\bar{y} = c + \int (a \cdot \bar{f} + b) d\mu \quad (2)$$

Choquet Integral is defined as the sum of combination of weights w_i and function f of criteria c_i. We can also write the integral as follows, $\bar{y}$ is an objective value in interval (fuzzy values), $\bar{f}$ is function value in interval, μ is weight, a is scaling parameter vector $(a_1, a_2, \dots, a_n)$, b is shift parameter vector $(b_1, b_2, \dots, b_n)$, and c is adjusting constant.

Linear pattern model has also already been developed in Inverse Function Neural Network using linear function as activation function [10] and has been applied in image object pattern recognition for static images [11]. Application in the static image showed capability to model fluctuation of the pixel values. The inverse function Neural Network calculates the weight changes of the neural network nodes using formulae as follows:

$$\sum_{j=0}^{n} w_{i,j,k+1,t} \cdot x_{i,t} = C_t \sum_{j=0}^{n} w_{i,j,k,t} \cdot x_{i,t} \quad (3)$$

$$C_t = \frac{g^{-1}(y_{j,t} - p)}{g^{-1}(y_{j,t} - Err_{k,t})} \quad (4)$$

, where $w_{i,j,k,t}$ is the connection weight between nodes, and C_t is the ratio of weight changes for the training set t.

3. Methodology

With a study case of face recognition problem, we use the structural and morphological method in object details description. We formulate the face object details using a multiple linear regression function describing the neighborhood grey-level pixel values in the object details. We compare the function curve patterns in the different face pose picture samples in the database of faces. We use ORL database of faces: http://www.cl.cam.ac.uk/research/dtg/attarchive/pub/data/att_faces.zip. For example, we model the eyebrows using neighborhood grey-level pixel values in a box area in the face images. We take a training sample from one pose image, and compare with the other poses of the same person image database. Example of poses are showed in Figure 1. For curve patterns fitting, we use various error evaluations including standard deviation of curve difference, Mahalanobis distance [14], and chi-square for Local Binary Pattern method (LBP) [13], all are applied in the eyebrows pixel area of an image compared to the other images of face poses. The result of error evaluation is described in Table 1.

Figure 1 shows example of different poses of a person from the face database. We use one pose as a training sample to compare with other poses to learn visual characteristics of the face element details, for example we use the eyebrows. For comparison, we expect to get knowledge about similarities between the visual characteristics of a face element detail in one pose which is also expected to be found related to the same face element detail in the other poses. The face element details, in this case we use the eyebrows, visually represented by the grey-level pixel values, in related pixel area of image. Fluctuation of grey-level values is then modeled using the Multiple Linear Regression function curve in Choquet Integral.

We use curve of simple neighborhood grey-level pixel values as a representation of the eyebrows face element which is showed in Figure 2. Local Binary Pattern method is also already introduced by [13] to represent face as a whole but not yet in detail of smaller elements. We also use LBP to represent the detail element of face, and further compared with simple cumulative neighborhood grey-level pixel values to represent the detail element.

In the field of face recognition, Local Binary Pattern (LBP) method is useful for eliminating difficulty due to lighting variation which affects grey-level values [9]. The LBP formula is described as follows [13]:

$$LBP(x_c, y_c) = \sum_{n=0}^{7} 2^n s(i_n - i_c) \quad (5)$$

where n runs over the 8 neighbors of the central pixel c, i_c and i_n are the gray-level values at c and n, and $s(u)$ is 1 if $u \geq 0$ and 0 otherwise. We also use simple neighborhood grey-level adapted from the LBP formula without the difference calculation between i_c and i_n.

Besides the chi-square for error evaluation, we also use Mahalanobis distance [14] to measure similarities between curves. Problem occurs when curves similarity measured by the smallest values of chi-square and Mahalanobis distance do not show similar object element in the curve pattern detection, then we have to define contiguous uniform deviation algorithm to improve the error evaluation to accommodate different curves with larger values chi-square and Mahalanobis distance which are actually represent the same element of face in different angle of pose images.

Table 1 shows curves pattern detection using LBP method and simple neighborhood grey-level values combined with chi-square, Mahalanobis distance, and simple standard deviation (without contiguous algorithm) for error evaluation. These results show big distances (position difference) between the targeted (actual) eyebrows position in the images and the detected positions which do not point toward the actual eyebrows position in the images. Smallest distance or difference cannot be used in error evaluation criteria because the face movement in different poses of head turning angle. The object movement makes the object detail curve values change into a slightly bigger distance from the beginning pose.

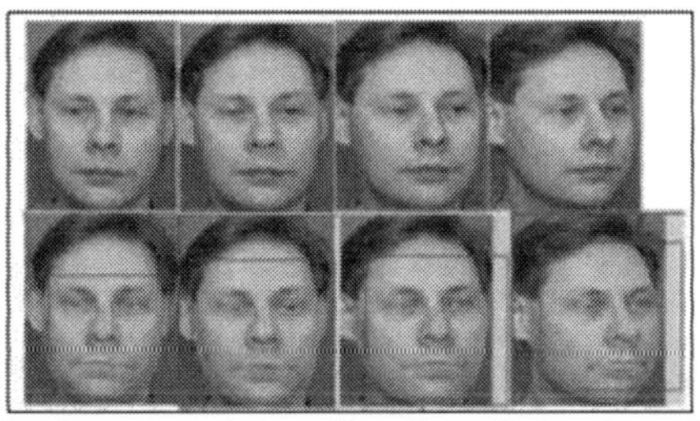

Figure 1. Example of poses variation: from front-facing (most left side pictures) until turning into certain angles in facial landmarks detection [1].

IOP Conf. Series: Journal of Physics: Conf. Series **801** (2017) 012046 doi:10.1088/1742-6596/801/1/012046

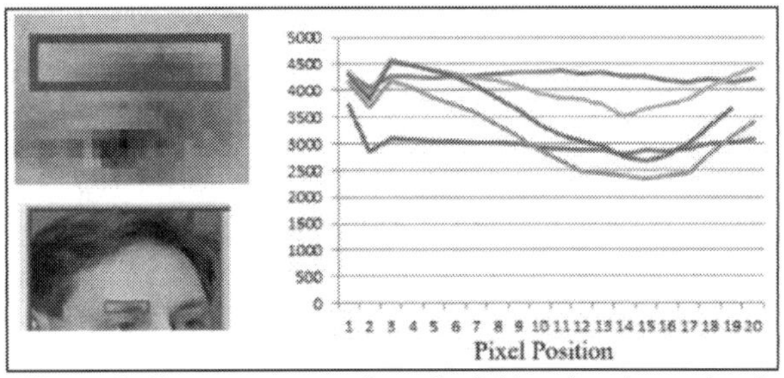

Figure 2. Curve representation of neighbourhood grey-level pixel values in each row of the eyebrows box area as training samples of face element details.

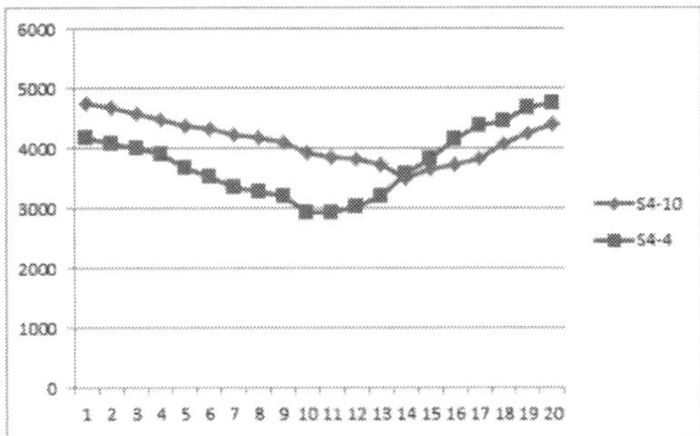

Figure 3. Neighborhood pixel values in a middle row of the eyebrows box area of one pose image compared to another pose image.

Table 1. Comparison of various methods and error evaluations on study case images.

		LBP Chi-Square		Neighbourhood Mahalanobis		Neighbourhood Std.Deviation	
ID Image	**Target position (*x*,*y*)**	**Detected position (*x*,*y*)**	**Position Differenc e (*dx*,*dy*)**	**Detected position (*x*,*y*)**	**Position Difference (*dx*,*dy*)**	**Detected position (*x*,*y*)**	**Position Difference (*dx*,*dy*)**
S4-4	(22,44)	(26,84)	(4,42)	(65,70)	(43,28)	(23,58)	(1,14)
S4-5	(29,43)	(44,85)	(15,42)	(22,83)	(-7,40)	(40,64)	(11,21)
S1-2	(25,35)	(66,33)	(41,-2)	(37,55)	(12,20)	(12,104)	(-13,69)

The contiguous uniform deviation algorithm is based on values of curve difference in the same range when the test curve is compared to the reference curve. A range of values is used because it is allowed to include bigger values of error (in chi-square or Mahalanobis distance) and to avoid larger curve difference (optimized difference). Difference values between curves should be spread uniformly or consistently along the curves, so that they do not fluctuate much. The formulation of contiguous uniform deviation of curve difference is defined as follows:

$$\Delta_1 \approx \Delta_2 \approx \Delta_k \approx \ldots \approx \Delta_n \tag{6}$$

$$\nabla_t = \begin{cases} |\Delta_t - \Delta_k| < \varepsilon_1 \;\; or \\ \varepsilon_2 < \Delta_t - \Delta_k < \varepsilon_3 \end{cases} \tag{7}$$

$$C_1 < \sigma = \left(\frac{1}{n} \sum_{t=1}^{n} (\Delta_t - \mu)^2 \right)^{\frac{1}{2}} < C_2 \tag{8}$$

Basically we expect the values of curve point differences Δ_k to be relatively almost equal each other along both compared curves. However, Δ_k values can change and the sign changes as well (positive or negative). Then first we need to check the vote of positive of negative sign, which sign is vote the most. The next step is defining the acceptance range ∇_t as the difference between Δ_k values which are expected to be small, and depend on the sign vote. Beside the acceptance range values of ∇_t, we also should define the standard deviation between Δ_k values with mean μ, and the standard deviation are expected to be in a range between certain values C_1 and C_2.

4. Results and Discussions

We apply the contiguous uniform deviation into face element recognition problems with a study case of the eyebrows localization, and give result in ability to localize the actual eyebrows position in the image. Table 2 shows a sequence of pixel position where the standard deviation values contiguously are in a range between 200 and 300, with μ is set to 249. The values of μ, C_1 and C_2 are set by experiments. It can be seen that position (24,44) has the lowest standard deviation value in the sequence, which has only 2 columns difference from actual position of the eyebrows in (22,44). However, there are still possibilities of false localization, and need combination of other rules of detection, for example combined with localization samples collected in facial landmarks detection [1].

Table 2. An example of a detection using pixel value standard deviation of differences in a range between 200 and 300 at each pixel position and mean of differences 249.

Position	Std.Deviation	In Range? (Y/N)
(19,44)	286	Y
(20,44)	225	Y
(21,44)	255	Y
(22,44)	298	Y
(23,44)	271	Y
(24,44)	249	Y
(25,44)	263	Y
(26,44)	331	N
(27,44)	409	N
(28,44)	476	N

We conclude that multiple linear regressions can be used to model visual characteristics of object details in grey-level pixel values, and can be used as a representation of object details as indicated in object detail detection algorithm constructed with contiguous uniform deviation error evaluation in image object detail detection. This highly detailed representation has more information in describing object details which have not been addressed by existing methods. However, more criteria of curve fitting should be investigated further to increase accuracy of localization of object details.

Acknowledgments

The author wish to thank The Indonesian Ministry of Research, Technology, and Higher Education in supporting this work through Indonesian Institute of Sciences and Institute of Technology Bandung research and education programs.

References

[1] Kazemi V and Sullivan J 2014 One millisecond face alignment with an ensemble of regression trees *IEEE Conference on Computer Vision and Pattern Recognition* pp 1867-1874

[2] Zafeiriou S, Zhang C, and Zhang Z 2015 A survey on face detection in the wild: Past, present and future (*Comput. Vis. Image Underst.*) vol 138 pp 1–24

[3] Antanas L, Frasconi P, Costa F, Tuytelaars T, and De Raedt L 2012 A relational kernel-based framework for hierarchical image understanding *Int. Conf. on Structural, Syntactic, and Statistical Pattern Recognition* pp 171-180

[4] Antanas L, van Otterlo M, Oramas M J, Tuytelaars T, and De Raedt L 2012 A relational distance-based framework for hierarchical image understanding *Int. Conf. on Pattern Recognition Applications and Methods* vol 2 pp 206–218

[5] Olszewski R T, Maxion R, Siewiorek D, and Banks D 2001 *Generalized Feature Extraction for Structural Pattern Recognition in Time-Series Data* (Pittsburgh PA: Carnegie Mellon University) pp 1-26

[6] Martinez G E, Mendoza O, Castro J R, and Rodriguez-Diaz A 2015 Response integration in modular neural networks using Choquet Integral with interval type 2 Sugeno measures *Fuzzy Inf. Processing Society World Conf. on Soft Comput.* (Redmond WA, USA: IEEE) pp 1-6

[7] Mendoza O, Melin P, and Licea G 2009 A hybrid approach for image recognition combining type-2 fuzzy logic, modular neural networks and the Sugeno integral *Inf. Sci. (Ny*: Elsevier*)* vol 179 no 13 pp 2078–2101

[8] Torres-Huitzil C 2014 Fast hardware architecture for grey-level image morphology with flat structuring elements (*IET Image Process.*) vol 8 no 2 pp 112–121

[9] Yang R, Wang Z, Heng P A, and Leung KS 2008 Fuzzified Choquet Integral with a fuzzy-valued integrand and its application on temperature prediction (*IEEE Trans. Syst. Man. Cybern. B. Cybern*) vol 38 no 2 pp 367–80

[10] Andriana D 2013 Linear function and inverse function with weight ratio for improving learning speed of multi-layer perceptrons feed-forward neural networks *Proc. IC3INA* pp 255-259

[11] D. Andriana, "Contiguous uniform deviation for artificial neural network pattern recognition," *Advanced Science Lett.*, vol. 21, no. 2, pp. 189-191, Feb. 2015.

[12] A. F. Tehrani, M. Strickert, and E. Hüllermeier, "The Choquet kernel for monotone data," *Esann*, no. April, pp. 23–25, 2014.

[13] X. Tan, and B. Triggs Enhanced local texture feature sets for face recognition under difficult lighting conditions (*IEEE Trans. on Image Processing*) vol 19 pp 1635–650

[14] Sobieranski AC, Chiarella VF, Alexandre EB, Linhares RTF, Comunello E, and von Wangenheim A 2014 Color skin segmentation based on non-linear distance metrics *Progress in Pattern Recog.,Image Analysis, Comp. Vis.,and App.: Proc.CIARP,* (Mexico: Springer) ed EB-Corrochano and E Hancock pp 143-150

Implementation of The LDA Algorithm for Online Validation Based on Face Recognition

Z Zainuddin[1] and A S Laswi[2]

[1]Department of Teknik Elektro Universitas Hasanuddin, Makassar, Sulawesi Selatan, Indonesia
[2]Department of Manajemen Informatika AMIK Ibnu Khaldun, Palopo, Sulawesi Selatan, Indonesia

Email : [1]zainuddinzahir@gmail.com, [2]aishiyah.rustam@gmail.com

Abstract. This paper report work in implementation of computer vision application in face recognition to the on-line validation for distance learning. Face recognition is chosen among many other alternatives of validation because its robustness. The problem with basic validation such as password cannot validate the student in distance learning. This cannot be accepted especially is distance examination. Face recognition algorithm used in this research is Linear Discriminant Analysis (LDA). By using this algorithm, the system capable of recognize the authorized persons about 93% and reject the unauthorized persons 100%.

1. Introduction

In some of last decades, the use of biometric security system was widely discussed in the world of computer vision and pattern recognition research. Face recognition is one of the biometric technology has now been applied for many applications in the fields of security, such as *Access security system, Authentication system,* and as a tool in tracking criminals [1]. High security system is one of the reasons the biometric continue to be developed, where the workings of the biometrics cannot be imitated and transferred to the second party. Some government agencies now select system security by using face recognition biometric.

Distance educations require authentication systems in order to find out the validity of their participants. Distance education itself is a system of education where teaching and learning among teachers and learners are not limited by space and time [2]. Some contents of e-learning and distance learning, such as [3][4], should be limited to the only member of the distance learning group. Authentication using face recognition can be a solution to overcome the anxiety of the Government in implementing a system of distance education.

Face recognition system offers some algorithms in the process of face identification. Linear projection method is one of the most commonly used algorithm in the application of face recognition system [5]. The process of this method is by projecting the image pixels into a matrix. Then, matrix is converted to a single-dimensional as vector data to obtain the eigen value and covariance value. This method is applied on the two Principal Component Analysis algorithms (PCA) and Linear Discriminant Analysis (LDA). The purpose of PCA and LDA methods solely to reduce the dimensions of high dimensional space into a low-dimensional space.

In this research, the education is based on distance online system. The proposed security system used the student ID number as user and face recognition as password to access the system. The algorithm used in this research is the LDA.

2. Face recognition and Algorithms

Face recognition is a complex problem in the verification process and identification of a face through the digital images or video frames [6]. The process of face recognition in a digital image or video frame is not an easy job. There are some challenges in the process of face identification on an image as follows [7]: intensity, pose, face rotation, and occlusion.

To overcome difficulties at the identification process of face image, it should take the right algorithm. In general, there are two techniques that are proposed, namely the approach of local features and holistic approach [8]. But lately, the development of new technique in the form of hybrid approach has already been done, i.e. a merger between engineering approach of holistic and local features.

2.1 Principal Component Analysis

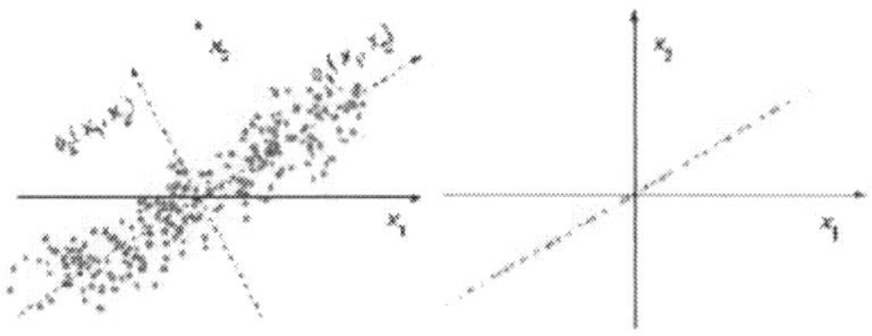

Figure 1. PCA Base (left) and PCA reduction to 1 d (Right) [9]

PCA is a standard technique for high-dimensional data visualization and for pre-processing of data [9], [10]. PCA can reduce data dimension without losing relevant information. Extracting the space component of the lower dimensions includes the highest variant. Figure 1 shows the data reduction process of the PCA.

2.2 Linear Discriminant Analysis

Like PCA, LDA is also an algorithm which aims to reduce dimensions. LDA finds an optimal projection so that it can project the input data on space with smaller dimensions where all patterns can be separately maximized [11][12]. Separation of the LDA will maximize the dissemination of data-input between different classes and also minimizing the spread of input in the same class. The difference between the class represented by the matrix of the Sb and the difference in class is represented by the matrix Sw [13].

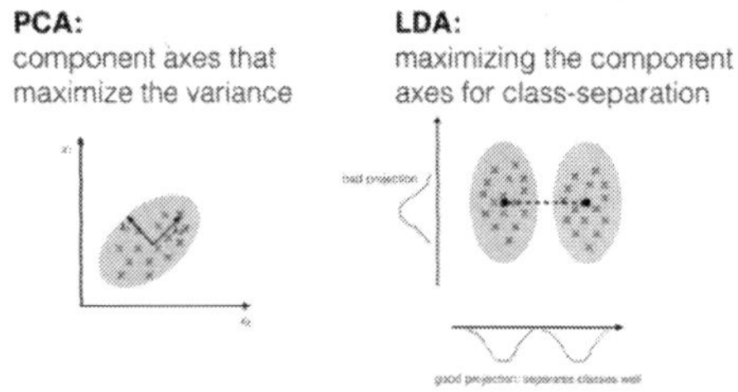

Figure 2. The difference of covariance between PCA and LDA [7]

Weight matrix Transpose PCA is the input for the LDA, this shows clearly that most of the algorithm LDA must run view steps through the PCA algorithm. The following algorithm LDA [10]:

- Vector data input from transpose weight PCA become input to the LDA:
  $X = [x_{11}, x_{12}, ..., x_{1n}, ..., x_{21}, x_{22}, ..., x_{2n}, ..., X\ x_{m1,\ m2,\ ...,\ mn}\ X]$.

- Calculates the spread between classes (S_b) and grade (S):

$$S_w = \sum_{j=1}^{n}\sum_{i=1}^{Ni} N_i\,(x_i^j - \Psi_j)(x_i^j - \Psi_j)^T$$

$$S_b = \sum_{i=1}^{n} N_i\,(\psi_i - \psi)(\psi_i - \psi)^T$$

Description:
N_i = The number of data training in class i
n = The number of different class
Ψ = The Mean total of vector to a sample to the i
ψ_i = The Mean of the sample on to the i
x_i^j = vector of PCA from pictures to i and results from classroom to j

- Calculate the value of the Covariant LDA with formula $C = S_b S_w^{-1}$
- Then calculate the value of the Eigen vector and eigen value equations:

$S_b\,\psi_i = \lambda_i\,S_w\,\psi_i$
Description:
λ_i = Value of the eigen value LDA
When the eigenvalues are lined up in ascending order, a matrix consisting of eigenvectors q becomes EQ [13].
$v_{LDA} = [\,\psi_1, \psi_2, \ldots\ldots \psi_q]$
Description:
v_{LDA}= Value of the eigen vector LDA

- Calculating Feature Matrix equation with LDA:

$$f = \sum_{n=1}^{N} (\mathrm{P} - \psi)^T x\, v$$

Description:
f = the value of the features of the PCA

3. The proposed Research Method

This research used a holistic approach with LDA algorithm. The system is tested using 10 dataset classes. Each class consists of 8 image training. Following are the steps in the process of face detection:

- Enrollment

At this point students enrolled in Distance Education website. This image will be stored in MySQL database as the training image. Figure 3 shows one of the training picture stored in Mysql.

Figure 3. Image training

- Face detection

  The following are the steps: Convert to grayscale image, Generate histogram of grayscale image, Equalize the image, Generate histogram of equalized image, Remove noise from an image, then the image classification of skin. Figure 4 and 5 show the process enhacement and face detection.

Figure 4. Process enhacement to grayscale image

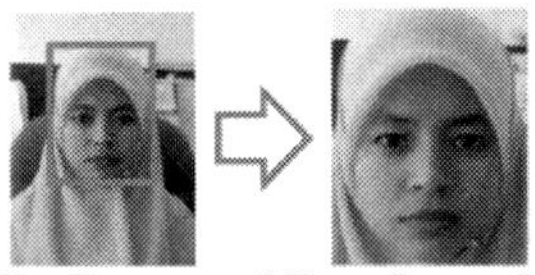

Figure 5. The Process of Face Detection

The process of face detection algorithm using viola jones but this research not emphasis on discussion of this algorithm.

- Training Image

  This stage uses LDA algorithm to train images stored in database after previous process. Figure 6 shows the results of LDA training.

Figure 6. Result of LDA training

- Face Recognition

  LDA algorithm becomes the final determinant in defining the face is recognizable or not in this process. Figure 7 shows the whole system.

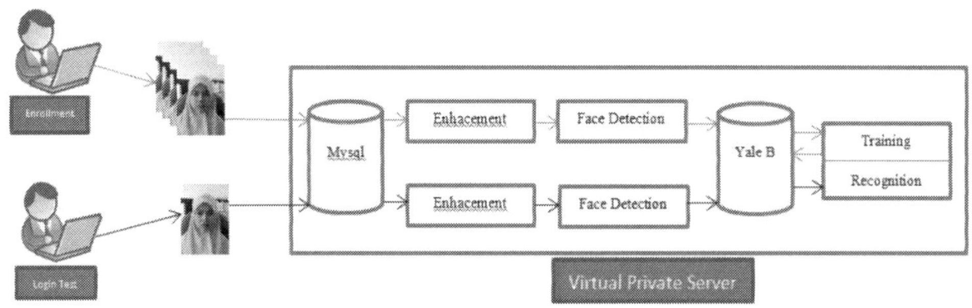

Figure 7. Online Distance Education System

4. Results and Discussion

This system was built using two different programming languages, on the side of the client/participant using the PHP programming language and on the server side using Java. System testing is performed with 10 participants, and 8 pictures for each of them. Each participant performs login tests to the system as much as 10 times to find out how it performs.

Table 1 shows the results of the testing participants who have already registered in the system. The results show that the system can work well when a face is unregistered. Average percentage indicates the level of accuracy as much as 93%. Error system shows the percentage of 7% at some login tests. e.

Table 1. The test results using the data training 8 per class for participants who enrolled in a Distance Education system.

No.	The initials	The Number of Test	To be known	Not Recognized	The accuracy of the %	Error %
1	User 1	10	9	1	90%	10%
2	User 2	10	10	0	100%	0%
3	User 3	10	10	0	100%	0%
4	User 4	10	8	2	80%	20%
5	User 5	10	9	1	90%	10%
6	User 6	10	9	1	90%	10%
7	User 7	10	10	0	100%	0%
8	User 8	10	10	0	100%	0%
9	User 9	10	9	1	90%	10%
10	User 10	10	9	1	90%	10%
		Average			93%	7%

In addition to conducting trials on the participants who have registered in the database, testing is also done on a user who is not listed in the database. Table 2 shows the percentage of 100%, where the face who tries to login and not listed on the database cannot access the system.

Table 2. The test results for a user who is not enrolled in a Distance Education system

No.	The initials	The Number of Test	To be known	Not Recognized	The accuracy of the %	Error %
1	User 1	10	0	10	100%	0%
2	User 2	10	0	10	100%	0%
3	User 3	10	0	10	100%	0%
4	User 4	10	0	10	100%	0%
5	User 5	10	0	10	100%	0%
6	User 6	10	0	10	100%	0%
7	User 7	10	0	10	100%	0%
8	User 8	10	0	10	100%	0%
9	User 9	10	0	10	100%	0%
10	User 10	10	0	10	100%	0%
		Average			100%	0%

5. Conclusion

From the results of this research it can be concluded that the online system could run on face authentication system using LDA algorithm. Some failure in the system are results of this holistic approach is still caused by several factors such as how big a difference the intensity of light received by each of the pictures, poses, and other additional attributes.

Bibliography

[1] Archana T., "Face Recognition : A Template Based Approach," *International Conference on Green Computing and Internet of Things, ICGCIoT 2015*, pp. 966–969, 2015.

[2] Peraturan menteri pendidikan dan kebudayaan republik Indonesia No. 24 Tahun 2012 Tentang Penyelenggaraan Pendidikan Jarak Jauh Pada Pendidikan Tinggi. 2012. Jakarta.

[3] Zainuddin Z, Manullang, E.V., "E-learning concept design of rijndael encryption process ," *Proceeding of 2013 IEEE International Conference on Teaching, Assessment and Learning for Engineering, TALE 2013*, 2013.

[4] Zainuddin Z, Ihsan I P, "Modelling of child psychology tests based game learning," *Proceeding of 2013 IEEE International Conference on Teaching, Assessment and Learning for Engineering, TALE 2013*, 2013.

[5] Li C., Liu J., Wang A., and Li K., "Matrix reduction based on generalized PCA method in face recognition," *Proc. - 2014 Int. Conf. Digit. Home, ICDH 2014*, no. 2, pp. 35–38, 2014.

[6] Saini R. and Narinder R., "Comparison of Various Biometric Methods," *Int. J. Eng. Sci. Technol.*, vol. 2, no. I, pp. 24–30, 2014.

[7] Bansal A., Mehta K., and Arora S., "Face Recognition Using PCA and LDA Algorithm," *2012 Second Int. Conf. Adv. Comput. Commun. Technol.*, pp. 251–254, 2012.

[8] Agrawal S, and Khatri P, "Facial expression detection techniques: Based on Viola and Jones algorithm and principal component analysis," *Int. Conf. Adv. Comput. Commun. Technol. ACCT*, vol. 2015–April, pp. 108–112, 2015.

[9] Wagh P, Chaudhari J, Thakare R., Patil S., "Attendance System based on Face Recognition using Eigen face and PCA Algorithms." *IEEE International Conference on Green Computing and Internet of Things, ICGCIoT 2015*, pp. 303–308, 2015.

[10] Tripathi C. and Singh K. P., "A new method for face recognition with fewer features under illumination and expression variations," *2012 19th Int. Conf. High Perform. Comput. HiPC 2012*, 2012.

[11] Elmaghraby A. S. and Reese K., "Performance Improvement of Face Recognition using Multispectral Images and Stereo Images," pp. 280–285, 2012.

[12] Bing Y, Ping C, and Pchenmailhzzjcn E B L, "A New LDA-based Method for Face Recognition," vol. 0, no. c, pp. 0–3, 2002.

[13] Song Y J, Kim Y G, Kim N, and Ahn J H, "Face Recognition Using Both Geometric Features and PCA/LDA," *Sixth Int. Conf. Adv. Lang. Process. Web Inf. Technol. (ALPIT 2007)*, pp. 248–252, 2007.

IOP Conf. Series: Journal of Physics: Conf. Series **801** (2017) 012048 doi:10.1088/1742-6596/801/1/012048

Traffic Congestion Model: Challenges and Opportunities

M.D. Enjat Munajat[1], Rinaldi Munir[2] Dwi H. Widyantoro[3],
School of Electrical Engineering and Informatics Institute of Technology Bandung,
Email: [1]mdenjatm@unpad.ac.id, [2]rinaldi@informatika.org, [3]dwi@stei.itb.ac.id,

Abstract. This paper discusses models to detect traffic congestion using two approaches: road detection and moving object detection. There are two methods proposed to detect roads and moving objects. The methods involve the detection of both moving and non-moving objects. The system presented in this paper is able to detect roads as well as moving objects. However, further development is needed to detect extreme road curves. The system presented in this paper is relatively inexpensive since it only uses one camera and capable of capturing satisfactorily detailed images.

1. Introduction

Traffic congestion model becomes an important issue, because a good model can effectively and efficiently determine the level of congestion, and inform the public as the basis for determining the best direction. Numerous methods have been developed to monitor roadways with the ultimate aim of detecting traffic congestion. Systems developed for this purpose almost always uses 3 (three) parameters, namely (1) vehicle detection, (2) vehicle tracking, (3) vehicle classification [2]. Potential issues arising are the resolution of details and accuracy of reading roads and vehicles. Road detection systems tend to have issues monitoring curved road sections. Vehicle detection, on the other hand, tends to work reliable if there are no obstructions present. However, such systems tend to be unreliable if view of the tracked object is partially obstructed. These shortfalls are inherent challenges in studies on road monitoring and vehicle tracking. This paper presents several road monitoring and vehicle detection methods, which are not susceptible to such issues.

2. Related Works

2.1. Vehicle Detection

Jun-Wei Hsieh et al. [2] conducted a study on a road with moderate traffic density. Their study found that the method they used were able to detect vehicles with an average accuracy of 80%. They [2] then developed their model further so that it could classify vehicular traffic. The model [2] developed improved detection accuracy to 91% even in conditions susceptible to occlusions. The process implemented in [2] was to reconstruct images using the Gaussian Mixture Model by adding boundaries onto each lane (Figure 1(a)). This boundary then removed vehicle occlusions caused by shadows to improve precision. In 1997, Beymer et al. [1] have conducted a similar study in vehicle detection. They [1] claimed that their method was, at the time, the best. Their method [1] of vehicle detection uses static cameras connected to a network of Texas Instruments C40 DSPs (Digital Signal Processor). Beymer et al. accumulated approximately 44 hours of traffic footage.

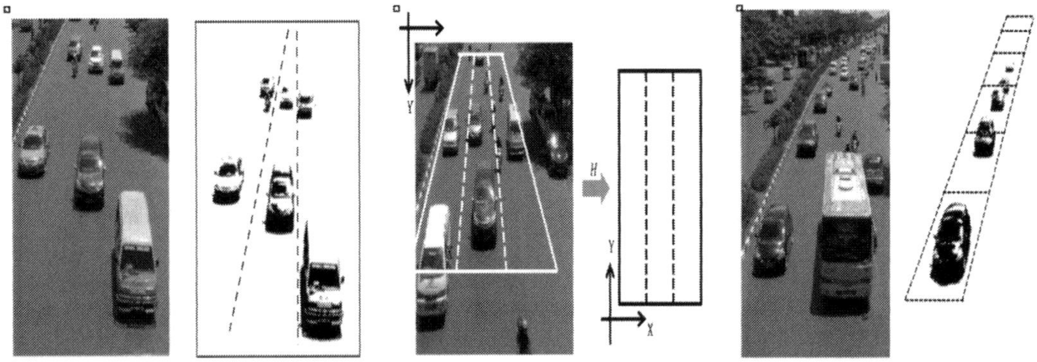

(a) (b) (c)

Figure 1. (a) Boundary of each lane is reflected by virtual lines; (b) H Projection transformation (Homography) to map coordinates (x,y) from the actual image to virtual image (X,Y); (c) Traffic condition and sequence of each area unit and corresponding entropy [6]

Another similar study was conducted by W.-L. Hsu et al. [3]. They proposed a vision-based traffic monitoring system that can perform real-time traffic parameter extraction. Entropy measurement is used as the underlying approach to calculate traffic flow and vehicle speed. The model they [3] developed drew a given background area into boundary blocks. The blocks were then calibrated. Thus, if a background area is filled, the foreground area of each video frame is read and number of vehicles passing calculated.

2.2. Vehicle Tracking

Shunsuke Kamijo et al. [4] looked into traffic monitoring and accident detection at intersections. Although [4] also incorporated object detection into their study, attention is given to the tracking process. Shunsuke [4] adopted the Spatio Temporal-Markov Random Fields (ST-MRF) based on the Hidden Markov Model (HMM). Images processed using the method are divided into 8×8 pixel blocks. The study was conducted at a busy intersection for 25 minutes. An updated model was proposed by David A. Ross et al. [5]. The algorithm [5] developed used incremental updates to an eigenbasis which has learned off-line from a set of training images. The algorithm then marked target particles in the image frame and predicts the object's location based on earlier learning. Tracking was done in varying weather and lightning conditions. The results showed that the algorithm was sufficiently robust to track vehicles. The algorithm also updated appearances of new object on the system reliably.

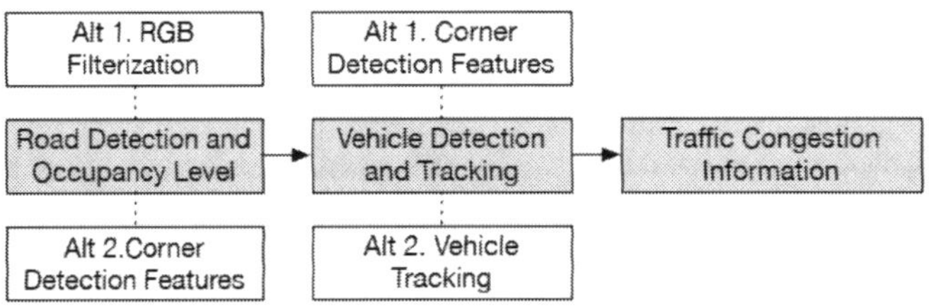

Figure 2. Proposed Model Flowchart

3. Proposed Methods

Following the methods described above, we propose a method to measure traffic congestion using a number of features we have developed, as seen in the flowchart in Figure 2. We have developed several alternatives (Alt), which we will describe in the following section.

4. Analysis

4.1. Road Detection and Occupancy Level

4.1.1. Alternative 1: RGB Filterization

The process begins with a road frame with no vehicles to identify the background using a traffic monitoring camera (Figure 3(a)).

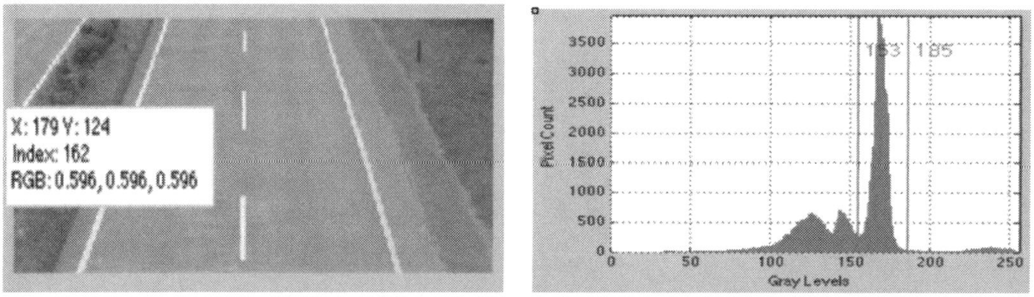

(a) (b)

Figure 3. (a) Red channel removed from RGB image; (b) Red channel histogram. [7]

Red, green, and blue channels of the grey-scale of the clean frame is then analysed. The result is shown in Figure 3(b). Identical analysis is also done to the green and blue channels. The peaks in the histogram indicate roads while the valleys indicate lane markings. The RGB histogram information is combined; consideration is only given to the region between peaks. This creates a better colour image, as shown in Figure 4.

(a) (b) (c)

Figure 4. Process (a) and (b) Noise Removal; (c) Masking [7]

Afterwards, the system is tested on a straight road (Figure 5(a)) and curved road (Figure 5(b)). The analysis shows that the system is sufficiently robust and able to differentiate between road and non-road areas. The lines on both sides of the road demarcate roads and non-roads by colour alone.

4.1.2. Alternative 2: Corner Detection Features

For the corner adjacent feature, we use a recording of a main road in Bandung, Indonesia. The recording is analysed frame-by-frame (Figure 6). The resulting recording is then converted to grayscale. By using the thresholding feature, the image is then converted to black and white (B/W) to enhance image boundaries. We then assign a thresholds value of about 30 for linking threshold and 50 for upper threshold. Images which have been subjected to the thresholding (B/W) processing is then detected for corners using the Kanade-Lucas-Tomasi (KLT) method. Detected corners are shown in Figure 6(b). Corner-detected images are then masked with the original image (Figure 6(c)) to clearly show the marked corners. The system then records the x and y positions of each readable corner points and mapped onto the original colour image. Here, the system detects corner positions. Each frame may have large numbers of marked corners (denoted by the "+" at each detected corner). Readable corners are stored in a database in one minute recording increments.

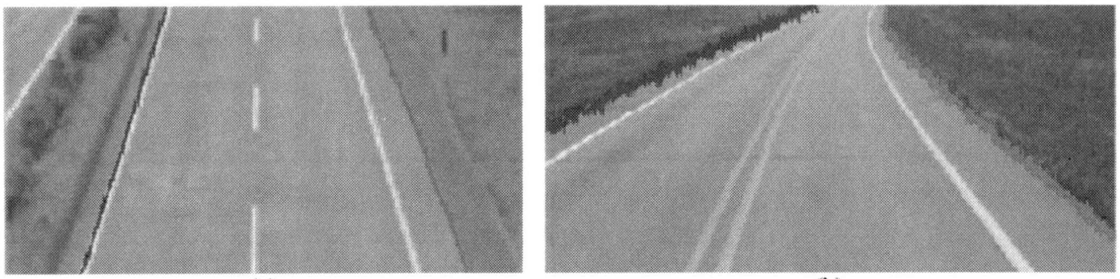

(a) (b)

Figure 5. (a) System testing on a straight road; (b) System testing on curved road [7]

The system reads the nodes of corners of each moving object in each pixel. We call this as corner (vertices), as shown in Figure 6(b). The corner is then stored in one minute recording increments until the road area is covered by enough corners. Following the reading and recording of corners, the system will conduct a Convex-Hull process using Melkman's algorithm to show the masking of the primary road. The convex shape is dynamic, depending on the number of outermost readable corners. In this case, the road may be widened to improve the accuracy of road boundary mapping.

(a) (b) (c)

Figure 6. System testing: (a) Traffic congestion; (b) corner reading; (vertices), marked by the '+' signs on objects' corners

4.2. Vehicle Detection and Tracking

Vehicle detection follows similar principles as that of road detection. However, the process is somewhat more complex due to the corner point database and a point's relative position to its neighbours. The first one-third of the frame is the initiation phase, where vehicular objects are recognised by comparing against the corner database established earlier. The remaining two-thirds of the frame is used for tracking. The system considers a set of corners (in xy position) as a vehicle or single object if their relative distance are constant. Conversely, if the corner distances are not constant, they are not from one single object. The system is able to determine the relationship among different objects in each frame and are able to record the relationship between identical objects. The condition of each object can then be discriminated into one single object.

(a) (b) (c) (d) (e)

Figure 7. System display using the proposed method: (a) Untouched image; (b) Thresholding process; (c) Corner detection; (d) Connecting corner points to form lines; (e) Vehicle demarcation using Convex-Hull algorithm

After establishing a single object, the system then tracks the moving object to obtain its average movement. Figure 7 shows that the proposed method is capable of discriminating two adjacent objects. This is largely due to the extremely robust adjacent corner and line algorithm in recognising the unity of an object. The proximity pattern is also implemented in measuring the consistency of movements of the points and lines in the corner marking process. In the event of two vehicles in very close proximity, the system disregards the generally inconsistent delta length of the line. An inconsistent delta length shows that there are two different objects between two vehicles, denoting that there are two objects. By doing so, the system makes its corner and line grouping calculations consistent. Each grouping marks every distinct object. The Convex-Hull algorithm clearly shows the limit between the two lines (Figure 7(e)). The method is significantly more robust compared to other similar methods, such as background subtraction. Figure 8 shows a comparison between the method proposed in this paper and the commonly implemented background subtraction method.

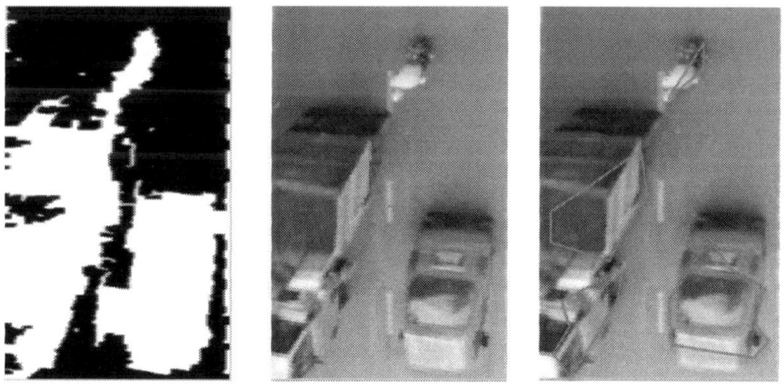

(a) (b) (c)

Figure 8. Comparison between: (a) Object thresholding; (b) background subtraction; (c) proposed method

Having identified all objects in the image, the next process is tracking vehicles to identify their speeds. It is possible to calculate the distance travelled by an object by calculating its distance range – distance between the start and end points. A series of distance range of each object can then be used to calculate average distance, from which object speed can be calculated. The average traffic speed can then be calculated from a number of individual object speeds.

5. Discussion

The system proposed in this paper requires further development to detect extreme road corners (or twisting roads). Vehicle detection using the adjacent corner method requires tuning to detect details on a given vehicle. Currently, the detection algorithm can only detect certain corners of vehicles, thus the polygonal outlines.

6. Conclusion

In detecting road condition, RGB histogram filtering is able to detect road corners. It should be noted that there are levels of road corners. In addition, road markings significantly affect detection area and speed. If road markings are absent, demarcation of roads will be difficult. Moreover, road detection using the adjacent corner method is unreliable if there are corners in the road. The adjacent corner method is a superior method of detecting vehicles compared to other similar methods. However, the method may be unreliable in capturing complete physical details of vehicles as it can mark only certain corner points. The system can easily identify and calculate traffic speed and volume. The keystone of this process is in the initiation phase (first one-third of each frame). If the system fails this phase, it will almost certainly fail to track vehicles and their speed.

Acknowledgment

This paper is funded by the Decentralization Grant awarded by the Directorate of Higher Education (DIKTI) and PUPT Scheme Grant, 2016.

References

[1] David Beymer, Philip McLauchlan, Benn Coifman, and Jitendra Malik, " A Real-time Computer Vision System for Measuring Traffic Parameters, "*1063-6919/97 $10.00 0 1997 IEEE*
[2] Jun-Wei Hsieh, Shih-Hao Yu, Yung-Sheng Chen, and Wen-Fong Hu, "Automatic Traffic Surveillance System for Vehicle Tracking and Classification, "*IEEE TRANSACTIONS ON INTELLIGENT TRANSPORTATION SYSTEMS, VOL. 7, NO. 2, JUNE 2006*
[3] W.-L. Hsu, H.-Y.M. Liao, B.-S. Jeng and K.-C. Fan, "Real-time traffic parameter extraction using entropy," *IEE Proc.-Vis. Image Signal Process., Vol. 151, No. 3, June 2004*
[4] Shunsuke Kamijo, Yasuyuki Matsushita, Katsushi Ikeuchi, and Masao Sakauchi, "Traffic Monitoring and Accident Detection at Intersections, "*IEEE TRANSACTIONS ON INTELLIGENT TRANSPORTATION SYSTEMS, VOL. 1, NO. 2, JUNE 2000*
[5] David A. Ross, Jongwoo Lim, Ruei-Sung Lin, and Ming-Hsuan Yang, "Incremental Learning for Robust Visual Tracking, "*Honda Research Institute, USA Motorola Labs*
[6] D. H. Widyantoro and M. D. Enjat Munajat, "Fuzzy traffic congestion model based on speed and density of vehicle," *Advanced Informatics: Concept, Theory and Application (ICAICTA), 2014 International Conference of,* Bandung, 2014, pp. 321-325. http://ieeexplore.ieee.org/stamp/stamp.jsp?tp=&arnumber=7005962&isnumber=7005902
[7] M. D. E. Munajat, D. H. Widyantoro and R. Munir, "Road detection system based on RGB histogram filterization and boundary classifier," *2015 International Conference on Advanced Computer Science and Information Systems (ICACSIS)*, Depok, 2015, pp. 195-200. http://ieeexplore.ieee.org/stamp/stamp.jsp?tp=&arnumber=7415163&isnumber=7415142

Height Measuring System On Video Using Otsu Method

C L M Sandy [1] and R Meiyanti [2]

University of Sumatera Utara, Medan, Indonesia

Email: likasandy3@gmail.com[1] and rinimeiyanti522@gmail.com[2]

Abstract. A measurement of height is comparing the value of the magnitude of an object with a standard measuring tool. The problems that exist in the measurement are still the use of a simple apparatus in which one of them is by using a meter. This method requires a relatively long time. To overcome these problems, this research aims to create software with image processing that is used for the measurement of height. And subsequent that image is tested, where the object captured by the video camera can be known so that the height of the object can be measured using the learning method of Otsu. The system was built using Delphi 7 of Vision Lab VCL 4.5 component. To increase the quality of work of the system in future research, the developed system can be combined with other methods.

1. Introduction

Image segmentation based on thresholding is one of the oldest and powerful technique, since the threshold value divides the pixels in such a way that pixels having intensity value les than threshold belongs to one class while pixels whose intensity value is greater than threshold belongs to another class. [1] Measurement is an activity that compares a quantity measured by the measuring instrument used as a unit. In this system of measurement of height, the image processing is able to perform recognition of the object and is capable of processing the measurement of height so that the correct parameter value is obtained. In this system the authors use the method of Otsu, which aims to separate objects from the background that comes naturally. Algorithms have been because, in principle, a video is a row of images that can be moved so as to form a picture of the state or event that is not silent (real time), which is the absolute nature that must be maintained.

2. Theoritical Basis

2.1 Image Processing

The image processing is the manipulation of the image to be, especially with the use of a computer, the images that have better quality. Digital image processing refers to the processing of two-dimensional images using a computer. In a broader context, digital image processing refers to the processing of any two-dimensional data. Digital image is an array containing the values of real or complex which is represented by a specific bit stream.

2.2 *Grey-Scale*

Gray-scale image is the image using shades of gray which is a combination of black and white. In the image processing, color images are often converted first into gray-scale image. The images received from the video must be converted to the Gray scale image because though the images are in gray matter the Red Green Blue components are not visible in the image. [2] Then, through the gray-scale image processing is carried out, for example, to obtain the texture of objects. To convert the RGB image into a gray-scale image, we use the following formula:

$$bit_{RGB}(x,y) = 0{,}11 \times bit_R(x,y) + 0{,}59 \times bit_G(x,y) + 0{,}3 \times bit_B(x,y)$$

where:

$bit_R(x,y)$ = The value of the intensity of the red color

$bit_G(x,y)$ = The value of the intensity of the green color

$bit_B(x, y)$ = The value of the intensity of the blue color

2.3 Video Image

The image of the video is a collection of images that moved along a specific time duration. The movement the images contained in the video to form a dynamic shift so easily trick the human eye. The movement or rapid change is a challenge for us to implement the right or efficient algorithm in a situation of rapid and dynamic. In recent years, hardware technologies and standards activities have matured to the point that it is becoming feasible to transmit, store, process, and view video signals that are stored in digital formats, and to share video signals between different platforms and application areas. [3] This is a natural evolution, since temporal change, which is usually associated with motion of some type, is often the most important property of a visual signal.

2.4 *Otsu*

Otsu method is calculating the threshold value T automatically based on the input image. The approach used by the Otsu method is by performing discriminant analysis that is by determining a variable that can distinguish between two or more groups that arise naturally. Otsu's thresholding method involves iterating through all the possible threshold values and calculating a measure of spread for the pixel levels each side of the threshold, i.e. the pixels that either falls in foreground or background. The aim is to find the threshold value where the sum of foreground and background spreads is at its minimum. [4] The threshold operation is regarded as the partitioning of the pixels of an image into two classes C0 and C1 (e.g., objects and background) at grey-level t, i.e., C0 = {0, 1, 2,..., t} and C1 = {t + 1, t +2,....L-1}. let σ2w σ2B and σ2T be the within-class variance, between-class variance, and the total variance, respectively. [5] Discriminant analysis will maximize these variables in order to separate objects from the background.

Thresholding technique segment scalar images by generating a binary partitioning of the image intensities. A thresholding technique tries to find an intensity value, called the threshold, which separates the desired classes. The segmentation is then obtained by grouping all pixels with intensity greater than the threshold into one class, and all other pixels into another class. [6]

The probability for pixel i given by:

$$p_i = n_{i\,/N} \tag{1}$$

where:
ni = the number of pixels by the number of grayish i
N = the sum of all pixels in the image
Pi = result

In Otsu method, the threshold that minimizes intra-class variance is determined, which is defined as the weighted sum of the variance of two classes:

$$\sigma_w^2(t) = w_1(t)\sigma_1^2(t) + w_2(t)\sigma_2^2(t) \tag{2}$$

The weights wi is the probability of two classes separated by a threshold t and σ_i ^ 2 variations of these classes.

Otsu shows that minimizes intra-class variance is the same as maximizing inter-class variance.

$$\sigma_b^2(t) = \sigma^2 - \sigma_w^2(t) = w_1(t)w_2(t)[\mu_1(t) - \mu_2(t)]^{\,2} \tag{3}$$

stated in class probabilities wi and class mean μ_i. Probability class wi (t) is calculated from the histogram as follows:

$$w_i\,(t) = \textstyle\sum_0^t p(i) \tag{4}$$

while the average class μ_1 (t) is:

$$\mu_1(t) = \left[\sum_0^t p(i)\,x(i)\right]/w_1 \quad (5)$$

Where x (i) is the value in the middle i histogram. Similarly, to calculate w_2 (t) and μ_2 on the right side of the histogram larger than t. Probability of class and class facility can be calculated iteratively. This idea resulted in an effective algorithm.

The optimum threshold value can be obtained in two ways. The first way is carried out by minimizing WCV. The second way is implemented by maximizing the BCV.

3. The Scheme of System

The scheme of the height measurement system developed in this study is illustrated in Figure 3.1.

Source: Grey-Scale Otsu Height measurement results: 155 cm

Figure 3.1. The scheme of height measurement system in video

3.1 *Grey-Scale*

Flowchart for gray-scale processes are developed based on Figure 3.2:

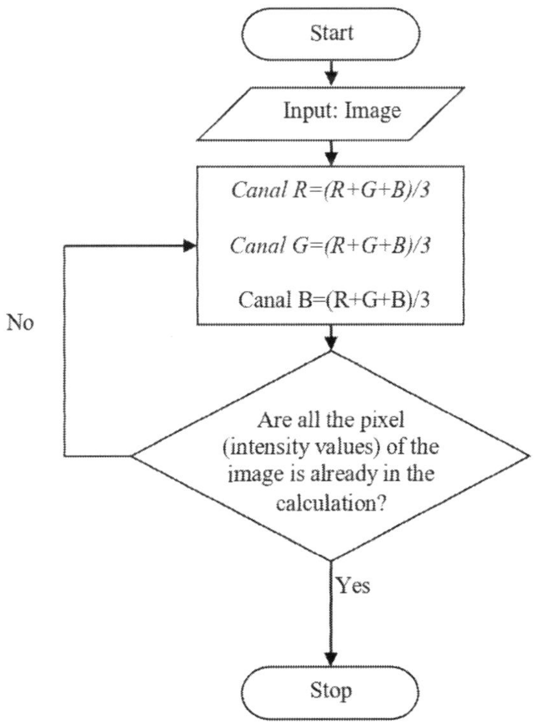

Figure 3.2. Flowchart of the *grey-scale* process

3.2 *Otsu*

Flowchart for the process of Otsu in the system is developed based on Figure 3.3:

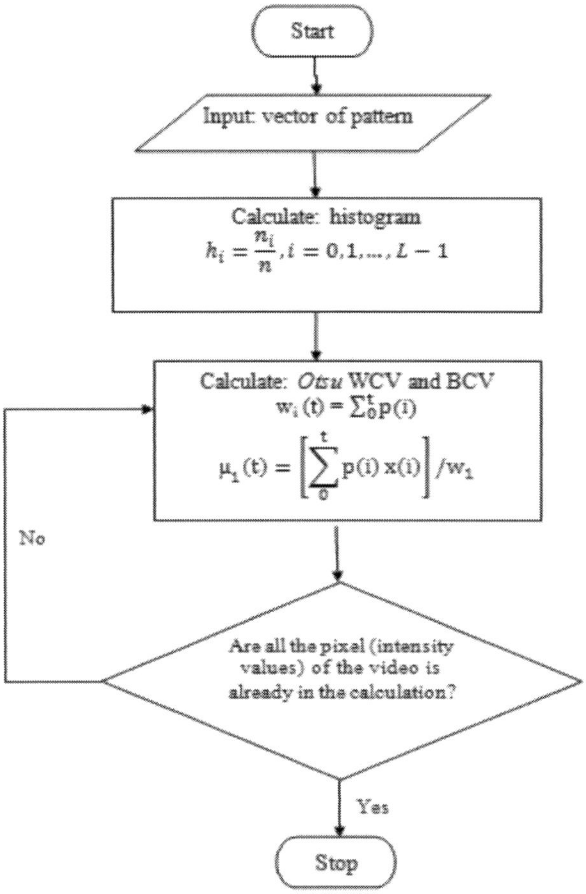

Figure 3.3. Flowchart of the Otsu Process

4. *User Interfaces*

4.1 View of the Main Form

Before the operation, the program will display a page like this:

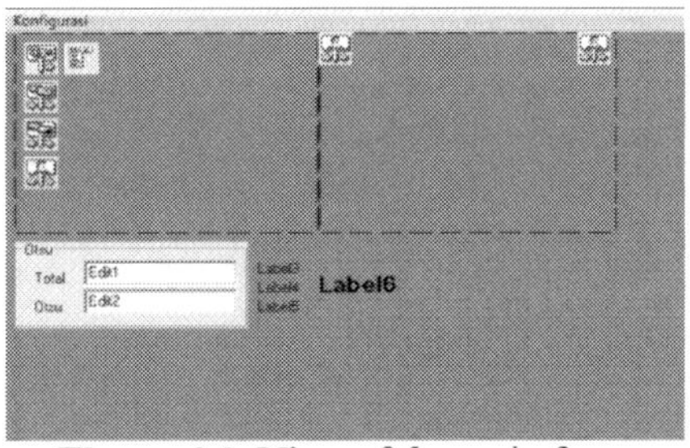

Figure 4.1. View of the main form

4.2 View of the *Otsu* form
Once executed, the program will display a page like this

Figure 4.2. View of the *Otsu* form

4.3 The test results
This form is intended for testing of Measurement of Height, here is a view of test results:

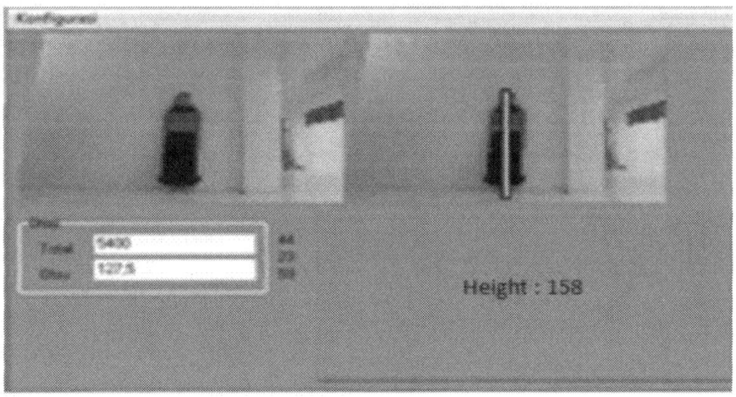

Figure 4.3 The test results of measurement of height

5.1 Conclusions
From the results of research and discussion undertaken conclusions can be drawn as follows:

1. Excellence of method of Otsu is to maximize the compatibility of a threshold so that the object can be separated from the background. All this is achieved by selecting the threshold value which gives the best class division for all the existing pixels in the image.

2. To get the end result is precise and accurate, then the condition or environment of the image catcher must be considered. The conditions referred to, among other things, are light, place, distance, and the position of the object of image.
3. Fairly evenly distributed light makes Otsu method is good enough to use. Light in the image is not always smooth, the meaning is that the existing light, has a different intensity depending on the spread of light, if the light is fairly evenly distributed then the intensity is not too bright and not too dark so the test in the system can work well.
4. Height measurement system is not able to work well if the region, in which the object to be measured, contains many other objects, because the measuring ruler can be moved to the place of other objects. To simplify and prevent this, you can choose the right place while doing the test so that the system can work well.

References

[1] Muzamil B 2014 Digital Image Processing *International Journal of Scientific & Technology Research* vol 3 Issue 1 ISSN 2277-8616
[2] Asole N 2012 Brain Tumor Detection Using Digital Image Processing Based On Soft Computing *Journal of Signal and Image Processing* ISSN: 0976-8882 & E-Issn: 0976-8890 vol 3 Issue 3 pp 102-105
[3] Prasad B 2012 An Efficient Medical Image Segmentation Using Conventional OTSU Method *International Journal Of Advanced Science And Technology* vol 38
[4] Choudhary K and Pote T 2012 Enhanced Identification of Malarial Infected Objects using Otsu Algorithm from Thin Smear Digital Images *International Journal of Latest Research in Science and Technology* ISSN (Online):2278-5299 vol 1 Issue 2 Page No 159-163 July-August ISSN (Online):2278-5299
[5] Bovik Al Handbook of Image Am Video Processing Austin, 2000 Isbn 0-12-119790-5

IOP Conf. Series: Journal of Physics: Conf. Series **801** (2017) 012050 doi:10.1088/1742-6596/801/1/012050

Image Classification of Ribbed Smoked Sheet using Learning Vector Quantization

R F Rahmat[1], A F Pulungan[1], S Faza[1], and R Budiarto[2]

[1] *Department of Information Technology, Faculty of Computer Science and Information Technology, University of Sumatera Utara, Medan, Indonesia*
[2] *Department of Information System, College of Computer Science and Information Technology, Albaha University, Saudi Arabia*

Email: romi.fadillah@usu.ac.id, annisa_fadhillah@students.usu.ac.id, sharfina.faza@gmail.com, rahmat@bu.edu.sa

Abstract. Natural rubber is an important export commodity in Indonesia, which can be a major contributor to national economic development. One type of rubber used as rubber material exports is Ribbed Smoked Sheet (RSS). The quantity of RSS exports depends on the quality of RSS. RSS rubber quality has been assigned in SNI 06-001-1987 and the International Standards of Quality and Packing for Natural Rubber Grades (The Green Book). The determination of RSS quality is also known as the sorting process. In the rubber factories, the sorting process is still done manually by looking and detecting at the levels of air bubbles on the surface of the rubber sheet by naked eyes so that the result is subjective and not so good. Therefore, a method is required to classify RSS rubber automatically and precisely. We propose some image processing techniques for the pre-processing, zoning method for feature extraction and Learning Vector Quantization (LVQ) method for classifying RSS rubber into two grades, namely RSS1 and RSS3. We used 120 RSS images as training dataset and 60 RSS images as testing dataset. The result shows that our proposed method can give 89% of accuracy and the best perform epoch is in the fifteenth epoch.

1. Introduction

The rubber industry, as a reliable earner of foreign exchange, has always been regarded as a major contributor to national economic development [1]. Natural rubber is an important export commodity for Indonesia. Approximately 1.3 million farm households rely on rubber cultivation and provide 75% of the national production [2]. One type of rubber processed into export material is ribbed smoked sheet (RSS).

RSS has a quality that has been assigned by the International Standards of Quality and Packing for Natural Rubber Grades (The Green Book) and SNI 06-001-1987. Every quality of RSS has different prices. To determine the quality of processed rubber RSS, they do the sorting for processing before the cutting process. The sorting is a process of classifying the quality of rubber sheet RSS. Until now, the sorting process is still done manually by direct observation of the rubber surface. Manual observation is usually to look at the physical defects on the surface, dirt, uneven colour, fungus, air bubbles and sticky so that the results of sorting is merely subjective [3]. Therefore, a system is required a system able to classify RSS through its image.

Several studies on the classification of RSS quality has been done by some previous researchers including on detecting air bubbles in RSS where the air bubble is one of the factors that affects the quality of the RSS. In 2011, Umyai et al. studied about Bubble detection in RSS based on the fractal dimension in 500 RSS image and obtained 98% success rate for classifying the presence or absence of bubbles in RSS. However, this study did not discuss about the classification of RSS rubber quality [4]. Another method, k-Means Clustering and Euclidean Distance, was used by Pornpanomchai et al. for classifying RSS from colour identification into five qualities, RSS1, RSS2, RSS3, RSS4, and RSS5 and their proposed method can obtain 80.90% of accuracy with 10.88 seconds of average time per RSS image [5].

Due to the constraints in classifying the quality of rubber, a method is required to classify it based on the number of bubbles in a rubber sheet RSS. A method often used in the classification is the Learning Vector Quantization (LVQ). LVQ is a classification method based on the model of Kohonen known as Network Self-Organizing Map (SOM). However LVQ is different from SOM, SOM is unsupervised learning, while LVQ is a supervised learning with a simple algorithm architecture that consists of only one input layer and output layer [6].

Several studies have been done to resolve the problems in the RSS. Ahmad, et al. analysed the characteristics of the colour of the rubber surface using RGB and HIS colour model and also texture characteristics. The conclusion shows that the colour's feature can be used for classifying the quality of RSS type 1 rubber, while it is not effective for determining the bound of RSS type 2. Component blue in RGB model can be used to classify the quality of the RSS with quite high compatibility, but a consistent and better result is obtained by using HIS colour model with criteria $H \leq 28$ and $I \geq 220$ for RSS1, $H \geq 68$ and $S \leq 73$ for RSS3 and else for RSS2. In their conclusion, it generated accuracy rate of 86% for RSS1, 77.5% for RSS2, and 95% for RSS3 [3].

Moreover, another related research was conducted using image processing techniques and did four different experiments of lighting and resolution while taking the RSS image. Experiment 1: down lighting with the resolution of 192 x 144; Experiment 2: up lighting with the resolution of 192 x 144; Experiment 3: down lighting with the resolution of 341 x 256; and experiment 4: up lighting with the resolution of 341 x 256. The used parameter of RSS image includes an index colour RGB, component HIS colour and texture components of the image. The results of experiment 1 showed that only blue colour index parameters that can be used as a sorting parameter with the value of the upper limit and lower limit of 0.2921 at 0.2843. The results of experiment 2 showed that the index parameters of red, green, blue and saturation can be used as a parameter sorting with an upper limit and lower limit of each Red (0.4143, 0.3914), Green (0.3321, 0.3258), Blue (0.2743, 0.2574) and Saturation (95.76). The results of experiment 3 showed that only blue colour index parameters that can be used as a sorting parameter with the value of the upper limit and lower limit of 0.2929 at 0.2852. The results of experiment 4 showed that the index parameters of red, green, blue and saturation can be used as a parameter sorting with an upper limit and lower limit of each Red (0.4168, 0.3927), Green (0.3305, 0.3241), Blue (0.2740, 0.2570) and Saturation (96.77). From the result of comparison in the experiments, they concluded that the percentage of success for classifying the RSS rubber quality 341 x 256 resolutions is higher than 192 x 144 resolution [7].

Related research was done by Prabpal, et al. in 2014 which also classified RSS rubber quality using ANN method and generated 90% of accuracy for 100 samples of rubber RSS image divided into four levels, namely A (Very Good), B (good), C (Medium), and D (Poor) [8].

In this research, we propose a system that is able to classify the quality of the RSS by using Learning Vector Quantization (LVQ) to help the classification of RSS quality accurately. The rest of the paper will be organized as follows: Section II describes the proposed method. Results and discussions are presented in Section III. Section IV provides summary and suggestions for future research.

2. Research methodology

In this section, we will discuss about our proposed method for classifying RSS. There are three main steps in this research, namely preprocessing, feature extraction and classification. In pre-processing step, there are several morphological operations, such as resizing, gray scaling and thresholding, to make the image be easily processed in the next steps. Next, we use zoning method for feature extraction and Learning Vector Quantization (LVQ) for classification. The general architecture of our proposed method is shown in figure 1.

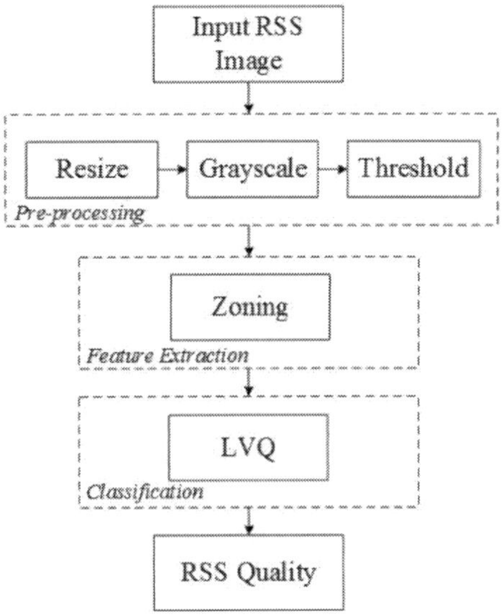

Figure 1. General Architecture

2.1. Image Acquisition
RSS images, used as an input in this research, had been captured in the same distance and density, so the classification result will be obtained more accurately.

2.2. Pre-processing

2.2.1. Resizing. It aims to process and only focus on the surface area of bubble on RSS rubber. We resize all the acquired images into 1024 x 1024 pixels.

2.2.2. Grayscaling. It aims to separate the shadow from the original colour in RSS image. In this step, the RSS colour image, which has RGB components, will be converted to grayscale image. The conversion from RGB image into grayscale image can be done using equation (1). Figure 2 shows the result of conversion of RSS colour image into grayscale image.

$$Gray = (R + G + B)/3 \quad (1)$$

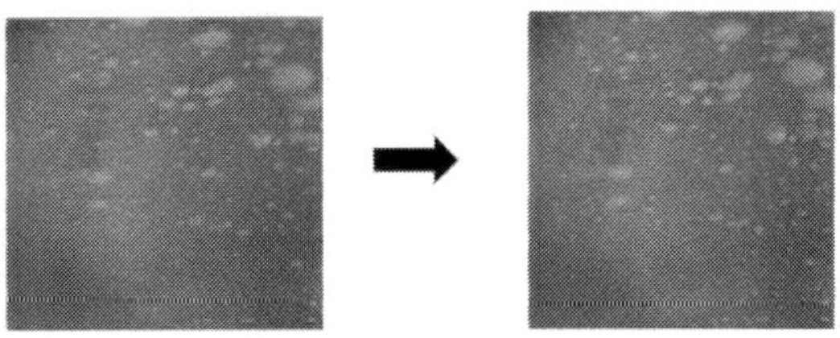

Figure 2. Grayscale Image Processing

2.2.3. Thresholding. After getting the grayscale image, it is converted into a binary image. In this step, the value of each pixel grayscale image will be replaced with 0 (black) or 1 (white). If the pixel value is less than the threshold value, it will be replaced with black and if the pixel value is more than the threshold value, it will be replaced with white (1). The threshold value can be obtained from the equation (2), where T = threshold value, f_{max} = maximum pixel value and f_{min} = minimum pixel value. The result of thresholding the grayscale RSS image is shown in figure 3.

$$T = \frac{f_{max} + f_{min}}{2} \tag{2}$$

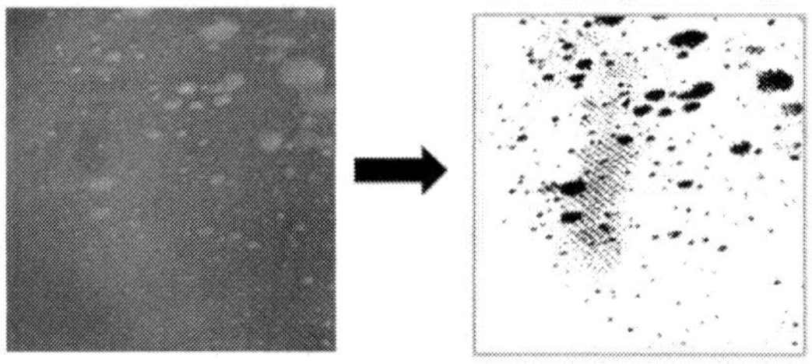

Figure 3. Threshold Image Processing

2.3. Feature Extraction

Feature extraction aims to get a unique feature to get the image characteristics. In this research, we use zoning method to extract the feature of RSS image before classifying it. The image will be divided into 11 columns and 11 rows to obtain 121 zones representing 121 features. Each of zone consists of 100 pixels. The illustration of zoning method is shown as figure 4 and will be describe in several steps below.

Z1	Z2	Z3	Z4	Z5	Z6	Z7	Z8	Z9	Z10	Z11
Z12	Z13	Z14	Z15	Z16	Z17	Z18	Z19	Z20	Z21	Z22
Z23	Z24	Z25	Z26	Z27	Z28	Z29	Z30	Z31	Z32	Z33
Z34	Z35	Z36	Z37	Z38	Z39	Z40	Z41	Z42	Z43	Z44
Z45	Z46	Z47	Z48	Z49	Z50	Z51	Z52	Z53	Z54	Z55
Z56	Z57	Z58	Z59	Z60	Z61	Z62	Z63	Z64	Z65	Z66
Z67	Z68	Z69	Z70	Z71	Z72	Z73	Z74	Z75	Z76	Z77
Z78	Z79	Z80	Z81	Z82	Z83	Z84	Z85	Z86	Z87	Z88
Z89	Z90	Z91	Z92	Z93	Z94	Z95	Z96	Z97	Z98	Z99
Z100	Z101	Z102	Z103	Z104	Z105	Z106	Z107	Z108	Z109	Z110
Z111	Z112	Z113	Z114	Z115	Z116	Z117	Z118	Z119	Z120	Z121

11 rows

11 columns

Figure 4. Zoning Image Processing

Step 1 : Count the number of black pixels for each zone (Z1-Z121).
The number of black pixels every zone, among others:
Z1=20, Z10=80, Z20=40, Z50=50

Step 2 : Determine the zone that has the number of highest black pixel.
Zone with the highest black pixel value is Z10 (80).

Step 3 : Calculate the feature value of each zone (Z1-Z121) using equation (3), where n is zone number

$$feature\ value\ Z_n = \frac{Z_n}{Z_{highest}} \tag{3}$$

From equation (3), it is obtained:
Z1 = 20/80 = 0.25
Z10 = 80/80 = 1
Z20 = 40/80 = 0.5
Z50 = 50/80 = 0.625

Step 4 : Round off the feature value of each zone using equation (4). It aims to generate the binary image, used as an input for the next step. Figure 5 shows the sample result of feature extraction using zoning method, which generates 121 features as an input for classification.

$$\begin{aligned} &if\ Z_n < 0.5, then\ Z_n = 0, \\ &else\ Z_n = 1 \end{aligned} \quad (4)$$

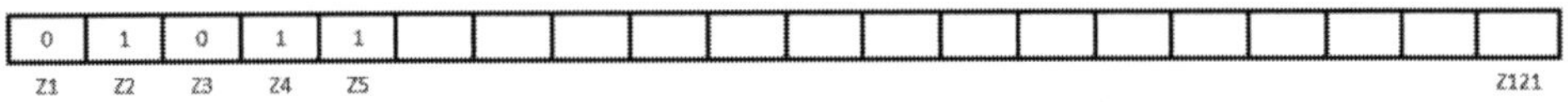

Figure 5. Feature Extraction Value

2.4. Classification
In this research, we use Learning Vector Quantization (LVQ) for classifying the RSS rubber into two grades, namely RSS1 and RSS3. LVQ is used for classifying pattern into several classes by searching the final weight by calculating the distance between the data and the final weight. The general architecture of LVQ is shown as Figure 6.

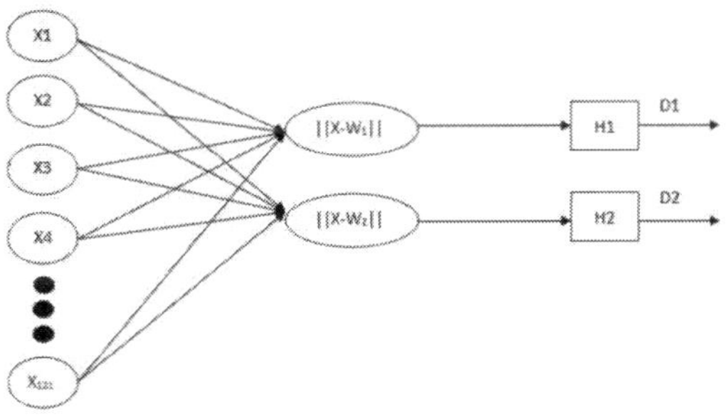

Figure 6. General Architecture LVQ

Where:
- $X_1, X_2, X_3, \ldots, X_{121}$ = Input Layer
- $||X-W_1||, ||X-W_2||$ = Distance Weighting
- H1, H2 = Output Layer
- D1, D2 = Output Value
- W1, W2.. W121 = Initialization Value

The parameters required in LVQ algorithms include:
1. Learning rate (α) is the value of training level. If α is too large then the algorithm becomes unstable and too small then the required processing time becomes longer. The value of α is in the range $0 < \alpha < 1$.
2. Decline Learning rate (Dec α) is decreasing the level of training. Decline of Learning rate is performed after the iterations on each data and will be performed on the next iteration.
3. Minimum Learning rate (Min α) is the level of training that is still allowed.

4. Maximum Epoch (MaxEpoch) is the maximum number of iterations may be performed during the training process. While the iteration has reached the maximum number of iterations, the iteration will be terminated.

2.4.1. Training Process. In the training process, LVQ will process the input vector to 121 features with their feature class description. Then the vector will calculate the distance of all represented vectors for the existing classes by using Euclidean Distance. These following steps will describe the implementation process of LVQ in the training process.

Step 1 : Initialization of initial weight, maximum iteration, minimum error and Learning rate to get the final result accurately.

Step 2 : Initialize input value and input target value. Input value is obtained from the feature extraction. The level target has been determined for each input value (RSS1 or RSS3).

Step 3 : Initial condition, epoch = 0 and error = 1. In this step, the weight calculation starts from the initial conditions of the epoch to specified maximum epoch.

Step 4 : If the epoch is smaller than the maximum epoch or learning rate > minimum epoch (*Eps*), then count each weight in an input and set the weight with the smallest value as the shortest distance on the weight

Step 5 : Compare the target class with weight. If the target class are same as the weight, then update the weight using equation (5)

$$W_{j(1)} = W_{j(0)} + \alpha \left[x - W_{j(0)} \right] \tag{5}$$

if the target class and weight are different, then update the value using equation (6)

$$W_{j(1)} = W_{j(0)} - \alpha \left[x - W_{j(0)} \right] \tag{6}$$

Step 6 : Doing the same calculation as initial input on each input using updated weights

Step 7 : Decrease the value of learning rate and do iterations for each epoch until reaching maximum epoch or learning rate value approaches *Eps* value.

Step 8 : If the iteration of epoch ends, set the new weight in the last epoch as the weight used on the testing process.

2.4.2. Testing Process. In the testing process, the input data will be classified in the same way as the training process, calculating the value of each weight to the input and select the smallest distance on both weights. The values of the smallest distance will represent the class on the input.

3. Result and discussion

3.1. Training

The training dataset consists of 120 image RSS (60 images of RSS1 and 60 images of RSS3). The input data is RSS image with size of 1024 x 1024 and stored on a local hard drive. In training phase, the classification using LVQ can work well on the 15th epoch. The feature value of training process will be stored to be used as an input in the testing phase to get the proper weight. The parameter value used in the training RSS rubber image will be mentioned in table 1.

Table 1. Training parameters.

Learning rate **(α)**	0.05
Decrease *Learning rate* (Dec α)	0.1
Minimum *Learning rate* (Min α)	0.0001
Maximum *Epoch* (Max*Epoch*)	20
Learning Algorithm	*LVQ*

3.2. Testing

In the testing process, the used data as an input consists of 60 RSS images (30 images of RSS1 and 30 images of RSS3). Table 2 shows the testing results.

Table 2. Testing results.

No	Rubber Name	Weight 1	Weight 2	System result	Manual	True/ False
1	Karet4.jpg	7.115	4.837	RSS3	RSS1	False
2	Karet12.jpg	5.456	4.95	RSS3	RSS1	False
3	Karet39.jpg	3.129	5.4	RSS1	RSS1	True
4	Karet40.jpg	1.782	5.433	RSS1	RSS1	True
5	Kare41.jpg	1.782	5.433	RSS1	RSS1	True
6	Karet42.jpg	1.782	5.433	RSS1	RSS1	True
7	Karet43.jpg	2.673	5.569	RSS1	RSS1	True
8	Karet44.jpg	3.498	5.449	RSS1	RSS1	True
9	Karet45.jpg	1.782	5.433	RSS1	RSS1	True
10	Karet46.jpg	2.311	5.411	RSS1	RSS1	True
11	Karet47.jpg	5.271	5.582	RSS1	RSS1	True
12	Karet48.jpg	1.782	5.433	RSS1	RSS1	True
13	Karet49.jpg	3.247	5.616	RSS1	RSS1	True
14	Karet59.jpg	8.146	5.577	RSS3	RSS3	True
15	Karet70.jpg	8.055	5.505	RSS3	RSS3	True
16	Karet73.jpg	8.763	5.782	RSS3	RSS3	True
17	Karet76.jpg	5.491	5.404	RSS3	RSS3	True
18	Karet77.jpg	9.056	6.473	RSS3	RSS3	True
19	Karet78.jpg	7.914	6.63	RSS3	RSS3	True
20	Karet80.jpg	7.008	6.359	RSS3	RSS3	True

From the testing results, we can conclude the average accuracy of our proposed method for classifying RSS rubber is 90% with 54 correct result of 60 image, as formulated using equation (7).

$$Accuracy = \frac{number\ of\ correct\ result}{total\ of\ RSS\ Image}\ x\ 100\ \% \qquad (7)$$

Table 3. The result of validation.

Training	Accuracy
1	90 %
2	88.33%
3	85%
4	90%
5	88.33%
6	90%
7	88.33%
8	85%
9	90%
10	90%
Average accuracy	89 %

To validate the truth of the accuracy, we also did 10 times of testing with 60 test data with the random training data sequence for each testing process. Table 3 shows the results of the validation.

4. Conclusion

Based on testing result of classifying RSS rubber quality, our proposed method can obtain accuracy of 90%. However, the pre-processing is still not able to remove a motif found in the image perfectly and handle the color condition in the image, so that it affects the feature value used as input value. And based on the training data, LVQ can work well in the testing process in the 15th epoch.

Further research can be developed by comparing other methods to obtain higher accuracy for classifying RSS rubber image. Another research possibly done is classifying rubber RSS image based on color to determine it matured rubber or not.

5. References

[1] Slocomb M 2010 The Privatisation of Cambodia's Rubber Industry *Cambodia's Economic Transformation* 94

[2] Wulan Y C, Budidarsono S and Joshi L 2008 Economic analysis of improved smallholder rubber agroforestry systems in West Kalimantan, Indonesia-implications for rubber development *Sustainable sloping lands and watershed management conference Luang Prabang, Lao PDR*

[3] Ahmad U, Riadi F and Subrata I D M 2006 Pemeriksaan Mutu Karet RSS menggunakan Pengolahan Citra (RSS Rubber Quality Inspection using Image Processing) *Jurnal Keteknikan Pertanian* **20** 265-274

[4] Umyai P, Kumhom P and Chamnongthai K 2011 Air bubbles detecting on ribbed smoked sheet based on fractal dimension *Int. Symp. on Intelligent Signal Processing and Communication System (ISPACS)*

[5] Pornpanomchai C and Chantharangsikul N 2010 Ribbed Smoked Sheet Grading System (RSSGS) *Int. Conf. on Electronic and Information Engineering (ICEIE)* pp 128-132

[6] Azara M, Fatayer T and Halees A E 2012 Arabic text classification using Learning Vector Quantization *The 8th Int. Conf. Informatics and System (INFOS2012)* pp 40-44

[7] Kurniawan R 2003 *Kajian Karakteristik Mutu Karet Alam Olahan Jenis RSS Dengan Teknik Pengolahan Citra (Study of Quality Characteristics of Processed Ribbed Smoked Sheet) Natural Rubber RSS using Image Processing Techniques)* Thesis Institut Pertanian Bogor

[8] Prabpal P, Manasri T and Sonasang S 2014 The classify of rubber sheet quality by image processing with artificial neural network *Int. Science, Social Science, Engineering and Energy Conf. (I-SEEC 2014)* pp 1-6

IOP Conf. Series: Journal of Physics: Conf. Series **801** (2017) 012051 doi:10.1088/1742-6596/801/1/012051

Tropical Timber Identification using Backpropagation Neural Network

B Siregar[1], U Andayani[1], N Fatihah[1], L Hakim[2], F Fahmi[3]

[1] Faculty of Computer Science and Information Technology, University of Sumatera Utara, Medan-Indonesia
[2] Faculty of Forestry, University of Sumatera Utara, Medan-Indonesia
[3] Dept. Electrical Engineering, University of Sumatera Utara, Medan-Indonesia

Email: fahmimn@usu.ac.id

***Abstract*—** Each and every type of wood has different characteristics. Identifying the type of wood properly is important, especially for industries that need to know the type of timber specifically. However, it requires expertise in identifying the type of wood and only limited experts available. In addition, the manual identification even by experts is rather inefficient because it requires a lot of time and possibility of human errors. To overcome these problems, a digital image based method to identify the type of timber automatically is needed. In this study, backpropagation neural network is used as artificial intelligence component. Several stages were developed: a microscope image acquisition, pre-processing, feature extraction using gray level co-occurrence matrix and normalization of data extraction using decimal scaling features. The results showed that the proposed method was able to identify the timber with an accuracy of 94%.

1. Introduction

Wood is one of the commodities that can be used for multiple purposes, such as building materials, furniture, etc. Each type of wood has different characteristics so that it can affect the quality and price of the timber. Accordingly, the identification of the type of wood is important, especially for industries who need to know the type of timber precisely, so as to enable the parties concerned may use the appropriate timber. There are two ways that can be done to identify the type of wood that is by observing the general nature and properties of wood anatomy [1]. General nature of the timber that the physical properties of wood directly without the aid of a magnifying shadow. The physical properties of wood including color, texture, grain direction, image, weight, impression felt, the circle grows, smell and so forth. The nature of anatomy include the shape, structure and size of cells that can be observed with the aid of a magnifying shadow.

In identifying the type of wood there are obstacles that not everyone has the knowledge to be able to identify the type of wood properly. Identifying the type of wood to observe the general nature or anatomical properties of wood can be done by people who are experts and have a lot of experience. However, there is very limited experts available. In addition, the identification manually conducted by expert inefficient because it requires a lot of time.

Backpropagation is a learning algorithm on artificial neural network that can be done for pattern recognition. Neural network itself usually used in most image based analysis such as image reading [2], image differentiation [3] or in biometric [4]. Backpropagation can create balance in the network to train a network to be able to recognize a pattern in the training and can provide the right response to the input pattern similar to the pattern used on the training [5]. Risaldi et al. (2014) using the method of backpropagation to classify the quality of coconut timber by comparing the propagation neural network algorithms and libSVM [6].

In 2014, Mohan S. conduct research to identify the wood in India [7]. Wood image taken using a digital camera with high resolution. Image taken is the outer surface of the bark. This study uses feature extraction Grey Level Co-Occurrence Matrix to acquire the image of wood. Gunawan et al. (2011) conducted a study identifying timber in order to develop a system that can classify four types of timber traded in Indonesia [8]. Identification of timber is made using a microscopic image of wood. Image taken using a microscopic camera. The method used for image classification of wood is the support vector machine. While the method for image feature extraction is two-dimensional principal component analysis (2D-PCA).

In this study, backpropagation neural network was developed as a specific identification method for tropical timber identification.

2. Method and Material

Methods in this study consisted of several stages as follows: image acquisition using a microscope, pre-stage image processing consists of scaling and gray scaling; feature extraction stage of each image with a value of 5 Haralick features of Grey Level Co-Occurrence Matrix; normalization of the extracted features using decimal scaling; and image classification using artificial neural network backpropagation.

The general architecture of the research methodology can be seen in Figure 1.

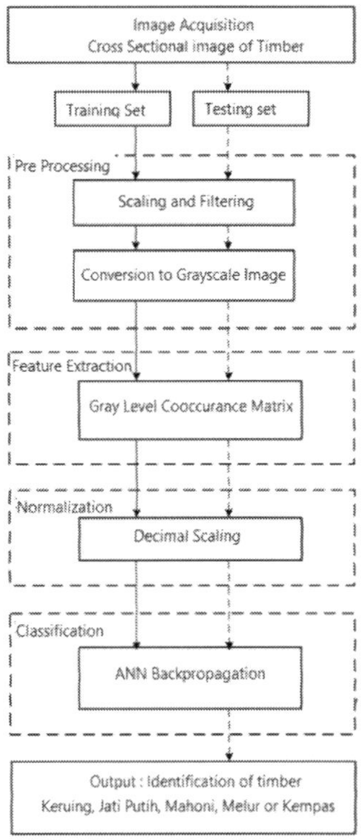

Figure 1. General Architecture

2.1 Image Acquisition

The acquisition of a cross-sectional image of tropical timber was taken using Digital Microscope Camera AxioCam with ERC 5s with a magnification of the microscope is 1:25 time and intensity of light is at 3200 K.

2.2 Data used

Type of wood used in this study is wood keruing oil (Dipterocarpus eurynchus), teak white (Gmelina arborea), mahogany (Swietenia mahagoni), wood jessamine (Dacrydium elatum) and wood kempas (Koompassia malaccensis) which can be seen in Figure 2.

Figure 2. (a) keruing minyak (b) Jati putih (c) Mahoni (d) Melur (e) Kempas

All types of wood in Figure 2 is made in the form of wooden blocks each of 5 pieces. The timber will be placed on top of the microscope to image is taken in the form of a block of wood measuring about 1x1x1 cm. This research will be done taking as many as 10 pieces of wood image on each block of wood so that each type of wood has 50 pieces of image data.

The image that has been collected will be divided into two groups of data that training data and test data. Training data amounted to 80% of the overall image data while the test data accounts for 20% of the overall image data so that in this research training data for each type of timber amounted to 40 units, while the test data for each type of timber amounted to 10 pieces.

2.3 Pre-Processing

Image pre-processing stage was done to produce a better image to be processed to the next stage of feature extraction phase. In this study, pre-processing the image taken was image scaling and the conversion of gray level image (gray scaling).

2.4 Feature Extraction

In this study, the method used for extraction of image features was the Gray Level Co-Occurrence Matrix (GLCM). GLCM is one of the second-order texture analysis methods. This method was introduced by Haralick et al in 1973 [13]. GLCM represents the relationship of two neighboring pixels in which two pixels associated has a certain intensity of gray and has a certain distance and direction between the two.

Calculation of statistical features can be made after the co-occurrence matrix had already been already normalized. In this study, some quantitative features of statistics obtained from the co-occurrence matrix were: Energy, Entropy, Contrast, Inverse Difference Moment, and Correlation.

2.5 Data Normalization

Normalization is a technique of pre-processing of data by transforming the value attribute of a dataset in a certain range, for example between 0 and 1. The normalization can be used in problems such as data classification and clustering neural network [11].

Data were normalized in advance before classification using backpropagation. In this study, normalization techniques used were decimal scaling. Decimal scaling was done by moving the decimal point of the value of an attribute. The size of the decimal point shifting was determined by the maximum absolute value of an attribute.

2.6 Classification Using Backpropagation Neural Network
Backpropagation is one method of supervised artificial neural network (supervised learning). Backpropagation requires the target as a reference in the training process, where the purpose of propagation is set so that the value of the error in the network becomes increasingly smaller, or in other words make the output value close to the target. The results of the training backpropagation is a set of weight value, where the final weight is a representation of knowledge of the learning process that carried out.

Backpropagation method consists of two phases: the propagation phase (feed forward) and backward direction phase (backward). This method modifies the value of weights in phase backward direction (backward) by using the error output. Advanced propagation phase (feed forward) must be passed first in order to be able to obtain the value of the error.

Determination of network architecture is done by determining the number of neurons in the input layer, hidden layer and output layer. In this study, the number of input neuron was determined by the number of feature extraction results (input neurons 20). While the number output neurons was 5, which is determined based on the target output of the number of types of wood that are classified.

3. Results

Tests were conducted to determine the ability of a propagation method in identifying five types of wood. The system's ability to identify types of wood depends on the backpropagation training process for generating weight that will be used in the testing phase. Parameters used in the backpropagation training phase can be seen in Table 1.

Table 1. Parameters of ANN *backpropagation*

No.	Parameter Backpropagation	Setting Value
1.	Hidden Neuron	40
2.	Activation Function	Sigmoid Binary
3.	Maximum Epoch	2000
4.	Minimum Error	0.1
5.	Learning Rate	0.8

The result of identifying the type of timber based on the final weight can be seen in Table 2 where the test images from every kind of timber was used.

Table 2. Test Result

No.	Image	Desired Output	Actual Output
1.	keruing-test (1).jpeg	Keruing	Keruing
2.	keruing-test (2).jpeg	Keruing	Keruing
3.	keruing-test (3).jpeg	Keruing	Keruing
4.	keruing-test (4).jpeg	Keruing	Keruing
5.	keruing-test (5).jpeg	Keruing	Keruing
6.	keruing-test (6).jpeg	Keruing	Keruing
7.	keruing-test (7).jpeg	Keruing	Keruing
8.	keruing-test (8).jpeg	Keruing	Keruing
9.	keruing-test (9).jpeg	Keruing	Keruing
10.	keruing-test (10).jpeg	Keruing	Keruing
11.	Jati-Putih-test (1).jpeg	Jati Putih	Kempas
12.	Jati-Putih-test (2).jpeg	Jati Putih	Jati Putih
13.	Jati-Putih-test (3).jpeg	Jati Putih	Jati Putih
14.	Jati-Putih-test (4).jpeg	Jati Putih	Jati Putih

15.	Jati-Putih-test (5).jpeg	Jati Putih	Jati Putih
16.	Jati-Putih-test (6).jpeg	Jati Putih	Jati Putih
17.	Jati-Putih-test (7).jpeg	Jati Putih	Jati Putih
18.	Jati-Putih-test (8).jpeg	Jati Putih	Jati Putih
19.	Jati-Putih-test (9).jpeg	Jati Putih	Jati Putih
20.	Jati-Putih-test (10).jpeg	Jati Putih	Jati Putih
21.	mahoni-test (1).jpeg	Mahoni	Mahoni
22.	mahoni-test (2).jpeg	Mahoni	Mahoni
23.	mahoni-test (3).jpeg	Mahoni	Mahoni
24.	mahoni-test (4).jpeg	Mahoni	Mahoni
25.	mahoni-test (5).jpeg	Mahoni	Mahoni
26.	mahoni-test (6).jpeg	Mahoni	Mahoni
27.	mahoni-test (7).jpeg	Mahoni	Mahoni
28.	mahoni-test (8).jpeg	Mahoni	Mahoni
29.	mahoni-test (9).jpeg	Mahoni	Mahoni
30.	mahoni-test (10).jpeg	Mahoni	Mahoni
31.	Melur-test (1).jpeg	Melur	Melur
32.	Melur-test (2).jpeg	Melur	Melur
33.	Melur-test (3).jpeg	Melur	Melur
34.	Melur-test (4).jpeg	Melur	Melur
35.	Melur-test (5).jpeg	Melur	Melur
36.	Melur-test (6).jpeg	Melur	Melur
37.	Melur-test (7).jpeg	Melur	Melur
38	Melur-test (8).jpeg	Melur	Melur
39.	Melur-test (9).jpeg	Melur	Melur
40.	Melur-test (10).jpeg	Melur	Melur
41.	Kempas-test (1).jpeg	Kempas	Jati Putih
42.	Kempas-test (2).jpeg	Kempas	Kempas
43.	Kempas-test (3).jpeg	Kempas	Kempas
44.	Kempas-test (4).jpeg	Kempas	Kempas
45.	Kempas-test (5).jpeg	Kempas	Kempas
46.	Kempas-test (6).jpeg	Kempas	Kempas
47.	Kempas-test (7).jpeg	Kempas	Kempas
48.	Kempas-test (8).jpeg	Kempas	Kempas
49.	Kempas-test (9).jpeg	Kempas	Kempas
50.	Kempas-test (10).jpeg	Kempas	Jati Putih

To calculate the accuracy of the test, the equation used is the following equation:

$$\text{Accuracy} = \frac{\text{Jumlah data uji yang benar}}{\text{Jumlah data uji keseluruhan}} \text{ x } 100$$

The accuracy of the test results in Table 2 can be seen in Table 3.

Table 3. Accuracy of system

No	Timber Type	Correct Identification	Accuracy
1.	Keruing	10	100%
2.	Jati Putih	9	90%
3.	Mahoni	10	100%
4.	Melur	10	100%
5.	Kempas	8	80%

$$\text{Average Accuracy} = \frac{(10 + 9 + 10 + 10 + 8)}{50} \text{ x } 100 \quad = 94\%$$

For further testing, we simulated the maximum parameter selection epoch in the training process. Epoch maximum parameter selection was done by the experiment performed using the same initial weight. The test results of 10 images per type of wood can be seen in Figure 3.

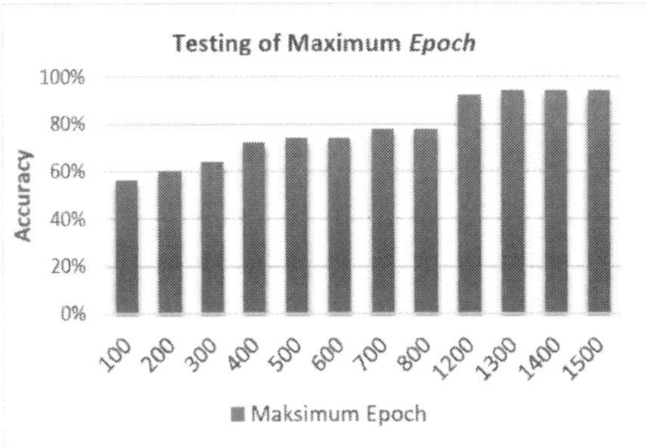

Figure 3. Testing of maximum *epoch*

Figure 3 shows the maximum epoch, the number of actual output in accordance with the desired output per type of wood as well as accuracy. The test results shows that the maximum epoch to 1300, the accuracy has reached 94%.

4. Discussion
Based on the tests performed, the identification of the type of wood can be done by using artificial neural networks backpropagation as a method of classification of types of wood according to predetermined targets with accuracy 94%. Propagation parameters such as the maximum value of the epoch has an influence on the accuracy of the test results. In testing the propagation parameters, the test gives an accuracy of 94% at the maximum value exceeds epoch 1300. The larger the value, the accuracy is increased. In a subsequent study is expected to develop by combining other methods to improve accuracy.

5. Conclusion
We have successfully develop a digital image based method to identify the type of timber automatically. In this study, backpropagation neural network is used as artificial intelligence component. Several stages were developed: a microscope image acquisition, pre-processing, feature extraction using gray level co-occurrence matrix and normalization of data extraction using decimal scaling features. The results showed that the proposed method was able to identify the timber with an accuracy of 94%

References
[1] Mandang YL and Pandit IKN 1997 *Pedoman Identifikasi Jenis Kayu di Lapangan.* Seri Manual. PROSEA: Bogor
[2] Triantoro T, Batubara FR and Fahmi F 2014 Image based water gauge reading developed with ANN Kohonen Proceedings of 2014 *Int. Conf. on Electrical Engineering and Computer Science, ICEECS Bali*
[3] Muhammadiy W and Fahmi F 2016 Mobile Application to differentiate flesh meat between beef and work *The 2016 IEEE International Conference on Computational Intelligence and Cybernetics (*CYBERNETICSCOM*)* Makassar
[4] Siswanto A, Tarigan P and Fahmi F 2013 Design of contactless hand biometric system with relative geometric parameters *Proc. of 2013 3rd Int. Conf. on Instrumentation, Communications,*

Information Technol., and Biomedical Engineering: Science and Technol. for Improvement of Health, Safety, and Environ., ICICI-BME Bandung

[5] Listia R and Harjoko A 2014 Klasifikasi massa pada citra mammogram berdasarkan grey level cooccurence matrix (GLCM). *Indonesian Journal of Computing and Cybernetics Systems* **8(1):** 59-68.

[6] Risaldi M, Purwanto and Himawan H 2014 Klasifikasi kualitas kayu kelapa menggunakan algoritma *neural network backpropagation. Jurnal Teknologi Informasi* **10(1)**: 69-78.

[7] Mohan SK, Venkatachalaphaty K and Sudhakar P 2014 An intelligent recognition system for identification of wood species. *Journal of Computer Science* **10(7)**: 1231-1237.

[8] Gunawan AAGR, Nurdiati S and Arkeman Y 2011. Identifikasi jenis kayu menggunakan *support vector machine based on image data. Jurnal Ilmu Komputer Agri-Informatika* **3(1)**: 1-8.

[9] Gadkari D 2004 Image quality analysis using GLCM. *Thesis under publication. University of Central Florida.Yang, X.S.* 2009.

[10] Ferguson JR 2007 Using the grey-level-co-occurrence matrix to segment and classify radar imagery. *Thesis under publication*. University of Nevada.

[11] Kadir A, Nugroho LE, Susanto A and Santosa PI 2011 Neural network application on foliage plant identification. Int. J. of Computer Applications **29(9)**: 15-22.

[12] Atomi WH 2012 The effect of data preprocessing on the performance of artificial neural networks techniques for classification problems. *Thesis under publication. University Tun Hussein Onn Malaysia.*

[13] Purnamasari RW 2013 Implementasi jaringan syaraf tiruan *backpropagation* sebagai sistem deteksi penyakit *tuberculosis* (TBC). *Thesis under publication. Universitas Negeri Semarang.*

IOP Conf. Series: Journal of Physics: Conf. Series **801** (2017) 012052 doi:10.1088/1742-6596/801/1/012052

The Role of Social Media User Experience as a Mediator for Understanding Social Media User Behavior in Indonesia's Museum Industry: Literature Review

A M Sundjaja[1]; F LumanGaol[2]; T Budiarti[3]; B S Abbas[3]; S B Abdinagoro[3]; H Ongowarsito[1]

[1] School of Information System, Bina Nusantara University, Indonesia
[2]Binus Graduate Program, Bina Nusantara University, Indonesia
[3] Doctoral of Research in Management, Bina Nusantara University, Indonesia

Abstract. Social media has changed the interaction between the customer and the business, social media has proven to provide new opportunities in facilitating access to information, efficiency and ease of interaction between customers and businesses that are distributed geographically dispersed. Ease of interaction to improve access to information about products, services, and prices have proven to have a positive impact for consumers. The purpose of this article is to develop a conceptual model to test the effect of user motivation, user expectations, and online community involvement to the intention of behavior that is mediated by the use of social media museum experience. This article is a literature study on exploration of social media user experiences museum in Indonesia. Authors searched and examined 85 articles from google scholar with the following keywords: motivation, expectations, online communities, user experience, social media, Technology Acceptance Model, Experiential Marketing, Uses and Gratification Theory. Proposed data collection techniques are literature study, survey and observation. The sample used in this research is 400 respondents of social media users that follow the social media managed by Indonesia's museum. The sampling technique are systematic sampling. We use Structural Equation Model with AMOS for analyze the data.

1. Introduction

Based on [1], social media today have 2.3 Billion active users and accessed by 1.9 Billion through mobile devices in the world, Facebook is the social media that has the most active users. The market share of the social media in Indonesia is 30% of the total population of social media users in the world and access their social media during 2.9 hours per day. So we can conclude that social media including Facebook, Google+, Twitter, etc., are very important to the company's strategy, products, and services. In general, a company must prepare a huge budget for advertising in conventional media but the presence of internet offers free advertising with measurable and extraordinary impact. Marketing communications should be integrated in order to integrate the customer database and how to deliver a consistent experience to customers through communication. Ease of interaction for improving the access to information about products, services, and prices have proven provide a positive impact for consumers. It can be concluded that social media has changed the way of communication between the customer and the business, social media has proven to provide new opportunities in facilitating access to information, efficiency and ease of interaction between customers and businesses that located in the geography diversity.

Social media offer new ways for people to develop and maintain social networks, create relationships, share information, create and modify the content, and social change in the internet. Based on the observations, people who want to visit the museum will search or obtain the information about the museum via the Internet (search engines or social media), newspapers, television, radio, and schools. Museum managers have come to realize these opportunities so that they began to adopt social media to provide information and museum services, activities, exhibitions, and collections so that the public can know the existence of the museum. Through social media managed by the museum, museum managers share experiences visiting guests during their museum visit through media images and video. In addition, people who have visited the museum and have a social media can share their experience while visiting the museum that will increase awareness of colleagues who become friends in social media.

Research on the factors that affect the experience of use of social media to the intention of behaviour has been done on the tourism industry [1]–[5]. The method used in previous study is quantitative analysis using Structured Equation Model and the sample of 400 respondents. [2] conducted a study to examine the factors that influence the intention to share information using social media in the United States and declared that the perceived ease of use and confidence in the integrity of influencing consumer behavior and travel. [4] conducted a study to test the conceptual model to learn why the tourists in the United States share the experience of their travel by using social media and stated that the success of the desire of tourists to share the experience of travel affected by the use of social media in planning travel, experience with social media, and the perception of pleasure. [5] conducted a study to explore the tourist behavior in social media in the United States and stated that there is influence between engagement on social media against the intentions of visitors to re-visit. [3] conducted a study to analyze the factors that influence the intention of consumers to follow the advice presented in the community of online travel in Spain and stated that the intention rating is positively influenced by the trust given by the people who give advice through an online community, trust in the community online travel advice and perceived usefulness has a direct influence on attitudes toward suggestions. Based on earlier research, experience of the use of social media users is affected by user motivation [5]–[9], user expectations [2], [10]–[14], and online community involvement [3], [15], [16].

With the adoption of social media technologies by the manager of the museum to introduce the museum to the public that this is an interesting phenomenon to be study. Museum managers need to understand the expectation and needs of the community in using social media properly to the public awareness about the museum. Based on the model of the Technology Acceptance Model as a theory of technology adoption, User and Gratification Theory as a theory of motivation users, and Strategic Experiential Marketing as the theory of the visitor experience, this study is expected to clarify whether experience of the use of social media use in the museum to become a mediator between the motivation of the user, expectations users, and online community involvement to the behaviour intention.

This paper objective is to review the previous studies to develop a conceptual model of the role of social media usage experience the museum in Indonesia as a moderator between the user motivation, user expectations, and community engagement to the behaviour intention. This paper implication is to be a reference, research road map, and encourage new ideas for future research regarding the experience of social media use in industrial museum in Indonesia.

2. Methods

The emergence of the idea of research in this area the author has a hobby in photography and involve in a cultural community. The author observes that the majority of social media owned by the museum management in Indonesia is not managed properly. Therefore, the authors conducted a literature search to develop a conceptual framework about the role of social media usage experience at the museum. To be able to manage a good social media, the author felt the need for a comprehensive research on what is the needs and the expectations of the community. Author search for some articles with keywords, such as user motivation, user expectations, involvement of online communities, social media usage experience, and behavioral intentions. The author uses Google Scholar to find articles that discuss about these variables. In literature search, the authors use a variety of keywords in English and Indonesian, as follows: social media usage, social media adoption, social media involvement, social media experience, user motivation, user expectation, community involvement, user behavior. The authors found thirty-four (34) articles used to understand the application of social media in museums and twenty-one (21) articles used to develop a conceptual framework. This article is divided into several sections as follows: (1) Discussion of the main theories that are used to develop a conceptual framework. (2) Development of a conceptual model and (3) The study design.

3. Results and Discussions

In the first section of this article, the author presents how the Technology Acceptance Model, Strategic Experiential Marketing, User and Gratification Theory as the main theory to develop a conceptual framework. Technology Acceptance Model (TAM) built by [6] to explain and predict individual adoption to information technology. The model was constructed by adopting the Theory of Reasoned Action (TRA) developed by Fishbein and Ajzen in 1975 that describes the behaviour of a person through their intentions in approving or rejecting behaviour, the intention is determined by the individual attitudes toward behaviour and social norms. [6] suggests the perception of usefulness, perceived ease of use is the main factors that affect the adoption of technology. [7] proposed several constructs to enhance the use of TAM in the context of Facebook, these constructs are Critical Mass (CM), Capacity (CP), and Perceived Playfulness (PP). [8] proposed a Strategic Experiential Modules that aims to help managers develop some type of customer experience for their customers. [8] identified five modules in Experiential Marketing, namely: think experience, related experience, sense experience, act experience, and feel experience. [9] stated that in the aspect of social media user experience, sense experience and act experience aren't relevant research themes for social media users can only interact socially and share knowledge among users. The community is the involvement of a group of people sharing a particular interest [10]. The concept of online community developed in the early advent of the Internet, when a computer network can connect the machine as well as the group of people who are connected with the aid of a machine into a social network. The online community members want to connect globally with hopes for the emergence of friendship, sharing information, and social support from their homes and offices. Relationships are created based on the common interests between members of a group to achieve common goals and supported by a technology can form an online community [11]. [12] identified four types of consumer needs can be met by the online community: community that facilitates the purchase and sale transactions and delivery of information related to the process, communities of interest involving a higher level of communication among members about a topic, community facilitating the emergence of a new story, a personality or a new environment, a community that facilitates the opportunity for users to share a life experience. [13] classifies online community based interactivity, focus and cohesion among members. Interactivity refers to the degree to which members of the community to interact and communicate. Focus refers to the goal of a community or the interests of members. Coherence refers to the degree of community members feel closeness to one another. Uses and Gratification Theory originally used to explain how the origin of the psychological and social needs of users who raises certain expectations in the use of a media [14]. One basic premise of the use of Theory of Use and Gratification is that users will search for the best media among its competitors to meet the needs and leads to the best gratuities [15]. Currently users are actively using the media, smart, and motivated so that the Theory of Use and Gratification has focused on what users are doing on the media than what influence or effect of the media towards the user [14]. In the thought of people actively selecting and using the media to respond to a need basis to examine the satisfaction derived from a medium. For example, if a user has a need to seek a new atmosphere, there is a special media are available to meet those needs satisfactorily. [14] proposed five (5) categories in the Theory of Use and Gratification, namely cognitive, affective, integrative personal, social interaction, and release of tension.

In the second section of this article, the author proposes a conceptual framework to examine whether the role of Social Media Usage experience as a mediator between User Expectations, The user Motivation, Engagement Online Communities against the intension of Conduct. Figure 1 is a conceptual framework developed by the authors.

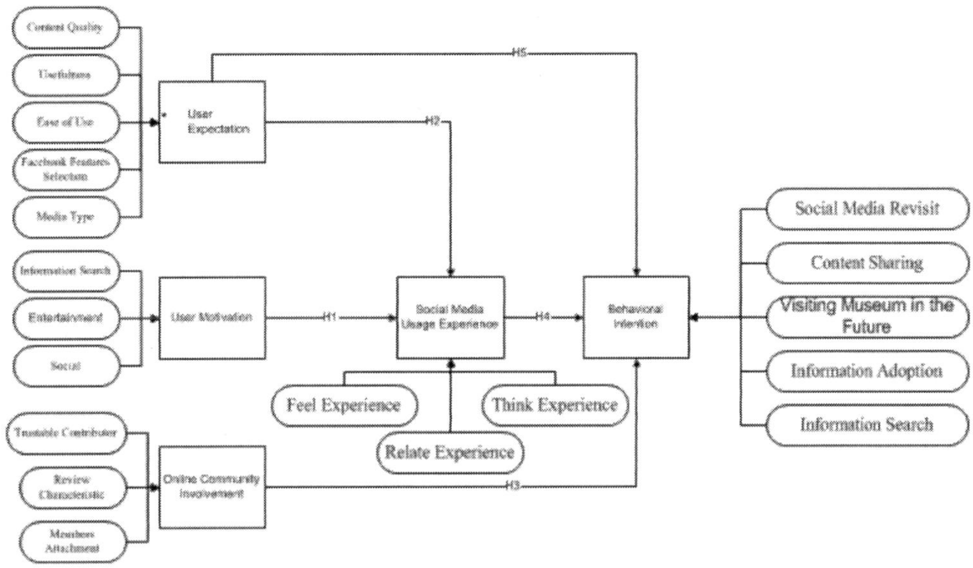

Figure 1. Conceptual framework

The first hypothesis is there is influence between motivations of users to experience the use of social media. Motivational influence users to experience social media users have been done by previous researchers but aspects studied still are gaps and inconsistencies [5], [9], [24]–[26]. Consumer behavior in using social media in general merely consume and participate and contribute only a small portion of content [26]. Motivation users in the use of social media are processing the information, entertainment activities, and social connections [9], [24], [26]

The second hypothesis is that there are expectations of users to experience the effect of the use of social media. Influence user expectations of the experience of the use of social media has been done by previous researchers but aspects studied there is still [3]–[5], [9], [25], [27], [28]. In previous studies, user expectations have been measured by the perceived ease of use [2], [24], [29], perceived usefulness [30], quality of content [27], [29], [31], media type, and features provided by social media [32]. Social media user experience can be interpreted as a form of user activity in a subject of discussion, offline events and activities held by the manager if users like the Facebook fan page and considers it to entertain them [24].

The third hypothesis is that there is an online community involvement influences the behavior intention. Research on the effects of online community engagement to the intention of behavior has been studied by several researchers previously [3], [33]–[36]. [34] stated that community leaders need a framework to manage discussions and activities that can be trusted by the community, where the two components to analyze this framework is a contributor and trustworthy content.

The fourth hypothesis is there is the influence of social media usage experience to the intention of behavior. Research on the influence of social media usage experience to the intention of behavior has been studied by several researchers previously [2]–[5].

The fifth hypothesis is there is influence between user expectations of the behavior intention. User expectations influence the behavior intention has been done by previous researchers but aspects studied there is still [2], [24], [27], [29]–[31], [37].

The last section in this article is a discussion of the research design. Data collection techniques used in this research is literature study, survey and observation. Selection of the sample is determined based on the selected population; in this case are social media users on the museum in Indonesia. Components of the study population is not infinite where the data source to obtain the number of members of the population media users can't be identify certainty. Therefore, the number of samples used in this study was 500 respondents. The sampling technique used in this research is non-random sampling technique aims samples (purposive sampling). The analytical method used in this research is to use Structural Equation Model.

4. Conclusions

The presence of social media has changed consumer behavior, the manager of the museum, and the community so that it can encourage research in the field of social media adoption. A literature study conducted by the authors of the thirty-four (34) article on the research and synthesize twenty-one (21) of articles to produce a conceptual model of the role of experience using social media as a mediator between the motivation of the user, user expectations, and community engagement online to the intention behavior. The articles used in the development of this model are obtained from a variety of relationships between constructs. The findings of this literature study is a new understanding of the viewpoint of social media usage experience at the museum in defining the needs and desires of social media users or visitors of the museum. The significant contribution of this work is the author developed a model to better understand the behavior of comprehension regarding the use of social media. There are several limitations to this study. First, the limited number of articles about the involvement of the online community to experience the use of social media so it is necessary to study in more depth. Second, research in the field of social media is still at an early stage and began to appear articles empirically so that research in this area can be considered to be further developed. Third, the literature search conducted has similar terminology, such as online communities, blogs, web 2.0, and social networking sites. So it takes exploration in the use of other terms that are relevant to social media, especially the emergence of technological development of new social media.

5. References

[1] Oktadiana H, and Kurnia A, "How Customers Choose Hotels," *Binus Bus. Rev.*, vol. 2, no. 1, p. 510, May 2011.

[2] Bilgihan A, Barreda A, Okumus F, and Nusair K, "Consumer perception of knowledge-sharing in travel-related Online Social Networks," *Tour. Manag.*, vol. 52, pp. 287–296, 2016.

[3] Casaló L V, Flavián C, and Guinalíu M, "Understanding the intention to follow the advice obtained in an online travel community," *Comput. Human Behav.*, vol. 27, no. 2, pp. 622–633, Mar. 2011.

[4] Kang M, and Schuett M, "Determinants of Sharing Travel Experiences in Social Media," *J. Travel Tour. Mark.*, vol. 30, no. 1, p. 107, 2013.

[5] Leung X Y and Bai B, "How Motivation, Opportunity, and Ability Impact Travelers' Social Media Involvement and Revisit Intention," *J. Travel Tour. Mark.*, vol. 30, no. February 2015, pp. 58–77, 2013.

[6] Davis F D, "Perceived Usefulness, Perceived Ease of Use, and User Acceptance of Information Technology," *MIS Q.*, vol. 13, no. 3, p. 319, Sep. 1989.

[7] Rauniar R, Rawski G, Yang J, and Johnson B, "Technology Acceptance Model (TAM) and Social Media Usage: an Empirical Study on Facebook," *J. Enterp. Inf. Manag.*, vol. 27, no. 1, pp. 6–30, Feb. 2014.

[8] Schmitt B, "Experiential Marketing," *J. Mark. Manag.*, vol. 15, no. 1–3, pp. 53–67, 1999.

[9] Chen G, Yang S, and Tang S, "Sense of virtual community and knowledge contribution in a P3 virtual community: Motivation and experience," *Internet Res.*, vol. 23, no. 1, pp. 4–26, Jan. 2013.

[10] Rothaermel F T and Sugiyama S, "Virtual internet communities and commercial success: individual and community-level theory grounded in the atypical case of TimeZone.com," *J. Manage.*, vol. 27, no. 3, pp. 297–312, Jun. 2001.

[11] Wellman B, Salaff J, Dimitrova D, Garton L, Gulia M, and Haythornthwaite C, "Computer Networks as Social Networks: Collaborative Work, Telework, and Virtual," *Source Annu. Rev. Sociol.*, vol. 22, no. January 2016, pp. 213–238, Aug. 1996.

[12] Armstrong A and Hagel III J, "Creating Value in the Network Economy," in *Creating value in the network economy*, D. Tapscott, Ed. Boston, MA, USA: Harvard Business School Press, 1999, pp. 173–185.

[13] Figallo C, *Hosting Web Communities: Building Relationships, Increasing Customer Loyalty, and Maintaining a Competitive Edge*. Wiley, 1998.

[14] Katz E, Haas H, and Gurevitch M, "On the Use of the Mass Media for Important Things," *Am. Sociol. Rev.*, vol. 38, no. 2, p. 164, Apr. 1973.

[15] Lariscy R W, Tinkham S F, and Sweetser K D, "Kids These Days: Examining Differences in Political Uses and Gratifications, Internet Political Participation, Political Information Efficacy, and Cynicism on the Basis of Age," *Am. Behav. Sci.*, vol. 55, no. 6, pp. 749–764, Jun. 2011.

[16] Charlesworth A, *An Introduction to Social Media marketing*. London [England]: Routledge, 2015.

[17] Tuten T, *Advertising 2.0: Social Media Marketing in a Web 2.0 World: Social Media Marketing in a Web 2.0 World*. Praeger, 2008.

[18] Proctor N, "Digital: Museum as Platform, Curator as Champion, in the Age of Social Media," *Curator Museum J.*, vol. 53, no. 1, pp. 35–43, 2010.

[19] Fletcher A and Lee M J, "Current Social Media Uses and Evaluations in American Museums," *Museum Manag. Curatorsh.*, vol. 27, no. 5, pp. 505–521, 2012.

[20] Charitonos K, Blake C, Scanlon E, and Jones A, "Museum learning via social and mobile technologies: (How) can online interactions enhance the visitor experience?," *Br. J. Educ. Technol.*, vol. 43, no. 5, pp. 802–819, 2012.

[21] Harahap K A E, "Analisis Pengaruh Kualitas Pelayanan, Harga, dan Promosi Terhadap Kepuasan Pengunjung di Museum Ronggowarsito Semarang," 2014.

[22] Jannah D A M, Andriani N, and Arief M, "Pengaruh Strategi Experiential Marketing Terhadap Kepuasan Pengunjung Museum Sepuluh Nopember Surabaya," *J. Stud. Manaj. Dan Bisnis*, vol. 1, no. 1, pp. 53–64, 2014.

[23] Sundjaja A M, Simamora B H, and Sundjaja A M, "An Adoption of Social Media for Marketing and Education Tools at Museum Industry," *Adv. Sci. Lett.*, vol. 21, no. 4, pp. 1028–1030, 2015.

[24] Chen H, Papazafeiropoulou A, Duan Y, and Chen T K, "The Antecedents And Outcomes Of Brand Experience On The Social Networking Site," in *ECIS 2013 Proceedings*, 2013, p. Paper 10.

[25] Hau Y S, and Kim Y G, "Why Would Online Gamers Share Their Innovation-Conducive Knowledge in the Online Game User Community? Integrating Individual Motivations and Social Capital Perspectives," *Comput. Human Behav.*, vol. 27, no. 2, pp. 956–970, Mar. 2011.

[26] Heinonen K, "Consumer Activity in Social Media: Managerial Approaches to Consumers' Social Media Behavior," *J. Consum. Behav.*, vol. 10, pp. 356–364, 2011.

[27] Chung N and Koo C, "The Use of Social Media in Travel Information Search," *Telemat. Informatics*, vol. 32, no. 2, pp. 215–229, May 2015.

[28] Lee W, Xiong L, and Hu C, "The effect of Facebook users' arousal and valence on intention to go to the festival: Applying an extension of the technology acceptance model," *Int. J. Hosp. Manag.*, vol. 31, no. 3, pp. 819–827, 2012.

[29] Kwon O and Wen Y, "An Empirical Study of the Factors Affecting Social Network Service Use," *Comput. Human Behav.*, vol. 26, no. 2, pp. 254–263, Mar. 2010.

[30] Sheng C W and Chen M C, "A Study of Experience Expectations of Museum Visitors," *Tour. Manag.*, vol. 33, no. 1, pp. 53–60, 2012.

[31] Filieri R and McLeay F, "E-WOM and Accommodation: An Analysis of the Factors That Influence Travelers' Adoption of Information from Online Reviews," *J. Travel Res.*, vol. 53, no. 1, pp. 44–57, Jan. 2014.

[32] Hsu Y L, "Facebook as international eMarketing strategy of Taiwan hotels," *Int. J. Hosp. Manag.*, vol. 31, no. 3, pp. 972–980, 2012.

[33] Kavoura A, and Stavrianea A, "Following and Belonging to an Online Travel Community in Social Media, its Shared Characteristics and Gender Differences," *Procedia - Soc. Behav. Sci.*, vol. 175, pp. 515–521, Feb. 2015.

[34] Shneiderman B, "Building Trusted Social Media Communities: A Research Roadmap for Promoting Credible Content," in *Roles, Trust, and Reputation in Social Media Knowledge Markets*, Springer Cham Heidelberg New York Dordrecht London: Springer International Publishing, 2015, pp. 35–43.

[35] Zhou T, "Understanding Online Community User Participation: a Social Influence Perspective," *Internet Res.*, vol. 21, no. 1, pp. 67–81, Jan. 2011.

[36] Muchardie B G, Yudiana N H, and Gunawan A, "Effect of Social Media Marketing on Customer Engagement and its Impact on Brand Loyalty in Caring Colours Cosmetics, Martha Tilaar," *Binus Bus. Rev.*, vol. 7, no. 1, p. 74, May 2016.

[37] Hsu M H, Tien S W, Lin H C, and Chang C M, "Understanding the Roles of Cultural Differences and Socio-Economic Status in Social Media Continuance Intention," *Inf. Technol. People*, vol. 28, no. 1, pp. 224–241, Mar. 2015.

IOP Conf. Series: Journal of Physics: Conf. Series **801** (2017) 012053 doi:10.1088/1742-6596/801/1/012053

KINECTATION (Kinect for Presentation): Control Presentation with Interactive Board and Record Presentation with Live Capture Tools

Rhio Sutoyo, Herriyandi, Tri Fennia Lesmana, Edy Susanto

Bina Nusantara University, Jl. K. H. Syahdan No. 9, Kemanggisan / Palmerah, Jakarta Barat, 11480, Indonesia

Email : rsutoyo@binus.edu

Abstract. Presentation is one the most common activity performed in various fields of work (e.g. lecturer, employee, manager, etc.). The purpose of presentation is to demonstrate or introduce presenters' idea to the attendees. Within the given time and specific place, presenters must transfer their knowledge and leave great impression for their audience. Generally, presenters use several handy tools such as mouse, presenter, and webcam to help them to navigate their slides. Nevertheless, some of these tools have several constraints and limitations such as not portable and does not support multimedia. In this research, we develop an application that assist presenters to control their presentation materials by using Microsoft KINECT. In this research, we manipulate colour image, image depth, and the skeleton of the presenters captured by the KINECT. Then, we show the post-process image results into the projector screen. The KINECT has more useful than other tools because it supports video and audio recording. Moreover, it also able to capture presenters' movement that can be used as an input to interact and manipulate the content (i.e. by touching the projection wall). Not only this application provides an alternative in controlling presentation activity, but it also makes the presentation more efficient and attractive.

1. Introduction

In various type of profession, presentation has become one of the most common activity. Presentation is performed with the help of computer technology. The purpose of presentation is to share knowledges, results, or ideas to other people, generally within the same room. For example, an advertising company try to present their work results to please their employer, or a lecturer in a university teach his/her student with PowerPoint slideshow.

To start presentation activity, presenters must have knowledge or material that they want to share. The act of presentation generally uses computer, mouse, keyboard, and projector. Those tools will be set in a specific room for presentation. To navigate their materials, presenters can press space on the keyboard or right-click on the mouse. Furthermore, they sometimes could also bring a presenter or a pointer to help them navigate the slides. To write notes, presenters usually will be provided white-board and colored-marker.

Those above solutions for presentation has been done for ages and there are not so much different in doing presentation. Nevertheless, some of these tools have several constraints and limitations such as not portable and does not support multimedia. For example, by using a mouse or a keyboard, the presentation must back and forth from the main stage and the computer just to move to the next slide. That problem could be solved by using a presenter, but a presenter could be missing and has limited battery. Moreover, those solutions cannot record (i.e. video and audio) the activity of presentation.

With the objective of giving better solution for presentation activity, we design and create a smart-board application that can assist act of presentation. Smart board is an interactive board that allows the presenter to display and manipulate presentation's content to the board directly with the touch [1]. Smart board was first introduced in 1991 by SMART Technologies Company. Users can replace the functions of the monitor into the board or vice versa. Moreover, users can use their finger or an electronic pen (stylus) to interact directly with the board [2]. The result of this research is an application that can help the presenter or teacher to control the presentation of content, add annotations on the screen, and control the mouse pointer using the touch on the board projector. Furthermore, this application can also be used to record video and audio of the presentation activity.

This application was built by using KINECT from Microsoft. Introduced as the complimentary device for XBOX game console, KINECT is a gaming sensor to detect people and the movements of the people in the game situation (e.g. running, jumping, dodging, etc.). Nowadays, KINECT's impact has extended beyond the gaming industry [3]. It helps in the area of education [4] [5], healthcare [6], indoor mapping [7], and robot [8]. Those are some examples of KINECT Effect in various wok area.

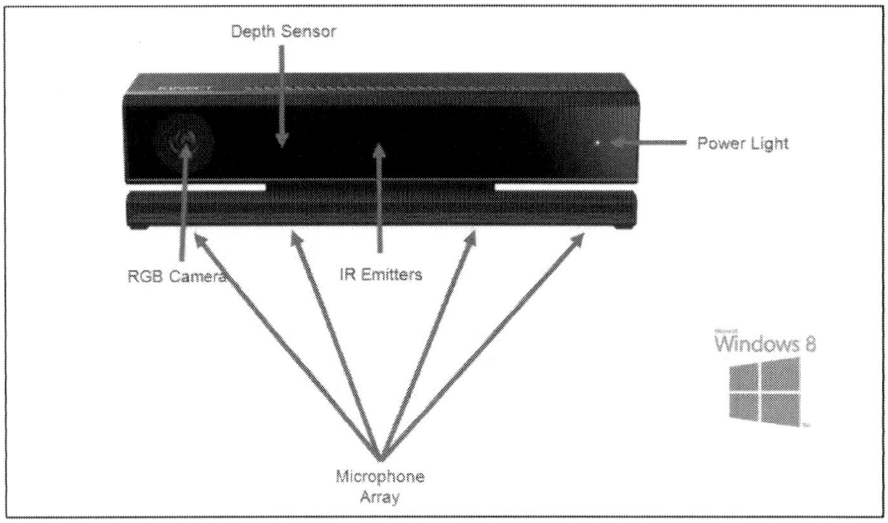

Figure 1 - Microsoft KINECT

The KINECT is used to detect the gesture of the presenters or the users. The gesture recognition that is done in here is optical camera-based methods [9]. Moreover, the presenters provide input on a computer that is currently active, and then run a process such as controlling a mouse pointer and navigate the presentation's content. Gesture will be used as input to navigate presentations that simplify the presenter to navigate. The input will be wireless and ease the input from the presenters. Moreover, the KINECT is always connected to the electricity, thus it will always have batteries.

The use of KINECT is chosen because KINECT has decent camera sensor and built-in microphone as shown in the Figure 1. Thus, our application could utilize those features to navigate and record presentation within the same application. Compared with webcam, it only provides a camera feature. There is also a tool called leap motion. However, this tool is limited as only a hand-detection sensor. We believe, this application can increase efficiency and effectiveness of the presentation. It is because the slide navigation, annotation process, and record of presentation can be done within the same application. After the presentation, the annotated slides or materials and videos can be shared to the viewers for them as a tutorial or review material

2. Related Works

Based on our observation, this type of research has by done by several researchers which are: Beeboard: From WiiMote to Interactive Whiteboard Application for Classroom [10] and The Implementation of Hand Detection and Recognition to Help Presentation Processes [11].

The prior research utilizes the WiiMote controller from Nintendo Wii Game Console for its sensor. The Wii Remote, as shown in the Figure 2, will be projected to the screen and searching for the infra-red pen tools. The presenter uses the infra-red pen tools, as shown in the Figure 3, to point to the projection screen and make selection with it. BeeBoard provides features for video recording along with the sound of the presenter, but required an additional tool in the form of a microphone and had to use infrared pen tools to be able to navigate.

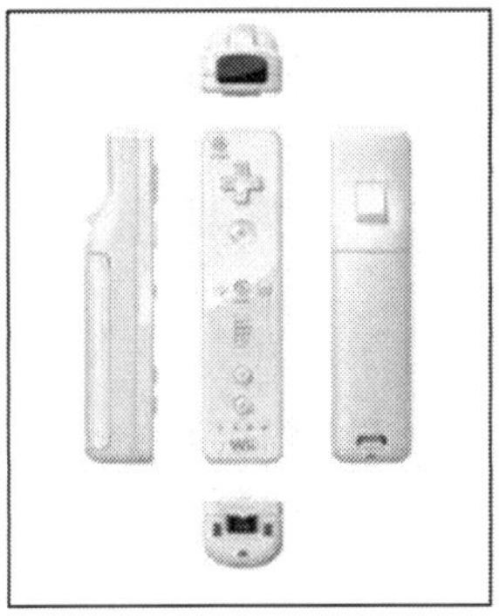

Figure 2 - Nintendo WiiMote

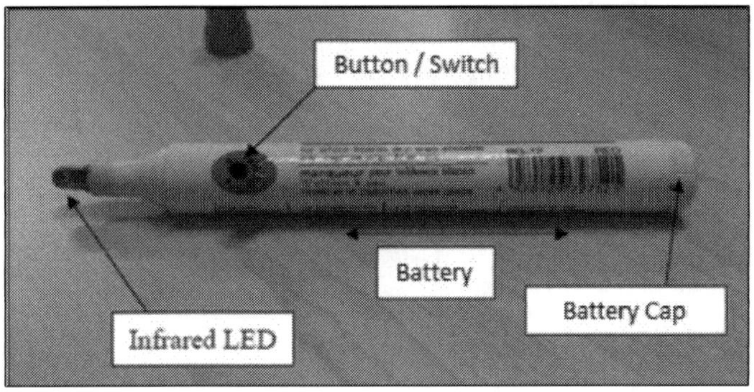

Figure 3 - Infra-Red Pen Tools

The latter research implements hand detection and recognition to help presentation by using a pattern of hand to control a PowerPoint slide as start slide show and slide navigation. It uses a webcam as the sensor to detect hands and fingers. The signal of the finger will become the input for the system and will be do pre-set actions (e.g. next slide, previous slide, etc.).

3. System Overview

To build this application, we have designed the architecture of the system. The design can be seen in the Figure 4. The design is used to show overall process of the system. The computer is connected with a KINECT and a projector. The application run in the computer and communicate with the KINECT and the projector. Gesture sensors on the KINECT will be activated to detect movement made by the presenter. The projector, as usual, will display computer screen to the display board.

Presenter give input to the computer by touching the board and those gestures will be captured by the KINECT sensor. The detected input then will be used as an action on the application. The application will perform the action and will be shown in the display by projector (e.g. give annotation to slides, next slide, etc.). The gesture will be similar with manipulating content on the tablet. We choose this approach because users are already familiar with it. The gestures are: touch and drag left for next slide, touch and drag right for previous slide, touch and drag up to show menu, and touch and drag down to show PowerPoint overview. The presenters could also record their presentation activities by simply touch the record button in the application. The KINECT will start recording and the results could be later use for reviews or documentations.

With this application, the presenters can do several functions which are: calibrate screen, pencil mode, navigate cursor, annotate slides, undo, and redo annotation, draw simple figure (i.e. circle, triangle, and square), navigate PowerPoint, erase canvas, open PowerPoint slide, register PowerPoint slide to the presentation list, start presentation, and record video and audio.

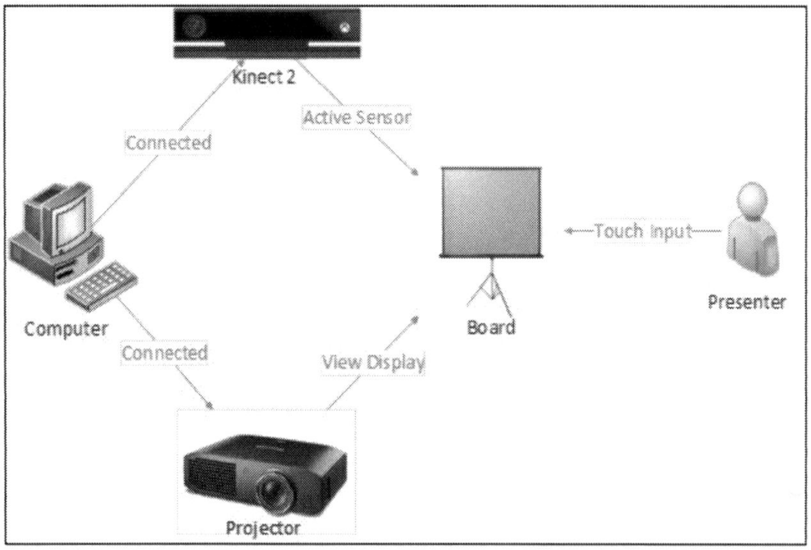

Figure 4 - KINECTATION's System Architecture

As we have said before, this application detects the presenter gesture and transform it to become input for the system. To detect the gesture, screen calibration must be done in the first step. The Figure 5 below explained the detailed steps for calibration:

4. Result and Discussions

We implement our system in the room of laboratory at Bina Nusantara University. The computer has dual-core, 2.66-Ghz and 2 GB of RAM. Moreover, the operating system should be minimum Windows 8 and 64-bit (x64). After the system is ready to be use, we conduct user evaluation that invites 35 people. The respondents work as lecturer and assistant laboratory. The evaluation consists of 18 questions that measure the accuracy, the ease to use, and user interface. The system can be seen in the Figure 6.

For the presentation navigation by using KINECTATION, 82.4% of respondent say it is easy to do. But, 17.7% found it is harder to navigate by using the application because they are not familiar with the input gestures. Moreover, 100% respondents agree that the recording materials from the presentation are useful for reviews and tutorials. For the annotation feature for the slide, 24% people says that the feature is not useful, while the rest find it quite useful. In the user interface, 74% respondent already familiar with the input and the menu. Nevertheless, 26% respondent says it is difficult and confusing.

In short, 30% of respondent is very satisfied with the result of the application, 62% of respondent is quite satisfied, but think that the application could be more improved. Then, 8% of respondent do not think that this application is suitable for presentation tools. With these results, we believe KINECT could be used for presentation tools in the near future. Not only it will ease the presentation system and more interactive, but it could directly record the presentation and make it as review materials.

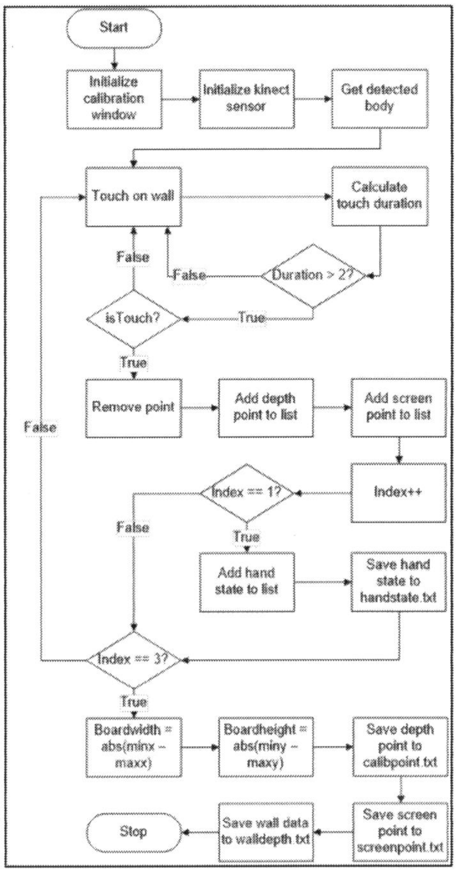

Figure 5 - Screen Calibration Process

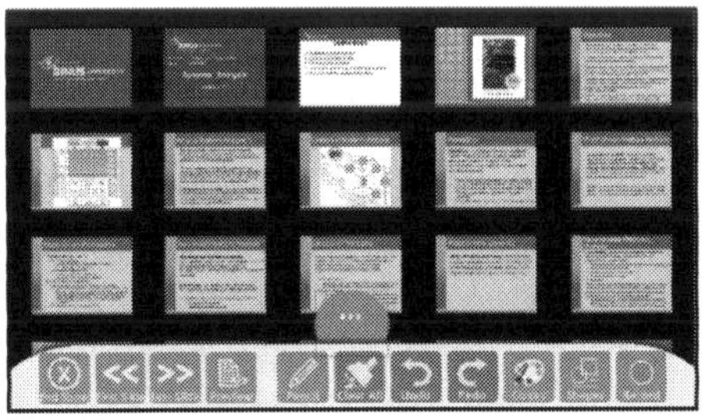

Figure 6 - KINECTATION Application

Figure 7 - KINECTATION Application

5. Conclusion and Future Works

In this paper, we present KINECTATION (Kinect Presentation) application, a tool to help presenters navigate their presentation materials. Moreover, KINECTATION can also be used to record audio and video of the presentation. We also have shown the system architecture and calibration step for our system. Moreover, the user evaluation shows positive results that says 92% of the respondents are satisfied. Thus, we believe KINECT is a promising tool to be integrated into the presentation room.

In the future works, we hope we could bring voice commands as additional features for our application. Moreover, we also plan to add feature to save annotation result as separated images (not saved in the slides).

References

[1] Nejem K M, and Muhanna W, "The Effect of Using Smart Board on Mathematics Achievement and Retention of Seventh Grade Students," *Int. J. Educ.*, vol. 6, no. 4, p. 107, 2014.

[2] Khamis A H A, and Al-Faki I M, "Difficulties Facing Teachers in Using Interactive Whiteboards in Their Classes," *Am. Int. J. Soc. Sci.*, vol. 3, no. 2, pp. 136–158, 2014.

[3] Zhang Z, "Microsoft kinect sensor and its effect," *IEEE Multimed.*, vol. 19, no. 2, pp. 4–10, 2012.

[4] Hsu H J, "The potential of Kinect in Education," *Int. J. Inf. Educ. Technol.*, vol. 1, no. 5, pp. 365–370, 2011.

[5] Holmes K, "Planning to teach with digital tools: Introducing the interactive whiteboard to pre-service secondary mathematics teachers," *Australas. J. Educ. Technol.*, vol. 25, no. 3, pp. 351–365, 2009.

[6] Webster D, and Celik O, "Systematic review of Kinect applications in elderly care and stroke rehabilitation.," *J. Neuroeng. Rehabil.*, vol. 11, p. 108, 2014.

[7] Khoshelham K, and Elberink S O, "Accuracy and resolution of kinect depth data for indoor mapping applications," *Sensors*, vol. 12, no. 2, pp. 1437–1454, 2012.

[8] Du G, Zhang P, Mai J, and Li Z, "Markerless Kinect-based hand tracking for robot teleoperation," *Int. J. Adv. Robot. Syst.*, vol. 9, pp. 1–10, 2012.

[9] Jiang F, Zhang S, Wu S, Gao Y, and Zhao D, "Multi-layered Gesture Recognition with Kinect," *J. Mach. Learn. Res.*, vol. 16, pp. 227–254, 2015.

[10] Yulianto B, and Dewi L C, "Beeboard: From wiimote to interactive whiteboard application for classroom," *J. Comput. Sci.*, vol. 10, no. 10, pp. 2117–2123, 2014.

[11] Sutoyo R, Prayoga B, Fifilia, Suryani D, and Shodiq M, "The Implementation of Hand Detection and Recognition to Help Presentation Processes," in *Procedia Computer Science*, 2015, vol. 59, pp. 550–558.

File Type Identification of File Fragments using Longest Common Subsequence (LCS)

R F Rahmat[1], F Nicholas[1], S Purnamawati[1], O S Sitompul[1]

[1]*Department of Information Technology, Faculty of Computer Science and Information Technology, University of Sumatera Utara, Medan, Indonesia*

Email: romi.fadillah@usu.ac.id, filbertnicholas@students.usu.ac.id, sarah_purnamawati@usu.ac.id, opim@usu.ac.id

Abstract. Computer forensic analyst is a person in charge of investigation and evidence tracking. In certain cases, the file needed to be presented as digital evidence was deleted. It is difficult to reconstruct the file, because it often lost its header and cannot be identified while being restored. Therefore, a method is required for identifying the file type of file fragments. In this research, we propose Longest Common Subsequences that consists of three steps, namely training, testing and validation, to identify the file type from file fragments. From all testing results, we can conlude that our proposed method works well and achieves 92.91% of accuracy to identify the file type of file fragment for three data types.

1. Introduction

Digital evidence, such as file or document is so important for any cases. The computer forensic analyst, who investigates and tracks the digital evidence, sometimes has difficulty for finding it due to deletion file or file damage. When restoring the file, the header file could possibly lost, so the type of file cannot be identified and difficult to reconstruct the file.

The file type identification of file fragments is such a complicated case that there are various file type, such as block of text ASCII, JPEG (image), or the complex one like Portable Document Format (PDF) [1]. Longest Common Subsequences, used for finding the longest sequence, is a classic method used for solving problems, such as comparing two files or bioinformatics. It also has the strengthness in coding optimization. Generally, header and footer in a file is rarely change, so it can reduce the required memory for comparison matrix.

There are several previous research related to classifying the file type of file fragment. In 2008, Calhoun and Coles compared two methods, namely Fisher's linear discriminant and Longest Common Subsequences. They concluded that the Longest Common Subsequences is not as fast as the linear discriminant method, but it could generate higher accuracy for predicting the type of file fragment [2]. Another related research is done by Sencar and Memon in 2009, who identified and recover JPEG file with missing fragments [3]. They used Huffman code tables to identify file fragments. In their conclusion, they stated that their proposed method to identify the next fragment of a file is more efficient than random decoding approach.

In 2010, Axelsson classified 28 different file types using the combination of Normalised Compression Distance (NCD) and K-Nearest Neighbour with 31541 of total number of file fragments [4]. Their proposed method is simple and easily implemented, but it could not give higher accuracy than random classification. Another method used for classifying the file fragments is NLP (Natural

Language Processing) techniques and Support Vector Machine conducted by Fitzgerald, et al. in 2012 [5]. They represented each file fragment with a feature vector that consist of the unigram and bigram counts of bytes in the fragment, along with several other statistical measurements (such as entropy). They classified 24 file type and generated an average prediction accuracy of 49.1%.

In this research, we propose Longest Common Subsequences to identify the file type of file fragments. The rest of the paper will be organized as follows: Section II describes the proposed method. Results and discussions are presented in Section III. Section IV provides summary and suggestions for future research.

2. Research methodology

Our proposed method consists of three main phases, namely training, testing and validation. In the training phase, the used data is document files collected before, while in the testing phase, the used data is fragment files generated from other document files with various formats by randomly overriding some hex number in the file and renaming the file extension to .tmp, so that it becomes unreadable. Figure 1, figure 2 and figure 3 respectively shows the general architecture for each phase.

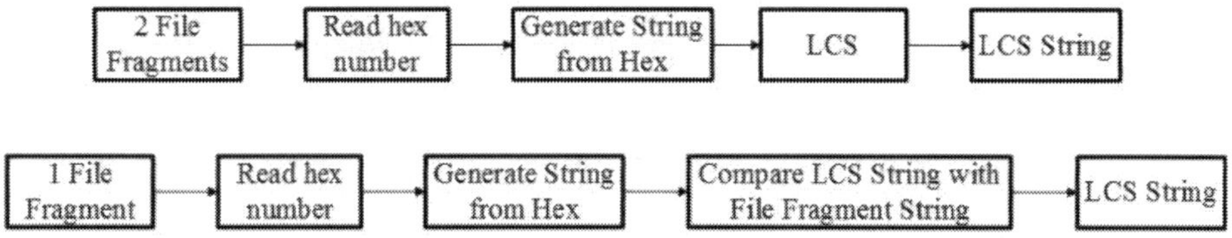

Figure 1. General architecture of training phase

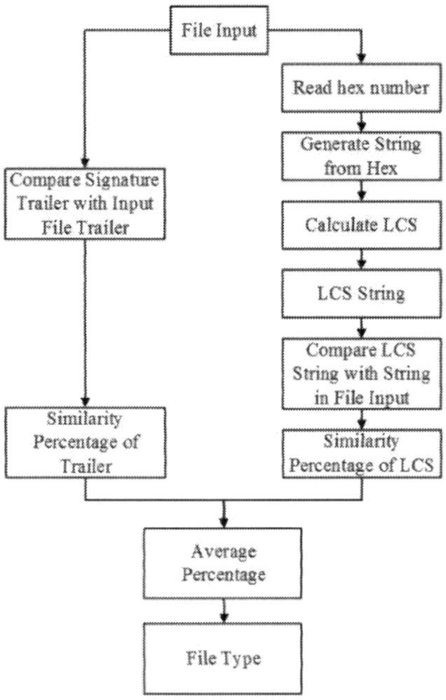

Figure 2. General architecture of testing phase

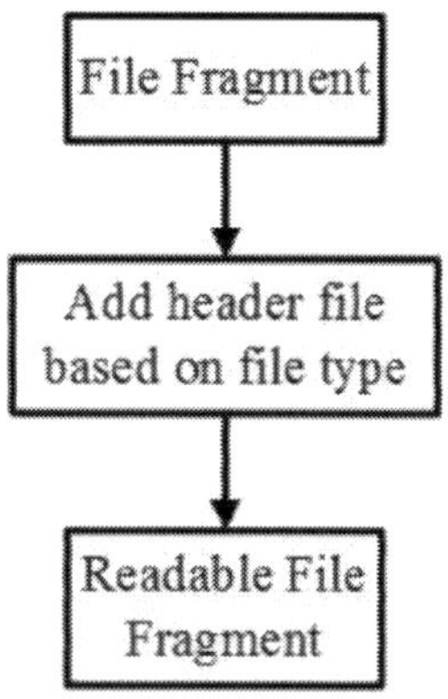

Figure 3. General architecture of validation phase

In the training phase, LCS will be calculated from each of file type, and then it will be stored in the external file. After that, the calculation of LCS will be repeated for each of file type with the external file and the training result will be used in testing phase. In the testing phase, the system will be tested to identify the file type from the file fragments and the accuracy will be calculated. In the validation phase, the system will add the header file based on the result of file type identification. If the file is accessible, then the validation is successful. However, sometimes the result will be not accessible or detected as a corrupt file because too many deleted hex numbers. Each steps of the general architecture will be clearly explained in the next subsection.

2.1. Dataset

The used data in this research especially for training and testing phase is government document files randomly collected and available at http://digitalcorpora.org/corpora/files. The data is divided into two parts namely training data and testing data. The calculation of LCS will be conducted using training data and then it will be used as reference for the comparison with the testing data. The testing data is file fragments generated from the collected data file. It is used for generating the file type.

There are three types of file used in this research, namely Portable Document Format (.pdf), Rich Text Format (.rtf) and Word Document (.doc). The total number of collected file is 615 files (300 files for training phase, 75 intact files and 240 files fragment used in testing phase). The fragment files will be categorized based on the percentage of fragmentation, represents how big the file size overwritten or fragmented by other file. The system will be tested to identify the intact file for each type of file. Table 1 and table 2 respectively show the file specification used for training and testing.

Table 1. File specification for training data.

File Type	File Number	Min. File Size (byte)	Max. File Size (byte)
PDF	100	150,735	594,339
RTF	100	6,876	48,258,055
DOC	100	762,880	1,502,208

Table 2. File specification for testing data.

File Type	File Number	Min. File Size (byte)	Max. File Size (byte)
PDF	25	75,509	178,805
RTF	25	255,469	437,361
DOC	25	93,696	109,568
PDF *fragment*	80	167,770	204,648
RTF *fragment*	80	252,133	575,967
DOC *fragment*	80	107,520	167,936

2.2. Read Hex Number and Generate String

The hex number from input file will be read and then changed into string. The result of string will be used for the calculation of LCS. The process of reading hex number and generating string will be describe in this following pseudocode.

```
int count
char temp
check hex number from file with StringBuffer
StringBuffer result = new StringBuffer(" ")
while (true){
if count >= file length
        break
else
        read hex number
        store hex number at temp
        count++
        result.append(temp)
return result.toString()
```

2.3. Longest Common Subsequences Algorithm

The calculation of LCS aims to get the longest sequence used for testing phase and percentage of similarity between file fragment and initial files (.pdf, .rtf and .doc). It will be describe in these following steps with a sample of input BDCABA and ABCBDAB.

Step 1 : Calculate the length of both string input.
Length of character BDCABA = 6 characters
Length of character ABCBDAB = 7 characters

Step 2 : Provide a table for the comparison based on the length of string input and 1 column addition for initial value (0). Table 1 shows the illustration of LCS comparison table.

Step 3 : Compare the strings and input the result using formula as shown in table 4 (if the strings equal) and table 5 (if the strings not equal). If the strings equal, then the input value will be a+1, while if the strings do not equal, then the input will be the highest value between the strings.

Table 1. LCS comparison table.

		B	D	C	A	B	A
	0	0	0	0	0	0	0
A	0						
B	0						
C	0						
B	0						
D	0						
A	0						
B	0						

Table 4. Result of LCS comparison

		B	D	C	A	B	A
	0	0	0	0	0	0	0
A	0	0	0	0	1	1	1
B	0	1	1	1	1	2	2
C	0	1	1	2	2	2	2
B	0	1	1	2	2	3	3
D	0	1	2	2	2	3	3
A	0	1	2	2	3	3	4
B	0	1	2	2	3	4	4

Table 2. Formula when string matches.

	X	
	a	**B**
Y	**c**	*a + 1*

Table 3. Formula when string not matches.

	X	
	a	**c**
Y	**b**	*d*

Step 4 : The highest value from LCS comparison table represents the longest sequence. In this case, the longest is 4 characters. Backtracking is conducted in this process to get LCS.

Step 5 : While backtracking, it starts from placing pointer in the bottom-right cell and will end if the pointer get the 0 value. The pointer will move up and left consistently. If the pointer pass the different value, for example, the cell with value of 4 to the cell with value of 3, the system will store the string then head to the next cell diagonally. It will be clearly explained in this following pseudocode.

```
Initialize a two-dimension-array i*j
Currently i equal to 8 and j equal to 7
While i and j not equal to 0
        If Array[i][j] equal to Array[i-1][j-1]+1
                lcs equal to char from Array[i][j] + lcs
                Array[i][j] go diagonal
        If Array[i-1][j] more than or equal to Array[i][j-1]
                go up
        If Array[i][j] equal to Array[i][j-1]{
```

```
        go left
    Return lcs
```

2.4. Average Percentage Calculation

In this research, LCS will generate the average percentage of similarity between file fragments and initial file and percentage of similarity between trailer file fragments and trailer initial file, for each type of file. The average of similarity percentage of PDF, RTF and DOC file can be calculated respectively using equation 1 until equation 3.

$$\left(Percentage_{PDF} + Percentage_{trailer\_PDF}\right)/\ 2 \quad (1)$$

$$\left(Percentage_{RTF} + Percentage_{trailer\_RTF}\right)/\ 2 \quad (2)$$

$$\left(Percentage_{DOC} + Percentage_{trailer\_DOC}\right)/\ 2 \quad (3)$$

2.5. Validation

In the last main phase, validation, an empty file will be created and then add the header file based on the identification result. After that, hex number from file fragment will be read and added to the empty file before. The file extension will be changed based on identification result. This following pseudocode ill describe the process of validation.

```
Create new text file
Read identification result
If result == pdf
        read pdf header hex bytes
        add pdf hex bytes to new text file
        read file fragment hex bytes
        add hex bytes to new text file with pdf header
        change text file extension to .pdf
else if result == rtf
        read rtf header hex bytes
        add rtf hex bytes to new text file
        read file fragment hex bytes
        add hex bytes to new text file with rtf header
        change text file extension to .rtf
else if result == doc
        read doc header hex bytes
        add doc hex bytes to new text file
        read file fragment hex bytes
        add hex bytes to new text file with doc header
        change text file extension to .doc
else
        cannot be verified
```

3. Result and discussion

3.1. Training Result

Figure 4, figure 5 and figure 6 respectively shows the training result of LCS for PDF, RTF and DOC file. The figure show the training result of LCS as follows: 25 characters for PDF file, 116 characters for RTF file and 13 characters for DOC file. The result of training is a longest sequence that will be compared with the hex number of initial file in the testing phase to get the percentage of similarity.

IOP Conf. Series: Journal of Physics: Conf. Series **801** (2017) 012054 doi:10.1088/1742-6596/801/1/012054

```
50 44 46 2d  31 2e 25 0d  0a 30 6f 62  6a 3c 3c 2f   PDF-1.%..0obj<</
65 30 30 ..  .. .. .. ..  .. .. .. ..  .. .. .. ..   e00.............
.. .. .. ..  .. .. .. ..  .. .. .. ..  .. .. .. ..   ................
```

Figure 4. Result LCS training for PDF file

```
5c 72 74 66  31 5c 61 6e  73 69 5c 64  65 66 6c 61   \rtf1\ansi\defla
6e 67 31 30  33 33 5c 64  65 66 66 30  7b 5c 66 6f   ng1033\deff0{\fo
6e 74 74 62  6c 7b 5c 66  30 5c 66 5c  66 63 68 61   nttbl{\f0\f\fcha
72 73 65 74  30 5c 66 70  72 71 32 7b  5c 5c 65 30   rset0\fprq2{\\e0
65 65 7d 7b  5c 5c 5c 63  61 72 73 65  74 5c 7b 5c   ee}{\\\carset\{\
5c 6e 20 5c  72 73 5c 72  61 5c 5c 5c  61 73 5c 70   \n \rs\ra\\\as\p
72 20 5c 5c  5c 73 5c 72  61 5c 5c 5c  5c 5c 5c 73   r \\\s\ra\\\\\\s
5c 20 5c 5c  5c 5c 5c ..  .. .. .. ..  .. .. .. ..   \ \\\\\.........
```

Figure 5. Result LCS training for RTF file

```
11 81 7d 1a  3e 03 09 06  10 01 5c 09  04 .. .. ..   .± .>.....\.....
.. .. .. ..  .. .. .. ..  .. .. .. ..  .. .. .. ..   ................
.. .. .. ..  .. .. .. ..  .. .. .. ..  .. .. .. ..   ................
```

Figure 6. Result LCS training for DOC file

3.2. *Testing Result*

This phase aims to calculate the similarity between the result of LCS in training phase and file fragment. Both files will be compared by the sequence of file and trailer file using Longest Common Subsequences method. The result of testing phase is shown as table 7.

Table 7. LCS Comparison of initial file and file fragment.

File Type	Percentage of Fragment	Number of Files	Similarity Percentage	
			Lowest	Highest
PDF	1-10%	20	23%	65%
	11-30%	20	23%	23%
	31-50%	20	61.5%	61.5%
	>50%	20	65%	65%
RTF	1-10%	20	11%	98%
	11-30%	20	15%	15%
	31-50%	20	15%	16%
	>50%	20	16%	16%
DOC	1-10%	20	84%	84%
	11-30%	20	69%	84%
	31-50%	20	15%	15%
	>50%	20	84%	84%

Table 8. LCS Comparison of initial file trailer and file fragment trailer.

File Type	Percentage of Fragment	Number of Files	Similarity Percentage	
			Lowest	Highest
PDF	1-10%	20	57%	100%
	11-30%	20	71%	100%
	31-50%	20	57%	100%
	>50%	20	57%	100%
RTF	1-10%	20	85%	100%
	11-30%	20	85%	100%
	31-50%	20	85%	100%
	>50%	20	0%	100%
DOC	1-10%	20	85%	90%
	11-30%	20	85%	90%
	31-50%	20	85%	90%
	>50%	20	85%	90%

The testing also aims to determine whether the file type can be detected only by testing the similarity of the training result of LCS and file fragment. The low similarity percentage shows that the most sequence is in the header file. Therefore, the file fragment will be difficult to detect if doing only one-time of testing LCS.

While testing the trailer of file fragment, the result shows that if all of file fragment is not overwritten, then the hex number in the trailer of file fragment is also not overwritten. The similarity of signature trailer file and file fragment trailer will also be identified and the result is shown as table 8. The highest percentage of similarity can be seen in the trailer files. Therefore, the first pointer in identifying the file type of file fragment is by identifying the similarity of trailer from the file fragment.

In this phase, we also compare the hex number of file fragment with the training result of LCS. If the similarity percentage of trailer file is equal or more than 70% (average percentage from lowest percentage in table 8), then the system will compare the hex number and calculate the percentage of similarity. If the similarity percentage of trailer file is less than 70%, then the system will recognize the file as unidentified file (not include of .pdf, .rtf or .doc file). After that, the average of similarity percentage will be generated from the similarity percentage of LCS trailer file and the similarity percentage of LCS hex number. If the average of similarity percentage is equal or more than 50% (average percentage from lowest percentage in table 7 and table 8), then the system can identify the file type of the file.

Before testing the file fragment, the system will test on the intact file to identify the file type and the result is shown in table 9. From the testing result of file type identification for the intact file, we can get 97.3% average of accuracy, so it will be continued for file type identification of file fragment. Table 10 shows the testing result of file type identification of file fragment.

Table 9. Testing result of file type identification of intact file.

File Type	Number of Files	Accuracy
PDF	25	96%
RTF	25	100%
DOC	25	96%

Table 10. Testing result of file type identification of file fragment dataset.

File Type	Fragmented	Number of Files	Accuracy
PDF	1-10%	20	75%
	11-30%	20	70%
	31-50%	20	95%
	>50%	20	95%
RTF	1-10%	20	100%
	11-30%	20	100%
	31-50%	20	100%
	>50%	20	100%
DOC	1-10%	20	95%
	11-30%	20	100%
	31-50%	20	85%
	>50%	20	100%
Avg. Percentage of Accuracy			**92.91%**

It was estimated that the bigger the size of overwritten file, the lower the generated accuracy. However, from table 10, we can conclude that the proposed method can generate higher accuracy, especially for .doc and .rtf file. It is caused by the various used file for overwriting the file fragment, so that the file overwriting the file fragment with high percentage of fragmentation has similar file structure, for example, .doc has similar file structure with .ppt or .xlc file.

The identification result shows that the system give lower accuracy while identifying the PDF file type than the two other types. It is not able to identify 5 files of 0-10% fragmented files, 6 files of 11-30% fragmented files, 1 file of 31-50% fragmented files and 1 file of >50% fragmented files. Figure 7, figure 8 and figure 9 shows respectively the result of file type identification for PDF, RTF and DOC file fragment. From the testing result, LCS can identify the file type of PDF, RTF and DOC file fragment with 92.91% of average accuracy from 240 file fragments.

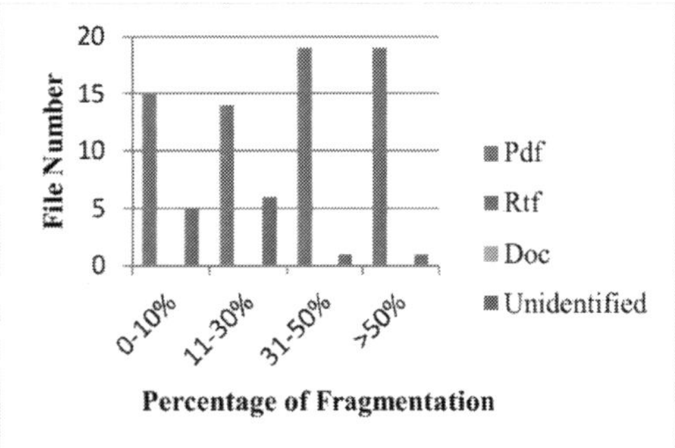

Figure 7. Graphic of identification result for PDF file fragment

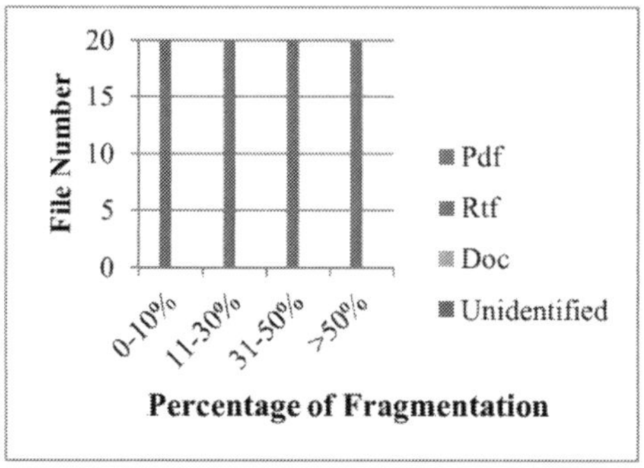

Figure 8. Graphic of identification result for RTF file fragment

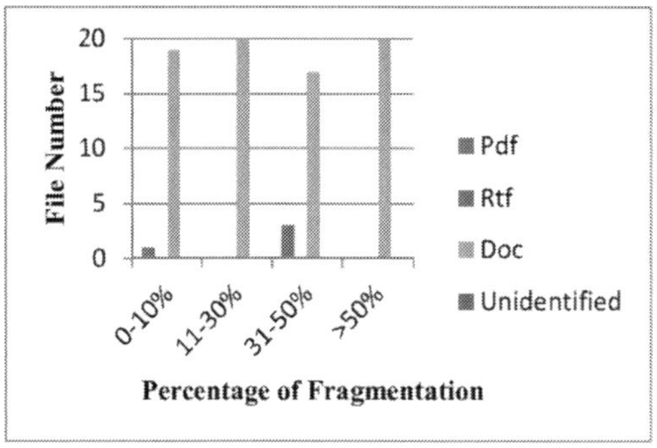

Figure 9. Graphic of identification result for DOC file fragment

In the testing phase, we also try to identify another file type besides pdf, rtf and doc file using the proposed method. In this research, we use 25 HTML files with specification described in table 11. It will be compared with LCS PDF, RTF and DOC file by its hex number. The result of HTML file identification is shown as table 12. The files are identified as PDF file (22 files) and also unidentified (3 files). It is caused that the proposed method cannot identify the file type of file fragment that has not been trained yet. Therefore, we also try to train and test the HTML file type and the result of identification is shown in table 13.

Table 11. HTML File Specification

Number of Files	25
Min. File Size (byte)	44,214
Max. File Size (byte)	53,481

Table 12. HTML File Identification

Number of HTML File	**Identified As**			
	PDF	**RTF**	**DOC**	**Unidentified**
25 file	22 file	0 file	0 file	3 file

Table 13. Identification result of HTML file after training

Number of HTML File	**Accuracy**	**Highest Percentage of Similarity**	**Lowest Percentage of Similarity**
25 file	100%	97,7%	97,7%

The identification result shows that the proposed method can identify the trained file type. It can identify 25 HTML files with 97.7% of accuracy.

3.3. Validation
In validation phase, header file will be added to the file fragment with the lowest percentage of fragmentation. If the file is accessible, then the validation is successful. However, sometimes the result will be not accessible or detected as a corrupt file because too many deleted hex numbers. If the validation is successful, then it will continue for the higher percentage of fragmentation. The result of validation in this research is shown in table 14.

Table 14. Validation result

File Type	Percantage of Fragmentation	Number of File	Corrupt File	Accessible File
PDF	0-10%	20	20	0
RTF	0-10%	20	20	0
DOC	0-10%	20	20	0

In table 14, it is shown that all fragment files is still corrupt despite the addition of the header. It is caused by the lack of effectiveness of the proposed method or the overwritten main component of file.

4. Conclusion
From all result of phases for identifying file type using Longest Common Subsequences, there are several generated conclusions. First, the proposed method can generate the accuracy up to 83.75% for PDF file fragment, 100% for RTF file fragment and 95% for DOC file fragment. The generated average accuracy for file type identification is 92.91%, so we can conclude that the proposed method, Longest Common Subsequences, can identify the type of file, such as pdf, rtf and doc files. It also can identify the file fragment with partly overwritten hex number, because the method still needs the hex number of trailer file for identification. The success rate depends on the data type overwriting the file that will be identified.

However, while testing another file type, the method should train it first. In addition, to get the appropriate output, every sequence from training result should be tested first, so it requires longer time of processing.

For further research, Byte Frequency Distribution (BFD) can be used for identification method for identifying more types of file. Another method can be used for detecting and distinguishing between intact file and file fragment.

5. References
[1] Roussev V and Garfinkel S L 2009 File fragment classification-the case for specialized approaches *Fourth International IEEE Workshop on Systematic Approaches to Digital Forensic Engineering (SADFE'09)* 3-14 (IEEE)

[2] Calhoun W C and Coles D 2008 Predicting the Types of File Fragments *Digital Investigation* **5** S14-S20

[3] Sencar H T and Memon N 2009 Identification and recovery of JPEG files with missing fragments *Digital Investigation* **6** S88-S98

[4] Axelsson S 2010 Using normalized compression distance for classifying file fragments *International Conference on Availability, Reliability, and Security (ARES'10)* 641-646 (IEEE)

[5] Fitzgerald S, Mathews G, Morris C and Zhulyn O 2012 Using NLP techniques for file fragment classification *Digital Investigation* **9** S44-S49

IOP Conf. Series: Journal of Physics: Conf. Series **801** (2017) 012055 doi:10.1088/1742-6596/801/1/012055

Cloud Based Drive Forensic and DDoS Analysis on Seafile as Case Study

R B Bahaweres[1], N B Santoso[2], A S Ningsih[3],

[1,2,3]Electrical Engineering Dept., Universitas Mercu Buana, Jakarta, Indonesia
[1]Informatics Dept., Faculty of Science & Tech., UIN Jakarta, Indonesia

Email: [1]rizalbroer@ieee.org, [2]benisantoso72@gmail.com, [3]sn.anggun@yahoo.com

Abstract. The rapid development of Internet due to increasing data rates through both broadband cable networks and 4G wireless mobile, make everyone easily connected to the internet. Storages as Services (StaaS) is more popular and many users want to store their data in one place so that whenever they need they can easily access anywhere, any place and anytime in the cloud. The use of the service makes it vulnerable to use by someone to commit a crime or can do Denial of Service (DoS) on cloud storage services. The criminals can use the cloud storage services to store, upload and download illegal file or document to the cloud storage. In this study, we try to implement a private cloud storage using Seafile on Raspberry Pi and perform simulations in Local Area Network and Wi-Fi environment to analyze forensically to discover or open a criminal act can be traced and proved forensically. Also, we can identify, collect and analyze the artifact of server and client, such as a registry of the desktop client, the file system, the log of seafile, the cache of the browser, and database forensic.

1. Introduction

Cloud computing is an emerging storage service in IT because with cloud computing we can increase storage capacity without the cost of buying new infrastructure, training new personnel, or licensing new software[1]. Service providers usually provide free services up to a certain capacity before required to pay[2].Although with all the convenience provided, there are still many companies are hesitant and cautious storing their data to public cloud storage services[3].

Based on Gartner [4] there are seven security issues for cloud computing. Due to the above considerations, we chose to use personal cloud storage seafile. Seafile is hosting file system software. Files are stored on a central server and can be synchronized with a personal computer and a mobile device through a client seafile. Files can also be accessed with server web interface.

Seafile function is similar to other popular services like ownCloud, Dropbox and Google Drive, but seafile is a free and open source, so users can host their own servers without limits imposed on the storage space or client connections, it only depends on the size of the storages[5]. In order for these technological advances, we chose the Raspberry Pi[6], [7] as a model of a private cloud storage infrastructure based on Seafile.

There are 5 (five) component of Seafile [16]. They consist of Seahub (django) as the web frontend, seafile server (seaf-server) as data service daemon, handles raw file upload, download and synchronization, ccnet server (ccnet-server) as RPC service daemon to enable communication among the other components, FileServer: handles raw file upload/download functions for Seahub, and Controller: monitors ccnet and Seafile daemons.The architecture of Seafile application can be seen as the following figure 1.

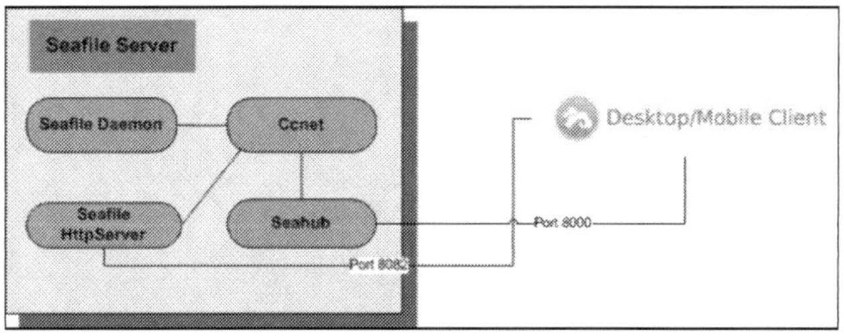

Figure 1. The Architecture of Seafile Application.

The framework of computer forensics used in this paper consists of identification and preservation, collection, examination and analysis, reporting and presentation[1], [8–10]. The diagram of framework forensics can be seen in figure 2.

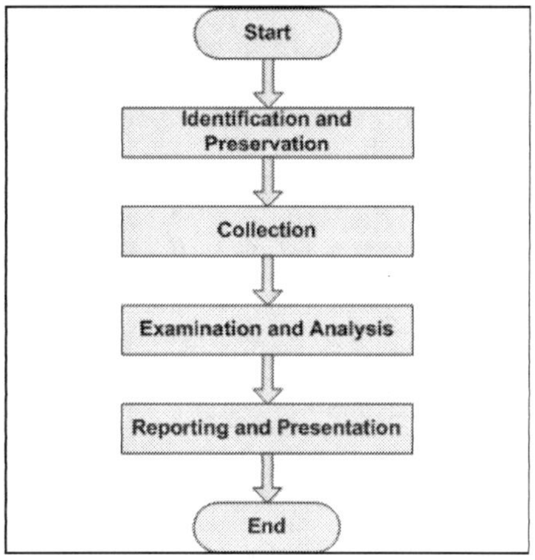

Figure 2. The Framework of Computer Forensics.

The digital artifact reference as the term is widely used in computer forensics. Content types of digital artifact include text, audio, video, image, animation or a combination. The digital artifact result from hardware malfunction, software malfunction, compression, aliasing, rolling shutter and error diffusion.[17]

Denial of Service is one type of attack on Server or Computer that makes the target crash or hangs so that the resources of the server cannot be accessed by the end user. The Distributed Denial of Services (DDoS) is a coordinated attack on the availability of services of a given target system that is launched indirectly through many compromised computing systems.

The goal of the DoS/DDoS attack : bandwidth depletion, cut off the connection between servers, prevent the victim using the services of system or resources depletion and devastate the system[11], [12]. From the security approach, the internal user can do DoS attack from internal local area network or Wi-Fi Environment. The schema of DDoS as following figure 3.

IOP Conf. Series: Journal of Physics: Conf. Series **801** (2017) 012055 doi:10.1088/1742-6596/801/1/012055

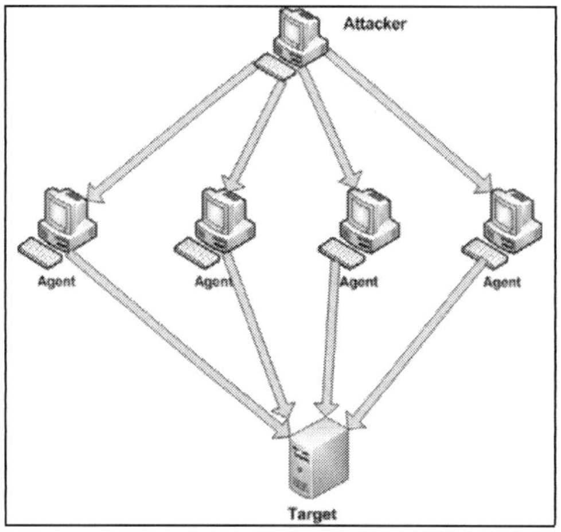

Figure 3. The Schema of DDoS Attack.

2. Related Works

Ben Martini, et. al focus on upon client and server artifacts on ownCloud cloud storage[1]. Darren Quick, et. al, discuss data remnant on end user devices using Dropbox in Windows 7 and Apple iPhone[9] and Google Drive Analysis[8]. Chen Long and Zhang Qing proposes a method to investigate and analyze the artifacts for reconstructing the event of user's activities on 360 Cloud service and Baidu Cloud service[13]. Kayvan Atefi, et. al, try to find a number of artifacts and make some measurements of CPU load on Seafile cloud storage[5]. Rizal BroerBahaweres, et. al, focus on denial of service on ownCloud personal storage[12]. Our research focuses on client and server artifact and DoS analysis on Seafile cloud storage.

Table 1. Matrix Related Research.

No	Journal Title	Author	Server Artifact	Client Artifact	Cloud Storage	DoS
1	Forensic Analysis to China's Cloud Storage Services	Chen Long, et.al		√	√	
2	Cloud storage forensics: ownCloud as a case study	Ben Martini, et.al	√	√	√	
3	Dropbox analysis: Data remnants on user machines	Darren Quick, et.al		√	√	
4	Google Drive: Forensic Analysis of Cloud Storage Data Remnants	Darren Quick, et.al		√	√	
5	Building a Private Cloud Computing and the analysis against DoS (Denial of Service) attacks: a Case study at SMKN 6 Jakarta	R.B. Bahaweres, et.al		√	√	√
6	Cloud Based Drive Forensic & DOS Analysis on Seafile as a Case Study	N. B. Santoso, et.al	√	√	√	√

3. Methodology

This study tries to install and implement a personal cloud-based storage based on Seafile. Installation and configuration Seafile on Raspberry Pi[14]. After the system is built, we install Seafile client on a computer running windows 7 and Android mobile phone.[15] The framework of research methodology as seen the following picture:

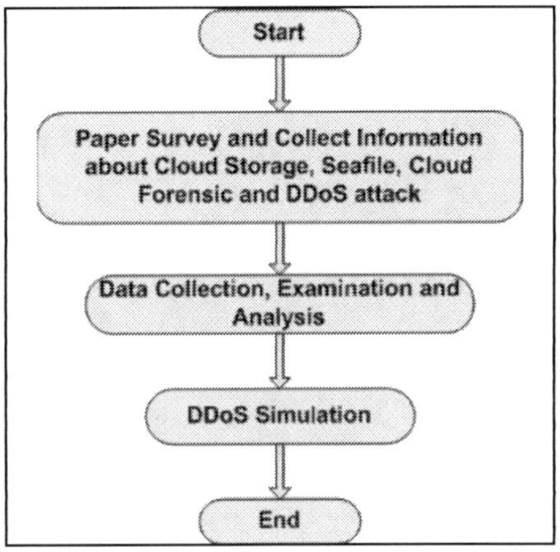

Figure 4. The Framework of Research Methodology.

The identification, preservation, and collection were performed as follow:

- The artifact in a browser at windows client.
- The artifact in seafile windows client apps and seafile android client's apps.
- Collectlog on Seafile Personal Cloud Storage such as access log, Syslog, etc.
- Collect database and registry artifact at Seafile and Windows 7 desktop client.
- Throughput with DDoS and without DDoS simulation on seafile cloud storage with 5PC sending packets simultaneously.

The cloud storage specification:

- Raspberry Pi 2, 900MHz quad-core ARM Cortex-A7 CPU
- Memory 1GB
- External SanDisk 64GB
- Network 10/100 Mbps
- Operating System : Raspbian

The cloud storage system consists of 1 (one) Seafile app and 1 (one) external disk attach to Raspberry Pi. And the simulation of DoS and client artifact analysis, we use 2 (two) notebook, 2 (two) server and 2 Android Smartphone. The architecture of Seafile cloud storage can be seen at following figure 5:

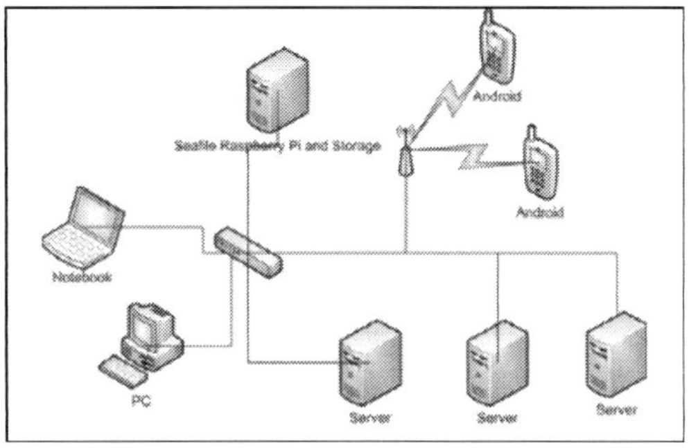

Figure 5. The Architecture of Private Seafile Cloud Storage.

4. Result and analysis

We try to present the overview of Seafile filesystem and configuration. To know depth forensic of Seafile, we must know all about Seafile architecture, from the filesystem, log file, database and configuration file. Here the overview about file system of seafile application [16]:

a. File system of Seafile server: /home/pi, seafile-server-6.0.3, seafile-data (seafile configuration and data), seahub-data (seahub data), ccnet (ccnet configuration and data), seahub.db (sqlite3 database used by seahub), seahub_settings.py (optional config file for seahub).
b. Configuration file of seafile server: ccnet.conf, seafile.conf, seafdav.conf, seahub_settings.py, and seafevents.conf (professional edition only).
c. Log files of seafile server: seafile.log (logs of Seafile server), controller.log (logs of Controller), seahub_django_request.log (logs of Seahub), seahub.log (logs from Django framework and emails sending), and ccnet log: logs/ccnet.log (logs for internal RPC, not useful).
d. Firewall setting:
   By default, open 2 ports, 8000 and 8082, in the firewall settings.
   If run Seafile behind Nginx/Apache with HTTPS, only need to open ports 443.
e. Security of seafile server:
   Seafile provides a feature called the encrypted library to protect the privacy. The file encryption/decryption is performed on client-side when using the desktop client for file synchronization. The password of an encrypted library is not stored on the server. Even the system admin of the server can't view the file contents, however they can view the metadata which is currently not encrypted. The metadata includes the complete list of directory and file names, every files size, the history of editors, when, and what byte ranges were altered.
   After we know filesystem and architecture of Seafile, we can separate forensic into some of forensic methods, as follow:
f. File system and Registry Forensic
   In Seafile desktop client, we can find the file system and registry artifact in a certain folder. We can see the following figure.

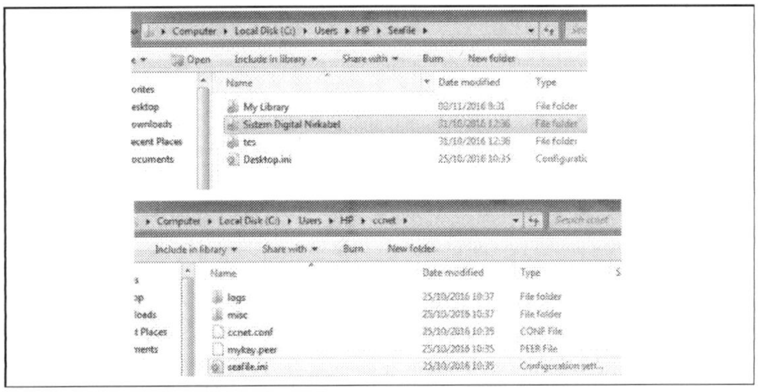

Figure 6. The Artifact of Seafile Desktop Client.

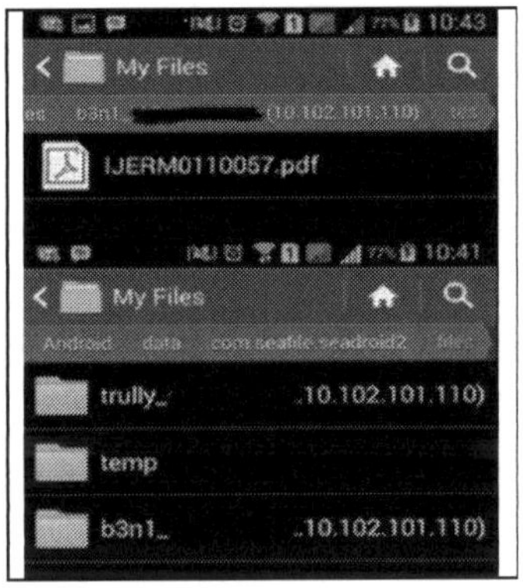

Figure 7. The Artifact of Seafile Android Client.

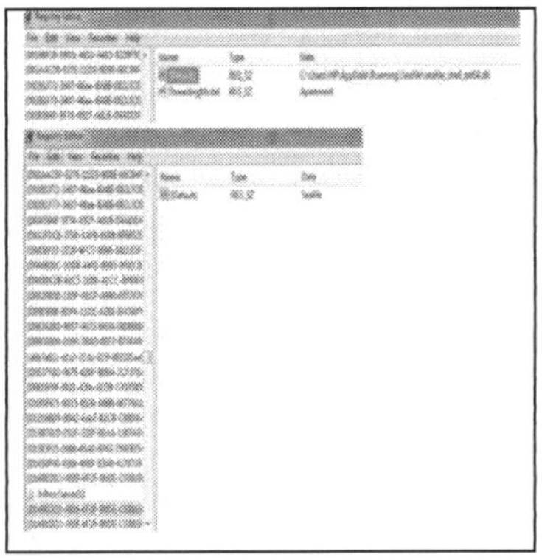

Figure 8. Registry Artifact of Seafile Desktop Client.

g. Database Forensic
With DB Browser for SQLite, we can view and display artifact of user data, like a platform that used by the user, device name, last accessed, the last login, created by, etc

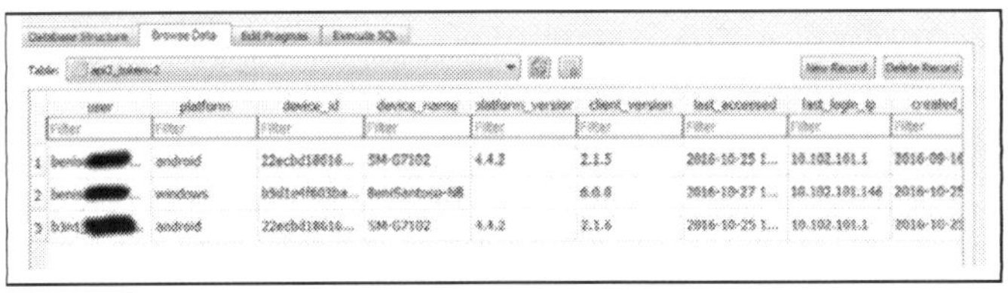

Figure 9. DB Browser for SQLite.

h. Browser Forensics
With Chrome Forensic tools, we can capture and analyze data that store the following SQLite Files in google chrome browser:

- History

- Cookies
- Web Logins
- Archived History (Web History and search terms)
- Bookmarks
- File locations : C:\Users\USERNAME\AppData\Local\Google\Chrome\User Data\Default

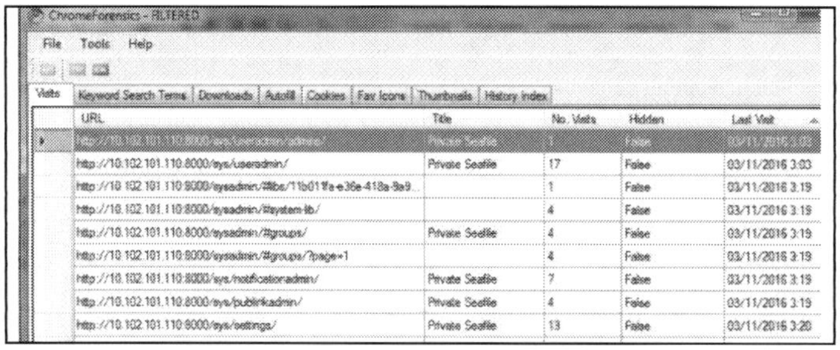

Figure 10. Chrome Forensics Tools.

i. Throughput calculation with or without DDoS attack Simulation
The researchers use Wireshark tool to see throughput when uploading files to Seafile Cloud Storage. The researchers used three scenarios to analyze throughput. The uploaded size file has 60.9MB, we make 3 scenarios of DDoS simulation, the first scenario, the file is uploaded without DDoS attack and throughput is captured with Wireshark tool. The second scenario, the file is uploaded simultaneously with DDoS attack by 5 computers with 30000 bytes, the third scenario, the file is uploaded simultaneously with DDoS attackby 5 computers with 50000 bytes.The result of throughput rate as seen by the following table 2.

Tabel 2. Throughput Calculation without DDoS and with DDoS Simulation.

No	Trial	Throughput without DDoS (MBit/s)	Throughput DDoS 30000 bytes (Mbit/s)	Throughput DDoS 50000 bytes(Mbit/s)
1	N1	9.784	7.472	4.615
2	N2	5.486	4.344	2.634
3	N3	10.842	8.425	8.084
4	N4	6.355	8.249	9.166
5	N5	8.455	9.024	9.188
6	N6	8.487	6.602	6.225
7	N7	11.202	8.618	8.013
8	N8	7.486	6.137	5.967
9	N9	11.174	6.014	6.027
10	N10	8.972	7.151	7.229
Average		8.8243	7.2036	6.7148

With DoS attacks influence the reduction in throughput rate, though the cloud services still able to running well, but this should be a concern so it does not become an obstacle later.The throughput rate has reduced to 81.63% with DDoS (30000 bytes), and 76.09% with DDoS (50000 bytes).

5. Conclusion and Future Works

This research demonstrated that private cloud Seafile provides a significant number of useful artifacts for forensic investigators. And provides a discussion on private cloud forensics from both client and server artifact, database forensics, and simulation of DDoS attack against private cloud services itself. The Seafile has a default security feature such as the encrypted library.

For future works, we need to protecting and securing the Seafile cloud storage from DDoS with making clustering Seafile applications, both in the application, database, Memcached enabled and storage clustering. And protect the cloud storage with firewall and bandwidth management.

Acknowledgment

Many thanks to all colleagues and family to support us to finish this paper completely.

References

[1] Martini B and Choo K K R 2013 Cloud storage forensics: ownCloud as a case study *Digital Investigation* vol **10** no 4 pp 287–299

[2] Drago I, Bocchi E, Mellia M, Slatman H and Pras A 2013 Benchmarking personal cloud storage *in Proceedings of the 2013 conference on Internet measurement conference* pp 205–212

[3] So K 2011 Cloud computing security issues and challenges *International Journal of Computer Networks* vol **3** no 5

[4] Singh V 2014 Well-known Gartner's Seven Security Issues Which Cloud Clients Should Advert

[5] Atefi K, Yahya S and Atefi A 2014 A survey on digital forensics investigation of Seafile as a cloud storage *International Journal of Engineering Research And Management (IJERM)* vol **1**

[6] Maksimovi M, Vujovi V, Davidovi N, Milosevi V and Perisi B 2014 Raspberry Pi as Internet of things hardware: performances and constraints *Design issues* vol **3** p 8

[7] Severance C 2013 Eben upton: Raspberry pi Computer vol **46** no 10 pp 14–16

[8] Quick D and Choo K K R 2014 Google drive: forensic analysis of data remnants *Journal of Network and Computer Applications* vol **40** pp 179–193

[9] Quick D and Choo K K R 2013 Dropbox analysis: Data remnants on user machines *Digital Investigation* vol **10** no 1 pp 3–18

[10] Roussev V, Barreto A and Ahmed I 2016 Forensic acquisition of cloud drives *arXiv preprint arXiv:1603.06542*

[11] Chauhan K and Prasad V 2015 Distributed Denial of Service (DDoS) Attack Techniques and Prevention on Cloud Environment *International Journal of Innovations & Advancement in Computer Science* pp 210–215

[12] Bahaweres R B, Syarif J and Alaydrus M 2016 Building a Private Cloud Computing and the analysis against DoS (Denial of Service) attacks: Case study at SMKN 6 Jakarta The 4th International Conference on Information Technology for Cyber and IT Service Management

[13] Long C and Qing Z 2015 Forensic Analysis to China's Cloud Storage Services *International Journal of Machine Learning and Computing* vol **5** no 6 p 467

[14] Chung H, Park J, Lee S and Kang C 2012 Digital forensic investigation of cloud storage services *Digital investigation* vol **9** no 2 pp 81–95

[15] Gracia-Tinedo R, Artigas M S, Moreno-Martinez A, Cotes C and Lopez P G 2013 Actively measuring personal cloud storage *Proceedings of the Cloud* pp 301–308

[16] https://manual.seafile.com

[17] https://en.wikipedia.org/wiki/Digital_artifact

IOP Conf. Series: Journal of Physics: Conf. Series **801** (2017) 012056 doi:10.1088/1742-6596/801/1/012056

Comparison of methods for localizing the source position of deauthentication attacks on WAP 802.11n using Chanalyzer and Wi-Spy 2.4x

R B Bahaweres[1], S Mokoginta [2], M Alaydrus[3],

[1,2,3]Electrical Engineering Dept., Universitas Mercu Buana, Jakarta, Indonesia
[1]Informatics Dept., Faculty of Science & Tech., UIN Jakarta, Indonesia

Email: rizalbroer@ieee.org, stenly_m@live.com, mudrikalaydrus@mercubuana.ac.id

Abstract. This paper describes a comparison of three methods used to locate the position of the source of deauthentication attacks on Wi-Fi using Chanalyzer, and Wi-Spy 2.4x adapter. The three methods are wardriving, absorption and trilateration. The position of constant deauthentication attacks is more easily analyzed compared to that of random attacks. Signal propagation may provide a comparison between signal strength and distance which makes the position of attackers more easily located. The results are shown on the chart patterns generated from the Received Signal Strength Indicator (RSS). And it is proven that these three methods can be used to localize the position of attackers, and can be recommended for use in the environment of organizations using Wi-Fi.

1. Introduction

Models of communication as a means of sending and receiving information have become very popular in the today's world. One of the popular models of communication is a wireless mode[1]. IEEE 802.11 Standard is a standard for Wi-Fi[2]. Nowadays, almost all computer devices, including personal computers, tablets, or mobile phones, are already equipped with a Wi-Fi device. That at any time it can use a Wi-Fi network as an alternative mode of communication to transmit and receive information.

Some types of interference that could affect Wi-Fi are among others in the forms of signal interference or attacks from attackers. The easiest way to attack Wi-Fi networks is by using the first method, namely wardriving which is aimed to find and mark the location and the status of a Wi-Fi network. There are three ways to find a Wi-Fi network: active scanning, monitor-mode sniffing, and traffic analysis to look for the AP (Access Point)[4].

The current security standard for Wi-Fi networks keeps developing starting from the security which uses WEP up to WPA2 personal and enterprise for encryption on the Wi-Fi AP to restrict the use of special authorized AP[5]. However, the security protocol of the WPA is not a guarantee that our Wi-Fi network is safe from attacks, as currently there are numerous free tools and also self-made tools that can be used for password recovery[4], [6]. One of the many reasons used by attackers for doing attacks on Wi-Fi networks is just to have fun[4]. Even, they could do it for something more harmful such as to steal data and information from devices connected via a Wi-Fi AP. One of the most well-known attacks is the man-in-the-middle attack. This attack uses the Denial-of-Service method, and creates a fake access point to trick the station so that the station moves toward the connection of the fake access point created by the hacker. Deauthentication attacks can be divided into 2 targets:

a) Attacks on AP
b) Attacks on stations

In this paper, the author discuss a form of interference that is deauthentication attacks, and tracks the position of the source of deauthentication attacks using a spectrum analyzer, namely Chanalyzer. And Wi-Spy 2.4x adapter against constant deauthentication attacks on a Wireless Access Point (WAP).

IOP Conf. Series: Journal of Physics: Conf. Series **801** (2017) 012056 doi:10.1088/1742-6596/801/1/012056

2. Related Works

There are some previous research papers or articles on deauthentication attack techniques and localization methods used as references in this paper.

The characteristics of omni-directional antenna, having a 3600 pattern of signal distribution, and the range depends on the environmental conditions that can cause loss of Wi-Fi signals especially in an indoor environment. Such loss of Wi-Fi signals could be due to an existing material such as furniture, walls and even humans, with the level of losses ranging from low to very high, from wooden doors to metals[8].

Deauthentication attacks are undetected because they are performed by taking advantage of the process of authentication and deauthentication on Wi-Fi networks, where the control frame is a 'Clear to Send' frame on the frame management, which means that it is unencrypted. This creates a loophole and is taken advantage of by the attacker to cause interference on Wi-Fi networks. These attacks also cause traffic drops[7], [9].

To detect deauthentication attacks, many methods can be used and one of them is by using the Packet Delivery Ratio (PDR) technique. With this technique, the patterns of the signal that are generated can be distinguished, in that when the signal gets closer to the attacker, the shape of the gradient gets more circular, and vice versa[10], [11].

Another method to determine the location that also uses the Wi-Fi network is the trilateration method, with calculations based on the log-normal path loss model to estimate the distance between the sending device and receiving device [12], [13].

3. Localization method

There are 3 methods performed in this paper to determine the location of attackers, all of which use Chanalyzer ver. 5.8.5.23 and Wi-Spy 2.4x as tools to find the location of the attacker. Since deauthentication attacks are performed in a constant manner, Chanalyzer can easily analyze the signal patterns obtained. The three methods are as follows.

A. Wardrive Method

This method is performed by comparing the signal patterns captured by Chanalyzer when moving closer to the Wireless Access Point, to the Station and to the Attacker.

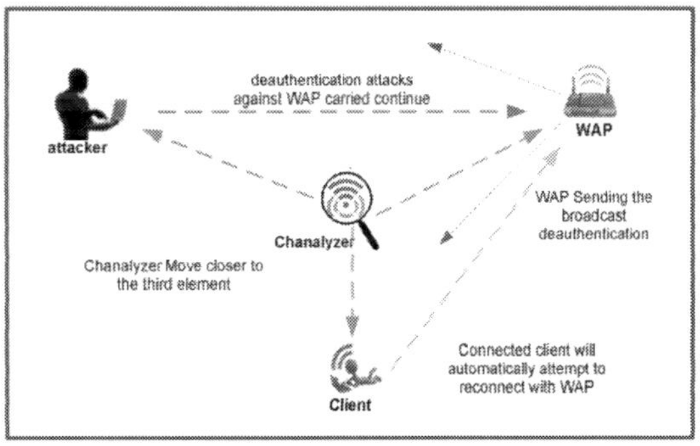

Figure 1. WarDrive Method

Based on the information obtained from Chanalyzer, the signal patterns that generated can be determining the position of the source of attack.

B. Absorption method

This method is performed by putting a barrier between the Chanalyzer and the WAP, the Station and the Attacker when the attack is performed, in order to get the information that the shape of the amplitude signal pattern change received is from the attacker.

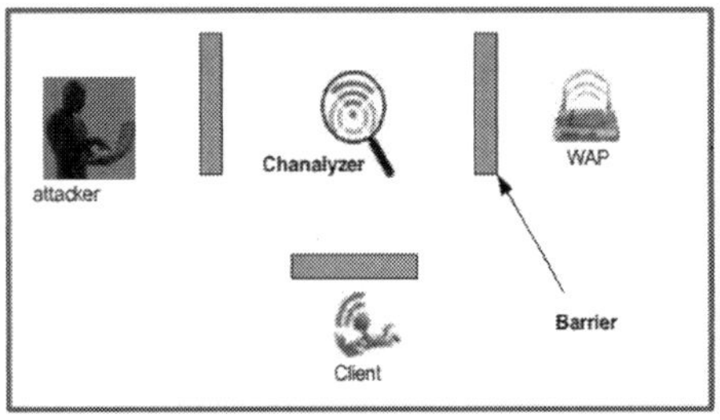

Figure 2. Absorption method

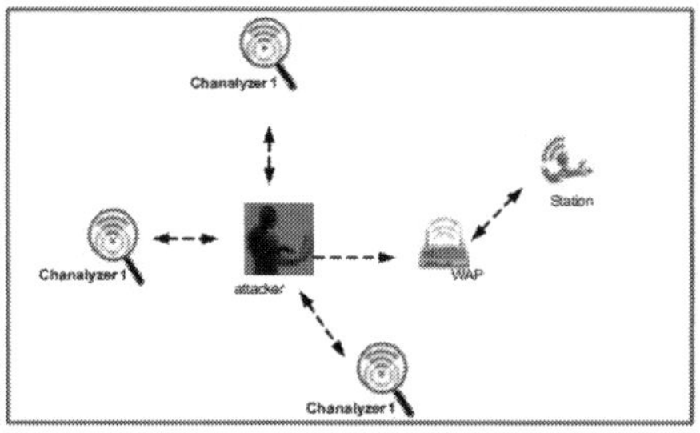

Figure 3. Trilateration Method

$$R(dB) = 10 \log (s2/s1) \quad (1)$$

This method actually takes advantage of the weakness of radio waves, in that there may be a reduction in the power received at the receiving antenna, which happens when the radio waves pass through a barrier. The power received after the waves pass through the barrier will decrease because the energy has been absorbed by the barrier. The amount of attenuation value depends on the type of material used and the distance between the barrier and the receiving antenna. Attenuation values vary depending on the material of the barrier.

C. Trilateration Method

The trilateration method is one of the localization methods, where trilateration refers to a process of determining the absolute location or a relative point using a circular, spherical or triangular geometry. This is done by placing Chanalyzer at 3 different positions to capture the attack signal from the attacker, and then making a comparison between the estimated distances and the RSSI values, in order to determine the position of the attacker

In performing the three methods mentioned above, in this study the author used the calibration technique in the comparison between the distance and RSSI, with a maximum distance of 15m. The purpose of this calibration is to make the determination of the position easier.

D. Experimental Environment

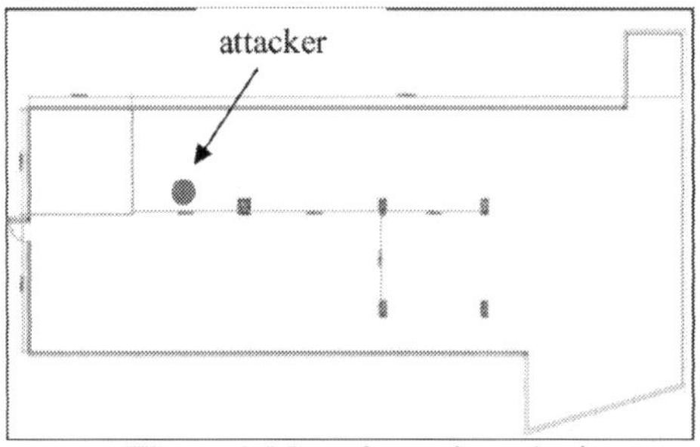

Figure 4. Map of experiment's site

E. Measurement tools

- Machine laptop Dell Latitude E6410 : kali linux 3.18.0-kali3-686-pae, wifi adapter Alfa network AWUS036NHA antena 5dBi and wireshark

- Machine laptop HP Pavilion dm4 : chanalyzer version 5.8.5.23 dan wi-spy 2.4x
- Machine Laptop IBM T430 as Station
- Wireless Access Point Cisco CVR100W
- Lumia 710 as WAP

F. Deauthentication attack

The deauthentication attack was performed as follows:

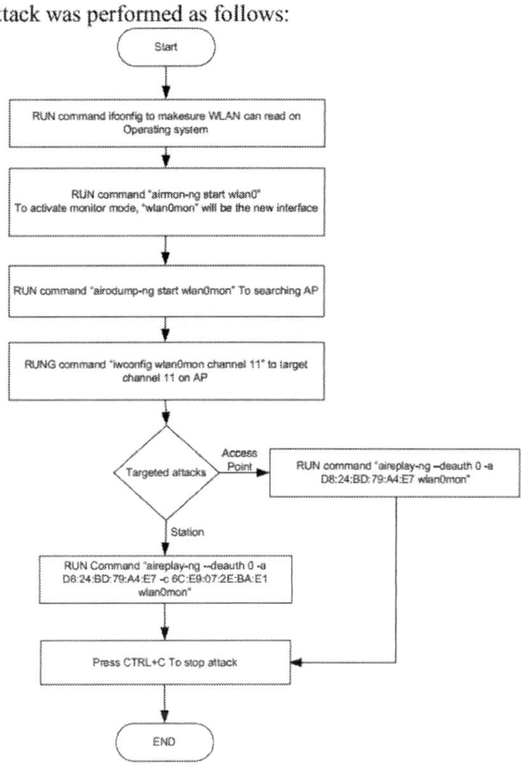

Figure 5. Deauthentication attacks' flow

4. Result

The three methods mentioned above basically have the same principles, in that they are performed by comparing the RSS values received with the distance. The information obtained by using a different antenna would of course be different. On the table v how the three methods are compared to one another. The following figures and table show the differences in the shapes of the patterns on Chanalyzer

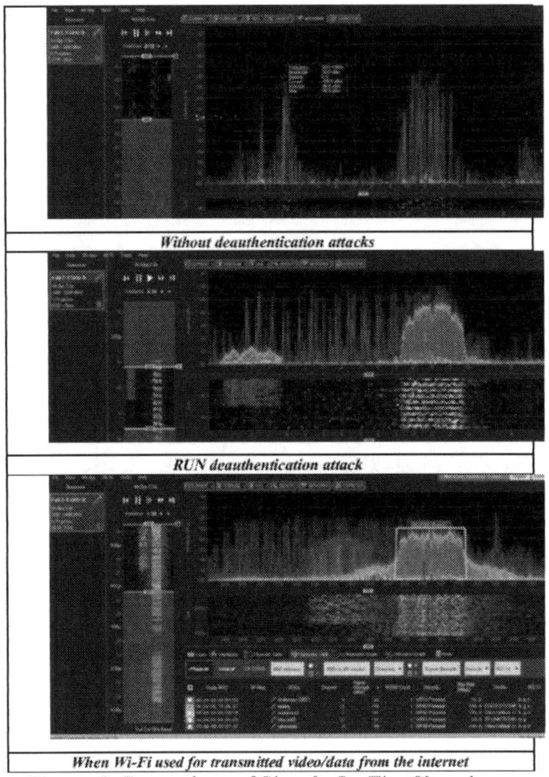

Figure 6. Comparison of Signals On The Chanalyzer

The differences in the amplitude signal patterns generated by Chanalyzer are apparent, i.e. the pattern shown when there is no deauthentication attack and that during the deauthentication attack. This attack is performed using channel 11

The calculation of the amount of power received by Chanalyzer using the path loss calculation formula is as follows:

$$P_{rx} = P_{tx}G_{rx}G_{tx}(\lambda / (4\pi d)) \qquad (2)$$

When testing was performed without using a formula, the results obtained are as follows. Testing was done indoors and the distances tested were from 0 to 15 meters. The value taken is the maximum value achieved by Chanalyzer. Following is the calibration data of each distance tested.

TABLE I. CALIBRATION RESULT

Distance (m)	Calibration (dBm)
0	-39.3333
1	-36.1667
2	-38.6667
3	-41.3333
4	-45
5	-42.6667
6	-47.3333
7	-49.3333
8	-46.8333

Distance (m)	Calibration (dBm)
9	-49.8333
10	-51.8333
11	56.8333
12	-54.6667
13	-52
14	-52.6667
15	-57.6667

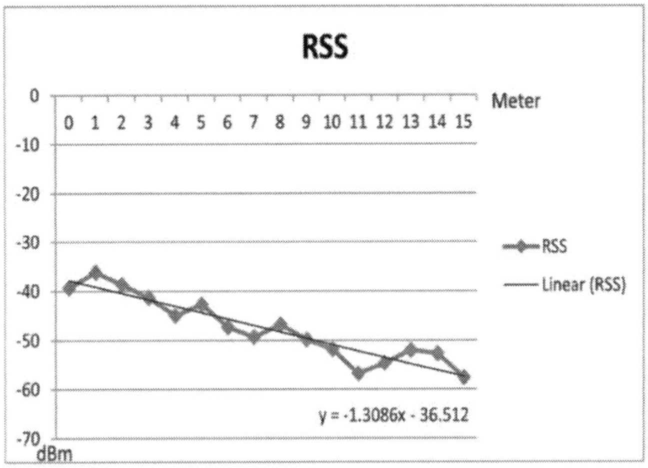

Figure 7 Graphics of RSS Calibration and linear gradient

From the graphics on figure 7, we can see a trend that the greater the distance between the attacker and the target value decreases. From the calibration results above all we need to do is testing the three methods as follows

A. Wardriving method

Using the wardriving method, on figure 8 shows the change in the signal generated. Chanalyzer moves toward the attacker, and then the data when Chanalyzer is far from and moves closer to the attacker are compared.

TABLE III. RESULT OF WARDRIVING EXPERIMENT

Distance (meter)	RSS(dBm)
7	-55
0	-36

IOP Conf. Series: Journal of Physics: Conf. Series **801** (2017) 012056 doi:10.1088/1742-6596/801/1/012056

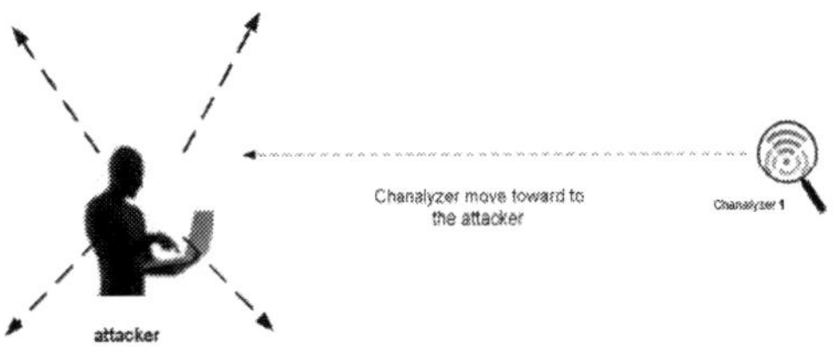

Figure 8 Method of wardriving experiments

In first position, chanalyzer is close to an access point. But being away from the attacker, there is a difference in the amplitude of the signal patterns received by chanalyzer. When chanalyzer are away, RSS value received by chanalyzer on the power of attacker was small. The test result are show on the table III. The closer chanalyzer to the attacker, the bigger the RSS value received by Chanalyzer.

B. Absorption method

By taking advantage of the weakness of Wi-Fi networks, we can determine the direction from which the source of attack comes by placing a barrier in front of the Wi-Spy 2.4x antenna facing the direction of the attacker, as in the following result.

The barrier must be placed near the position of the Chanalyzer's antenna, so that the differences in the RSS values received can be seen. In this testing, the barrier technique is done by placing a hand facing the direction of the attacker. When the barrier is placed on the other side of the antenna, there is no decrease in the RSS value. This indicates that the power received by Chanalyzer comes from the attacker. The difference in the attenuation values above is as follows:

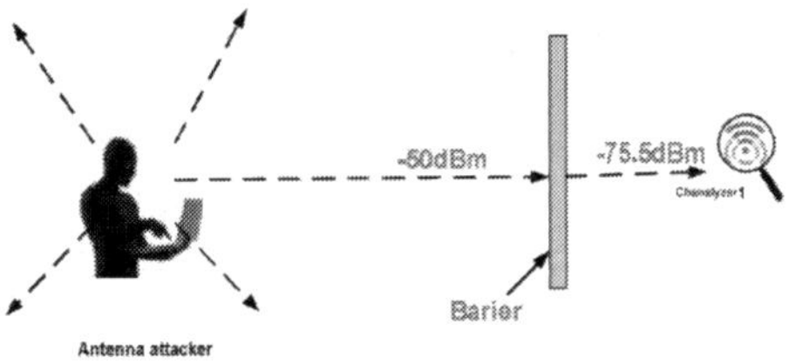

Figure 9 Absorption method

Thus, with this technique we can estimate the direction and the position of the source of the attack if compared to the calibration value, which is 7-9 meters from Chanalyzer.

C. Trilateration method

Using the trilateration method, which is performed by positioning Chanalyzer in three different locations, the following result is obtained:

TABLE IV. TRILATERATION RESULT

Position	dBm	distance (meter)
Position 1	-47	6 - 9
position 2	-48	6 -9
position 3	-53	13

Using the trilateration technique, it is easier to determine the location of the attacker.

5. Conclusion and Future Works

The three methods mentioned above basically have the same principles, in that they are performed by comparing the RSS values received with the distance. The information obtained by using a different antenna would be have a different result. Table IV shows that the three methods are compared to one another.

IOP Conf. Series: Journal of Physics: Conf. Series **801** (2017) 012056 doi:10.1088/1742-6596/801/1/012056

TABLE IV. CONCLUSION

No	Wardriving method	Absorption method	Trilateration method
1	Probability of success 1:8	Probability of success 1:8	Probability of success 1:1
2	more effort is needed because we need to walk all around the area	less effort than wardriving method, because we don't have to walk around the area to be able to determine the attacker location	3 site are needed to installation of satellite
3	need 1 chanalyzer + 1 wi-spy 2.4x adapter	need 1 chanalyzer + 1 wi-spy 2.4x adapter	need 3 chanalyzer + 3 wi-spy 2.4x adapter

The source of the attacker check in the busy area within a wifi, and the antenna from attacker use the direction antenna are our future work.

Acknowledgment
Many thanks to all colleagues and family to support us to finish this paper completely.

References

[1] A. Bawiskar and B. B. Meshram, "Survey of Attacks on Wireless Network," *Int. J. Innov. Res. Comput. Commun. Eng.*, vol. 1, no. 1, pp. 90–100, 2013.

[2] Ieee, *IEEE Standard for Information technology--Telecommunications and information exchange between systems Local and metropolitan area networks--Specific requirements Part 11: Wireless LAN Medium Access Control (MAC) and Physical Layer (PHY) Specifications*, vol. 2012, no. March. 2012.

[3] F. Project, "802.11 Jamming Using Consumer WiFi Cards," 2014.

[4] H. Kumar, D. Sarma, and a. Kar, "Security threats in wireless sensor networks," *IEEE Aerosp. Electron. Syst. Mag.*, vol. 23, no. June, pp. 21–23, 2008.

[5] A. H. Lashkari, M. I. R. Mohammad, S. Danesh, and B. Samadi, "A Survey on Wireless Security protocols (WEP , WPA and WPA2 / 802 . 11i)," *Comput. Sci. Inf. Technol. 2009. ICCSIT 2009. 2nd IEEE Int. Conf.*, no. Iv, pp. 48–52, 2009.

[6] L. Butti and J. Tinnès, "Discovering and exploiting 802.11 wireless driver vulnerabilities," *J. Comput. Virol.*, vol. 4, no. 1, pp. 25–37, 2008.

[7] R. Cheema, "Deauthentication / Disassociation Attack : Implementation and Security in Wireless Mesh Networks," vol. 23, no. 7, pp. 7–15, 2011.

[8] K. Ogunjemilua, J. N. Davies, R. Picking, and V. Grout, "An Investigation into Signal Strength of 802.11n WLAN," no. November, pp. 191–204, 2009.

[9] T. Farooq, D. Jones, and M. Merabti, "MAC Layer DoS Attacks in IEEE 802.11 Networks," ... *Telecommun. Netw. ...*, 2010.

[10] K. Pelechrinis, I. Koutsopoulos, I. Broustis, and S. V. Krishnamurthy, "Lightweight Jammer Localization in Wireless Networks: System Design and Implementation," *Glob. Telecommun. Conf. 2009. GLOBECOM 2009. IEEE*, 2009.

[11] N. Sufyan, N. Saqib, and M. Zia, "Detection of jamming attacks in 802.11b wireless networks," *EURASIP J. Wirel. Commun. Netw.*, vol. 2013, no. 1, p. 208, 2013.

[12] S. N. Makarov, "Antenna and EM Modeling with MATLAB -." pp. 1–284, 2002.

[13] W. Stallings, *Wireless Communications and Networks*. 2002.

IOP Conf. Series: Journal of Physics: Conf. Series **801** (2017) 012057 doi:10.1088/1742-6596/801/1/012057

Enhancing Honeypot Deception Capability Through Network Service Fingerprinting

R N Dahbul[1] , C Lim[2] , J Purnama[3]

Department of Information Technology, Swiss German University, Edutown BSD City, Tangerang, 15339 Indonesia.

E-mail: [1]`rasyid.dahbul [at] student.sgu.ac.id`, [2]`charles.lim [at] sgu.ac.id`, [3]`james.purnama [at] sgu.ac.id`

Abstract. Honeypot is designed to lure attackers away from the computer resources the attackers are trying to compromise. In addition, honeypot also tracks attacker's activities and helps researchers learn about their attack patterns. However, honeypot can also be identified by attackers using various fingerprinting methods. In this research, we use threat modeling to identify potential threats that reveal its existence which made honeypot ineffective. Various countermeasures are discussed and the proposed countermeasures have proved effective to enhance the deception capability of the honeypots we tested.

1. Introduction

Internet security threats continue to rise at an alarming rate as more and more system and application vulnerabilities continue to increase [1]. These threats are contributed by common Internet services such as Email & Web services and by the rise of mobile devices and Internet of Things usage as well, as demonstrated by the recent Distributed Denial of Service (DDoS) attack [2]. Attackers persist to find new ways to probe, attack and compromise systems and/or applications they are targeting. Firewalls and Intrusion Detection Systems (IDS) have always been used intensively to defense against known Internet security threats and attacks to the systems connected to the Internet. However, these security protection does not protect against unknown threats, that over the time may cause even larger damage to the system or system owner.

Honeypot provides an exceptional way to detect these unknown threats, which may include possible intrusion by attackers [3]. It can gather information useful attack patterns, which may help system administrator to learn in hardening systems to provide better security defense. With the increased usage of honeypots [4, 5], attackers also prepare their ways to defeat them. Attacker often uses fingerprinting techniques to probe system's profiles including system version, open services and their vulnerabilities. Using the same techniques, honeypot system could also be identified and thus make the system useless. In the worst case, the attacker can create a false intrusion to the honeypot causing disturbance to system.

The objective of this research is to explore various fingerprinting techniques at different Open Systems Interconnection (OSI) layers to detect the presence of honeypots. We use information gathered using these techniques as the new threat sources to our threat model, used to analyze these new threats. The new countermeasures are designed and applied to the existing honeypot systems. Security experts are then invited to review and validate the applied countermeasures.

2. Literature Review

2.1. Cyber Deception

Deception is an approach to deceive or mislead users into believing that the given information is beneficial to the user, while at the same time protecting the real assets and the owner's intention [6]. The idea of deception is not new, but with the more sophisticated attacks may bypass the common protection at the perimeter such as Firewall and IDS. The use of Cyber deception, hence, engages attackers using the useless resources and spend most of their time with the provided resources [7]. For the deception to be effective, there are 6 principles of deception to be taken into consideration [6]:

- **Focus** - The deception must have a target and it must have an effect on the target's decision. It can be targeted against groups or individuals.
- **Objective** - The deception must cause the target decision's to be affected and take suggested actions, not just observing the trap.
- **Centralized planning** - The operation must be centrally planned and move the unity to achieve a certain objective.
- **Security** - The creators of the deceptive system must halt any knowledges that may alert the target's of the trap.
- **Timeliness** - The deception requires patience, it needs careful timing and synchronizing.
- **Integration** - The deception needs to cooperate every system that it supports.

The goal is to have the attacker to think that they are in a real system in order for the operation to successfully continue. For cyber deception, there are many ways to deceive attackers, one of the most popular tools to achieve this purpose is honeypot [6].

2.2. Honeypot

Honeypot is essentially a decoy system that stores many "valuable" information that attackers are looking for and lures them to stay engaged in the system [8, 9]. Honeypot does not block the attackers to explore the system, instead, it simulates network services to interact with the attackers and record all the activities in the system. Since honeypot is intentionally placed in the network, any connections to the honeypot is considered as suspicious [3]. The lesson learned from monitoring attack patterns and distribution of malware during the attack can be very useful to provide real countermeasures.

Even though honeypot has the above mentioned advantages, like any other systems, the attacker can eventually find out the system is a honeypot. Once identified, the attacker will try to compromise the system and use it as a stepping stone to the next system.

3. Threat Modeling

As discussed, one of the counter deception strategy is to use fingerprinting techniques [10] to identify the existence of honeypots. Threat modeling is a process to analyze the security and vulnerabilities of an application [11] and it is commonly used to analyze security issues of an application or network services by identifying system's vulnerabilities [12, 13]. In our research, Threat modeling is used to enumerate all possible security threats by fingerprinting honeypots, as shown on Table 3.

There are 3 models of direct attacks to the honeypots [14, 15]: flooding the honeypots by sending garbage packets, compromising the honeypot to become a bot and using the honeypot as a spying platform. These attacks can be performed once the honeypot is detected through fingerprinting. To counter these threats, we need to evaluate each of the fingerprinting results for each OSI layers' parameters, including TTL (Time-to-live) data in layer 3, TCP or UDP ports on layer 4 and so on.

Table 1. Honeypots Threat Modeling

Attacks	**Attacks Description**	**Root Causes**	**Attacks Surface**	**Possible Mitigation**
Poisoning	Flooding the Honeypots	Honeypot fingerprinting	Multiple connections with same sources with no limit	Using LaBrea Tarpit
Compromising	Staging Internal Attacks	Honeypot fingerprinting	Honeypots that allow any activity entering or leaving the honeypot	Using Honeywall
Learning	Gaining private information	Honeypot fingerprinting	The honeypots are allowed to access the root system	Limit privileges

4. Honeypots Fingerprinting

Attackers fingerprint their targets before committing any intrusion; identifying their target makes it easier to infiltrate using known vulnerabilities. If the honeypots are detected with these attackers, it will be deemed useless as it needs interaction from attackers. Attackers often use port scanning tools such as NMAP or others to detect running network services. In our research, we setup 2 system: the honeypot system and the real system (such as real http server or ssh server) and fingerprinting is performed on both systems. During both system fingerprinting, we are looking for specific information, in particular information from layer 4 and 7 of OSI layers, on each network services, such as network application banner, application version, network protocols, etc. The fingerprinting results become the inputs to our threat modeling, the enhancement is done by comparing the result of real system and honeypot system. The honeypots used in this research are: HoneyD, Dionaea, Kippo, and Glastopf [16]. We chose to install these honeypots on a virtual machine for ease of portability and maintenance. Table 2 describes the overview of the honeypot weaknesses, divided into honeypot groups and focus. Based on these weaknesses, various enhancement is designed and applied.

5. Honeypots Enhancements

In this section, each of the enhancements is being implemented on the relevant honeypots.

5.1. Open Ports Fix

Only open ports that a real server would open will quickly fix the problem. However, opening port should be done carefully since opening port, such as port 17300 on the system may indicate the presence of Kuang2 worm. Opening well-known ports such as port 21, 80 and 143 will make it less suspicious for hackers to attack.

5.2. HoneyD IIS Directory Traversal Exploit Fix
To fix the Microsoft Internet Information Server (IIS) problem, the fix should be done carefully. The problem with the Honeyd IIS scripts is the timestamp itself, hence it is important to modify the timestamp similar to the one of a real server. The modification should also include folder name to make the honeyd appearance looks more closely to the real IIS.

5.3. HoneyD HTTP Service Fix
To fix the outdated and invalid replies of honeyd's HTTP service, we need to find which script that honeyd uses to emulate the HTTP service. The script that honeyd uses to emulate the HTTP service is named "apache.sh" on the scripts folder (`/usr/share/honeyd/scripts/unix/linux/suse8.0`). After full examination of the scripts, there is a problem with the script, it does not implement "POST", "PATCH", "DELETE", "COPY", "LINK", "UNLINK" and "PURGE". This implementation provides an easy way to fingerprint the system as honeyd by the attacker. Comparing and examining the response of the real Apache version 1.3.23, we can recreate the replies on the HoneyD HTTP Services.

5.4. Dionaea FTP Service Fix
The naming of the FTP service of Dionaea is different from any other real services. To fix this problem, changing the configurations into a more believable services. For example, changing the FTP banner into something else is a start to enhance the honeypot from fingerprinting activities. To make it hidden from service scan, we changed the "Welcome to the ftp service" header in the configuration file into a real Microsoft FTP service header, i.e. "Microsoft FTP Service."

5.5. Dionaea SMBD Fix
NMAP scan results revealed that Dionaea's SMBD service uses particular names on Workgroup and Homeuser names. If the server has a domain name of WORKGROUP and the computer name of HOMEUSER-XXXXXX, where X is any random character, then it is detected as a Dionaea SMBD service. To evade this fingerprint, modification should be performed on the Dionaea configuration file. For example, we change the name OemDomainName into "MIDOMINIO" and the ServerName into "EQIUPO-TEST."

5.6. Dionaea MS-SQL Service Fix
NMAP detects MS-SQL fingerprint by checking the response of a TDS package that the Dionaea responds. Dionaea sends this response to any client when they are connecting to its database. NMAP detects it by comparing the value of the in the field named "Token Type" during the login process into the Dionaea's MS-SQL service. The modification could be performed on the file of the MS-SQL service script is named "mssql.py" in the scripts folder. Changing the "`r.VersionToken.TokenType`" value from "0x00" to "0xAA" will fix the problem.

5.7. Kippo Regex Ping Fix
To fix the problem, the file that emulates the ping function need to be located. It was found that the ping function is emulated by the script named "ping.py". To repair the issue, several lines of codes need to be modified to restrict the ping command only works within the set of network groups. If the target ping is not in the network group, then it is seen as "Unknown host." After the modification, pinging using an impossible IPP address such as "999.999.999.999" returns with an error [17].

Table 2. Honeypot Weaknesses and Enhancements

Honeypot	Layer	Focus	Vulnerabilities	Enhancements
	4	Open Ports	Suspicious open ports	Opening & Configuring essential ports
HoneyD	7	HTTP	Invalid replies	Configuring scripts
	7	IIS	Suspicious timestamps	Fixing timestamps
	4	Open Ports	Suspicious open ports	Opening & Configuring essential ports
	7	FTP	Detected by port scan	Configuring scripts
Dionaea	7	MySQL	Detected by port scan	Configuring scripts
	7	SMBD	Detected by port scan	Configuring scripts
Kippo	7	SSH	Commands not working properly	Configuring scripts
Glastopf	7	HTTP	Invalid error reply	Configuring scripts

5.8. Glastopf LFI Vulnerability Fix

At first, if any hacker tries to access a folder that is not in the vulnerability group, it will reply "Warning: include(vars1.php)" error, which is a static error message made by the developer. To workaround this fingerprinting exploit, it can be changed to a "Permission Denied" error. It does make more sense to do this, rather just throw a warning error.

Table 2 summarizes the honeypots' weaknesses and their enhancements.

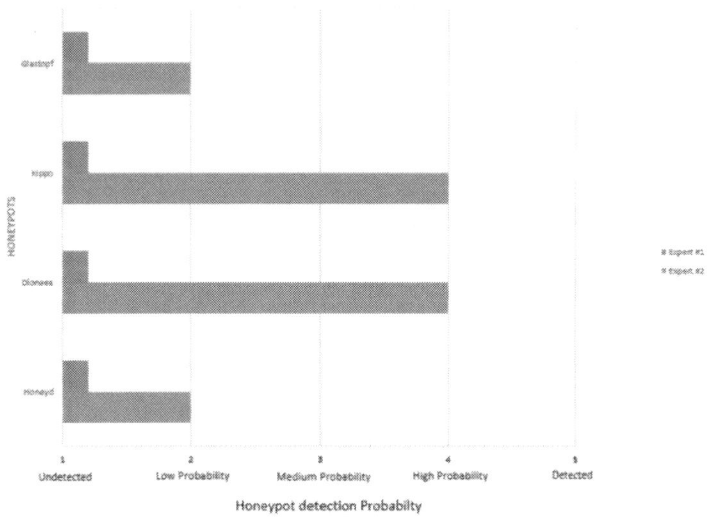

Figure 1. Expert Review results

6. Expert Review

To validate our enhancement results, security experts are invited to review the enhancements. Both of the experts are security professional pen-testers and consultants to various industries, including Banking, Telecommunications and other Enterprises. Security experts were given the freedom to use any security tools to perform fingerprinting on the honeypots. The tests attempted by 2 different experts are using black-box testing method, in which the experts are not given any information about the targets. The result of the tests (detection probability) are divided into 5 categories, i.e. "Undetected", "Low probability", "Medium probability", "High probability" and "Detected". The results show that Glastopf and HoneyD honeypots (with enhancements) were detected with "Low Probability", which indicates that the experts have difficulty in detecting them as honeypots. The other honeypots (with enhancements), Kippo and Dionaea, are detected with "High Probability", which indicates that there are some fingerprinting weaknesses that are need to be addressed. Figure 1 summarizes the security expert reviews on the enhancement of the honeypots.

7. Conclusion

Honeypots being used as a cyber deception tool will continue to evolve. Attackers can detect the presence of honeypot by "fingerprinting" them and inspect the result carefully on different OSI layers. The fingerprinting results of the existing system may indicate whether the existing system is a honeypot, hence, the honeypot should be "hardened" to reduce the chance of the honeypots being discovered by attackers. Threat modeling is then performed to analyze the potential security threats of the fingerprinting results. To enhance its deception capability, system configuration and custom scripts are used. We have successfully enhanced the deception capability of the honeypots to avoid the honeypots being identified by the attackers. Current limitation of our research is still focusing on layer 3, 4 and 7. Our future works may include enhancing the deception capability the system by fingerprinting layer 1 to check for clock skew of the system [18].

References

[1] Rubino D Symantec internet security threat report 2015 URL `https://www.symantec.com/content/en/us/enterprise/othe_resources/21347933_GA_RPT-internet-security-threat-report-volume-20-2015.pdf`
[2] Security K 2016 Hacked cameras, dvrs powered todays massive internet outage URL `https://krebsonsecurity.com/2016/10/hacked-cameras-dvrs-powered-todays-massive-internet-outage/`
[3] Spitzner L 2003 *Honeypots: tracking hackers* vol 1 (Addison-Wesley Reading)
[4] Nemade S, Darekar M M A and Bachhav M J 2015 *Journal of Multidisciplinary Studies* **1**
[5] Kao C N, Chang Y C, Huang N F, Liao I J, Liu R T, Hung H W *et al.* 2015 A predictive zero-day network defense using long-term port-scan recording *Communications and Network Security (CNS), 2015 IEEE Conference on* (IEEE) pp 695–696
[6] Bodmer S, Kilger M and Carpenter 2012 *Reverse Deception: Organized Cyber Threat Counter-Exploitation* (McGraw Hill Professional)
[7] Rowe N C 2008 *Cyber Warfare and Cyber Terrorism* 97–104
[8] Provos N *et al.* 2004 A virtual honeypot framework. *USENIX Security Symposium* vol 173 pp 1–14
[9] Sokol P, Kleinova L and Husak M 2015 Study of attack using honeypots and honeynets lessons learned from time-oriented visualization *EUROCON 2015-International Conference* (IEEE) pp 1–6
[10] Aguirre-Anaya E, Gallegos-Garcia G, Luna N S and Vargas L A V 2014 *International Journal of Electrical and Computer Engineering* **4** 848
[11] Shostack A 2014 *Threat modeling: Designing for security* (John Wiley & Sons)
[12] Potteiger B, Martins G and Koutsoukos X 2016 Software and attack centric integrated threat modeling for quantitative risk assessment *Symposium and Bootcamp on the Science of Security* (ACM) pp 99–108
[13] Chang J S, Jeon Y H, Sim S and Kang A N 2015 *International Journal of Distributed Sensor Networks* **2015**
[14] Piller K and Wolfgarten S 2004 *Ernst&Young Risk Advisory Services ppt. Retrieved on November* **16** 2005
[15] Krawetz N 2004 *IEEE Security & Privacy* **2** 76–79
[16] Gorzelak K, Grudziecki T, Jacewicz P, Jaroszewski P, Juszczyk L and Kijewski P 2012 Proactive detection of security incidents ii - honeypots Tech. rep. Tech. Rep., ENISA
[17] Oosterhof M 2015 Kippo github pull URL `https://github.com/desaster/kippo/pull/148/commits/19241a374dec3e8d198e7d679ed4e5c6cefe1e2c`
[18] Kohno T, Broido A and Claffy K C 2005 *Dependable and Secure Computing, IEEE Transactions on* **2** 93–108

IOP Conf. Series: Journal of Physics: Conf. Series **801** (2017) 012058 doi:10.1088/1742-6596/801/1/012058

Mal-Xtract: Hidden Code Extraction using Memory Analysis

Charles Lim[1] , Yohanes Syailendra Kotualubun[2] , Suryadi[3], Kalamullah Ramli[4]

[1,2]Department of Information Technology, Swiss German University, Edutown BSD City, Tangerang, 15339 Indonesia
[3]Department of Mathematics, Universitas Indonesia, Kampus UI, Depok 16424, Indonesia
[4]Department of Electrical Engineering, Universitas Indonesia, Kampus UI, Depok 16424, Indonesia

E-mail: [1]`charles.lim [at] sgu.ac.id`, [2]`yohannes.kotualubun [at] student.sgu.ac.id`, [3]`yadi.mt [at] sci.ui.ac.id`,[4]`kalamullah.ramli [at] ui.ac.id`

Abstract. Software packer has been used effectively to hide the original code inside a binary executable, making it more difficult for existing signature based anti malware software to detect malicious code inside the executable. A new method of written and rewritten memory section is introduced to to detect the exact end time of unpacking routine and extract original code from packed binary executable using Memory Analysis running in an software emulated environment. Our experiment results show that at least 97% of the original code from the various binary executable packed with different software packers could be extracted. The proposed method has also been successfully extracted hidden code from recent malware family samples.

1. Introduction

Static Analysis through reverse engineering is one of the main techniques to analyze assembly codes inside a binary executable, especially for malicious software (malware). To prevent this, malware authors commonly use various obfuscation techniques to protect and hide their original malicious codes from being detected. With the rate of new malware now is more than 200 malware every second [1], malware analyst faces even more challenges to be able to analyze the huge volume of malware. Nevertheless, the new malware often is a "recycle" version of existing version, added with various obfuscation techniques to create a new signature to avoid from being detected. Obfuscation techniques include source code obfuscation [2, 3], Control-flow obfuscation [4], Instruction Emulation [5], and binary code packing [6]. Recent study [7] indicated up to 80% of detected malware is packed. Ugarte [8] found that common packers will run their unpacking routine and reveal their original body at some time in physical memory. First, packers write their original body in some memory sections and jump to their Original Entry Point, which contain the newly written memory section. Based on this characteristic, we define some pre-defined memory section size that each packer will write into it and use these pre-defined values as the threshold for memory section. With the increased use of software packer in malware, malware detection becomes less effective [9]. We believe that our method can help to increase the malware detection rate by unpacking the encryption and obfuscation from the packed malware.

In this experiment, we propose a method to detect the end of unpacking routine from common packer that runs inside QEMU [10]. We present a method called Mal-Xtract, a memory access analysis technique that determine the threshold of written section size to be used in extracting the original hidden-body by dumping the whole physical memory and extract its unpacked code. The contributions of this paper are as follows:

(i) We develop a memory analysis technique to determine the end of unpacking routine. We propose a method to detect the end of memory write section before program counter start to execute the original body of packed executable and to identify the memory address that has been written by the executable, and mark it as newly written or rewritten. Our method relies on the number memory section written or rewritten by the executable, the end of unpacking routine is indicated by the last written memory section.

(ii) We identify the threshold of written memory section size for every unpacking routine from the software packer. Normal distribution probability is used to determine dump sample probability threshold of written memory section.

(iii) We propose a method to validate the extracted code from physical memory dump with the original code. We use the similarity percentage to measure the percentage unpacked code from the binary executable under investigation.

2. Related Works

To avoid from being detected by anti-virus software, malware authors obfuscate the binary code using various software packers. By making the malicious code "stealth" from the presence of anti-virus, malware can accomplish the tasks that author has designed for. Rad et al [11] argued that several reasons to use "stealth" techniques in malware. i.e. (i) to have invincibility from all malware analysts (ii) to prevent it from reverse-engineered, (iii) to extend the lifetime of the malware, and (iv) to prevent it being modified or tampered.

There are several type of binary encryption such as oligomorphism [12], polymorphism [13, 14, 15], metamorphic encryption [16], and also a new encryption approach, mimimorphism [17]. According to Sun et al [18], a packer is an executable program that compresses or encrypts the original body of executable file. The key method in extracting a hidden-code from packed executable is to find the exact time when the unpacking routine is completed. The idea is, when a packed executable being run, all of its instructions and data will be loaded into memory. All of the unpacking process will be done at the memory level, with the original body instructions will also be shown after all unpacking process is completed.

PolyUnpack [19] compared the instruction from the static analysis with the state instruction from memory that can be used to acquire the original body of unpacked code by creating the complete executable version of the hidden-code extracted. Renovo [20] focused on the memory write instruction execution at memory level. They used 'shadow memory' to taint all the newly written memory section as 'dirty'. When the instruction pointer jumped to the 'dirty' memory, then the executable dump procedure is called. OmniUnpack [21] observed the execution tracking and using a page-level protection mechanism that can monitor both user and kernel space to flag when the code is executed from a page that was modified. Eureka [22] developed a kernel driver that performs a memory dump of the executable and disassemble it whenever the malicious payload has been fully unpacked. Rotalume [5], focused on the emulator based packer, and identified the fundamental characteristic of emulation decoding: iterative main-loop, bytecode fetches using virtual program counter and opcodes monitoring based on the bytecode fetches to create the control flow graph from each emulation packer. Our previous study, Mal-unpack [23], described method of using Binary Instrumentation to extract instruction trace of the running executables at the user space. Some drawbacks from these previous studies is that all of the unpacking engine only can apply for [Type I] up to [Type IV] from Ugarte's Packer classifications [8]. They cannot detect the modern anti analysis technique, such as metamorphic obfuscator and emulation based packer that can translate or re encrypt the memory section after its executions as mentioned by Ugarte et al [8] as [type V] and [Type VI] packers classification. In this paper, Our Mal-Xtract manages to extract the hidden body from [type V] and [Type VI] packers classification, as defined by Ugarte et al, by tracking memory access (read, write and rewrite) for memory write section size. Comparison between our approach, Mal-Xtract and other methodologies is shown in the table 1.

Table 1. Mal-Xtract compared wiht other framework

Key Items	**Poly Unpack**	**Renovo**	**Eureka**	**mal-unpack**	**Mal-Xtract**
Memory Access Space	User Space	User and Kernel Space	User space	User Space	User Space and Kernel Space
Features	Instruction Trace	Runtime Memory Instruction	Runtime System Call API	Instruction Trace	Memory Access Write & Read
Tracking End of Unpacking Methods	Comparing the static instruction with runtime memory instruction	Detecting the memory write instruction (mov %eax, [%edi] and push %eax)	Detecting process start and termination System Call API and Code/data Ratio	Detecting the memory write and jump Program counter	Using memory Analysis to find the end of unpacking routine using memory write section threshold

There are several techniques can be used to perform dynamic binary analysis such as system emulator, system call hooking, dynamic taint analysis, dynamic binary instrumentation and system emulation. Tools such as cuckoo sandbox [24] uses Out-VM Application Programming Interface (API) Hooking as its main analysis. One of the major drawback of hooking technique is that it runs on a user memory space and can not monitor kernel memory space. Another dynamic analysis technique, dynamic taint analysis, e.g. used in Panorama [25], tracks tainted data flow and program behavior. Unfortunately, taint tracking method is not strong enough when dealing with unpacking routine since the analysis need to go deeper into memory and instruction level access. Eureka [22] developed a custom kernel driver to inspect all instruction pointer. Our previous work, mal-unpack [23] uses binary instrumentation to observe the instruction trace at the user memory space only and can not trace kernel memory space. Dynamic binary analysis using system emulation, i.e. QEMU [26], has been applied by various researchers, includes Renovo [20], Ether [27], CWSandbox [28] and Lindorfer et al [29]. One of the major advantage is that QEMU perform instruction translation that enables more detail information tracking regarding instruction, control flow and register content. In this research, we use PANDA [30], an open source Platform for Architecture-Neutral Dynamic Analysis that built upon QEMU [26], a whole system emulator, in which the analysis has access to all data and the executing code in the guest. In addition, PANDA is also capable of recording and replaying executions - enabling iterative, deep the whole system analysis.

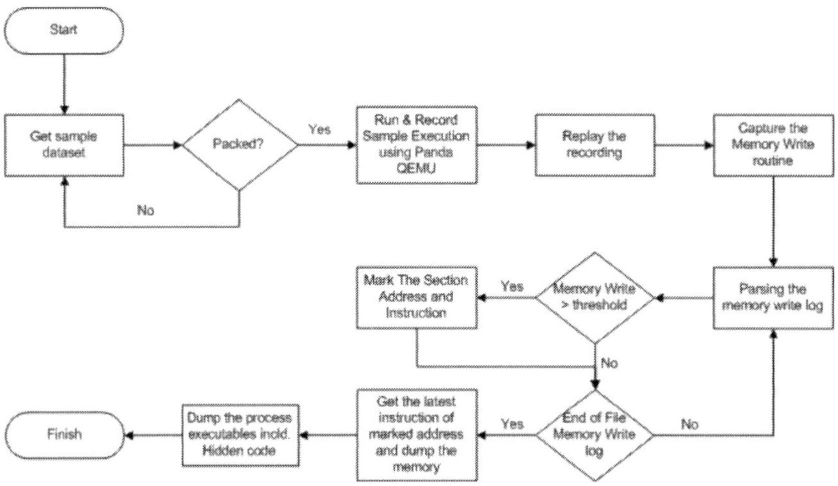

Figure 1. Mal-Xtract Framework

3. Mal-Xtract

In this paper, we develop Mal-Xtract, that monitors memory write access both user and kernel memory space for the running binary executable under PANDA [30]. Figure 1 shows the Mal-Xtract framework that allows us to track physical memory access of the running binary executable. The goal is to uncover an indicator of full or partial original hidden-code on the physical memory has been written. To do this, PANDA runs the binary executable and record whole system emulation in a log file [31].

To track memory written by the binary executable, the result recording log of PANDA is replayed with the tracking of memory write and its adjacent memory address selected. All written address and its adjacent memory address will be marked as memory section written by the running binary executable. Since there is no ground truth about how many written bytes in memory that can determine the memory section, we use an empirical approach, assuming the normal distribution of each dump of memory, to extract the revealed code from the memory space. The number of instruction count is recorded during each dump and based on the experiments an upper value of number of instructions is set as the threshold value. Section 3.1 discusses the process of determining memory section threshold.

Once the physical memory dump is available, Volatility memory forensic tool is used to extract the binary executable and the result is validated. This binary executable resulting from memory dump cannot be executed directly in the Windows environment due to missing or damaged Import Address table (IAT). Most software packers manipulate import address table to protect its code obfuscation and to perform anti-analysis. The pseudo code of the proposed Mal-Xtract is listed in Algorithm 1.

3.1. Finding Written Memory Section Threshold

To determine written memory section threshold, we track the memory address for writing, especially in contiguous manner, e.g. starting at memory address `0x00100101` and next memory address would be `0x00100102, 0x00100103`, etc. Whenever the memory address written started to jump into different memory address such as`0x70089030`, the memory section size is counted and the memory section is marked. Each memory write/rewrite section are stored in a log and marked the above threshold section as instruction section. The advantage of this method is that it does not depend on Virtual Address Descriptor (VAD) section, but on the actual memory

IOP Conf. Series: Journal of Physics: Conf. Series **801** (2017) 012058 doi:10.1088/1742-6596/801/1/012058

Algorithm 1 Memory Analysis Algorithm

```
procedure MAIN MEMORY ANALYSIS
    Perform Recording from the Executables
    Replay the Recording using ASIDSTORY            ▷ Get PID from the executables
    Replay the Recording using BUFMON      ▷ Get all Memory Write/Read Log from PID
            ▷ Parsing Raw Memory Write/Read Log to Memory Address Write / Rewritten
    i = 1
    while Line(i) <> End Of File Memory Logs do
        if Address(i) == Written Address then
            AddressWrittenCount + 1
        else
            New Written Address    ▷ add New Written Address into Memory Address List
        i ++
            ▷ Parsing the Memory Address List to memory section by capturing all adjacent
  Address
    j = 1.
    while Address(j) <> End Of Memory Address List do
        if Address(j + 1) == Address(j) then
            Address Section Size + 1
        else
            New Written Memory Section (Section Size)
        j ++
                          ▷ Parsing the Memory Section to the Memory section Threshold
    k = 1.
    while Memory Section(k) <> End Of Memory Address Section List do
        if Memory Section(k) > Memory Dump Threshold then
            Mark the Memory Section as Memory Dump Threshold
        j ++
                                  ▷ Parsing The percentage of each dump threshold
    for each T ∈ DumpThresholdS do
        Get Instruction Count from the latest Memory Section of Threshold T
        Percentage of Threshold T = Instruction Count (T) / Instruction Total of Recording
```

write. According to Sharif et al [22], any memory section could contain data or code, and when unpacking process occurred, the packer will unpack the original instruction and data to the physical memory [8]. To determine the threshold of memory section, an empirical approach is used, assuming normal distribution statistics of all memory dumps of our experiments. In our early experiments, written memory section tends to be larger and gets smaller as the unpacking routine completes it tasks. Figure 2 shows the dump performed in the range of threshold from 50,000 bytes to 100 bytes per written memory section. Thus, a minimum of 100 bytes of memory section is used as the lower level threshold value since all of the software packers investigated in our early experiments are already unpacked at this value. On the other hand, memory section of size 50,000 bytes are chosen to the upper level threshold value for any memory sections found above 50,000 bytes or above. For example, for some software packers, there is a write up to 200k bytes for 1 memory section, and they are considered as part of 50,000 byte range.

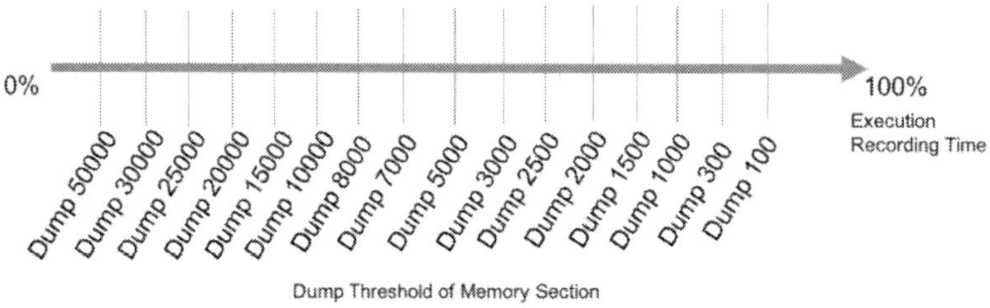

Figure 2. Dump Threshold of Written Memory Section

3.2. Hidden-Code Extraction Validation

To validate the code extracted from each dump of the running executable, process dump is performed after each dump completed, using Volatility [32] with procdump feature. Since the executables extracted from this process which include slack space between the Portable Executable (PE) sections are not page-aligned, further extraction is performed to cover all the expanded section written after the executables run. While Renovo [20] extract only from memory sections that had been expanded from the executables, our extraction method ensure ensures a more complete code coverage from the memory dump. The binary executables resulted from the dump are further processed using IDA Pro [33], a debugger and static analysis tool, to provide an assembly ASM file. The ASM file, which consists of all the code and data from the sections of dumped executables, are then parsed using python script to obtain all the mnemonic instructions or x86 Opcode Instructions [34]. Only the mnemonic instructions are retrieved since the data and assembly variables part are already translated for every packer as shown in figure 4 and 5. Figure 4 shows the comparison the API-string translation from the dumped packed executable (left) and the original executable (right) while figure 5 presents assembly variable translation from the dumped packed executable (left) and the original executable (right).

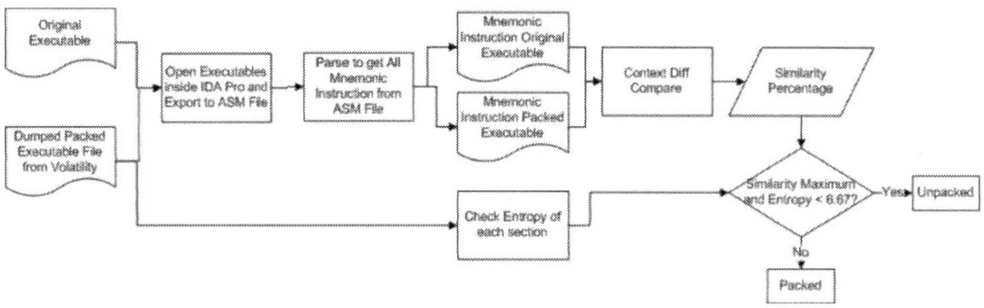

Figure 3. Validation Method

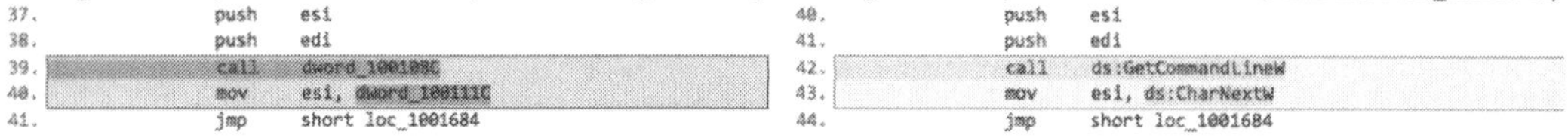

Figure 4. API String Translation (Left: Dumped Packed Executable, Right: Original Executable)

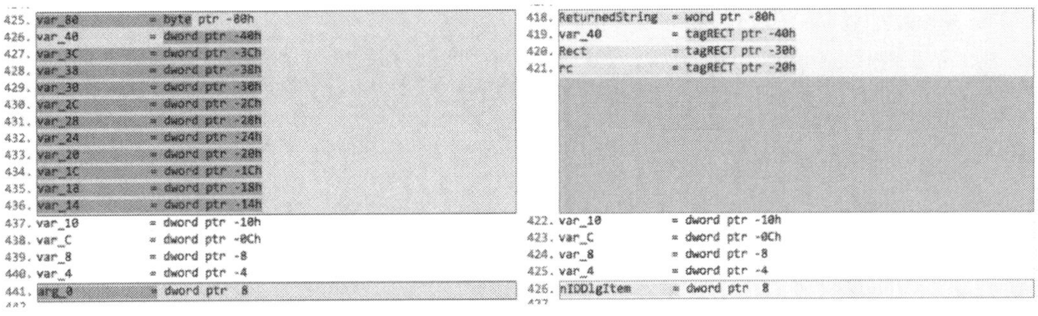

Figure 5. Assembly Variable Translation (Left: Dumped Packed Executable, Right: Original Executable)

The same process is also applied to the original binary executable (no software packer is applied), the parsed instructions from original binary executable is then compared using context diff comparison technique, which first align the whole line of each executable's parsed ASM file and then search for diff line for each aligned line. All the diff line are totaled and similarity percentage is then calculated using the following equation 1:

$$Similarity = \frac{\sum Diff}{\sum Line} * 100\% \tag{1}$$

Other than similarity percentage, file entropy [35], is used to determine whether the binary executable is packed. Thus the combination of similarity percentage value and file entropy are used in our research whether the binary executable file extracted is packed.

4. Experiment Data

A similar sample binary executables as [36], two generic Windows applications in Windows 7 (`calc.exe` and `notepad.exe`) and one custom standalone program (`helloworld.exe`) that print string in the command prompt, are used in this research. These binary executables are packed with a collection of third party binary packers, both free and commercial packers, i.e. UPX [37], PECompact2 [38], FSG [39], Armadillo [40], Molebox [41], WInUPack [42], Themida [43] and VMProtect [44]. We also evaluate our method on 20 malware samples, obtained from from VirusShare and MALHEUR [45] dataset, which contain of 4 different malware families, Lipler (2009), domaiq (2014), Sality (2015), and Uparte (2015). For all malware samples, only packed malware samples are selected and used in this research, standard file entropy [35] is used to check for packed executables.

5. Experiment Setup

All experiments are performed using an Apple Macintosh i7 2.7 Ghz with 16 GB RAM Mac OS X 10.12, VMWare Fusion 8.0.2 that runs Ubuntu 14.04 with 6 GB RAM and 4 Core Processor. PANDA QEMU is installed inside Ubuntu VM and runs Windows 7 32bit SP1 with 2 GB RAM for its QEMU VM Emulation and all samples executed inside the Windows 7 OS. All the sample binary executables also gathered from Windows 7 32bit SP1 Portable Executables. We are using Windows XP SP2 VM for running IDA PRO to extract ASM file from extracted binary executables.

IOP Conf. Series: Journal of Physics: Conf. Series **801** (2017) 012058 doi:10.1088/1742-6596/801/1/012058

6. Evaluation and Discussion

For all the memory dumps, the lower and upper memory section threshold values of 100 bytes and 50,000 bytes respectively, are used in this research. In addition, based on Lyda [35], file entropy threshold value of 6.67 is used to indicate whether a file is packed. Table 2 shows the memory dump results of various packed sample binary executables.

Even tough Armadillo v8.6, categorized as Type VI packer [8], implements numerous anti-reverse-engineering technique, our method could extract the whole body from original body in the Dump 5000 with similarity of 99.6%. Similarly, for UPX and WinUpack as Type I packer, FSG, PECompact2 and Molebox that categorized as Type III packer can be extracted with 99% similarity. Thus, for all type of common software packer unpack their original body into physical memory at the threshold of 5000 bytes as the memory write section threshold that indicate the end of unpacking routine before it continue to execute the unpacked instructions.

Table 2. Memory Dump of Various Packed Sample Binary Executables Result

	Notepad.exe			Calc.exe			helloworld.exe		
Packers	*D*	*E*	*S*	*D*	*E*	*S*	*D*	*E*	*S*
Armadillo	5000	5.96	99.6%	5000	5.61	98.8%	5000	5.72	99%
FSG	50000	4.8	99.8%	50000	5.8	99.8%	50000	5.1	99.8%
Molebox	15000	6.2	99.6%	25000	6.2	99.8%	15000	6.0	99%
UPX	25000	3.7	99.6%	25000	5.3	99.8%	25000	5.2	99.1%
PECompact2	8000	3.56	99.6%	25000	6.12	99.8%	25000	5.96	99.2%
WinUpack	50000	3.5	99.6%	50000	4.5	99.8%	50000	4.2	99.6%
Themida	50000	7.6	97.1%	50000	7.91	97.4%	50000	7.3	97.5%
VMProtect	50000	6.2	97.5%	50000	6.1	97.2%	50000	6.3	97.3%

Remark:

D Memory Section Threshold
E Max Entropy from Executable Sections
S Similarity Percentage

For software packer that uses virtualization which implement advanced code obfuscation using emulation and bytecode translation [46], our method successfully extract 97% the original body at dump 50,000 bytes. From the statistical view in Figure 6, UPX, FSG, PE2 and WinUpack, reached maximum similarity at dump 50,000, while some other packers only reach the same similarity level at 5000 bytes. This is due to UPX, FSG, PE2 and WinUpack are categorized as compression packers which don't have any code protection feature. However, memory dump with the lower memory section value than 50,000 bytes, the similarity was significantly decreased to 89% for VMProtect and no executables could be extracted for Themida packed executables. In fact, in our experiments, no file extraction is possible for memory dump with the lower memory section value at 30,000 bytes. Themida starts to apply its memory protection features from memory dump 30,000 bytes to memory dump 100 bytes. Based on this observation, we can conclude that Themida and VMProtect write all original body in physical memory before any emulation and anti analysis routine are applied. In addition, there are also differences between all the unpacked dump on each executable due to the complexity of instructions for each binary executable; the more complex binary executable, the more more protection, encryption and obfuscation in the code and memory level.

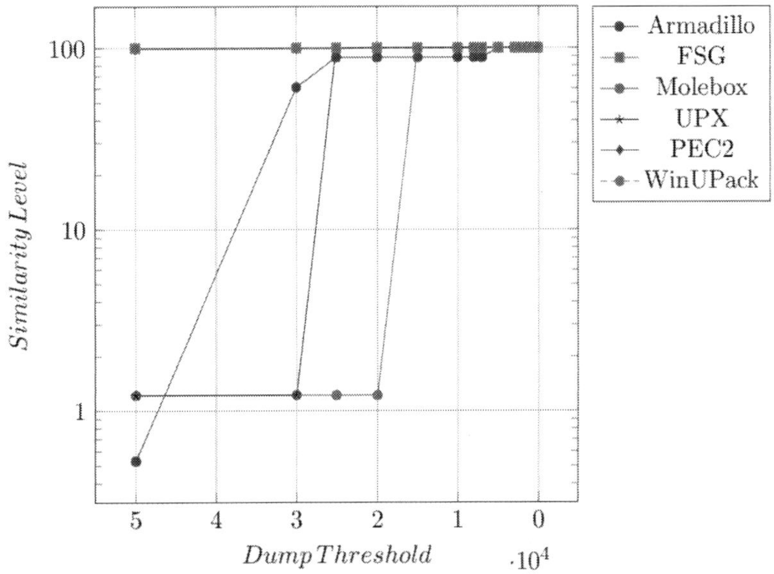

Figure 6. Similarity Level for Dumped Packed Executables

Figure 7 shows the memory section distribution for all binary executable, Benign (not packed) and Packed binary executables, used in our research. Based on the graph in figure 7, there is a large memory section access (write and rewrite) at memory dump of 100 bytes, bu we have either the value of 0 or 1 for memory section access at the memory dump of 50,000 bytes. It is also interesting to note that all binary executables, including benign binary executable, demonstrate a large memory section usage at the memory dump 100 bytes. Thus, for all software packers the unpacking process range from memory dump 50,000 bytes up to memory dump 100 bytes, and the unpacked binary executable (both from benign and packed binary executable) run normally when the memory section usage hits 100 bytes.

Table 3 shows the result of our evaluation of the malware samples using Mal-Xtract framework. Static analysis tools such as PEStudio [47] is then used to gather all static information from fiel extracted from memory dump of running malware binary samples. Due to the lack of ground truth to validate the unpacked malware samples, a combination of parameters are used in this research, i.e. Indicator of Compromise (IoC) [48], Blacklisted string and Suspicious API Import as comparative features from before malware unpacking process and after malware unpacking itself. The results show that our proposed method successfully unpack all the packed malware indicated by much lower entropy value after unpacked [35].

During the unpacking process, malware samples evaluated exhibit a larger body size before and after the process, as the binary executable writes a new section or rewrite the existing section. All the comparative features show a positive results except for Upatre family. On Upatre family no obfuscation is found, i.e. on the suspicious string or API, but a larger size for each unpacked dump malware can be observed. This type of packer can be correlated with Ugarte [8] as Packer type I,II, and III. On the other hand, Obfuscator such as Armadillo and Molebox, which can be correlated with Ugarte [8] as packer type IV, V, and VI. In the malware experiments, Lipler, Domoiq and Sality malware family demonstrated obfuscation functionality, the malware samples have a lot of suspicious strings and API calls based on Indicator of Compromise static analysis.

The limitation of dump executable resulted from Mal-Xtract is not a runnable executable inside Windows Operating System, due to the Import Address Table that already tampered by

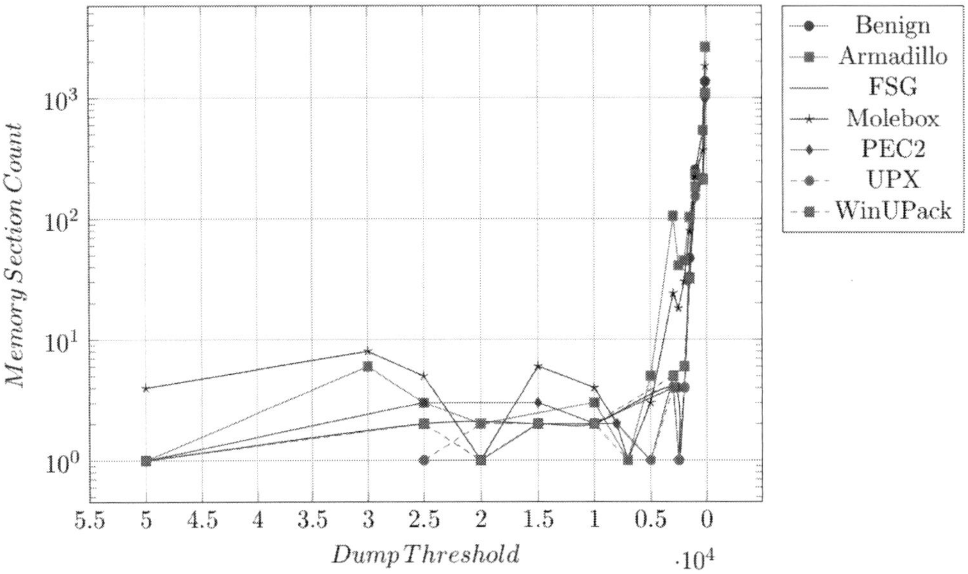

Figure 7. Memory Section Distribution for Packers

Table 3. Malware Unpacking Result

		Indicator of Compromise		Suspicious API Import		Blacklist String	
Families	*Result*	*Before*	*After*	*Before*	*After*	*Before*	*After*
Lipler	Full	5	7	49	109	57	218
Domaiq	Full	15	10	72	180	227	568
Upatre	Full	7	6	52	52	90	82
Sality	Full	5	19	1	1	3	30

the packer. There are also huge overhead in the analysis logs because we are using recording and memory access capture.

7. Conclusion

Mal-Xtract, a memory analysis based on system emulation to extract hidden-body code has been presented. Our experiments shows that memory analysis technique is effective to handle common [Type I] to [Type IV] software packers [8]. Depending on type of packers, memory write section may differ: memory write section for most software packers can be found at the threshold of dump 5,000 bytes and memory write section for advanced software packer that utilizes virtualization is found at dump 50,000 bytes. Our experiment results show that at least 97% of the original code from the various binary executable packed with different software packers could be extracted. Our proposed method has also proved to be successful in extracting hidden code from recent malware family samples tested in this research. In the future, we are planning to include instruction and register tracing as part of the features to determine the completion of unpacking routine.

References

[1] AVTEST Av test - the independent it security institute URL https://www.av-test.org/en/statistics/malware/

[2] Ceccato M, Di Penta M, Falcarin P, Ricca F, Torchiano M and Tonella P 2014 *Empirical Software Engineering* **19** 1040–1074 URL https://dx.doi.org/10.1007/s10664-013-9248-x

[3] Viticchié A, Regano L, Torchiano M, Basile C, Ceccato M, Tonella P and Tiella R 2016 Assessment of source code obfuscation techniques *In Proceedings of the 16th IEEE International Working Conference on Source Code Analysis and Manipulation* (IEEE Computer Society) pp 11–20 URL http://dx.doi.org.ezproxy.biblio.polito.it/10.1109/SCAM.2016.17

[4] Balachandran V and Emmanuel S 2013 *IEEE Transactions on Information Forensics and Security* **8** 669–681 URL https://dx.doi.org/10.1109/TIFS.2013.2250964

[5] Sharif M, Lanzi A, Giffin J and Lee W 2009 Automatic reverse engineering of malware emulators *2009 30th IEEE Symposium on Security and Privacy* (IEEE) pp 94–109 URL http://dx.doi.org/10.1109/SP.2009.27

[6] Kim M J, Lee J Y, Chang H Y, Cho S, Park Y, Park M and Wilsey P A 2010 Design and performance evaluation of binary code packing for protecting embedded software against reverse engineering *2010 13th IEEE International Symposium on Object/Component/Service-Oriented Real-Time Distributed Computing* (IEEE) pp 80–86 URL http://dx.doi.org/10.1109/ISORC.2010.23

[7] McAfeeLabs 2015 *McAfee Labs* URL http://www.mcafee.com/sg/resources/white-papers/wp-good-bad-the-unknown.pdf

[8] Ugarte Pedrero X, Balzarotti D, Santos I and Bringas P G 2015 SoK: Deep packer inspection: A longitudinal study of the complexity of run-time packers *SSP 2015, IEEE Symposium on Security and Privacy, May 18-20, 2015, San Jose, CA, USA* (San Jose, UNITED STATES) URL http://dx.doi.org/10.1109/SP.2015.46

[9] Moser A, Kruegel C and Kirda E 2007 Limits of static analysis for malware detection *Computer security applications conference, 2007. ACSAC 2007. Twenty-third annual* (IEEE) pp 421–430

[10] QEMU Qemu - open source processor emulator URL http://wiki.qemu.org/Main_Page

[11] Rad B B, Masrom M and Ibrahim S 2012 *International Journal of Computer Science and Network Security* **12** 74–83 URL http://paper.ijcsns.org/07_book/201208/20120813.pdf

[12] Aycock J 2006 *Computer viruses and malware* vol 22 (Springer Science & Business Media) URL http://dx.doi.org/10.1007/0-387-34188-9

[13] O'Kane P, Sezer S and McLaughlin K 2011 *Security & Privacy, IEEE* **9** 41–47 URL http://dx.doi.org/10.1109/MSP.2011.98

[14] Szor P 2005 *The art of computer virus research and defense* (Pearson Education)

[15] Li X, Loh P K and Tan F 2011 Mechanisms of polymorphic and metamorphic viruses *Intelligence and Security Informatics Conference (EISIC), 2011 European* (IEEE) pp 149–154 URL http://dx.doi.org/10.1109/EISIC.2011.77

[16] Christodorescu M and Jha S 2003 Static analysis of executables to detect malicious patterns *USENIX Security Symposium*

[17] Wu Z, Gianvecchio S, Xie M and Wang H 2010 Mimimorphism: A new approach to binary code obfuscation *Proceedings of the 17th ACM conference on Computer and communications security* (ACM) pp 536–546 URL https://doi.org/10.1145/1866307.1866368

[18] Sun L, Versteeg S, Boztaş S and Yann T 2010 Pattern recognition techniques for the classification of malware packers *Information security and privacy* (Springer) pp 370–390 URL http://dx.doi.org/10.1007/978-3-642-14081-5_23

[19] Royal P, Halpin M, Dagon D, Edmonds R and Lee W 2006 Polyunpack: Automating the hidden-code extraction of unpack-executing malware *Computer Security Applications Conference, 2006. ACSAC'06. 22nd Annual* (IEEE) pp 289–300 URL http://dx.doi.org/10.1109/ACSAC.2006.38

[20] Kang M G, Poosankam P and Yin H 2007 Renovo: A hidden code extractor for packed executables *Proceedings of the 2007 ACM workshop on Recurring malcode* (ACM) pp 46–53 URL https://doi.org/10.1145/1314389.1314399

[21] Martignoni L, Christodorescu M and Jha S 2007 Omniunpack: Fast, generic, and safe unpacking of malware *Computer Security Applications Conference, 2007. ACSAC 2007. Twenty-Third Annual* (IEEE) pp 431–441 URL http://dx.doi.org/10.1109/ACSAC.2007.15

[22] Sharif M, Yegneswaran V, Saidi H, Porras P and Lee W 2008 Eureka: A framework for enabling static malware analysis *Proceedings of the 13th European Symposium on Research in Computer Security: Computer Security* (Springer-Verlag) pp 481–500 URL http://dx.doi.org/10.1007/978-3-540-88313-5_31

[23] Lim C, Sulistyan D Y, Ramli K *et al.* 2015 *Advanced Science Letters* **21** 3333–3336 URL https://doi.org/10.1166/asl.2015.6487

[24] Oktavianto D and Muhardianto I 2013 *Cuckoo Malware Analysis* (Packt Publishing Ltd)
[25] Yin H, Song D, Egele M, Kruegel C and Kirda E 2007 Panorama: capturing system-wide information flow for malware detection and analysis *Proceedings of the 14th ACM conference on Computer and communications security* (ACM) pp 116–127 URL https://doi.org/10.1145/1315245.1315261
[26] Bartholomew D 2006 *Linux Journal* **2006** 3
[27] Dinaburg A, Royal P, Sharif M and Lee W 2008 Ether: malware analysis via hardware virtualization extensions *Proceedings of the 15th ACM conference on Computer and communications security* (ACM) pp 51–62 URL https://doi.org/10.1145/1455770.1455779
[28] Willems C, Holz T and Freiling F 2007 *IEEE Security and Privacy* **5** 32–39 URL https://doi.org/10.1109/MSP.2007.45
[29] Lindorfer M, Kolbitsch C and Comparetti P M 2011 Detecting environment-sensitive malware *International Workshop on Recent Advances in Intrusion Detection* (Springer) pp 338–357 URL https://dx.doi.org/10.1007/978-3-642-23644-0_18
[30] Dolan-Gavitt B F, Hodosh J, Hulin P, Leek T and Whelan R 2014 Repeatable reverse engineering for the greater good with panda Tech. rep. Department of Computer Science, Columbia University URL http://dx.doi.org/10.7916/D8WM1C1P
[31] PANDA Panda - open-source architecture-neural dynamic analysis URL https://github.com/panda-re/panda
[32] Volatility: open source memory forensics URL http://www.volatilityfoundation.org
[33] Ida is a windows, linux or mac os x hosted multi-processor disassembler and debugger URL https://www.hex-rays.com/products/ida/
[34] x86 opcode and instruction reference URL http://ref.x86asm.net/coder32.html
[35] Lyda R and Hamrock J 2007 *IEEE Security & Privacy* **5** 40–45 URL https://doi.org/10.1109/MSP.2007.48
[36] Böhne L 2008 *Pandoras bochs: Automatic unpacking of malware* Ph.D. thesis University of Mannheim URL http://2014.hack.lu/archive/2009/PandorasBochs.pdf
[37] Upx, a free and open source, cross-platform runtime packer URL http://upx.sourceforge.net/
[38] Pecompact2, windows executable compressor URL http://pecompact2.software.informer.com
[39] Fsg 2.0, f[ast] s[mall] g[ood] perfect compressor for executable files URL http://www.downloadpcsoft.com/Windows/Development/Other/FSG_24767.html
[40] Armadillo, overlays packer and obfuscator URL http://the-armadillo-software-protection-system.software.informer.com
[41] Molebox, a free executable compression and encryptor URL https://molebox.en.softonic.com
[42] Winupack, a freeware runtime packer URL http://www.softpedia.com/get/PORTABLE-SOFTWARE/Compression-Tools/Windows-Portable-Applications-Portable-WinUpack.shtml
[43] Themida, a commercial executable protector URL http://www.oreans.com/
[44] Vmprotect, a commercial executable virtualization URL http://vmpsoft.com/
[45] Rieck K, Trinius P, Willems C and Holz T 2011 *Journal of Computer Security* **19** URL https://dx.doi.org/10.3233/JCS-2010-0410
[46] Rolles R 2009 Unpacking virtualization obfuscators *3rd USENIX Workshop on Offensive Technologies.(WOOT)* URL http://dl.acm.org/citation.cfm?id=1855877
[47] Pestudio URL https://www.winitor.com
[48] Cohen M, Bilby D and Caronni G 2011 *digital investigation* **8** S101–S110

IOP Conf. Series: Journal of Physics: Conf. Series **801** (2017) 012059 doi:10.1088/1742-6596/801/1/012059

Combination of Huffman Coding Compression Algorithm and Least Significant Bit Method for Image Hiding

D Rachmawati[1*], A Amalia[1], and J Surya[1]

[1]Department of Computer Science, Faculty of Computer Science and Information Technology, University of Sumatera Utara, Jl. Universitas No. 9-A, Medan 20155, Indonesia.

*Email: dian.rachmawati@usu.ac.id, amalia@usu.ac.id, johansurya20871@gmail.com

Abstract. Least Significant Bit method can be used to hide image data into another image, but it need 8 bytes of data to store 1 bit of data, therefore, to store an image with red, green and blue channel. It would require 24 times the size of the image itself. To solve the size problem in hiding data, we implement and experiment with Huffman Coding compression algorithm to reduce the size of the image data needed in C#. Our experimental results are as follows. First, the less colour value variation, Huffman Coding algorithm will give a better compressed size. Second, the average file size change in this research is 29.17%.

1. Introduction

Steganography is one of the most powerful techniques to conceal the existence of hidden secret data inside a cover object [1]. Least Significant Bit steganography is one such technique in which least significant bit of pixels of the image is replaced with data bits. This approach has the advantage that it is simplest one to understand, easy to implement and results in stego-images that contain embedded data as hidden [2]. Images created from pixels i.e. If any pixel created by using these three colors red, green and blue are called as RGB. Each color of a pixel is one byte information that shows the density of that color. So if only last layer of information is used, then the last bits of the pixels has to be changed, in other hands we have 3 bits in each pixel so we have 3*height*width bits memory to write our information [5].

Compression refers to reducing the quantity of data used to represent a file, image or video content without excessively reducing the quality of the original data. It also reduces the number of bits required to store and/or transmit digital media [3]. Image compression can be bifurcated as lossy and lossless compression. Lossy compression as the name implies results in loss of some information. The compressed image is similar to the original uncompressed image but not identical to the previous as in the process of compression some information regarding the image has been lost. They are generally suited for photographs. The most common example of lossy compression is JPEG. Lossless compression compresses the image by encoding all the information from the original file, so when the image is decompressed, it will be exactly identical to the original image. Examples of lossless image compression are PNG and GIF. (GIF only allows 8-bit images). When to use a certain image compression format really depends on what is being compressed [6]. The purpose for image compression is to reduce the amount of data required for representing sampled digital images and therefore reduce the cost for storage and transmission [4].

Huffman coding is a loseless data compression technique. Huffman coding is based on the frequency of occurrence of a data item i.e. pixel in images. The technique is to use a lower number of bits to encode the data in to binary codes that occurs more frequently [3].

2. Method

The experiments are conducted on the Windows 7 Notebook which has Intel Core i3 processor with 32-bit architecture and 2048MB RAM. The development environment being used for coding C# scripts is SharpDevelop.

The image hiding process began with the secret image being compressed before insert into cover image. The compression process will create the Huffman Tree at first.

Steps for creating Huffman Tree:

1. Get all the frequency of all color value from the secret image.
2. Select the two lowest frequency color value from it and treat it as the first node of Huffman Tree.
3. Create parent node from those two nodes and count the sum of their frequency.
4. Remove those two nodes and replace with the parent node. It will be used to create the tree.
5. Do the steps above continuously until there is only one node left.
6. Every node that place at left branch will be given 0 value and the right branch will be given 1 value.
7. Read from the root of the tree to all nodes by checking their branch.

The Huffman Code is created from the Huffman Tree and will be use to encode all the secret image data. This compression result will be divided into 3 parts. The first part is the bits that contain information about the image size, the Huffman Code bit length variation which will be used to read the code length and its frequency that will be used to read the Huffman Code table in the second part. The second part is the bits that contain the pixel value and its Huffman Code. This part will be used to decode all the compressed data in the third part. The third part is the bits that contain all the image pixels value that have been encoded with Huffman Code. Those merged part will be insert into cover image by using Least Significant Bit method to produce stego-image. Because the Least Significant Bit method only replace the last bit, then the steps to this method is as follow:

1. If the bit value that need to store is 1 and the color value modulo 2 is 0, then increase the color value by 1
2. If the bit value that need to store is 0 and the color value modulo 2 is 1, then decrease the color value by 1
3. If the bit value that need to store is same with the color value modulo 2, then skip to the next bit that need to store.

The image recovery process began with the stego-image being extracted bit by bit by using Least Significant Bit method. In each extraction process, the system will check the extracted bits for decompression. If one of the decompression process failed, then the whole recovery process will be failed. If the decompression process has produced an image, the recovery process will be success and sending the decompressed image as the result.

3. Results and Discussions

The results of the experiments of each set are presented in Table 1 and Table 2 as follows.

Table 1. The Compression result for 5 sample images.

Image	Color Value Variation (0-255)	Original (Bit)	Compressed (Bit)	Compressed Ratio (%)	Compression Time (ms)
All Black.bmp	1	2400	365	15.21	1
BlackWhite.bmp	2	2400	374	15.58	1
BlackWhiteGray.bmp	3	2400	503	20.96	1
_sicon.bmp	195	24576	17878	72.75	3
woman.bmp	199	98304	28826	29.32	9

In Table 1, it can be seen that the average compressed ratio is 30.76% and the average compression time is 3ms.

Table 2. File size changes for 10 sample images.

Image	File Size	File Size (After inserted)	Change (%)
All Black.bmp	376 bytes	454 bytes	20.74
BlackWhite.bmp	376 bytes	454 bytes	20.74
BlackWhiteGray.bmp	376 bytes	454 bytes	20.74
fivepointstar.bmp	3.05 KB	4.05 KB	32.79
Airplane.bmp	6.80 KB	8.86 KB	30.29
woman.bmp	12.0 KB	16 KB	33.33
_sicon.bmp	3.05 KB	4.05 KB	32.79
brick.bmp	175 KB	234 KB	33.71
Core2Quad.bmp	670 KB	893 KB	33.28
Intel-X38.bmp	597 KB	796 KB	33.33

In Table 2, it can be seen that the average file size change is 29.17%. The file size change showed in Figure 1.

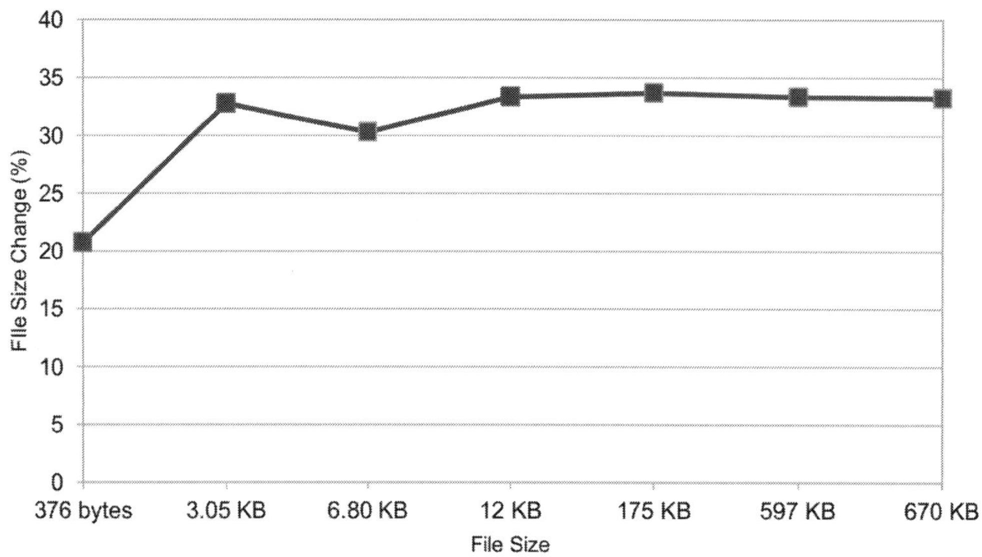

Figure 1. File Size Change

4. Conclusions

The conclusion in this research are as follows. First, the less the color value variation, Huffman Coding algorithm will give a better compressed size. Second, the average file size change in this research is 29.17%.

Acknowledgments
The authors gratefully acknowledge that the present research is supported by Lembaga Penelitian Universitas Sumatera Utara.

References

[1] Champakamala B S, Padmini K and Radhika D K Least Significant Bit algorithm for image steganography *International Journal of Advance Computer Technology (IJACT)* **3** (4) 34-38

[2] Verma V, Poonam and Chawla R 2014 An Enhanced Least Significant Bit Steganography Method Using Midpoint Circle Approach *International Conference on Communication and Signal Processing* 105 -108

[3] Sharma M 2010 Compression Using Huffman Coding *IJCSNS International Journal of Computer Science and Network Security* **10** (5) 133 – 141

[4] Pujar J H and Kadlaskar L M 2010 A New Lossless method of Image Compression and Decompression Using Huffman Coding Techniques *Journal of Theoretical and Applied Information Technology* 18-22

[5] Patel K, Utareja S, Gupta H 2013 Information Hiding Using Least Significant Bit Steganography and Blowfish Algorithm *International Journal of Computer Applications* **63** (13) 24 – 28

[6] Mathur MK, Loonker S, Saxena D 2012 Lossless Huffman Coding Technique For Image Compression And Reconstruction Using Binary Trees *International Journal of Computer Technology and Applications* **3** (1) 76 – 79

IOP Conf. Series: Journal of Physics: Conf. Series **801** (2017) 012060 doi:10.1088/1742-6596/801/1/012060

Online trust building through third party trust transfer and third party protection

Wanda Wandoko[1], Bahtiar Saleh Abbas[2], Dyah Budiastuti[3], Raymond Kosala[4]

[1,3]School of Business and Management, Bina Nusantara University, Indonesia
[2]Industrial Engineering, Faculty of Engineering, Bina Nusantara University, Indonesia
[4]Binus Business School, Bina Nusantara University, Indonesia

Email: wwandoko@binus.ac.id

Abstract. The primary objective of this research is to develop an online trust building mechanism for SME (Small Medium Enterprise). Trust is very important in e-commerce. The nature of online shopping has a greater uncertainty than offline shopping. Seeing as there is an uncertainty that can produce risks, a prospective buyer's trust is needed. A lot of people's unwillingness to shop online is caused by their lack of trust toward e-commerce. E-commerce is said to be one of the ways for SME to compete with bigger companies. However, building trust requires immense time and cost. SME with limited resources may experience difficulties in building trust just with their own resources. Base on literature research that needs to be validated in next research, we found that trust can be built through trust transfer from the reputable and well-known trust-mark issuer, and third party protection such as escrow account service and credit card issuer.
Keywords—Trust, Risk, Trust-mark, Third party protection, e-commerce

1. Introduction

Indonesia is a big country, 3rd largest in population and 16th largest GDP in the world, with GDP 888,838 million USD on 2014. With Internet penetration reaching about 71.2 million people in 2013 and expected to reach 100 million people on 2016. According to Google survey, 1 in 2 Indonesian people who have internet access has an intention to shop online within 12 months[1]. The potential market of Indonesian e-commerce is 8 billion USD in 2013 and expected to be 25 billion USD in 2016[1]. Several multinational companies have already put their feet in Indonesia. The growth of e-commerce in Indonesia does not only attract multinational enterprises, but also unethical ones. This calls for a more serious attention in e-commerce users' protection[2]. Indonesia e-commerce market still has a lot of space to improve. The percentage of Indonesian online shopper compare to Indonesia people who have access to the internet is only 27% in 2014[3], while the world already reaches 38% in 2011 and expected to be 45% in 2015. The nature of the online transaction, where there is no physical contact and there is a temporal and spatial separation between the vendor and the consumer, creates more uncertainty[4][5]. In the e-commerce environment, the transaction is done without physical contact, thus buyers are forced to place their trust based on the information that are displayed on the website [6]. It makes buyers insecure; this condition makes buyers' uncertainties bigger than when purchasing in traditional commerce environment [7][8]. These uncertainties may lead to risk, and most people are risk-averse[9]. Several empiric researches show that a lot of consumer refrain from online shopping because of the risk it exposes [10][3]. Trust helps a consumer to reduce their perception of risk [11][12]. Several surveys show that trust is the biggest factor for e-commerce buyers to make a transaction. The absence of trust may stunt the growth of e-commerce [13][2]. This is consistent with TNS's survey result, which says that 40% of people who don't shop online do so because of their distrust [1]. SME in Indonesia has an important role by having more than 50% share in PDB. E-commerce is also considered as a substantial opportunity for SME to compete with enterprise because of wider market reach, more efficient market, and cheaper transaction cost [14][15][16]. However, there is an important issue in e-commerce, which is consumer's trust[8]. Building trust is a job that needs a big effort in term of resources and money, for a long period [17]. This might not be easy for SME. Other than doing it by their own resources, trust can also be built by collaborating with the third party through certification or endorsement.

2. LITERATURE REVIEW

2.1. E-commerce

E-commerce is a business transaction that transfers the ownership of the good or services through an electronic medium such as internet, telephone, television, or fax. There are stages in e-commerce transaction: 1) The Order Placement, 2) The Payment and the 3) delivery of the good or services. A transaction can be defined as e-commerce transaction if the transaction does at least the first phase through electronic medium. There are several types of e-commerce that involving consumer as a buyer as follow: 1) B2C (Business to Consumer) and 2) C2C (Consumer to Consumer). E-commerce also can be categories by the market or the place of transaction, like: Online Marketplace, Classified Ads, Daily Deals, Online Retail and social media such as Instagram and Facebook[18].

2.2. Online Trust

Trust has been studied by multi-discipline for years, but until now there is no single scientific definition that generally accepted[19][8]. One of the trust definition that is widely accepted: Trust is a psychological condition where one party is willing to be vulnerable based on a positive expectation of the intention of the other party[19]. In the e-commerce context, trust or online trust can be defined as the willingness of the consumer to depend on the vendor. The willingness is based on the perception that the vendor will fulfill or deliver the commitment[20]. In e-commerce, trust plays more significant role compared to brick and mortar shop or to traditional market which has a face to face relationship because there is no physical contact in e-commerce, either with the vendor, goods or money. The consumer and the vendor can be separated by a few km to thousands km, it causes the consumer to have no possibility to use social cues to assess the vendor[21][22]. According to Mayer and McKnight, trust is a very important role to make a consumer be willing to take a risk where there are uncertainty, interdependence, and fear of opportunism[12]. Trust involves a calculation process to calculate the gain and the loss for when the trusted party delivers the commitment or otherwise[23]. To build the trust, the buyer needs analytical skill to predict the behavior of the other party. There are several mechanisms in building online trust, which are: Institution-based, Knowledge-based mechanism dan Transfer [11][24].

2.3. Institution-based Trust

Institution-based trust is the perception or beliefs that a structural condition exists to mitigate the risk and increase the possibility for the transaction to succeed[25][21][26]. In the e-commerce environment where the uncertainty is bigger compared to face to face, trust is fundamental in building and retaining inter-organizational relationship [7]. Even though institution-based trust is an important factor to drive purchase intention in an online environment, the trust itself is not enough to make a consumer do an online purchase[12]. These vendor-specific mechanisms are useful in building trust and eliminating risks inherent with transaction related to specific vendor[26]. Institution-based trust is built by 1) Structure Assurance and 2) Situational Normality [27][25][20]. Structure assurance can be defined as a belief that a protection mechanism from the institution is in place to ensure the condition are conducive for a transaction to succeed, such as warranty, contract, rule, process, procedure and system [25][11]. Trust that is earned through structure assurance is usually obtained by: 1) Transferences, 2) Guarantees, 3) Signaling [12]. Structure Assurance sources can be : Trust Mark (Dependent Vendor), Escrow Account (Independent Vendor), bank as credit card issuer (Independent vendor) [12][26]. Situational Normality can be defined as a belief based on general belief that the e-commerce environment and the vendor are normal, favorable and appropriate to gain success[11][28].

2.4. Perceived Risk

People tend to give a bigger value to something that is already certain compared to something that is not certain yet[29]. Most people are risk-averse, they will choose a less gain but with 100% probability, compared to bigger gain but with uncertainty[12][29]. For example, people will choose to gain 3,000 but sure, compared to having a probability of 80% to gain 4,000[29]. Risk-averse people will reject the opportunity to get 200 or lose 100 with the equal probability [30]. Risk can be defined as a situation where the result of a decision involves uncertainty. This happens because of the decision maker's incapability to determine end result that depends on the upcoming situation. A decision with countable probability is still considered as a risk because the result is still uncertain[31][32]. Shopping has risks, especially online shopping because the buyer doesn't have a complete information to make a decision and cannot perfectly predict the impact of the decision, whether it's disappointing or otherwise[22][32]. According to Hong and Cha (2013), problems that may arise and thought by the consumer while they are online shopping can be called as online shopping risk perception, which henceforth will be called as perceived risk[32]. The source of risk in e-commerce or online risk is the uncertainty of e-commerce environment, such as internet technology and uncertain seller's character. Risk does not only affect trust, but it also directly affects buying intention. if a consumer think that an online retail shop has a high-risk degree, then the consumer will reduce risk or risk reduction, such as canceling the order or reduce purchase[33]. There are several categories of online shopping risk, the categories are:[34][35][31]

- Product/Performance risk: Potential loss caused by the product is not as expected, in term of performance, color or size.
- Financial Risk: Potential loss caused by the good is not delivered, the bad after sales, bad warranty.
- Time Risk :Potential loss caused by the order, payment, and good delivery process take longer time,
- Confidential Risk: Potential loss caused by the lost control of personal dan financial data.
- Psychological Risk, Potential loss caused by un an easy feeling or anxiety about the online purchase that might not end as expected or meet her/his goals. Psychological risk also caused by possibility financial loss caused by the credit card information is stolen.
- Delivery Risk: Most of the online shopping good are delivered by courier. Except for digital thing such as mp3, ticket, software etc that usually the buyer can get/download the good right away. For nondigital thing there is a potential risk the good is damage when they delivered, the delivery takes longer time than expected, the delivery cannot be tracked etc.

2.4.1. Trust Mark

Trust Mark is an integrity certification logo that is displayed on the website. The purpose is to show the visitor of the website that the institution in which the web-site belongs to is eligible to conduct trade transaction through electronic, having followed and passed a series of assessment and audit from authorized organization. With the label, mark, or logo of trust mark embedded on the institution website, it means that the institution has proved that they have a certain standard. For example, can be one of them or a combination of them, but not limited to: computer security, data security, the business model, financial stability, after sales support[2][36]. There are different names for the similar concept of trust mark. For example, Web Assurance Seals, Third-Party Seal. One of trust mark goal is to ensure e-commerce customer is protected by e-commerce institution, through securing the website in processing the transaction, guarding personal and financial information from a hacker, virus, and other means of electronic data exchange[2][37]. Trust mark can increase the trust, especially on the small medium enterprise where their brands are not well known[2]. Displaying the trust mark is one of the ways to increase consumer's trust in doing online transaction. If an institution is displaying the e-commerce trust mark on their website, we can expect that the institution has the integrity to respect the law, good business behavior, and other standards [2][25][38]. Every trust mark issuer represents their own standard. There are several categories in trust mark: a) Transaction Security assurance (e.g., VeriSign), b) consumer privacy assurance (e.g.,TRUSTe), and c) transaction integrity assurance (e.g., BBBOnLineReliability)[39].The several reasons consumer look for the trust mark when shopping online[2]:

- Give the consumer confidence in shopping with the vendor
- The vendor complies with the consumer legislation
- The vendor has secure payment methods
- The trust mark may provide help when a problem occurs.

2.4.2. Trust Transfer

Trust Transfer is a process in which an object that has been proven to be reliable gives the effect of trust on the other objects that are not too famous or have a little experience [23][27][40]. Trust can be transferred from different kind of sources. The source of trust can be a place, industry, associations or a person. Trust transfer comes from one entity to another entity that perceived as having a relationship [16]. The major use of the trust transfer is to gain trust from a third party for the object of trust based on third party's previous experience with an object that has association with the object of trust, rather than to gain it based on direct experience with the object of trust or parties that will be trusted [41]. Some examples of the trust transfer: According to Johnson and Kaye (2004) and Liu (2004), a traditional vendor who has a physical store can be perceived as more credible than the one who has an online store only[42]. When they see the trust-mark logo on the online store, and their trust toward the online store is increased, this process called trust transferred[43].

2.5. Online Payment Provider Protection (OPPP).
There are several online payment providers which give the extra protection in the online shopping environment such as:

2.5.1. Escrow Service
Escrow service is one of dispute avoidance system, that prevent the seller to get the money before buyer confirmation on good receiving and prevent the buyer don't pay when the seller sent the good because the seller only sent the good after he/she has a notification from the escrow service provider (or the escrow) that the buyer already deposits the money into escrow account. Escrow service cause extra cost for the fee [44]. In addition, the consumer has several options to use other payment services to make online purchases which have a similar degree of protection such as Paypal. Paypal extends the protection more than the common escrow services. That usually ended when the buyer already confirm that the good already received with the condition as agreed. Paypal extends the protection period until 180 days after the purchase[45][44]. Some escrow service providers have an ADR (Additional Dispute Resolution) to help its customer. The escrow is not only helped its customer by facilitates communications between the buyer and seller. But if the negotiation between the seller and the buyer can not produce an agreed solution, the escrow will make a decision as online arbitrator. The ADR provided by Paypal able to decrease 50% complain to the seller, and reduce 20% chargeback (retracts the money from the seller account and transfers back to the buyer account) to the seller[44].

2.5.2. Credit card protection
3D Secure already implemented by several credit card companies like Visa (Verified by Visa), Master Card (MasterCard SecureCode), JCB International (J/Secure) ect., to give another layer of protection on online payment using a credit card. We can see their logo on the online shop that working with them. 3D Secure has OTP (one-time password) and 'Phrase' to authentic the seller and the buyer. With OTP, the consumer needs to key in the OTP that sent from the credit card company to their phone cell. And the consumer also will be able to verify whether the online shop is the real one by the phrase that shown on the online shop. The phrase that displayed in the online shop should be the same with the one that the consumer registered on the credit card issuer, usually bank. Several numbers weakness of 3D secure:

- 3D Secure are only give the protection on the online shop that working with them, while the consumer needs the protection on all online shop.
- There is no law to enforce online shop to use 3D Secure in only shop payment using a credit card.

In Indonesia, there is no law that particularly to protect credit cardholder from the fraudulent transaction. But some credit card issuer already have a policy regarding this issue, for example Visa has a policy called Zero Liability, that guarantee the credit card holder won't be held any responsibility for transactions that not authorized by her/him [46].

2.6. Purchase Intention
Purchase mostly preceded by purchase intention. Even though not all purchase intention will be converted to real purchase[47]. Purchase intention is not perfect to portray the real purchase because it can be changed by the time. But the purchase intention variable still reliable to represent the real purchase, and still use by researcher widely [48]. Without real purchase, both the seller and the consumer can't get the full benefit of e-commerce and the technology behind them. The consumer can't get the benefit that been offered by e-commerce such as a lower price and the convenient[22]. And the seller can't have the more efficient market, and cheaper transaction cost[14].

3. Model and Hypotheses Development
Trust in general is built slowly, and requires a big amount of resources and time[17][49]. One of the methods that relatively fast to build trust is trust transfer method, which recently received great attention[41]. Trust transfer can be done by using the trust-mark from the well-known trust-mark issuer. Trust-mark that more widely used, will be more effective[50]. There are several studies that show the trust-mark able to increase the online trust, especially for small or not very famous online retailer[50][51]. And also able to reduce online risks[50][52][53]. Hence, we hypothesized:

H1. Trust-mark is positively related to online trust.
H2. Trust-mark is negatively related to perceive risk.

Risk in a simple way can be formulated into the consequences or potential loss multiply by the occurrence probability[54]. So to reduce the risk, we can reduce one of the two components, in this case is the potential loss. In the online retail environment, there are some institutions that are independent of online retailer that helps protect consumers from potential losses caused by the online transaction, for example credit card issuer and escrow service[12]. With the protection, the online risk can be expected to be lower. This protection also able to increase the online trust[12][55] . Hence, we hypothesized:

H3. Online payment provider protection is positively related to online trust.
H4. Online payment provider protection is negatively related to perceive risk.

Shopping online has a greater risk compared to traditional shopping. Because of the risk, trust is necessary to make a customer want to shop online [21][12]. Trust and risk perception is an important psychological factor in influencing consumer shopping behavior in shopping online [56]. To participate in online shopping requires complex calculations relating to the behavior of sellers and internet security, trust can reduce complexity and reduce the threat of a possible seller of doing things that are beyond his expectations[32][57]. Trust in online shopping site has an important role in online shopping, because it is likely customers will not shop if the customer does not trust the online shopping site[32][58]. Trust is a strong indication of the intention to buy [57]. It is supported by several empirical studies that show trust has a positive influence on buying intention[58][59]. Hence, we hypothesized:

H5. online trust is positively related to purchase intention.
H6. Risk is negatively related to Purchase Intention.

There are several studies with different results according to the trust mark, some of them show the positive relation between trust mark and the trust, on the other hand, the trust mark does not have any significant impact on trust or on purchase intention[39][50][60]. The negative result might lead by the consumer didn't know what the trust-mark is. For example, the survey conduct by The European Consumer Centre's Network in 19 European countries[2] shows only 37% consumers look for a trust mark when they are shopping, but it maybe because only 49% percentage of people know what a trust mark is, and only 34% know of any trust mark. Other research shows that trust-mark with high-dissemination are more effective[50]. Same result on escrow services, the main cause of escrow services is not used is the lack of awareness (70%)[26]. And according to trust transfer theory, trust is transferred from a more trusted and known institution to less known institution[23][27]. Hence, we hypothesized:

H7. Institution familiarity is moderates relationship between trust-mark and online trust
H8. Institution familiarity is moderates relationship between trust-mark and perceived risk
H9. Institution familiarity is moderates relationship between OPPP and online trust
H10. Institution familiarity is moderates relationship between OPPP and perceived risk

Based on literature review and hypotheses above, we propose the model:

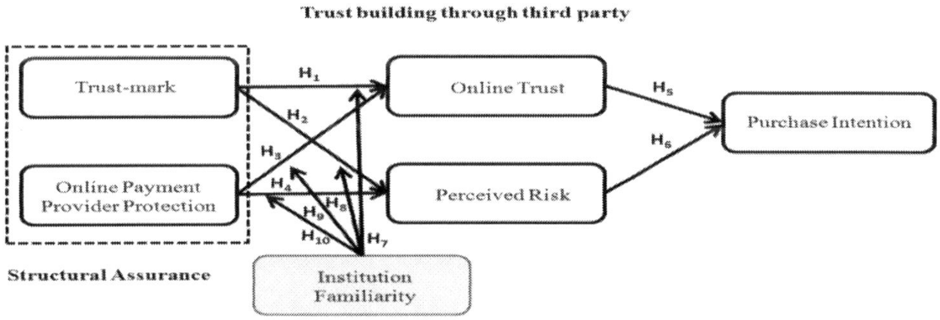

Fig. 1. The Proposed Model

4. CONCLUSION AND FUTURE WORK

Trust is an important key in e-commerce, including in Indonesian market. Some Indonesian people didn't want to do online shopping, mostly because there are risks that inherent in online shopping. But building trust is cost time and money, while SME has more limited resources and time. Choosing the third party is not an easy job, most of the third parties that already have a reputation usually demanding better condition and more money for the fee compare to the less known third party. But regarding trust theory and several researches, more reputable and well-known institution will give better impact in trust transfer.

This research proposed online trust building based on structure assurance, which is a third party institution that consists of trust-mark and online payment services protection, that already in the market and ready to use. Those source of trust needs relative small initial investment and relative faster, more effective on a small company that not really famous. The proposed model will be tested empirically in Indonesia market. There are a few open research related to this study such as: e-commerce risk ramification to differentiate the source of risk, other trust building that suitable for different market, e-commerce risk reduction to increase the purchase intention.

References

[1] Google, "Online Shopper Study : IDEA Roundtable," Jakarta, 2014.
[2] The European Consumer Centres' Network, "Trust marks report 2013 : " Can I trust the trust mark ?"," 2013.
[3] Asosiasi Penyelenggara Jasa Internet Indonesia and Puskakom UI, *Profil Pengguna Internet indonesia 2014*, Pertama. Jakarta, 2015.
[4] X. Li, G. Rong, and J. B. Thatcher, "DO we trust the technology? People? Or both? Ruminations on technology trust," *15th Am. Conf. Inf. Syst. 2009, AMCIS 2009*, vol. 6, pp. 3813–3822, 2009.
[5] I. L. Wu, "The antecedents of customer satisfaction and its link to complaint intentions in online shopping: An integration of justice, technology, and trust," *Int. J. Inf. Manage.*, vol. 33, no. 1, pp. 166–176, 2013.
[6] K. C. Lee, S. Lee, and Y. Hwang, "The impact of hyperlink affordance, psychological reactance, and perceived business tie on trust transfer," *Comput. Human Behav.*, vol. 30, pp. 110–120, 2014.
[7] P. A. Pavlou, Yao-Hua Tan, and D. Gefen, "The transitional role of institutional trust in online interorganizational relationships," in *Hawaii International Conference on System Sciences*, 2003, no. C, p. 10.
[8] S.-J. Yoon, "The antecedents and consequences of trust in onlin-purchase decisions," *J. Interact. Mark.*, vol. 16, pp. 47–63, 2002.
[9] Y. Y. Liu, J. C. Nacher, T. Ochiai, M. Martino, and Y. Altshuler, "Prospect theory for online financial trading," *PLoS One*, vol. 9, no. 10, p. 8, 2014.
[10] Erwan Teguh, L. Lauwira, C. Peter P. Sutedja, and D. Sumartono, "The Dawn of an e-commerce age," Jakarta, 2015.
[11] D. H. McKnight, V. Choudhury, and C. Kacmar, "Developing and validating trust measures for e-commerce: An integrative typology," *Inf. Syst. Res.*, vol. 13, no. 3, pp. 334–359, 2002.
[12] Y. Fang, I. Qureshi, H. Sun, P. McCole, E. Ramsey, and K. H. Lim, "Trust, Satisfaction, And Online Repurchase Intention: The Moderating Role of Perceived Effectiveness of E-commerce Institutional," *MISQ*, vol. 38, no. 2, pp. 407–427, 2014.
[13] K. Ozpolat and W. Jank, "Getting the most out of third party trust seals: An empirical analysis," *Decis. Support Syst.*, vol. 73, pp. 47–56, 2015.
[14] United Nations and UN, "Information Economy Report 2015: Unlocking the Potential of E-commerce for Developing Countries," Geneva, 2015.
[15] S. L. Jarvenpaa, N. Tractinsky, and L. Saarinen, "Consumer trust in an internet store," *J. Comput. Mediat. Commun.*, vol. 5, pp. 1–35, 2000.
[16] K. J. Stewart, "Trust Transfer on the World Wide Web," *Organ. Sci.*, vol. 14, no. 1, pp. 5–17, 2003.
[17] M. M. M. Head and K. Hassanein, "Trust in e-commerce: evaluating the impact of third-party seals," *Q. J. Electron. Commer.*, vol. 3, pp. 307–325, 2002.
[18] C. S. P. Ng, "Intention to purchase on social commerce websites across cultures: A cross-regional study," *Inf. Manag.*, vol. 50, no. 8, pp. 609–620, 2013.
[19] D. M. Rousseau, S. B. Sitkin, R. S. Burt, and C. Camerer, "Not so different after all: A cross discipline view of trust," *Acad. Manag. Rev.*, vol. 23, no. 3, pp. 393–404, Jul. 1998.
[20] W. Wang and R. Chen, "How to Retain Your Customers : The Impact of Consumer Trust and

Commitment in E- marketplaces Commitment in E-marketplaces," *Elev. Wuhan Int. Conf. E-bus.*, pp. 133–143, 2012.
[21] P. Pavlou, Y.-H. Tan, and D. Gefen, "Institutional Trust and Familiarity in Online Interorganizational Relationships," *Proc. Elev. Eur. Conf. Inf. Syst.*, vol. 16, no. 1, p. 21, 2003.
[22] C. M. Chiu, E. T. G. Wang, Y. H. Fang, and H. Y. Huang, "Understanding customers' repeat purchase intentions in B2C e-commerce: The roles of utilitarian value, hedonic value and perceived risk," *Inf. Syst. J.*, vol. 24, no. 1, pp. 85–114, 2014.
[23] P. M. Doney and J. P. Cannon, "An Examination of the Nature of Trust in Buyer-Seller Relationships," *J. Mark.*, vol. 61, no. April, p. 35, Apr. 1997.
[24] N. Wang, X. L. Shen, and Y. Sun, "Transition of electronic word-of-mouth services from web to mobile context: A trust transfer perspective," *Decis. Support Syst.*, vol. 54, no. 3, pp. 1394–1403, 2013.
[25] D. H. Mcknight and N. L. Chervany, "What Trust Means in E-Commerce Customer Relationships: An Interdisciplinary Conceptual Typology.," *Int. J. Electron. Commer.*, vol. 6, no. 2, pp. 35–59, 2001.
[26] P. A. Pavlou and D. Gefen, "Building effective online marketplaces with institution-based trust," *Inf. Syst. Res.*, vol. 15, no. 1, pp. 37–59, 2004.
[27] Q. Yang, L. Huang, and Y. Xu, "Role of Trust Transfer in E-Commerce Acceptance," *Tsinghua Sci. Technol.*, vol. 13, no. 3, pp. 279–286, 2008.
[28] W. Wang, R. Chen, Carol Xiaojuan Ou, and F. Liu, "Information Systems: Transforming the Future," *Australas. Conf. Inf. Syst.*, no. December, p. 11, 2013.
[29] D. Kahneman and A. Tversky, "Econometrica , 47(2) ," *Econometrica*, vol. 47, no. March, pp. 263–291, 1979.
[30] M. Rabin, "Risk aversion and expected utility theory: A calibration theorem," *Econometrica*, vol. 68, no. 5, pp. 1281–1292, 2000.
[31] L. Zheng and M. Favier, "Consumer Perceived Risk and Risk Relievers in E-Shopping," *J. Electron. Commer. Res.*, vol. 13, no. 3, pp. 255–274, 2012.
[32] I. B. Hong and H. S. Cha, "The mediating role of consumer trust in an online merchant in predicting purchase intention," *Int. J. Inf. Manage.*, vol. 33, no. 6, pp. 927–939, 2013.
[33] A. Leeraphong and A. Mardjo, "Trust and Risk in Purchase Intention through Online Social Network: A Focus Group Study of Facebook in Thailand," *J. Econ. Bus. Manag.*, vol. 1, no. 4, pp. 314–318, 2013.
[34] Y. Lu, "The establishment of perceived risk and trust in e-commerce," *Proc. 2012 2nd Int. Conf. Bus. Comput. Glob. Informatiz. BCGIN 2012*, pp. 229–232, 2012.
[35] M. V. Nepomuceno, M. Laroche, and M. O. Richard, "How to reduce perceived risk when buying online: The interactions between intangibility, product knowledge, brand familiarity, privacy and security concerns," *J. Retail. Consum. Serv.*, vol. 21, no. 4, pp. 619–629, 2014.
[36] L. A. R. Remotti, A. F. van Veenstra, M. van Lieshout, L. Kool, G. Rumpf, B. Ipektsidis, and T. Damvakeraki, "EU online Trustmarks Building Digital Confidence in Europe (Final Report)," 2012.
[37] W. Koo, "Online Store Atmospherics : Development of a Gender-Neutral Measure," University of Tennessee, 2014.
[38] CaseTrust, "Webfront (E-commerce)," 2016. [Online]. Available: https://www.casetrust.org.sg/accreditation-detail.aspx?id=9. [Accessed: 14-Apr-2016].
[39] X. Hu, G. Wu, Y. Wu, and H. Zhang, "The effects of Web assurance seals on consumers' initial trust in an online vendor: A functional perspective," *Decis. Support Syst.*, vol. 48, no. 2, pp. 407–418, 2010.
[40] G. W. Bock, J. Lee, H. H. Kuan, and J. H. Kim, "The progression of online trust in the multi-channel retailer context and the role of product uncertainty," *Decis. Support Syst.*, vol. 53, no. 1, pp. 97–107, 2012.
[41] B. L. Delgado-Márquez, N. E. Hurtado-Torres, and J. A. Aragón-Correa, "The dynamic nature

of trust transfer: Measurement and the influence of reciprocity," *Decis. Support Syst.*, vol. 54, no. 1, pp. 226–234, 2012.

[42] S. C. Wingreen, S. L. Baglione, and G. R. Storholm, "Transfer of electronic commerce trust between brick-and-mortar and online business environments," *Australas. Conf. Inf. Syst.*, vol. 23, no. Gefen 2002, pp. 1–12, 2012.

[43] P. Jiang, D. B. Jones, and S. Javie, "How third-party certification programs relate to consumer trust in online transactions: An exploratory study," *Psychol. Mark.*, vol. 25, no. 9, pp. 839–858, Sep. 2008.

[44] P. Cortés, *Online Dispute Resolution for Consumers in the European Union*. New York: Routledge, 2010.

[45] PayPal, "PayPal User Agreement," 2015. [Online]. Available: https://www.paypal.com/us/webapps/mpp/ua/useragreement-full. [Accessed: 16-Mar-2016].

[46] Visa, "Visa's Zero Liability Policy," 2016. [Online]. Available: https://www.visa.com/chip/personal/security/zero-liability.jsp. [Accessed: 01-Aug-2016].

[47] A. Osman and L. Y. Jin, "Key Drivers of Purchase Intention Among Undergraduate Students," *Int. J. Econ. Commer. Manag.*, vol. II, no. 11, pp. 1–11, 2014.

[48] E. Risen and L. Risen, "The Use of Intent Scale Translations to Predict Purchase Interest," 2008.

[49] M. K. Chang, W. Cheung, and M. Tang, "Building trust online: Interactions among trust building mechanisms," *Inf. Manag.*, vol. 50, no. 7, pp. 439–445, 2013.

[50] S. Van Baal, "Not all seals are equal: An experimental investigation of the effect of third-party seals on purchase probability in electronic commerce," *Electron. Commer. Res.*, pp. 177–206, 2015.

[51] H. Zhang, "Trust-Promoting Seals in Electronic Markets: Impact on Online Shopping Decisions," *J. Inf. Technol. theory Appl.*, vol. 6, no. 4, pp. 29–40, 2005.

[52] M. Salehan, D. J. Kim, and J.-N. Lee, "Antecedents , Processes and Consequences of Web Assurance Seals : A Meta-Analysis Approach," *PACIS Proc.*, p. 12, 2015.

[53] S. E. Kaplan and R. J. Nieschwietz, "A Web assurance services model of trust for B2C e-commerce," *Int. J. Account. Inf. Syst.*, vol. 4, no. 2, pp. 95–114, 2003.

[54] Bilal M. Ayyub, *Risk Analysis in Engineering and Economics*, Second Edi. Boca Raton, 2014.

[55] X. Chen, Q. Huang, D. Robert, and Z. Hua, "The Moderating Effects of Contextual Factors on a Buyer's Trust in E-commerce Platform and Sellers," *PACIS 2014 Proc.*, p. 16, 2014.

[56] J. Mou, "Trust and risk in consumer acceptance of e-services," Witwatersrand, 2015.

[57] B. Lu and W. Fan, "Social Presence, Trust, and Social Commerce Purchase Intention: An Empirical Research," *Pacific Asia Conf. Inf. Syst. – PACIS*, p. 17, 2014.

[58] E. Bonsón Ponte, E. Carvajal-Trujillo, and T. Escobar-Rodríguez, "Influence of trust and perceived value on the intention to purchase travel online: Integrating the effects of assurance on trust antecedents," *Tour. Manag.*, vol. 47, pp. 286–302, 2015.

[59] M. H. Hsu, C. M. Chang, and L. W. Chuang, "Understanding the determinants of online repeat purchase intention and moderating role of habit: The case of online group-buying in Taiwan," *Int. J. Inf. Manage.*, vol. 35, no. 1, pp. 45–56, 2015.

[60] K. Kim and J. Kim, "Third-party Privacy Certification as an Online Advertising Strategy: An Investigation of the Factors Affecting the Relationship between Third-party Certification and Initial Trust," *J. Interact. Mark.*, vol. 25, no. 3, pp. 145–158, 2011.

IOP Conf. Series: Journal of Physics: Conf. Series **801** (2017) 012061 doi:10.1088/1742-6596/801/1/012061

Analysis and Implementation of Graph Clustering for Digital News Using Star Clustering Algorithm

A B Ahdi[1], K R SW[1], A Herdiani[1]

Jl. Telekomunikasi Bandung 40287, West Java, Indonesia

E-mail: aufaba@yahoo.co.id, bagindokemas@telkomuniversity.ac.id, anisaherdiani@telkomuniversity.ac.id

Abstract. Since Web 2.0 notion emerged and is used extensively by many services in the Internet, we see an unprecedented proliferation of digital news. Those digital news is very rich in term of content and link to other news/sources but lack of category information. This make the user could not easily identify or grouping all the news that they read into set of groups. Naturally, digital news are linked data because every digital new has relation/connection with other digital news/resources. The most appropriate model for linked data is graph model. Graph model is suitable for this purpose due its flexibility in describing relation and its easy-to-understand visualization. To handle the grouping issue, we use graph clustering approach. There are many graph clustering algorithm available, such as MST Clustering, Chameleon, Makarov Clustering and Star Clustering. From all of these options, we choose Star Clustering because this algorithm is more easy-to-understand, more accurate, efficient and guarantee the quality of clusters results. In this research, we investigate the accuracy of the cluster results by comparing it with expert judgement. We got quite high accuracy level, which is 80.98% and for the cluster quality, we got promising result which is 62.87%.

1. Introduction

The abundant information in Internet make difficult for user to find related news about specific topic without entering precise and detailed keywords in the search engine. To overcome this problem, we need a news clustering system which can grouping news based on the content and the characteristic of the news automatically. And for data modelling purpose, graph model is the righteous option because it can represent the news and the relation among the news into nodes and edges.

A grouping or clustering is a process in which objects are classified into groups called clusters [1]. The goal of clustering is to divide the data into specific clusters where each cluster has a certain resemblance or relation [2]. There are many methods of graph clustering such as MST Clustering, Chameleon, Makarov Clustering, and Star Clustering. We consider Clustering Star as the most appropriate, because it is easy to understand, accurate, efficient and guarantee the quality of the resulting cluster [3].

2. Theoretical Foundation

2.1 Graph Theory

Systematically graph G is defined as the set of pairs (V, E) where V is the set of nodes is not empty v1, v2, v3, ..., vn. While E is the set of edges connected between two nodes e1, e2, e3, ..., en So it can be written G = (V, E). Degree (degree) of a node is the number of side

by side with that node is denoted by d (v). Examples of modeling graph can be seen in Figure 1.

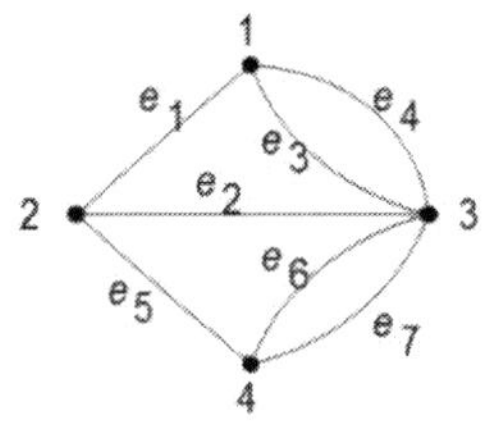

Figure 1. Example of graph

Each node can be connected by an edge and a node can be connected to more than 1 node like example in figure 2 and figure 3.

Figure 2. Node and edge

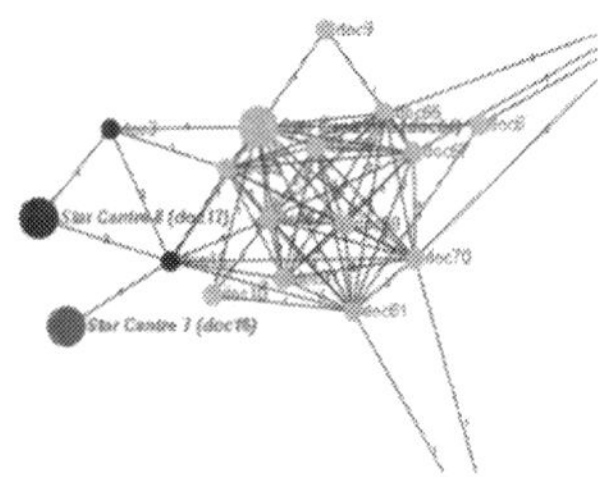

Figure 3. Node and edge

2.2 Graph Clustering

Graph Clustering is a method of grouping a graph into several groups / clusters based on particular relevance. Making the cluster is done by cutting or discard useless edges by certain sizes [3]. Based on the literature, the clusters in graph is called community clusters [1]. The application of clustering on a graph unlike the usual clustering and has its own methods such as MST Clustering [3], Chameleon [3], Makarov Clustering [3], and Star Clustering [3].

2.3 Star Clustering

Star clustering algorithm is widely used because it is quite easy to understand, simple, and can produce precise cluster more efficiently. A star cluster in the clustering consisted of 1 centre node that is called star centre and other nodes that are connected to the star centre which is called the satellites [4]. Each satellite node may be connected to more than 1 star centre. Star clustering algorithm use user-defined threshold for creating clusters where the edge of value below the threshold will be removed from the graph. The example of star clustering graph can be seen in figure 3. Following are the detailed of star clustering algorithm [4]:

- For each node, count the number of degree and sort the graphs according the number of degree (from the most into the least)
- Next, choose a node as the star centre by conducting examination towards each node that has the most number of degree and does not connected with other star centres.
- Repeat the previous process until we obtain all-star centre from the graph database.
- Lastly, create the clusters on the graph. Definition of a cluster in the star clustering algorithm is a collection of a star centre and its satellite nodes. If a node is connected to more than one satellite star centre node, then they will be joined with the star cluster centre which has the most weight towards the satellite node.

2.4 Cosine Similarity

Cosine Similarity is a calculation method to determine the weight / value relationship between two objects based on the cosine angle between these two objects [5]. This method is often used in text mining process. Cosine similarity formula is as follow:

$$\textit{Cosine Similarity}\ (\mathrm{d1,d2}) = \frac{dot\ product(d1,d2)}{||d1||\ X\ ||d2||}$$

Figure 4. Cosine Similarity Equation

D1 and D2 are the two document that will be determined. If two vectors have similar values which is approaching 1, it can be said there are similarities between the two vectors. To calculate the value of cosine similarity, we must first calculate the value of TF, IDF, and TFxIDF.

3. System Description and Testing Scenario

3.1. System Description

In this research, we will construct graph clustering system by utilizing Star Clustering algorithm, where each node in the graph represents a digital news in Indonesian language. The overview of System Description could be seen in Figure 4: Following are the explanation of Figure 4:

- The format of dataset is JSON. And the digital news is in Indonesian language. The words in dataset is processed so that become form of tokens that can be calculated using the cosine similarity relation.

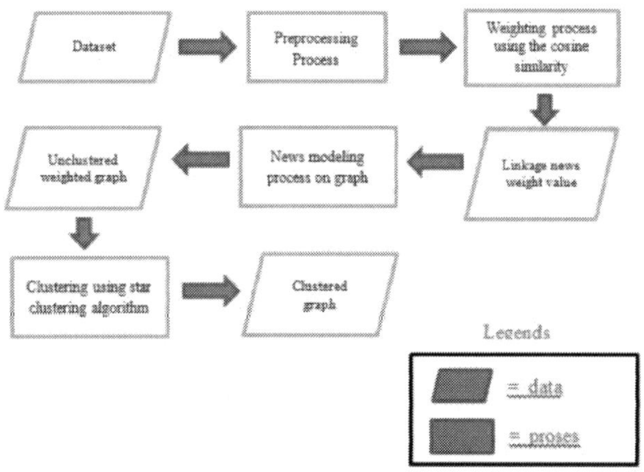

Figure 5. Overview of System Description

- Those token then processed using cosine similarity formula. This process produce weight values inter-document based on the relationship between news.
- The next process is modelling the news data into a graph that has not been clustered yet. Each news will become a node and the cosine similarity weight become the weight of the edge.
- Cluster the graph database using star clustering algorithm, forming a collection of news based linkages between news.

3.2. Testing Scenario
Following are the testing scenario that we use in our research:

- To know the effect of the number of news and threshold towards the value of intra cluster and inter cluster, we will perform several experiments to test the accuracy of intra cluster. Experiments were performed with four number of news that is 223, 167, 112, 56, and three kinds of threshold value which is 0.05, 0.1, and 0.15.
- To measure the accuracy of the clustering result, we will compare the clusters which is produced by our system with the grouping/clustering that is performed by an expert in Indonesian Language. Expert who perform clustering in this test is a journalists / private media reporter who graduates study program of Communication Studies, University of Indonesia. We calculate the accuracy by using F-measure, precision and recall for each cluster that is produced.

4. Discussion
4.1 Results from first scenario

Based on the experimental results, we got the percentage of 62.87% for the main data test which consist of 202 nodes and the threshold value that we use is 0.1. The experiment results could be seen in table 1. Following are our analysis toward the results:

- In general, the greater the amount of news, the less accurate the intra-cluster value comparison will be and vice versa, but there is a difference in the 0.15 threshold.
- The number of nodes and the number of clusters affect the accuracy of the results. If the number of nodes and clusters increase, then the accuracy value will also increase and if the number of nodes and clusters become smaller, then the accuracy value will also be smaller.
- At 0.15 threshold occurs different result because if the number of news abounded, the number of clusters which are formed are increasing rapidly with fewer nodes.
- At 0:15 threshold, we got good and consistent intra-cluster value which is above 70% for 112 nodes of the news.
- The threshold value has an effect to intra-cluster accuracy where the greater the threshold value, the greater accuracy value will be.

Table 4.1 Result of first scenario

Number of Node	Threshold	Number of Cluster	Node with greater intracluster value	Node wih greater intercluster value	Good Cluster Accuracy
54	0.05	10	43	12	79.62%
102	0.05	14	67	35	65.68%
150	0.05	11	72	78	48%
202	0.05	19	72	130	35.64%
54	0.1	16	42	12	77.778%
102	0.1	17	71	31	69.608%
150	0.1	25	107	43	71.333%
202	0.1	33	127	75	62.87129%
54	0.15	23	36	18	66.67%
102	0.15	28	76	26	74.5%
150	0.15	41	114	36	76%
202	0.15	54	153	49	75.742%
252	0.15	75	182	70	72.22%
297	0.15	84	215	82	72.39%

4.2 Results from second scenario

In the second scenario, we compare the cluster results that is produced by the system with the cluster results from the expert. The average accuracy value that we got is 80.98%. See Table 4.2 for more complete result.

Table 2. Result of second scenario

Cluster Name	Node Expert	Node Sistem	Node Relevant	Recall	Precision	F Measure
Cluster 1	3	6	3	100%	50%	67%
Cluster 2	3	3	3	100%	100%	100%
Cluster 3	8	10	7	87.50%	70%	78%
Cluster 4	2	2	2	100%	100%	100%
Cluster 5	2	2	2	100%	100%	100%
Cluster 6	18	17	16	88.89%	94.12%	91%
Cluster 7	1	1	1	100%	100%	100%
Cluster 8	1	1	1	100%	100%	100%
Cluster 9	3	7	3	100%	42.86%	60%
Cluster 10	2	2	2	100%	100%	100%
Cluster 11	21	20	20	95.22%	100%	98%
Cluster 12	3	1	1	33.33%	100%	50%
Cluster 13	4	3	3	75%	100%	86%
Cluster 14	18	15	13	72.22%	86.67%	79%
Cluster 15	32	32	31	96.88%	96.88%	97%
Cluster 16	15	9	9	60%	100%	75%
Cluster 17	1	4	1	100%	25%	40%
Cluster 18	10	10	10	100%	100%	100%
Cluster 19	3	3	3	100.00%	100%	100%
Cluster 20	5	0	0	0%	0%	0%
Cluster 21	1	2	1	100%	50%	67%
Cluster 22	10	6	5	50%	83.33%	62%
Cluster 23	8	7	7	87.50%	100%	93%
Cluster 24	2	2	2	100%	100%	100%
Cluster 25	9	9	7	77.78%	77.78%	78%
Cluster 26	1	3	1	100%	33.33%	50%
Cluster 27	2	2	2	100%	100%	100%
Cluster 28	2	2	2	100%	100%	100%
Cluster 29	5	4	4	80%	100%	89%
Cluster 30	2	3	2	100%	66.67%	80%
Cluster 31	2	2	2	100%	100%	100%
Cluster 32	2	1	1	50%	100%	67%
Cluster 33	1	2	1	100%	50%	67%

Based on the generated number of clusters, the expert produced 33 clusters while the system generates 37 clusters, with different contents. Differences clusters results that occur due to some node that is associated based on common words, but they had different main topic, such as for example on the type of news" kriminal" where there are many types of different clusters, but there are several nodes belonging to the wrong cluster because containing more same word with the star centre on another cluster than the star centre cluster it should be. This error occurs because many news containing more word like "korban" in comparison with its own special topics such as "mesum", "jambret", "curanmor", etc. Another factor that affects the value of accuracy is the imperfect pre-processing process. This cause imperfect test data, and in return the clustering result become not 100% accurate.

5. Conclusion

- Digital news can be modelled using graph. This model is easy to understand and we could see the relationship between the news by making edge between nodes to illustrate the interconnectedness of news and rated the weight of each node.
- The star clustering algorithm can be applied in the news grouping case based on the linkages of the news content. This approach defines the node with biggest degree as the star centres and then grouped clusters based by linkages with its star centre.
- The star clustering algorithm can perform grouping of digital news based on linkage of the news content with easy to understand, fast, effective, has a pretty good degree of accuracy (62.87% accuracy for comparison of intra-cluster and 80.98% for the comparison with the clustering by expert)
- Grouping news using graph clustering is easy to understand because it produces good clusters visualization, so that the observer can see the linkage between clusters more easily by the colouring between nodes that have the same cluster.
- Results of star clustering algorithm relies heavily on the star centre that it formed
- The clusters that is formed in this research is not on the basis of topics commonly found on the grouping of digital news in the news media, but by the words that frequently appear on the news.

6. Acknowledgement

The research presented in this paper conducted by the School of Computing, Telkom University. This research was funded by research grant "Hibah Bersaing 2016". We thank to the Ministry of Research Technology and Higher Education of the Republic of Indonesia (RISTEKDIKTI) for contributing to the success of this research.

7. References

[1] Feldman, Ronen, and James S, 2007. The Text Mining Handbook *"Advanced Approaches in Analyzing Unstructed Data"*. University Press, Cambridge
[2] Schaeffer, Satu E, 2007. *"Graph Clustering"*. Laboratory for Theoretical Computer Science, Helsinki University of Technology, Finland
[3] Wijaya, Derry T, 2008. Graph, Clustering and Applications. Depatment of Computer Science National University of Singapore, Singapore.
[4] Aslam, J A, Pelekhov K, and Rus D, 2004. The Star Clustering Algorithm for Information Organization.
[5] Sree, K. S. (2012). Clustering Based on Cosine Similarity Measure.

IOP Conf. Series: Journal of Physics: Conf. Series **801** (2017) 012062 doi:10.1088/1742-6596/801/1/012062

Digital News Graph Clustering using Chinese Whispers Algorithm

M F E Pratama, R S W Kemas and H Anisa

School of Computing, Telkom University, Jl. Telekomunikasi Bandung 40287, West Java, Indonesia

E-mail: fitrah.eka@gmail.com, bagindokemas@telkomuniversity.ac.id, anisaherdiani@telkomuniversity.ac.id

Abstract. As the exponential growth of news creation on the internet, the amount of digital news has reached out billion numbers. Digital news is naturally linked each other but it needs to be grouped so that user can easily classify the news that they read. Graph is the most suitable data model to represent digital news since its can describing relation in easy and flexible manner. Thus, to overcome grouping problems, in this paper we using Chinese Whispers Algorithm as the graph clustering approach. We choose Chinese Whisper Algorithm based on consideration that the algorithm is able to make clusters from a big graph data with a relatively fast process [8], that appropriate with the characteristics of digital news. In this research, we examine the graph quality by comparing intra and inter-cluster weights of every node. This scenario gives us a quite high result that 95% of nodes have intra-cluster weight higher than its inter-cluster weight. We also investigate the graph accuracy by comparing the cluster results with expert judgement. As the result, the average accuracy of digital news graph clustering using Chinese Whisper algorithm is 80%.

1. Introduction

Nowadays digital news is considered as a common thing for people in around the world. The online audience engaged with newspaper content reached a new peak in August 2015, totaling 179.3 million adult unique visitors. The count is a 10% increase in adult unique visitors measured by comScore since August 2014, and is double the growth rate for the Internet overall (5%). The phenomenon happens due to the ease of online news media to spreading the information, its flexible form, timeless accessibility, and borderless accessibility [1].

Digital news data that has been saved in the data storage has reached out ten billion to hundred billions of news. The data is so big that we need a simple representation to make data analytics easier. Graph model then implemented to represent the digital news because the model is more understandable and in line with human's way of thinking. Besides, graph model can handle a huge scale of data and has semi structured characteristics [2][3] that comply with digital news data. In this graph, every node represents a news, and edges represent news' connectivity based on its content.

News grouping based on its similarity or connectivity is necessary to help the readers found related articles, but it is hard to be performed if there are no explicit tags on the news or articles. To solve this problem, we use clustering approach. Clustering aims to make groups based on the element similarities [4][5]. Since we use graph to represent the news, we will perform a graph clustering approach. Graph clustering is a process of nodes grouping, so that the node that have similar characteristics will be grouped in the same unit.

There are quite much graph clustering algorithm such as MST clustering, Chameleon, Makarov Clustering, and Star Clustering [7]. In this paper we use Chinese Whispers Algorithm based on consideration that the algorithm is able to make clusters from a big graph data with a relatively fast process [8], that appropriate with the characteristics of digital news.

2. Theoretical Foundation

2.1. Graph Theory

Theoretically, a graph consists of a group of vertex (V) and edge (E), and symbolized by G = (V, E). According to the edge direction of a graph, there are directed and undirected graph. Direct graph has a vertex used as a start and an end [9] meanwhile undirected graph has no distinction between the two vertex associated with each edge. In other words, undirected graph has bidirectional edges. Example of simple undirected graph can be seen on Fig. 1.

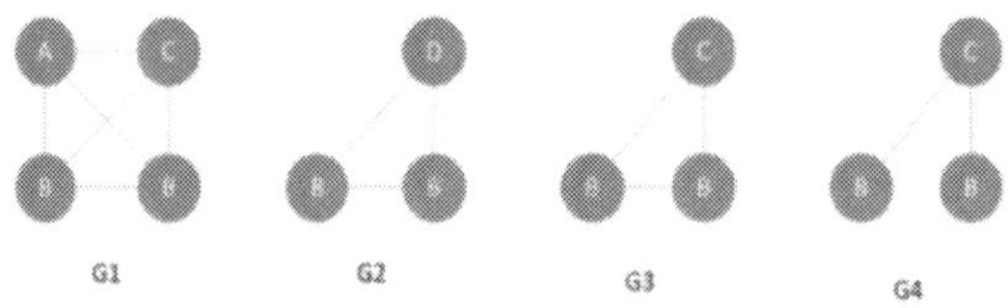

Fig. 1. Example of simple undirected graph

Besides directed and undirected graph, there is also exist the term of weighted graph. Weighted graph is a graph that has numerical label associated with its edge, called the weight of edge [8]. Fig 2 is an example of weighted graph.

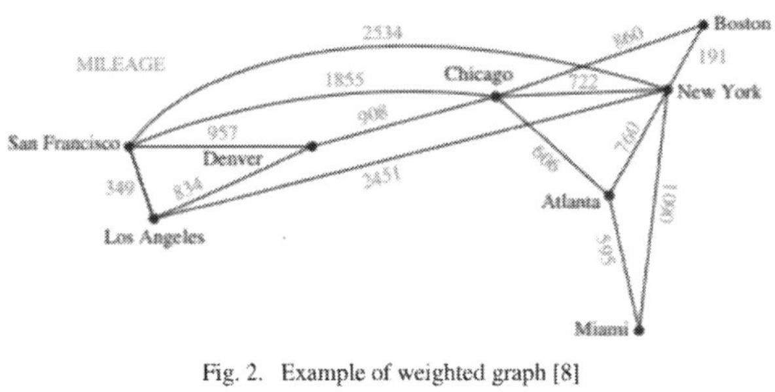

Fig. 2. Example of weighted graph [8]

2.2. Graph Clustering

Graph clustering is a process of node grouping so that every similar node gets into the same group. Some of researchers were also defined that graph clustering is an effort to grouping nodes in a graph so that every node in the same group have a high value of intra-connectivity and every node in different group have a low value of interconnectivity [6].

Graph clustering has been widely used in biology field especially in gen study, VLSI chip design, social networks, and others [10].

2.3. Chinese Whisper Algorithm

Chinese Whispers (CW) is an easy and effective algorithm in grouping/clustering a weighted, undirected graphs. The idea of CW came from a traditional children's game, where every child has to spreads words or sentence by whispering it to another child consecutively. Several steps in CW algorithm based on [8] is as follow:

a. Every node in the graph is divided in different cluster. One node in one cluster.

b. Point one node randomly, then take that node into the strongest class in the neighborhood. This is the class whose total edge weights to the current node is maximal.
c. Repeat the process b for every node in the graph.
d. In case of multiple strongest classes, one is chosen randomly.
e. Regions of the same class stabilize during the iteration and grow until they reach the border of a stable region of another class. Note that classes are updated immediately: a node can obtain classes from the neighborhood that were introduced there in the same iteration.

2.4. Cosine Similarity

Cosine similarity is a method that often used in the text mining especially in the information retrieval and also clustering as well for measuring text similarity [11]. Cosine similarity measures the corner size between document vector (point (ax, bx)) and (point (ay, by). Every vector that represents every word in every documents then compared and formed in a triangle. If there are two identical documents, it will form a zero degree (0°) angel so that the similarity's value is one (1); if two document are not identical at all, the angle that will be formed is 90 degree ($90^{o)}$ so that the similarity's value is zero (0) [12]. The formula of Cosine Similarity is as follow [13][14]:

$$Cosine\ Similarity\ (\delta 1, \delta 2) = \frac{dot\ product(d1,d2)}{||d1||\ X\ ||d2||} \quad (1)$$

To prepare data for cosine similarity calculation, we have to do the term weighting by finding out the value of TF (Term Frequency), IDF (inverse-document frequency), and TF • IDF. Fig 3 represent the process for preparing data for Cosine Similarity calculation.

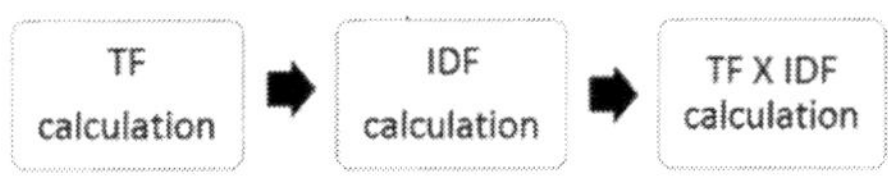

Fig. 3. Process before calculating the Cosine Similarity.

TF is weight of a term that occurs in a document, while IDF is a way of diminishes the weight of terms that occur very frequently in the document set and increases the weight of terms that occur rarely.

The TF formula is as follow:

$$\text{TF}\ (t_k, d_j) = \text{f}(t_k, d_j) \quad (2)$$

With (t_k, d_j) is the number of times that term *k* occurs in document *j*.

The IDF formula is as follow:

$$\text{IDF}(t_k) = 1 + \log \frac{N}{df\ (t)} \quad (3)$$

With *N* is total number of documents in the corpus and *df(t)* is total number of documents in the corpus where the term *t* appears.

The TF • IDF formula is as follow:

$$\text{TF} \bullet \text{IDF}\ (t_k, d_j) = \text{TF}\ (t_k, d_j) \bullet \text{IDF}(t_k) \quad (4)$$

3. System Design

In this chapter we proposed the methodology we used to implement Chinese Whisper Algorithm to cluster the Indonesian digital news. Fig 4 depicting the system design.

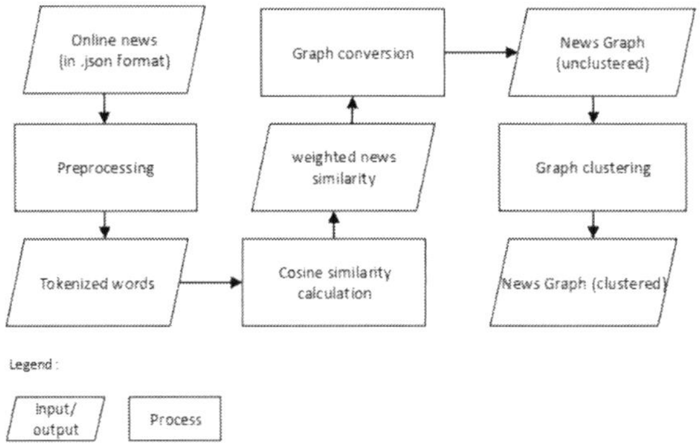

Fig. 4. System Design

Following is the description of the system design:

a. The first step is to prepare data by data preprocessing that consist of tokenization, case folding, stop word removal, and stemming. The raw data is digital news in .json format.
b. Preprocessed data is the input of the term weighting step (TF, IDF, TF • IDF) and cosine similarity calculation.
c. The results value of Cosine Similarity calculation is approximately between 0 and 1. To simplify the number, we normalized the value by multiplying the value by 10 and making it integers. Then, we defined that two documents are related if the value of similarity between 1 - 9.
d. After we know which documents are related one another, we convert the documents into graph. Every node in the graph gives information about the news such as author, title, date of released, content of news, and others. Meanwhile the value on the edges represents the similarity values.
e. All nodes then ready to be clustered using Chinese Whisper Algorithm.

4. Experiment

To measure whether the methodology satisfy our goal, we already conduct some experiments that follow the methodology we have proposed on the previous section.

Datasets. We use 258 Indonesian digital news captured from http://news.okezone.com in .json formats. One example of the news format can be seen in Fig. 5.

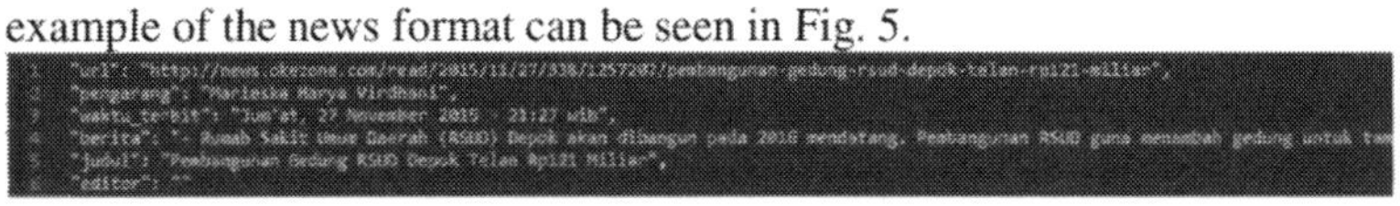

Fig. 5. Digital news on .json format

Scenarios. (1) In order to measure the graph quality, we examine the edge's weight between node in the same cluster (intra-cluster value), and edge's weight between node in different cluster (inter-cluster value). (2) In order to measure the accuracy of clustering result, we compare the clusters with news' group that manually grouping by expert. The expert is a Bahasa Indonesia teacher, graduated from Indonesian Language and Literature Department.

To get a stable shape of clusters, we perform six (6) iterations of graph clustering based on steps on Chinese Whisper Algorithm.

5. Results and Discussion

5.1. Scenario 1

The result of the first iteration clusters shows that only 82,69% of its nodes have intra-cluster values higher than its inter-cluster values. It happened because the clusters shape was still not stable. The second iteration shows the promising improvement percentage that is 92.79%. On the second iteration, the clusters are more stable than the previous one.

On the third and fourth iteration, we got even better percentage results that is more than 95%, but there are several nodes that it's inter-cluster value higher than its intra-cluster because the cluster is still not stable enough. Meanwhile, for the 5th and 6th iteration, the percentage is also more than 95%, but the clusters have been stable because no nodes have higher point of inter-cluster value. The reason why we didn't achieve 100% is because some nodes has the same value of inter-cluster and intra-cluster. Fig 6 illustrate the results of scenario 1.

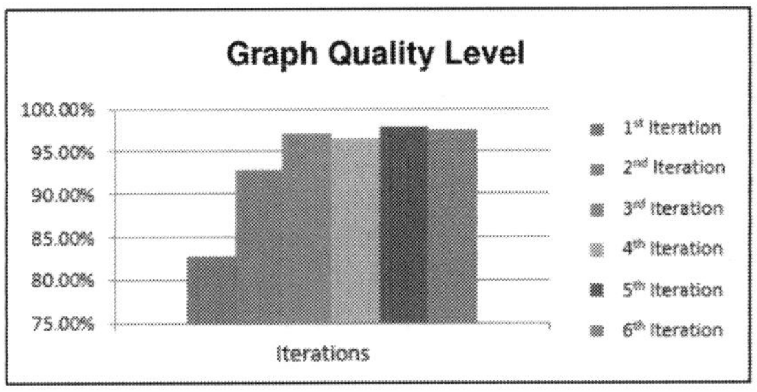

Fig. 6. Result of scenario test 1

5.2. Scenario 2

To perform scenario 2, we asked the expert to grouping all digital news on the dataset. As the result, there are 22 groups of news. Then we compare the result with clusters that we get after the 6th iteration from previous scenario. Table 1 shows the comparison results.

Table 1. Comparison Result

Experiment	Number of clusters	Number of correct documents clustered	Number of incorrect documents clustered	Accuracy
1	19	167	41	80.3%
2	18	158	50	76.0%
3	19	169	39	81.3%
4	19	167	41	80.3%
5	20	171	46	82.2%

Inconsistency of accuracy happens because every result of the clustering has different clusters result even though the total number of the clusters was similar. The main factor that differ one clustering result to another is the random phase. For example, experiment 1 and 3 both have 19 number of clusters, but the element of every cluster are different. That is the reason why level of accuracy is also different.

Another factor causing the accuracy not maximum was because of something not worked really well on the preprocessing phase. We investigated that the stemming process was not 100% correct. Thus it influenced the term weighting and overall clustering process. But after all, the average 80% accuracy is quite promising.

6. Conclusion

Transforming digital news into graph can answer the problem of news grouping. Every news document represented by node, meanwhile the similarity between document represented by the weight of the edges. In this research, we use Cosine Similarity to perform graph clustering and we get quite high result on graph quality (95%) and clustering accuracy (80%).

Acknowledgements
This study is one of research roadmap of the Faculty of Informatics Telkom University, and has been funded by Ministry of Research, Technology and Higher Education of the Republic of Indonesia, through "Hibah Bersaing" grant.

References

[1] Husnul K, "Analisis faktor-faktor website quality dalam menggunakan situs berita digital kompas.com 2013". Bandung: FEB Universitas Telkom, 2013.

[2] Isjhar K, "Analisis performansi metode graph decomposition index pada graph database". Bandung, Indonesia: Fakultas Informatika Telkom University, 2014.

[3] Fahri F, Erwin Y H, and Ika N D, NoSQL: Latar Belakang, Konsep, dan Kritik. Semarang, Indonesia: Fakultas Ilmu Komputer, Universitas Dian Nuswantoro, Semarang, 2012.

[4] Yang Z, Hong C, and Jeffery X Y, Graph Clustering Based on Structural/Attribute Similarities. Hong Kong: Engineering Management University of Hong Kong.

[5] Satu Elisa S, Graph CLustering. Helsinki, Finland: Laboratory for Theoretical Computer Science, Helsinki University of Technology, 2007.

[6] Reena M, Shashwat S, Deepak A, and Mohit K, An Effective Comparison of Graph Clustering Algorithms via Random Graphs. India: Department of Computer Science and Engineering Amity University, 2011.

[7] Derry T W, Graph, Clustering and Applications. Singapore: Depatment of Computer Science National University of Singapore, 2008.

[8] Chris B, Chinese Whispers - an Efficient Graph Clustering Algorithmand its Application to Natural Language Processing Problems. Leipzig, Germany: University of Leipzig, NLP Department.

[9] Kenneth H R, Discrete Mathematics and Its Applications, Seventh Edition ed. New York: McGraw-Hill, 2007.

[10] Mousumi D and Shukla K K, Characteristic of Restricted Neighbourhood Search Algorithm and Markov Clustering on Modified Power-Law Distributor. India: Departement of Computer Engineering, 2012.

[11] Bhanu P A. and Venkata G R S., Space and Cosine Similarity measures for Text Document Clustering. Hyderabad, India: Vardhaman College of Engineering, 2013.

[12] Radiant V I, Adelia , Mewati A, and Alexander R, Implementasi Cosine Similarity dan Algoritma Smith-Waterman untuk Mendeteksi Kemiripan Teks. Bandung, Indonesia: Fakultas Teknologi Informasi Universitas Kristen Maranatha, 2014.

[13] Jana V, (2013, Oktober) Seeking Wisdom. [Digital]. https://janav.wordpress.com/2013/10/27/tf-idf-and-cosine-similarity/

[14] Jon B, Event-Centric Clustering of News Articles. Uppsala, Sweden: Department of Information Technology, Uppsala University, 2013.

Frequency Analysis of Chaotic Flow in Transition to Turbulence in Taylor-Couette System with Small Aspect Ratio

Tatsuya Fujii[1], Yoshihiko Oishi[1], Hideki Kawai[1], Hiroshige Kikura[2], Riky Stepanus Situmorang[1], Himsar Ambarita[3].
[1]Muroran Institute of Technology, Mizumoto-cho 27-1, Muroran, Hokkaido, Japan 050-8585
[2]Tokyo Institute of Technology, Ookayama 2-12-1, Meguro-ku, Tokyo, Japan 152-8550
[3]University of Sumatera Utara, Jl. Almamater Kampus USU Medan 20155, Indonesia

E-mail: hdkawai0@mmm.muroran-it.ac.jp

Abstract. Taylor-Couette flow with small aspect ratio has characteristics such as the different vortex structure, because of a boundary layer of the upper and lower wall and the acceleration of the inner cylinder. In this study, the mechanism of Taylor-Couette system with the small aspect ratio is measured and analyzed by using an ultrasound measurement and a numerical simulation. The process of transition to turbulent flow is observed by using a spectra analysis in a radial and an axial direction. The experimental and numerical results confirmed the characteristics of the broadband component in Taylor-Couette system.

Key Words: Taylor-Couette flow, Small aspect ratio, Chaotic wavy vortex flow, Numerical simulation

1. Introduction.

Many researchers have been studying about the experiments and theoretic of Taylor-Couette flow (TCF), since it can be reproduced under the most fundamental conditions to understand the turbulent transition process of a fluid. In the previous studies, effect of the aspect ratio (Γ) which defined with the clearance between coaxial cylinder and the working fluid height was considered. Benjamin et. al. [1] built up the fundamental research with small aspect ratio of TCF. It is also expected as a stirring flow with the less local shear. This flow system can be used in blood filtration and bioreactor microorganism culturing devices. Therefore it is suggested to be compact due to the circumstances such as the place where it will be applied, so it is important to elucidate the flow problem with small aspect ratio in TCF. In the study of infinite cylinder TCF, Fenstermacher et al. [2] and David Andereck et al. [3] report the frequency analysis of the TCF velocity, and reveal the TCF transition with the changing of Reynolds number (Reynoldss number : *Re*). In the process transition from Taylor vortex flow (TVF), Wave vortex flow (WVF), Modulate wavy vortex flow (MWV) to turbulence are ordered to the Reynolds number. Especially in the turbulent state, existence of broadband component: B has been reported. This is also studied by Takeda et. al. [4][5], they report that there is a stage called chaotic wavy vortex flow (CWV) before the transition to turbulence, which has the B component and the fundamental frequencies.

In this research, we are studying a normal 2-cell mode with the aspect ratio is 3 and stable behaviour caused of the upper and lower boundary conditions. Nakamura et.al. [6] report that TCF with small aspect ratio has instability such as branches of the number of cells comparing to TCF with large aspect ratio. And it has been reported that the CWV state continues relatively even at large Reynolds number under this equipment condition. According to Kageyama et. al. [7], by numerical simulation and experiment of TCF with small aspect ratio, it is stated that the critical Reynolds number which divides laminar flow and turbulent flow is $Re_{crit} \fallingdotseq 5\times10^5$. Our consideration is the fixed upper and lower boundaries that prevent turbulence of TCF. The purpose of calculating the velocity field of TCF by numerical simulation is to estimate the cause of modulation of TCF by deriving the radial velocity component that is difficult to be measured by experiment.

2. Methodology

2.1. Experiment Method to Measure Velocity Field Using Ultrasound Doppler Velocity Profiler (UVP).

The experimental apparatus was filled with glycerin aqueous solution (68% $C_3H_8O_3$, Wako Pure Chemical Ltd.) and 80 μm nylon particles were added as the tracer. The inner cylinder was rotated to generate TCF in the normal 2-cell mode. An ultrasonic transduce (TDX) was installed 7 mm from the surface of the inner cylinder to measure the velocity in the *z* axial direction. We selected to install it at 7 mm from the surface of the inner cylinder because at that condition the characteristic of the cell type vortex of TCF appeared well.

Inner cylinder radius is R_{in}=50 mm, Outer cylinder radius is R_{out}=75 mm, inner and outer cylinder clearance is d=25 mm, inner and outer cylindrical height is H=75 mm. It is defined that radius ratio is $\eta = R_{in} / R_{out}$, aspect ratio is $\Gamma = H / d$, and the Reynolds number is $Re=dW_{in}/v$, with assumption that the inner cylinder speed is W_{in}, v is the viscosity of the glycerin aqueous solution. This experimental condition is shown in table 1.

Table 1 Experimental conditions

Boundary condition		Fixed
Cell mode		Normal-2cell
Aspect ratio	Γ	3
Radius ratio	η	0.667
Reynolds number	Re	500-4000

2.2. Numerical Simulation Model and Condition.

The numerical simulation model is shown in Fig. 1. The radial direction is *r* (the direction of the outer cylinder is positive from the inner cylinder), the axial direction is *z* (the direction from the lower boundary end to the upper boundary end is positive) and the circumferential direction is set to θ (CCW rotation is positive). The analysis conditions are shown in table 2.

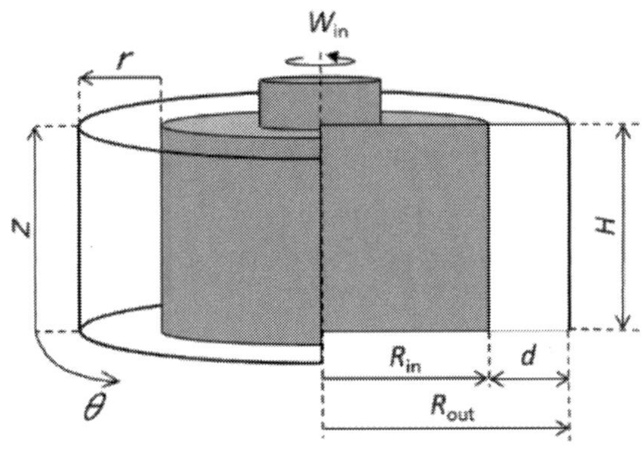

Fig.1 Model of numerical simulation

Table 2 Numerical simulation conditions

Boundary condition		Non-Slip
Cell mode		Normal-2cell
Aspect ratio	Γ	3
Radius ratio	η	0.667
Reynolds Number	Re	500-4000
Grid num.	$r\times z\times\theta$	50×150×160
Time step	Δt	0.01

2.3. Numerical Simulation Method.

Numerical simulation method used to solve the Navier-Stokes equation by setting the fluid inside the model as an incompressible viscous fluid. In addition, the Navier-Strokes equations represent the length as the inner cylinder radius: R_{in}, and speed as the inner cylinder circumferential velocity: W_{in}, and normalize all the parameters. As an example, the radial component of the normalized cylindrical coordinate system in Navier-Stokes equation is shown in equation (1) and the continuous equation is shown in equation (2)

$$\frac{\partial u}{\partial t}+u\frac{\partial u}{\partial r}+\frac{w}{r}\frac{\partial u}{\partial \theta}-\frac{w^2}{r}+v\frac{\partial u}{\partial z}=-\frac{\partial p}{\partial r} \\ =-\frac{\partial p}{\partial r}+\frac{1}{Re}\left(\frac{\partial^2 u}{\partial r^2}+\frac{1}{r}\frac{\partial u}{\partial r}-\frac{u}{r^2}+\frac{1}{r^2}\frac{\partial^2 u}{\partial \theta^2}-\frac{1}{r^2}\frac{\partial w}{\partial \theta}+\frac{\partial^2 u}{\partial z^2}\right) \quad \text{....(1)}$$

$$\frac{\partial u}{\partial r}+\frac{u}{r}+\frac{\partial v}{\partial z}+\frac{1}{r}\frac{\partial w}{\partial \theta}=0 \quad \text{....(2)}$$

For the analysis algorithm, SMAC method (Simplified Marker and Cell Method) is used. This scheme is explained by the radial expression. Regardless of the known velocity field satisfies the continuous equation or not, predictive value of $\tilde{u}$ is obtained by the Euler explicit method as shown in Equation (3). n in the equation refer to the time.

$$\tilde{u} \approx u^n + \Delta t\left(-u^n\frac{\partial u^n}{\partial r^n} - \frac{w^n}{r^n}\frac{\partial u^n}{\partial \theta^n} + \frac{w^{n2}}{r^n} - v^n\frac{\partial u^n}{\partial z^n} - \frac{\partial p^n}{\partial r^n} + \frac{1}{Re}\left(\frac{\partial^2 u^n}{\partial r^{n2}} + \frac{1}{r^n}\frac{\partial u^n}{\partial r^n} - \frac{u^n}{r^{n2}} + \frac{1}{r^{n2}}\frac{\partial^2 u^n}{\partial \theta^{n2}} - \frac{2}{r^{n2}}\frac{\partial w^n}{\partial \theta^n} + \frac{\partial^2 u^n}{\partial z^{n2}}\right)\right) \quad ...(3)$$

However, this predicted value does not satisfy the continuous equation. Therefore introduce a vortex-less velocity field u' and obtain a corrected speed u^{n+1} so as to satisfy the continuous equation from the equation 4.

$$u^{n+1} = \tilde{u} + u' \quad ...(4)$$

In addition, scalar potential: φ exists since u' is not vortexed. Therefore, the composite speed V^{n+1} of u^{n+1}, v^{n+1}, and w^{n+1} that omit the intermediate process are expressed as equation (5), when divergence take place and converges to 0.

$$divV^{n+1} = div(\tilde{V}^n + grad\varphi) = 0 \quad ...(5)$$

From the equation (5), Poisson equation for the scalar potential φ is described as the following equation

$$\frac{1}{r}\frac{\partial}{\partial r}\left(r\frac{\partial \varphi}{\partial r}\right) + \frac{1}{r^2}\frac{\partial^2 \varphi}{\partial \theta^2} + \frac{\partial^2 \varphi}{\partial z^2} = -\left(\frac{1}{r}\frac{\partial}{\partial r}(r\tilde{u}) + \frac{1}{r}\frac{\partial \tilde{w}}{\partial \theta} + \frac{\partial \tilde{v}}{\partial z}\right) \quad ...(6)$$

By calculating this Poisson equation iteratively, the vortex-less velocity field is obtained from the scalar potential, and the velocity at the next time is obtained from the predicted value and the modified value. According to the program in this study, the iteration number is set to 10 for the convenience of shortening calculation time, and the 10th calculated value is taken as a solution regardless of the continuous equation is satisfied or not. The sufficiently small value of the data is used to the continuous formula. For discretization, Adams-Bashforth method is used for time progression with the second order accuracy, the central difference with the fourth order accuracy is used for convection term in the space, and central difference with the second order accuracy is used for viscosity term and pressure term as explained respectively in equation (7), (8) and (9).

$$u^{n+1} = u^n - \Delta t\frac{p_{i-1}{}^n - p_i{}^n}{\Delta r} + \Delta t\frac{3(A_r + B_r)^n - (A_r + B_r)^{n-1}}{2} \quad ...(7)$$

$$f_i' = \frac{-f_{i-1/2} + f_{i+1/2}}{\Delta} \quad ...(8)$$

$$f_j' = \frac{f_{j-3/2} - 27f_{j-1/2} + 27f_{j+1/2} - f_{j+3/2}}{24\Delta} \quad ...(9)$$

2.4. Frequency Analysis.

For the frequency analysis, in order to discriminate the transition process of TCF from the frequency with respect to the flow velocity, Fast Fourier Transform (FFT) is applied to the velocity data obtained by UVP and numerical simulation. Then the oscillation frequency and power spectral density (PSD) are obtained at each Reynolds number. PSD is the amplitude obtained by FFT squared and divided by frequency resolution. The number of data is 2048, and the frequency is normalized by the rotation speed of the inner cylinder. In order to confirm the accuracy of numerical simulation, FFT is applied to the axial flow velocity value on the same line as on the measurement line in UVP. Also, there are two evaluation methods. The first method is to evaluate the change in the vibration state accompanying the increase in the Reynolds number by taking the average of the vibration frequencies on the measurement or calculation line. The second method is by taking the vibration frequency of each channel in the axial direction, the vibration state at each position in the axial direction is evaluated in detail. The results are summarized in a color plot that showing the intensity of the PSD.

3. Results and discussion

Fig. 2 shows the PSD distributions map of the frequency components on the axial velocity in space. Fig. 2-(a) is the experimental data with the UVP method and Fig. 2-(b) is the numerical. The vertical axis indicates the normalized frequency f^* and the horizontal axis is the Reynolds number (*Re*). Red color shows the higher PSD value and the blue one is the lower. In Fig.2-(a), a sharp frequency around f^*= 2.1 can be observed continuously from *Re*=700. This is the flow transition from TVF to WVF. However, this continuous line is disappeared at *Re* =1000, then at the same time, new three types of the sharp peaks lines are appeared at f^*=1.8, 1.7 and around 0 respectively. This means another flow transition has been occurred from the WVF to the MWV around *Re*=1000. The first peak which was observed at the beginning of the MWV keeps its existence clearly until *Re*=4000. This phenomenon is observed in both the experimental and the numerical results, which can be assumed as the fundamental frequency in the flow.

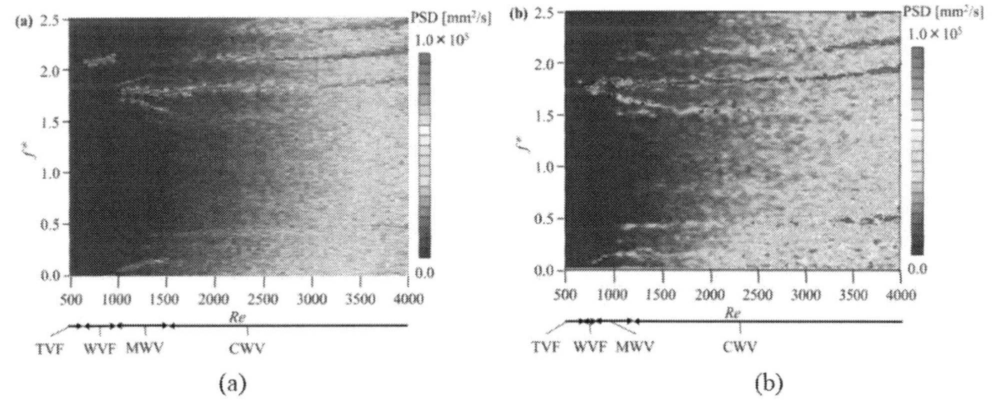

Fig. 2 Variation of PSD result of the velocity in z direction on Re - f^* plane. (a) The velocity measured by experiment and (b) the velocity calculated by numerical analysis.

Fenstermacher et.al. [2] and Takeda et.al. [4] found a broadband component called B component that was observed in the lower frequency area of MWV. This B component is gradually powered up with increasing the Reynolds number. In this study, the similar B component was observed in the small aspect ratio of Γ=3 in both of the experimental and the numerical results. In this experiment, the PSD value of the B component in the lower frequency is gradually increased and widely distributed after *Re*=3000

However the beginning of the B component is anomalous compared with the previous studies in Fenstermacher's. In Γ=3, the two fundamental frequencies are much closer in each other where the MWV mode starts, which results in the cause of the growl in lower frequency and it is clearly observed in both Figs. 2-(a) and (b). In the previous studies, some discussions are made that the B component is caused from the growl generated in the MWV mode. In the case of Γ=3, we found the pair of the fundamental frequencies in MWV just start from the closer frequency values and thus our experiment is easier in estimating the lower frequency growl. In fact from Figs. 2-(a) and (b), the growl generates between the difference of the upper two fundamental frequencies and exists in the lower part of frequency area. But there can be seen a jumping from the growl line at the starting point of the B component with increasing the Reynolds number, which can be certificated in both the experimental and numerical results. It could be discussed that the growl in the lower part of frequency is not causing the B-component. The both results in the experimental and the numerical frequencies correspond well qualitatively until *Re*=4000 including the starting point of the B component. The flow seems to transfer the CWV around *Re*=3000 where the PSD of the B component is much powered up and widely distributed in the lower frequencies, but still we can see the MWV mode in the upper frequencies.

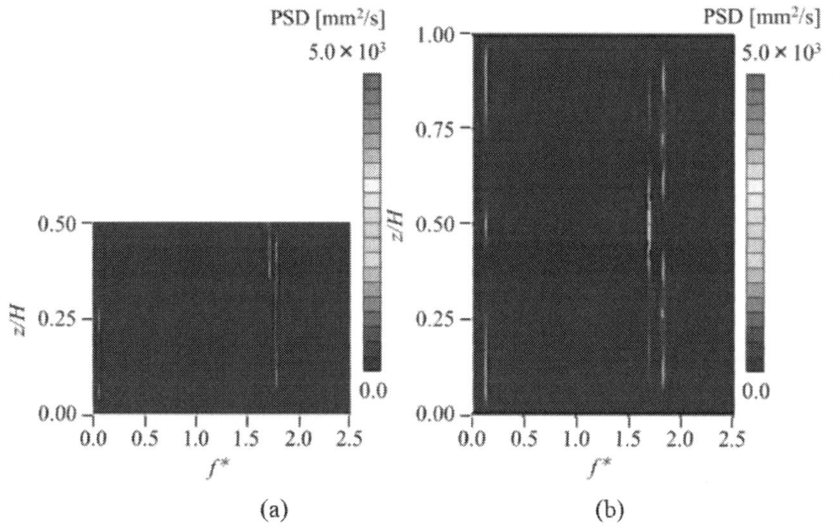

Fig. 3 Variation of power spectra of (a) the velocity that measured by experiment and (b) the velocity that calculated by numerical analysis in z direction at Re = 1000

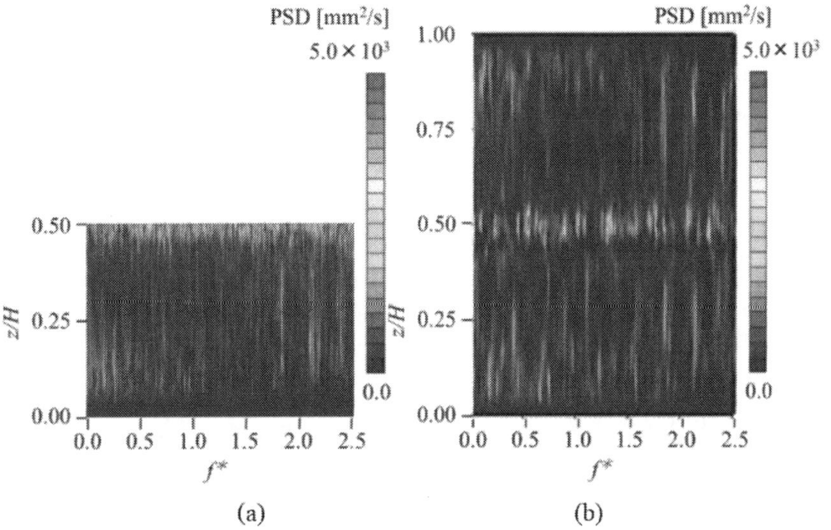

Fig. 4 Variation of power spectra of (a) the velocity that measured by experiment and (b) the velocity that calculated by numerical analysis in z direction at Re = 3000.

Figs.3 and 4 show the colored PSD map with the dimensionless z axial coordinate and the normalized frequency from the axial velocity components. The vertical is the z direction and the horizontal is the frequency. In both figure, (a) is from the experimental data with the UVP method, and (b) is the numerical, however the experimental maps are shown in the half part of the space region as Z/H=0~0.50 where the error from the near field of the ultrasound is less influenced

At Re=1000, three peaks of the frequencies are also clearly observed in Fig. 3 with the distribution of z coordinate, which are well corresponded to Fig. 2. The peak of f^*=1.7 is seen in the middle area of z direction and the peak of f^*=1.8 is in a whole z direction except for the middle area. From this state, the frequency of f^*=1.7 is mainly from the outer flow oscillation and the other frequency of f^*=1.8 is the self-oscillation in each vortex cell. f^*=0.1 is the beat frequency caused by the fundamental frequency between f^*=1.7 and 1.8. In comparison with the both experimental and numerical, the frequency distributions in space are also in good agreement with each other. In Fig. 4 at Re=3000, a large peak of the frequency is observed in the middle of z direction and the spectrum value is more increased than Re=1000, especially in the lower frequency region around f^*=0~0.5.

4. Conclusion

Measurement experiment and numerical simulation were carried out with small aspect ratio TCF with Γ=3 and the following findings were obtained from the frequency analysis of the axial velocity components of TCF.

- Although the broadband (B) component spreads in a high Reynolds number, the fundamental frequencies maintain until Re=4000, and at Re=4000 or less, the CWV state continues with undeveloped turbulence. The power spectrum of each channel with respect to the axial direction quantitatively agreed with the results of experiment and numerical simulation at Re=1000 and 3000.

- The broadband component which occurs from upper and lower boundary develops in between the vortex cells. The transition to turbulence of TCF with small aspect ratio is affected by the boundaries and the effect of boundaries is a factor to maintain a chaotic flow state.

An acknowledgment
This research was conducted with the grant of JSPS Grant 26420096.

References

[1] Benjamin, T. B., 1978, Bifurcation Phenomena in Steady Flows of a Viscous Fluid. II. Experiment, *Proc. R. Soc. London*, Ser. A, 359, pp. 27-43

[2] Fenstermacher, P. R., and Swinney, H. L., 1979, Dynamical instabilities and the Transition to chaotic Taylor vortex flow, *J. Fluid Mech.* Vol.94 part 1, pp. 103-128

[3] David, C. A., Liu, S. S., Swinney, H. L., 1986, Flow regimes in a circular Couette system with independently rotating cylinders, *J. Fluid Mech*, Vol. 164, pp. 155-183

[4] Takeda, Y., 1999,Quasi-periodic state and transition to turbulence in a rotating Couette system, *J. Fluid Mech.* , Vol. 389, pp. 81-99

[5] Takeda, Y., Fischer, W. E., 1993, Experimental observation of the quasiperiodic medes in a rotating Couette system, *Phys. Rev. E* Vol.47, No.6, pp. 4130-4134

[6] Nakamura, I., Toya, Y., 1994, Configurational Condition of Mode of Multiple Taylor Vortex Flow (In the case of a symmetric condition), *Transactions of JSME Series B*, 60-571,pp. 1-7

[7] Kageyama, A., Ji, H., Goodman, J., Chen, F., Shoshan, E., 2004, Numerical and Experimental Investigation of Circulation in Short Cylinders, *Journal of the Physical Society of Japan* Vol. 73, No. 9, September, pp. 2424-2437

Table Extraction from Web Pages Using Conditional Random Fields to Extract Toponym Related Data

Hayyu' Luthfi Hanifah[1,2], **Saiful Akbar**[1,3]

[1]School of Electrical Engineering and Informatics, Bandung Institute of Technology

[2]hayyuhanifah52@gmail.com
[3]saiful@informatika.org

Abstract. Table is one of the ways to visualize information on web pages. The abundant number of web pages that compose the World Wide Web has been the motivation of information extraction and information retrieval research, including the research for table extraction. Besides, there is a need for a system which is designed to specifically handle location-related information. Based on this background, this research is conducted to provide a way to extract location-related data from web tables so that it can be used in the development of Geographic Information Retrieval (GIR) system. The location-related data will be identified by the toponym (location name). In this research, a rule-based approach with gazetteer is used to recognize toponym from web table. Meanwhile, to extract data from a table, a combination of rule-based approach and statistical-based approach is used. On the statistical-based approach, Conditional Random Fields (CRF) model is used to understand the schema of the table. The result of table extraction is presented on JSON format. If a web table contains toponym, a field will be added on the JSON document to store the toponym values. This field can be used to index the table data in accordance to the toponym, which then can be used in the development of GIR system.

Keywords - table extraction, toponym, GIR, CRF, gazetteer

1. Introduction

Nowadays, many people fulfill their needs of information by browsing on the internet. There are billions web pages composing the World Wide Web [1]. On these web pages, a lot of information is presented by some visualization techniques, such as text, image, table, audio, and video. The aim of certain visualization technique used on a web page is to ensure people can understand the information easily without considering computer's ability to understand the information. With abundant number of web pages, there is a need for an automatic system (computer-aided) to help people find the needed information effectively and efficiently.

Among several visualization techniques mentioned above, table has attracted researchers to find methods for automatically extracting data from it. Tables use visual structure besides textual form to deliver information. This combination differs data extraction from web tables from data extraction from articles (textual form only). Some researches on table extractions are [2], [3], [4], [5], and [6].

One type of information on the internet is information about certain place. This information is usually identified by the name of the place (toponym). On [7] and [8], it is stated that approximately

12%-15% of search engine queries contain toponym. This becomes a background for research on Geographic Information Retrieval (GIR) [9] focusing on extracting and indexing documents (i.e. web pages) so that people can easily get geo-referenced information from them. In indexing process, toponym recognition is done to identify location-related information in documents. Some researches about toponym recognition are [10] and [11].

In information extraction field, it is common to use some statistical models. These models can be used for some tasks, such as sequence labeling. Some examples of statistical models are Hidden Markov Model (HMM), Maximum Entropy Markov Model (MEMM), and Conditional Random Fields (CRF). Among those models, CRF outperforms the others. So, in this research, CRF will be used on the process of table extraction from web pages.

This paper is organized as follows. Section 1 introduces the underlying problems of this research. Section 2 summarizes some works related to this research. Section 3 explains the proposed method to detect toponym from web table. Section 4 describes the experiment and evaluation for the proposed method. And section 5 states the conclusion of this research.

2. Related Works

2.1. Table Extraction from Web Pages

There are some researches on web table extraction so far. Generally, those researches use two approaches: rule-based approaches and machine learning approaches.

[2] and [3] attempt to recognize HTML tables using rule-based approach. They construct a rule set to detect tables on web pages containing worth extracted information (genuine tables). This rule set will eliminate HTML tables that are used for formatting/layouting purposes, not for displaying relational data. There are rules about number of rows, number of columns, and number of some HTML elements, such as images, forms, and links in this rule set. In [3], Penn, et al add a rule which specifies that genuine HTML table must not contain other HTML table.

[5] and [6] try to extract data from web tables with different approach. They use CRF model to recognize table schema before extracting data from those tables. Table schema can be understood by knowing the role of each row on a table. [6] categorizes rows on a table as title rows, header rows, group-header rows, data rows, aggregate rows, non-relational rows, and blank rows as showed on Figure 1. To build a CRF model, some features are extracted from each row on the table. After that, each row is labeled coresponding to its role. Then, this feature set and labels are used for training CRF model.

Provinsi	2010	2011	2012	Header
Sumatera				Group Header
Jambi	234	135	235	Data row
...	...	...	...	Data row
TOTAL	1300	1213	1700	Agregat row
Jawa				Group Header
Banten	543	236	435	Data row
...	...	...	...	Data row
TOTAL	1793	1982	2000	Agregat row

Fig. 1. Example of table with labeled rows

2.2. Toponym Recognition

In toponym recognition field, many researches are conducted in attempt to recognize toponym from textual corpus (set of sentences). [10] uses a rule-based approach to detect toponym on Italian text while [11] combines rule-based and statistical approach to recognize toponym in social media.

The rules which are used on [10] take advantage of some prepositions (e.g. 'from', 'to') and adjectives (e.g. 'near') to detect toponym in a sentence. Then, a dictionary is used to ensure whether the word/phrase after those prpositions/adjectives is a toponym.

In [11], toponym recognition is done by the help of CRF model. Although the use of CRF model is considered as statistical approach, Sagcan & Kagaroz use a rule (i.e. regex) as a feature to build the model. So, it can be said that this is a hybrid approach.

2.3. Conditional Random Fields

Conditional Random Fields (CRF) is one of probabilistic model which is often used on sequence labeling tasks (e.g. part-of-speech tagging, named entity recognition, etc). Compared to other probabilistic models, CRF is better than HMM because of its ability to take more complex feature set on model building [12] and it can handle the label bias problem which may occur on MEMM.

There are some terms related to CRF model: observation, state, and transition. Observation is something that can be observed from the corpus, something that can be used to formulate the feature set. State is the label and transition is the change from a state to another state.

CRF model tries to find the most probable label sequence for a given observation. To do so, CRF model calculates the probabilities for each lable sequence. There are two components affecting this calculation: feature functions and weights of feature functions. The feature function has four parameters, i.e. current state, previous state, observation sequence, and current position. The weight of feature function will be updated on each iteration to get the best model (model which can predict the label sequence with high accuracy).

Mathematically, CRF model building complies a formula as shown in Equation (1).

$$p(y_{1:N}|x_{1:N}) = \frac{1}{Y} exp(\sum_{n=1}^{N} \sum_{i=1}^{F} \lambda_i f_i(y_{n-1}, y_n, x_{1:N}, n)) \quad (1)$$

In that equation, f denotes the feature functions, λ is the weight of the feature function, y is the state, x is the observation, and n is the sequence number.

3. Toponym Recognition from Web Table

The most common way to create a HTML table is by using table tag (`<table>`). Nevertheless, not all HTML elements with table tag are genuine tables. Table tag is often used for formatting/layouting purposes on web pages (non-genuine table). As stated in the previous section, some rules can be applied to distinguish genuine tables from non-genuine tables. Nevertheless, HTML element that complies those rules is not necessarily a valid relational table. A table will be considered as a valid relational table if it consists of at least a header row and some data rows. Therefore, the schema of a table must be recognized before we extract data from the table.

Meanwhile, the problem we face on toponym recognition from web tables is that table does not usually use full sentence for cell contents. In the other research, toponym recognition is usually done by using the features of adjacent words on a sentence. For web tables, the role of adjacent words on toponym recognition can be replaced by column names, i.e. words/phrases on the header rows.

The proposed method to recognize toponym from HTML table is described in Figure 2 and the following subsections.

Fig. 2. Diagram of the proposed method

3.1. Table Detection

The first step in table recognition is to find HTML element with table tag and complies these rules (adapted from [2] and [3]):

- It does not contain other HTML table. In other words, it must be the innermost HTML table element.
- It has at least 3 non-empty rows and 2 non-empty columns, according to Drasden Web Table Corpus (DWTC) framework. To ensure the majority of rows do not only consist of a cell, a rule regarding the ratio of cell number and row number is added. If the ratio is more than 1.5, the table will be considered as genuine table.

- It does not contain too many images, forms, and links. The maximum limit for images and forms is less than the number of non-empty rows on the table. And the maximum limit for links is less than 50% from the table content.

The limit values in those rules are determined after doing some observations on web table in some websites.

3.2. Schema Recognition

Table schema recognition in this research is adapted from the proposed method in [6]. In brief, the steps of schema recognition are:

- Extract the attributes of each cell on a row of the table.
- Combine those attributes to be the features of the row.
- Assign a label for the row corresponding to its role on the table (title row, header row, data row, group header row, aggregate row, non-relational row, or blank row).

The cell attributes used in this research involve attributes related to cell layout and cell content. Here is the list of those attributes:

- IsMerged, whether the cell has rowspan or colspan attribute.
- IsHead, whether the cell is an HTML element with head tag (<th>) .
- IsEmpty, whether the cell is empty.
- IsShortText, whether the cell content is a short text (less than 80 characters long).
- IsLongText, whether the cell content is a long text (more than or equal to 80 characters long).
- IsNumeric, whether the cell content is a numeric string.

After extracting these boolean-valued attributes, row features are constructed by applying logarithmic binning [6] to those attributes.

To build a CRF model, all rows in some HTML table elements from table detection module are labeled manually. Figure 3 shows an example of row features and labels for a table that is used as data train. This CRF model will be used to predict the label of each row on a table before the table is parsed to a more structured form.

	IsMerged	IsHead	IsEmpty	IsShortText	IsLongText	IsNumeric	B	Label
Row 1	1.0	0.0	1.0	1.0	0.0	0.0	2.0	T
Row 2	0.0	0.0	0.0	0.0	0.0	0.0	4.0	B
Row 3	0.0	0.0	2.0	1.0	0.0	0.0	4.0	H
Row 4	0.0	0.0	2.0	2.0	0.0	3.0	4.0	D
Row 5	0.0	0.0	1.0	3.0	0.0	3.0	4.0	D
Row 6	0.0	0.0	1.0	3.0	0.0	3.0	4.0	D
Row 7	0.0	0.0	1.0	3.0	0.0	3.0	4.0	D
Row 8	0.0	0.0	0.0	0.0	0.0	0.0	4.0	B
Row 9	1.0	0.0	2.0	1.0	0.0	1.0	2.0	N
Row 10	1.0	0.0	1.0	0.0	1.0	0.0	2.0	N

Fig. 3. Example of row features and row labels

3.3. Parsing

As table extraction method proposed on [2], before parsing, a web table is transformed to matrix first. This can be done by duplicating cells which have `rowspan` or `colspan` attributes. After that, each row is parsed corresponding to its label.

In this research, JSON is chosen as the format of data structure to represent the result of table extraction. The reason behind this choice lies in the flexibility of JSON usage for information retrieval system. Figure 4 shows the format of table extraction result.

```
"table_notes": "...",
"table_title": "...",
"table_data": {
  "group_header1": {
    "row-1": {
      "key1": "value1",
      "key2": "value2"
    },
    "row-2": {
      "key1": "value1",
      "key2": "value2"
    },
    "_aggregate": {
      "key1": "agregat_value1",
      "key2": "agregat_value2"
    }
  },
```

Fig. 4. JSON format of table extraction result

There are three main parts in the JSON: table_notes, table_title, and table_data. Value for table_notes is taken from the content of non-relational rows on a table. Value for table_title is the content of title rows on a table. And values for table_data are composed from the content of group header, header, data, and aggregate rows. Contents of header rows will be the keys on each row ("key1" and "key2" in Figure 4) while contents of data rows will be the value of the corresponding key.

3.4. Toponym Recognition

Toponym recognition from the result of table extraction begins with an examination on table header. This aims to find table column presuably containing toponyms. Such column can be identified by some toponym indicator words, e.g. 'province', 'city', 'country', etc.

If the table has column which presumably contains toponym, all contents of that column will be checked whether they meet the following rules:

- The word/phrase must be an alphabet string (not alphanumeric or numeric string)
- The length of the word/phrase is less than 200 characters.

If those rules are met, each word/phrase in the column will be checked whether it is registered in the gazetteer or not. In this research, the gazetteer is constructed from location names registered in GeoNames. Because of the abundant number of toponyms in GeoNames, a tool having feature for text searching is used. We choose Elasticsearh over other tools because its text search feature supports fuzzy matching. The requirement for fuzzy matching is caused by the possibility of different spellings between toponym on a web page and toponym on the gazetteer (although they refer to the same location).

If the table does not have column which presumably contains toponym, then each cell content on the header rows which is an alphabet string and less than 200 characters in length will be checked on the gazetteer. This step is to handle a table which has toponym on its header.

4. Experiments and Evaluations

There are two experiments in this research: experiment of CRF model building and experiment of gazetteer usage to recognize toponym on table. The building of CRF model is done several times with different combinations of parameters to obtain the best model. Meanwhile, the experiment of gazetteer usage is done by trying some settings on Elasticsearch to obtain the best configuration which can recognize toponym with high accuracy.

In addition to those experiments, this section will explain the tests performed on table detection module and parsing module.

4.1. CRF Model Building

The scenario for CRF model building experiment is described as follows:

- This experiment uses two types of dataset. The first dataset contains both valid tables and invalid tables. The second dataset only contains valid tables (tables which have at least header and data rows).
- The maximum numbers of iterations in the model building are 100, 200, 250, 300, 350, 400, and 500.
- In the model building process, it can be set whether it will take unseen transitions into account or not. Unseen transition is the transition between states/labels that does not exist in the dataset.

CRF models from this experiment will be evaluated by per label accuracy (especially H and D label accuray) and by full table accuracy. Full table accuracy is associated with the ability of CRF model to predict the label of each row on a table precisely.

Table I describes the statistics of dataset for this experiment.

TABLE I. STATISTICS OF THE DATASET

Dataset	**Table**	**Labels**						
		T	**H**	**G**	**D**	**A**	**N**	**B**
Valid + invalid tables (mixed)	160	24	80	9	1243	27	101	120
Valid tables	145	18	56	16	781	13	69	96

From this experiment, it can be concluded that the best CRF model is obtained from the model building process with 200-300 iterations and without taking unseen transitions into account. Per label F1-score for this CRF model reaches 0.88. In terms of dataset usage, CRF model which is built by a mixed dataset will yield a better full table accuracy when it is used to predict invalid tables.

4.2. Toponym Recognition using Gazetteer

There are three fields on GeoNames which are used to store the names of a location, i.e. name, asciiname, and alternatenames. Name and asciiname fields contain a toponym. Meanwhile alternatenames field may contain more than one toponym.

To check whether a word/phrase is registered in the gazetteer, the text search feature in Elasticsearch is used. In this feature, there are some configurations which can be set to obtain the optimum look up results. Generally, these configurations are related to mapping the data (index structure) and query structures used for searching toponym on Elasticsearch.

There are three kinds of mapping scenarios in this experiment:

- Mapping-1 indexes the three toponym fields from GeoNames into three different fields on Elasticsearch. For name and asciiname fields, not_analyzed index type is used. This type will index the words on a field without analyzing (e.g. do tokenization) those words.
- Mapping-2 indexes the three toponym fields into three different fields and uses `analyzed` index type for all fields.
- Mapping-3 indexes the three toponym fields into one field and uses analyzed index type for the field.

Meanwhile the query structures used in this experiment are the combination of match query, bool query, and fuzziness parameter provided by Elasticsearch. Generally, there are two types of queries, i.e. match (or multi_match) queries and bool queries. The difference lays in the score calculation. Bool query calculates the score in accordance with the type of its subqueries (`must`, `should`, or `must_not`).

From this experiment, the best configuration for toponym recognition using gazetteer is to use bool query for searching on Mapping-1. This configuration can recognize the toponym on the web tables with decent accuracy. The bool query has `must` subquery and `should` subquery. Combination of these subqueries can be used to do fuzzy text searching but it will give a higher score for toponym which exactly matched to the query.

4.3. Table Detector Testing

The rules which are mentioned in section III.A are effective enough to detect genuine web tables. From 3684 HTML elements with table tag on BPS web pages (http://bps.go.id), there are 942 elements that comply those rules. The remainder 2742 elements are not detected as genuine tables and will not be processed any further. These non-genuine tables are HTML elements with tabel tag which are used for form layouting or to display a list of images and links.

4.4. Parser Testing

Before parsing a web table to JSON format, the validity of the table is checked first. A table must have at least header and data rows. The validity of a table is checked using a Finite State Machine (FSM) as shown in Figure 5.

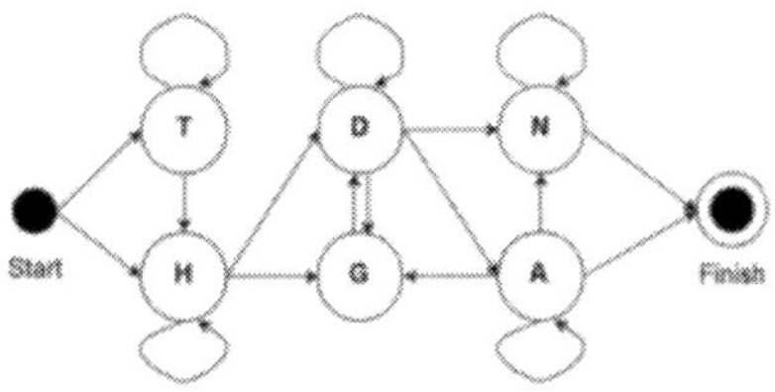

Fig. 5. FSM to check table validity

From 160 manually labeled web tables, there are 20 tables which cannot reach the finish state of this FSM (e.g. tables which only contain non-relational rows, tables that do not have header rows).

From this testing, it is also known that there is a small amount of valid tables which are not parsed to JSON perfectly (corresponding to the designed format described in section III.C). This mostly because of the vagueness of data group boundaries (on a table that has group header or aggregate rows).

5. Conclusion

Table extraction from web pages can be done by combining the rule-based approach and statistical-based approach. Rule-based approach is used to distinguish genuine tables from non-genuine tables. Meanwhile the statistical-based approach is used to recognize the schema of a table before that table is parsed to a more structured format. Table schema recognition can be done by using CRF model to predict the role of each row on a table. In this experiment, the best CRF model yields 0.88 F1-score.

Toponym recognition on the extracted data from a table can be done by implementing a rule-based approach and using a gazetteer. The first step to recognize toponym on a table is to check the table header. If there is a column presumably containing toponym, the values of that column will be checked in the gazetteer. If such column cannot be detected from the table header, then an examination is conducted to detect toponym in the header itself. After the toponym on a table can be recognized, the data of that table can be indexed by this toponym and can be used in future development of GIR system.

This research only tries to handle web tables which are made by using HTML table tag. It is possible that a web table is not made by using HTML table tag, even though it is rare. Hence, it will be better if the rules and features, which are used on table extraction, are defined more generally in the future works so that they can detect and extract all kinds of web tables. Besides, an experiment that involves many more toponyms is needed to test the toponym recognition module (this research uses location names on Indonesia only).

References

[1] D. Yadav. (2010). Users Search Trends on WWW and Their Analysis. Proceeding IITM '10 Proc. First Int. Conf. Intell. Interact. Technol. Multimed. , pp. 59-66.
[2] H. Chen, S. Tsai, J. Tsai. (2000). Mining Tables from Large Scale HTML Texts. COLING'00 Proceedings of the 18th conference on Computational linguistics – Volume 1, page 166-172.
[3] G. Penn, J. Hu, H. Luo, R. McDonald. (2001). Flexible Web Document Analysis for Delivery to

Narrow Bandwidth Devices. *Sixth International Conference on Document Analysis and Recognition Proceedings.*

[4] Y. Wang, J. Hu. (2002). A Machine Learning Based Approach for Table Detection on The Web. *WWW '02 Proceedings of the 11 th international conference on World Wide Web.*

[5] D. Pinto, A. McCallum, X. Wei, W.B. Croft. (2003). Table Extraction Using Conditional Random Fields. SIGIR '03 Proceedings of the 26th annual international ACM SIGIR conference on Research and development in information retrieval.

[6] M.D. Adelfio, & H. Samet. (2013). Schema Extraction for Tabular Data on the Web. *Proceedings of the VLDB Endowment, vol. 6 issue 6, page 421-432.*

[7] M. Himmelstein, "Local Search : The Internet Is the Yellow Pages," Computer (Long. Beach. Calif). , vol. 38, no. 2, pp. 26–34, 2005.

[8] M. Sanderson, W. Bank, and J. Kohler, "Analyzing geographic queries," in SIGIR Workshop on Geographic Information Retrieval., 2004, vol. 2, pp. 8–10.

[9] A. Dewandaru, I.S. Suwardi, and S. Akbar. (2015). Evaluation on Geospatial Information Extraction, Retrieval Mining Thematic Maps from Web Source. *Proceedings International Conference on Information and Communication Technology (ICoICT),* 283-288. Nusa Dua.

[10] D. Caruso, R. Giunta, D. Messina, G. Pappalardo, E. Tramontana. (2015). Rule-based Location Extraction from Italian Unstructured Text. *WOA, volume 1382 of CEUR Workshop Proceedings, page 59-64.*

[11] M. Sagcan, P. Karagoz. (2015). Toponym Recognition in Social Media for Estimating the Locations of Events. *IEEE International Conference on Data Mining Workshop.*

[12] X. Zhu. (2007). CS838-1 Advanced NLP: Conditional Random Fields. *Technical report, The University of Wisconsin Madison.*

A Framework for Sentiment Analysis Implementation of Indonesian Language Tweet on Twitter

Asniar[1] and B R Aditya[2]

[1,2] School of Applied Science, Telkom University, Jl. Telekomunikasi No.1, Bandung, Indonesia

Email: [1]asniar@telkomuniversity.ac.id, [2]bayu@tass.telkomuniversity.ac.id

Abstract. Sentiment analysis is the process of understanding, extracting, and processing the textual data automatically to obtain information. Sentiment analysis can be used to see opinion on an issue and identify a response to something. Millions of digital data are still not used to be able to provide any information that has usefulness, especially for government. Sentiment analysis in government is used to monitor the work programs of the government such as the Government of Bandung City through social media data. The analysis can be used quickly as a tool to see the public response to the work programs, so the next strategic steps can be taken. This paper adopts Support Vector Machine as a supervised algorithm for sentiment analysis. It presents a framework for sentiment analysis implementation of Indonesian language tweet on twitter for Work Programs of Government of Bandung City. The results of this paper can be a reference for decision making in local government.

1. Introduction

In the digital era, people have a variety of social media accounts such as Facebook, Twitter, Instagram, Path, Pinterest, Blogs, and other social media. Every time these people put the idea to share on social media or leave a comment. With the social media, people as if finding media to express opinions more easily and more freely. Besides that, people prefer to express what they think in social media rather than express directly to the object or the person.

Along with the widespread use of social media, the data become more and more unsystematic. Sometimes the writing is shared socially mediated consists of only one word that is difficult to understand or not rare that someone makes writing that does not mean anything. Millions of digital data are scattered and are still not used to be able to provide any information that has value to the government or the company.

Sentiment analysis that is part of the opinion mining [4], is the process of understanding, extracting and processing the textual data automatically to get information [2]. Conducted to see opinion on an issue, or it can also be used to identify the tendency of things in the market [1]. Sentiment analysis in this research is the process of textual document classification into two classes, positive and negative sentiment. The magnitude of the effect and benefits of sentiment analysis, leading research or applications on the rapidly growing sentiment analysis, even in America, approximately 20-30 companies that focus on sentiment analysis service [4]. Basically, sentiment analysis is a classification, but the reality is not as easy as usual because the classification process is related to the use of language. In which there is ambiguity in the use of words, the absence of intonation in a text, and the development of language itself [3].

The interaction between leaders and members is a necessity in order to create synergy and can produce a business or product that fits the vision and mission. For example in this case is the interaction between the Municipal Government, specifically in this case the mayor of officials and citizens. There are many ways to interact with the mayor, through a forum, official letters, letters of complaint, demonstrations, Facebook pages, and even Twitter. Interaction with twitter carried out by one mayor in Indonesia, namely Mr. Ridwan Kamil who is the mayor of Bandung City. Mr. Ridwan Kamil almost every day using Twitter as a means to interact with citizens. His twitter account is @ridwankamil. According to the observations, Mr Ridwan Kamil receives a response from the public ranging from around dawn to midnight. Tweet from Mr. Ridwan Kamil usually contains about socialization of program of the government (local government) and response about complaints from the public.

Sentiment analysis can be performed to monitor Bandung City Government's Programs which quickly can be used as a tool to see the public response to that program so that the next strategic steps can be conducted. Based on this background, it is critical to conduct research about sentiment analysis towards Indonesian language documents of Bandung City Government's Programs.

Researches about sentiment analysis of tweet have been conducted by many researchers such as "Twitter as a Corpus for Sentiment Analysis and Opinion Mining" [8] which provides a method of data collection. Furthermore, the research entitled "Analisis Sentimen Pada Dokumen Berbahasa Indonesia dengan Pendekatan Support Vector Machine" [6] is added language detection process to get data in Indonesian language and to produce result that the classification process that is supervised by an algorithm using Support Vector Machine (SVM) has level of accuracy that is better than the Naïve Bayes. Unfortunately, no study has tried to make the implementation of these two studies in the form of software tools that can automatically perform sentiment analysis of Indonesian tweet to see public opinion on Bandung City Government's Program. However, twitter is already used by Mayor of Bandung City to interact with the community. So it is very important software tools that can rapidly process and analyze data in social media tweet related to Bandung City Government's Programs. By doing so, Mayor of Bandung City can make policie regarding to its work program quickly.

Through this study, researcher can propose a framework of implementation of sentiment analysis of Indonesian language tweet on twitter for Work Programs of Bandung Government so that it can become a reference in the creation of applications sentiment analysis for Indonesian language tweet for work program of Bandung City.

In summary, the contribution of this work is described as follows Opinion Lexicon, Sentiment Analysis, and Proposed Framework for Implementation of Sentiment Analyisis of Indonesian Language Tweet on Twitter.

2. Opinion Lexicon

In the process of sentiment analysis, document is identified whether it is an opinion (sentiment contained in the document). To get the sentiment, the document should be analyzed so that it can be concluded whether it contains sentiment or not. Opinion lexicon is a collection of words containing sentiment. Examples of words containing opinions with positive sentiment are "cantik" , "indah", "bagus" and "luar biasa". While examples of words that contain the negative sentiment are "jelek", "lemah" and "payah" [4].

In practice, the use of word having opinion consists of two kinds [2] :

2.1. Direct Opinion

Direct opinion is a direct expression of the sentiments of the target object such as product, topic or person in a document.

2.2. Comparative Opinion

Comparative opinion is to compare the similarity or difference of more than one object.

The technique of collecting sentiment words can use dictionary-based approach [4]. The strategy used is listing words that are previously known their sentiment, then conducting a query to the dictionary to get the equation of words (synonyms) or opposite (anonymous). Query results are then used as input parameters for the next query, until no more new word again. For English dictionary, the commonly used dictionary is WordNet, which has already contained full lexical English. The weakness of this technique is not able to get a word in accordance with a particular domain. To eliminate it, then a technique based on corpus is used in which word identification is done on a corpus with certain existing domain. The problem that arises is there is no corpus with a complete vocabulary for all languages.

3. Sentiment Analysis

Sentiment analysis is the process of understanding, extracting and processing the textual data automatically to get the sentiment information contained in an opinion sentence [2]. Sentiment analysis is one part of the opinion mining [4]. Sentiment analysis is conducted to see opinions towards a problem or an object by a person to view negative or positive [3]. One example of the use of sentiment analysis in the real world is the identification of market trend and market opinion to an object.

The magnitude of the effect and benefits of sentiment analysis led to the rapid growth of research and applications of sentiment analysis. Even in America, there are about 20-30 companies that focus on sentiment analysis service [4]. Pang et al in 2002 classified the review of the film on the level of a document that has positive or negative opinion by using supervised learning techniques. A set of movie reviews that had previously been determined to be either positive or negative is used as training data for a machine learning algorithm that is already there. Accuracy obtained ranged from 72% to 83% [1].

Outline techniques for performing sentiment analysis is divided into two [5]:

3.1. Using Engineering Symbols

In this technique, analysis of every word in the document and extracting relationships to get the sentiment are conducted. To use this technique, the meaning of each word must be known and the rules of words in a sentence varies depending on the language used. Examples of the use of symbols techniques such as by using a dictionary as a base, where a document is declared as a representation of the word. Each word is determined its sentiments then the use of certain functions (sum, average, weight etc.) is used to get the sentiment of a document. The use of this technique is highly dependent on the language used, because the rules of words is different in languages. Even though the standard rule has been formulated in language, in fact, it cannot be applied completely. It is associated with the language itself that evolves.

3.2. Using Machine Learning Techniques

In this technique the features used are unigram and n-gram [8]. In unigram features, words and symbols in the document are represented into vector shapes, and each word or symbol is counted as a single feature. Whereas the n-gram features use syntactic, semantic, link-based and part of speech. In this case the features taken are not word by word but by whole document, in which the relationship of each word in a document is analyzed previously to obtain the relation among words. The words that make up relation are taken into a feature classification. Machine learning technique emphasizes the use of statistic to process the texts. The examples of the use of machine learning techniques are the use of Super Vector Machine (SVM), Naïve Bayes, Maximum Entropy, Centroid Classifier and K-nearest Neighbor.

4. Implementation of Sentiment Analysis of Indonesian Language Tweet on Twitter

To implement sentiment analysis of Indonesian language tweet on twitter for work programs of Bandung City Government, it is necessary to develop the device application with the proposed framework includes the following steps as shown in figure 1.

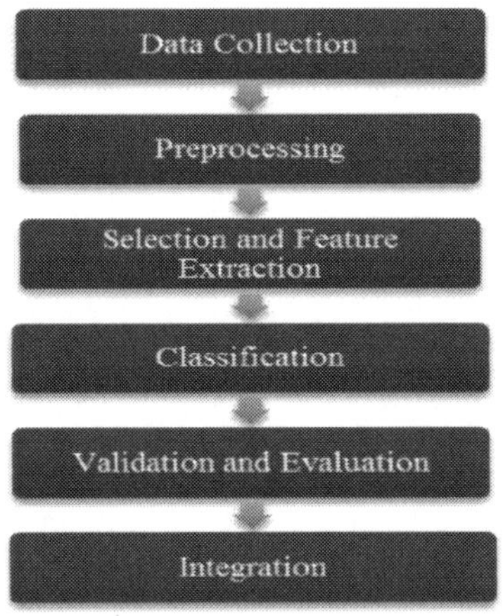

Figure 1. Proposed Framework for Sentiment Analysis Implementation of Indonesian Language Tweet on Twitter for Work Programs of Bandung City Government.

4.1. Data Collection

At this stage, software implementation will be made for data collection. The data used come from Twitter, a social networking site. Topics are limited regarding the comments about the Government's Work Programs in the City of Bandung in era of Mr. Ridwan Kamil. Methods of data collection used refers to the study "Twitter as a Corpus for Sentiment Analysis and Opinion Mining" [8] and "Analisis Sentimen Pada Dokumen Berbahasa Indonesia dengan Pendekatan Support Vector Machine". [6] For the process of crawling, it uses the keyword of several Special Work Program, such as: the city park, culinary night, etc.

4.2. Preprocessing

At this stage, software implementation for preprocessing will be made. The purpose of preprocessing is to change the unstructured data into structured data.

The Methods that will be implemented at this stage are as follows [6]:

4.2.1. Cleansing. Cleansing is the process of cleaning of the document that contains words that are not required to reduce noise. The word omitted is HTML code, keyword, emotion code, hashtag (#), username (@ username), url (http://situs.com), and email (nama@situs.com).

4.2.2. Case folding. Case folding is the uniformity of letter as well as the elimination of the numbers and punctuation. In this case, only latin letters between a through z are used.

4.2.3. Parsing. Parsing is the process of breaking document into a word. This is according to the features used, namely unigram.

4.3. Selection and Feature Extraction

At this stage, software implementation will be made for the selection and feature extraction. The purpose of the selection and feature extraction is to create and classify data set which contains sentiment words.

The Methods that will be implemented at this stage are as follows [6]:

4.3.1. Part of Speech (POS). Part of Speech (POS) tagger is the process of giving a class to the word. Class of word selected is the word that contains many sentiments. Determination of the class of words is based on Kamus Besar Bahasa Indonesia (KBBI).

4.3.2. Stemming. Stemming aims to reduce variations of words that have the same basic word. As in the POS Tagger processing, stemming processing is also done based on KBBI.

4.4. Classification

At this stage, the implementation of the software will be made to do the weighting and classification. The purpose of weighting and classification is to determine the number of frequency of occurrences of positive sentiment and negative sentiment. The weighting processing is carried out by using Feature Term Frequency (TF), Feature Term Presence (TP), and the Term Frequency-Inverse Document Frequency (TF IDF). The classification processing uses supervised method with Support Vector Machine (SVM) algorithm because it has level of accuracy that is better than the Naïve Bayes according to the result of research "Analisis Sentimen Pada Dokumen Berbahasa Indonesia Dengan Pendekatan Support Vector Machine" [6].

Basically, SVM method classify data into two classes by constructing the N-dimensional hyper plane [10]. SVM uses g(x) as the discriminate function where the formula of g(x) is as follows: [11],

$$g(x) = w^T f(x) + b \tag{1}$$

where w is the weights vector, b is bias and f(x) is nonlinear mapping from input space to hogh-dimensional feature space. The parameters w and b are learned automatically on the training dataset following the principle of maximized margin using the following formula :

$$Min\ ½\ w^T w + C \sum_{i=1}^{N} ci \tag{2}$$

where N is the slack variables and C is the penalty coefficient.

4.5. Validation and Evaluation

At this stage the software implementation will be created to perform the validation and evaluation by calculating the accuracy with coincidence matrikx showing by Table 1[12].

Table 1. Coincidence Matrix [12]

		True Class	
		Positive	Negative
Predicted Class	Positive	True Positive Count (TP)	False Positive Count (FP)
	Negative	False Negative Count (FN)	True Negative Count (TN)

Table 1 shows the number of rows test data predicted with the true and false classification. Thus, the accuracy of classification can be obtained from the sum of true positive and true negative divided by the total, with the following formula:

$$Accuracy\ (A) = \frac{TP+TN}{(TP+TN+FP+FN)} \quad (3)$$

4.6. Integration
This stage is the integration of the results of the implementation of the previous stages which is the developing of software integration tools from the stages of the data collection, preprocessing, feature selection and extraction, classification, validation and evaluation.

5. Conclusion
This research resulted the proposed framework for implementation of Sentiment Analysis of Indonesian Language Tweet on Twitter for Work Programs of the Bandung City Government. To implement this proposed framework, the development of software applications for each steps, from Data Collection stages, Preprocessing, Selection and Feature Extraction, Classification, Validation and Evaluation, and Integration is required.

Finally, This proposed framework is expected to be developed in the future research to make the application of implementation of Sentiment Analysis of Indonesian Language Tweet on Twitter for Work Programs of the Bandung City Government.

References
[1] Pang B, Lee L and Vaithyanathan S 2002 Thumbs up? Sentiment Classification using Machine Learning *in Proc. of the ACL-02 conf. on Empirical methods in natural language processing* vol 10 pp 79-86 (USA: Morristown, NJ)
[2] Pang B and Lee L 2008 Opinion Mining and Sentiment Analysis, Foundations and Trends *Information Retrieval* vol 2 no. Issue 1-2 pp 1-135
[3] Liu B 2010 Sentiment Analysis: A Multi-Faceted Problem *IEEE Intelligent Systems*
[4] Liu B 2010 Sentiment Analysis and Subjectivity *Handbook of Natural Language Processing*
[5] Boiy E, Hens P, Deschacht K, Moens M F 2007 Automatic sentiment analysis in on-line text *Proc. of the 11th Intl. Conf. on Electronic Publishing (ELPUB 2007)* (Austria: Vienna)
[6] Nur M Y and Santika D 2011 Analisis Sentimen Pada Dokumen Berbahasa Indonesia Dengan Pendekatan Support Vector Machine (Indonesia : KNSI 2011)
[7] O'Keefe T and Koprinska I 2009 Feature Selection and Weighting Methods in Sentiment Analysis *Proc. of the 14th Australasian Document Computing Symp.*
[8] Pak A and Paroubek P 2010 Twitter as a Corpus for Sentiment Analysis and Opinion Mining *Proc. of the Seventh conf. on Intl. Language Resources and Evaluation (LREC'10)* pp 1320-1326
[9] Naradhipa A R and Purwarianti Ayu 2011 Sentiment Classification for Indonesian Message in Social Media *Intl. Conf. on Electrical Engineering and Informatics* (Indonesia : Bandung)
[10] Vapnik V 2000 The Nature of Statistical Learning Theory *Springer-Verlag* pp 863-884
[11] Yang Y and Liu X 1999 A re-examination of text categorization methods *Proc. of ACM SIGIR Conf. on Research and Development in Information Retrieval (SIGIR)* (USA : New York) pp 42–49
[12] Olson D L and Delen D 2008 Advanced Data Mining Techniques, 1st ed. Heidelberg (Berlin: Springer)

Set of Frequent Word Item sets as Feature Representation for Text with Indonesian Slang

Dian Sa'adillah Maylawati[1,3], **G. A. Putri Saptawati**[2,4]
[1]Departement of Informatics, State Islamic University of Bandung, Indonesia
[2]Informatics, Bandung Institute of Technology, Indonesia
[3]Email: `diansm@uinsgd.ac.id`
[4]Email: `putri@informatika.org`

Abstract. Indonesian slang are commonly used in social media. Due to their unstructured syntax, it is difficult to extract their features based on Indonesian grammar for text mining. To do so, we propose Set of Frequent Word Item sets (SFWI) as text representation which is considered match for Indonesian slang. Besides, SFWI is able to keep the meaning of Indonesian slang with regard to the order of appearance sentence. We use FP-Growth algorithm with adding separation sentence function into the algorithm to extract the feature of SFWI. The experiments is done with text data from social media such as Facebook, Twitter, and personal website. The result of experiments shows that Indonesian slang were more correctly interpreted based on SFWI.

1. Introduction

Social media acts as a medium for the emergence of a variety of slang, including Indonesian slang. The use of Indonesian slang results in difficulties for social media mining since it has unstructruced syntax, and since there is no complete and official source of Indonesian slang [9]. Moreover, Indonesian slang is increasingly deviate from the rules, it called 'Alay' language. There have been available text mining research, none has pre-processing phase for handling Indonesian slang.

Compared with bag of words, Multiple of words is considered as good text representation [5] [8] since it can semantically maintain the meaning of sentence by maintaining the relationship between words, phrases, clauses and sentences [1] [5] [10]. Frequent Word Sequence (FWS) [2] [4] [10] and Set of Frequent Word Sequence (HFWS) [10] which are text features based on sequential pattern, have showed their ability in semantically maintaining document meaning. However, they work well for formal text (documents). Meanwhile, Frequent Word Item sets (FWI) which is feature based on frequent pattern [2] [3], is not work well for formal text [10]. FWI does not consider the sequence words (phrases, clauses, moreover sentence) to maintain the text meaning [2] [10]. Therefore, frequent pattern can be used as a structured representation of the text that is not formal.

In this paper, we propose Set of Frequent Word Item sets (SFWI) which based on frequent pattern as text features for Indonesian slang. In description above, FWI can not maintain the text document meaning well because FWI does not consider sequence of words. But intead, in SFWI, the sequence of sentence is payed, because the sentence contain the meaning of text. In our work, feature of SFWI was produced by Frequent Pattern Growth (FP-Growth) algorithm [7] with adding separation sentence function into algorithm.

2. Features Extraction for text with Indonesian slang

We begin this section by explaining the general concept of SFWI and how it is used as features for text with Indonesian slang. Then, we explain how to prepare text with Indonesian slang before conducting feature extraction based on SFWI.

2.1. SFWI representation for text with Indonesian slang

FWI is a structured representation for the text which are text based on frequent pattern [2] [10]. FWI has the following syntax structure: <(w1,w2),(w3,w4),…>, where (w1,w2) is FWI i, (w3,w4) is FWI i+1, etc. All items in FWI always appear together, but do not mean consecutive. For example, (*cerita, seru*) will be treated the same as (*seru, cerita*).

To overcome the problem of unstructured syntax in Indonesian slang, we propose to use Set of FWI (SFWI) instead of FWI. SFWI representation has the following syntax structure: << (w1, w2), (w3, w4)>, <(w5, w6) >>, where (w1, w2) is FWI i, (w3, w4) is FWI i + 1, (w5, w6) is FWI i + 2. Each item in FWI does not mean consecutive, instead it means that all items always appear together. <(w1, w2), (w3, w4)> is called SFWI i with two FWI in sentence i, means that FWI i, FWI i + 1 and FWI i + 2 are consecutive. Moreover, <(w5, w6)> is SFWI i + 1 with one FWI in sentence i + 1. Sentence i and Sentence i+1 are consecutive too. For example, <<(*drama, korea, hipnotis, gue*), (*asli, seru, abis*))>>, <(*baper, parah*)>> is not equal to <<(*baper, parah*)>, <(*drama, korea, hipnotis, gue*), (*asli, seru, abis*)>>. But equal to <<(*gue, hipnotis, drama, korea*), (*seru, abis, asli*)>, <(*parah, baper*)>>. It means that the appearance of FWI in SFWI is consecutive depend on the sequence of sentence in text.

As mentioned before, Indonesian slang does not have structured syntax by which their meaning is not represented by the sequence of words, but by the sequence of sentence. Accordingly, SFWI is more applicable compared with FWS and more applicable than FWI because using sequence of sentence for maintain the meaning of text.

2.2. Handling Indonesian slang before the establishment SFWI

Prior to the extraction of SFWI we need to address some Indonesian slang as follow:

- Slang words which are exist in the slang dictionary will be recognized. If slang word is not exist in the slang dictionary, SFWI will still recognize the slang word if it appears frequently. For example, from 10 documents with minimum frequency is 5, we get SFWI <<(*doi, gantung, gue*)>, <(*galau, tingkat, dewa*)>> with frequency is 6. So, SFWI is still able to keep the meaning from those 10 documents.
- Slang that change the prefixes and suffixes that are not match with Indonesian (eg. the prefix 'me-' into 'nge-', the suffix '-kan' to be '-in', for example "mengerjakan" to be "ngerjain". For handling this problem, the word will be changed to its basic word with stemming process. To do stemming process, it is prepared stem word dictionary. We used Porter algorithm for stemming with several modifications. Such as checking the word which is the stem word. Stemming process is only done for the words with affixes. Next, we add prefixes and suffixes which are appeared in Indonesian slang.
- There is a case in Indonesian slang that the slang word is the same as Indonesian word but has different meaning. For example the word "secara", in Indonesian slang it means "confirmation" but in Indonesian it means "in a way". Another example is the word "cumi", in Indonesian it means "squid as animal" but in Indonesian slang it means "mockery". Therefore, SFWI cannot represent the proper meaning of sentences if they use some Indonesian words for expressing different meaning in Indonesian slang. The word "cumi" in Indonesian slang and the word "cumi" in Indonesian will be considered as the same word by SFWI.
- In Indonesian slang, there is sound of word which is usually used conversation is used to be word. For example, the word "lihat", in Indonesian slang is written "liat" (use the sound of the word). The problem is in Indonesian, "liat" it means "something soft, flexible" but in Indonesian slang it

means "lihat (see)". If the sound of word which is written as word is the same as Indonesian word, then SFWI cannot differentiate them.

- Since Indonesian slang is common to shorten words such as "makan" to be "mkn", "mkan", "makn". So, it is necessary to translate them into the complete words before extracting SFWI. This is a part of Natural Language Processing (NLP) that recognize an unfamiliar words or containing words or structures that have not been seen before. As consequences, it is required to recognize a complete collection of abbreviations for all possibilities of shortened words. SFWI will be able to keep the meaning as long as word acronym has been changed into the complete word.
- Excessive use of punctuation marks can be removed. Whilst the punctuation which is not appropriate will affect the process of separating sentences. The affect is the different SFWI with the same item. But as long as the word or sentence is frequent, the meaning is kept in SFWI.
- Slang for excessive or "Alay" and slang that mix with foreign or local language are only a few in this works, so they are not considered to be frequent. For handling foreign of local language, of course we need dictionary of all foreign and local language to translate them to Indonesian. Then, for "Alay" language which is mix character numbers and letter still difficult to distinguish. If all number characters are changed to be letter, such as "4" to "a", "5" to "s", "2" to "z", "3" to "e", "13" to "b", "12" to "r", and so on, for example "6u3 131n6un9 54m4 tul154n 6u3 53nd1121" to be "gue ieingung sama tulisan gue sendiizi" or "gue bingung sama tulisan gue sendiri". However, because of "Alay" language is often used by particular group such as teenager, it is not considered to be frequent.

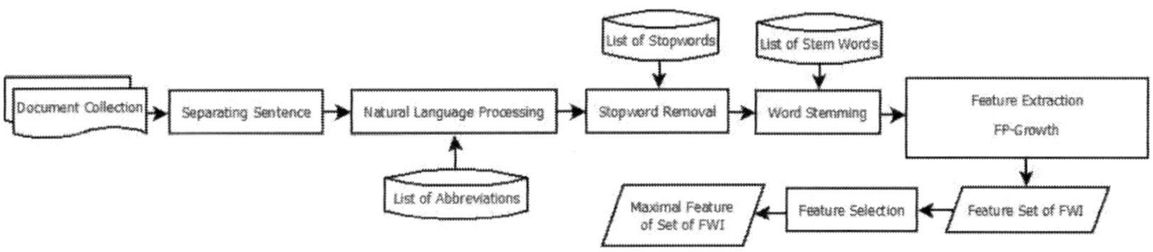

Figure 1. Process of creating the SFWI representation

2.3. The process of producing SFWI feature

This process starts with separating sentence. Followed by NLP tackling the characteristics of slang which use abbreviations. Next processes are eliminating stop words and word stemming. Finally, the process of generating and selecting SFWI features are conducted. Figure 1 describes the overall process in more detail. Detail of each phase for the process of extracting SFWI feature will describe as follow (the example available in Indonesian language in Figure 2):

- First, collection of text document will be separated by sentence in accordance with the SFWI's structure. SFWI looks carefully at sequence of sentence, because a sentence has a complete meaning. Sentence separating process is done with simple algorithm, over the contents of document is read, if it finds the end of the sentence punctuation such as point (.), exclamation (!), or question (?), then it will be separated as different sentence.
- After separating sentence process, the content of document will be checked weather it contains a natural language such as abbreviation word. If it finds, it will be changed to be an original form. We used abbreviation dictionary which save abbreviation and its original form. Content of abbreviation dictionary such as *ak* to *aku*, *rame* to *ramai*, *lby* to *lebay*, *cys* to *ciyus*, *jys* to *jayus*.
- Next, the stop words will be removed. If the content of document contains stop word which exist in stop words dictionary, it will be removed. Stop words in Indonesian, including slang, such as *apakah, apalagi, adalah, atau, ataukah, ataupun, bagai, bagaikan, sebagai, sebagainya, bagaimana, dan, yang, ini, itu, loe, gw, deh*, etc

IOP Conf. Series: Journal of Physics: Conf. Series **801** (2017) 012066 doi:10.1088/1742-6596/801/1/012066

- Then, after removing the stop words, the words that contain an affixes (prefix or suffix) will be stemmed. We used Porter algorithm which is modified. We added stem word checking process at the first step. It the word is the stem word that exist in the stem dictionary, then it will not be stemmed. We also changed the prefix and suffix in accordance with the Indonesian grammar and Indonesian slang. Stem words in Indonesian, including slang, such as *belajar, sapu, mampu, ada, keles, bingit, rempong, lol, unyu, kepo, php, geje, gokil*, etc.
- Next step is feature extraction. This step is produced SFWI features from document text. We used FP-Growth for create SFWI. FP-Growth was selected because it produce frequent item set which is accordance with FWI representation. Frequent item sets which is produced by FP-Growth is the same as FWI's structure that it is not represent sequence of words. Just need to add the separating sentence process before use FP-Growth so that SFWI can be created.
- The last step is feature selection for selecting redundant features. For example if an item set has frequented, automatically sub item set has frequented too. We used maximum approach for this work. A feature f is called maximum if no more feature f' which sub feature of feature f.

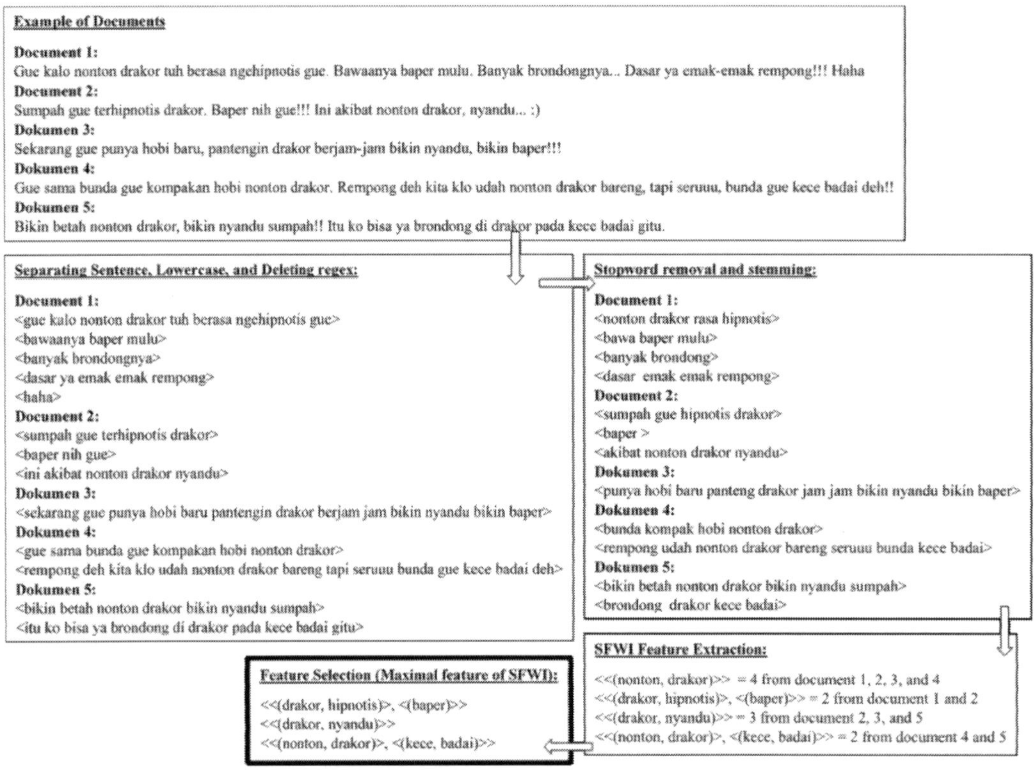

Figure 2. Example of producing SFWI with 40% minimum support, 1 maximum gap, and 1 maximum length

3. The experiment and result

In accordance with the needs of research, text documents taken from tweet of Twitter, some of the blogs that use Indonesian slang and Facebook. Text documents are grouped into directories according to the topic. SFWI representations are stored in a file. We build the text preprocessing library for the needs of this experiment. We also use Sequential Pettern Mining Framework (SPMF) Library for FP-Growth Algorithm [6]. The experiments were conducted based on several test scenarios as follows:

- Examine the meaning of Indonesian slang by SFWI representation. Text data is taken from Facebook (168 of data, with five category, among others: "galau", "hiburan", "gaul", "statustereliye", and "cinta"), Twitter (681 of data, with five category, among others:“kampanye”,“tahunbaruislam”, “kabinetkerja”, “gayahidup”, and “artiskorea”), and form Blog (191 of data, with five category, among others: “anakgaul”,“cirihamilmuda”, “masalahrumahtangga”, ”“nasehatcinta”, and “pengobatantradisional”). This test aims to see accuracy of keeping slang meaning by SFWI representation. The combination of parameters entered for this test is the support of at least 10-100%, maximal gap 5 and 10, and a minimum length of 1, 2, and 3. The test that produces SFWI accordance with the input data topics rated 1 and the results are not appropriate topics or do not produce SFWI rated 0.

- Seeing the comparison of accuracy between SFWI representation and FWI representation in maintaining the meaning of Indonesian slang. The same test scenario for examining the meaning of Indonesian slang by SFWI representation will be used for FWI representation.

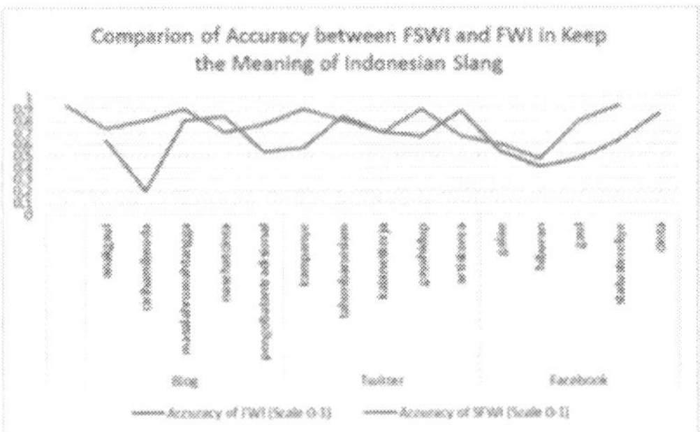

Figure 3. Comparison of Accuracy between SFWI and FWI

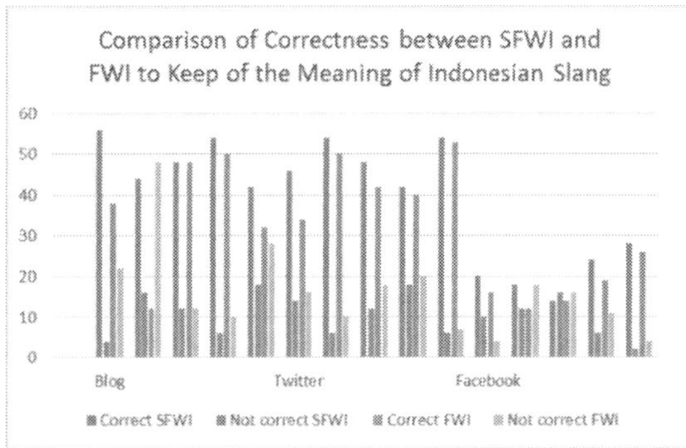

Figure 4. Comparison of Correctness between SFWI and FWI

Table 1. Result of keep the meaning of slang between Set of FWI and FWI representation

No	Dataset	Tot. File	Tot. Test	Correct SFWI		Not correct SFWI		Accuracy (Scale 0-1)		Time Average (ms)	
				SF WI	FWI	SF WI	FWI	SF WI	FWI	SFWI	FWI
1	anakgaul	72	60	56	38	4	22	0,93	0,63	15536	9125
2	cirihamilmuda	28	60	44	12	16	48	0,73	0,2	3092	2311
3	masalahrumahtangga	24	60	48	48	12	12	0,8	0,8	3773	2429
4	nasehatcinta	44	60	54	50	6	10	0,9	0,83	55161	14696
5	pengobatantradisional	23	60	42	32	18	28	0,7	0,53	1864	3206
6	kampanye	450	60	46	34	14	16	0,77	0,57	8363	2363
7	tahunbaruislam	100	60	54	50	6	10	0,9	0,83	1497	2369
8	kabinetkerja	41	60	48	42	12	18	0,8	0,7	1471	1652
9	gayahidup	23	60	42	40	18	20	0,7	0,67	1157	657
10	artiskorea	67	60	54	53	6	7	0,9	0,88	1531	778
11	galau	25	30	20	16	10	4	0,67	0,53	2649	3001
12	hiburan	22	30	18	12	12	18	0,6	0,4	2429	987
13	gaul	41	30	14	14	16	16	0,47	0,47	2468	2409
14	statustereliye	50	30	24	19	6	11	0,8	0,63	2575	2095
15	cinta	30	30	28	26	2	4	0,93	0,87	4807	4632
Total Time Average of SFWI : 7225,411922; Total Time Average of FWI : 5014,5890574											
Total Accuracy Average of SFWI : 0,773; Total Accuracy Average of FWI : 0,6367											

The evaluation of the results can be summarized as follows:

SFWI representation can keep the meaning of slang with an average percentage of accuracy is 77.33% (contained in Table 1), while FWI representation has 63.67% average percentage of accuracy (contained in Table 2). Even though in the case of the same slang word but the meaning is different,

SFWI, moreover FWI, cannot tell the difference, this experiment proves that SFWI pretty well in keeping the meaning of slang compared with FWI (Figure 3 and Figure 4). Only one case from all test case that FWI has the same total correctness and same accuracy. Several of the experiment show that SFWI has an accuracy that does not differ much with FWI, but most of them has a higher accuracy. And also the experiment proves that pay attension to the sequence of sentence can keep the meaning of unstructure language such as Indonesian slang well. The time process to extract FWI representation is faster than SFWI because there is no sentence separation process.

4. Conclusion

- In accordance with the results of the analysis and experiment, SFWI representation can be used for keeping the meaning of unstructured language such as Indonesian slang. SFWI has a structure that similar to structure of slang that it is not pay attention to sequence of words. However, SFWI is still pay attention to sequence of sentence which contain of meaning. Even though, SFWI not be able to keep the meaning of slang that contains a lot of meaning. Such as the same word with different meaning. The processes before producing SFWI, from separating sentence until stemming can affect the result of SFWI. Including completeness of stop words, abbreviation, and stem words which are owned.
- Testing and evaluation for keeping slang meaning with FWI representation use a collection of documents from Facebook, Twitter, and Blog. The percentage of accuracy of SFWI is 77.33%. This is higher than FWI that has the precentage of accuracy 63.67%. This accuracy is obtained based on the comparison of the correct and uncorrect value.
- For the future work, the accuracy of SFWI can be improved by increasing vocabulary of slang and abbreviation, because slang language always increases and a lot of new vocabulary. The FP-Growth algorithm which is used to extract SFWI can be replaced by another frequent pattern algorithm. And also the problem of SFWI which can not keep the meaning of word that contains a lot of meaning must be resolved to keeping the meaning better.

5. References

[1] Adityarani, Kania. 2009. *Aplicating CFWS* Algorithm for Document Grouping with Text representation Set of Frequent Word Sequence and Text Representation Qord Frequent itemset.* Informatic, Bandung Institute of Technology.

[2] Agrawal, Rakesh, et al. 1999. *Mapping Words, Phrases using Sequential Pattern to Find User Specific Trends in a Text Database – Patent Document.* International Business Machines Corporation, Armonk, NY.

[3] Agrawal, Rakesh. Ramakrishnan Srikant. 1995. *Mining Sequential Patterns.* IBM Research Center,

[4] Ahonen, Helena. 1999. *Knowledge Discovery in Document by Extracting Frequent Word Sequence.* Library Trends. Vol. 48, No. 1, Summer 1999, pp. 160-181.

[5] Doucet, Antoine. 2005. *Advance Document Description, a Sequential Approach.* Accademic Dissertation, Departement of Computer Science, Series of Publication A. Report A-2005-2, University of Helsinki, Finland.

[6] Fournier-Voger, Philippe. 2008-2016. *Sequential Pattern Mining Framework* (SPMF) *Data Mining Software. Tools.* Available in: http://www.philippe-fournier-viger.com/spmf

[7] Han, Jiawei. Kamber,Micheline. 2006. *Data Mining: Concepts and Techniques Second Edition.* Simon Fraser University. Morgan Kaufmann Publisher.

[8] Hoonlor, Apirak. July, 2011. *Sequential Patterns and Temporal Patterns for Text Mining.* A Thesis Submitted to Graduate Faculty of Rensselaer Polytechnic Institute in Partial Fulfillment of the Requirement for Degree of Doctof og Philisophy. Major Subject: Computer Science. NewYork.

[9] Muliady, Wahyu. 2012. *Generating Indonesian Slang Lexicons from Twitter.* International Conference on Uncertainty Reasoning and Knowledge Engineering.

[10] Nalistia, Rindhu Astri. 2008. *Feature-Based Clustering for documents grouping with the Structured Representation of Text.* Informatic, Bandung Institute of Technology.

Comparison between BIDE, PrefixSpan, and TRuleGrowth for Mining of Indonesian Text

Dian Sa'adillah Maylawati[1,2], **Mohamad Irfan**[1,3] **and Wildan Budiawan Zulfikar**[1,4]
[1]Departement of Informatics, State Islamic University of Bandung, Indonesia
[2]Email: diansm@uinsgd.ac.id
[3]Email: irfan.bahaf@uinsgd.ac.id
[4]Email: wildan.budiawan.z@uinsgd.ac.id

Abstract. Mining proscess for Indonesian language still be an interesting research. Multiple of words representation was claimed can keep the meaning of text better than bag of words. In this paper, we compare several sequential pattern algortihm, among others BIDE (BI-Directional Extention), PrefixSpan, and TRuleGrowth. All of those algorithm produce frequent word sequence to keep the meaning of text. However, the experiment result, with 14.006 of Indonesian tweet from Twitter, shows that BIDE can produce more efficient frequent word sequence than PrefixSpan and TRuleGrowth without missing the meaning of text. Then, the average of time process of PrefixSpan is faster than BIDE and TRuleGrowth. In the other hand, PrefixSpan and TRuleGrowth is more efficient in using memory than BIDE.

1. Introduction

Indonesia is one of the biggest social media user, such as Facebook, Twitter, Instagram, Path, and so on. Mining Indonesian text from social media is interesting things. To get the meaning of text, we have to keep the meaning well. Moreover, Indonesian language has a unique grammer, such as many form to use the affixes that is combined in words. And also, in social media, many Indonesian slang words or it is called "bahasa alay" which should be more attention so that the meaning is manitained.

Compared with bag of words, multiple of words is considered as good text representation [6][11]. As we know, the meaning of text is stored in complete sentence, minimal own the subject and predicate. Frequent word sequence [3][5][13] and Set of frequent word sequence [13] are text features based on sequential pattern which is proved can maintain the meaning of text well. Because of sequential pattern semantically can maintain the meaning of text by maintaining the relationship between words, phrases, clauses and sentences [1][6][13]. Sequential pattern mining which is proposed by Agrawal and Srikant [2][4] is one of the most popular technique to find relationship between event in large database [8][9][15]. Sequential pattern mining will discover the frequent events or elements or subsequence in sequence database [2][4][8][9][10][14][16]. Sequential pattern is called frequent when it meets the minimum support or threshold that is given. Sequential pattern for text has a words as an items, and it is called sequence of word. Frequent word sequence and set of frequent word sequence is sequence of word [13].

Many previous study about efficient algorithm of sequential mining, among others BIDE (BI-Directional Extention) [16], PrefixSpan [14], and TRuleGrowth [8]. In this paper, we compare BIDE, PrefixSpan, and TRuleGrowth algorithm to discover sequence of word from Indonesian text. The form of sequence of words from BIDE, PrefixSpan, and TRuleGrowth is frequent word sequence. The different between frequent word sequence and set of frequent word sequence is frequent word sequence does not pay attention to sequence of sentence [13]. Otherwise, set of frequent word sequence pay attention to sequence of sentence [13].

In this paper, we compare BIDE, PrefixSpan, and TRuleGrowth algorithm because the algorithms have similarity to growth the pattern but different in use approach to process and efficiency. We compare about the efficiency and the ability of those algorithm to discovered frequent word sequence and to keep the meaning of Indonesian text.

2. Text preprocessing for Indonesian text

Text is an unstructure data that need special treatment prior to the mining process. Preprocessing is an important process that can affect the next process to get the best result, essecially for text. Bad text preprocessing will exacerbate the meaning of the text custody.

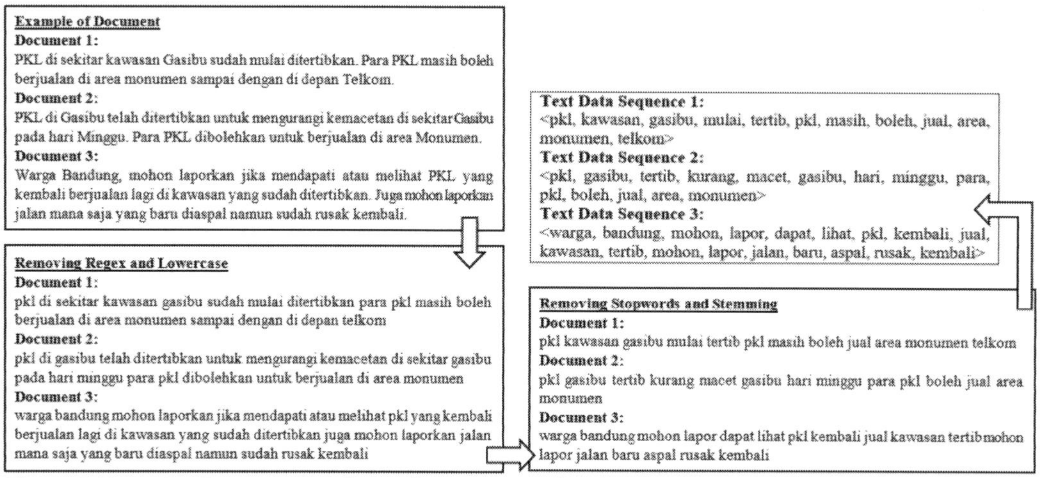

Figure 1. Example of text preprocessing until get sequence data for Indonesian text.

For example, the meaning of text is contained in verb, if there is no stemming process for verb with affixes, the meaning of text will be lost. Because of many slang words in Indonesian text from social media, there is several process that is customized for Indonesian text case [11], among others:

- All of non letter character or regular expression will be deleted from text. Because non letter character such as punctuation must be in every text and it will make frequent. Then, the words will be changed to the same lowercase form.
- The next process is going to check natural language in text. Especially in social media, most of the text contains natural language, among others abbeviation words and slang. Abbreviation word will be changed to be the original form. We used abbreviation dictionary which save abbreviation and the original form, including abbreviation for slang words. For example, *bc* to be *baca*, *sk* to be *suka*, *cnt* to be *cinta*, *cnta* to be *cinta, hdup* to be *hidup*, and so on.
- Stop words removal will be done after all of words have changed to the original form. All of Indonesian slang words is also removed, such as *gue, gw, loe, ini, itu*, and so on.
- After stop words removal, the next important process is stemming. The stemming process is crucial, because verb contain the meaning of text. We use Porter Algorithm for stemming process which has been customized for Indonesian language and Indonesian slang. For example *nyuci* tu be *cuci*, *memasak* to be *masak*, *ngerjain* to be *kerja, belajar* to be *ajar*, *bayarin* to be *bayar*, and so on.
- After cleaning process, from remove regular expression until create text data sequence, next BIDE, PrefixSpan, and RuleGrowth algorithm will discovered the sequence of words from sequence database of the text. For example the ready text after text prepocessing described in Figure 1 with Indonesian text.

3. Discovering Frequent Word Sequence with PrefixSpan

PrefixSpan or Prefix-projected Sequential Pattern Mining is a sequential pattern mining algorithm that adopts a divide-and-conquer and pattern growth principle for efficient mininig of sequential pattern in large sequence database [10][14]. PrefixSpan offers ordered growth and reduce projected database, better than a priori-based algorithm, such as GSP, FreeSpan, and SPADE. In the a priori-based algorithm, a huge set of candidate sequence could be generated in a large sequence database, do multiple scans of database in mining, and generate a combinatorially explosive number of candidate when mining long sequential pattern [2][4][14].

PrefixSpan introduce the concept of prefix and suffix where prefix suppose all the item within an element are listed alphabetically. While, sequence s' is called the suffix of sequence s with regards to prefix or consecutive after prefix. Sequential pattern would be extracted locally from projected database through process of finding frequent items that appended to the prefix [14][15]. The algorithm of PrefixSpan is described below in paper [14][15]. The item of sequential pattern for text mining is the word. For example, from Figure 1, the text document collection will be a sequence database and if the minimum support for sequential pattern is 2, then the result of sequence of word in Table 1 that present with Indonesian text.

Table 1. An example of sequence pattern from Figure 1 which is produced by PrefixSpan

Sequential Pattern	Freq.	Sequential Pattern	Freq.
<pkl>	3	<jual, area, monumen>	2
<pkl, tertib>	3	<area>	2
<pkl, jual>	3	<gasibu>	2
<pkl, kawasan>	2	<gasibu, jual>	2
<pkl, gasibu, jual, area>	2	<tertib, area, monumen>	2
<pkl, gasibu, tertib, jual>	2	<gasibu, jual, area, monumen>	2

4. Discovering Frequent Word Sequence with BIDE

BIDE or BI-Directional Extention is a sequential pattern mining algorithm without candidate maintenance for mining frequent closed sequence [10][16]. Another frequent closed itemset mining algorithms need to maintain the set of already mined closed sequence candidates that cause poor scalability in the number of frequent closed pattern. Because a large number of frequent closed pattern will occupy much memory and lead to large search space for the closure checking of new pattern, specially when the support threshold is low or the pattern become long.

BIDE do not need closure checking to keep track of any single candidate for new pattern [16]. BIDE propose a deep search space pruning method with the *BackScan* pruning method and *ScanSkip* optimization techniques. The algorithm of BIDE is described in reference number [16].

A frequent closed sequence is said to be close if there is no more subsequence s' from sequence s and s' has the same minimum support threshold as s. Based on the definition, from Table 2, frequent closed sequences that are produced among others <pkl> = 3; <pkl, tertib> = 3; <pkl, jual> = 3; <pkl, gasibu, jual, area, monumen> = 2; <pkl, kawasan, tertib> = 2; <pkl, pkl, boleh, jual, area, monumen> = 2. The number of feature of frequent closed sequence less than classic sequential pattern, but will not eliminate the meaning of text because the meaning of subsequence saved in supersequence.

5. Discovering Frequent Word Sequence with TRuleGrowth

TRuleGrowth is a modified version of RuleGrowth algorithm that accept a sliding-window constraint [8] for mining sequential rules common to several sequences uses a pattern-growth approach for discovering sequential rules [8][9]. Pattern-growth approach can be much more efficient and scalable. Sequence database S is a set of sequences s, S={s1, s2…sn}, and a set of items I={i1, i2,…in}, where each sequence *s* is an ordered list of itemsets *s*={X1, X2, … Xn} such that X1, X2, …Xn $\subseteq$ I . A sequential rule X⇒Y is defined as a relationship between two itemsets X, Y $\subseteq$ I such that X$\cap$Y = Ø and X, Y are not empty [8][9]. If the items of X occur in some itemsets of a sequence, then the items in Y will occur in some itemsets afterward in the same sequence as long as there is no ordering restiction between items in X and between items in Y. And also that items from X and items from Y do not need to appear in a same itemset in a sequence. The algorithm of RuleGrowth and TRuleGrowth is described in reference number [8].

TRuleGrowth same as RuleGrowth has two measure for the rule, the first measure is sequential rule support that defined as sup(X⇒Y) where denotes the number of sequence from sequence database where all the items of X appear before all items of Y. The second measure is sequential confidence that defined as conf(X⇒Y) where denotes the number of sequences that contains the items of X or the ratio of the number of Y to the number of X in all sequence in sequence database. From the example text data sequence in Figure 1 for 50% of minimum support and minimum confident will produce a sequence rule such as in Table 2 that present with Indonesian text.

Table 2. An example of sequence rule by TRuleGrowth

Rule	Supp.	Conf.	Rule	Supp.	Conf.
<pkl>⇒<kawasan>	0.67	0.67	<tertib>⇒<area, monumen>	0.67	0.67
<pkl, gasibu>⇒<tertib, jual>	0.67	1	<jual, area>⇒<monumen>	0.67	1
<pkl>⇒<tertib>	1	1	...	...	...

6. The Experiment and Result

Table 3. An example of sequence rule by PrefixSpan

No	Label	Number of File	No	Label	Number of File	No	Label	Number of File
1	Agama	427	8	Fashion	191	15	Kampanye	450
2	Bencana	170	9	Film	1011	16	Keadilan	580
3	Bisnis	752	10	Gaul	437	17	Keamanan	68
4	Bola	1557	11	Hobi	211	18	Kehidupan	432
5	Cantik	1378	12	Hukum	398	19	Kejahatan	342
6	Cinta	4322	13	Ibadah	495	20	Komputer	362
7	Ekonomi	226	14	Idola	197			

In accordance with the needs of research, text documents taken from 14.006 of Indonesian tweet of Twitter. Text documents are grouped into 20 directories according to the topic that labeled with Indonesian text (explained in Table 3). We use the text preprocessing library for Indonesian slang [12] and Sequential Pettern Mining Framework (SPMF) Library for BIDE, PrefixSpan, and TRuleGrowth algorithm for this experiment [7].

The experiments were conducted based on several test scenarios as follows:

- Examine the meaning of Indonesian language with sequence of word that produced by BIDE, PrefixSpan, and TRuleGrowth algorithm. And also comparing the accuracy of sequence of word between BIDE, PrefixSpan, and TRuleGrowth algorithm. We test the algorithms with several various of minimum support, among others 10%, 25%, and 50% of minimum support value. And for RuleGrowth, we use 10%, 25%, and 50% for minimum confident and 10 for size of window-sliding. We will get a number of sequence of word features from each algorithms that keep the meaning of text.
- Comparing the time process (in milisecond) and memory usage (in megabyte) between BIDE, PrefixSpan, and TRuleGrowth in discovered the sequence of word.

The evaluation of the results can be summarized as follows:

- PrefixSpan and BIDE algorithm are better in keep the meaning of text than TRuleGrowth algorithm. It is shown in Figure 8, Figure 9, and Figure 10. Surely, the number of sequence of words or frequent word sequence that are produced depend on contain of text and text preprocessing. But, the experiment shows that with same minimum support, PrefixSpan and BIDE can produce frequent word sequence well. Instead BIDE can produce more efficient the number of frequent word sequence than PrefixSpan without losing the meaning of text. In other hand, TRuleGrowth can not find frequent word sequence that BIDE and PreffixSpan found. It shows in 25% and 50% minimum support, TRuleGrowth can not find frequent of words from the most of categories, only in "Idola" and "Kampanye" with bigger number of frequent word sequence than PrefixSpan. This causes the meaning of text is missing. This result is caused by dependence on minimum confident and window-sliding constraint in TRuleGrowth.

- The average of time process in Figure 2, Figure 3, and Figure 4 for producing frequent word sequence shows that PrefixSpan is the fastest. The average time process of PrefixSpan is 336,883 ms, BIDE is 3391,05 ms, and TRuleGrowth is 4057,37 ms. BIDE have higher time process because BIDE finds frequent closed sequence for efficiency number of feature. While, the average of time process between PrefixSpan is not much different from TRuleGrowth, but TRuleGrowth produces fewer frequent word sequence with longer time process than PrefixSpan and BIDE. Moreover, TRuleGrowth do not produce frequent word sequence mostly.
- For memory usage experiment, we find that PrefixSpan and TRuleGrowth are more efficient than BIDE. Not only from the average of memory usage but also TRuleGrowth and PrefixSpan is not use memory when do not fine the feature or frequent word sequence. It is different than BIDE that still use the memory though do not find the sequence of words. The average of memory usage of PrefixSpan is 3,603 mb, BIDE is 4,587 mb, and TRuleGrowth is 1,733 mb.

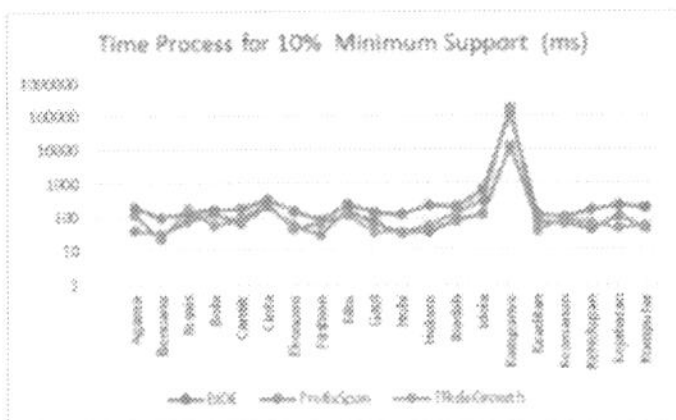

Figure 2. Result of Time Process for 10% Minimum Support

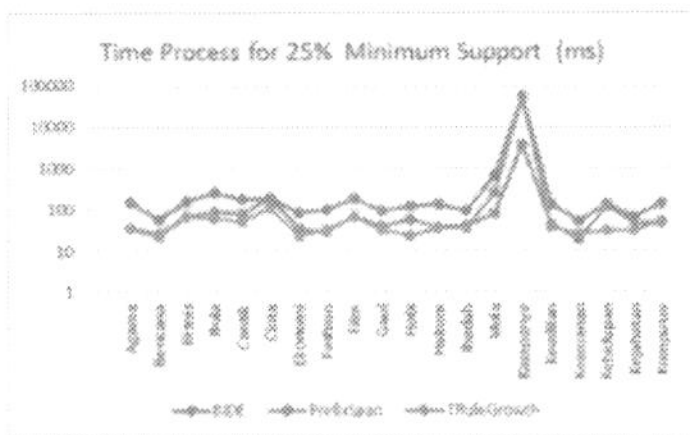

Figure 3. Result of Time Process for 25% Minimum Support

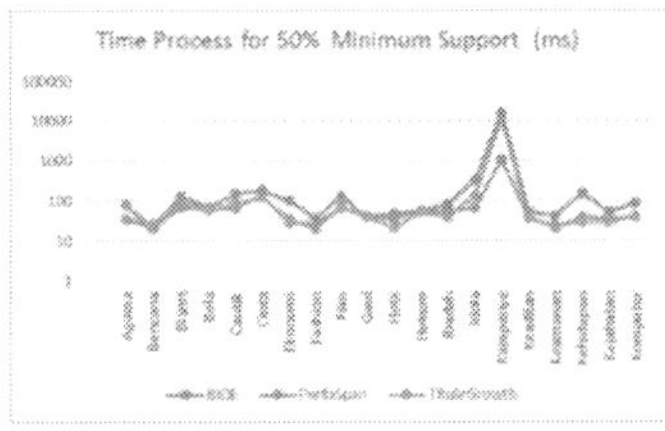

Figure 4. Result of Time Process for 50% Minimum Support

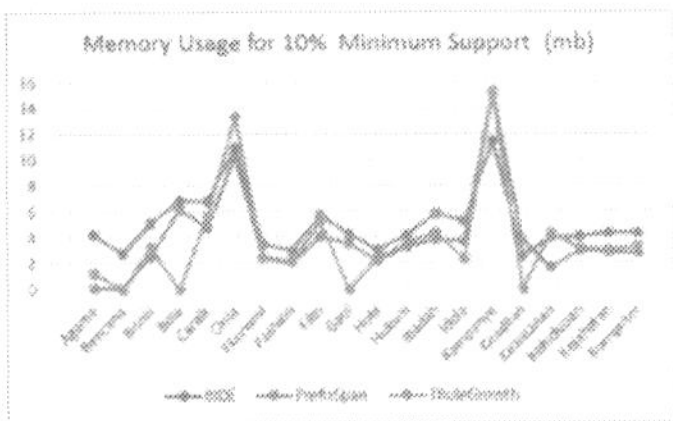

Figure 5. Result of Memory Usage for 10% Minimum Support

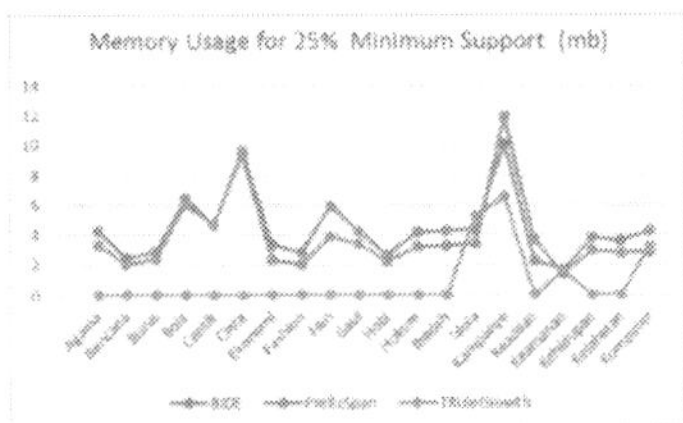

Figure 6. Result of Memory Usage for 25% Minimum Support

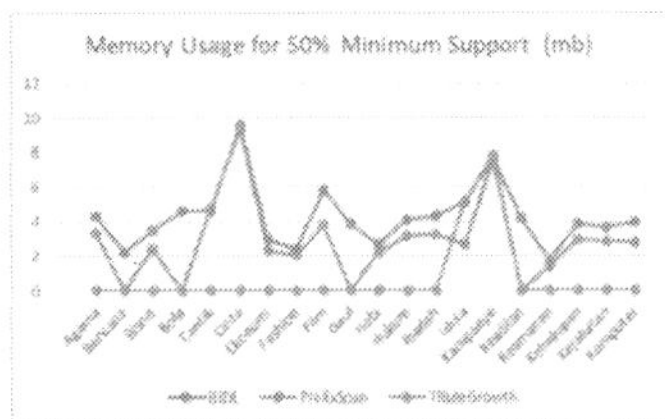

Figure 7. Result of Memory Usage for 50% Minimum Support

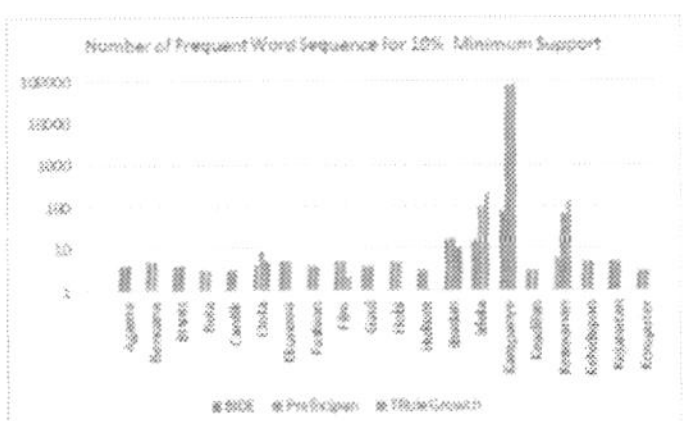

Figure 8. Result of Number of Word Sequence for 10% Minimum Support

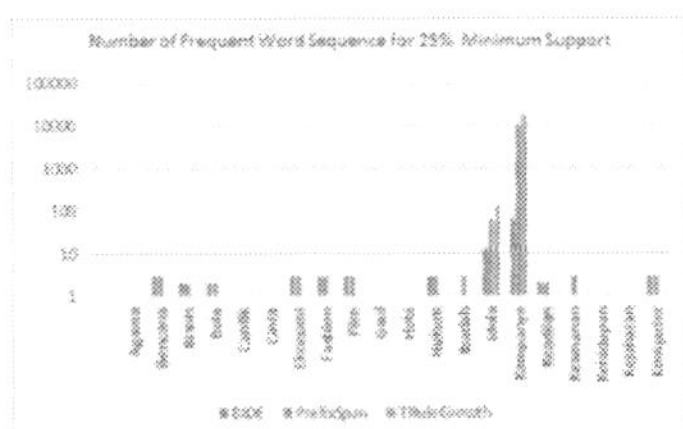

Figure 9. Result of Number of Word Sequence for 25% Minimum Support

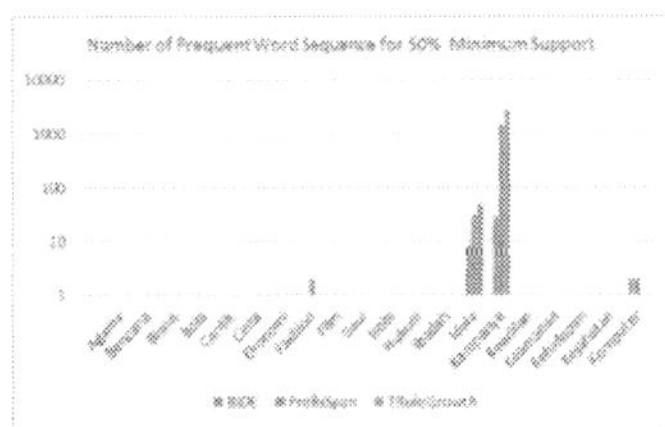

Figure 10. Result of Number of Word Sequence for 50% Minimum Support

7. Conclusion

- There is many sequential pattern algorithm that used for sequential pattern mining. This research do comparison between PrefixSPan, BIDE, and TRuleGrowth algorithm for mining Indonesian text with frequent word sequence feature. The experiment with 14.006 tweets from Twitter shows that PrefixSpan and BIDE are better in keep the meaning of text than TRuleGrowth. Moreover, BIDE produces more efficient sequence of word with frequent closed sequence without missing the meaning of text. It was different than TRuleGrowth that do not produces features mostly.
- The average of time process of PrefixSpan shows that PrefixSpan is fastest. It is because BIDE must find frequent closed sequence for efficient feature and TRuleGrowth must have more time for satisfy minimum confident and window-sliding constraint. However, for the memory usage, PrefixSpan and TRuleGrowth are more efficient than BIDE because PrefixSpan and TRuleGrowth do not use memory when do not find the feature. It is different than BIDE that keep using memory though the feature is not found.
- For discussion and future work, we have to try do the experiment with larger or scalable or big text data for PrefixSpan, BIDE, and TRuleGrowth. Because BIDE and TRuleGrowth is match for scalable data. It does not rule out the possibility for comparing another sequential pattern or frequent pattern algorithm for Indonesian text.

8. References

[1] Adityarani, Kania. 2009. *Aplicating CFWS* Algorithm for Document Grouping with Text representation Set of Frequent Word Sequence and Text Representation Qord Frequent itemset.* Informatic, Bandung Institute of Technology.

[2] Agrawal, Rakesh. Srikant, Ramakrishnan. 1994. *Fast Algorithm for Mining Association Rules.* Proceeding of the 20th VLDB Converence. Santiago, Chile.

[3] Agrawal, Rakesh, et al. 1999. *Mapping Words, Phrases using Sequential Pattern to Find User Specific Trends in a Text Database – Patent Document.* International Business Machines Corporation, NY.

[4] Agrawal, Rakesh. Srikant, Ramakrishnan. 1995. *Mining Sequential Patterns.* IBM Research Center.

[5] Ahonen, Helena. 1999. *Knowledge Discovery in Document by Extracting Frequent Word Sequence.* Library Trends. Vol. 48, No. 1, Summer 1999, pp. 160-181.

[6] Doucet, Antoine. 2005. *Advance Document Description, a Sequential Approach.* Accademic Dissertation, Departement of Computer Science, Series of Publication A. Report A-2005-2, University of Helsinki.

[7] Fournier-Voger, Philippe. 2008-2016. *Sequential Pattern Mining Framework* (SPMF) *Data Mining Software Tools.* Available in: http://www.philippe-fournier-viger.com/spmf

[8] Fournier-Viger, Philippe. Wu, Cheng-Wei. Tseng, Vincent S. Cao, Longbing. Nkmbou, Roger. 19th of February 2015. *Mining Partially-Ordered Sequential Rules Common to Multiple Sequences.* IEEE Transactions on Knowledge and Data Engineering.

[9] Fournier-Viger, Philippe. Nkambou, Roger. Tseng, Vincent Shin-Mu. 2011. *RuleGrowth: Mining Sequential Rules Common to Several Sequence by Pattern Growth..* SAC'11 Proceedings of the 2011 ACM Symposium on Applied Computing, pages 956-961.

[10] Han, Jiawei. Kamber, Micheline. 2006. *Data Mining: Concepts and Techniques Second Edition.* Simon Fraser University. Morgan Kaufmann Publisher.

[11] Hoonlor, Apirak. July 2011. *Sequential Patterns and Temporal Patterns for Text Mining.* A Thesis Submitted to Graduate Faculty of Rensselaer Polytechnic Institute in Partial Fulfillment of the Requirement for Degree of Doctof og Philisophy. Major Subject: Computer Science, NewYork.

[12] Maylawati, Dian Sa'adillah. 2011. "Development of Pre-processing Library with Set of Frequent Word Item Set Representation for Text Mining Application. Case Study: Indonesian's Slang Language". Master of Informatic Programs. STEI, Bandung Institute of Technology

[13] Nalistia, Rindhu Astri. 2008. *Feature-Based Clustering for documents grouping with the Structured Representation of Text.* Informatic, Bandung Institute of Technology.

[14] Pei, Jian. Han, Jiawei, et.al. October, 2004. *Mining Sequential Patterns by Pattern Growth: The PrefixSpan Approach.* IEEE Transaction on Knowledge and Data Engineering, Vol. 16, No. 10.

[15] Sabrina, Puspita Nurul. Saptawati, G.A. Putri. November, 2015. *Multiple MapReduce and Derivative Projected Database: New Approach for Supporting PrefixSpan Scalability.* International Conference on Data and Software Engineering (ICoDSE)

[16] Wang, Jianyong. Han, Jiawei. 2004. *BIDE: Efficient Mining of Frequent Closed Sequence.* ICDE '04 of 20th Interntional Conference on Data Engineering.

IOP Conf. Series: Journal of Physics: Conf. Series **801** (2017) 012068 doi:10.1088/1742-6596/801/1/012068

Keyword and Event Extraction for Thematic Map Retrieval from Indonesian Online News Site

A Dewandaru, I Supriana and S Akbar
School of Electrical Engineering and Informatics
Institut Teknologi Bandung, Indonesia

Abstract. Online news sites provide great deal of information that may be extracted and mined to create Geographical Information Retrieval (GIR) system that provides thematic map based on user query. The task requires extracting generic events and its attributes from the news corpus. This event extraction requires the keywords and topics that can help increase the accuracy of the extraction and retrieval process. We prepare a large online news corpus and compared the Latent Dirichlet Allocation (LDA) and CBOW and skip-gram model to help providing base thematic keywords which can assist the extraction process on the intended GIR system. LDA is better in terms of semantic relatedness and the CBOW and skip-gram is useful for providing semantic similarity.

1. Introduction

The staggering growth of internet data is followed by the dominance of textual retrieval system to aid search. However, the extraction and retrieval of thematic maps from web corpus is an interesting task that is still not well explored [1]. The purpose of the task is providing infographic in form of thematic geographic map that is relevant to the user's query. In other words, it is an automated Geographical Information Retrieval (GIR) system that processes data from web and serves geographical content to users. Essentially it is a novel type of search engine, in which –instead of returning set of ranked documents– it returns a thematic map relevant to the query. The basic definition of many GIR follows the generic architecture and components illustrated in Figure 1.

On the diagram, the Extraction process (inside the IE Process box) deals with identifying and filtering information from the source web documents (such as named entity), notably handling the various formats and ambiguities, and also tackle the complexity of mostly less structured free text in form of natural languages. Even if we restrict the source to news sites (as the main focus of this paper) there are still a lot of open problems in the extraction process.

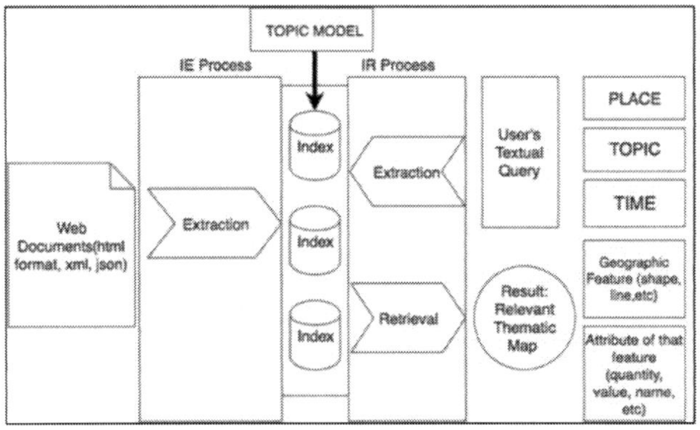

Figure 1. GIR System Architecture for Thematic Map Extraction & Retrieval from Web

Taking it to the greater level of difficulty, the task pretty similar to the machine reading or text understanding endeavour, exemplified by works [2,3]. However, we would like to underline the key scopes here that: 1) the text needs to have geographic reference(s) and 2) the result is a thematic map of a geographic scope that's relevant, which need not to provide direct answer from the user's query as in QA based system. The paper will explain the target output and how to deal with less-structured input in news documents. As later will be shown, we found that topic modeling will help in finding important keywords for extraction.

As stated in the earlier part, we would like to aim in creating GIR system that is able to output thematic map in response to user's query as in the example below (**Figure 2**). The map was published by Indonesian formal body (BNPB) and is projecting flood data in Jakarta in particular time range. The map was color-coded, which was further explained by a brief legend: blue area indicates a flooded area, while the light yellow showed clear area.

Such thematic maps were (typically) prepared manually using help of GIS by a formal or institutional body which has predefined editorial process and standards. We are aiming to create similar output automatically, tapping the techniques available in Machine Learning and Text Mining. However, instead of having a structured input such as database systems, we are interested in utilizing the world wide web as a huge and recent source of data and information. The inputs are taken from web documents instead of sampling from the field reports. Hence, the challenge is that it must be able to extract key information from the various types of web documents that is served in various formats.

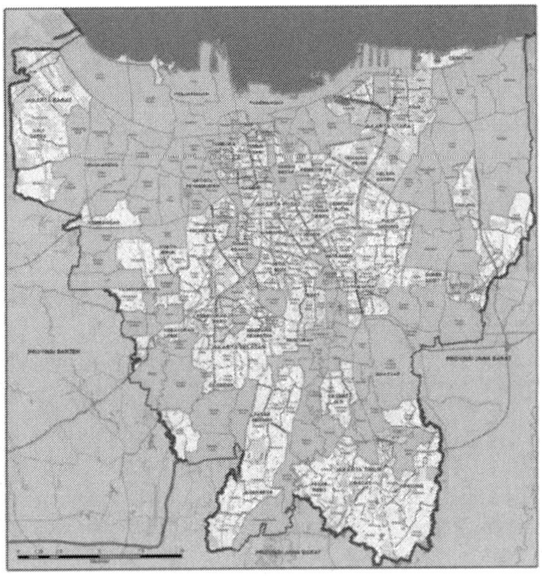

Figure 2. Thematic Map of Jakarta Flood 18th Jan 16

2. System inputs

There are myriad types of information and structure in the web documents that is suitable for thematic map extraction, i.e. such as formal body that publishes statistics regularly, but it may not match the timely update and coverage offered by social media and online news sites. However, we think that social media typically are not accurate enough for this purpose. Thus, in this paper we start with focusing on the (rather in the middle/balanced) news sites, especially in terms of the update frequency and trustworthiness due to a specific editorial process [4].

Table 1. Web data assumptions

Aspects	Social Media	Formal Body	News Site	Wiki
Update frequency	Fast	Slow	Medium	Slow
Information that can be mined	Events, preferences	Aggregate/ Instance statistics	Events	Events, Aggregate Statistics
Coverage	Broad	Narrow	Medium	Broad
Trustworthiness	Low	High	Medium	High
Format	Natural Language	Tabular HTML, DIVs	Natural Language	Natural Language, Tabular HTML
Example Sites	Facebook, Twitter	CIA World's Factbook	DailyMail, CNN	Wikipedia. org

Let us take a look on a sample news document (Figure 3) offered by the online news sites to get more details into its nature. Typically, news articles are about events and its attribute. By simply reading at the text, and by making use of the important keywords, it is easy for average adult human to infer the 5W1H (what, where, when, who, why, how) attributes of an event presented in the article. However, it turns out to be a challenging task and still open problem for machine learning as we will shortly.

Dark sky, heavy rain and turbulent wind come across Jakarta. Sheds of water gushed out, as had happened in Kuningan area. Even when the water depth level reaches 30 cm, drivers were still trying to get across.

The report from detik.com, the rain started since 1:40 PM local time, Friday (29 February 2008) making a 30 cm watersheds in front of KPK building, situated at Jl HR Rasuna Said, Jakarta.

This had caused the traffic jam to direction of the Kuningan area. Drivers were slowing down when trying to get through. Meanwhile in the reverse side to Halimun, the flood caused vehicle queuing as far as 500 meters.

Figure 3. Translated Sample News Article (from http://news.detik.com/berita/902053/hujan deras-dan-angin-kencang-kuningan-banjir-30-cm)

3. Identifying keywords to extract events

An obvious problem that we face is that it is common to have multiple subtopics within a single news article. In the example given (Figure 3), the first portion (roughly first and second paragraph) was talking about the flood event. The second portion was about the traffic jam as the implication of the flood. A proper event extraction by a human annotator for that single article should give result similar to the following table:

Table 2. Extracted Values form Single Document Sample

Topics #	Variable	Value
1	Event	Flood
	Time	29/2/2008
	Place	Jakarta, Kuningan, Jl. HR Rasuna Said
	Depth	30 cm
2	Event	Traffic Jam
	Time	29/2/2008
	Place	Jakarta, Kuningan, Halimun
	Queue Length	500 m

Note that the example presented is only a simple case of a news event and with a specific domain (i.e. disaster news). With a specific domain, an information extraction wrapper or supervised wrapper induction system is easier to craft manually or trained, notably by identifying keywords term which plays some semantic role in the typical sentence. For example, Tanev et. al [5] are using set of keywords (kill, shoot) to help construct handcrafted rules which will handle with specific domain of disaster and violence.

Similarly, [6] had prepared manually annotated corpus in Bahasa Indonesia to train some supervised machine learning classifiers (Adaboost.M1 and C4.5) to perform event extraction in output form of 5W1H. The learning is based on a limited number of only 90 annotated news articles. We think that it is not possible to cover a huge repertoire of articles in the real data without training pretty large portion of a manually annotated samples (the training would usually assign the statistical significance of some keywords).

In the work of [7], the identification of key terms are done automatically using TF-IDF metric that is monitored daily with some simple rules: 1) it has to be present in more than two documents and 2) it is ranked via additional "novelty" metric that takes care of the currency of the term (we compare this in modern parlance as "trending topic" concept). However, the method only extracted hot terms as the main topic from the news article; whereas in our case we need to get down to the attributes keyword (such as queue, water depth, etc.) and its relevant value (500 m, 30 cm).

In a real world scenario –e.g. in the context of a typical news sites– we are facing multi-faceted, intertwined and yet multi-topics (hundreds) each with its distinct sets of keywords that arise from multiple domains of news articles to extract from, and the number of tokens exposed is pretty huge (in hundred-thousands tokens, as will be confirmed later on based on our crawled data). Hence it is not a simple task, and requires great deal of manual work if done solely using supervised algorithms.

Recall that the GIR described in the introduction needs to have a great emphasis in the ability of returning relevant maps pertaining to user's query. Other GIR had specified different query formats. As an example, the SPIRIT project [8] for example had offered the query format as triplet of theme (e.g. schools), spatial relationship (such as near, within, etc.) and location string. Note the theme that is being queried as important keyword in the search query there.

In our designated system we wanted to be able to extract such thematic information, and even provide ability to extract intangible object (not just some building or premises, but such as events as in floods, earthquake, or spatially related info such as GDP statistics) and enable users to query these information needs (e.g. floods in Jakarta 2014, or World's GDP 2015). The triplet of query format would form as <theme><location><time> in an unspecified sequence. The examples of queries are given in **Table 3**.

Table 3. Example of queries and expected output

Query example	Output
"Flood Jakarta 2014"	Summarized map of flood in Jakarta within 2014
"World GDP 2015"	Map of world countries' growth domestic products statistics
"Earthquake Asia 2010"	Map of earthquake centers that happens in 2010 in Asia Pacific.

To tackle this problem, we suggest the usage of unsupervised algorithms to assists us in learning or extracting important topics keywords that would eventually useful in extracting proper values based on the theme or topics. Probabilistic topic modeling framework falls exactly as a good option for this.

4. Topic modeling for keyword extraction

Topic modeling is a task of capturing hidden relations between terms and finding clusters terms in the corpus. It can be traced from a very simple TFIDF metric, which can be used to gauge term significance by accounting on the term frequency and the term specificity. Deerwester et. al [9] developed Latent Semantic Indexing (LSI), which tries to capture hidden pattern not directly visible in the co-occurrence of words using Singular Value Decomposition (SVD) applied to term-document matrix. Later this work was developed further and extended by basing it under probabilistic framework called Probabilistic Latent Semantic Indexing (PLSI) by Hoffman [10]. This provided an aspect model and was a significant upgrade over the former.

The seminal Latent Dirichlet Allocation (LDA) by Blei et.al. [11] supposed to fix the overfitting problem and the need to linearly grow the parameter as the number of documents increase in PLSI. Unlike some traditional IR clustering approach (e.g. agglomerative clustering or k-means), Both LDA and PLSI does a soft-clustering in which they do not form an absolute partition. These architectures offer a mixture model whereby a words formed from a mixture of term components (we can imagine a cluster of words) that represents topic. It models the possibility that a document consisted of several topics at the same time.

In a mixture model, an element can be a member of more than one cluster with varying degree of membership [12]. In the case of LDA and PLSI, a topic is nothing more than virtual "set" of terms, of which each term is having degree of membership to any topic to all other topics. Thus a term belongs to more than one topic, forming many to many relationships.

As an example, we can model topics of two different events (flood and quake) as a mixture components of terms, each having an attached probability value indicating how likely it forms that topic. Note that "victim" term are both highly probable member of Flood and Quake topic. Any terms in the processed corpus are actually having a probability to any topic, but for practical purpose some cut-off values are being used to prevent low probability terms to appear.

We believed that if these clusters of important keywords discovered accurately, it can be a very useful information that can assists many tasks such as classification, extraction and retrieval. We will show later on the results found are pretty consistent with our expectation.

5. Latent Dirichlet Allocation

The LDA model is a widely topic modeling framework and is based on a sound probabilistic graphical model framework that itself belongs to a family of Bayesian Network. Its use on GIR literature is still minimal however, and we are interested in using it due to its ability to extract keywords and topics at once.

In this model, the probability assumptions is encoded in form of directed acyclic graph[13], where each node represents a random variable which can be observable or hidden. A hidden or latent node is not observable directly but can often be estimated from the visible ones. The directed edge of this graph represents dependencies between random variables (parent nodes has directed edge to its child nodes, the child is conditionally dependent on its parents).

To illustrate better, Figure 4 shows LDA model which is presented in a plate notation (or often called hierarchical model). Plate (drawn as rectangular boxes) notation is a visual extension which provides easy way to visualize duplication of nodes semantics. N and M in the graph are the quantifier variable, indicating how many nodes are duplicated inside the plate box. Thus, Z and W will be duplicated a number of N times.

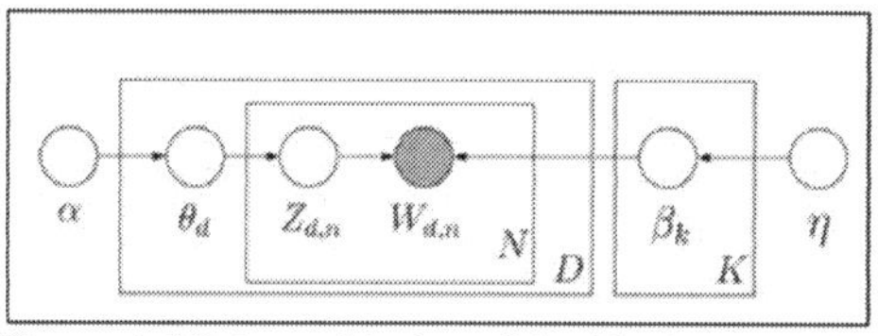

Figure 4. LDA graphical model with plate notation

From the graphic, we can also say that the nodes α, β, θ, Z, W are random variables that is related one another via the edge annotation. For example, W is conditioned (depends) on Z and so forth. W is the only random variables that is observable (shaded). In the LDA case it is the word inside documents formed from N words.

As with many Bayesian inference models, LDA is a generative models, in which it tries to construct a joint probability distribution[14] that models the entire corpus and able to "generate" new documents. From the model standpoint it is simply multiplication result of each instantiated node given its parent's value. It can be used to create a new document that is having similar distribution as the learned ones. Note that LDA uses bag-of-words unigram assumptions, thus it does not care about the sequence of words as it uses bag of words representations.

$$P_B(X_1, X_2, \ldots, X_n) = \prod_{i=1}^{n} P_B(X_i|\pi_i)$$

Figure 5. Joint Probability Distribution of A General Bayesian Network

Using the formula from the generative model, and by observing the training documents, the parameters to the network will be approximated or estimated. One problem is that it is often the case that the formula for the equation requires calculating integral that is intractable to compute and needs to be simplified further. Parameter can be used by some methods. Researchers use Gibbs Sampling (based on stochastic simulation) or some Expectation Maximization (EM) algorithm (including Variational). Both EM and Variational are deterministic and are not based on stochastic simulations. These learned parameters will determine topic structure: the terms and its memberships within the topics.

A final note on these method is that it requires a great deal of data to train with. They are unsupervised however, so minimum human participation is anticipated. So we are going to proceed by describing the acquisition of the training data from two of the biggest Indonesian news sites, which should produce a good enough corpus for these model.

6. Skip Gram and Continuous Bag of Words Models (Word2Vec)

The recent phenomena of massive generation of text via the internet motivates models and algorithms to be scalable, faster and better in handling data, at even more than hundreds of million of tokens. Also, the performance improvements of machine learning inference techniques enable more complex model to be trained within reasonable time.

The language model used within LDA context is the unigram known as bag of words. This model discards the information sequence that may arise within document. It is simpler and faster to train and handle, but the prediction performance or the quality of the model may suffer if compared to more complex model.

Mikolov et.al [15] introduced a novel approach which seen by many as promising direction in language modelling. The skip-gram model proposed is an extension of n-gram, which, instead of modelling a contiguous series of tokens in the model, it allows one or more "skips" between the token involved. It is also introduced a novel way to measure the "cleverness" of the algorithm in noticing patterns such as answering question like this: "Jakarta is to Indonesia is like Berlin to Germany".

7. Experiment

We are going to describe main steps in this experiment. The manual annotation of the corpus is not yet used but article tagging and application has been used as proof of concept. The main objective is to check whether we can use keywords learned from the unsupervised learning to serve in topic extraction and retrieval. The following methods are done:

1. Obtain good amount of news corpus in Bahasa Indonesia.
2. Unsupervised training using LDA and Word2Vec algorithm from the entire set of documents within the corpus.
3. Analysis the keyword against sample documents.
4. Analysis the similarity of sample keywords.

#554:
banjir, air, sungai, warga, hujan, cm, tergenang, rumah, surut, deras, jalan, ketinggian, wilayah, pompa, meter, terendam, kawasan, menggenangi, lokasi, meluap, airnya, tanggul, aliran, mengguyur, titik, mencapai, daerah, akibat, luapan

#517:
kendaraan, tol, kemacetan, arah, jalan, macet, lintas, arus, jalur, lancar, pengendara, melintas, padat ,jam, roda, kilometer, mobil, petugas, antrean, pintu, masuk, titik, kawasan, motor ,bergerak, lokasi, alternatif, ditutup, pagi

#209:
gempa, bumi, mengguncang, tsunami, berkekuatan, kerusakan, kilometer, akibat, warga, barat ,korban, petang, wilayah, rumah, peringatan, pantai, pusat, bencana, orang, jiwa, dampak, laporan, tewas, stroke, sebelah, berolahraga, dilaporkan, dilansir, tersumbat, terserang

Figure 6. Snippet of LDA topic model learned, topic #554 (flood) and #517 (traffic jam) and #209 (earthquake)

The first target of news corpus was not directly available, so it was necessary to create such. We crawled news articles from www.detik.com and www.kompas.com, both were the leading news portal in Indonesia at the moment. We assume they hosted a great number of articles and having a relatively good editorial quality. Both sites are having archive index page which sort articles based on date of publication.

The crawling was done within a two level scopes: the index and the detail article. Scraping had been done for the detail article page based on CSS selectors (manually crafted wrappers) which has been adapted for both sites.

The crawling process was directed to gather title, content, and the URLs of the articles from January 2013 to May 2016. It resulted 271.979 articles in Bahasa Indonesia (Indonesian) from www.detik.com and 330.249 articles from www.kompas.com. The result of this crawling was cleaned from HTML tags and irrelevant scripts. We did not apply stemming as we would like to learn as much word forms as possible from the corpus. A stop words filtering was applied based on simple dictionary in Bahasa Indonesia. We also applied standard TF-IDF transformation to downplay the less specific but often appearing terms. Total unique terms count extracted for www.detik.com reaches 168,865 tokens.

Afterwards, the LDA model was constructed and the unsupervised learning method to extract the topics were executed based on the cleaned corpus. The training is done using online LDA algorithm by [16] via gensim python library[17]. We were playing with several combinations of number of topics ($k = 100, 200, 500, 600$) and number of terms per topic ($t = 10, 20, 30$) before settling to $k = 600$ and $t = 30$ which we think best capturing the size of the data.

The Word2Vec family model was trained using both skip-gram and continuous bag of words (CBOW) model using line-by-line training sequence. The maximum distance between current and predicted word is set to 5 and the dimensionality of the feature vector is 400. The Word2Vec training process took some minutes and is much faster compared to the LDA.

8. Results and Discussion

8.1 LDA

We had trained a model of LDA with the obtained corpus in an unsupervised manner. We manually picked three notable clusters (topics) out of 500 described on Figure 6 to serve as example and applied for one article. With that configuration, the LDA seems to be able to cluster semantically related terms pretty well and suited to our task of keyword extraction. When this result was applied on real articles example cited above (Figure 7 was the real article whose translation is presented on Figure 3), we found that many important keywords to be extracted from page already described by the clusters' term keyword. This serve as proof-of-concept of the utility of using topic modeling approach towards thematic map extraction GIR system.

The experiment showed some weaknesses however, in that we need to arbitrarily picked number of topics (repeated trials). Bayesian nonparametric techniques such as HDP [18] are able to infer the number of clusters automatically. Also, LDA is agnostic about geospatial data and temporal data. However some specialization on LDA in this direction has been made, e.g. [19][20] with Spatial LDA and 3S-LDA to incorporate geospatial and temporal data. LDA also does not regard sequential pattern nor works in the sentence level. It only views document as bag of words, not accounting sequence as in n-gram models. It helps in simplifying the model assumption but we think that it is very crucial in order to do proper extraction to reveal hidden patterns in sentence or sequence level as in Word2Vec.

Langit kelam, *hujan* deras dan angin kencang melanda Jakarta. Genangan *air* langsung bermunculan, seperti halnya di <u>*kawasan*</u> Kuningan. Meski *ketinggian air* mencapai **30** *cm*, pengendara tetap mencoba melintasinya.

Pantauan detikcom, *hujan* yang turun sejak pukul 13.40 WIB, Jumat (29/2/2008), membuat genangan air **30** *cm* di **depan kantor KPK, Jl HR Rasuna Said, Jakarta.**

Akibatnya <u>*arus*</u> lalu <u>*lintas*</u> mengarah ke Kuningan tersendat. <u>*Pengendara*</u> memperlambat laju <u>*kendaraan*</u> saat hendak melintasi genangan *air*. Sedangkan <u>*arah*</u> sebaliknya ke Halimun, *banjir* menyebabkan <u>*antrean*</u> <u>*kendaraan*</u> mencapai **500** *meter*.

Sementara <u>*pengendara*</u> <u>*roda*</u> dua memilih berteduh di halte-halte bus. Pengojek payung berwarna-warni pun laris manis.

Figure 7. Projection of the Extracted Keywords on A News Article from Topic #554 (flood) and Topic #517 (traffic jam). Bold terms indicate the values that should be extracted with regard to the event. Italicized keywords are related to flood and underlined words are related with traffic jam topic.

8.2 Word2Vec

The skip-gram and CBOW model offered interesting word similarity result. The example of this can be seen on **Table 4**. We noticed that LDA results are focused towards semantic relatedness while Word2Vec are more on semantic similarity. The Word2Vec result cannot be immediately used as keywords for extraction, however we are interested in its ability to produce synsets, hyponyms, and hypernims.

Table 4. Similarity of "Flood" (Ind. "Banjir") (same word forms are omitted)

Word	Score
Cileuncang	0.668
Bandang	0.651
Kekeringan	0.611
Terendam	0.609
Rob	0.596
Longsor	0.584
Genangan	0.576
Luapan	0.559
Hujan	0.520
Plengsengan	0.503
Bencana	0.494

9. Conclusions and Future Works

Topic modeling framework with its unsupervised algorithms can be used to assist Event extraction process by providing priors over some important keywords. We plan to employ the set of keywords learned to assists the in-page information extraction process by commencing a web browser extension module development. The planned algorithm for extraction falls under the family of discriminative type (e.g. sequential Conditional Random Field), which will be able to make use of neighbourhood term observation data. For the topic and keyword extraction we plan to adapt the Bayesian approach of LDA, combined with the skip gram approach to fit better with online news domain in terms of tackling more on geospatial and temporal data.

REFERENCES

[1] Dewandaru A, Supriana I and Akbar S 2015 Evaluation on Geospatial Information Extraction and Retrieval : Mining Thematic Maps from Web Source *Information and Communication Technology (ICoICT), 2015 3rd International Conference on* (IEEE) pp 283–8
[2] Etzioni O, Banko M, Soderland S and Weld D S 2008 Open information extraction from the web *Commun. ACM* 51 68
[3] Etzioni O, Popescu A, Weld D S, Downey D and Yates A 2004 Web-Scale Information Extraction in KnowItAll (Preliminary Results) *Proceedings of the 13th international conference on World Wide Web Pages* pp 100–10
[4] Haristya S, Suwana F and Kurniana I 2012 THE CREDIBILITY OF NEWS PORTAL IN INDONESIA : *J. Commun. Stud.* 5 1–17
[5] Tanev H, Piskorski J and Atkinson M 2008 Real-time news event extraction for global crisis monitoring *Lecture Notes in Computer Science (including subseries Lecture Notes in Artificial Intelligence and Lecture Notes in Bioinformatics)* vol 5039 LNCS pp 207–18
[6] Khodra M L 2015 Event Extraction on Indonesian News Article Using Multiclass Categorization 1–5
[7] Liu M, Liu Y, Xiang L, Chen X and Yang Q 2008 Extracting key entities and significant events from online daily news *Lect. Notes Comput. Sci. (including Subser. Lect. Notes Artif. Intell. Lect. Notes Bioinformatics)* 5326 LNCS 201–9
[8] Purves R S, Clough P and Jones C B 2007 The Design and Implementation of SPIRIT : a Spatially- Aware Search Engine for Information Retrieval on the Internet *Int. J. Geogr. Inf. Sci.* 717–45
[9] Deerwester S, Dumais S T, Furnas G W, Landauer T K and Harshman R 1990 Indexing by latent semantic analysis *J. Am. Soc. Inf. Sci.* 41 391–407
[10] Hofmann T 1999 Probabilistic latent semantic indexing *Acm Sigir* 50–7
[11] Blei D M, Ng A Y and Jordan M I 2003 Latent Dirichlet Allocation *J. Mach. Learn. Res.* 3 993–1022
[12] Song M, Wu Y B and Klinger K 2009 *Handbook of Research on Text and Web Mining Technologies* vol II (IGI Global)
[13] Koller D and Friedman N 2009 *Probabilistic Graphical Models: Principles and Techniques*
[14] Faltin F and Kenett R 2007 Bayesian Networks *Encycl. Stat. Qual. Reliab.* 1 4
[15] Mikolov T, Corrado G, Chen K and Dean J 2013 Efficient Estimation of Word Representations in Vector Space *Proc. Int. Conf. Learn. Represent. (ICLR 2013)* 1–12
[16] Hoffman M, Blei D and Bach F 2010 Online learning for latent dirichlet allocation *Nips* 1–9
[17] Řehůřek R 2011 Scalability of Semantic Analysis in Natural Language Processing 147
[18] Teh Y W, Jordan M, Beal M and Blei D 2006 Hierarchical Dirichlet Processes *J. Am. Stat. Assoc.* 101 1556–81
[19] Wang X and Grimson E 2008 Spatial Latent Dirichlet Allocation *Adv. Neural Inf. Process. Syst.* 1–8
[20] Pan C-C and Mitra P 2011 Event detection with spatial latent Dirichlet allocation *Proceeding 11th Annu. Int. ACM/IEEE Jt. Conf. Digit. Libr. - JCDL '11* 349

Detecting spam comments on Indonesia's Instagram posts

Ali Akbar Septiandri[1] and Okiriza Wibisono[2]

School of Informatics, The University of Edinburgh, 11 Crichton St, Edinburgh EH8 9LE, UK

E-mail: [1]`s1500216@sms.ed.ac.uk`, [2]`s1367442@sms.ed.ac.uk`

Abstract. In this paper we experimented with several feature sets for detecting spam comments in social media contents authored by Indonesian public figures. We define spam comments as comments which have promotional purposes (e.g. referring other users to products and services) and thus not related to the content to which the comments are posted. Three sets of features are evaluated for detecting spams: (1) hand-engineered features such as comment length, number of capital letters, and number of emojis, (2) keyword features such as whether the comment contains advertising words or product-related words, and (3) text features, namely, bag-of-words, TF-IDF, and fastText embeddings, each combined with latent semantic analysis. With 24,000 manually-annotated comments scraped from Instagram posts authored by more than 100 Indonesian public figures, we compared the performance of these feature sets and their combinations using 3 popular classification algorithms: Naïve Bayes, SVM, and XGBoost. We find that using all three feature sets (with fastText embedding for the text features) gave the best F_1-score of 0.9601 on a holdout dataset. More interestingly, fastText embedding combined with hand-engineered features (i.e. without keyword features) yield similar F_1-score of 0.9523, and McNemar's test failed to reject the hypothesis that the two results are not significantly different. This result is important as keyword features are largely dependent on the dataset and may not be as generalisable as the other feature sets when applied to new data. For future work, we hope to collect bigger and more diverse dataset of Indonesian spam comments, improve our model's performance and generalisability, and publish a programming package for others to reliably detect spam comments.

1. Introduction

As reported in [1], Instagram has 22 million monthly active users in Indonesia from 500 million users worldwide. With 95 million photos and videos posted every day, Instagram has obviously become a prominent photo and video sharing platform in the world. However, just like e-mails, the high number of activities mean that there is a chance that people could get more attention even when they post totally irrelevant comments which we also know as spam.

Indonesia, in particular, has this problem where some users put advertising on public figures' posts as comments. This problem has gone too far to the fact that the first 50 comments on a public figure's post could be all spam[1]. Like e-mail spam, we believe that these comments can be very annoying and should be automatically filtered.

To the best of our knowledge, we have not seen any investigation on Instagram's spam on Indonesian accounts before. The case might be platform specific, but the insight from this study

[1] e.g. username: @cita_citata, post_id: BIjHGG3BUEK

IOP Conf. Series: Journal of Physics: Conf. Series **801** (2017) 012069 doi:10.1088/1742-6596/801/1/012069

could hopefully be used to similar cases, e.g. spam comments on blogs or e-mail spam. This is because the machine learning algorithms we used in our study which have been proven to work well in general. Our focus is less on finding the algorithm that works best on our dataset and more on identifying the salient features to detect spam comments.

2. Related Work

While we might be able to detect spam to some extent by using origin-based filters (e.g. using IP or email address), in the case of spam email, content-based filtering is more common to do [3]. The same thing also applies to spam comments, as people can easily create new accounts for this spamming purpose. However, content-based filtering also comes with some problems. For instance, spammers could obfuscate their message (e.g. by writing "f r 3 3" instead of "free") to trick the filter [2].

Machine learning approach, which can automatically filter the spam by building adaptive model, has become more popular nowadays. Before sending the features to be learned by the algorithms, according to [2], the structure of a spam filter can be grouped into:

(i) tokenization, which extracts the words in the message body;
(ii) lemmatization, reducing words to their root forms;
(iii) stop-word removal, eliminating some words that often occur in many messages;
(iv) representation, which converts the set of words present in the message to a specific format required by the machine learning algorithm used.

Nevertheless, as it was also pointed out in the study, not all of the steps are mandatory.

As the name might suggest, words are the features in content-based spam filtering. The problem now is how to represent them properly for the algorithms. Some of the representation that has been used in content-based spam filtering are bag-of-words (BoW) [5], term frequency-inverse document frequency (TF-IDF) [6], and binary representation of word occurences [4]. In [4], it was also considered to use of upper case words as one of the features.

Since using BoW and TF-IDF can result in a sparse matrix, we can reduce the dimension by using Latent Semantic Analysis (LSA). As explained in [9], this method applies Singular Value Decomposition (SVD) to the matrix so that we can learn "expected contextual usage of words in passages of discourse." LSA also helps us to make the learning process faster.

In more recent work [7], we can see the improved version of skip-gram model [8], "where each word is represented as a bag-of-character n-grams." This method enriches the word vectors with subword information on rare words. It also has the advantage of not needing any preprocessing of the data.

3. Methodology

3.1. Features

We used several techniques for representing the comments as follows:

(i) Binary Bag-of-Words with LSA;
(ii) TF-IDF with LSA;
(iii) Word2Vec using skip-gram model.

The first two representations were made using the library provided in scikit-learn[2] [10]. We set the minimum document frequency (`min_df` parameter) to be 5. We then reduced the dimension of BoW and TF-IDF representations to 100 using LSA. The chosen dimension is the same as

[2] `http://scikit-learn.org/stable/`

the default output dimension in the Word2Vec library we used in this study, i.e. fastText[3] [7]. We did not apply any lemmatization or stop-word removal in our study.

The word vectors produced by Word2Vec are 100-dimension vectors for each word in the document. We then took the average of all word vectors that constitutes each document, yielding 100-dimension document vectors. These vectors can then be used as the features in our experiment.

These features alongside some numerical features that can be easily extracted from the text, e.g. number of tokens, number of upper case words, number of numerical characters, percentage of emoji, and the length of the text (see Table 1), were then passed onto a number of machine learning algorithms. We call these as *basic* features in our experiemnt. We also utilised some hand-engineered keyword patterns in our experiment (see Table 2), which later on in this paper are known as *keywords* features.

Table 1. List of basic features

Name	Description
`n_token`	Number of tokens
`n_capital`	Number of upper case words
`n_emoji`	Number of emojis
`n_unique_emoji`	Number of unique emojis
`n_number`	Number of numerical characters
`n_mention`	Number of mentions
`%_capital`	Percentage of upper case words
`%_number`	Percentage of numerical characters
`%_emoji`	Percentage of emojis
`%_unique_emoji`	Percentage of unique emojis
`log_char`	$log(number of characters)$
`has_phone_number`	Contains phone number
`has_bbm_pin`	Contains BlackBerry® Messenger PIN

3.2. Algorithms

In this study, we used Naïve Bayes (NB), Support Vector Machine (SVM) with RBF kernel, and XGBoost as the algorithms to classify the data. We based our selection of algorithms on a generic review of machine learning algorithms performance in [11] where SVM with RBF kernel and Random Forest turned out to be performing quite well in many cases. XGBoost [12] as a variation of decision trees method that shares some similarities with Random Forest, such as in the column sampling method and ensemble trees concept, became our choice of implementation because it has been proven to be successful in many machine learning competitions. On the other hand, we also incorporated the results from Naïve Bayes algorithm as the baseline for our study because of its simplicity and efficacy in practice [13].

We held out 20% of the dataset to be used as the test set later on in our study. We evaluated our models using cross-validation and concurrently tuned the hyperparameters. Using consistent random number generator, the hyperparameters we tested in our study can be seen in Table 3. The best models were then tested on the test set to know how well our model performs in general.

[3] `https://github.com/facebookresearch/fastText`

IOP Conf. Series: Journal of Physics: Conf. Series **801** (2017) 012069 doi:10.1088/1742-6596/801/1/012069

Table 2. Hand-engineered keyword patterns

<table>
<tr><th>Regular Expression</th></tr>
<tr><td>bbm|pin bb|line|whatsapp</td></tr>
<tr><td>(ch?ec?k|intip|liat|kepoin)(out)?((our|my))? (ig|insta|koleksi|profil)</td></tr>
<tr><td>cekidot</td></tr>
<tr><td>dada|herba|langsing|payudara|pemutih|penggemuk|peninggi|tahan lama|tinggi badan</td></tr>
<tr><td>efek samping</td></tr>
<tr><td>follow</td></tr>
<tr><td>free (delivery|ongkos|ongkir|pengiriman)|gratis|murah|promo|terjangkau</td></tr>
<tr><td>garansi|kualitas</td></tr>
<tr><td>impor</td></tr>
<tr><td>invit</td></tr>
<tr><td>jerawat</td></tr>
<tr><td>jual|sell</td></tr>
<tr><td>langganan</td></tr>
<tr><td>luar biasa</td></tr>
<tr><td>mampir</td></tr>
<tr><td>nyaranin</td></tr>
<tr><td>order</td></tr>
<tr><td>password</td></tr>
<tr><td>penghasilan</td></tr>
<tr><td>produk</td></tr>
<tr><td>s(o|u|ou)venir</td></tr>
<tr><td>stock|stok</td></tr>
<tr><td>yu+k</td></tr>
</table>

Table 3. List of hyperparameters

Classifier	Hyperparameters	Values tested
Naïve Bayes	`fit_prior`	[True, False]
SVM	`C`	$[2^{-5}, 2^{-3}, 2^{0}, 2^{3}, 2^{5}, 2^{7}, 2^{9}]$
	`gamma`	$[2^{-9}, 2^{-7}, 2^{-5}, 2^{-3}, 2^{0}, 2^{3}, 2^{5}, 2^{7}, 2^{9}]$
XGBoost	`n_estimators`	[10, 20, 50, 100]
	`subsample`	[0.5, 0.6, 0.7, 0.8, 0.9, 1.]
	`colsample_bytree`	[0.5, 0.6, 0.7, 0.8, 0.9, 1.]
	`reg_alpha`	[0., 0.001, 0.005, 0.01, 0.05, 0.1, 0.5, 1., 5., 10.]
	`reg_lambda`	[0., 0.001, 0.005, 0.01, 0.05, 0.1, 0.5, 1., 5., 10.]
	`scale_pos_weight`	[1., 2.28, 3.55, 4.83, 6.11, 7.38, 8.66, 9.94, 11.22, 12.49]

3.3. Evaluation Metrics
Since the distribution of the classes in the dataset is imbalanced, we used precision, recall, and F_1-score to evaluate our models. These metrics are formulated as:

$$\textbf{precision} = \frac{TP}{TP + FP} \tag{1}$$

$$\textbf{recall} = \frac{TP}{TP + FN} \tag{2}$$

$$F_1 = 2 \cdot \frac{\textbf{precision} \cdot \textbf{recall}}{\textbf{precision} + \textbf{recall}} \tag{3}$$

where TP, TN, FP, FN denote true positives, true negatives, false positives, and false negatives respectively.

In this study, we focused on getting the highest F_1-score. Therefore, most of the comparison would be based on F_1-score. We only provide the precision and recall for the best model achieved.

4. Experiments

4.1. Dataset

We collected 24,602 comments from 500 posts authored by 104 Indonesian public figures. We annotated those comments ourselves by looking at the original post to see the relevance of the comments. We focused on comments that are promoting websites or products to be categorised as spam. Eventually, we got 22,743 ham and 1,859 spam in the dataset.

4.2. Results

Our preliminary result from using only basic features as mentioned in Table 1 was promising. We got 0.7775 of F_1-score using XGBoost. We can see in Figure 1 that the logarithmic value of number of characters by itself can be a discerning feature.

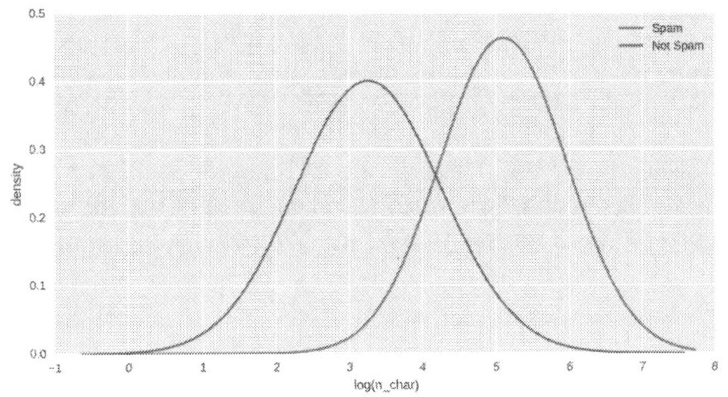

Figure 1. Difference in length on spam vs ham

We achieved an even better result using the features from hand-engineered keyword patterns (see Table 4). However, as the keywords can be altered in the future by the spammers to trick the classifiers, we cannot rely solely on this result. Thus, we continued our experiments with the representations mentioned in subsection 3.1.

By using only word vectors extracted from the text, we can see in Table 5 that the algorithms, with the exception of Naïve Bayes, managed to beat the corresponding algorithms with our hand-engineered keywords. Using word vectors from fastText resulted in the best models from three representations.

Our next effort was combining the features in Table 1 with the word vectors. We wanted to see how adding the basic features and our hand-engineered keyword patterns could improve the models. The result for this experiment can be seen in Table 6. Note that "+b" means we were

Table 4. F_1-scores of preliminary results

Classifier	Basic	Keywords	Basic + Keywords
Naïve Bayes	0.6667	0.8391	0.8772
SVM	0.7688	0.8726	0.9093
XGBoost	0.7775	0.8655	0.9089

Table 5. F_1-scores on word vectors

Classifier	BoW	TF-IDF	fastText
Naïve Bayes	0.5361	0.2385	0.8249
SVM	0.9074	0.9074	0.9398
XGBoost	0.9121	0.9089	0.9248

adding basic features, while "+k" means we were adding keyword patterns features when we trained the models.

Table 6. F_1-scores on combined features

Classifier	BoW+b	BoW+b+k	TF-IDF+b	TF-IDF+b+k	fastText+b	fastText+b+k
Naïve Bayes	0.5802	0.7366	0.7331	0.8818	0.8621	0.8801
SVM	0.9399	0.9309	0.9373	0.9349	0.9523	0.9601
XGBoost	0.9268	0.9381	0.9377	0.9436	0.9512	0.9512

We can see that adding our proposed basic features improved all the F_1-scores. However, adding keyword patterns to word vectors with basic features did not improve the performance significantly. We confirmed this by using McNemar's test defined as follows:

$$\chi^2 = \frac{(b-c)^2}{b+c} \tag{4}$$

where b and c denote the number of difference in class predictions (test 1 negative and test 2 positive, and vice-versa).

The p-value we got from McNemar's test on SVM classifiers is 0.22, which means the difference is not statistically significant ($p > 0.05$). This result indicated that using word vector representation with semantic analysis is sufficient to detect important features from the text. Having said that, the best F_1-score in our experiment was from adding basic and keyword patterns features to SVM with fastText: 0.9601.

5. Conclusions and Future Work

Our experiments show that employing fastText can produce robuster models. This technique can also minimise the time it takes for the feature extraction step. However, there is still a possibility to improve the document vectors by taking the maximum or signed maximum values for each dimension of the word vector instead of averaging them out.

This study also corroborates the thorough investigation in [11] where it is stated that SVM-RBF and Random Forest performs quite well on many cases. These algorithms combined with fastText and our proposed basic features and keyword patterns turned out to be the best models to identify spam comments. In our experiment, SVM and XGBoost got the best F_1-scores of 0.9601 and 0.9512 respectively.

References

[1] Fajrina H N 2016 Ada 22 Juta Pengguna Aktif Instagram dari Indonesia URL `http://www.cnnindonesia.com/teknologi/20160623112758-185-140353/ada-22-juta-pengguna-aktif-instagram-dari-indonesia/`
[2] Guzella T S and Caminhas W M 2009 *Expert Systems with Applications* **36** 10206–10222
[3] Nosseir A, Nagati K and Taj-Eddin I 2013 *IJCSI International Journal of Computer Science Issues* **10** 1694–0814
[4] Drucker H, Wu D and Vapnik V N 1999 *IEEE Transactions on Neural networks* **10** 1048–1054
[5] Metsis V, Androutsopoulos I and Paliouras G 2006 *CEAS* pp 27–28
[6] Joachims T 1996 A probabilistic analysis of the rocchio algorithm with tfidf for text categorization. Tech. rep. DTIC Document
[7] Bojanowski P, Grave E, Joulin A and Mikolov T 2016 *arXiv preprint arXiv:1607.04606*
[8] Mikolov T, Sutskever I, Chen K, Corrado G S and Dean J 2013 *Advances in neural information processing systems* pp 3111–3119
[9] Landauer T K, Foltz P W and Laham D 1998 *Discourse processes* **25** 259–284
[10] Pedregosa F, Varoquaux G, Gramfort A, Michel V, Thirion B, Grisel O, Blondel M, Prettenhofer P, Weiss R, Dubourg V, Vanderplas J, Passos A, Cournapeau D, Brucher M, Perrot M and Duchesnay E 2011 *Journal of Machine Learning Research* **12** 2825–2830
[11] Fernández-Delgado M, Cernadas E, Barro S and Amorim D 2014 *J. Mach. Learn. Res* **15** 3133–3181
[12] Chen T and Guestrin C 2016 *Proceedings of the 22nd ACM SIGKDD International Conference on Knowledge Discovery and Data Mining* KDD '16 (New York, NY, USA: ACM) pp 785–794 ISBN 978-1-4503-4232-2 URL `http://doi.acm.org/10.1145/2939672.2939785`
[13] Rish I 2001 *IJCAI 2001 workshop on empirical methods in artificial intelligence* vol 3 (IBM New York) pp 41–46

Rhetorical Sentence Categorization for Scientific Paper Using Word2Vec Semantic Representation

G H Rachman[1], M L Khodra[2], and D H Widyantoro[3]

[1, 2, 3]School of Electrical Engineering and Informatics, Institut Teknologi Bandung, Bandung, Indonesia

Email: [1]ghoziyahaitan@gmail.com, [2]masayu@stei.itb.ac.id, [3]dwi@stei.itb.ac.id

Abstract. One of some ways to summarize scientific papers is by employing rhetorical structure of sentences. Determining rhetorical sentence itself passes through the process of text categorization. In order to get good performance, some works in text categorization have been done by employing semantic similarity words. Therefore, this paper aims to present the rhetorical sentence categorization from scientific paper by using selected features, added previous label, and Word2Vec to capture semantic similarity words. Then, this paper shows the result of employing resampling for balancing the existing instances per class and combining resampling and Word2Vec representation itself. Every experiment is tested in two classifiers, namely IBk and J48 tree. It shows that the use of previous label, Word2Vec (Skip-Gram), and resampling improves performance. After doing all the experiments in the 10-fold cross-validation, the highest performance of F-measure is achieved 84.97% by combining Word2Vec (Skip-Gram), all features, and resampling.

Keywords. Rhetorical categorization; scientific paper; word2vec; feature extraction.

1. Introduction

Scientific papers become one kind of document that continues to increase in number along with more research conducted by academics and researchers themselves. They initially use, at least one, scientific paper to collect relevant information for their own argument in the research [1]. Abstracts of that papers can be read in advance to find out the summarized meaningful information [2]. However, abstracts does not always give all important information as needed by readers and does not expose the correlation of scientific papers at each other. Due to this condition, readers prefer summary of a collection of scientific papers in the form of outline on certain points. A scientific paper has rhetorical structure, in which every segment of a text (i.e. sentence) has a meaningful category in the body of each section [3]. The classified rhetorical sentences are easier to be structured into a summary as needed by readers [4].

Determining rhetorical sentence passes through the process of sentence categorization. This process produces the high dimensionality of feature space which represents text in the document by employing string to word vector [5]. Its vector representation can be formed into distributional semantic model by capturing the meaning of each existing words [6]. For example, because the word of "aim" and "purpose" has the similar meaning, these two words should also have high word similarity. Therefore, the more words with similar meaning on the scientific paper exist, the higher effect against performance of classification can be acquired. In this paper, we will conduct rhetorical sentence categorization on scientific paper by employing Word2Vec as a tool for computing word vector representation of semantic similarity words [6]. Since rhetorical dataset is an imbalanced dataset, this paper will also involve handling of imbalanced data set.

The rest of this paper is organized as follows. The next section provides the related works on rhetorical sentence categorization from scientific paper and implementation of word2vec semantic representation. Our method is explained in section 3. We also define various rhetorical categories and feature sets for classifying sentences. We describe our experiments in section 4. The results are analyzed in section 5. Finally, the conclusion and further work are described in section 6.

2. Related Works

Text categorization is conducted to assign label of a text based on its text features. Rhetorical sentence categorization is commonly used for scientific papers. Teufel and Moens [7] conducted paper summarization by first assign one of seven rhetorical categories to every sentence. These seven are "aim", "textual", "own", "background", "contrast", "basis", and "other". They [8] employed Naïve Bayes classifier and got the F-measure result from 26% to 86%. Then, its result has risen up until 92.93% after employing Maximum Entropy [9]. These seven categories next become 15 rhetorical categories, namely "aim", "nov_adv", "co_gro", "othr", "prev_own", "own_mthd", "own_fail", "own_res", "own_conc", "codi", "gap_weak", "antisupp", "support", "use", and "fut". Widyantoro, et al. [10] implemented these 15 categories added with "textual" and employed multi-heterogeneous classifier. Its average of F-measure result is about 25%. It still does not significantly show the high performance for this 15 rhetorical sentence categorization model. The main problem is imbalanced dataset and word variation in sentences.

Word2Vec is used to group words which have similar meaning into vector representation. This model is proposed by Mikolov [6] and released by Google in 2013. There are two architectures of Word2Vec, namely the continuous bag-of-words (CBOW) and skip-gram. The CBOW predicts the current word based on the context and the skip-gram predicts surrounding words given the current word [11].

Heffernan and Teufel [12] employed Word2Vec representation to identifying problem statement in scientific text. They used 18,753,472 sentences from a biomedical corpus based on all full-text Pubmed articles and then built model from the 200 semantically similar words to only "problem". Its result showed that Word2Vec model caused significant performance increase, because Word2Vec attributes have the greatest information gain compared the other features.

Rahmawati and Khodra [13] have employed Word2Vec representation in multi-label classification for Indonesian news article. They did experiment by using CBOW (once) and Skip-Gram in the vector length from 200 to 500. Its result showed the testing F-measure value from 76.73% to 81.63%. It is proven that Word2Vec has better performance than TF-IDF in their previous research. Putra and Khodra [14] also showed that text representation using semantic model has higher accuracy than lexicon model which does not consider about semantic meaning of words. They obtained the best accuracy by using ANN with word2vec CBOW at 82.94%.

3. Methodology

In this research, we will use the data set of scientific paper from [10]. Every sentence in these papers had been annotated into one of 16 rhetorical categories from Teufel, namely "aim", "nov_adv", "co_gro", "othr", "prev_own", "own_mthd", "own_fail", "own_res", "own_conc", "codi", "gap_weak", "antisupp", "support", "use", "fut", and "textual". The description of these categories is explained in [10]. The example of an annotated paper is shown in Figure 1.

We use sentence features i.e. content, absolute location, explicit structure, sentence length, citation and sequential label. These features, except previous label, are adapted from [7] and [10]. Its explanation can be seen in Table 1. Then this paper employs Word2Vec representation algorithm from Medallia[1] library. We use 75 scientific papers still from [10] to build the word vector itself. Actually that data set was taken from ACL Anthology Reference Corpus (ACL-ARC).

[1]https://github.com/medallia/Word2VecJava

```
<paper>
...
   <Section><Sentence><SectionTitle>Abstract</SectionTitle></Sentence>
   <P><Sentence><Aim>This paper aims to analyze word dependency structure in compound
   nouns appearing in Japanese newspaper articles.</Aim></Sentence> <Sentence><Othr>The
   analysis is a difficult problem because such compound nouns can be quite long, have
   no word boundaries between contained nouns, and often contain nnregistered words such
   as abbreviations.</Othr></Sentence> <Sentence><Gap_Weak>The non-segmentation property
   and unregistered words cause initial segmentation errors which result in erroneous
   analysis.</Gap_Weak></Sentence> <Sentence><Own_Mthd>This paper presents a corpus-
   based approach which scans a corpus with a set of pattern matchers and gathers co-
   occurrence examples to analyze compound nouns.</Own_Mthd></Sentence>
   <Sentence><Own_Mthd>It employs boot-strapping search to cope with unregistered words:
   if an unregistered word is lound in the process of searching the examples, it is
   recorded and invokes additional searches to gather the examples containing
   it.</Own_Mthd></Sentence> <Sentence><Own_Conc>This makes it possible to correct
   initial over-segmentation errors, and leads to higher accuracy.</Own_Conc></Sentence>
...
</paper>
```

Figure 1. Example of annotated paper

Table 1. Features and its description

Type	Name	Description	Values
Content	Cont-1	Occurrence of 10 significant terms of document using TF-IDF	1, 0
	Cont-2a	Incidence of words occurring in document title	1, 0
	Cont-3	Occurrence of 10 significant terms of abstract using TF-IDF	1, 0
Absolute location	Loc	Sentence position in document to 10 segments	A-J
Explicit structure	Struct-1	Sentence position within section	7 values
	Struct-2	Sentence position within paragraph	Initial, Medial, Final
	Struct-3	Prototypical type of section title	15 section or Non-Prototypical
Sentence length	Length	Whether sentence has longer than 15 words or not	1, 0
Citations	Cit-1	Occurrence of citation or self citation	Citation, Self-Citation, None
	Cit-2	Citation location in sentence	Beginning, Middle, End, or None
Sequential Label	PrevLabel	Previous label	Previous rhetoric category or "start" for first sentence in the document

In addition, resampling would be included in the experiment process for solving the imbalanced data [15]. Resampling is conducted by using WEKA. It will do oversampling for categories in smaller number and under-sampling for categories in higher number. So that every instances of categories is balanced in number.

Actually, our method is generally divided into four main processes. The first is preprocessing, and then constructing the vector representations of the vocabularies of data set by using Word2Vec. The second is producing all the features. The third is implementing resampling before building classification model. The last is we do 10-fold cross validation to know the F-measure detail of rhetorical category by using some classifiers, namely IBk and J48 in Weka.

We consider four preprocessing that consist of case folding, tokenization, stemming, and stop word elimination. These are done using Apache Lucene library and Weka. All of them were fixtures that would always be applied in all experiments. Then we use our annotated preprocessed sentences to output a model containing word vectors through Word2Vec representation (CBOW and Skip-Gram). We set its layer size into the vector length of 200. In the end, all classification models are produced from the last process.

4. Experiment

We conducted the rhetorical sentence categorization using the data set of scientific paper from [10]. This data has 75 annotated scientific papers that have been split into 10880 sentences. The number of sentences in every category before and after doing resampling can be seen in Table 2 and 3.

Table 2. The number of sentences in first eight rhetorical categories

Resampling	Aim	Nov_adv	Co_gro	Othr	Prev_own	Own_mthd	Own_fail	Own_res
Before	214	248	386	967	625	5325	70	419
After	663	631	689	724	691	673	718	678

Table 3. The number of sentences in second eight rhetorical categories

Resampling	Own_conc	Codi	Gap_weak	Antisupp	Support	Use	Fut	Textual
Before	581	111	367	60	392	426	154	535
After	673	632	692	709	649	734	667	657

In addition, there are five experiments that would be conducted. It is related to existence of previous label as sequential feature, resampling, and Word2Vec of CBOW and Skip-Gram. These are as follows:

1. *Baseline*. The types of feature to use are content, absolute location, explicit structure, sentence length and citation, without sequential label (previous label);
2. *Scenario 1*. In this experiment, all the features are used, including sequential label (previous label), to know whether previous label increases performance or not;
3. *Scenario 2*. In this experiment, scenario 1 will be conducted and completed by doing resampling. This method is used because [16] showed that resampling is the best method for solving imbalanced data set;
4. *Scenario 3*. This experiment will employ all the features and involve Word2Vec representation by using CBOW and Skip-Gram algorithm. For this scenario, resampling is not used, because this scenario is only to know the effect of CBOW and Skip-Gram against performance; and
5. *Scenario 4*. After knowing the best one of Word2Vec algorithm from comparing the result of scenario 3, this experiment will combine all the features, the best one of Word2Vec (Skip-Gram), and resampling.

For the last scenario, only Skip-Gram is used, because this kind of Word2Vec has better F-measure result than CBOW if compared with the result of first scenario. In addition, every scenario would be evaluated in two classifiers, namely IBk and J48 tree. Actually, determining classifiers for this research is based on [16] showing that IBk and J48 are the first two classifiers that have the highest performance in its last experiment. Then from all the experiments, we would know how some features have effect on the accuracy of F-measure in every rhetorical categorization.

5. Result

First, we built the word vectors from the data set which consist of 75 scientific papers that have been split into 10880 sentences. We extracted all the features without previous label. This was evaluated by two classifiers, namely IBk and J48 tree.

Table 4 shows the result of experiments. We can see that previous label and method of resampling has significantly increased the performance. In the first experiment (baseline), the F-measure average from using IBk classifier achieved 17.39% and J48 achieved 18.79%. Then these two performance have increased becoming 30.07% for IBk and 39.12% for J48 tree in the scenario 1 that involves previous label. Its increase is until about 18% from initial performance in baseline. It is proven that previous label has a large effect to raise the performance of classification as concluded in [16]. It is because previous label as sequential feature shows some patterns which always appear in rhetorical sentence categorization that are correlated with the target classes [17].

In fact, the data set we got has the imbalanced instances per category, so we have to make it balanced. Then, the third experiment (scenario 2) exists to know how much the effect of handling imbalanced data set, by using resampling in Weka, against the performance. Its result shows that the F-measure average from using IBk classifier achieved 79.91% and J48 achieved 72.1%. These two increases are significant from the result of scenario 1. However, there is a difference in this scenario

Table 4. The result of experiment

Category	Baseline		Scenario 1 (Baseline + Previous Label)		Scenario 2 (Resample)		Scenario 3 (CBOW or Skip-Gram)				Scenario 4 (Skip-Gram + Resample)	
	IBk	J48	IBk	J48	IBk	J48	IBk (CBOW)	J48 (CBOW)	IBk (Skip-Gram)	J48 (Skip-Gram)	IBk	J48
Othr	24.7	24.2	47	57.6	69	60.1	39.6	48.9	38.6	44.2	72.8	64.6
Own_mthd	69.3	71.7	75.9	79.6	47.1	42.3	71	77.3	71.7	76.1	53	49.9
Nov_adv	6	6.4	11.4	18.7	81.4	73.8	12.5	16.4	16.9	13.5	90.3	84.5
Gap_weak	12.3	7	23.5	28.5	79.7	63.7	21.9	19.6	23	20.4	83.8	75.7
Aim	18.7	18	28.3	34.6	85.2	77.5	28.2	27.5	34.7	23.9	92	85.4
Textual	14.3	16.7	22.3	26	68	55.5	19.1	32.3	28.3	29.4	79.5	72.1
Support	19	28.4	37	48	85	76	33.6	42.4	30.2	40.9	85.4	80.7
Use	6.9	6.9	27.3	45	81.5	73.7	26.2	37.1	28.5	35.2	84.9	80.6
Fut	29.9	36.4	36.7	42.2	89.3	85.5	28.2	32.6	30.6	31.3	93.9	91.4
Own_conc	18.8	16.9	30.4	40.6	73.3	60.5	27.9	34.2	36.8	29.2	80.8	71.6
Co_gro	24.3	25	42.6	52.5	80.2	71.8	39.6	45.1	38.9	40.8	87.5	79.3
Codi	2.9	9.9	11.8	18.3	89.7	86.6	7.1	14	8.4	9.9	95.7	92.5
Own_res	14.7	6.7	23.8	39.5	79.2	65.3	26.2	30.4	35.3	29.2	84.7	78.7
Own_fail	0	2.6	4.3	23.9	93.3	92.4	6.2	14.1	8.4	16.1	96.1	95.1
Prev_own	16.4	26.7	56.3	70.9	81.9	76.4	48.1	65.7	47.8	61.7	80.8	82.1
Antisupp	0	0	2.6	0	94.7	92.5	5.9	2.7	14.6	0	98.4	96.5
Average	**17.39**	**18.97**	**30.07**	**39.12**	**79.91**	**72.1**	**27.58**	**33.77**	**30.79**	**31.36**	**84.97**	**80.04**

result compared with the two before. In the first two experiment, J48 always has the higher performance, but in the third, IBk becomes the highest. This result is the same as concluded in [16] that resampling is more suitable for IBk classifier than J48 tree.

The next is we do the last two scenarios that related to Word2vec representation. In the scenario 3, we compare between using CBOW and Skip-Gram algorithm without doing resampling. Its result will be compared with the result of scenario 1, because both of them use all the extracted features. The most result from this scenario is not good. When using CBOW algorithm, the F-measure average from using IBk classifier achieved only 27.58% and J48 achieved 33.77%. If compared with the result of scenario 1, this result has decreased from 2.5% to 5.5%. It is different when using Skip-Gram. Although in J48 tree classifier, the F-measure has decreased becoming 31.36% more than when using CBOW, in IBk classifier it shows the contrary that using Skip-Gram is better than CBOW. Moreover the only one of all result in scenario 3 that increases from scenario 1 is acquired when using Skip-Gram and IBk classifier. In [13], Skip-Gram is indeed better than CBOW. It could be because Skip-Gram predicts surrounding words given the current word. In this research we check the similarity between words existing in sentences and words existing in vocabulary in Word2Vec representation. So that every word in sentence has weight of the semantic similarity between vocabulary clearly. It is different when using CBOW. It will predict the current word based on the context, in which sentence can be that context.

Therefore in the scenario 4, we only involve Skip-Gram algorithm then it is combined with resampling. This scenario is compared with scenario 2. Its result shows that the F-measure average from using IBk classifier achieved only 84.97% and J48 achieved 80.04%. After using resampling, the performance has always increased. It means that combination of Skip-Gram of Word2Vec representation and resampling give the higher performance for rhetorical sentence categorization. Resampling helps to solve the imbalanced data set and Skip-Gram Word2Vec to capture semantic similarity words in the data set.

6. Conclusion and Further Work

For the first, the implementation of previous label as sequential feature is used to catch class pattern which always appears that is correlated to the target class. So that it could be used to predict the next rhetorical category for sentences. For the second, the implementation of resample can get the performance up. It is because this method make the imbalanced data set in every class become balanced. For this research, we get the highest performance of using resampling when employing IBk classifier. It means that resampling is more suitable with IBk classifier than with J48 tree. Furthermore our work focused also in employing Word2Vec semantic representation. There are two architecture of Word2Vec, namely Skip-Gram and CBOW. Actually, after doing this research, we got that the performance of using Skip-Gram is better than CBOW. The last is implementation of combining Word2Vec and resampling can give the higher performance.

Actually, in this work we still do Word2Vec representation in the annotated data set itself and have not yet conducted testing and training. These two things can affect the performance whether it becomes better or not. In the further work, it will be considered to achieve good performance of rhetorical sentence categorization. Moreover the data set should be added more also to get higher performance.

Acknowledgement

This research was partially supported by the Master's Program toward Doctoral Degree for Excellent Graduate (*Program Pendidikan Magister Menuju Doktor untuk Sarjana Unggul*/PMDSU) from Kemenristekdikti Indonesia within research entitled "Automatic Guided Summarization".

References

[1] R A Schwegler and L K Shamoon 1982 *The Aims and Process of the Research Paper* (College English vol 44) no 8 pp 817-824
[2] H P Luhn 1958 *The Automatic Creation of Literature Abstracts* (IBM JOURNAL) pp 159-165
[3] M Taboada and W C Mann 2006 *Rhetorical structure theory: Looking back and moving ahead* (Discourse studies vol 8) no 3 pp 423-459
[4] M L Khodra, D H Widyantoro, E A Aziz, and B R Trilaksono 2012 *Automatic Tailored Multi-Paper Summarization based on Rhetorical Document Profile and Summary Specification* (Journal of ICT Research and Applications vol 6) no 3 pp 220-239
[5] Y Yang and J Pedersen 1997 *A Comparative Study on Feature Selection in Text Categorization, Int. Conf. on Machine Learning* (Nashville: ICML) pp 412-420
[6] X Rong 2014 word2vec Parameter Learning Explained (Cornell University Library) *Preprint* cs/1411.2738
[7] S Teufel 1999 *Argumentative Zoning: Information Extraction from Scientific Text* (Edinburgh: University of Edinburgh)
[8] S Teufel and M Moens 2002 *Summarizing scientific articles: experiments with relevance and rhetorical status* (Computational Linguistics - Summarization vol 28) issue 4 pp 409-445
[9] S Merity, T Murphy, and J Curran 2009 *Accurate Argumentative Zoning with Maximum Entropy models, Proc. of the 2009 Workshop on Text and Citation Analysis for Scholarly Digital Libraries* (Suntec: World Scientific Publishing Co Pte Ltd) pp 19-26
[10] D H Widyantoro, M L Khodra, B R Trilaksono and E A Aziz 2013 *A Multiclass-based Classification Strategy Sentence Categorization from Scientific Papers* (Journal of ICT Research and Applications vol 7) no 3 pp 235-249
[11] T Mikolov, K Chen, G Corrado, and J Dean 2013 Efficient estimation of word representations in vector space (Cornell University Library) *Preprint* cs/1301.3781
[12] K Heffernan and S Teufel 2016 *Identifying Problem Statements in Scientific Text, Foundations of the Language of Argumentation, Proc. of COMMA 2016 Workshop* (University of Potsdam) p 18
[13] D Rahmawati and M L Khodra 2016 *Word2vec Semantic Representation in Multilabel Classification for Indonesian News Article, Proc. of 2016 Int. Conf. on Advanced Informatics: Concepts, Theory and Application (ICAICTA)* (Penang)
[14] Y A Putra and M L Khodra 2016 *Deep Learning and Distributional Semantic Model for Indonesian Tweet Categorization, Proc. of 2016 International Conference on Data and Software Engineering (ICoDSE)* (Bali)
[15] N V Chawla, N Japkowicz and A Kotcz 2004 *Editorial: special issue on learning from imbalanced data sets* (ACM Sigkdd Explorations Newsletter vol 6) no 1 pp 1-6
[16] G H Rachman and M L Khodra 2016 *Automatic Rhetorical Sentence Categorization on Indonesian Meeting Minutes, Proc. of 2016 International Conference on Data and Software Engineering (ICoDSE)* (Bali)
[17] N Lesh, M J Zaki and M Oglhara 2000 *Scalable feature mining for sequential data* (IEEE Intelligent Systems and their Applications vol 15) no 2 pp 48-56

IOP Conf. Series: Journal of Physics: Conf. Series **801** (2017) 012071 doi:10.1088/1742-6596/801/1/012071

An Automatic Labeling of K-means Clusters based on Chi-Square Value

[1*]**R Kusumaningrum**, [2*]**Farikhin**

[*]Master Program of Information System, School of Postgraduate Studies
[2]Department of Informatics, Faculty of Science and Mathematics
[3]Department of Mathematics, Faculty of Science and Mathematics,
Diponegoro University

E-mail: [1]retno@live.undip.ac.id, [2]farikhin.math.undip@gmail.com

Abstract. Automatic labeling methods in text clustering are widely implemented. However, there are limited studies in automatic cluster labeling for numeric data points. Therefore, the aim of this study is to develop a novel automatic cluster labeling of numeric data points that utilize analysis of Chi-Square test as its cluster label. We performed K-means clustering as a clustering method and disparity of Health Human Resources as a case study. The result shows that the accuracy of cluster labeling is about 89.14%.

1. Introduction

Clustering is an unsupervised method that is widely used in several domains to categorize data into some clusters based on its similarity. In image domain, clustering can be implemented as a method in image segmentation [1], images database categorization [2], image quality verification [3], etc. Moreover, clustering are also widely implemented in text domain, such as [4],[5],[6], etc.

In general, the studies of clustering method only implement related method for grouping the data into several clusters without annotations (automatic labeling) against the resulted clusters. Some of studies in text domain have been implemented cluster labeling, such as Wikipedia-based cluster labeling [7], hierarchical cluster labeling [8], [9], etc. However, there are limited studies for automatic cluster labeling in other domains. The existing cluster labeling for numeric data points is cluster labeling method for Support Vector Clustering (SVC) which is developed based on some invariant topological properties of a trained kernel radius function [10]. Its labels are support vectors, data points, and stable equilibrium points.

On the other hand, Chi-Square test is statistical methods that can be used detect different conditions between actual condition and ideal condition. Therefore, we proposed a novel method for cluster labeling of numeric data points that utilize analysis of Chi-Square test as its cluster label. In this study we performed K-means clustering as a clustering method and a disparity of Health Human Resources (HHR) as a case study.

2. Related Works

As clustering is an unsupervised method which categorizes data on some clusters based on its similarity, then there is no label for each resulted cluster. This leads to the emergence of challenges in the cluster labeled according to the data characteristics. The cluster label has been widely implemented in text data domain. However, it still very rarely implemented in numeric data. One of implemented methods in text data domain is chi-square method. It is used to test each word at each node in a hierarchy starting at the root and recursively moving down the hierarchy [11]. Since chi-square method can be used to find the association between two variables, i.e. observed value and expected value, then it is possible to be implemented in numeric data for cluster labeling.

2.1. Clustering

There are several methods of clustering that are widely implemented such as K-means Clustering, K-Medoids, Hierarchical Clustering, DB Scan, etc. In this study, we apply K-means Clustering as its simplicity. We implemented K-means Clustering with the following specification, namely a standard *Euclidean distance* as similarity measure and Rule of Thumbs formula as a method for determining the number of cluster based on the following equation:

$$K \approx \sqrt{\frac{n}{2}} \tag{1}$$

Where K is defined as the number of cluster and n is the number of data.

The algorithm of K-means Clustering is as follow [12]:

1. Select k initial cluster centroid
2. Iteratively refining them as follows:
   a. Each instance d_i is assigned to its closest cluster centroid
   b. Each cluster centroid C_j is updated to be the mean of its constituent instances
3. Stop the iteration when there is no further change in assignment of instances to each cluster.

2.2. Cluster Homogeneity

Cluster homogeneity can be determined based on the silhouette coefficient value that can be obtained through the stages as follows [13]:

1. Calculate the average distance from a document supposes i with all other documents that are located in one cluster

$$a_i = \frac{1}{|A|-1}\sum_{j\in A, i\neq j} d(i,j) \tag{2}$$

   Where $|A|$ is defined as the number of data in Cluster A, $d(i,j)$ is defined the distance between i^{th} document and j^{th} document, and (i,j) is defined as the index of document
2. Calculate the average distance from i^{th} document with all the documents in the cluster and take the smallest value

$$d(i,C) = \frac{1}{|A|}\sum_{j\in C} d(i,j) \tag{3}$$

   Where $d(i,C)$ is defined the distance between i^{th} document over all object in other cluster, in which $A \neq C$
3. The value of the Silhouette coefficient is

$$s(i) = \frac{b(i)-a(i)}{\max\big(a(i),b(i)\big)} \tag{4}$$

2.3. The Standard Analysis of Requirement Status of HHR

According to the Regulation Number 32 of 1996, health professionals is everyone who devoted themselves in the field of health and has the knowledge and/or skills through education in the field of health, which requires the authority to do the efforts of health in specific types. Based on the Regulation of the Minister of Health Republic of Indonesia, Number 33 of 2015 on Preparations Guidelines of Health Human Resource Requirement Planning, which is included into HHR, i.e. (i) health care practitioner includes general practitioner, dentist, and specialists; (ii) nursing includes nurses, dentist nurses, and midwives; (iii) pharmacy includes pharmacies and assistant pharmacists; (iv) public health officer; (v) sanitarian, (vi) nutrition experts, and (vii) therapists both physical and medical.

One of methods that can be applied to plan the HHR requirement is "population ratio" method, namely the ratio of health professionals to the total population for a region. The ratio of the standard population is determined based on population per regency, population growth per regency as well as a number of health professionals.

On the other hand, a disparity analysis can be implemented to determine the appropriateness between the availability of HHR with "standards of the population ratio". A Chi-Square analysis is a method that can be used to perform the disparity analysis with the following formula:

$$X^2 = \sum \frac{(F_o - F_h)^2}{F_h} \tag{5}$$

Where X^2 is *Chi-square* value, F_o is observed value, and F_h as predicted value

The interpretation value of the Chi-Square is based on the comparison between the values of arithmetic Chi-Square with the value of Chi-Square table. The greater value of arithmetic Chi-Square means there is a significant difference between the existence of health professionals with the normative needs of health care professionals, and it's applied on the contrary. Subsequently, this framework applied as a method for cluster labels.

3. Research Procedures

Illustration of the research procedure can be seen in Figure 4, whereas the details explanation about each step can be seen on the following subsection.

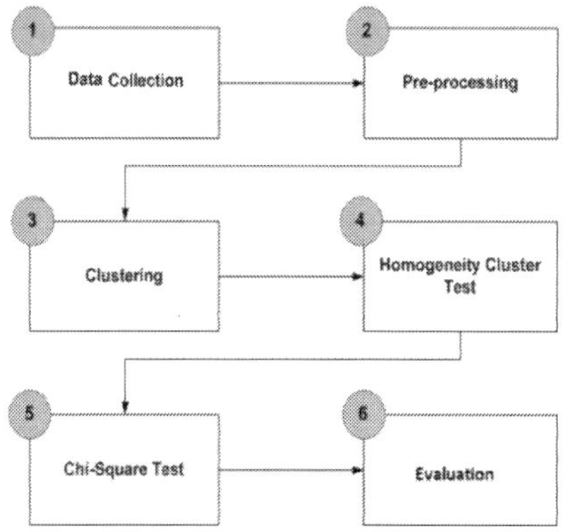

Figure 1. Research procedures

3.1. Data Collection

As explained before, the collected data including:

- Population data obtained from the Central Bureau of Statistics, Central Java Province
- HHR data (doctors, nursing, midwives) are acquired through an online media, i.e. dashboard of Health Human Resources Information System version 2016. This system is published by the PPSDM, Ministry of Health Republic of Indonesia.

In addition, this research also uses the HRR Requirement ratio per 100,000 total populations as described in Table 1.

Table 1. HHR Requirement over Population Data

No	Type of Health Professional	HHR Requirement per 100.000 Total Populations
1	General Practitioner	40
2	Specialist	10
3	Dentist	12
4	Nurse	158
5	Midwife	100

IOP Conf. Series: Journal of Physics: Conf. Series **801** (2017) 012071 doi:10.1088/1742-6596/801/1/012071

3.2. Pre-processing
The pre-processing step is used to normalize the obtained data into range [0,1] by using min-max normalization. This step is important since the range of the obtained data is difference for each type and it would give an impact on the performance of clustering process, where the process of the clustering data was conducted based on the similarity level between the data.

3.3. Clustering
The clustering process will be applied with K-means clustering method, whereas the Euclidean distance applied as the measurement techniques of similarity level. The detail of clustering methods has been described in Subsection 2.1.

3.4. Cluster Homogeneity Test
The aims on performing the cluster homogeneity cluster are to define the quality and the strength of the cluster, how good an object placed in a cluster. The method for performing the cluster homogeneity test on this research is Silhouette coefficient. The detail of the homogeneity cluster test can be seen on the Subsection 2.2.

3.5. Chi-Square Test
As described previously Chi-square test is performed to know the requirements disparity of HHR in accordance with the Government Regulation No. 32 of 1996 compared with the real conditions in the field about the availability of HHR.

3.6. Evaluation
Evaluation was performed to evaluate the accuracy level of the Chi-Square based labeling.

4. Result and Analysis
The results from the clustering process using K-means Clustering can be seen in Figure 2 with the value of the centroid for each cluster can be seen in Table 2 and the number of the cluster based on the Rule of Thumbs formula is defined as 4 clusters.

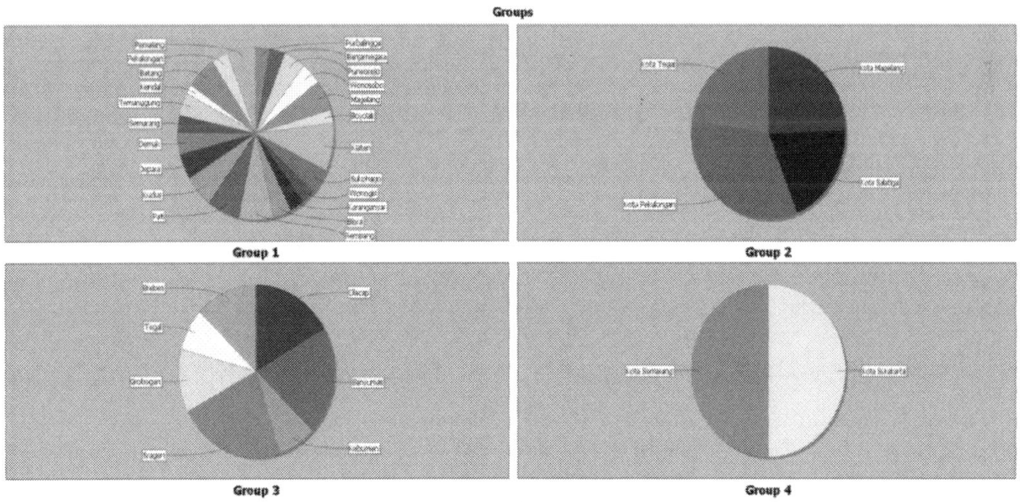

Figure 2. Clustering Result

Table 2. Last Centroid from *Clustering* Process

Cluster ID	The Number of Population	HHR Availability				
		General Physician	Specialist	Dentist	Nurse	Midwife
1	9363060	1160	990	310	6100	306
2	2101091	841	1231	231	5891	96
3	14143392	1482	1692	442	9512	662
4	10915493	3733	9283	1343	31043	404

As mention in the previous section, the cluster homogeneity test was performed based on Silhouette coefficient of about 0.541 which means the results of clustering have a good structure.

Subsequently, the calculation of Chi-Square Test was performed for automatic cluster labeling in general as well as the identification of HHR disparity status (the difference between the value of the availability and value of requirement) for regency in Central Java province in specific.

Table 3. *Chi-Square* Test

	Cluster ID	General Practitioner	Specialist	Dentist	Nurse	Midwife
F_o – Availability	1	1160	990	310	6100	306
	2	841	1231	231	5891	96
	3	1482	1692	442	9512	662
	4	3733	9283	1343	31043	404
F_o – Requirement	1	3746	937	1124	14794	9364
	2	841	211	253	3320	2102
	3	5658	1415	1698	22347	14144
	4	4367	1092	1310	17247	10916
Expected Value (F_h)	1	2453	963.5	717	10447	4835
	2	841	721	242	4605.5	1099
	3	3570	1553.5	1070	15929.5	7403
	4	4050	5187.5	1326.5	24145	5660
Chi-Square Value (Availability)	1	681.5528	0.728853	231.0307	1808.788	4242.366
	2	0	360.749	0.5	358.8123	915.3858
	3	1221.217	12.34776	368.5832	2585.411	6138.198
	4	24.8121	3233.373	0.205239	1970.694	4880.837
Chi-Square (Requirement)	1	681.5528	0.728853	231.0307	1808.788	4242.366
	2	0	360.749	0.5	358.8123	915.3858
	3	1221.217	12.34776	368.5832	2585.411	6138.198
	4	24.8121	3233.373	0.205239	1970.694	4880.837
Chi-Square	1	1363.11	1.46	462.06	3617.58	8484.73
	2	0	721.5	1	717.62	1830.77
	3	2442.43	24.7	737.17	5170.82	12276.4
	4	49.62	6466.75	0.41	3941.39	9761.67

Since we used 2 categories (availability and requirement) then the degree of freedom is *(2-1) = 1*. Based on its value and fault tolerance 0.5, thus the value of Chi-Square table is 3.841. Therefore, we perform the automatic labeling process based on the rule in eq. (6) and the result label in table 4.

$$disparity\_label = \begin{cases} Fair, if\ Arithmetic\ Chi - Square < 3.841 \\ High, otherwise \end{cases} \quad (6)$$

Table 4. General Label for Each *Cluster*

Cluster ID	General Practitioner	Specialist	Dentist	Nurse	Midwife
1	HIGH	FAIR	HIGH	HIGH	HIGH
2	FAIR	HIGH	FAIR	HIGH	HIGH
3	HIGH	HIGH	HIGH	HIGH	HIGH
4	HIGH	HIGH	FAIR	HIGH	HIGH

Based on Figure 2 and Table 4, there are four cities in Central Java Province that have good disparity label (fair status for both general practitioner and dentist availability), namely the Magelang City, Salatiga City, Pekalongan City, and Tegal City. Those cities are included in cluster 2. The final evaluation process was performed in order to measure the accuracy of the labeling process based on the ratio between the number of true predicted label and the total number of labels. The result shows that the labeling accuracy is about 89.14%.

5. Conclusions and Future Works

The automatic labeling of K-means cluster based on Chi-Square Tests was successfully applied with the accuracy is about 89.14%. Based on the resulted label, there are four cities in Central Java Province that have good disparity label (fair status for both general practitioner and dentist availability), namely the Magelang City, Salatiga City, Pekalongan City, and Tegal City.

The same strategy can be applied to all the regencies and cities in Indonesia and to assess the availability of other HHR's such as pharmacists, public health officer, sanitarian, etc. While to improve the accuracy of the value can be done by optimizing the K-means Clustering algorithm thus they have better Silhouette coefficient.

References

[1] Dhanachandra N, Manglem K and Chanu Y J 2015 Image Segmentation using K-means Clustering Algorithm and Subtractive Clustering Algorithm *Proc. of Eleventh Intl. Multi-Conf. on Information Processing-2015* pp 764–771

[2] Le Saux B and Boujemaa N 2002 Unsupervised Categorization for Image Database Overview *Proc. of Intl. Conf. on Visual Information System (VISUAL'2002)*

[3] Niemeijer M, Abràmoff M D and van Ginneken B 2006 Image structure clustering for image quality verification of color retina images in diabetic retinopathy screening *Med. Image Anal.* **10** no. 6 pp. 888–898

[4] Yin J and Wang J 2016 A Text Clustering Algorithm Using an Online Clustering Scheme for Initialization *Proceedings of the 22nd ACM SIGKDD International Conference on Knowledge Discovery and Data Mining*, 2016 pp. 1995–2004

[5] Xu T and Oard D W 2011 Wikipedia-based Topic Clustering for Microblogs *Proc. of the American Society for Information Science and Technology* pp. 1–10.

[6] Aggarwal C C, Zhao Y and Yu P S 2012 On Text Clustering with Side Information *Proc. of the 2012 IEEE 28th Intl. Conf. on Data Engineering* pp. 894–904

[7] Carmel D, Roitman H and Zwerdling N 2009 Enhancing Cluster Labeling Using Wikipedia in *Proc. of the 32nd Intl. ACM SIGIR Conf. on Research and Development in Information Retrieval* pp. 139–146.

[8] Treeratpituk P and Callan J. 2006 Automatically Labeling Hierarchical Clusters *Proc. of the 2006 Intl. Conf. on Digital Government Research* pp 167–176

[9] Moura M F and Rezende S O 2007 Choosing a Hierarchical Cluster Labelling Method for a Specific Domain Document Collection *New Trends in Artificial Intelligence* pp. 812–823

[10] Lee J and Lee D 2005 An Improved Cluster Labeling Method for Support Vector Clustering *IEEE Trans. Pattern Anal. Mach. Intell.* **27** 3 pp. 461–464

[11] Popescul A and Ungar L H 2000 Automatic Labeling of Document Clusters [Online] Available: http://www.cis.upenn.edu/~popescul/Publications/labeling_KDD00.pdf

[12] Wagstaff K, Cardie C, Rogers S and Schroedl S 2001 Constrained K-means Clustering with Background Knowledge *Proc. of the Eighteenth Intl. Conf. on Machine Learning* pp. 577–584

[13] Rousseeuw P J 1987 Silhouettes : a graphical aid to the interpretation and validation of cluster analysis *J. Comput. Appl. Math.* **20** pp. 53–65

Pre-processing Tasks in Indonesian Twitter Messages

A F Hidayatullah[1] and M R Ma'arif[2]

[1]Department of Informatics, Universitas Islam Indonesia
Jl. Kaliurang KM 14,5 Sleman, Yogyakarta, Indonesia
[2]Department of Informatics Management, STMIK Jend. Achmad Yani Yogyakarta
Jl. Ringroad Barat, Banyuraden, Gamping, Yogyakarta, Indonesia

E-mail: [1]fathan@uii.ac.id, [2]rifqi@stmikayani.ac.id

Abstract. Twitter text messages are very noisy. Moreover, tweet data are unstructured and complicated enough. The focus of this work is to investigate pre-processing technique for Twitter messages in Bahasa Indonesia. The main goal of this experiment is to clean the tweet data for further analysis. Thus, the objectives of this pre-processing task is simply removing all meaningless character and left valuable words. In this research, we divide our proposed pre-processing experiments into two parts. The first part is common pre-processing task. The second part is a specific pre-processing task for tweet data. From the experimental result we can conclude that by employing a specific pre-processing task related to tweet data characteristic we obtained more valuable result. The result obtained is better in terms of less meaningful word occurrence which is not significant in number comparing to the result obtained by just running common pre-processing tasks.

1. Introduction

Twitter is a microblogging system that allows its users to post about their activities through short message called a tweet. Many tweets are posted by people around the globe. Thus, the amount of tweets is increasing steadily. From those data, people can get some useful information by classifying tweets into some categories. For example, we can classify tweets based on sentiments, contents, opinions, news, topics, etc.

However, like other social media text message, Twitter text messages are very noisy. Moreover, tweet data are unstructured and complicated enough. These are happened because there is no regulation for users to write tweet. So, they can post their tweet by ignoring grammar, spelling, etc. Therefore, Twitter messages usually contain misspelled words, abbreviations, and other bad language forms. In addition, tweets also contain symbols, link, emoticons, etc. To extract information from tweets, we need to transform those inappropriate words into standard form. According to those problems, pre-processing tasks are very important and critical in text mining [1]. Pre-processing aims to eliminate noises from the text data.

This paper addresses the issue of text pre-processing steps in Bahasa Indonesia, particularly in Twitter text messages. In addition, we will conduct two different tasks in pre-processing tasks. The first section is a common pre-processing task and the second one is a particular task for Twitter messages.

The rest of this paper is organized as follows. Section 2 presents some related works. Section 3 describes the experiment about tweet pre-processing. Section 4 explains the result and discussion of this research. Finally, section 5 presents the conclusion and future work of this research.

2. Related Works

Some previous researchers have been discussed about text pre-processing. Hemalatha et al. [2] has conducted the research about the pre-processing techniques to perform sentiment analysis in Twitter. The pre-processing steps proposed are tokenization, stemming, removing URLs, filtering, removing special characters and removing retweets. Sun et al. [3] proposed the pre-processing text on online financial text corpora. Six processing steps conducted in that research are URL and number removal; abbreviation extending; additional punctuation and lengthening words extraction and replacement before tokenization; negation identification and handling; POS tagging and removal of pronouns, prepositions and conjunctions, and punctuations; and lemmatization.

Duwairi and El-Orfali [4] presented several pre-processing and feature representation strategies in Arabic text. Some pre-processing tasks such as stemming, feature correlation, and n-gram models were experimented to investigate the effects of the accuracy in Arabic sentiment analysis. Rushdi-Saleh et al. [5] applied different pre-processing tasks on movie reviews collected from different web pages and blogs in Arabic. They proposed several pre-processing techniques including stop word elimination, stemming and n-grams generation for unigrams, bigrams and trigram.

In case of text classification, stop words removal step in pre-processing influence essentially toward the accuracy and classification performance [6] [7]. Torunoglu, et al. [1] also investigate the impact of stop words in pre-processing methods in text classification on Turkish texts. According to their experiments, the research concluded stop words only give little impact. Moreover, stemming has not significantly affect the Turkish text classification.

3. Pre-processing Experiments

In this part we divide our proposed pre-processing experiments into two parts. The first part is common pre-processing task which widely used for typical text mining job, and the second part is a specific pre-processing task for tweet data corresponding to the tweet data characteristics.

3.1. Common Pre-processing Task

3.1.1. Removing symbols, numbers, ASCII strings, and punctuations. Twitter messages usually contain symbols, numbers, and punctuations. All of these will be removed using regular expression syntax.

3.1.2. Tokenization. Tokenization task aims to divide sentence into some parts called token. The token can be formed in words, phrases or the other meaningful elements. This task is performed by using word_tokenize function from nltk.tokenize library.

3.1.3. Case folding. Case folding is the process to convert words into the same form, for instance lowercase or uppercase. In this step, we transform all words into lower case using Python string lower method.

3.1.4. Stemming. Stemming is the process to obtain the base or root of word by omitting affixes and suffixes. This research utilizes Sastrawi Python library to alleviate inflected words in Bahasa Indonesia to their base form. The algorithm for stemming in Sastrawi library is based on Nazief-Adriani algorithm.

3.1.5. Stopword removal. Stopword removal eliminates the common and frequent words which do not have the significant influence in the sentence. In this pre-processing task, we remove the stop word in the Twitter message according our stop word lists containing stop words in Bahasa Indonesia such as 'dan' (and), 'atau' (or), etc. This step is conducted by importing our stop words list from nltk.corpus library.

3.2. Specific Pre-processing Task

In typical short message text like Twitter, the messages have a wide range of quality ranging from high quality well defined text to meaningless strings. The variations raised due to typos, ad hoc abbreviations, phonetic substitutions, ungrammatical structures and emoticons. Hence the pre-processing task for tweet data cannot simply relies on common pre-processing task describes on previous sub section. On this following sub section, we have defined four characteristic of Twitter messages and our method to handle those kinds of texts.

3.2.1. Special Symbols on Twitter. Twitter has special symbols in its message such as hashtag (#), username (@username), and retweet (RT). These characters will be removed in this task. However, our method will only remove the hashtag symbol and leaving the word because the hashtag symbols usually followed by word or phrase which represent the discussed topic.

3.2.2. Emoticons Handling. Emoticons will be converted into their represented word. This work grouped seven emoticon types such as smile *(emot-senyum)*, laugh *(emot-tawa)*, love *(emot-cinta)*, sad *(emot-sedih)*, wink *(emot-kedip)*, cry *(emot-tangis)*, and stick out tongue *(emot-ejek)*. This agglomeration created based on the emoticons that found in our dataset. The snippet code of emoticon conversion can be seen in Figure 1.

```
('emot-senyum', [':-)', ':)', '(:', '(-:', '^^', '^-^', '^_^'] )     ,\
('emot-tawa',   [':-D', ':D', 'X-D', 'XD', 'xD', ] )     ,\
('emot-cinta',  ['<3', ':\*', ] )    ,\
('emot-kedip',  [';-)', ';)', ';-D', ';D', '(;', '(-;', ] ) ,\
('emot-sedih',  [':-(', ':(', '(:', '(-:', ] )  ,\
('emot-tangis', [':,(', ':\'(', ':"(', ':(('] ) ,\
('emot-ejek',   [':p', ':P', ':-p', ':-P', '=p', '=P'] )     ,\
```

Figure 1. Snippet code of emoticon conversion

3.2.3. Removing URLs. Twitter messages usually contain URLs, for example http://www.ift.tt/1QBmUt3. The URLs are removed because we only focused on the words containing in the tweets.

3.2.4. Tweet Characteristics of Indonesian People. The occurrence of non-standard words in social media text including in Twitter messages is very high. Abbreviations or shorthand and miss spellings are the most common examples of non-standard words [8]. Moreover, there are some other non-standard words that usually found such as combining letter and number, lengthening words and writing message using slang words.

Based on previous research by Hidayatullah [9], there are some unique tweet characteristics which commonly posted by Indonesian which related to the appearance of non-standard word. The first characteristic is the Indonesian sometimes show their expression by lengthening words. They repeat the letter 'e' in the word *'hore'* which means hurray in English. Indonesian people sometimes write *'horeeeee'* to show happiness in their tweet. Because of this, the lengthening words should be normalized by omitting the excess letter based on the dictionary.

The second characteristic related to a non-standard word is abbreviation or shorthand. For instance, Indonesian usually write *'g', 'gk',* or *'tdk'* to express the word *'tidak'* which means no or not. In addition, Indonesian people also often to use slang words to express the word *'tidak'* by writing *'nggak'* or *'gak'*. In addition, people also usually combining between letters and number such as *'hati2'* (be careful) that should be *'hati-hati'* to repeat the word, *'se7'* which means agree.

We presented to normalize those non-standard word into appropriate word based on the dictionary in this pre-processing step. For the lengthening word problem, we use dictionary of Indonesian standard word which contain nearly 35.000 words to obtain the standard form of those text. For the abbreviation or shorthand, we build a corpus of commonly non-standard word used by Indonesian and the corresponding standard word for replacement when such word found within incoming tweets.

3.3. Pre-processing Steps

The two previous sub sections defined the common approach of text pre-processing and specific approach to twitter text processing consecutively. Each approach has its own sub task. Figure 2 depicts the overall steps and workflow regarding the tweet pre-processing which combines the sub tasks from both approaches. In figure 2, yellow box indicating common pre-processing step, while blue box indicating specific pre-processing step.

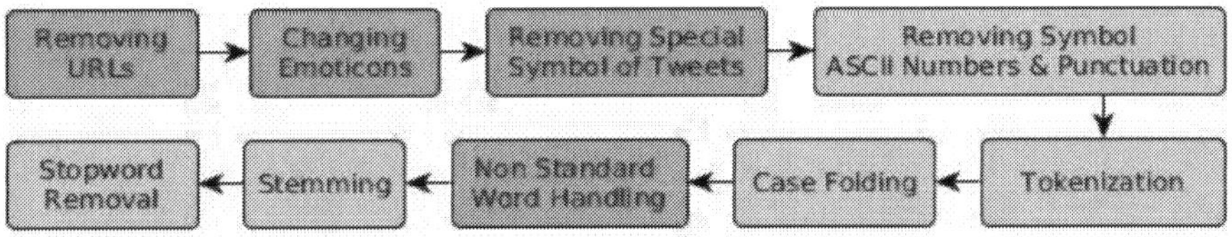

Figure 2. Pre-processing step

4. Result and Discussion

About 25,189 tweets have been obtained from Twitter using Twitter Search API v1.1. We have collected Twitter dataset by adding between our current dataset and our previous research [9] [10]. The main goal of this experiment is to clean the tweet data for further analysis. Thus, the objectives of this pre-processing task is simply removing all meaningless character and left valuable words.

Following the steps defined on sub section 3.3, we obtained the following result depicted as a word cloud in figure 3. In figure 3, we can figure out that most of words appeared on the visualization is a key term which build up the idea of the tweet. Even though there are left some of the meaningless word likes *"ya", "dr", "aja", "at"*, etc., the number of such words are not significant. In word cloud, the bigger the size of the word, they appeared more frequently on the dataset.

Figure 3. The result obtained by common and specific pre-processing

For a comparison purposes, we have visualized the result obtained by just employing the common pre-processing step in figure 4. In figure 4, there are exist more meaningless word in significant number. The words like "*ga*", and "*yg*" which is actually abbreviated stop word is still appeared on the dataset. Another meaningless term such as "*p*", "*d*", "*rt*", "*org*", "*udh*", and so on still also appeared in significant number.

Figure 4. The result obtained by just common pre-processing task

5. Conclusion and Future Work

In this paper, we have presented pre-processing tasks for Indonesian Twitter messages. In general, we can divide the pre-processing task into two categories. Firstly, there are common pre-processing tasks in text mining such as tokenization, case folding, removing punctuations, removing stopwords, and stemming. The second category are the specific task to handle special characteristics that frequently occurs in Twitter and social media text messages, for example removing Twitter symbols (e.g. hashtag, username, etc.), non-standard words normalization, and emoticon handling. From the experimental result, we can conclude that by employing a specific pre-processing task related to tweet data characteristic we obtained more valuable result. The result obtained is better in terms of less meaningful word occurrence which is not significant in number comparing to the result obtained by just running common pre-processing tasks.

For the future work, we can develop new algorithms to automatically recognize non-standard words in Bahasa Indonesia from social media text. In addition, we also can categorize those non-standard words into two classes such as meaningful words and meaningless words.

6. References

[1] Torunoğlu D, Çakırman E, Ganiz M C, Akyokuş S and Gürbüz M Z 2011 Analysis of Pre-processing Methods on Classification of Turkish Texts *Innovations in Intelligent Systems and Applications (INISTA), 2011 International Symposium on IEEE*

[2] Hemalatha I, Varma G P S and Govardhan A 2012 Pre-processing the informal text for efficient sentiment analysis *International Journal of Emerging Trends and Technology in Computer Science (IJETTCS)* vol 1 no 2 pp 58-61

[3] Sun F, Belatreche A, Coleman S, McGinnity T M and Li Y 2014 Pre-processing online financial text for sentiment classification: A natural language processing approach *2014 IEEE Conference on Computational Intelligence for Financial Engineering and Economics (CIFEr)* London

[4] Duwairi R and El-Orfali M 2013 A study of the effects of pre-processing strategies on sentiment analysis for Arabic text *Journal of Information Science* pp 1-14

[5] Rushdi-Saleh M, Martín-Valdivia M T, Ureña-López L A and Perea-Ortega J M 2011 OCA: Opinion corpus for Arabic *Journal of the American Society for Information Science and Technology* vol 62 no 10 pp 2045-2054

[6] Srividhya V and Anitha R 2010 Evaluating Pre-processing Techniques in Text Categorization *International Journal of Computer Science and Application* vol 47 no 11 pp 49-51

[7] Ghag K V and Shah K 2015 Comparative analysis of effect of stopwords removal on sentiment classification *International Conference on Computer, Communication and Control (IC4) 2015 IEEE* pp 1-6

[8] Baldwin T, de Marneffe M C, Han B, Kim Y B, Ritter A and Xu W 2015 Shared tasks of the 2015 workshop on noisy user-generated text: Twitter lexical normalization and named entity recognition *Proceedings of the ACL 2015 Workshop on Noisy User-generated Text*, Beijing, China

[9] Hidayatullah A F 2015 Language tweet characteristics of Indonesian citizens *International Conference on Science and Technology 2015, RMUTT*

[10] Hidayatullah A F, Ratnasari C I and Wisnugroho S 2016 Analysis of stemming influence on Indonesian Tweet sentiment analysis *TELKOMNIKA (Telecommunication Computing Electronics and Control),* vol 14 no 2

Latent Dirichlet Allocation (LDA) for Sentiment Analysis Toward Tourism Review in Indonesia

[1]I R Putri, [2]R Kusumaningrum

[1 2] Department of Informatics, Faculty of Science and Mathematics, Diponegoro University, Semarang

E-mail: [1]indiati@student.undip.ac.id, [2]retno@live.undip.ac.id,

Abstract. The tourism industry is one of foreign exchange sector, which has considerable potential development in Indonesia. Compared to other Southeast Asia countries such as Malaysia with 18 million tourists and Singapore 20 million tourists, Indonesia which is the largest Southeast Asia's country have failed to attract higher tourist numbers compared to its regional peers. Indonesia only managed to attract 8,8 million foreign tourists in 2013, with the value of foreign tourists each year which is likely to decrease. Apart from the infrastructure problems, marketing and managing also form of obstacles for tourism growth. An evaluation and self-analysis should be done by the stakeholder to respond toward this problem and capture opportunities that related to tourism satisfaction from tourists review. Recently, one of technology to answer this problem only relying on the subjective of statistical data which collected by voting or grading from user randomly. So the result is still not to be accountable. Thus, we proposed sentiment analysis with probabilistic topic model using Latent Dirichlet Allocation (LDA) method to be applied for reading general tendency from tourist review into certain topics that can be classified toward positive and negative sentiment.

1. Introduction

Nowadays, the tourism industry has experienced continued growth and deepening diversification to become one of the fastest growing foreign exchange sectors in the world that acknowledged by UN and other various international organizations such as World Bank and World Tourism Organization (WTO). These dynamics have turned tourism into a key driver for socio-economic progress in the world including Indonesia. In tourism industry, compared to other Southeast Asia countries such as Malaysia with 18 million tourists and Singapore 20 million tourists, Indonesia which is the Southeast Asia's largest country have failed to attract higher tourist numbers compared to its regional peers. Indonesia only managed to attract 8.8 million foreign tourists in 2013, with the value of foreign tourists each year which is likely to decrease such as in 2013 with 9,43% of foreign tourists every year but in 2014 only 7.19%. Meanwhile, if we compared to other countries, sceneries, and attractions that had been offered in Indonesia was not much different from other countries in South-East Asia. Apart from the infrastructure problems, marketing and managing also form of obstacles for tourism growth in Indonesia. Tourism management itself cannot be separated from the presence of data sources which manually managed by institutions that are directly related to the state of foreign tourists, i.e the Ministry of Manpower and Immigration Office, Central Bureau of Statistics (BPS), Bank Indonesia, Ministry of Culture and Tourism (Depbudpar), Department of Tourism regions (Disparda), and the Indonesian Hotel and Restaurant Association (IHRA). In general, the data sources used still relies on statistical data in the form of secondary data that have been most affected by the data collection period (periodic) so that its implementation would need a huge number of resources. The difficulties encountered when the statistics then combined or compared with other numerical statistics data from other countries, as the result given an imprecise number and increase the risk of human error.

One of the technologies that have been developed to address these challenges is an online survey ever conducted by several websites like www.indo.com. But the accuracy of the online survey is still not accountable cause the user or the survey participants can not precisely be known. Other technologies such as grading and rating which are widely implemented in many cases of the online survey also did not give the results that can be assessed objectively because it is only given a fixed value based on numerical user given input. In this case, online opinion or online review can be considerate as one of most appropriate media that can be assessed more objective to represent user satisfaction or user impression through text. Online review is one of an unstructured text form that can be used in extracting data and gain more information related in user impression using sentiment analysis as one of *Information Extraction* process. Sentiment analysis is an *Information Extraction* and *Natural Language Processing* task that aims to obtain writer's feelings expressed in positive or negative review by analyzing large numbers of documents as a corpus.

The purpose of this research is to study, implement and analyze the most suitable techniques in the case of sentiment analysis for tourism review in Indonesia. In this research approach we proposed an unsupervised technique using probabilistic topic models to classify online review based on the sentiment behind those reviews. Currently, probabilistic topics models became a model which most developed and applied in a variety of research and well known to produced a pretty good effect on various applications especially in the field of classification of text and Information Retrieval [1]. As an advantage of probabilistic topics models, large data corpus can be interpreted as latent semantic in more comprehensive and stable manner [2]. Probabilistic topics models consist of Latent Semantic Analysis (LSA), Probabilistic Latent Semantic Indexing (PLSA) and Latent Dirichlet Allocation (LDA). This research use LDA method to performed classification on online review for sentiment analysis, since the LDA method designed to address problems in the LSA and the PLSA. LDA is also known to have a good ability and stability in handling large data scale because it gives the parameters as a random variable [3][4]. The rest of the paper as follows: in section 2 the overview and literature survey of some techniques which are commonly used in sentiment analysis. Then in section 3 present the proposed sentiment analysis approach method, techniques. Section 4 present the experimental process, result, and analysis. Finally, the main conclutions and future directions of research are presented in section 5.

2. Literature Survey

Sentiment Analysis is a large growing field which first brought by Liu in 2012. Sentiment Analysis also known as Opinion Mining is a field of study that provides analysis of opinion, sentiment, evaluation, attitudes, and emotions against the entities such as product, service, organization, individual, problem, events, topics, and other attributes [5]. The idea of Sentiment analysis is to determine the attitude of a writer through online text data toward certain topic or the overall tonality of a document. Sentiment analysis has been handled as natural language processing task in many ranges of field area including hotel, restaurant, tourism, devices and other consumer or marketing area. In the field of tourism, a corpus for sentiment analysis used online text data through travel sites such as TripAdvisor, Expedia, and Booking.com[8].

Various supervised or data-driven techniques such as Naïve Bayes, Maximum Entropy, SVM has been implemented for sentiment analysis in microblog data like Twitter. However some of those techniques tend to gives a rigid result, and some only give the best result in a feature based.

In this study, we design sentiment analysis using probabilistic topic model as one of topic modeling approach which assumes a document is a distribution over a topic and topic is represent as the distribution over word [1]. Probabilistic topic model is one of Topic Modelling approach which provides convenience to automatically organize, search and summarize large scale data. This model also proved to be well implemented on text fields and retrieval of information [1]. Implementation of probabilistic topic models on document will generate a set of low-dimension polynomial distribution, and for each polynomial distribution then referred as topic. Topics will be used to capture information which related to each word in the document, so each topic can be generated and extracted into a semantic structure [2][6]. Recently one of probabilistic topic models that are very popular to use, and has more comprehensive assumptions on the generation of text compare to other methods is Latent Dirichlet Allocation (LDA)[6].

The development of LDA originally intended to address problems in PLSA (Probabilistic Latent Semantic Analysis). On PLSA, the input is only determined by one hyperparameter value symbolized as beta (β). While LDA perfected with the addition of hyperparameter alpha (α), then both values will generate a latent variable value which symbolized as theta (θ) and phi (φ). Therefore the number of parameters in LDA does not grow according to the size of train corpus.

LDA method is divided into two difference processes, i.e. LDA as generative process and LDA as inference process. LDA as inference process has been characterized by the availability of corpus data as observed variable to gain latent variable including the word distribution over topic (φ) and topic proportion for each document (θ). Otherwise, LDA as generative process implemented to create corpus data or a set of document from latent variable which already known. LDA as one of the probabilistic topic model that we implemented in this study applied as inference process by using a Gibbs sampling algorithm as approximate posterior inference algorithm.

3. Research Approach

3.1. Problem formulation

In this section, we present the problem formulation for the research approach. In a case of tourism in Indonesia, we propose a sentiment analysis that will be implemented to identify positive and negative sentiment given the user or visitors through the review on a travel site in other to gain the overall sentiment tendency toward tourism in Indonesia.

Sentiment analysis is composed of three major processes i.e. Subjectivity Classification, Orientation Detection, and Opinion Holder and Target Detection. The first process, Subjectivity Classification is applied when we want to determine whether a sentence is an opinion. For the second process, Orientation Detection is applied to determine whether the opinion is positive or negative. And the last process, Opinion Holder, and Target Detection are applied to determine who is opinionated and define the object of opinion [7]. Sentiment Analysis in this research implement orientation detection process, since we aim to identify positive and negative sentiment in tourism review.

The tourism review that we used as a corpus in this research is collected from TripAdvisor since these sites applied a systematic rule in publishing review so it will provide a better data as a corpus. Furthermore, the number of review's contributors that scattered in various countries in the world became a reasoning appeal for using TripAdvisor as a corpus data since we only used review in an English language.

According to the previous explanation, an appropriate technique to applied in this issue is Sentiment Analysis with orientation detection process in the document-level task. Thus, for each review given by user or visitors through travel site will be identified and classified as each one value of sentiment.

3.2. LDA for Sentiment Analysis

As mention before, this research implement LDA as inference process for sentiment analysis especially for document-level classification task. In the case of LDA for document-level classification task, there is two main processes with each difference data for those process, training data for training process and the testing data for testing process. Training and testing data is a result of the 10-fold cross validation process that implemented to divide the data in a balanced and objective manner. In advance of the main process, for both of training and testing data will be implement basic text processing i.e. tokenization, stopwords removal and stemming. Thus text processing aims to eliminate the words, symbols or characters were give insignificant meaning for information extraction.

Tokenization is a process to eliminating characters that likely insignificant for information extraction such as white space and punctuation marks. Afterwards, the second process is stopword removal, implemented to eliminating a word that likely insignificant for information extraction. In this research, there are 277 words in the English language that are considered as stop-words, which consist 55 slang words as an addition. The last text processing, stemming is a process to derive each word to their root form. This research implement porter's stemming for stemming process.

As the input of training process is Bag of Words (BoW) which consist a set of words token from all training document that has been processed through text processing. In training process, we implement LDA as inference process over the BoW to produce a classification model. In term of implementing LDA over the BoW, there is several parameter to be defined, there are number of iteration, number of topics, hyperparameter alpha (α) and beta (β). As the result of implementing LDA as inference process, there is topic distribution and topic proportion for each document and word distribution for each topic. Subsequently, we implement harmonic mean over topic proportion for each document to compute topic proportion for each class including positive and negative class since the aims of this classification process are to classify each document into two difference sentiment classes. Afterward, the classification model for LDA as inference process in document is value of topic proportion for each class, topic proportion for each document, word distribution for each topic and topic distribution, which are used in testing process.

The testing process used testing data, which consists an extracted word for each testing document. In testing process, we applied Kullback Leiber Divergence (KLD) to measures the similarity between two distribution, the topic proportion for each class from training process as defined as classification model and the topic proportion of testing data. Thus the smallest value of KLD indicated as the sentiment class of the document.

4. Experiment and Result

4.1. Experimental Setup

As mentioned in the section before, dataset we used as a corpus is collected from TripAdvisor tourist's reviews site which consists of 100 review in English language for two sentiment classes and each class consists 50 reviews. For dividing the dataset we applied the 10-folds cross-validation, thus one folds define as testing data and the rest folds as training data and it is performed repeatedly. Accordance with 10-folds cross-validation concept, for each fold will be assigned as testing data alternately so each data will be used as testing and training data. Hereafter, the accuracy value is measured for each fold and each average of accuracy over 10-folds and is computed to indicate performance for the classification model.

The experiment is implemented using Matlab programming in Windows 8.1 Pro – 64 bit, with the following hardware specification :

- Inter Core i7- 4710HQ CPU 3.50GHz
- 4 GB of memory (RAM)
- 1 TB of Hard Disk Drive

4.2. Scenario Experiment

In this study, we compare of several parameters in LDA as a combination experiment. The first parameter is number of iteration, N, with N = 9000, 10000, 30000, 50000, 70000, 90000, 100000, or 300000. The second, number of topics, T, with T = 5, 10, 15, 20, 25, 30, 35, 40, 45, 50, 55, 60, 65, 70, 75 or 90. Another parameter is hyperparameter alpha (α), α = 0.1 or 50/T (number of topics), which for all combination hyperparameter beta defined as 0.1.

Based on those parameter combinations, we conduct two scenario of experiments to obtain the best classification model for sentiment analysis. The aim of the first scenarios is to compare each combination and obtain the information of parameter affect for each accuracy in classification model. The second scenarion aims to identify the best parameter combination that given the highest accuracy value.

4.3. Experiment Result and Analysis

previous subsection, as the result of the first scenario showed in Figure 1. Figure 1(a) shows the comparison of accuracy value for each combination of number of iteration in two alpha. As seen in Figure 1 (a) we used eight parameter number of iteration as combination, while Figure 1 (b) shows the comparation of accuracy value for 16 parameter number of topics as combination in two alpha.

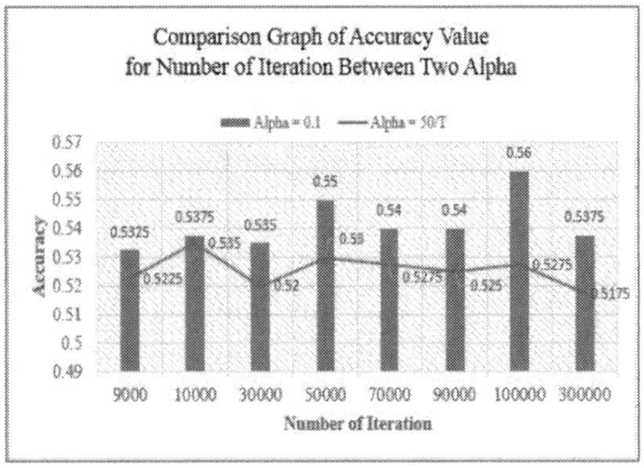

(a)

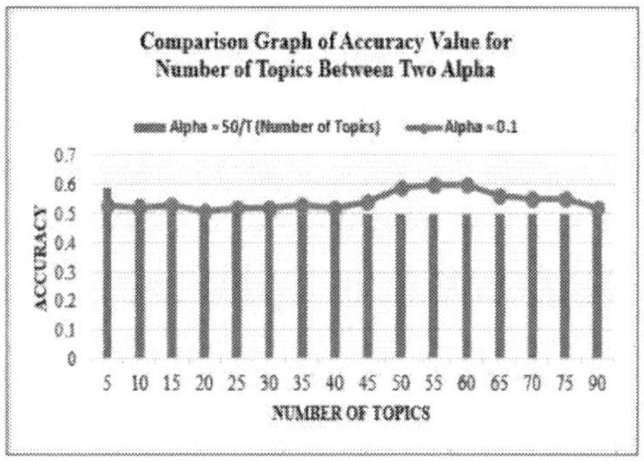

(b)

Figure 1. Comparation Graph of Accuracy Value for Number of Iteration (a) and Number of Topics (b) between two alpha.

Through comparison graph in Figure 1 (a) we can drawn a conclusion that the number of iterations is also one of parameters that play an important role to obtain a good accuracy values by satisfying the stationary distribution in model classification using LDA. Although in science there are no definite methods in determining the appropriate number of iterations to reached a good accuracy and meet the given distribution, but in LDA as inference process, the word distribution of each topic and proportion topics will be strongly influenced by the number of iterations value. As conclusion the influence of number of iterations on the value of accuracy in case of sentiment analysis using LDA can be define that when the number of iteration is too small will increase the risk of Gibbs sampling to give a false representation of the posterior distribution, while the number of iterations is too high will certainly pose a greater computational cost. As well as the number of iteration, the least amount of number of topics that did not coincide with the data and the value of hyperparameter alpha will obtain an overly broad interpretation and tend to pose a high ambiguity. Otherwise, The number of topics that are too high and did not coincide with the data and the value of hyperparameter alpha can create a very distinctive interpretation so it will be difficult to be categorized or interpreted in a certain topic. The comparison of word distribution over topic to present the ambiguity interpretation can be seen in Figure 2.

Figure 2 shows the comparation of word distribution over topic in two difference number of topic. In small amount of number of topics shows the ambiguity interpretation when "monkeys" were interpret in several time for difference topic. As seen in Figure 2 for higher number of topics, the word presented in more distinctive manner.

Figure 3 shows, higher value of hyperparameter alpha determines the interpretation level of words over topic for more evenly distributed. Thus, the higher value of alpha will reach more stable posterior distribution when the number of topic is used in smaller value. The opposite applies, smaller value of alpha will interpret each words over topics diversely so it required higher value of number of topic to meet the stable posterior distribution.

Number of Topics = 55			
TOPIC_1	0,01545	TOPIC_2	0,02113
W-1066:steal	0,01266	W-1223:up	0,04011
W-1238:view	0,01266	W-212:close	0,03129
W-83:attract	0,00663	W-404:first	0,01807
W-101:bad	0,00663	W-134:bit	0,01366
W-109:bananasoverall	0,00663	W-447:getting	0,01366
W-129:better	0,00663	W-590:jumped	0,01366
W-150:bought	0,00663	W-929:rip	0,01366
W-154:breathtaking	0,00663	W-89:avoid	0,00926
W-186:centre	0,00663	W-100:backpack	0,00926
W-381:fee	0,00663	W-345:everything	0,00926

Number of Topics = 5			
TOPIC_1	0,18435	TOPIC_2	0,25042
W-711:monkeys	0,03894	W-711:monkeys	0,03724
W-750:not	0,02730	W-415:forest	0,02676
W-884:rabies	0,02471	W-710:monkey	0,02676
W-135:bite	0,01824	W-750:not	0,02581
W-137:bitten	0,01695	W-593:just	0,01629

Figure 2. The comparation of word distribution over topic

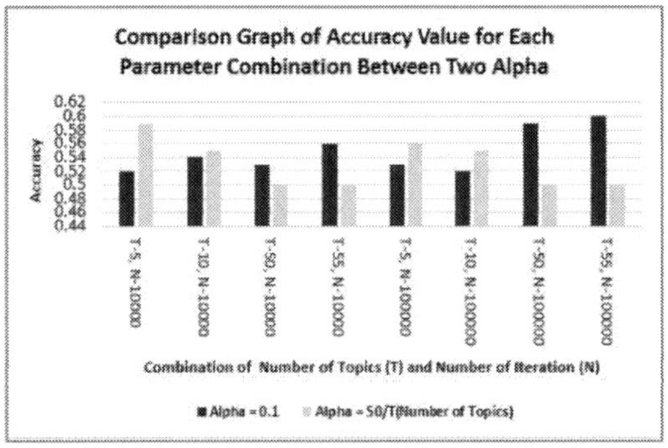

Figure 3. The Overall Comparation Graph of Accuracy Value for Each Parameter Combination Between Two Alpha

5. Conclusions

In this proposed study, we used Latent Dirichlet Allocation (LDA) as document classification method for sentiment analysis in TripAdvisor's review. The experiment was performed to evaluate the performance of the LDA-based document classification method in sentiment analysis by using 10-folds cross-validation techniques. As the result shows the best accuracy is about 60% as an average accuracy of all folds and the best accuracy is about 80% by fold 6 and 7.

Based on the experimental setup, a modification of each parameter induce different effects. A balance in each portion of the parameters is needed to obtain optimum levels of accuracy. The number of iterations or number of topics that are too small will certainly make the bias level of LDA going higher. Otherwise, the number of iterations that are too high will increase the cost computational, while the number of topics that are too high will lead LDA to difficult to interpreted the words in a certain topic.

In inversely related to the number of iterations and the number of topics, the value of hyperparameter alpha that are too small need to be balanced with the amount of the number of iterations and the number of topics to achieve the optimum accuracy. Where alpha> = 0.1 would be better used for the classification.

References

[1] Blei D 2012 Probabilistic topic models *Communications of the ACM* 55 77
[2] Arora S, Ge R and Moitra A 2012 Learning Topic Models -- Going beyond SVD *2012 IEEE 53rd Annual Symposium on Foundations of Computer Science* 2 1-10
[3] Girolami M and Kaban A 2003 On an equivalence between PLSI and LDA *Proceedings of the 26th annual international ACM SIGIR conference on Research and development in informaion retrieval - SIGIR '03* 433
[4] Lu Y, Mei Q and Zhai C 2010 Investigating task performance of probabilistic topic models: an empirical study of PLSA and LDA *Information Retrieval* 14 178-203
[5] Liu B 2012 Sentiment Analysis and Opinion Mining *Synthesis Lectures on Human Language Technologies* 5 1-167
[6] Liu Z and Li M 2013 *High performance latent dirichlet allocation for text mining* (London: Brunel University)
[7] Zagibalov T and Carroll J 2008 Automatic seed word selection for unsupervised sentiment classification of Chinese text *Proceeding of the 22nd International Conference on Computational Linguistics*
[8] Liu S, Law R, Rong J, Li G and Hall J 2013 Analyzing changes in hotel customers expectations by trip mode *International Journal of Hospitality Management* 34 359-371

IOP Conf. Series: Journal of Physics: Conf. Series **801** (2017) 012074 doi:10.1088/1742-6596/801/1/012074

Review assessment support in Open Journal System using TextRank

S R Manalu[1], Willy[1], A M Sundjaja[2], Noerlina[2]

[1] School of Computer Science, Bina Nusantara University, Jl. K. H. Syahdan No. 9, Kemanggisan, Palmerah, Jakarta, Indonesia
[2] School of Information System, Bina Nusantara University, Jl. K. H. Syahdan No. 9, Kemanggisan, Palmerah, Jakarta, Indonesia

E-mail: smanalu@binus.edu

Abstract. In this paper, a review assessment support in Open Journal System (OJS) using TextRank is proposed. OJS is an open-source journal management platform that provides a streamlined journal publishing workflow. TextRank is an unsupervised, graph-based ranking model commonly used as extractive auto summarization of text documents. This study applies the TextRank algorithm to summarize 50 article reviews from an OJS-based international journal. The resulting summaries are formed using the most representative sentences extracted from the reviews. The summaries are then used to help OJS editors in assessing a review's quality.

1. Introduction

Open Journal System (OJS) is an open-source journal management platform, created by the Public Knowledge Project in 2002 as part of the research program at the University of British Columbia, under the direction of John Willinsky [1]. OJS was designed to provide an easy to use, technical infrastructure of the entire journal publishing workflow, such as article submissions, reviewing to editing work, and online publication. In addition, OJS also demonstrate how the cost of journal publishing can be reduced significantly while maintaining a quality publication. This study focuses on improving the reviewing procedure of OJS. Particularly, it is an attempt to develop a decision support for the reviewer ratings feature in OJS.

In this study, we propose a summarization technique to support the review assessment procedure in OJS. To illustrate, when an article is submitted to OJS, the editor will take a look to see if it is suitable for the journal and if it is, the editor will proceed to assign a relevant reviewer to evaluate the paper. After the review process is completed, the editor will assign a rating to indicate the review's quality, based on various parameters defined by each journal. By employing the TextRank algorithm, an automatic summary of each review can be acquired to give the editor a quick grasp of their content.

2. Methodology

Text summarization is a process of creating a condensed form of text document, containing few important sentences that represent the whole document. TextRank [2] is an unsupervised, graph-based algorithm for automatic text summarization. It is an extractive summarization technique where several sentences will be extracted from the original document to create the summary, contrary to the abstractive summarization technique that produces a summary containing rephrased words or phrases, based on the concept or idea in the document. Currently, the extractive technique is the standard in auto summarization because it is easier to implement. TextRank uses graph-based algorithm to represent the text document by the calculating the importance of each sentence. The best scoring sentences will then be used to form the summary.

2.1. Similar methods

Some of the successful methods of extractive summarization uses supervised approach [3], by learning from a large number of labelled training data to determine how to make a good summary, such as [3,4]. However, the drawback of using such methods is the inability to adapt to a new domains or languages, as it needs an entirely new training data each time. TextRank uses an unsupervised approach where no training data is required to produce a summary, independent of the language used.

In a graph-based summarization, text is represented as a graph where nodes represent words or sentences, and edges represent the semantic relations between nodes. Graph based ranking model takes into account all the information within a graph, rather than relying only on each local node information [3]. One graph-based algorithm for summarization is LexRank [6], which uses cosine similarity and TF-IDF to calculate the weight of edges, applied to summarize multiple documents. In comparison, TextRank's measurement is based on the number of words two sentences have in common, applied to summarize single documents or to extract important keywords. While graph-based ranking model offers good result, the time it takes to complete the algorithm is proportional to the complexity of the graph, therefore it might not be a very efficient algorithm, timewise [7].

2.2. TextRank for sentence extraction

TextRank is a variant of Google's PageRank algorithm [8], which uses the concept of recommendation. When a node is connected to other node, it is said that the node is recommending that other node. As the number of recommendations a node has increases, the importance of that node also increases. In addition, the importance of each node also influences the importance of its recommendations. Formally described, the PageRank algorithm used to score of a node V_i is as follows:

$$S(V_i)=(1-d)+d*\sum_{j\in In(V_i)}\frac{1}{|Out(V_j)|}S(V_j) \tag{1}$$

where the damping factor d is set between 0 and 1, used to give a random probability of a node jumping to another random node in the graph. To illustrate, assume a random surfer that is given a random web page and keeps opening links on provided on the page, until the surfer becomes bored and decided to open a new random page. The algorithm takes into account the random factor by utilizing d, usually set to 0.85 according to Brin and Page [8]. At first, each node in the graph will be assigned the value of 1. Every node will then be computed iteratively using the PageRank algorithm until they reach convergence. Convergence will be achieved once the error rate of each node falls below a given threshold. The error rate ΔS can be approximated by calculating the difference between two successive score computation of a node V_i:

$$\Delta S=S^{k+1}(V_i)-S^k(V_i) \tag{2}$$

where $k+1$ is the current algorithm iteration. In this implementation, an undirected weighted graph is used. Unlike the PageRank algorithm, TextRank takes into account the strength of connections (edge) between each node. Therefore, a new formula that integrates weighted edges will be used, as follows:

$$WS(V_i)=(1-d)+d*\sum_{V_j\in In(V_i)}\frac{w_{ji}}{\sum_{V_k\in Out(Vj)} w_{jk}}WS(V_j) \tag{3}$$

To apply the graph-based ranking model in natural language sentence extraction, a graph representation of the text document will be built. The graph will have nodes containing text units and edges containing the relationship value between every text unit pairs. Depending on the implementation, text units could mean words, sentences, n-gram sentences, or others. The relationship between text units will also be dependent on the algorithm used to calculate it. In this study, sentence similarity is determined as the overlap of two sentences S_i and S_j, formally defined as follows:

$$Similarity(S_i, S_j) = \frac{\left|\{w_k \mid w_k \in S_i \,\&\, w_k \in Sj\}\right|}{\log(|S_i|) + \log(|S_j|)} \tag{4}$$

where a sentence is represented as n set of words: $S_i = w_1^j, w_2^j, \dots, w_n^j$. The end result is a complete graph with weighted edges and nodes, ready to be extracted to create the summary. One advantage of using a ranking model is the capability to freely adjust the summary length, as sentences used in the summary are taken from the ordered list of weight-sorted sentences. Followings are the core steps used to implement the TextRank algorithm:

- Preprocess the text document as clean text units and build the nodes of the graph using those units. In this case, each node contains one sentence.
- Decide the form of relationship between each node in the form of edges, and calculate each their weight. In this case, the weight is represented by the overlap between two sentences.
- Iterate the TextRank algorithm to calculate the weight of nodes recursively until convergence is achieved or until the iteration limit is passed.
- Extract the values of each node as sentences, sorted by their TextRank score.

3. Discussion & Evaluation

In this study we use a collection of 50 article reviews, taken with permission from Binus Business Review, an international journal hosted by the Research and Technology Transfer Office of Bina Nusantara University. Using Python as the programming language to implement the TextRank algorithm, a typical process of the summarization technique presented in Section 2 will be demonstrated. First, one of the review will be preprocessed to segmentate the review text into sentences, as shown in figure 1. Note that the number '0' is used to denote the first sentence.

[0] I am sorry that I could not say this paper is suitable for publishing at BBR.
[1] This paper does not have some components a journal should have, except literature review : I could not find Research problem, Research question, Variable and Hypothesis, Methodologies, Research Model, and also, Statistical approach etc
[2] You may need to provide some of components mentioned above
[3] About the purpose of this paper: I could not find the reason why the author want to draw a map of phenomena that occur in students who are taking classes entrepreneurship
[4] The authors may provide research problem, objective, and insight and plan to the problem-solving effort
[5] Even though this paper has literature reviews on variables, but I could not find operational definition on each variables, and theories to support each variables and measurement
[6] Author may describe operational definition on each variables and explain how to measure them
[7] And also, the author may provide a questionnaire with references to retain objectivity, and also to confirm and check how to measure variables
[8] Because this paper employed only simple average-value (MEAN) from specific group to draw a map of phenomena, there are two critical errors for the research
[9] In fact, using only simple average-value (MEAN) is not enough to estimate a characteristics of parameter without any statistics
[10] There is the error of generalization, because the authors collected some data from only one specific group (no other comparative groups) and used only simple MEAN from the group to explain and generalize some phenomena
[11] So, the results (the conclusion generalized) must be contaminated by some biased sample
[12] It is impossible to estimate some characteristics of social activities and to measure some phenomena, using only average-value (MEAN) from a specific group
[13] Because a sample is not population, and MEAN from a sample does not represent/estimate a parameter without any standard deviation and variance
[14] The author should do empirical approach to estimate a map of phenomena, using statistics such as mean, variance, standard deviation of a sample (specific group), etc.

Figure 1. Segmented review.

IOP Conf. Series: Journal of Physics: Conf. Series **801** (2017) 012074 doi:10.1088/1742-6596/801/1/012074

Next, a graph model will be built, where a node will be inserted to the graph for each sentence in the text. The nodes will be given an arbitrary initial score of 1. Each node will then be linked to each other. After that, the weight of each edge will be determined using equation (4). The more similar two sentences get, the higher the value that will be assigned to them. Edges with zero similarity value will be eliminated. The resulting graph is presented in figure 2.

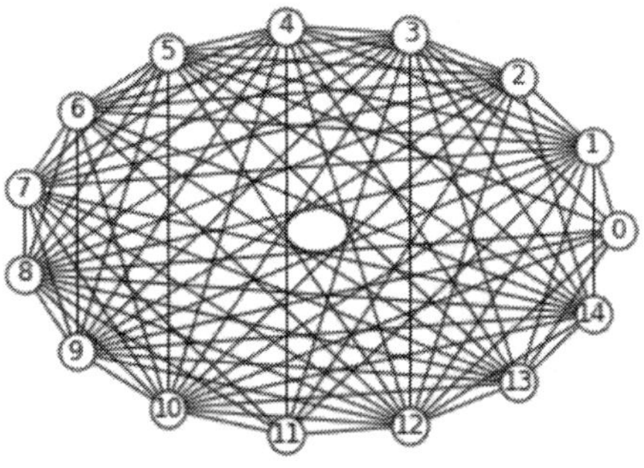

Figure 2. A highly connected graph consisting of nodes and edges. Some node pairs don't have an edge between them (0-14, 0-7, etc.) because their similarity value is zero. In total, 93 edges are built.

Once the graph is properly built, the score of each node can be calculated iteratively using equation (3), taking into account the TextRank score of each neighbour node and their similarity values. The more adjacent edges a node has, the higher the score of that node will be. In addition, the more high scoring neighbour a node has, the higher also the score of that node will be. The process ends when the error rate (2) falls below the threshold value (0.0001) as it achieves convergence. Figure 3 and figure 4 illustrate the iterative process.

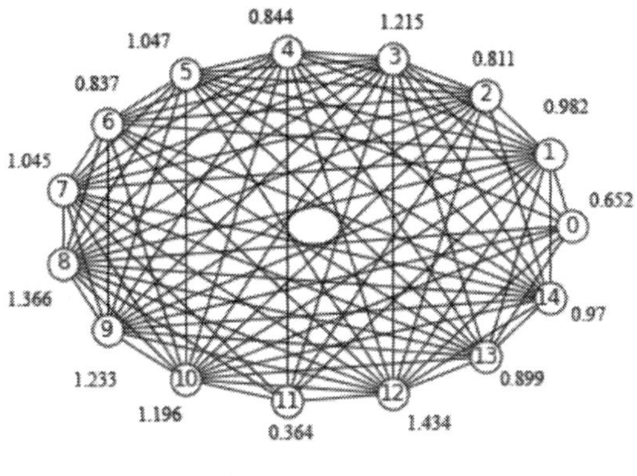

Figure 3. The graph after the second iteration of TextRank, consisting of nodes and their corresponding scores. The value was obtained after using equation (3) to calculate each node' value.

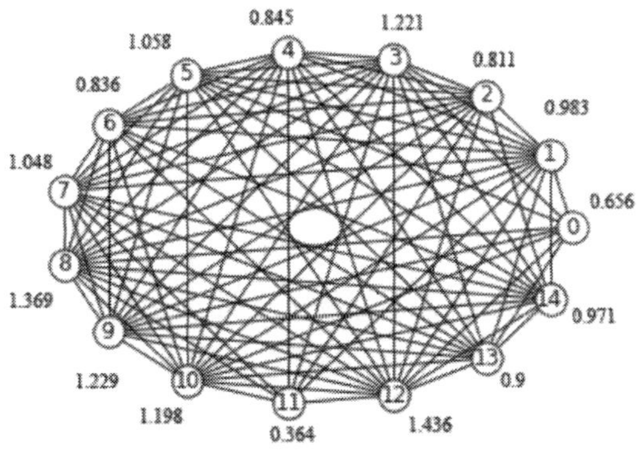

Figure 4. The graph after the third iteration of TextRank. The algorithm will keep repeating and readjusting the value of each node until convergence is achieved.

Figure 3 and figure 4 shows the score of each node during the second and third iterations, respectively. As seen above, the error rate of several nodes between the second and third iterations are still higher than the threshold value. In this case, once the error rate of each node fall below 0.0001, the algorithm will finish. Finally, to create the summary TextRank will extract the best scoring nodes from the graph. In this case, three sentences will be extracted. For this review a total of 12 iterations are run before the graph achieved convergence in Figure 5. The final result is presented in Figure 6.

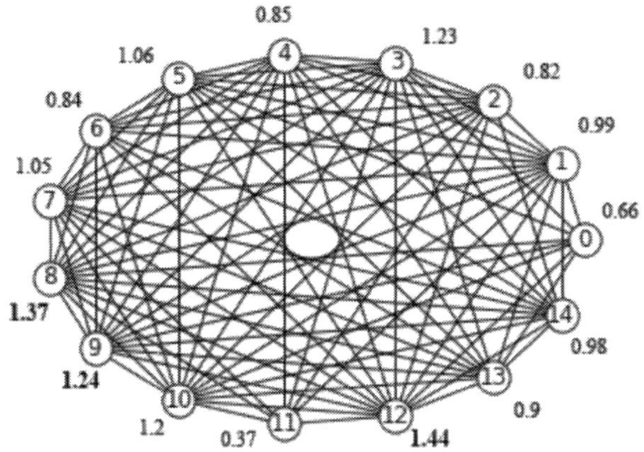

Figure 5. The final graph after the 12th TextRank iteration and reaching convergence. Node 8, 9, and 12 will be extracted to form the summary.

> Because this paper employed only simple average-value (MEAN) from specific group to draw a map of phenomena, there are two critical errors for the research. In fact, using only simple average-value (MEAN) is not enough to estimate a characteristics of parameter without any statistics. It is impossible to estimate some characteristics of social activities and to measure some phenomena, using only average-value (MEAN) from a specific group.

Figure 6. The final TextRank summary of the sample review.

After the required number of sentences is extracted, they will be presented in the summary with the same order they appeared in the original document, regardless of their TextRank score. The summary will then be ready to be read by the editor in charge of assessing the review, to provide a quick insight as to the quality of the reviews. The editor will then set the review's rating on a 1 to 5 scale. The review rating will impact a reviewer's total rating.

4. Conclusions

This paper proposes a way to develop a review assessment support for the OJS, by implementing TextRank to auto summarize reviews and giving editors a quick insight to the content of each review, thus helping them assess the reviews with less resources and time required. TextRank works by extracting the most representative sentences in a document, measured by the similarities of each sentences. TextRank is easy to implement as it doesn't require a lot of computational power nor a deep linguistic knowledge, allowing it to be used in other languages or document types.

This study suggests that TextRank as a summarization technique could be applied for various purposes. In our future work, we will develop additional review assessment features which will be used in conjunction with TextRank to provide automatic review score recommendations. We are also planning to develop more usability support for the OJS, whether it is to improve the reviewing process or other parts of the publishing procedure.

IOP Conf. Series: Journal of Physics: Conf. Series **801** (2017) 012074 doi:10.1088/1742-6596/801/1/012074

References

[1] Edgar B D and Willinsky J 2010 A survey of the scholarly journals using Open Journal Systems *Sch. Res. Commun.* **1** 1–22
[2] Mihalcea R and Tarau P 2004 TextRank: bringing order into texts *Proc. EMNLP* **85** 404–411
[3] Mihalcea R 2005 Language independent extractive summarization *Evaluation* 49–52
[4] Hirao T, Sasaki Y, Isozaki H and Maeda E 2002 Ntt's text summarization system for duc-2002 *Proc. Doc. Underst. Conf. 2002* 104–7
[5] Ko Y and Seo J 2008 An effective sentence-extraction technique using contextual information and statistical approaches for text summarization *Pattern Recognit. Lett.* **29** 1366–71
[6] Erkan G and Radev D R 2004 LexRank: Graph-based lexical centrality as salience in text summarization *J. Artif. Intell. Res.* **22** 457–79
[7] Ferreira R, De Souza Cabral L, Lins R D, Pereira E Silva G, Freitas F, Cavalcanti G D C, Lima R, Simske S J and Favaro L 2013 Assessing sentence scoring techniques for extractive text summarization *Expert Syst. Appl.* **40** 5755–64
[8] Brin S and Page L 1998 The anatomy of a large scale hypertextual Web search engine *Comput. Networks ISDN Syst.* **30** 107–17

Topic Identification and Categorization of Public Information in Community-Based Social Media

RP Kusumawardani and MH Basri

Department of Information Systems, Institut Teknologi Sepuluh Nopember (ITS)
Kampus ITS, Jl. Raya ITS, Keputih – Sukolilo
Surabaya, Indonesia

E-mail: renny.pradina@gmail.com

Abstract. This paper presents a work on a semi-supervised method for topic identification and classification of short texts in the social media, and its application on tweets containing dialogues in a large community of dwellers in a city, written mostly in Indonesian. These dialogues comprise a wealth of information about the city, shared in real-time. We found that despite the high irregularity of the language used, and the scarcity of suitable linguistic resources, a meaningful identification of topics could be performed by clustering the tweets using the K-Means algorithm. The resulting clusters are found to be robust enough to be the basis of a classification. On three grouping schemes derived from the clusters, we get accuracy of 95.52%, 95.51%, and 96.7 using linear SVMs, reflecting the applicability of applying this method for generating topic identification and classification on such data.

1. Introduction

An implication of the proliferation of the social media is its increasingly pertinent role in the society. Nowadays, not only is it used simply as a medium of self-expression and communicating with the members of one's circle of family and friends as it traditionally is, the social media has roles as divergent as a vehicle for the propagation of news, opinion, and lifestyles, coordinating efforts in disaster relief, to acting as means for organizing social and political uprisings.

This breadth of role of the social media is basically enabled due to its ability in assisting people in doing two things: to seek for information, and to share it. Due to its real-time nature and expansive network of members, people turn to social media to get and share information. As a result, the dialogues exchanged in the social media contain a wealth of information that is often the most up-to-date compared to more traditional information outlets.

However, the timeline format of the social media is often not the best for conveying and enabling its audience to get a better insight of the information content. In the case of Twitter as studied in this work, in order to get a thorough understanding on a topic, one must read all tweets in the stream of the account or search term of his interest, while it is very possible that not all of these are relevant, or that the tweet containing an important information occurring just before or shortly after the length of the stream that the person read.

Therefore, a further processing and organization of these social media posts is necessary. However, especially for Twitter, users often adopt linguistic styles that are not grammatically correct – rife with misspellings or deliberate omission of characters or words, and use of incorrect structures due to the restriction of only 140 characters allowed per tweet, and the informal nature of communication in Twitter. For these reasons, natural language processing techniques developed under the assumption of some regularity in the corpus language become unsuitable.

A method with minimal reliance to linguistic resources is thus the solution for this problem. The aim of this paper is to explore the use of such method by combining clustering for topic identification and classification techniques for organizing the tweets. In between these two stages, we perform an analysis to determine whether the clustering results in groups of tweets with meaningful content themes. To see if the groups are robust and retrievable, we then use these groups to generate labels and subsequently use these to create labeled data for the classification process.

In this paper, we are interested in applying such method to a dataset of tweets comprising dialogues in a community of city dwellers, sharing real-time information on topics such as traffic condition, latest news, and events in the city. Specifically, we use a set of tweets collected from a large community of citizens in the city of Surabaya, written mostly in Bahasa Indonesia. Using K-Means for the clustering, we develop three labeling schemes which consequently fed into Linear SVMs, resulting in accuracy of 95.52%, 95.51%, and 96.7%, suggesting that this minimal language-specific resource method could be applied successfully to organize information in public tweets, especially with regard to a specific community and regardless of the language. We also found that despite the high language irregularity, a relatively small number of tokens can be used as cue to the category of the tweets.

2. Related Work

Twitter as a means for sharing information has been widely studied from a diverge point of view; for example, for sharing health-related information [1, 2], its use in catastrophic disasters [3], and its roles in social unrest [4], among others. Its use as a probe of activities in an urban context has been discussed, e.g. by [5]. [6] discuss the use of Twitter for information sharing by and related to the police department of a big city. [7] works in the domain of e-Government, extracting complaints on public services.

Many of the previous work is related to the identification of generally trending topics in Twitter, for example works by [8, 9, 10, 11]. Others are aimed at specific topics, for example traffic information. [12] studied the extraction of traffic information in tweets in the Thailand language, with the first stage of the system filtering on words such as "traffic congestion" and "accident". Similar keyword-based filtering approach was taken by [13] for Japanese tweets. [14, 15, 16] uses hashtags or twitter accounts oriented for sharing traffic information as their data source, extracting traffic-related information using various methods, working on tweets in Bahasa Indonesia.

However, we note that in the case of urban dialogues in Twitter, the public information contained not only trending topics, but also seems to consist of multiple themes, which is quite stable from time to time. In this work we aim to investigate what these themes are, and whether they could be the basis of a classification system, which could then be used to organize these information, ultimately assisting people in getting the information relevant to them in real-time.

3. Data

The data used in this research is a five months' worth of tweets either mentioning or by the accounts @e100ss and @SapawargaSby. The first is the account of Suara Surabaya Radio, a radio station in Surabaya with a strong community actively engaging in the sharing of information on the city of Surabaya and its surrounding regions. The second account is the official account of the Government of the City of Surabaya. By using these two accounts as seeds, [17] collected a total of 30,339 messages, spanning from the beginning of September 2015 until early February 2016. Table 1 shows some examples of these tweets.

Table 1. Example tweets

No.	Message
1.	@e100ss mohon pihak terkait mengenai jalan mastrip yang banyak berlubang dan sangat berbahaya bagi roda dua pada saat musim hujan.
2.	@e100ss jln.girilaya ke arah dukuh kupang macet poooll...silahkan ambil jalan lain
3.	@e100ss agak kecewa sekali terhadap pegawai kecamatan thdp pelayanan.mengurus surat pengantar e-ktp
4.	RT @satpolppsby: Tim Odong Odong melepas baner yg salah penempatan di jl. Ngagel Jaya Selatan @e100ss @SapawargaSby

None of the messages in Table 1.can be considered to contain entirely correct Bahasa Indonesia. Each message contains multiple syntactic errors and variations in spelling. Especially, many abbreviations are employed on the basis of the assumed understanding of the reader. For example, the word 'jalan' – road – correctly spelled in tweet no. 1 is written as 'jln' in tweets no. 2, and as 'jl.' in tweet no. 4. Only in these four tweets we can find 'silakan' – please – written as 'silahkan', 'terhadap' – to – written as 'thdp', and 'banner' – banner – not even and Indonesian word and written as 'baner'. We can also find 'poooll' – not a word, just an emphasis, and employing repetition of characters, where there should have been none. Consequently, the number of token types become very large. More importantly, since many tokens appears very rarely or even only once, these could cause failure in finding a good generalization in the classification stage of this method, resulting in overfitting.

4. Token Variability and Minimum and Maximum Document Frequency

Tokenizing all 30,339 tweets in the dataset results in 28,201 distinct token types with a total number of tokens of 371,182. This results in a comparable ratio of distinct token type to number of tokens of 0.076 when compared to other corpora of Bahasa Indonesia. For example, the ratio of distinct token type to number of token in the Indonesian classic literature Sengsara Membawa Nikmat [18] is 0.09. It should be remembered, however, that considering that due to its shortness, the language in tweets are usually quite simple, such high variation of token types is more likely due to artefacts such as variations in spellings, both inadvertently or deliberately to reduce the number of characters necessary, rather than being actually informative.

Therefore, it is important to choose boundaries for the minimum document frequency appearance of tokens. If a token occurs only rarely, then it would not be something that reflects a category. However, if a tweet contains only rare tokens, it would have an empty representation. Obviously, the number of 'empty' tweets need to be limited. Using unigram + bigram as features, a min_df value of 20 was chosen, reducing the number of terms to 2052, while retaining 95% of the tweets; upon closer inspection, tweets with empty representation typically consist of hyperlinks, which are unique.

Conversely, we also might need to set a maximum limit for document frequency, since if a token appears in too many tweets, then we would not be able to infer the topic from it, thus rendering it useless in the topic identification in the clustering stage. However, the dataset shows that the highest document frequency for any token in the data is 3800, less than 13% of the number of documents, thus in this work we do not set a limit on maximum document frequency.

5. Topic Identification based on Clustering

The remaining 28,837 tweets are then clustered using the unigram and bigrams as features. Each feature is represented by its Tf-Idf value, the combination of term frequency (tf) and inverse document frequency (idf) [19, 20]. The tweets are primarily in Indonesian, but English words are occasionally also used. Therefore, the stopwords used are a mix of Indonesian [21] and English from the NLTK [22] used in this work.

For clustering the tweets, we use the K-means algorithm [23], a very popular algorithm for clustering due its speed and simplicity [24, 25]. Basically, it has a single parameter to set: k, the number of clusters to find. Using the elbow method [26] we plot the total sum of squared error (SSE) of the clustering against the k value used to create the clusters. However, as shown in Figure 1, the value of SSE is consistently gradually decreasing with the increase of k, thus indicating that the data does not have a natural set of grouping.

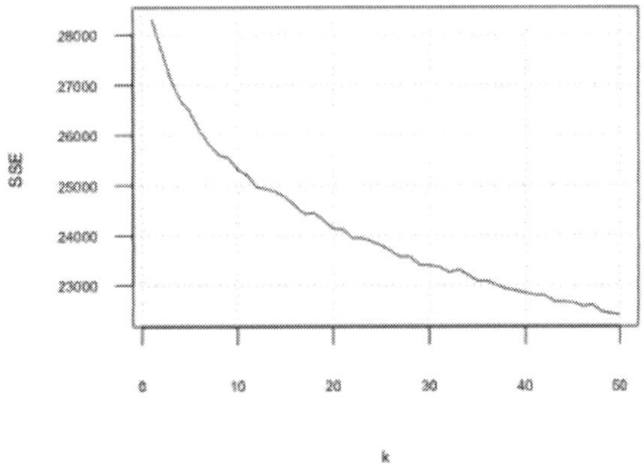

Figure 1. SSE of the clusters on different values of k

Therefore, we take the approach of creating a large number of clusters, then based on the most frequent words and the semantic content of each cluster, group together clusters that are deemed to be most similar, in accordance the merge-two-clusters post-processing approach of [27]. An initial k value of 45 was chosen and the result of topic analysis of these clusters is shown in Table 2.

Table 2. Resulting clusters

Clust #	Topic	# of members	Clust #	Topic	# of members	Clust #	Topic	# of members
1	traffic congestion	362	**16**	(spam)	138	**31**	traffic jam	339
2	traffic light	798	**17**	demonstration	276	**32**	loss	335
3	electricity	1041	**18**	to Surabaya	353	**33**	car	547
4	toll road	916	**19**	weather	780	**34**	Sidoarjo	532
5	weather	230	**20**	water	478	**35**	clear roads	600
6	greetings	409	**21**	clear roads	739	**36**	accident	256
7	Ahmad Yani	184	**22**	electricity	694	**37**	police	370
8	Surabaya	7210	**23**	(spam)	143	**38**	weather	198
9	Malang	821	**24**	lane	420	**39**	truck	708
10	electricity	446	**25**	government performance	433	**40**	clear roads	194
11	crowded	962	**26**	fire	314	**41**	congested	564
12	weather	374	**27**	bridge	231	**42**	child issues	218
13	motorcycle	460	**28**	Krian	212	**43**	weather	1268
14	congested	520	**29**	Gresik	315	**44**	economy	171
15	weather	394	**30**	traffic jam	1438	**45**	toll road	386

Note in Table 2 that some clusters have the same topic. We merge these and arrive at 31 topics. Looking at the content of the tweets belonging to each topic, it is obvious that some are the subtopics of larger topics. Therefore, we create three-layer hierarchy to group these topics from the most general to the more specific. This hierarchy is shown in Figure 2.

Figure 2 shows that a major theme of the dialogues around @e100ss and @SapawargaSby is about traffic. This is indeed the case, especially for @e100ss, since this is the kind of information most often shared by the community built around this account. In the topic of traffic, people may report about traffic condition in a particular area; whether there is traffic jam, congestion, or clear roads. @e100ss and @SapawargaSby are also often used by the public to voice their complaints on public services, especially on electricity and water. People also reports on general news and events, such as demonstrations, fire, and regional news.

6. Classifying the Tweets

The topics identified in Figure 2 gives us an idea of the types of public information shared in the tweets. We use the clustering results to create a dataset for training and testing classifiers based on the topics and topic hierarchy. This dataset is created by attaching labels to each tweet in a cluster with the topic category of that dataset.

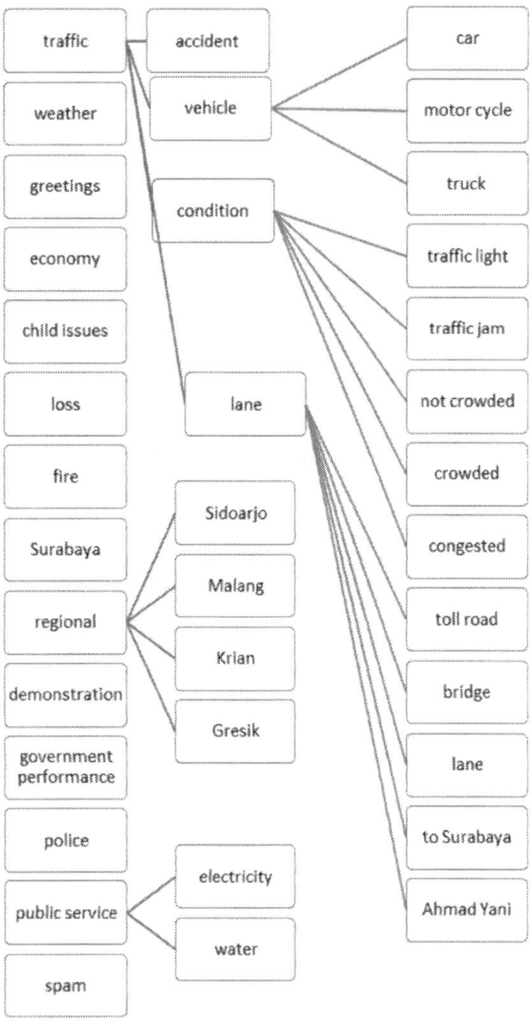

Figure 2. Topic hierarchy identified from the analysis of the clusters

As the topics form a hierarchy of different specificities, we created three schemes for classifying the data. The first scheme consists of all the topics found, thus there are 31 labels. In Figure 2, these are the result of following each topic in the first layer to the leaf in the last layer in the topic tree. In the second scheme, the labels are the topics up to the second layer of the tree, thus there are 21. Lastly, the third scheme consist of 14 topics in the first layer in Figure 3. These three schemes are used to create three datasets, as summarized in Table 3.

Table 3. Summary of the labelling schemes derived from the identified topics hierarchy

Data set	Topic Specificity	# of Topics	Example Labels
1	Up to layer 3	31	traffic_accident, traffic_vehicle_car, traffic_condition_congested, weather
2	Up to layer 2	21	traffic_accident, traffic_vehicle, traffic_condition, weather
3	Layer 1 only	14	traffic, weather, greetings, economy

For the classification, we explore the use of several algorithms popularly used for text classification: Naïve Bayes (see [28] for a thorough discussion), logistic regression [29], and SVMs [30] with RBF and linear kernels. The parameters used are the default parameters of the scikit-learn [31] implementation we use. In particular, the gamma value of the RBF kernel of the SVM is $1/(number\ of\ features)$, which in this case is $1/2052 = 4.8733 \times 10^{-4}$. The default value for the C parameter of the soft-margin SVM classifier [30] is 1.0.
Applying these algorithms to dataset 1 results in accuracy values shown in Figure 3. It shows that the best accuracy is 95.20%, given by SVM-linear, a stark contrast to Naïve Bayes with 32.13% and SVM-RBF with 24.48%. The performance of logistic regression of 94.10% is actually not too far behind. However, we choose SVM-linear because it has the soft-margin parameter C that could still be optimized. For datasets 2 and 3, we get similar accuracy profiles, thus we also choose SVM-linear for these datasets.

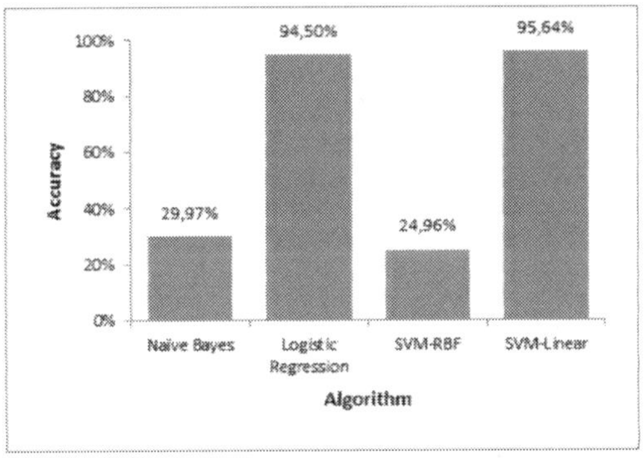

Figure 3. Accuracy of classifiers from dataset 1 automatically derived from topic labels on different algorithms

Subsequently, for each dataset, we optimize the C parameter of soft-margin SVM-linear to seek the best performance value possible. Figures 4, 5, and 6 plots the accuracy obtained by systematically varying the value of C using the grid search method [32]; each for datasets 1, 2, and 3, respectively. These figures shows that the categorization schemes, derived from topics identified from the clusters has successfully captured the themes in the public information shared by the community Twitter users revolving around the accounts of @e100ss and @SapawargaSby. Specifically, the highest performance of 96.97% is obtained with the third labeling scheme, occurring at the C value of 2; this is likely to do with the fact that this scheme has the least number of classes. The second best performance is from dataset 1, which is derived from the original clusters found: 95.92% at the C value of 5. Lastly, the second labeling scheme yields the lowest maximum accuracy of 95.51% at the C value of 2; a very good performance nevertheless.

In addition to accuracy, we calculated the precision, recall, and create the confusion matrix for the classifier, and these confirms that the performance of the classifier is indeed very satisfactory. Since all three levels categorizing data works very well, in practice, for a particular application, we could choose any labeling scheme out of the three best on the level of granularity of topic desired. We would also like to point out that since the method we use needs a minimal level of language-specific resources, it could easily be used for other languages as well.

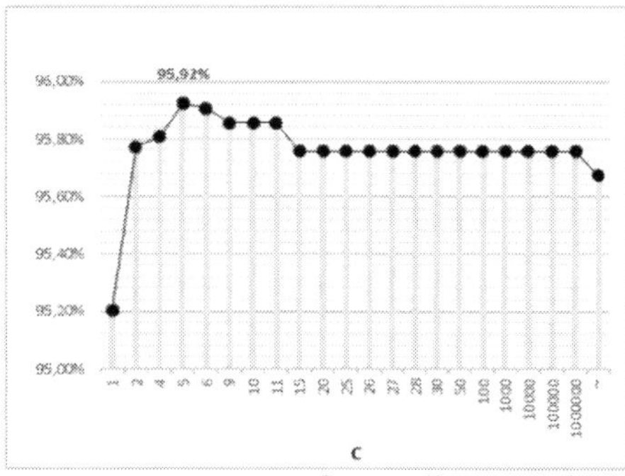

Figure 4. Accuracy of SVM-linear on dataset 1

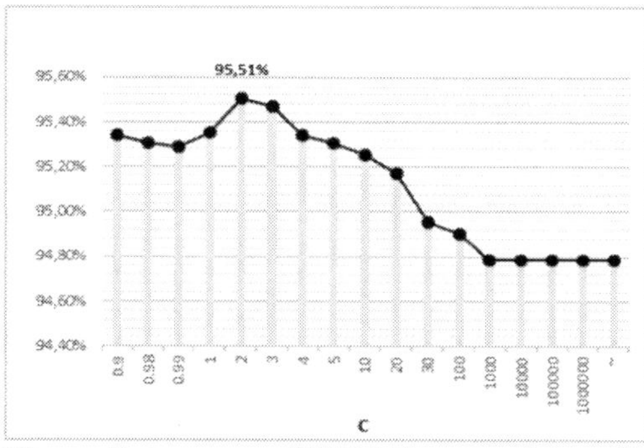

Figure 5. Accuracy of SVM-linear on dataset 2

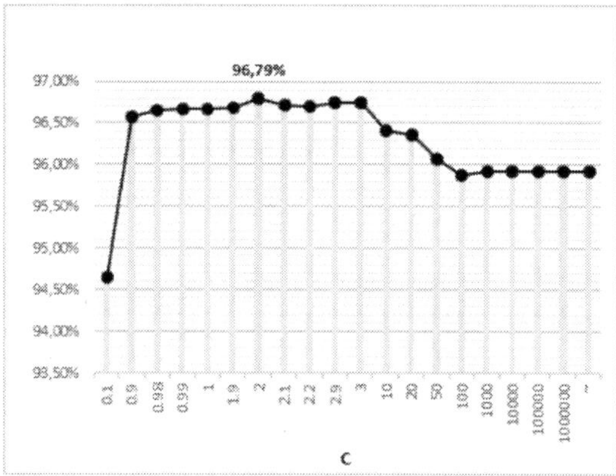

Figure 6. Accuracy of SVM-linear on dataset 3

7. Conclusion

In this paper, we described a work on a semi-supervised method for identifying topic and classifying tweets containing public information shared by citizens. We first cluster these tweets using the K-means algorithm using a large number of k, since the data does not seem to naturally have a value for k for which the clustering is optimal. We subsequently analyze the clusters to find the topics of each cluster. Clusters with similar topics are merged; furthermore, we found that some topics have hierarchical relations to each other. Based on this, we create three datasets which corresponds to three levels of granularity on the specificity of the topics, by automatically labeling each tweet with the word descriptive of the category of the cluster. Among the algorithms we explored for creating the classifier, SVM-linear gives the best performance. Optimizing for the soft-margin constant c, we get accuracy of more than 95% for all of the three levels of topic granularity, which shows that our method works very well in finding the topic and categorizing tweets containing public information shared over the social media.

The work in this paper is primarily aimed is to study whether it is possible to find a meaningful identification of topics of public information in the tweets and to retrieve an automatic categorization based on those topics. Having demonstrated that the tweets do have some identifiable internal organization of topics within them, it is interesting to explore the use of topic modelling with Latent Dirichlet Allocation [33]. LDA is a method which in recent years has been very popular for learning the latent topics in a corpus, using the assumption that the generation of words could be modelled by a Dirichlet distribution. One variation of this algorithm that might be relevant is its use with latent feature word representations [34], since it offers improvement in topic identification in cases where the documents are short.

References

[1] D. Scanfeld, V. Scanfeld and E. L. Larson, "Dissemination of health information through social networks: Twitter and antibiotics," *American journal of infection control,* vol. 38, no. 3, pp. 182-188, 2010.

[2] S. J. Sullivan, A. G. Schneiders, C.-W. Cheang, E. Kitto, H. Lee, J. Redhead, S. Ward, O. H. Ahmed and P. R. McCrory, "'What's happening?' A content analysis of concussion-related traffic on Twitter," *British Journal of Sports Medition ,* 2011.

[3] F. Toriumi, T. Sakaki, K. Shinoda, K. Kazama, S. Kurihara and I. Noda, "Information Sharing on Twitter During the 2011 Catastrophic Earthquake," in *22nd International Conference on World Wide Web*, Rio de Janeiro, 2013.

[4] E. Tonkin, H. D. Pfeiffer and G. Tourte, "Twitter, Information Sharing and the London Riots?," *Bulletin of the American Society for Information Science and Technology,* pp. 49-57, 2012.

[5] F. Kling and A. Pozdnoukhov, "When a city tells a story: urban topic analysis," in *20th International Conference on Advances in Geographic Information Systems*, Redondo Beach, 2012.

[6] T. Heverin and L. Zach, "Twitter for City Police Department Information Sharing," *ASIST 2010,* 2010.

[7] D. Anggareska and A. Purwarianti, "Information Extraction of Public Complaints on Twitter Text For Bandung Government," *2014 International Conference on Data and Software Engineering (ICODSE),* pp. 80-85, 2014.

[8] L. M. Aiello, G. Petkos, C. Martin, D. Corney, S. Papadopoulos, R. Skraba, A. Goker, Y. Kompatsiaris and A. Jaimes, "Sensing Trending Topics in Twitter," *IEEE Transactions on Multimedia,* vol. 15, no. 6, pp. 1268 - 1282, 2013.

[9] J. Benhardus and J. Kalita, "Streaming trend detection in Twitter," *International Journal of Web Based Communities,* vol. 9, no. 1, pp. 122-139, 2013.

[10] H. Rosa, J. P. Carvalho and F. Batista, "Detecting a Tweet's Topic within a Large Number of Portuguese Twitter Trends," *3rd Symposium on Languages, Applications and Technologies,* vol. 38, pp. 185-199, 201.

[11] E. Miller, K. Vodrahalli and A. Lee, "Estimating Trending Topics on Twitter with Small Subsets of the Total Data," 2015.

[12] N. Wanichayapong, W. Pruthipunyaskul, W. Pattara-Atikom and P. Chaovalit, "Social-based traffic information extraction and classification," in *11th International Conference on ITS Telecommunications*, Petersburg, 2011.

[13] T. Sakaki, Y. Matsuo, T. Yanagihara, N. P. Chandrasiri and K. Nawa, "Real-time event extraction for driving information from social sensors," *Proceedings of the 2012 IEEE International Conference on Cyber Technology in Automation, Control, and Intelligent Systems,* 2012.

[14] S. K. Endarnoto, S. Pradipta, A. S. Nugroho and J. Purnama, "Traffic Condition Information Extraction & Visualization from Social Media Twitter for Android Mobile Application," *2011 International Confrence on Electrical Engineering and Information,* 2011.

[15] M. Hasby and M. L. Khodra, " Optimal path finding based on traffic information extraction from Twitter," *In ICT for Smart Society (ICISS), 2013 International Conference,* pp. 1-5, 2013.

[16] A. Wibisono, I. Sina, M. A. Ihsannuddin, A. Hafizh, B. Hardjono, A. Nurhadiyatna, W. Jatmiko and P. Mursanto, "traffic intelligence system architecture based on social media information," *International Conference on Advanced Computer Science and Information Systems (ICACSIS),* pp. 25-30, 2012.

[17] D. F. Priyambada, N. Aini and R. Prasetianto, "Rancang Bangun Aplikasi Backend dan Web Services Penyampaian Laporan Masyarakat Berdasarkan Crowdsourcing melalui Jejaring Sosial Twitter," Juli 2016.

[18] T. S. Sati, Sengsara Membawa Nikmat, Jakarta: Balai Pustaka, 1929.

[19] H. P. Luhn, "a statistical approach to mechanize encoding and searching of literary information," *IBM Journal,* 1957.

[20] K. Spärck Jones, "A Statistical Interpretation of Term Specificity and Its Application in Retrieva," *Journal of Documentation,* vol. 28, pp. 11-21, 1972.

[21] D. H. Wahid, Februari 2016. [Online]. Available: https://github.com/masdevid/ID-Stopwords/blob/master/id.stopwords.02.01.2016.txt. [Accessed 18 August 2016].

[22] Bird, Steven and E. K. Edward Loper, Natural Language Processing with Python, O'Reilly Media Inc., 2009.

[23] S. P. Lloyd, "Least Squares Quantization in PCm," *IEEE Transactions on Information Theory,* vol. 28, no. 2, p. 129, 1982.

[24] D. Arthur and S. Vassilvitski, "How Slow is the k-Means Method?," in *Proceedings of the twenty-second annual symposium on Computational geometry*, Sedona, 2016.

[25] J. Issa, "Performance characterization and analysis for Hadoop K-means iteration," *Issa Journal of Cloud Computing,* 2016.

[26] R. L. Thorndike, "Who Belongs in The Family?," *Psychmetrika,* vol. 18, no. 4, p. 267, 1953.

[27] G. Ball and D. Hall, "A Clustering Technique for Summarizing Muktivariate Data," *Behavior Science,* vol. 12, pp. 123-155, 1967.

[28] D. D. Lewis, "Naive (Bayes) at Forty: The Independence Assumption in Information Retrieval," in *10th European Conference on Machine Learning*, London, 1998.

[29] D. R. Cox, "The Regression Analysis of Binary Sequences," *Journal of the Royal Statistical Society. Series B (Methodological),* vol. 20, no. 2, pp. 215-242, 1958.

[30] C. Cortes and V. Vapnik, "Support-Vector Networks," *Machine Learning,* vol. 20, no. 3, pp. 273-297, 1995.

[31] F. Pedregosa, G. Varoquaux, A. Gramfort, V. Michel, B. Thirion, O. Grisel, M. Blondel, P. Prettenhofer, R. Weiss and V. Dubourg, "Scikit-learn: Machine Learning in Python," *Journal of Machine Learning Research,* 2011.

[32] C.-W. Hsu, C.-C. Chang and a. C.-J. Lin, "A Practical Guide to Support Vector Classification," Taipe, 2016.

[33] D. M. Blei, A. Y. Ng and M. I. Jordan, "Latent Dirichlet Allocation," *Journal of Machine Learning Research 3,* pp. 993-1022, 2003.

[34] D. Q. Nguyen, R. Billingsley, L. Du and M. Johnson, "Improving Topic Models with Latent Feature Word Representations," *Transactions of the Association for Computational Linguistics, vol. 3,* p. 299–313, 2015.

IOP Conf. Series: Journal of Physics: Conf. Series **801** (2017) 012075 doi:10.1088/1742-6596/801/1/012076

Comparison of Document Index Graph Using TextRank and HITS Weighting Method in Automatic Text Summarization

Fadhlil Hadyan[1],, Shaufiah[2] and Moch. Arif Bijaksana[3]

School of Computing, Telkom University, Bandung, Indonesia

.

[1]fadhlilhadyan0@gmail.com, [2]shaufiah@telkomuniversity.ac.id, [3]arifbijaksana@telkomuniversity.ac.id

Abstract. Automatic summarization is a system that can help someone to take the core information of a long text instantly. The system can help by summarizing text automatically. there's Already many summarization systems that have been developed at this time but there are still many problems in those system. In this final task proposed summarization method using document index graph. This method utilizes the PageRank and HITS formula used to assess the web page, adapted to make an assessment of words in the sentences in a text document. The expected outcome of this final task is a system that can do summarization of a single document, by utilizing document index graph with TextRank and HITS to improve the quality of the summary results automatically.

1. Introduction

The digital era enables a lot of people to obtain article or news very easily through internet using their computer like pc or tablet [1]. Especially, we use these devices constantly in our entire day and a lot of article is available in the internet. This results in a large number of text data generated every day. although we have a large number of text, utilization of this data is very low, especially in a long text. It's not easy to obtain information from a long article with a short time, thus we need to take more time to read the text. It is necessary to automatically extract core information from an article, since it will reduce the time we need to obtain the information.

Automated text summarization aims to generate summary from a document or new in this research, reducing length of a news while keeping the same core information as the original news. The goal of document summarization is to expressing the original document with fewer words [2]. There are two types of document summarization, which is extractive and abstractive [3]. Extractive based summarization identifies and pick important sentence from document and combine the sentences into summary. Abstractive summarization attempt to generating new sentence, but currently this type off summarization is still difficult to implement. Majority of automated summarization system is still using extractive method [4]. This research is going to develop an extractive automated summarization system using document index graph.

Document index graph is a method to represent document as graph. With this representation a sentence will not be scored based on words. Sentence scoring is based on similarity of phrase [5]. Since phrase will be represented by connected nodes in graph. This ability to represent phrase is the advantages of document index graph. Graph scoring algorithm, TextRank and HITS, will be used to score the graph produced by DIG method. this scored nodes will be used to score sentence and identify important sentence.

2. Theoretical Review

Sentence scoring is an attempt to identify important sentences of a document. Sentence scoring can be divided into two process, graph construction and graph scoring. Graph scoring can utilize several algorithms like HITS or lexrank. This paper is using TextRank and HITS to score graph constructed by document index graph. Sentence score will determine by the score of each word contain in that sentence. Sentence with high score will be included as a result summary. Graph construction is using document index graph algorithm to create graph from word in a document.

2.1. Document index graph

Document index graph constructing graph from words inside a document, stopword will not be include in graph construction. DIG is a directed and weighted graph consist of nodes and edges. Nodes represent word in document, where edges represent occurrence of two words in the document [6]. As a weighted graph, each edges have its own weight. Edges weight is the frequency of occurrence of the two words in sequence.

In single document summarization, graph is constructed for each document. Since we assume there is no correlation between document we construct graph for single document. In each document index graph iterate through each word. The construction of document index graph is following this two rules:

- If a word never occurred before, then create new node (node A) and draw edge from node B (node in previous iteration) to node A, if node A is not the first word in the sentence. the weight of new edge is 1.
- If a word has occurred before, find the node that containing current word (node A) and draw edge from node B (node in previous iteration) to node A, if node A is not the first word in the sentence. If node A and B already connected by an edge, increment its edge by 1.

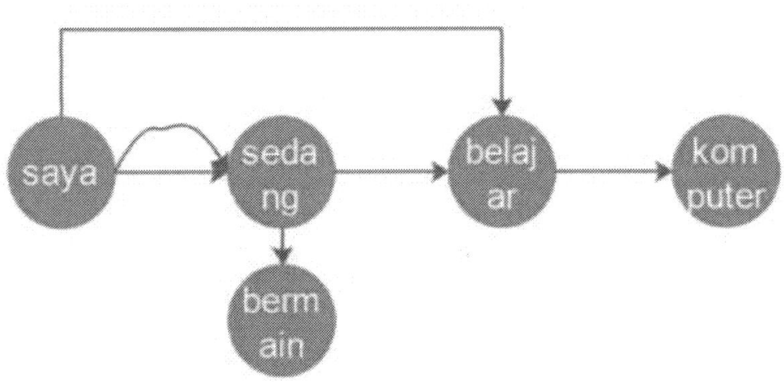

Figure 1. Example of DIG result

Document index graph method is able to score sentence based on words or phrase inside the sentence [7]. In figure 1, phrase is represented by two connected nodes with high edge weight. A phrase is two nodes that occurred in sequence multiple times. with this representation, phrase can be stored inside graph alongside all other word, thus there's no need to store phrase information separately and reduce resource requirement [5].

2.2. *TextRank*

TextRank is a graph scoring algorithm, TextRank attempt to determine score for all nodes based on information contained in the graph itself [8]. TextRank algorithm in based on web-site scoring algorithm PageRank, the original page rank is used to handle unweighted graph. since TextRank is going to score weighted graph, the original formula is changed to enable TextRank handles weighted graph.

Formula (1) [8] is an adapted PageRank formula to integrate edges weight.

$$PR(A) = (1-d) + d\sum_{i=1}^{n}(\frac{PR(Ti)}{C(Ti)}) \quad (1)$$

- PR(x) = weight of nodes
  n = number of nodes in graph
- C(x) = number of out edge from node x
- D = damping factor, a value between 0 and 1, dampening factor is usually set to 0.85 [9].

In formula (1), score of PR is determined by other PR score, thus PR score need to be initialize with a score, the initial score can be any number except for 1, since PR score will converge to 1 [10]. Number of iteration needed to complete TextRank algorithm is define by a convergence threshold, the threshold used in this paper is 0.0001 [8]. TextRank algorithm will stop after error is lower than the threshold, error in the difference between score of a same node in different iteration [7].

formula (2) [7] describe how error will be calculated:

$$e = PR^{i+1} - PR^{i} \quad (2)$$

2.3. *HITS*

HITS or hyperlinked induced topic search is another graph scoring algorithm, this algorithm using two score, authority and hubs, instead of a single score like in TextRank [11]. HITS algorithm will determine nodes wit high authority (node with large number of income edges) and node with high hub (node with large number of outgoing edges). There are two different formulas that will used to calculate each score.

Formula (3) and (4) [11] is HITS formula used to score graph:

$$HITS_A(V_i) = \sum_{V_j \in In(V_i)} HITS_H(V_j) \quad (3)$$

$$HITS_H(V_i) = \sum_{V_j \in In(V_i)} HITS_A(V_j) \tag{4}$$

- $HITS_A(x)$ = nodes authority score
- $HITS_H(x)$ = nodes hub score

HITS is an iterative algorithm formula (3) and (4) will use in sequence, each node will have starting assigned value. Similar to TextRank, HITS calculate each score until convergence below a certain threshold [12].

2.4. Rouge-n

The output result of a system need to be evaluate to analyse the system quality and improvement that could be made. ROUGE or recall-oriented understudy for gisting evaluation is an evaluation method that measure text similarity based on n-gram similarity between text [13]. ROUGE is a unit to compared similarity between system generated summary and reference summary or the gold standard. Formula (3) [13] describe how ROUGE calculate recall value of system generated summary by number of n-gram found in both summary (system and reference) divided by total number of n-gram in reference summary.

$$ROUGE - N = \frac{\sum_{C \in RSS} \sum_{gram_n \in C} Count_{match}(gram_n)}{\sum_{C \in RSS} \sum_{gram_n} Count(gram_n)} \tag{5}$$

3. System Architecture

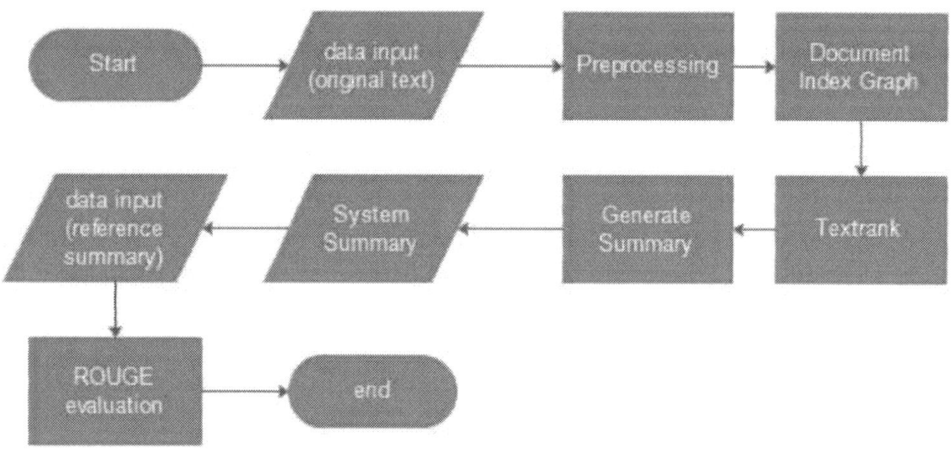

Figure 2. System architecture

News dataset will process using python programming language, and evaluate using ROUGE toolkit. News file will first be preprocessed to clean text before transformed into graph. the news file will split into list of sentence and each sentence will have tokenized in to list of words. Stopword removal applied in the list to remove common words that not significance in news (e.g. a, is, it, etc.). stemming

IOP Conf. Series: Journal of Physics: Conf. Series **801** (2017) 012075 doi:10.1088/1742-6596/801/1/012076

process will remove suffix (e.g. -ing, -ly, -er) to transform word to its root. The result of stemming will allow different word with same root score the same.

Preprocessed text will be transformed into graph. each sentence will iterate, thus last word in the sentence will not connected to first word in the next sentence. this process allows possibility of dangling nodes, which is a node that doesn't have out edge. TextRank algorithm cannot handle dangling nodes by itself, thus an extra process is needed to handle this condition. Dangling nodes can be handled by connecting it to all other nodes in graph [14]. The result graph will be transformed again into adjacency matrix to further scored using TextRank algorithm. Scored word then use to score each sentence. summary will generate by picking sentence with highest score.

The system generated summary will evaluated using rouge-2. system summary will compare to reference summary from news. Average recall of each summary will become the evaluation score of the system.

4. Testing and Analysis

4.1. Dataset

The dataset used in this research from MultiLing 2015 also used in [15], this research use 30 document with Indonesian language from the dataset. The original and reference text is stored in txt format in two separate folders. original and reference document using a same file name with slight difference after the filename, '_body.txt' for original text and '_summary.txt' for reference summary. Each reference summary has average range at 218.33 words.

4.2. Evaluation Result and Analysis

The system will be evaluate using rouge-1, also used in [15], which resulting recall, precision and f-measure. Evaluation will perform on 3 scenarios with 3 different compression rate, 30%, 40%, and 50%. The result of ROUGE-1 evaluation on three scenarios is show on table1.

Table 2. TextRank Evaluation result.

Summary Length	Recall	Precision	F-measure
50	0.074069	0.332153	0.120741
100	0.147472	0.327876	0.202573
150	0.209757	0.313522	0.250186
200	0.265075	0.297331	0.279062
250	0.309068	0.277654	0.291242

Table 3. HITS Evaluation result.

Summary Length	Recall	Precision	F-measure
50	0.078226	0.349438	0.127522
100	0.14939	0.331628	0.205178
150	0.217959	0.320518	0.258361
200	0.273328	0.302407	0.28592
250	0.322984	0.287407	0.302845

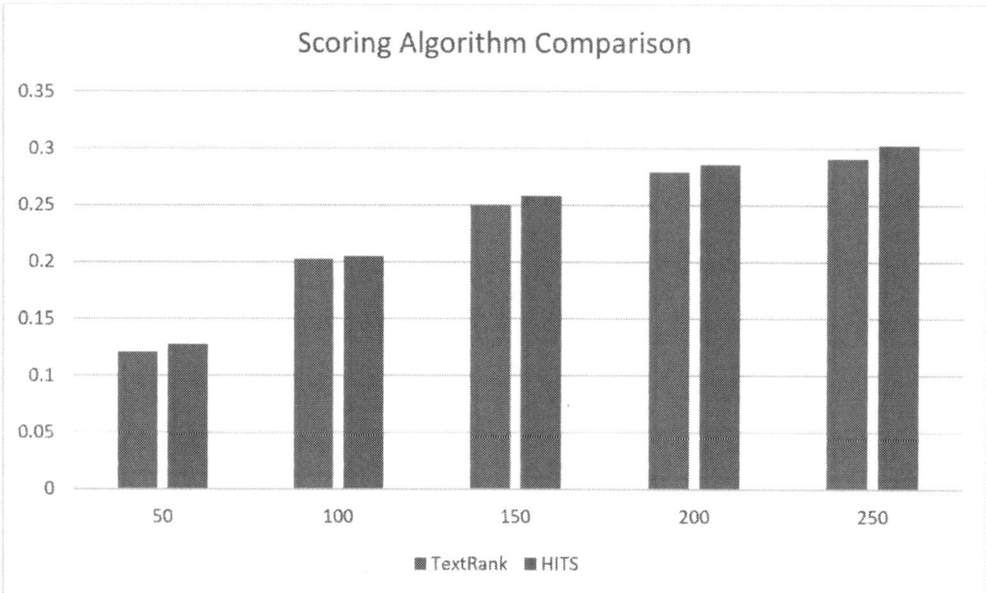

Figure 3. Scoring Algorithm Comparison

Based from table-2 and table-3, f-measure of the system is increasing as the number of words, the result is consistent between two systems. While recall following f-measure trend, precision have the opposite trend. Precision of the systems decrease as the length of the summary increased, this follow the usual precision-recall trend. The highest f-measure from both comes from summary with 250 word. The overall result is pretty low compared to the system mentioned in [15]. The system developed in this research is general text summarization system, while the reference summary is topic-focused summaries. The best f-measure score coms from HITS system with f-measure of 0.302845, slightly higher that TextRank best result of 0.291242.

Figure 3 show the comparison between TextRank and HITS on all summary length. The trend consist HITS have the highest score in all summary length, but not a large difference. The smallest difference is in 100 words summary with difference under 0.003.

5. Conclusion

From this research result, it can be concluded:

1. Summarization using document index graph with TextRank and HITS resulting f-measure score of 0.291242 and 0.302845 respectively. Furthermore, HITS algorithm resulting a slightly higher score than TextRank.
2. The average f-measure score of summary highly dependent on the length of the summary.

6. References

[1] K. LU and J. HOLCOMB, "Digital News Audience: Fact Sheet," Pew Research Center, 15 june 2016. [Online]. Available: http://www.journalism.org/2016/06/15/digital-news-audience-fact-sheet/. [Accessed 11 november 2016].

[2] L. Marina and L. Matk, "Graph-Based Keyword Extraction for Single-Document Summarization," *Proceedings of the Workshop on Multi-source Multilingual Information Extraction and Summarization,* p. 8, 2008.

[3] E. Hovy and C.-y. Lin, "AUTOMATED TEXT SUMMARIZATION," p. 18, 1998.

[4] G. R. Radev, E. Hovy and K. Mckeown, "Introduction to the special issue on summarization," *Journal Computational Linguistics - Summarization,* vol. 28, no. 4, pp. 399-408 , 2002.

[5] K. M. Hammouda and M. S. Kamel, "Efficient hrase-Based Document Indexing for Web Document Clustering," p. 18, 2004.

[6] Q. Zhou, L. Sun and J.-Y. Nie, "IS_SUM: A Multi-Document Summarizer based on Document Index," 2005.

[7] M. a. Jinnah, "Analisis dan Implementasi Peringkasan Teks Otomatis Dokumen Tunggal Berbahasa Indonesia Menggunakan Document Index Graph," 2011.

[8] R. Mihalcea and P. Tarau, "TextRank: Bringing Order int Texts," *Association for Computational Linguistics,* 2004.

[9] S. Brin and L. Page, "The Anatomy of a Large-Scale Hypertextual Web Search Engine," *Seventh International World-Wide Web Conference,* 1998.

[10] I. Rogers, "The Google Pagerank Algorithm," IPR Computing Ltd, [Online]. Available: http://www.cs.princeton.edu/~chazelle/courses/BIB/pagerank.htm. [Accessed 11 june 2016].

[11] J. M. Kleinberg, "A uthoritat ive Sources in a H y p erlinked Env ironment," *Journal of the ACM (JACM),* vol. 46, pp. 604-632, 1999.

[12] R. Mihalcea and P. Tarau, "A language independent algorithm for single and multiple document summarization," 2005.

[13] C.-Y. Lin, "ROUGE: A Package for Automatic Evaluation of Summaries," *Text summarization branches out: Proceedings of the ACL-04 workshop,* vol. 8, 2004.

[14] E. Yan and Y. Ding, "The effects of dangling nodes on citation networks," 2011.

[15] M. Vicente, O. Alcon and E. Lloret, "The University of Alicante at MultiLing 2015: approach, results and further insights," *16th Annual Meeting of the Special Interest Group on Discourse and Dialogue,* 2015.

Question analysis for Indonesian comparative question

A Saelan[1], A Purwarianti[2], D H Widyantoro[3]

School of Electrical Engineering and Informatics, Bandung Institute of Technology
Jl. Ganesa No. 10, Bandung, Indonesia

[1]athiasaelan@yahoo.co.id, [2]ayu@stei.itb.ac.id, [3]dwi@stei.itb.ac.id

Abstract. Information seeking is one of human needs today. Comparing things using search engine surely take more times than search only one thing. In this paper, we analyzed comparative questions for comparative question answering system. Comparative question is a question that comparing two or more entities. We grouped comparative questions into 5 types: selection between mentioned entities, selection between unmentioned entities, selection between any entity, comparison, and yes or no question. Then we extracted 4 types of information from comparative questions: entity, aspect, comparison, and constraint. We built classifiers for classification task and information extraction task. Features used for classification task are bag of words, whether for information extraction, we used lexical, 2 previous and following words lexical, and previous label as features. We tried 2 scenarios: classification first and extraction first. For classification first, we used classification result as a feature for extraction. Otherwise, for extraction first, we used extraction result as features for classification. We found that the result would be better if we do extraction first before classification. For the extraction task, classification using SMO gave the best result (88.78%), while for classification, it is better to use naïve bayes (82.35%).

1. Introduction

Human demand on information seeking is increasing progressively. More people used search engine to seek for information, whether for educational purpose, or even to obtain information about goods to be bought. One of the active researches on this field is question answering. With question answering system, people can get the answer of a question directly, rather than read all of the passage retrieved by search engine.

One purpose of information seeking, especially when looking something to be bought, is to compare several objects. If we used search engine, we have to search each object manually and read some information that we want to compare. This is clearly more inconvenient than searching for only one thing.

In this research, we will develop a question answering system for answering comparative questions. This type of question answering certainly required a different question analysis compared to common factoid question answering system. This paper will explain question analysis part of comparative question answering system. This will be done by classification and information extraction technique. The answer finder component will be done in the following step of the research

Research of question answering for Bahasa Indonesia have been done before, including for factoid questions [1][2][3], non-factoid questions [2][4], and list factoid questions [5]. Some research focused on the question analysis including [6], which is an Indonesian question analysis for cross language question answering. Meanwhile, research in [7] focused on question analysis for complex question, which is a question that can be decomposed into several simple questions. However, as far as we know, there is no research for comparative question answering yet.

Next part of this paper will continue as follows. Section 2 explains some related researches on information extraction from comparative sentences. Section 3 contains question analysis for Indonesian comparative question. Section 4 explains experiment and the result, while the last will conclude the result obtained from this research.

2. Related work

Several researches had studied information extraction techniques for analyzing comparative sentences. Some of the research analyzed comparative sentences for sentiment analysis [8][9][10]. Information extraction in [8] was used for extracting 3 comparative elements, that is subject entity, comparative predicate, and object entity. In [9], extracted information including 2 compared entities, relation word, and feature that is compared. Research in [10] studied how to determine preferred entities among the 2 compared entities that had been extracted in [9].

In other work, [11] also focused on comparative question. On that research, comparative questions, that was collected online, was used to get entities that is compared, or in other word comparable entities. Those comparable entities could be used for several purposes, such as recommendation systems.

3. Comparative question analysis

Comparative question is a question that comparing two or more entities, for example "*bagusan mana Galaxy Grand 2 sama lenovo p70?*" (English: "which one is better, Galaxy Grand 2 or lenovo p70?"). Unlike [11], we tried not to limit comparative question as questions that explicitly mentioned the compared entities. For example, "*saingan berat alcatel flash 2 ini apa ya?*" (English: "what is the competitor of alcatel flash 2?") mentioned just one entity. However that sentence was a comparative question because it compared the entity to other entities. Furthermore, the sentence "*Uang 1.6 jt bisa dpt hp sprti apa?*" (English: "what kind of hand phone can be bought with 1.6 million rupiahs?") was not mentioning any entities, but it was a comparative question because it was intended to compare all hand phones that could be bought with a certain amount of money.

For analyzing comparative questions, first we classified questions based on their type and source of the answer. Based on analysis of collected question examples, we classified comparative questions into 5 categories, which will be discussed in section 3.2. Before that, we will explain types of information that need to be extracted in section 3.1.

3.1. Information need to be extracted

There are 4 types of information to be extracted from a comparative question, including entity, aspect, comparison, and constraint. Those information were considered to be enough to represent a comparative question. For example, the sentence "*xiaomi yg kualitasnya setara sama redmi note 3 yg support micro sd tipe apa ya?*" (English: "which xiaomi smartphone that have similar quality with redmi note 3 and have micro sd support?"). Extracted information from that sentence are:

- Entity, i.e. the compared object. In that sentence, "*redmi note 3*".
- Aspect, i.e. the thing that is compared. In that sentence, "*quality*".
- Comparison. In that sentence, "*similar*".
- Constraint. In that sentence two constraints are mentioned, "*xiaomi*" and "*micro sd support*". The constraint "*xiaomi*" might be easily confused for an entity, but "*xiaomi*" here is considered as constraint because the expected answer for that question is limited to the "*xiaomi*" branded handphone.

Based on those information, we could understand what was asked, that is to search for an object with "similar" "quality" with "redmi note 3", "xiaomi" brand, and have "micro sd support". Even though so, this four type of information is not always exist in a question. For example, the sentence "*Uang 1.6 jt bisa dpt hp sprti apa?*" (English: "what kind of handphone can be bought with 1.6 million rupiahs?") only contain the constraint. This depends on the type of the question, which will be explained in the next section.

3.2. Question types

Based on type of answer, questions can be grouped into 3, which is question answered with entity (selection), comparison, and yes/no question. Question answered with entity can be further divide into 3 types based on source of the answer which is from entities mentioned in the question, from entities not mentioned in the question, and from any entity, whether it was mentioned or not. So we have 5 question types:

1. Selection between mentioned entities (mentioned)
2. Selection between unmentioned entities (unmentioned)
3. Selection between any entities (any)
4. Comparison
5. Yes or no

In the first type, selection between mentioned entities, the expected answer is one of the mentioned entities, for example "bagusan mana Galaxy Grand 2 sama lenovo p70?" (English: "which is better, Galaxy Grand 2 or lenovo p70?"). This type of question definitely contains entity. Furthermore, the question could contain aspect, for example in the question "kalau Kamarenya bagusan mana dengan Sony 13 mp dengan Asus 13 mp" (English: "how about the camera? Which one is better compared to Sony 13mp and Asus 13mp"), the aspect is camera. If there was no information about aspect, we could assume that the comparison made is general. While for the comparison, majority of the question asked about better entity among all mentioned entities.

For the unmentioned type, the question expects an answer from unmentioned entity. There are two majority questions in this type. The first asks about similar entity with a certain constraint, for example "xiaomi yg kualitasnya setara sama redmi note 3 yg support micro sd tipe apa ya?" (English: "which xiaomi smartphone that have similar quality with redmi note 3 and have micro sd support?"). The other one asked for entities that had some better qualitiy to the mentioned, for example "Stuart Hughes iPhone 5 Black Diamond masih biasa aja, ada yg lbih mahal ga?" (English: "Stuart Hughes iPhone 5 Black Diamond is just so-so, isn't there anything more expensive?". This type of question can contain all kinds of information, including entity, comparison, aspect, and constraint, like the first example. But the mandatory thing here is entity and comparison, for example the entity in the second example is "*Hughes iPhone 5 Black Diamond*" and the comparison is "*more expensive*".

Answer expected from the third question type is any entity, mentioned or not, as long as it meets the constraints and comparisons. However, most of this type of question was not explicitly mention the comparison, but it could be assumed that people will look for for the best thing with certain constraints. For example, in the question "Uang 1.6 jt bisa dpt hp sprti apa?" (English: "what kind of handphone can be bought with 1.6 million rupiahs?"), the asker was certainly looking for the best handphone that can be bought with a certain amount of money and was not looking for the worse even though the price is far below that mentioned amount.

Questions in comparison type requires an answer in the form of comparison between two or more entities. For example, "pngen tw bedanya redmi note 2 prime sama yg redmi note 2 biasa apa ya?" (English: "what is the difference between redmi note 2 prime and the ordinary redmi note 2?"). Unlike selection between mentioned entities, this type of question doesn't require any selected entity in the answer, just the comparison. Information contained in this type of question usually is just some compared entities. However, it is possible that there is only 1 entity, means to compare it to another entities generally. Furthermore, the question can also contain aspect, which means comparing only the selected aspect.

The last one is yes or no questions. Answer expected from this type of question is whether yes or no. For example, "semuanya sama ya dengan xiaomi redmi 2? hanya beda di ram dan internal aja kan?" (English: "it is the same as xiaomi redmi 2, right? The only difference is the ram and the internal, right?"). This type of question was more varied. Some asked the truth of a comparative sentence between some entities and with a comparative word, which would be similar to "selection between mentioned entities". While the other question asked whether there was exists some entity with specific constraints, and will easily confused with comparison class.

4. Experiment

This experiment was aimed to find out which scenario and algorithm is suitable for comparative question analysis. In this experiment, we used Weka as machine learning tools. We compared 2 scenario. First one, we did classification first and then extraction. And the last one we did extraction first, and then classification.

4.1. Experiment Data

Data was collected from several Indonesian gadget review sites. We have collected comments from 6 sites, that is www.detekno.com, www.ulasgadget.com, hargahpxiaomi.com, ulashape.com, www.begawei.com, and hpsaja.com. Gadget review sites were chosen because we can find comparative questions quite easily from its comments. Besides, we add some questions from a general forum, kaskus.co.id, because we found that gadget review site does not have many question classified as "selection from other entities".

Table 1. Statistics of classification data

Question Class	Data Size	Training	Testing
Selection between mentioned entities	106	85	21
Selection between unmentioned entities	11	9	2
Selection between any entity	15	12	3
Comparison	14	11	3
Yes-no	26	21	5
Total	172	138	34

Table 2. Statistics of extraction data

Token Class	Data Size	Training	Testing
other	1145	912	233
entity-b	304	240	64
entity-i	685	540	145
aspect-b	48	36	12
aspect-i	16	16	0
comparison-b	140	111	29
comparison-i	29	25	4
constraint-b	74	60	14
constraint-i	116	100	16
Total	2557	2040	517

IOP Conf. Series: Journal of Physics: Conf. Series **801** (2017) 012077 doi:10.1088/1742-6596/801/1/012077

Before doing the experiment, the data was preprocessed manually. We selected only comparative questions from the data. Some data that have ellipsis needs to be completed by adding entity in appropriate place according to the review title. Finally, we collected 172 comparative questions with 2557 token. Table 1 shows data for question classification, and table 2 shows data for extraction.

4.2. Experiment Result

Features that was used for classification task is bag of words. For extraction, we used 6 features, that is lexical, 2 previous words lexical, 2 following words lexical, and previous label. For the first experiment, that is classification first, extraction feature was added by classification result, which is question class. While the second experiment, extraction result was added as classification feature, which is number of entities, aspects, comparisons, and constraints in the sentence.

We compared 4 machine learning algorithms for each experiment, including naïve bayes (NB), SMO, j48, and random forest (RF) that is provided by Weka. The experiment result is shown in table 3.

Table 3. Experiment Result

Scenario	Task	Accuracy			
		NB	SMO	J48	RF
Classification first	classification	73.53%	73.53%	70.59%	67.65%
	extraction	84.33%	89.75%	84.72%	77.56%
Extraction first	classification	82.35%	76.47%	79.41%	76.47%
	extraction	84.53%	88.78%	84.72%	83.37%

From the experiment, we can see that for extraction task, there was no significant difference between the two scenarios, except for random forest classification. But for classification, the second experiment gave better result. So we could say that the extraction first scenario will gave better result generally.

For classification task, naïve bayes gave better result. Most of the prediction error occurred in yes/no class that is classified as mentioned class. This is consistent with the analysis in section 3.2.5, a question that asked the truth of a comparative sentence between some entities would be similar to "selection between mentioned entities". This encouraged us to do some further analysis for this type of question in the future research. In the best case of the extraction task, the one that use naïve bayes classifier, we found out that from 5 yes/no question, 3 question was misclassified as mentioned. SMO was better for classifying yes/no questions, which was only 2 questions that misclassified. But SMO was not really good compared to naïve bayes, especially for unmentioned and comparison class.

For extraction task, we found that SMO gave better result. The worst error occurred in aspect-b class and constraint-b class. For aspect-b class, only 1 from 12 training data was classified correctly. While for constraint-b, only 5 from 14 that were classified correctly. Most of the misclassified data were classified into other class. We guessed that aspect and constraint did not give good result because of inconsistent annotation.

5. Conclusion

We have analyzed some Indonesian comparative question and tried to make automatization of question analysis for comparative question. We grouped comparative questions into 5 types, which is selection between mentioned entities, selection between unmentioned entities, selection between any entity, comparison, and yes or no question. Then we extracted 4 types of information from comparative questions, which is entity, aspect, comparison, and constraint. Experiment showed that it is better to do the extraction first, and then the classification task used the extraction result as features. For extraction task, classification using SMO gave the best result (88.78%), while for classification, it is better to use naïve bayes (82.35%). For future research, we can further analyzing about yes/no questions. Besides, we will continue this research by developing answer finder for comparative questions.

Acknowledgement
This research is partially supported by "Penelitian Pendidikan Magister menuju Doktor untuk Sarjana Unggul (Kemenristekdikti)" (Master's Program toward Doctoral Degree for Excellent Graduate) within research "Pengembangan Chatbot Berbahasa Indonesia" (Indonesian Chatbot Development).

References

[1] Purwarianti A, Tsuchiya M and Nakagawa S 2007 A Machine Learning Approach for Indonesian Question Answering System *Proc. Artificial Intelligence and Applications (12-14 February 2007, Innsbruck, Austria)* (ACTA Press) pp 573-578
[2] Zulen A A and Purwarianti A 2011 Study and Implementation of Monolingual Approach on Indonesian Question Answering for Factoid and Non-Factoid Question *Proc. PACLIC 25 (16-18 December 2011, Singapore)* (Waseda University) pp 622-631
[3] Zulen A A and Purwarianti A 2013 Using Phrase-Based Approach in Machine Learning Based Factoid Indonesian Question Answering *Proc. CISAK (7 July 2013, Daejon, South Korea)*
[4] Purwarianti A and Yusliani N 2010 Study of Non Factoid Question Answering for Bahasa Indonesia *Proc. ICACSIS 2010 (20-23 November 2010, Bali, Indonesia)*
[5] Prabawa A E and Purwarianti A 2011 Indonesian List Question Answering System. *Proc. KNIF 2011 (23 November 2011, Bandung, Indonesia)*
[6] Faruqi M I and Purwarianti A 2011 An Indonesian Question Analyzer to Enhance the Performance of Indonesian-English CLQA *Proc. ICEEI 2011 (17-19 July 2011, Bandung, Indonesia)* (IEEE)
[7] Basuki S and Purwarianti A 2016 Statistical-based approach for Indonesian complex factoid question decomposition *IJEEI* **8(2)** 356
[8] Yang S and Ko Y 2011 Extracting Comparative Entities and Predicates from Texts Using Comparative Type Classification *Proc. 49th Annual Meeting of the Association for Computational Linguistics: Human Language Technologies (19-24 June 2011, Portland, Oregon)* vol 1 (ACL) pp 1636-1644
[9] Jindal N and Liu B 2006 Mining Comparative Sentences and Relations *Proc. AAAI 2006 (16-20 July 2006, Boston, Massachusetts)* **22** (Menlo Park, California: AAAI Press) pp 1331-1336
[10] Ganapathibhotla M and Liu B 2008 Mining Opinions in Comparative Sentences *Proc. 22nd International Conference on Computational Linguistics (18-22 August 2008, Manchester, UK)* vol 1 (ACL) pp 241-248
[11] Li S, Lin C Y, Song Y I and Li Z 2013 Comparable Entity Mining from Comparative Questions *IEEE Transactions On Knowledge And Data Engineering* **25(7)** pp 1498-1509

Modelling and Simulation of Search Engine

Mahyuddin K M Nasution[1*]

[1]Information Technology, Fasilkom-TI, Universitas Sumatera Utara, Padang Bulan USU Medan 20155, Indonesia.

[*]Email: mahyuddin@usu.ac.id

Abstract. The best tool currently used to access information is a search engine. Meanwhile, the information space has its own behaviour. Systematically, an information space needs to be familiarized with mathematics so easily we identify the characteristics associated with it. This paper reveal some characteristics of search engine based on a model of document collection, which are then estimated the impact on the feasibility of information. We reveal some of characteristics of search engine on the lemma and theorem about singleton and doubleton, then computes statistically characteristic as simulating the possibility of using search engine. In this case, Google and Yahoo. There are differences in the behaviour of both search engines, although in theory based on the concept of documents collection.

1. Introduction

To access or search for information in an information space or system, we need tools [1]. One of tools is the search engine, we know as a software system [2, 3]. In general, for helping to know and understand a system, we use the model to assemble it such that mathematically a model can represent the search engine [4]. Whereas, simulation can used for estimating the effect of search engine model on the information space or system [5].

There are many different search engines. The search engine that arises naturally with the database or search engine that grew up with the web (web search engine) [6, 7]. Dealing with the complexity of information, the search engines helpless and disappear, the search engine shifts to meet the capabilities required, or the search engines changed clothes and present be new. Therefore, all this will affect access to information in space. In this case, the mathematical principle is not only used to systematize, but it serves to optimize the creation of a search engine on information space. This paper aimed to express the characteristics of search engine based on the constraints in the information space.

2. Basic Concept and Motivation

Suppose we denote the information space or system such as Ω [8]. The information space contain the groups of documents or D [9]. Each group of documents consist of documents d_j whereby there a word w, i.e. the basic unit of discrete data, defined to be an item from a vocabulary indexed by $\{1,\ldots,K\}$, $w_k = 1$ if k in K or $w_k = 0$ otherwise [10, 11]. Next, we define the terms related to the word.

Definition 1. A term t_x coincide with at least one or more words, i.e. $t_x = (w_l|l=1,\ldots,L)$, $k \leq l$, l is a number of parameters representing words w, l is the number of vocabularies in t_x, $|t_x| = l$ is the size of t_x.

Suppose that we have a term, that is a person name t_l = 'Mahyuddin Khairuddin Matyuso Nasution', or $\{w_1,w_2,w_3,w_4\}$ = {"Mahyuddin","Khairuddin","Matyuso","Nasution"} as a set of words. We obtain the power of sets $\{\{\},\{w_1\},\{w_2\},\{w_3\},\{w_4\},\{w_1,w_2\},\{w_1,w_3\},\{w_1,w_4\},\{w_2,w_3\},\{w_2,w_4\},\{w_3,w_4\},\{w_1,w_2,w_3\},\{w_1,w_2,w_4\},\{w_2,w_3,w_4\},\{w_1,w_2,w_3,w_4\}\} = \{t^{2^{\wedge}\{k\}-1}\}$, note: 2^{k} is called *k*-th power of 2. In this case we have $|\{t^{2^{\wedge}\{k\}-1}\}| = 15$ and $l = k$. Therefore, probability of t_l or $p(t_l) = 1/(2^l-1)$. Suppose the vector space of t_l is $\{t^{2^{\wedge}\{l\}-1}\}$, we have a design for searching information in the search space by a software system as the search engine. We express it as follows [12, 13].

Definition 2. Suppose Ω is a set of documents indexed by search engine, i.e., a set consists of the ordered pair of the terms t_{li} and documents d_{lj}, or (t_{li},d_{li}), $i=1,\ldots,I$, $j=1,\ldots,J$. The relation table is two columns t_l and d_l as a representation of search engine whereby $\Omega_l = \{(t_l,d_l)_{ij}\}$ is a subset of Ω. The size of Ω is denoted by $|\Omega|$.

Definition 3. Let t_l is a search term and q is a query, then t_l in q for t_l in d_l, d_l in Ω.

In logical implication, Definition 3 express that a document is relevant to a query if it implies the query, that is if d=>q is true or d=>t_l is true for all d in Ω: $(d \Rightarrow t_l) = 1$. Thus, the degree of d=>q measured by $P(d$=>$q)$. Therefore there are the uniform mass probability function for Ω, i.e.

$$P : \Omega \rightarrow [0,1] \tag{1}$$

where $\Sigma_\Omega P(d) = 1$.

Definition 4. Suppose t_x is a search term or t_x in S whereby S is a set of singleton search terms of search engine. A vector space Ω_x, be a subset of Ω, is a singleton search engine event (singleton) of documents that contain an occurrence of t_x in d_x.

The same meaning of Ω_x as subset of Ω is if d=>t_x has true value, or $\Omega_x(t_x)\approx 1$ if t_x is true at d in Ω or $\Omega_x(t_x) \approx 0$ otherwise, and the cardinality of Ω_x be $|\Omega_x| = \Sigma_\Omega(\Omega_x(t_x)\approx 1)$. In other word, each document that is indexed by search engine contains at least one occurrence about the search term. In degree of uncertainty of d=>t_x on d=>q means that

$$P(\Omega_x) = P(\Omega_x(t_x)\approx 1) = \Sigma_\Omega(\Omega_x(t_x)\approx 1)/|\Omega| = |\Omega_x|/|\Omega|. \tag{2}$$

However, if search term in pattern, like t_x = "Mahyuddin Khairuddin Matyuso Nasution", then a different result appears. In other words, Ω_{xp}("t_x")=1 if t_x is true at d in Ω exactly or Ω_{xp}("t_x")= 0 otherwise, and the cardinality of Ω_x be $|\Omega_{xp}| = \Sigma_\Omega(\Omega_{xp}$("$t_x$")=1). In this case, each document that is indexed by search engine contains at least one occurrence of a search term. In degree of uncertainty of d=>"t_x" on d=>q is

$$P(\Omega_{xp}) = P(\Omega_{xp}(\text{"}t_x\text{"})=1) = \Sigma_\Omega(\Omega_{xp}(\text{"}t_x\text{"})=1)/|\Omega| = |\Omega_{xp}|/|\Omega|. \tag{2}$$

Thus $|\Omega_{xp}|/|\Omega| \leq |\Omega_x|/|\Omega|$, so $|\Omega_{xp}| \leq |\Omega_x|$ or Ω_{xp} is a subset of Ω_x.

Let t_x and t_y are two different search terms. If $t_x = t_y$, $t_x \neq t_y$, or $|t_x|<|t_y|$, then Ω_{xp} be a subset of Ω_x or Ω_{yp} be a subset of Ω_y or Ω_{xp} be a subset of Ω_y or Ω_{yp} be a subset of Ω_y.

3. Adaptive Approach to Model

Let t_x and t_y are search terms, refer to the definitions above, will be revealed some characteristics related to the search engine as a system. All characteristics derived from the adaptation formula that build model of the problem completion relating to the possible the results of the search engine. Some of the adaptive characteristics are as follows [12, 13, 14].

Lemma 1. If $t_x \neq t_y$ and $t_x \cap t_y = \phi$, then $|\Omega_x \cap \Omega_y|=0$ and $|\Omega_x \cup \Omega_y|=|\Omega_x|+|\Omega_y|$ where Ω_x and Ω_y are subsets of Ω.

Proof. $t_x \neq t_y$ and $t_x \cap t_y = \phi$ mean that for all w_x in t_x all w_x not in t_y and for all w_y in t_y all w_y not in t_x, then for all w_x in d_x all w_x not in d_y and for all w_y in d_y all w_y not in d_x such that $t_x \cup t_y = t_y \cup t_x$ and $d_x \cup d_y = d_y \cup d_x$.

Therefore, $\Omega_x=\{(t_x,d_x)\}$ and $\Omega_y=\{(t_y,d_y)\}$ are two independent events from queries, or t_x and t_y are true at d in Ω, respectively. In this case, $\Omega_x \cap \Omega_y=\phi$. In other words, $\{(t_x,d_x)\} \cup \{(t_y,d_y)\} = \Omega_x \cup \Omega_y$. Therefore, we have

$$|\Omega_x \cap \Omega_y| = 0 \quad (3)$$

and

$$|\Omega_x \cup \Omega_y| = |\Omega_x| + |\Omega_y| \quad (4)$$

Lemma 2. If $t_x \neq t_y$ $t_x \cap t_y \neq \phi$ and $|t_y|<|t_x|$, then $|\Omega_x|=|\Omega_x|+|\Omega_y|$ where Ω_x and Ω_y are subsets of Ω.
Proof. Based on assumption, we have for all w_y in t_y all w_y in t_x, but there are w_x in t_x whereby w_x not in t_y such that $t_x \cap t_y=t_y$ and $t_x \cup t_y=t_x$. Similar concept, for all w_y in t_y all w_y in d_y and because all w_y in t_x we conclude that w_y also in d_x, but there are w_x in t_x and x_x in d_x whereby w_x not in t_y such that w_x not in d_y. Thus, if $t_x \cap t_y=t_y$ then $d_x \cap d_y=d_y$ and if $t_x \cup t_y=t_x$ then $d_x \cup d_y=d_x$. Therefore, $\Omega_x = \{(t_x,d_x)\} = \{(t_x \cup t_y, d_x \cup d_y)\} = \{(t_x,d_x) \cup (t_y,d_y)\} = \{(t_x,d_x)\} \cup \{(t_y,d_y)\} = \Omega_x \cup \Omega_y$. In other words,

$$|\Omega_x| = |\Omega_x| + |\Omega_y| \quad (5)$$

Proposition 1. For search terms $t_z \neq \ldots \neq t_y \neq t_x$ and $|t_z| < \ldots < |t_y| < |t_x|$, then $|\Omega_x| = |\Omega_x| + |\Omega_y| + \ldots + |\Omega_z|$, where $\Omega_z, \ldots, \Omega_y, \Omega_x$ are subsets of Ω.
Proof. Based on generalization of Equation (3) and Equation (4), we derive $|\Omega_x| = |\Omega_x \cup \Omega_y| = |\Omega_x|+|\Omega_y| = |\Omega_x|+|\Omega_y \cup \ldots| = |\Omega_x|+|\Omega_y|+\ldots = |\Omega_x|+|\Omega_y|+|\ldots \cup \Omega_z|$, and

$$|\Omega_x| = |\Omega_x| + |\Omega_y| + \ldots + |\Omega_z| \quad (6)$$

Lemma 3. If $t_x \neq t_y$ $t_x \cap t_y=\phi$ and $d_x \cap d_y \neq \phi$, then $|\Omega_x| \approx |\Omega_y|$, Ω_x and Ω_y are subsets of Ω.
Proof. $t_x \neq t_y$ $t_x \cap t_y=\phi$ and $d_x \cap d_y \neq \phi$ mean that for all w_x in t_x all w_x not in t_y and for all w_y in t_y all w_y not in t_x then $t_x \cup t_y = t_y \cup t_x$, but for all w_x in d_x there are w_x in d_y and for all w_y in d_y there are w_y in d_x also, then $d_x \cap d_y = d_x = d_y$ and $d_x \cup d_y = d_y \cup d_x = d_x = d_y$. Or for $\Omega_x = \{(t_x,d_x)\}$ and $\Omega_y = \{(t_y,d_y)\}$ we obtain $\Omega_x \cap \Omega_y = \{(t_x,d_x)\} \cap \{(t_y,d_y)\} = \{(t_x,d_y)\} \cap \{(t_y,d_y)\} = \{(t_y,d_y)\} \cap \{(t_y,d_y)\}$ or $\Omega_x \cap \Omega_y = \Omega_y \cap \Omega_y$ or

$$\Omega_x \cap \Omega_y = \Omega_y. \quad (7)$$

Similarly,

$$\Omega_x \cap \Omega_y = \Omega_x. \quad (8)$$

In other words, $\Omega_x = \{(t_x,d_x)\} = \{(t_x,d_x \cup d_y)\} = \{(t_x,d_x) \cup (t_x,d_y)\} = \{(t_x,d_x) \cup (t_y,d_y)\} = \{(t_y,d_x) \cup (t_y,d_y)\} = \{(t_y,d_x \cup d_y)\} = \{(t_y,d_y)\} = \Omega_y$. Therefore, based on Equation (7) and Equation (8), we have $|\Omega_x| \approx |\Omega_y|$.

Definition 5. Suppose t_x and t_y are two different search terms. Let $t_x \neq t_y$, t_x and t_y in S, where S is a set of singleton search term of search engine. A doubleton search term is $D = \{\{t_x,t_y\}: t_x,t_y \text{ in } S\}$ whereby the vector space of doubleton search term denoted by $\Omega_x \cap \Omega_y$ is a doubleton search engine event of documents that contain a co-occurrence of t_x and t_y such that t_x,t_y in d_x and t_x,t_y in d_y whereby Ω_x, Ω_y, $\Omega_x \cap \Omega_y$ are subsets of Ω.

Theorem 1. Suppose t_x and t_y are two different search terms. $\Omega_x \cap \Omega_y$ is a doubleton search engine event for t_x and t_y whereby Ω_x and Ω_y are subsets of Ω, then $|\Omega_x \cap \Omega_y| \leq |\Omega_x| \leq |\Omega|$ and $|\Omega_x \cap \Omega_y| \leq |\Omega_y| \leq |\Omega|$.
Proof. Based on set theory $\Omega_x \cap \Omega_y$ be subset of Ω_x and $\Omega_x \cap \Omega_y$ be subset of Ω_y, thus $|\Omega_x \cap \Omega_y| < |\Omega_x|$ and $|\Omega_x \cap \Omega_y| < |\Omega_y|$. While based on Lemma 3 we have $|\Omega_x \cap \Omega_y| = |\Omega_x|$ and $|\Omega_x \cap \Omega_y| = |\Omega_y|$.
For $\Omega_x = \{(t_x,d_x)\}$, we have $\{(t_x \cap t_y, d_x \cap d_y)\} = \{(t_x,d_x) \cap (t_y,d_y)\} = \{(t_x,d_x)\} \cap \{(t_y,d_y)\} = \Omega_x \cap \Omega_y$ if $t_x \neq t_y$ and $|t_x|<|t_y|$. Thus $|\Omega_x \cap \Omega_y| = |\Omega_x|$. Similarly, we obtain $\Omega_y = \Omega_x \cap \Omega_y$, so $|\Omega_x \cap \Omega_y| = |\Omega_y|$. Based on Definition 5 $D = \{\{t_x,t_y\}, t_x,t_y \text{ in } S\}$, or $\{t_x,t_y\} = \{(t_x,d_x),(t_y,d_y)\} = \{(t_x,d_x) \cap (t_y,d_y)\} = \{(t_x,d_x) \cap (t_y,d_y)\} = \{(t_x \cap t_y), (d_x \cap d_y)\}$

$= \{(t_x,t_y),(d_x,d_y)\}$, for $\{t_x,t_y\} = \Omega_x \cap \Omega_y$, we get $\Omega_x \cap \Omega_y = \Omega_x$ and $\Omega_x \cap \Omega_y = \Omega_y$. In other words, $\{t_x,t_y\} = \{(t_x,d_x),(t_y,d_x)\} = \{(t_x,d_x)\},\{(t_y,d_y)\}$, for $\{t_x,t_y\} = \Omega_x \cap \Omega_y$ we get $\Omega_x \cap \Omega_y = \Omega_x \cap \Omega_y$, $\Omega_x \cap \Omega_y$ is a subset of Ω_x or Ω_y. If the comma logically means "and" in set theory it means an intersection.
Therefore, $|\Omega_x \cap \Omega_y| \leq |\Omega_x| \leq |\Omega|$ and $|\Omega_x \cap \Omega_y| \leq |\Omega_y| \leq |\Omega|$ for all search terms t_x and t_y.

Corollary 1. If t_x and t_y are the different search terms, then $|\Omega_x \cap \Omega_y| = |\Omega_x \cap \Omega_y| + |\Omega_x \cap \Omega_x| + |\Omega_y \cap \Omega_y|$.
Proof. As the direct or indirect consequence of Proposition 1 and Theorem 1.

4. A Selective Approach as Simulation

The purpose of simulation, in this case, is to construct an approach for selecting the documents in information space or for disclosing the information in the repository [15]. As an experiment to collect data, which is to select n objects from the community. For example, we collect data from the academic community of Faculty of Medicine University of Sumatera Utara (USU), i.e. $n = 51$ academic actors, or in a list is A = {`Abdul Majid, Abdul Rachman Saragih, Abdul Rasyid, Abdullah Afif Siregar, Achsanuddin Hanafie, Adi Kusuma Aman, Alfred C. Satyo, Askaroellah Aboet, Atan Baas Sinuhaji, Ayodhia Pitaloka Pasaribu, Aznan Lelo, Bachtiar Surya, Budi R. Hadibroto, Chairuddin Panusunan Lubis, Chairul Yoel, Darwin Dalimunthe, Daulat Hasiholan Sibuea, Delfi Lutan, Delfitri Munir, Erwin Dharma Kadar`}. Among the names of actors as the term, two different terms t_x and t_y have several options that correspond to words of each name, such as mutual, including, or intersection. Therefore, each term has the opportunity to be placed in the position of a particular index. The position of each term in the search engines for example based on the selected collection of a number of documents related to the term.

Table 1. Experiment design for simulation of search engine

Actor Name (A)	Medium of randomness test		Search engine as test simulation			Search engine as comparative simulation		
	1	2	$\|\Omega_x\|$	$\|\Omega_{"x"}\|$	$\|\Omega_x \cap \Omega_y\|$	$\|\Omega_x\|$	$\|\Omega_{"x"}\|$	$\|\Omega_x \cap \Omega_y\|$
a	0 or 1	0 or 1	0 or 1	0 or 1	0 or 1	0 or 1	0 or 1	0 or 1
...	...	...	...	...	...	...	...	...
b	...	...	...	...	...	...	...	...
...	...	...	...	...	...	...	...	...
...	...	...	...	...	...	...	...	...
...	...	...	...	...	...	...	...	...
z	...	....	...	...	...	...	...	...
Average	pr or lc	a or n	avg_1	avg_2	avg_3	avg_4	Avg_5	avg_6
n_1	n_1(pr)	n_1(a)	$n_{11}(\geq)$	$n_{12}(\geq)$	$n_{13}(\geq)$	$n_{14}(\geq)$	$n_{15}(\geq)$	$n_{16}(\geq)$
n_2	n_2(lc)	n_2(n)	$n_{21}(<)$	$n_{22}(<)$	$n_{23}(<)$	$n_{24}(<)$	$n_{25}(<)$	$n_{26}(<)$
Run (r)	r_1	r_2	r_3	r_4	r_5	r_6	r_7	r_8
μ_r	μ_{r1}	μ_{r2}	μ_{r3}	μ_{r4}	μ_{r5}	μ_{r6}	μ_{r7}	μ_{r8}
σ_r	σ_{r1}	σ_{r2}	σ_{r3}	σ_{r4}	σ_{r5}	σ_{r6}	σ_{r7}	σ_{r8}

In the sample that can represent population, we develop a table of information as experiment design for providing data, Table 1. Data that reveal characteristics of a search engine. In the table, the first column is the actor's names alphabetically ordered. The second column contains academic level: It is used to test whether the sample is random, the academic level as medium of randomness test (mrt). The third column involves data of scientific publications indexed by Scopus whereby the actor consists of two categories: the author or not, data of scientific publications as the comparative mrt. It is intended to support the randomness test of sample. The next columns contain the list of singletons respective to t_x and t_x in quotes, and a list of doubletons of t_x and t_y (singleton with keyword). In this case, we ensure that the singletons also are random.

In general, the information space consisting of documents viewed as the population. Statistically, the population is random, and it was tested whether the characteristics also lowered to the sample, so that any measurement about sample describe population. We seperate the sample into two categories: number of first categories

$$n_1 = \sum_{i=1...n} a_{i1} \quad (1)$$

or

$$n_2 = \sum_{i=1...n} a_{i2} \quad (2)$$

whereby a_{i1} is elements of A that meet first category and a_{i2} is elements of A that meet second category. While run (r) is how many times the category change in the sample. Thus, the average of run is

$$\mu_r = (2n_1n_2/(n_1+n_2))+1 \quad (3)$$

and the variance of run is

$$\sigma_r^2 = ((2n_1n_2(2n_1n_2-n_1-n_2))/((n_1+n_2)^2(n_1+n_2-1)))^{1/2}. \quad (4)$$

Then, we have Z_{count} as follows

$$Z_{count} = (r - \mu_r)/\sigma_r \quad (5)$$

for hypotheses used are as follows: H_0: the data sequence is random, and H_1: the data sequence is not random. For academic level as category: professor (pr) or lecturer (lc), we have $n_1 = 34$ and $n_2 = 17$. By using Equations (3), (4), and (5), we obtain $\mu_r = 23.67$, $\sigma_r = 0.93$ and $Z_{count} = -1.79$, and for $\alpha = 0.05$ we obtain $Z_{\alpha=0.025}$=1.96 $\leq Z_{count} \leq Z_{=0.025}$ = 1.96, and because r is located between the critical value then the decision is received H_0. Seen from the publication of scientific papers indexed by Scopus: author (a) or not (n), we have the similar conditions such that the sequence of data is random.

Furthermore, to test the randomness perfectly, tested independence of two data space by using chi-square (χ^2). Suppose the data space (ds) is presented in matrix form as follows,

$$ds = \begin{vmatrix} x_{11} & x_{12} & \dots & x_{1n} \\ x_{21} & x_{22} & \dots & x_{2n} \\ \dots & \dots & \dots & \dots \\ x_{m1} & x_{m2} & \dots & x_{mn} \end{vmatrix}$$

Amount of data x_{ij} is S_{ij} as follows

$$S_{ij} = \Sigma_{i=1,...,n;j=1,...,m}\, x_{ij}\,. \quad (6)$$

So that we can calculate the expectations of each data as follows

$$\begin{aligned} e_{11} &= (\Sigma_{i=1,j=1,...,m}\, x_{ij})(\Sigma_{i=1,...,n,j=1}\, x_{ij})/S_{ij} \\ e_{12} &= (\Sigma_{i=1,j=1,...,m}\, x_{ij})(\Sigma_{i=1,...,n,j=2}\, x_{ij})/S_{ij} \\ &\dots \\ e_{mn} &= (\Sigma_{i=n,j=1,...,m}\, x_{ij})(\Sigma_{i=1,...,n,j=m}\, x_{ij})/S_{ij} \end{aligned} \quad (7)$$

and we have a matrix of expectations as follows

$$es = \begin{vmatrix} e_{11} & e_{12} & \dots & e_{1n} \\ e_{21} & e_{22} & \dots & e_{2n} \\ \dots & \dots & \dots & \dots \\ e_{m1} & e_{m2} & \dots & e_{mn} \end{vmatrix}$$

Amount of data e_{ij} is E_{ij} as follows

$$E_{ij} = \Sigma_{i=1,...,n,j=1,...,m}\, e_{ij} \tag{8}$$

Then, we have

$$\chi^2 = \Sigma_{i=1,...,n,j=1,...,m}\, (x_{ij}\text{-}e_{ij})/e_{ij} \tag{9}$$

with degree of freedom (df) is (m-1)(n-1). For example, among 51 actor names we have x_{11} = 34 professors, x_{21} = 17 lectures, x_{12} = 17 authors, and x_{22} = 34 non-authors. Based on Eq. (7) we can calculate their expectations, i.e. $e_{11} = e_{12} = e_{21} = e_{22} = 25.5$, and based on Eq. (9) we obtain $\chi^2 = 11.33$ for test statistic T as chi-squared distribution with (m-1)(n-1) = (2-1)(2-1) = 1 degree of freedom and the acceptance region for T with a significance level of 5% is 3.841, then rejects the null hypothesis of independence because $\chi^2 > 3.841$. This tell us there is a relationship between type of academic level and authors.

In reveal characteristics of search engine based on a model, we conduct an experiment about singleton and doubleton of Google search engine as test simulation and of Yahoo search engine as comparative simulation as follows.

1. Randomness test: Calculate the randomness test for t_x, $t_{\text{"}x\text{"}}$, and t_x,t_y by completing the computations as follows:
   a. For t_x, we have the amount of 51 $|\Omega_x|$ from Google search engine, that is 2635514 with average (avg) = 51676.75. Number of $|\Omega_x|$ greater than or equal to avg is 10 $|\Omega_x|$ while number of $|\Omega_x|$ less than avg is 41. By using Eq. (3), Eq. (4), and Eq. (5), Z_{count}=-6.54 < $Z_{\alpha=0.025}$=-1.96. Therefore, reject H_0 and 51 singletons of t_x from Google search engine is not random.
   b. However, the amount of 51 $|\Omega_{\text{"}x\text{"}}|$ from Google search engine, that is 1095045 with average (avg) = 21471.47. Number of $|\Omega_{\text{"}x\text{"}}|$ greater than or equal to avg is 3 $|\Omega_{\text{"}x\text{"}}|$ while number of $|\Omega_{\text{"}x\text{"}}|$ less than avg is 48. By using the similar equations, Z_{count}=-1.50 > $Z_{\alpha=0.025}$=-1.96, and H_0 accepted whereby 51 singletons of $t_{\text{"}x\text{"}}$ from Google search engine is random.
   c. For doubleton t_x,t_y whereby t_y = "Universitas Sumatera Utara" as a keyword, we have amount of 51 $|\Omega_x \cap \Omega_y|$ from Google search engine, i.e 61092 with avg = 1197.88. Number of $|\Omega_x \cap \Omega_y|$ greater than or equal to avg is 15 and number of $|\Omega_x \cap \Omega_y|$ less than avg is 36. With that, we obtain Z_{count}=-1.31 > $Z_{\alpha=0.025}$=-1.96 based on Eq. (3), Eq. (4) and Eq. (5), and H_0 accepted, thus 51 doubletons of t_x,t_y from Google search engine is random.
   d. Whereas, for t_x by using Yahoo search engine, we have the amount of 51 $|\Omega_x|$ is 2365061 with avg is 46373.76. So $n_1(pr)$ = 5 and n_2(lc) = 46. Z_{count}=-0.03 > $Z_{\alpha=0.025}$=-1.96. On that basis, H_0 accepted, thus 51 singletons of t_x from Yahoo search engine is random.
   e. Simlarly for $t_{\text{"}x\text{"}}$, the amount of 51 $|\Omega_{\text{"}x\text{"}}|$ from Yahoo search engine, that is 395815 with average (avg) = 7761.08. Number of $|\Omega_{\text{"}x\text{"}}|$ greater than or equal to avg is 3 $|\Omega_{\text{"}x\text{"}}|$ while number of $|\Omega_{\text{"}x\text{"}}|$ less than avg is 48. By using the similar equations, Z_{count}=-1.50 > $Z_{\alpha=0.025}$=-1.96, and H_0 accepted whereby 51 singletons of $t_{\text{"}x\text{"}}$ from Yahoo search engine is random.
   f. For doubleton t_x,t_y whereby t_y = "Universitas Sumatera Utara" as a keyword, we have amount of 51 $|\Omega_x \cap \Omega_y|$ from Yahoo search engine, i.e 15361 with avg = 301.19. Number of $|\Omega_x \cap \Omega_y|$ greater than or equal to avg is 12 and number of $|\Omega_x \cap \Omega_y|$ less than avg is 39. With that, we obtain Z_{count}=-0.42 > $Z_{\alpha=0.025}$=-1.96 based on Eq. (3), Eq. (4) and Eq. (5), and H_0 accepted, thus 51 doubletons of t_x,t_y from Yahoo search engine is random.

2. Independence test: For a contingency table has m rows and n columns, a test of independency that null and alternative hypotheses are:

H_0: The two or more categorical variables are independent.
H_1: The two or more categorical variables are related.

Table 2. Samples and categories

Categories	Samples					
	Google search engine			Yahoo search engine		
	$\|\Omega_x\|$	$\|\Omega_{"x"}\|$	$\|\Omega_x \cap \Omega_y\|$	$\|\Omega_x\|$	$\|\Omega_{"x"}\|$	$\|\Omega_x \cap \Omega_y\|$
n_1	10	3	15	5	3	12
n_2	41	48	36	46	48	39

a. First, we test the independence $|\Omega_x|$ of Google and $|\Omega_x|$ of Yahoo. By using Eq. (6), Eq. (7), Eq. (8), and Eq. (9) toward $n_1(|\Omega_x|)$ and $n_2(|\Omega_x|)$ see Table 2, we obtain $\chi^2 = 1.95 < 3.84$ with $df = 1$, and H_0 accepted for $\alpha = 0.05$. Thus two samples are independent.
b. Second, we test the independence $|\Omega_{"x"}|$ of Google and $|\Omega_{"x"}|$ of Yahoo. By using similar equations against $n_1(|\Omega_{"x"}|)$ and $n_2(|\Omega_{"x"}|)$ see Table 2, we have obtain $\chi^2 = 0.00 < 3.84$ with $df = 1$, and H_0 accepted for $\alpha = 0.05$. Thus two samples are independent.
c. Third, we test the independence $|\Omega_x \cap \Omega_y|$ of Google and $|\Omega_x \cap \Omega_y|$ of Yahoo. By using similar equations with $n_1(|\Omega_x \cap \Omega_y|)$ and $n_2(|\Omega_x \cap \Omega_y|)$ see Table, we get value of $\chi^2 = 0.45 < 3.84$ for $df = 1$, and H_0 accepted for $\alpha = 0.05$. Therefore, two samples are independent.
d. For getting behavior of $|\Omega_x|$, $|\Omega_{"x"}|$, and $|\Omega_x \cap \Omega_y|$, we test independence among singletons and doubleton of Google search engine. By using Eq. (6), Eq. (7), Eq. (8), and Eq. (9) for $n_1(|\Omega_x|)$, $n_1(|\Omega_{"x"}|)$, $n_1(|\Omega_x \cap \Omega_y|)$, $n_2(|\Omega_x|)$, $n_2(|\Omega_{"x"}|)$, and $n_2(|\Omega_x \cap \Omega_y|)$ see Table 2, we obtain $\chi^2 = 9.53 > 7.82$ with $df = 3$, and H_0 rejected for $\alpha = 0.05$. Therefore, three samples of Google search engine are dependent.
e. In contrast to that, we test independence among singletons and doubletons of Yahoo search engine. Based on similar concept, we obtain $\chi^2 = 7.71 < 7.82$ with $df = 3$, and H_0 accepted for $\alpha = 0.05$. Therefore, three samples of Yahoo search engine are independent.
f. Therefore, for all characteristics in Table, based on Eq. (6), Eq. (7), Eq. (8), and Eq. (9), the $\chi^2 = 18.98$ greater than 12.59 for $df = 6$ and $\alpha = 0.05$ such that H_0 rejected. Therefore, all the data as a whole is dependent.

In general, a collection of documents in information space and indexed by a system be random, see randomness test (1a, 1c, 1d, 1e, and 1f), and information space Ω has a normal distribution, where Eq. (1) be the uniform mass probability function. A row of data in A is random with a confidence level of 95%.

Although the same characters can be derived based on set theory, but singleton from different search engines are not interdependent. So the information presented freely with each other, caused by each search engine has its own potential and capabilities. There are different potential between Google search engine and Yahoo search engine. In Google search engine, the singletons and doubleton are dependent. Whereas in Yahoo search engine, the singleton and doubleton are independent. Therefore, an information space such as system have information tied to each other, but in different sub-systems can be built mutually bound: Google search engine and Yahoo search engine, for example, as different subsystems.

5. Conclusion

To model and simulate the search engines has been developed the adaptive and selective approach. Adaptive approach produced some formal characteristics while the selective approach generates the characteristic in reality. Both reveal the possibility of the differences about the information presented by the search engine although they has same basic concept. For example, the Google and Yahoo search engines show the different behavior. Further research will reveal some other formulation and characteristic of search engine.

References

[1] Fredrik K, Andersson, and S D Silvestrov 2008 The mathematics of internet search engines *Acta Appl Math* **104**.
[2] M Haman, K Lakhotia, J Singer, D. R. White, and S Yoo 2013 Cloud engineering is search based software engineering too *The Journal of Systems and Software* **86**.
[3] R Roj 2014 A comparison of three design tree based search algorithms for the detection of engineering parts constructed with CATIA V5 in large databases *Journal of Computational Design and Engineering* **1(3)**.
[4] A Anagnostopoulos, A Broder, and K Punera 2008 Effective and efficient classification on a search-engine model *Knowl Inf Syst* **16**.
[5] G Meghabghab and A Kandel 2004 Stochastic simulations of web search engines: RBF versus second-order regression models *Information Sciences* **159**.
[6] M Song, I-Y Song and P P Chen 2004 Design and development of a cross search engine for multiple heterogeneous database using UML and design patterns *Information System Frontiers* **6**.
[7] S Agrawal, K Chakrabarti, S Chaudhuri, V Ganti, A C Konig, and D Xin 2009 Exploiting web search engines to search structured databases *WWW*, Madrid, Spain, ACM.
[8] M Tvarozek and M Bielikova 2007 Adaptive faceted browser for navigation in open information spaces *WWW* Banff, Alberta, Canada ACM.
[9] J G Davis, E Subrahmanian, S Kanda, H Granger, M Collins and A W Westerberg 2001 Creating shared information spaces to support collaborative design work *Information Systems Frontiers* **3(3)**.
[10] M K M Nasution and S A Noah 2012 Information retrieval model: A social network extraction perspective *IEEE International Conference on Information Retrieval & Knowledge Management*.
[11] M K M Nasution 2014 New method for extracting keyword for the social actor *Intelligent Information and Database Systems* **LNAI 8397**.
[12] M K M Nasution 2012 Simple search engine model: Adaptive properties *Cornell University Library*.
[13] M K M Nasution 2012 Simple search engine model: Adaptive properties for doubleton *Cornell University Library*.
[14] M K M Nasution 2011 Kolmogorov complexity: Clustering and similarity *Bulletin of Mathematics* **3(1)**.
[15] M K M Nasution 2013 Simple search engine model: Selective properties *Cornell University Library*.

Large-Scale Graph Processing Analysis using Supercomputer Cluster

Alfrido Vildario[1], Fitriyani[2], Galih Nugraha Nurkahfi[3]

[1,2,3]Department of Computational Science, Telkom University, Bandung, West Java, Indonesia

[1]bukan.rido@gmail.com, [2]fitriyani@telkomuniversity.ac.id, [3]galihnugrahanurkahfi@gmail.com

Abstract. Graph implementation is widely use in various sector such as automotive, traffic, image processing and many more. They produce graph in large-scale dimension, cause the processing need long computational time and high specification resources. This research addressed the analysis of implementation large-scale graph using supercomputer cluster. We impelemented graph processing by using Breadth-First Search (BFS) algorithm with single destination shortest path problem. Parallel BFS implementation with Message Passing Interface (MPI) used supercomputer cluster at High Performance Computing Laboratory Computational Science Telkom University and Stanford Large Network Dataset Collection. The result showed that the implementation give the speed up averages more than 30 times and eficiency almost 90%.

Keywords: Large-scale Graph, BFS, Shortest Path, Parallel

1. Introduction

Large web, huge database, social network, and many scientific problems has produce large-scale graph [1,12,13]. The graph is used on various sector. Such as GPS, social media, network system, traffic, search engine and many more[5]. This is growing and more widely used. Large-scale graph processing become very interesting to analyze and important to solved.

As these problems grow in scale, parallel computing resources are required to meet their computational needs. In the earlier times, supercomputers and mainframe had dominated the world of computing. These are very exotic machines with an integrated technologies inside them, and the scale at which they are built are very high, up to thousands of processors make up a single machine. This is where high performance computing emerges as a solution to provide such needs. Supercomputers are expensive and poor scalability. High performance computing (HPC) offer a scalable, cost-efficient, and easily-maintained resources to conduct a very large computational process. HPC system generally takes one of two paths: cluster computing, and grid computing. Cluster computing is a large number of dedicated computers which are connected in a very close proximity, managed in a centralized way, and have a very high speed interconnect. Among many important graph algorithms, graph traversal and shortest path routing are frequently used to solve other graph problems. For example, shortest path algorithm is an important tool to find betweenanes centrality, a measure of influence in social network.

Graph500 benchmark is the new standard for supercomputer benchmark beside Top500. This benchmark use Breadth-first search (BFS) for searching graph's vertices. The BFS do searching that start from root vertex and continue searching for their neighbor vertices until find the path [2].

Previous studies had implemented BFS on Graphics Processing Unit (GPU)[10] and processing large graph on cloud [3]. In this research we implemented BFS algorithm on cluster supercomputer, with given problem single destination shortest path. Hope the result in this research is provide knowledge about processing large scale graph with breadth-first search algorithm given problem single destination shortest path and show the performance of the supercomputer cluster.

2. Literature Review

2.1. *Graph Adjacency List*

Adjacency list is a list from set of couple vertex in graph which between the couple vertex is the edge[4]. The example of the representation is given below :
A Adjacent to $B = (A,B)$

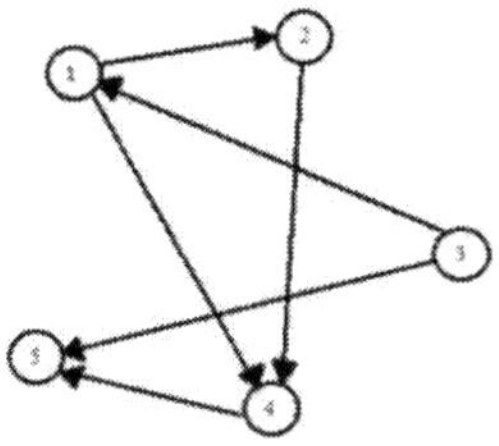

Figure 1. Graph *G*

Given directed graph *G* as **Figure 1** above, the adjacency list form on that graph *G* is : (1,2), (1,4), (2,4), (3,5), (4,5).

2.2. *Breadh-first Search Algorithm*

The BFS is searching algorithm for tree or graph. Start from the root (starting vertex) and search their neighbour from that vertices before continue to the another neighbour on another level[2]. This algorithm used in many applications that using in graph modelling for searching purpose, example the shortest path problem. For example, the BFS algorithm for graph *G* at **Figure 1**. The searching start form vertices that neighbor with vertex 1, they are vertices 4, 3 and 2. Then the seraching continue to neighbor of vertices 4, 3 and 2. Base on the grap, vertex 4 neighbor with vertex 5 and vertex 3 neighbor with vertex 5. So vertex 5 become the next and last vertex. Final result of traversal graph sequentially are vertices 1, 4, 3, 2, 5.

2.3. *Parallel processing*

Parallel computation is a technique for doing computation simultaneously by using many computer, processor or thread which are available[7]. Usually it is needed when the problem is large, either the data or the computing process [5,7]. In computer parallel machine, consist of the node for solve the computation. Head node is for give the instruction computation to compute and compute node for execution received from head node.
Message Passing Interface (MPI) is the system message-passing for parallel computation architecture. It support many programming languange such as C or C++[7].
Speedup is how faster execution of that parallel program. Speedup is widely used for show the performance after the resource is increase or changed. Efficiency is a measure of how much of your available power is being used. The speedup and efficiency can be count using $S_p = T_s / T_p$ and $E_p = S_p / p$. Where S_p = Speedup in p processor or thread, E_p= Efficiency in p processor or thread, T_s= Sequential execution time, T_p= Parallel execution time with p processor or thread, p = amount of using processor or thread.

3. Parallel large-scale graph processing

In this research, we implemented BFS algorithm that given problem single destination shortest path on Supercomputer Cluster at HPC Lab Telkom University using MPI.

3.1. Computer Cluster

A computer cluster consists of a set loosely or tightly connected computers that work together with goal that computation can be more faster and efficient rather than single compute. The component of the cluster usually connected to each other through fast local area network (LAN) and each node running its own instance of an operating system [5].

The Supercomputer Cluster of HPC Lab is consists of three nodes, one head node and two compute nodes. Head node specification is inter Xeon E5-606 @2.13 GHz, 24 GB RAM 1333 Mhz DDR3, SSD 250 GB. Each compute node have dual intel Xeon E5-2670v3 12 core (@2.3GHz), 128 GB RAM DD4, SSD 250 GB with accelator intel Xeon Phi 3120 and Xeon Phi 31 SP. All of node is connected with InfiniBand cable and use FDR-IB SwitchX Mellanox SX6005 Switch. Below is the topology of Supercomputer Cluster.

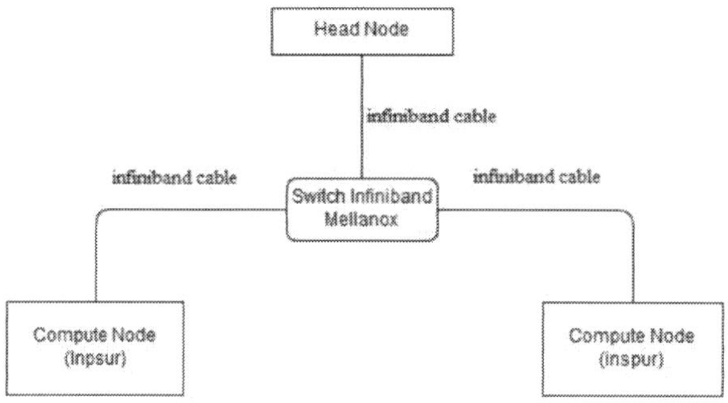

Figure. 2 : Supercomputer Cluster Topology HPC Lab

3.1.1. Graph Dataset.

Graph dataset adjacency list to simulate BFS algorithm on supercomputer cluster were obtained from Standford Large Network Dataset Collection[11]. The data type is directed and unweighted. Data is contain of set couple between vertex, which the vertex is showed by number. Number on the first column is starting vertex and on the other column is goal vertex. It means the vertex on first column have a directed edge to vertex on second column. For example, vertex 0 in first column and vertex 1 on second column, it means that vertex 0 have and directed edge to vertex 1.

We chose some of Gnutella dataset. They are p2p -Gnutella04, p2p -Gnutella05, p2p -Gnutella08, and p2p -Gnutella09. Gnutella dataset is a sequence of snapshots of the Gnutella peer-to-peer file sharing network from August 2002. Nodes represent hosts in the Gnutella network topology and edges represent connections between the Gnutella hosts. **Table 1** below describes the Gnutella dataset.

Table 1. Adjacency List Graph Dataset

Data name	Data type	Number of vertices	Number of Edges

p2p -Gnutella08	Directed, Unweighted	6301	20777
p2p -Gnutella09	Directed, Unweighted	8114	26013
p2p -Gnutella05	Directed, Unweighted	8846	31839
p2p -Gnutella04	Directed, Unweighted	10876	39994

4. Result and Discussion

BFS algorithm implemented in serial and parallel form. The parallel algorithm was designed as parallel data model. The data distributed to all processor by head node and each compute node doing computation by using this part of data. Thenn compute node send the part of result to head node, so the head note compile all part of result become the final result. The serial algorithm implemented on single super computer that consist of 48 threads as explained before. And the parallel algorithm were implemented on super computer cluster with various thread. **Table 2** show the speed up.

Table 2 Speed Up

Sum of threads	p2p-gnutella08	p2p-gnutella09	p2p-gnutella05	p2p-gnutella04
4	3,98	3,64	3,97	3,61
16	15,47	14,84	15,54	15,04
32	28,75	25,42	28,80	23,32
48	42,36	44,22	46,39	31,75
64	62,12	60,44	60,29	44,77
80	75,06	78,73	78,14	59,86
96	94,56	90,62	91,19	80,63

Base on **Table 1**, the highest speed up of p2p-gnutella08, p2p-gnutella09, p2p-gnutella05, and p2p-gnutella04 dataset sequentially are 94.56, 90.62, 91.19, and 80.63. The greater number of threads, the higher the speed up. And the greater number of graph edges and vertices, the lower the speed up. But, when we compare the speed up of p2p-gnutella09 and p2p-gnutella05, p-gnutella09 get less speed up than p2p-gnutella05 even the number of vertices is almost the same. It's because the number of edges from p2p-gnutella09 is less rather than p2p-gnutella05. So the number of edges more influential the computational time than vertices. Data p2p-gnutella04 as the largest graph gives the least speed up See figure 3 to show the comparison of speed up clearly.

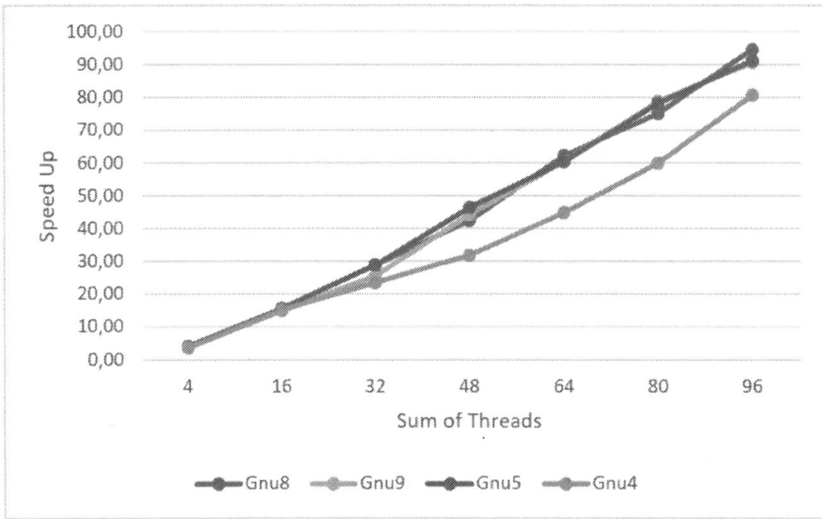

Figure 3 Speed Up

The result of the efficiency in all data is quite fluctuating. On data p2p-gnutella04 the best efficiency is on 8 thread with 95.43% and the worst is on 40 thread with 66.03%. On data p2p-gnutella05 and p2p-gnutella08 the best is on 4 thread with 99.33% and 99.57% efficiency and the worst on 8 thread with 88.76% in p2p-gnutella05 and 48 thread with 88.24% in p2p-gnutella08. On data p2p-gnutella09 the best on 96 thread with 94.39% efficiency and the worst is on 8 thread with 74.68%. See Figure 4 below.

Figure 4 Efficiency

Fugure 4 indicates that speed up is not linear with efficiency. Efficiency depend on speed up and number of threads, number of threads refer to cost. So the effiency means the optimal speed up with make sense cost. For dataset p2p-gnutella08, p2p-gnutella09, p2p-gnutella05, the effeiciency approximately 80-99 %, and p2p-gnutella08 is 66-94% . It's show that, the supercomputer cluster is quitely enough to solve the parallel BFS algorithm. To optimize the use of supercomputer cluster we can use the larger dataset and improve the parallel algorithm.

IOP Conf. Series: Journal of Physics: Conf. Series **801** (2017) 012079 doi:10.1088/1742-6596/801/1/012079

References

[1] Ajwani D, et al, 2015 Processing Large Graphs: Representations, Storage, Systems and Algorithms, *Proceedings of the 24th International Conference on World Wide Web*

[2] Beamer S,. 2013 *Distributed Memory Breadt-First Search revisited: Enabling Bottom-Up Search* Berkeley, California. 1-2.

[3] Chen R, et al. 2012 Improving large graph processing on partitioned graphs in the cloud *Proceedings of the Third ACM Symposium on Cloud Computing* Article No. 3

[4] Bondy, J.A. and Murty, U.S.R. 1976. *Graph Theory With Applications* Ontario, Canada. Elsevier Science Publishing Co., Inc.

[5] Fitriyani, Izzatul U, *2013* A Research On Parallel Computing Using Ant Colony Optimization Algorithm To Distribute Jobs Among Processors In The System Environment, *Proceeding, The 7th International Conference on Information & Communication Technology and Systems (ICTS)*

[6] Fujisawa, Katsuki, et al. 2016 *Advanced Computing and Optimization Infrastructure for Extremely Large-Scale Graphs on Post Peta-Scale Supercomputers* Springer Japan. 1-13.

[7] Grama A, Gupta A, Karypis G, and Kumar V, *Introduction to Parallel Computing* 2nd Edition, Addison-Wesley, 2003.

[8] Kuck D, (1978). *Computers and Computations*, Vol 1. John Wiley & Sons, Inc. p. 12.

[9] Leskovec J,. *Stanford Large Network Dataset Collection.* "Internet networks dataset".

[10] Luo, Lijuan. Wong, Martin. Hwu, Wen-mei. 2010 An effefctive GPU implementation of breadth-first search. *Proceedings of the 47th Design Automation Conference* pp 52-55

[11] Malicevic, Jasmina., Roy, Amithaba, et al. 2014 Scale-up graph processing in the cloud: challenges and solutions, *Proceedings of the Fourth International Workshop on Cloud Data and Platforms*

[12] Ummah, Izzatul., Fitriyani, 2013 *Parallel Computing for Newton Raphson Method (Case Study: Stewart Platform)*, KNSI 2013

[13] U. Kang, C. Tsourakakis, A. P. Appel, C. Faloutsos, and J. Leskovec. 2008 HADI: Fast diameter estimation and mining in massive graphs with hadoop. *Technical Report CMU-ML-08-117*, CMU.

[14] U. Kang, C. E. Tsourakakis, and C. Faloutsos.2009 PEGASUS: A peta-scale graph mining system - implementation and observations. *ICDM*

[15] Murphy, Richard. 2010 *Introducing the Graph 500* Albuquerque. 1-2

[16] Sadashiv, Naidila, and SM Dilip Kumar 2011 Cluster, grid and cloud computing: A detailed comparison, *The 6th International Conference on Computer Science & Education (ICCSE)*,. IEEE.

[17] Sibaroni, Yuliant. Fitriyani., Nhita, Fhira. 2016.. The Optimal High Performance Computing Infrastructure for Solving High Complexity Problem, *Journal of Telkomnika*. Vol.14, No.4, December 2016 p 281

[18] Willmore, Frank. 2012 *Introduction to Parallel Computing* Texas : Texas Advanced Computing Center.

Performance of parallel computation using CUDA for solving the one-dimensional elasticity equations

J B B Darmawan[1] and S Mungkasi[2]

[1]Department of Informatics, Faculty of Science and Technology, Sanata Dharma University, Yogyakarta, Indonesia
[2]Department of Mathematics, Faculty of Science and Technology, Sanata Dharma University, Yogyakarta, Indonesia

E-mail: b.darmawan@usd.ac.id, sudi@usd.ac.id

Abstract. In this paper, we investigate the performance of parallel computation in solving the one-dimensional elasticity equations. Elasticity equations are usually implemented in engineering science. Solving these equations fast and efficiently is desired. Therefore, we propose the use of parallel computation. Our parallel computation uses CUDA of the NVIDIA. Our research results show that parallel computation using CUDA has a great advantage and is powerful when the computation is of large scale.

1. Introduction

A large number of real world problems can be modelled mathematically. Solutions to mathematical models are representatives of the solutions to the real problems. Some mathematical models are in the form of differential equations.

In this paper, we solve a system of partial differential equations. In particular, we consider the nonlinear elasticity equations. These equations were derived from engineering problems relating to elastic wave propagation through heterogeneous media [1].

Some work has been done previously regarding the nonlinear elasticity equations. Supriyadi and Mungkasi [2] solved the nonlinear elasticity equations in a sequential computation. Solving the problem in sequential is not appropriate for a large-scale problem, as the computation is tedious for the large-scale problem. Darmawan and Mungkasi [3] solved the nonlinear elasticity equations in a parallel computation using the MPI with a cluster of workstations applying MPI_Send and MPI_Recv techniques. However, parallel programming with MPI_Send and MPI_Recv techniques in the MPI standard is not appropriate to be used in the one-dimensional elasticity equations. This is because the total time needed in the data exchange is considered more dominant compared with the total amount of time to conduct the basic elasticity computation using the finite volume method [4].

Nowadays, Graphic Processing Units (GPUs) have better performance than CPUs in both floating point operation and memory bandwidth. One of computing platforms and programming models in GPUs is Compute Unified Device Architecture (CUDA) [5]. GPUs with CUDA offer high performance at a very low cost, and they can also be integrated into high performance computer systems [6-7]. This paper investigates if we obtain high speed in parallel computations using CUDA to solve elasticity problems.

This paper is organized as follows. We recall the mathematical model concerning elasticity problems and present the numerical scheme to solve the model in Section 2. We describe the numerical method in parallel to solve elasticity problems in Section 3. Computational results and discussion are provided in Section 4. Finally, we draw some concluding remarks in Section 5.

2. Mathematical model and numerical scheme

In this section, we present the mathematical model to be solved and the numerical scheme used to solve the model.

The nonlinear elasticity equations are given by

$$\varepsilon_t - u_x = 0, \tag{1}$$

$$u_t - \sigma(\varepsilon)_x = 0. \tag{2}$$

In this model, the free variables are time t and the space x. In addition, the notation $\varepsilon = \varepsilon(x,t)$ is the strain, $u = u(x,t)$ represents the velocity, and $\sigma(\varepsilon)$ denotes the stress. The space domain that we consider in this paper is $0 \le x \le 100$. The initial condition for our test problem is

$$\varepsilon(x,0) = 0, \tag{3}$$

$$u(x,0) = 0, \tag{4}$$

for all x. The boundary condition at $x = 0$ and $x = 100$ is

$$\varepsilon(0,t) = 0, \tag{5}$$

$$u(0,t) = \begin{cases} -0.4\left(1+\cos\left(\dfrac{(t-30)\pi}{30}\right)\right) & \text{if} \quad t \le 60, \\ 0 & \text{if} \quad t > 60, \end{cases} \tag{6}$$

$$\varepsilon(100,t) = 0, \tag{7}$$

$$u(100,t) = 0. \tag{8}$$

The system of equations (1)-(2) have the form of a conservation law

$$q_t + f(q)_x = 0, \tag{9}$$

where $q = q(x,t)$ is the conserved quantity and $f(q)$ is the flux function. One of numerical methods that can be used to solve conservation laws is the finite volume method. The finite volume method itself is conservative, as the numerical quantity is conserved at any time.

The finite volume scheme for equation (9) in the fully discrete version is

$$Q_i^{n+1} = Q_i^n - \frac{\Delta t}{\Delta x}(F_{i+1/2}^n - F_{i-1/2}^n), \tag{10}$$

where Q_i^n is the approximate quantity computed in the finite volume frame work at the i-th cell at the n-th time step. In addition, the variable $F_{i+1/2}^n$ denotes the approximate flux at the $(i+1/2)$ interface from the n-th time step to the $(n+1)$-th time step. The notations Δt and Δx are the time step and the spatial cell width respectively. We use a uniform time step as well as a uniform spatial cell width. All fluxes are computed using the Lax-Friedrichs formulation. Lax-Friedrichs fluxes for equation (9) relating the finite volume scheme (10) are given by

$$F^n_{i+\frac{1}{2}} = \frac{1}{2}[f(q^n_{i+1}) + f(q^n_i)] - \frac{\Delta x}{2\Delta t}(q^n_{i+1} - q^n_i), \tag{11}$$

and

$$F^n_{i-\frac{1}{2}} = \frac{1}{2}[f(q^n_i) + f(q^n_{i-1})] - \frac{\Delta x}{2\Delta t}(q^n_i - q^n_{i-1}). \tag{12}$$

3. Method for parallel computations

The method for our parallel computations is described as follows.

NVIDIA developed the CUDA programming model and computing platform to let programmers write scalable parallel codes [5]. CUDA is an extension of the C and C++ programming languages. The programmer writes a serial program that calls parallel kernels, which may be functions or programs. A kernel executes a set of parallel threads. The threads must be organized into hierarchy of grids of thread blocks. A thread block is a set of concurrent threads that share access to a memory space privately to the block and cooperate among themselves through barrier synchronization. A grid is a set of thread blocks that each of them may be executed independently in parallel [8]. In this work, we use CUDA on a Personal Computer and a GPU for parallel programming with the C programming language.

Foster [9] mentioned that the execution time (which is varied with problem size) as well as the efficiency (which is independent of problem size) can be the metrics to evaluate parallel algorithm performance. The relative speed up S_{relative} is calculated as

$$S_{\text{relative}} = \frac{T_1}{T_p}, \tag{13}$$

where the execution time T_1 is used on one processor and the time T_p is used on p processors. The factor by which execution time is reduced on p processors is actually the relative speedup. Relative efficiency E_{relative} is calculated as

$$E_{\text{relative}} = \frac{T_1}{p\,T_p}. \tag{14}$$

Notice that from equations (13)-(14), the relative speedup is in relation with the relative efficiency.

However, according to Tan [6] in GPU parallel computing the execution times are tested on hardware platforms with totally different architectures, namely, the CPU and the GPU. The efficiency is not as a useful metric as in CPU parallel analysis. Therefore, in this paper we evaluate computational results using the execution time and the speedup defined by equation (13) with a fixed p processor equal to the number of GPU cores plus the number of CPU cores.

4. Computational results

In this section we present main results of our research.

Computations are conducted in a Personal Computer with one Core 2 Quad processor with RAM 8 GB and a GPU using GT 730 from NVIDIA with 96 cores, memory 2 GB and maximum of 1024 threads per block. We use the global memory in the GPU to share data between threads. The operating system that we use is Windows 7 on 64 bit. The parallel computing environment is CUDA Toolkit 7.5.

Table 1. Total time (in seconds) and speedup for elasticity simulation scenarios with 1001 threads per block conducted in parallel.

Size of arrays (N)	Sequential execution time (T_1)	Parallel execution time ($T_{p=96\ GPU+1\ CPU}$)	Speedup ($S_{relative}$)
10001	1.36267	0.76433	0.56091
20001	2.00200	1.52400	0.76124
30001	2.33433	2.28767	0.98001
40001	2.86500	3.08367	1.07632
50001	3.42167	3.81200	1.11408
60001	3.92067	4.60200	1.17378
70001	4.46700	5.39700	1.20819
80001	5.01800	6.16633	1.22884
90001	5.54867	6.94733	1.25207
100001	6.13567	7.73200	1.26017
110001	6.33533	8.51267	1.34368
120001	6.88533	9.29733	1.35031
130001	7.43600	10.11400	1.36014
140001	7.97700	10.87833	1.36371
150001	8.38233	11.66367	1.39146
160001	8.82933	12.46967	1.41230

We have conducted 16 simulations of parallel computations and 16 simulations of sequential computations for running the program in order to solve the elasticity problem mentioned in Section 2. The elasticity problem is based on one-dimensional elasticity equations implemented in one-dimensional array. We use scenarios with increasing size of the arrays from 10001 to 160001 with the step is 10000. We choose 1001 threads per block and calculate the number of blocks to obtain the number of threads at least equal to the dimension of the arrays.

For a basic elasticity problem we record the total time for each simulation in Table 1. As shown in Figure 1 for the total time of both sequential and parallel executions, we observe that more number of size arrays leads to slower computation. For the size of arrays between 10001 and 20001, sequential execution is faster than parallel execution. Surprisingly for the size of arrays between 30001 and 40001 we observe that parallel execution exceeds sequential execution, and parallel execution time continue to increase linearly exceeding the sequential execution time. As shown in Figure 2, for the speedup less than 1, the speedup increases linearly and sharply. However, for the speedup more than 1, the speedup increases linearly and gradually. This indicates that more size of arrays will gain better speedup. The limitation is the number of memory available. It means that for one-dimensional elasticity computations, more size of arrays in the CUDA parallel programming will gain more speedup.

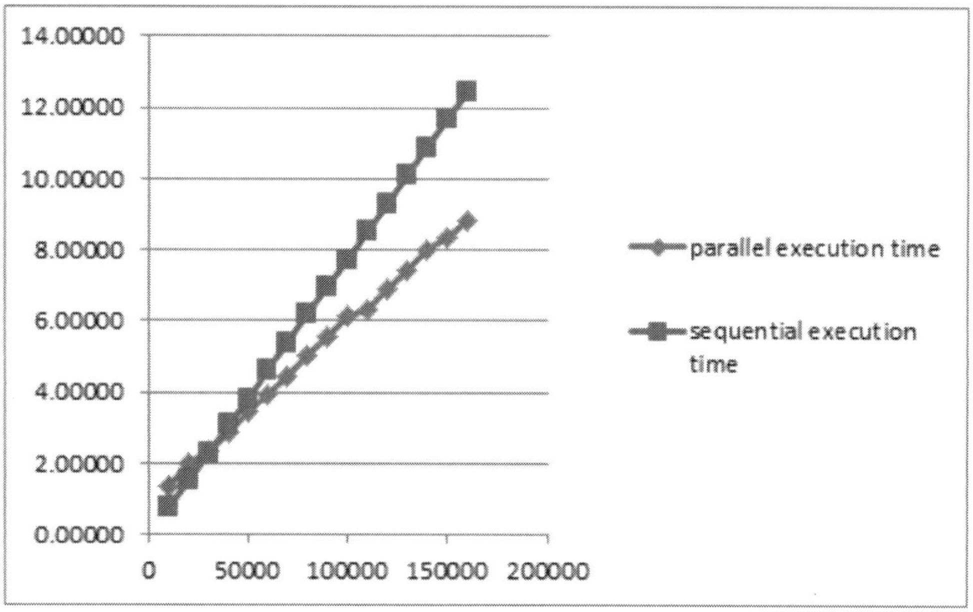

Figure 1. Total time elapsed in a elasticity simulation scenario for several computational settings. The horizontal axis is the size of arrays. The vertical axis is the total time used for computations.

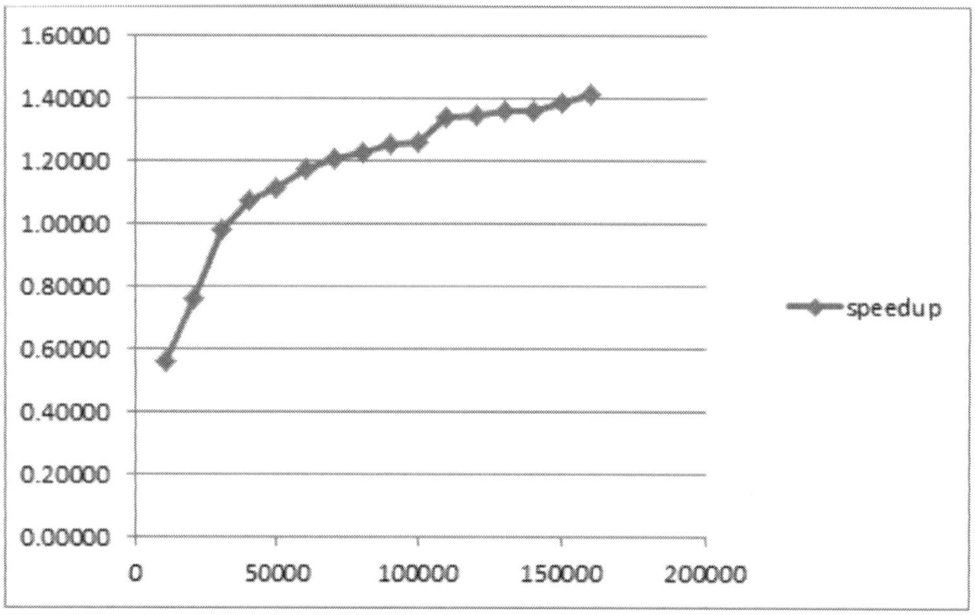

Figure 2. Speedup in an elasticity simulation scenario for several computational settings. The horizontal axis is the sizes of arrays. The vertical axis is the speedup.

5. Conclusion

We have simulated several scenarios of parallel computations for elasticity problems. We obtain that parallel programming with CUDA can be used to improve execution time of computation to solve the one-dimensional elasticity equations within an appropriate array dimension. The improvement of speedup can be obtained when the size of arrays are more than 40001 and it will continue to increase linearly until 160001. Without loss of generality, we recommend that the sizes of arrays in the one-dimensional elasticity computation must be appropriate to obtain improvement of speedup in parallel computations using CUDA.

Acknowledgments

This work was financially supported by the Institute for Research and Community Services of Sanata Dharma University (LPPM USD). The LPPM USD internal research grant year 2016 is gratefully acknowledged by both authors.

References

[1] LeVeque R J 2002 Finite-volume methods for non-linear elasticity in heterogeneous media *International Journal for Numerical Methods in Fluids* **40** 93

[2] Supriyadi B and Mungkasi S 2016 Finite volume numerical solvers for non-linear elasticity in heterogeneous media *International Journal for Multiscale Computational Engineering* **14** 479

[3] Darmawan J B B and Mungkasi S 2016 Parallel computations using a cluster of workstations to simulate elasticity problems *Journal of Physics: Conference Series* accepted

[4] Mungkasi S and Darmawan J B B 2015 Fast and efficient parallel computations using a cluster of workstations to simulate flood flows *Communications in Computer and Information Science* **516** 469

[5] NVIDIA Corp 2016 *CUDA C Programming Guide*. Available online (accessed on 21 November 2016) https:docs.nvidia.com/cuda/cuda-c-programming-guide

[6] Tan Y and Ding K 2016 A survey on GPU-based implementation of swarm intelligence algorithms *IEEE Transactions on Cybernetics* **46** 2028

[7] Coates A, Huval B, Wang T, Wu D J and Ng A Y 2013 Deep learning with COTS HPC systems *Proceedings of the International Conference on Machine Learning (ICML)* (Atlanta, GA, USA) 1337

[8] Nickolls J, Buck I, Garland M and Skadron K 2008 Scalable parallel programming with CUDA *ACM Queue* **6** 40

[9] Foster I 1995 *Designing and Building Parallel Programs: Concepts and Tools for Parallel Software Engineering* (Boston: Addison-Wesley)

IOP Conf. Series: Journal of Physics: Conf. Series **801** (2017) 012081 doi:10.1088/1742-6596/801/1/012081

Distributed Factorization Computation on Multiple Volunteered Mobile Resource to Break RSA Key

I Jaya[1], S M Hardi[2], J T Tarigan[3], E M Zamzami[4], P Sihombing[5]

Faculty of Computer Science and Information Technology,
University of Sumatera Utara

Email: ivanjaya@usu.ac.id, vani.hardi@usu.ac.id, jostarigan@usu.ac.id, elvi_zamzami@usu.ac.id, poltak@usu.ac.id

Abstract. Similar to common asymmetric encryption, RSA can be cracked by using a series mathematical calculation. The private key used to decrypt the massage can be computed using the public key. However, finding the private key may require a massive amount of calculation. In this paper, we propose a method to perform a distributed computing to calculate RSA's private key. The proposed method uses multiple volunteered mobile devices to contribute during the calculation process. Our objective is to demonstrate how the use of volunteered computing on mobile devices may be a feasible option to reduce the time required to break a weak RSA encryption and observe the behavior and running time of the application on mobile devices.

1. Introduction

1.1. Background

RSA [1] is one of the most common cryptosystem used in modern data communication. It is an asymmetric encryption that contains 2 keys: public key for encryption and private key for decryption. This capability makes asymmetric encryption is a better option than symmetric encryption, a method that require both party (sender and receiver) to share the same key. To generate these keys, it uses a series of modulus based calculation using prime numbers. This calculation can be reversed; hence it is possible to generate the private key based on the knowledge of public key. This process can be done by factoring the public key. However, most RSA based cryptosystem used a very large prime numbers to increase the complexity of the calculation process and reduce the risk of breaking the encryption. Hence, it is not possible to decompose these keys in a practical computing time using a common computer device.

The concept of using multiple hardware to solve a massive task, or mostly known as grid computing [14], has been explored for some time. A highly parallel task can be split into smaller tasks. By distributing them to multiple computers, we can reduce the computing time by performing those tasks concurrently. With a small modification on the algorithm, factorization key can be formed as a highly parallel task. It suits to be ran on a grid computing environment. However, building an infrastructure of dedicated computers to perform this computation may not be the best option for most cases. It can be expensive and redundant during idle times.

In this paper, we propose RSABreaker, an application that is capable to decompose RSA's key by using factorization on multiple volunteered mobile devices. We use Online Teamwork [10] [11] as a platform of our software. This platform allows users to let its Android based mobile device to donate it's processing power to a certain process. Our objective is to build an application to calculate RSA's private key by factoring its public key and allows the process to be ran on multiple mobile resources.

1.2. Related Works

The most straightforward method to break RSA key is to use factorization. However, there are a few works that attempted to find a better way to break the RSA key. In 1993, [19] created a trapdoor; a hardware that can easily factor RSA modulus. However, this option was denied by [20] that proposed a way to break the trapdoor and prevent the factorization attack. [15] proposes to break the key by using GPU to process the GDC computation. [16] [17] shares the same idea but with NVidias CUDA technology [18] which allows GPU to be used to perform generic computation instead of graphic computation.

The possibility of performing Grid Computing on mobile devices has been explored by many researchers. [5] gives a thorough survey regarding the implementation of grid computing on mobile devices. It summarized the positive and negative aspects of mobile grid computing and suggest that it is a possibility to that mobile devices may become the future platform of grid computing. Parmar et al [2] has suggested that while the performance is far below common desktop PC, the amount of available mobile devices makes it an option to perform grid computing.

Mobile OGSI.NET is an implementation example of Grid Computing on Mobile Devices [6]. This framework allows developers to create an application that will run on top of Mobile OGSI.NET platform. By using this platform, an application can run on mobile devices with .NET framework and be able to communicate to it's server by using HTTP request.

Some research focus on volunteered computing, a variant of grid computing where the client is capable to voluntarily donate its power. SETI@Home [4] and AVRF [3] proposed solutions to allow users voluntarily donate it's processing power to perform a task. These solutions give possibility to perform grid computing on non-dedicated hardware and gives control to the user to decide whether to collaborate on the grid or not. SETI@home targets personal computers as it's donor and use the donated processing power to calculate radio signals from space. AVRF focuses on building a grid computing framework that uses mobile devices with Android Operating System as its clients. Both solutions show a promising result on performing grid computing without accessing dedicated machines. Castro et al [13] uses volunteer computing approach to simulate electrophysiological models. [12] has done a thorough examination how manage and distribute task effectively in volunteer computing system. The paper explains important aspects to be considered in defining a task in volunteer computing, such as bandwidth and power usage.

2. RSA Encryption

RSA stands for Ron Rivest, Adi Shamir, and Leonard Adleman. First published in 1978 [1], this algorithm is one of the most commonly used method in modern cryptosystem. RSA is an asymmetric encryption: it relies on two keys (public key and private key) to perform the encryption process. Public key is publicly known and used by sender to encrypt the message, while private key is used to decrypt the message. These keys are generated by using a set of large random prime numbers. Given two prime numbers p and q, we calculate n as the modulus for the public and private key where it satisfies the rules $n = p * q$. We calculate Euler's Toitent $\varphi(n) = (p - 1) * (q - 1)$. This value is kept as part of the private key. The next step is to calculate e that satisfies two rules; $1 < e < \varphi(n)$ and the greatest common divisor of e and $\varphi(n)$ is 1 which makes e and $\varphi(n)$ coprime. We then determine d as $d = e^{-1}\ mod\ \varphi(n)$ which makes d a modular multiplicative inverse of $(e\ mod\ \varphi(n))$. The public key, which is used by the sender to encrypt the message, is (n, e) and the private key, which is used by receiver to decrypt the chipertext, is d. Given a message m, the encryption function is $c(m) = m^e\ \%\ n$. Conversely, to decrypt the encrypted message c, we can use this formula $m = c^d\ mod\ n$. If the value of p and q are available, the private key d can be computed easily. However, to obtain p and q from encryption key (n, e), we need to decompose n into p and q using factorization computation which may require a great amount of computation considering RSA cryptosystem relies on large numbers.

In this paper, we will use a simple factorization method to find the p and q values based on a known value n. Given a value n to be factorized, our method will look through all possible values from square root of n to 0 and find a value that evenly divides n. Since n is a product of multiplication two prime numbers, we will only look for odd values.

3. Application Architecture

Our RSABreaker application runs on top of Online Teamwork application. It consists of two part: the server and the client. The client is a Java based Android application with an objective to calculate factorization from a set of numbers at a certain range. When a user chooses to collaborate in the process, it will run the Online Teamwork application and pick the RSABreaker app on the menu. The Online Teamwork will then download the RSABreaker client application and automatically run the program. The client will communicate to server and the factorization process will be started as soon as the server send the parameter. The server is a PHP based application responsible to manage and distribute the task to clients and keep the result in a database. Figure 1 shows the architecture of RSA Breaker.

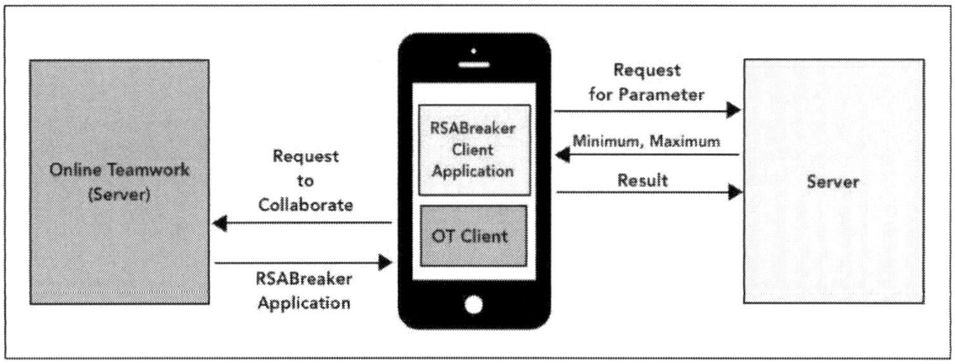

Figure 1. The Architecture of RSABreaker

Another important issue needed to be discussed is the definition of task. Determining the definition of task is important to keep the balance between bandwidth and processing power. Our objective is to split the factorization process into smaller tasks that are going to be distributed to multiple clients. While most of the operations are modulo calculations, it is important to notice that we are using big integer as our values. The clock cycle needed to run the modulo calculation using BigInteger [21] is longer than a standard 32-bit integer operation. Our current program defines each task as a set of 100,000 possible values from a certain maximum and minimum point. We only calculate the odd numbers; thus the range of maximum and minimum values is 200,000.

Communication between client and server is done via HTTP request. A PHP based server will then response with a proper task parameter. When a client sends a request for a task from the server, it will receive two values; maximum and minimum point. Client will then run the factorization process by dividing n to each values. When a new client request for another task, the server will look for the smallest minimum value that has been sent previously. This value is then set as the maximum value of the new task. The process is repeated until the minimum value is 0.

4. Implementation, Test, and Result

We conducted our test on 10 unique mobile devices with 4 different users. During the test, the user may use their mobile devices as usual while RSABreaker works on the background. In this test, we observe the behaviour and running time of RSABreaker while working on the background.
The specifications of the mobile devices are varied as listed in table 1.

Table 1. List of Mobile Devices used in test

Model	Specification (Processor, Memory)
Sony Xperia C3	Quad-core 1.2 GHz Cortex-A7, 1GB
Sony Xperia C5	Octa-core 1.7 GHz Cortex-A53, 2GB
Sony Xperia Z5	Octa-core (4x1.5 GHz Cortex-A53 & 4x2.0 GHz Cortex-A57), 3GB
Samsung Galaxy J5	Quad-core 1.2 GHz Cortex-A53, 1.5GB

To keep the running time practical for 4 devices, we use a 100 decimal digits RSA key instead of the standard 617 digits in 2048-bit RSA. Using this value will enable us to observe the practical running time when the calculation is processed on the mobile devices.

4.1. Application Behavior

In this test, we observe the Android processing power allocation based on the usage of the mobile device. The application uses AsyncTask [4] to run the factorization process, hence the user is able to use the mobile device as usual. However, we test the application with 2 modes: dedicated and normal use. In dedicated test, the user is not allowed to run other program except the Android OS built in process and installed service. The second mode is normal use, where the user is allowed to open a commonly used application such as social media and instant messaging. We took our sample from one of our client, Sony XPeria Z5. Figure 2 shows the processing power allocation during our test in both dedicated mode (left) and normal mode (right).

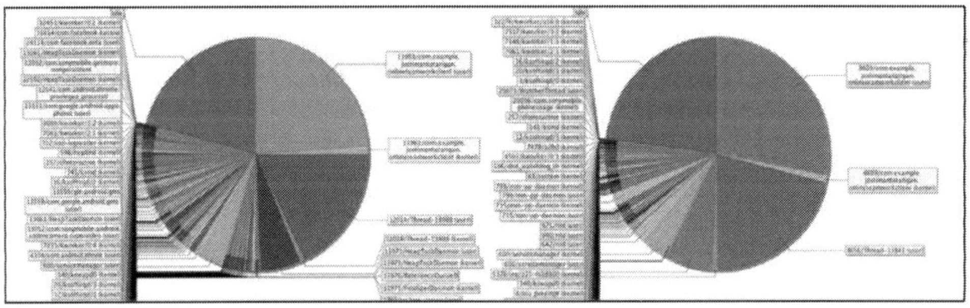

Figure 2. Processor Usage on XPeria Z5

During dedicated test, the application is able to grab approximately 50% of the processing power (light green, gold, and purple). We noticed there are 20% of idle processing power (dark green) that we were unable to use. Our current verdict that cause this behavior is the processor architecture of Z5 that split the processor into two parts (4 cores on each part)

4.2. Running Time

We also observe the running time of the application while trying to solve the 100 decimal digit key. Each task consists of 100,000 possible values (which makes the range of maximum and minimum value of each task is 200,000). Due to the high decimal digit used for the key, we were unable to observe the total time needed to find the key. However, our objective is to see the behavior of the running time of a client to complete a single task.

The observed client (XPeria Z5) was able to finish a single task in approximately 7 seconds during the dedicated mode and approximately 10 seconds in normal mode. The other clients, which have lower specifications than the XPeria Z5, were able to complete the calculation of a single task in 15-20 seconds average. We also observe as the value went lower, there were tendency that the running time was decreased slightly. This behavior is expected since BigInteger operators works faster with a smaller value.

5. Conclusion and Future Works

In this paper, we proposed a new way to break RSA key by using 4 volunteered mobile devices with various specifications. RSABreaker have successfully performed factorization in order to find the prime numbers that construct the public key. During test, we observed that each client is capable to finish a single task consists of 100,000 possible values in 10-15 seconds average. While we have not yet able to break the 50 decimal digit key, we were able to observe the behavior of using BigInteger operators and its performance in mobile devices.

Our current plan for the future works is to implement a faster algorithm to do the factorization. There are also a few methods available to break RSA key faster than factoring the public key. We also need to conduct a full test to find the total running time required to break the RSA key.

References

[1] R L Rivest, A Shamir, and L Adleman 1978 A method for obtaining digital signatures and public-key cryptosystems Communications of the ACM, vol. 21, pp. 120 – 126

[2] Parmar K B, Jani N N, Shrivastav P S, and Patel M H 2013 Mobile Grid Computing: Facts or Fantasy *International Journal of Multidisciplinary Sciences and Engineering*, vol. 4 no.1 p 26-33

[3] Arnold E 2011 AVRF: A Framework to Enable Distributed Computing Using Volunteered Mobile Resources, *Summer Research, University of Puget Sound*

[4] SETI@home: *http://setiathome.berkeley.edu/*

[5] Bichhawat A and Joshi R C, 2010 A Survey on Issues in Mobile Grid Computing *International Journal of Recent Trends in Engineering and Technology*, vol. 4, No. 2, p 15-19

[6] Chu D C and Humphrey M 2004 Mobile OGSI.NET: Grid Computing on Mobile Devices *Proceedings of the 5th IEEE/ACM International Workshop on Grid Computing*

[7] Purcell T J, Buck I, Mark W R, Hanrahan P 2002 Ray Tracing on Programmable Graphics Hardware, *Proceedings of the 29th Annual Conference on Computer Graphics and Interactive Tehcnique*, p 703-712

[8] Nah J, Kwon H, Kim D, Jeong C, Park J, Han T, Manocha D, and Park W 2014 Raycore: A Ray-Tracing Hardware Architecture for Mobile Devices *ACM Transaction on Graphic* vol 3 Article no. 162

[9] Sheharyar A and Bouhali O 2013 A Framework for Creating a Distributed Rendering Environment on the Compute Cluster *International Journal of Advanced Computer Science and Applications* vol 4 no. 6 pp 117-123

[10] Pinem R A, Tarigan J T, and Barus E S 2015 Perancangan Aplikasi Client Online Teamwork Berbasis Mobile, STMIK Kristen Neumann Indonesia

[11] Baloren C, Tarigan J T, and Barus E S 2015 Perancangan dan Pengembangan Aplikasi Server Online Teamwork, STMIK Kristen Neumann Indonesia

[12] Anderson D P, Korpela E, and Walton R 2006 High-performance Task Distribution for Volunteer Computing *e-Science and Grid Computing*

[13] Castro J, Monasterio V, Jesys 2016 Volunteer Computing Approach for the Collaborative Simulation of Electrophysiological Models *Enabling Technologies: Infrastructure for Collaborative Enterprises (WETICE)* pp 129-132

[14] Foster I and Kesselman C 1999 *The Grid: Blueprint for a New Computing Infrastructure* Morgan Kaufmann.
[15] Nakano K and Ito Y 2015 Bulk GCD Computation Using a GPU to Break Weak RSA Keys *IEEE International Parallel and Distributed Processing Symposium and Workshops* pp. 385-394
[16] Nakano K and Ito Y 2015 Bulk execution of Euclidian Algorithms on the CUDA-enabled GPU *International Journal of Networking and Computing* Volume 6 pp. 385-394
[17] Scharfglass K, Weng D, White J, Lupo C 2012 Breaking Weak 1024-bit RSA Keys with CUDA *Proceedings of 13th International Conference on Parallel and Distributed Computing Applications and Technology* pp. 207-212
[18] CUDA: *http://www.nvidia.com/object/cuda_home_new.html*
[19] Andersson R. 1993 Practical RSA Trapdoor *Electronics Letter* Vol. 29 pp. 995
[20] Kaliski Jun B S 1993 Anderson's RSA Trapdoor Can Be Broken *Electronics Letter* Vol. 29
[21] BigInteger: *https://docs.oracle.com/javase/7/docs/api/java/math/BigInteger.html*

IOP Conf. Series: Journal of Physics: Conf. Series **801** (2017) 012082 doi:10.1088/1742-6596/801/1/012082

Designing and Characterizing Periodic Image Monitoring Device for Remote Surveillance Purpose

Rahmad Fauzi, Suherman, Rhobby Maulana

Electrical Engineering Department, University of Sumatera Utara

Email : suherman@usu.ac.id

Abstract. The 433 MHz radio works in the Industrial, scientific and medical (ISM) frequency and gains popularity for various applications as it has low signal attenuation and insensitive transmission characteristics. This paper examines the radio for periodic monitoring image application. The assessment results shows that the radio link is able to deliver image with losses lower than 6% and period of image 40 s for 100 kbps bit rate.

1. Introduction

The 433 MHz frequency channel is part of the Industrial, scientific and medical (ISM) channel allocation which is license free (Hill R., 2014). The ISM 433 MHz signal propagation is low signal attenuation and not sensitive to metal and water reflection, which is suitable for low power transmission for few kilometers distant. Beside its low power consumption, the 433 MHz radio construction is relatively simple with small antenna requirement (Tuset-Peiró et al, 2013). The system has been used in many countries including Europe, America, China, and Japan which results high availability (Lin Z. et al, 2014).

There has been intensive works done by researchers in employing 433 MHz radio system. Li et al (2013) experimented 433 MHz radio propagation characteristics on greenhouse. The regression analysis results demonstrated the regression parameters and the transmitting power were quadratic relationship. It also shows that there is quadratic regression equation between the environmental factor n and transmitting power. The 433 MHz radio has also been used for biomedical applications (Kim S., 2013). The results show that the 433 MHz body sensor network (BSN) is comparable to IEEE 802.15.4 BSNs in the proximity of Bluetooth or WiFi networks. The radio frequency identification (RFID) can also take advantage of 433 MHz based wireless sensor communication (Weyn M., 2013). Considering the wide usages of 433 MHz radio, and due to the nature of insensitive signal reflection to metal and non-metal materials, this paper examines the 433 MHz radio system to be used for a radio link providing periodic image transmission for eruption image monitoring system. This paper focuses on 433 MHz channel capability assessment in delivering image packets.

2. Research Method

2.1 Hardware and software design

Figure 1 shows the radio design for periodic monitoring image delivery that consists of remote unit and the image receiver. Remote unit is built by using Raspberry Pi as the controller, 3DR 433 MHz transmitter as the radio link, power bank 3000 mAH for the power source and a camera for capturing image periodically. The receiver contains a 3DR 433 MHz radio rreceiver connected to a laptop.

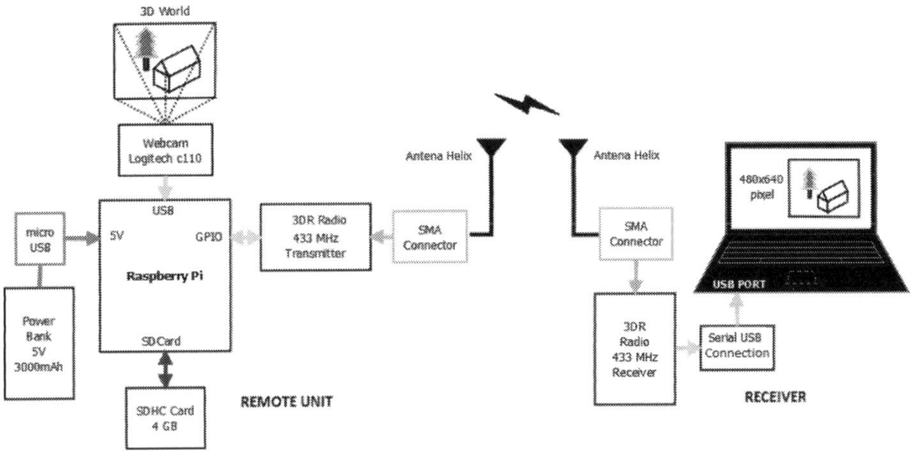

Figure 1. Radio design for image monitoring

Figure 2 shows the constructed radio system. Software included in image transfer is based on those of installed in devices. The image data is captured by using Real Term, serial capture software. Image is encoded to RGB format by using Matlab code.

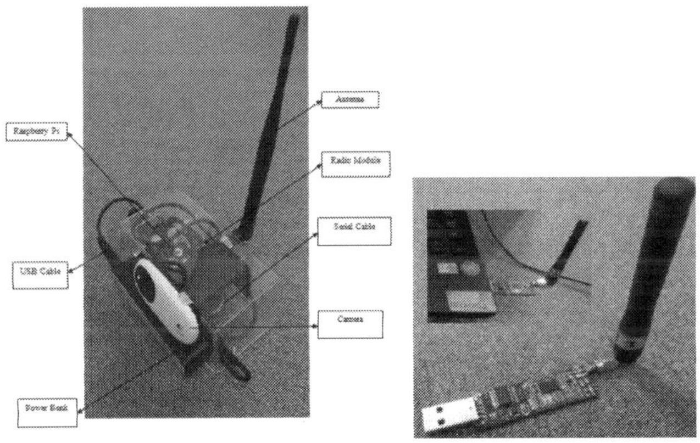

(a) Remote unit (b) Receiver unit

Figure 2. The evaluated 433 MHz radio system

2.2 Evaluation design

The 433 MHz radio evaluation for periodic image transmission involves device tests and image streaming tests. Device tests include remote unit battery test, camera test, connection test, data capture test, communication range test. The streaming tests measure loss and delay parameters.

Device tests are started from the hardware connections, Raspberry Pi operating system installation, radio configuration and connection establishment. Device tests quantification is represented by the remote powering measurement. Streaming tests include direct connection through the Ethernet, wireless connection through the 433 MHz radio, image capturing and image transmissions.

3. Evaluation Result

3.1 Remote Powering Measurement

There are three conditions measured in the experiments; standby, data transmission, and image transmission.

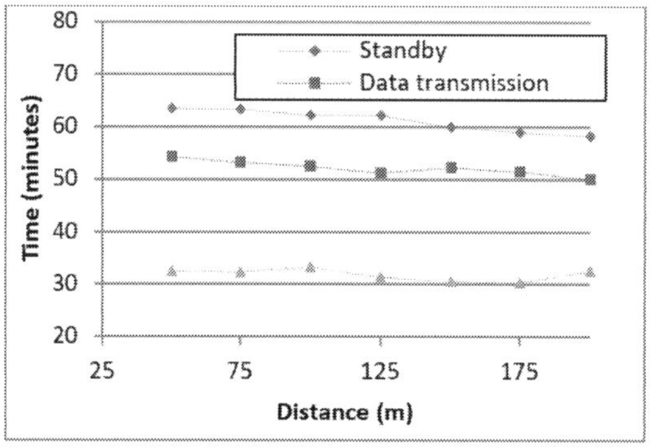

Figure 3. Battery lifetime

The battery lifetime decreases when the distance between receiver and remote is further (Figure 3). Three conditions show the same reduction pattern. However, image transmission consumes energy the most, while the standby power absorbs the least. Image transmission uses in average 48% higher than standby power and 37% more than data transmission. This data brings the fact that the remote monitoring unit should the worst condition (image transmission) and the distance when planning the power reliability for remote monitoring unit.

3.2 Direct Connection Test

Direct connection test is performed to find out the potential factors affecting the monitoring image transmission besides channel interferences. Therefore, direct connection is made by using the Ethernet cable from remote unit (Raspberry Pi port) to computer Ethernet port. The statistic results are plotted in Figure 4, which shows that most connections are successful, the main disturbance is the cable connection between modules (9.23%), and the unknown problem achieves 4.62%. The unknown problem may be caused by the problem within the receiver software.

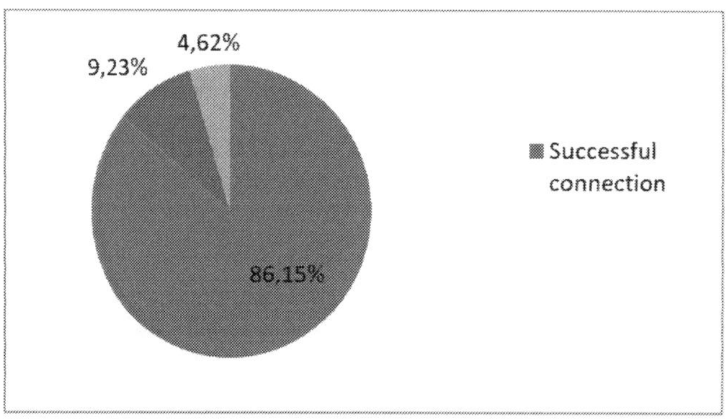

Figure 4. Direct connection test

3.3 Wireless Connection Test

Wireless connection uses the Universal Asynchronous Receiver Transmitter (UART) protocol through 3DR 433 MHz radio module. The experiment shows that the radio connection is sensitive to signal reception that results 30% connection failure (Figure 5). Therefore, sufficient signal reception should be ensured before the radio link is used.

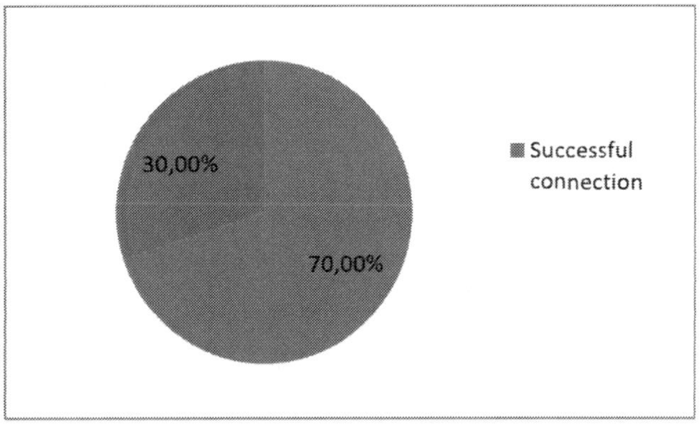

Figure 5. Wireless connection test

3.4 Image Transmission Test

During image capturing and transmission using JPEG format of 480x480 image resolution, it is obtained that packet loss increases to radio link distance (Figure 6). The packet loss increases in the second order of polynomial with R-square 0.549. The loss occurs from 2.3% to 5.2%.

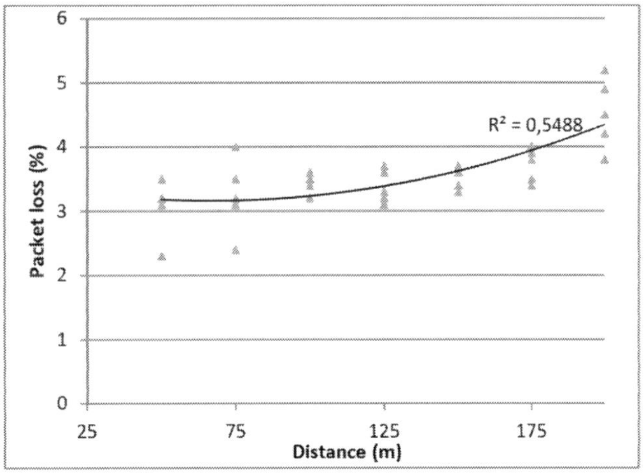

Figure 6. Image transmission test

By adjusting bit rate of 433 MHz radio to certain values, the required time for sending one image for 50 m distance is plotted in Figure 7.

Figure 7 shows for the given bit rate, the period of image capturing should be less than certain value. For instance, by using bit rate 100 kbps, a single 480x480 JPEG image requires 40 s to be sent completely. This means that for periodic monitoring application, the period of image capturing should not exceed 40 s. The higher the bit rate setting, the smaller the image capturing period. The bit rate setting is depending upon the 433 MHz radio type.

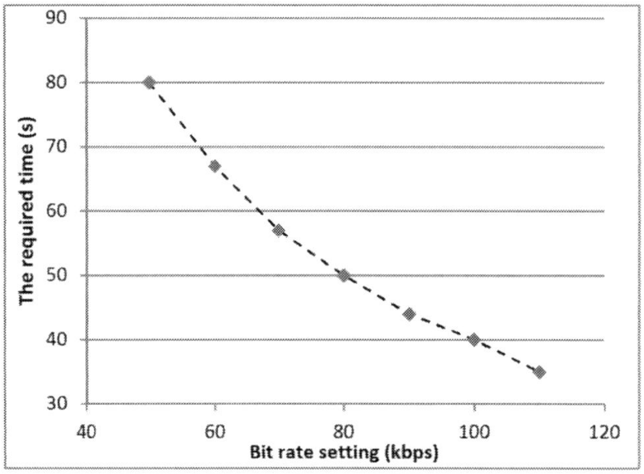

Figure 7. The time for sending a single image

4. Conclusion

This paper examines the use of 433 MHz radio for periodic image monitoring which covers device and connection tests. It is found that the link distance influences the amount of energy taken from battery, and image transmission energy consumption is 48% higher than standby power consumption. Further, for sufficient bandwidth and distance, the packet loss of image transmission is between 2.3% and 5.2%. The image-capturing period depends on radio bit rate, for 100 kbps bit rate, image capturing period should not less than 40 s.

Future works may deal with efforts for increasing the channel capacity as well as the frequency of image capturing.

5. References

[1] Tuset Peiró, P., Anglès Vazquez, A., López Vicario, J., & Vilajosana Guillén, X. (2014). On the suitability of the 433 MHz band for M2M low-power wireless communications: propagation aspects. Transactions on Emerging Telecommunications Technologies, 25(12), 1154-1168.

[2] Hill, R. (2014). Conclusions and Implications for National Legislators and Regulators. In The New International Telecommunication Regulations and the Internet (pp. 141-145). Springer Berlin Heidelberg.

[3] Lin, Z., Mak, P. I., & Martins, R. (2014, February). 9.4 A 0.5 V 1.15 mW 0.2 mm 2 Sub-GHz ZigBee receiver supporting 433/860/915/960MHz ISM bands with zero external components. In 2014 IEEE International Solid-State Circuits Conference Digest of Technical Papers (ISSCC) (pp. 164-165).

[4] Li, X., Zang, Y., Luo, X., Li, T., Liu, Y., & Kong, Q. (2013). Experiment of propagation characteristics based on 433MHz channel of WSN in orchid greenhouse. Transactions of the Chinese Society of Agricultural Engineering, 29(13), 182-189.

[5] Kim, S., Brendle, C., Lee, H. Y., Walter, M., Gloeggler, S., Krueger, S., & Leonhardt, S. (2013). Evaluation of a 433 MHz band body sensor network for biomedical applications. Sensors, 13(1), 898-917.

[6] Weyn, M., Ergeerts, G., Wante, L., Vercauteren, C., & Hellinckx, P. (2013). Survey of the DASH7 alliance protocol for 433 MHz wireless sensor communication. International Journal of Distributed Sensor Networks, 2013.

Ubiquitous learning model using interactive internet messenger group (IIMG) to improve engagement and behavior for smart campus

K Umam[1,2], S N S Mardi[1] and M Hariadi[1]

[1] Department of Electrical Engineering, Institut Teknologi Sepuluh Nopember (ITS) Surabaya, 60111, Indonesia.
[2] Walisongo State Islamic University, Semarang, 50185, Indonesia.

Email: Khothibul@gmail.com, smardi88@gmail.com, mochar@ee.its.ac.id

Abstract. The recent popularity of internet messenger based smartphone technologies has motivated some university lecturers to use them for educational activities. These technologies have enormous potential to enhance the teaching and ubiquitous learning experience for smart campus development. However, the design ubiquitous learning model using interactive internet messenger group (IIMG) and empirical evidence that would favor a broad application of mobile and ubiquitous learning in smart campus settings to improve engagement and behavior is still limited. In addition, the expectation that mobile learning could improve engagement and behavior on smart campus cannot be confirmed because the majority of the reviewed studies followed instructions paradigms. This article aims to present ubiquitous learning model design and showing learners' experiences in improved engagement and behavior using IIMG for learner–learner and learner–lecturer interactions. The method applied in this paper includes design process and quantitative analysis techniques, with the purpose of identifying scenarios of ubiquitous learning and realize the impressions of learners and lecturers about engagement and behavior aspect, and its contribution to learning.

1. Introduction

The idea of a smart campus today is incredibly powerful. Smart campus has been widely accepted and used in strategy of education reformation in several countries in the world for improving the quality of the campus learning environment[1]. With the adopted of technological development of smartphone technology, smarter campuses are built to benefit the lecturer and learner, manage the available resources and enhance user experience with proactive services anytime and anywhere [2].

Within the context of smart campus, technology of smartphone devices can be used to support teaching and learning in ubiquitous environment through Interactive internet messenger group (IIMG). Some examples of IIMG educational activities are content generating, interacting, collaborating and sharing [3][4], Content generating occurs when affective interaction and developing Question and Answer sessions. This application is designed to provide a platform for learners to share their knowledge, broadcast news, and asks for help [3][5][6]. Member are able to share or publish their work and ideas by message for others to view and download. For example, multimedia and documents files can be shared with message[3][7]. IIMG can be used by learners to collaboratively learn how to solve problems with members of a group, or to organize collaborative learning and study groups[3][4].

Many researchers have discussed and focused on the infrastructure required for constructing smart campuses and applications that related to learning at ubiquitous environment, in [8] have proposed a framework specification for u-learning in a smart campus model that integrates real world learning resources with a social network of campus. This model aims to embed of learners within a smart campus environment that provides context based personalized learning and interaction or feedback, in [5] has designed and implemented a prototype system based OnCampus scheme. The model developed in three functional modules, namely Group, Transaction, and Forum modules to provide services related to learning. The model OnCampus can significantly contribute in the following: through the creation of social environment based on interests mining, the equipping of educational guidance based on emotion analysis, as an information sharing platform, and by developing a secondary platform that aimed at the optimal allocation of campus resources, in [9] has proposed an adaptive contents delivery model for context aware ubiquitous learning according to three level service models, which create the adaptive contents for learners to get a all access in learning according to learners' interest and contexts. The evaluation of a Ubiquitous learning system shows that the learners may not only study in media on mobile device at anytime and at any place and also get a better learning experience.

The majority of mobile and u-learning model studies showed positive effects. However, the design ubiquitous learning model using interactive internet messenger group (IIMG) and empirical proof that would favor a broad application of mobile technology devices and ubiquitous learning in smart campus settings to improve engagement and behavior is still limited. In addition, the expectation that mobile learning could improve engagement and behavior on smart campus cannot be confirmed because the majority of the reviewed of studies followed instructions paradigms.

This paper aims to design and implement and showing empirical evidence about Ubiquitous learning model using interactive internet messenger group (IIMG) to improve engagement and behavior for learner–learner and learner–lecturer interactions on smart campus. The method applied in this paper includes model design process and quantitative analysis techniques, with the purpose of identifying scenario of the ubiquitous learning model using IIMG and realize the impressions of learners and teachers about the engagement and behavioral aspects, and its contribution to learning.

2. Literatur review

2.1. Ubiquitous Learning on Smart Campus

Education in a smart campus supported by mobile and smart technologies, making use of smart tools and mobile devices, can be considered smart education with ubiquitous environment. In this respect, we observe that novel technologies have been widely adopted in campus for provides connectivity between learners and their surrounding environments. For learners, learning goals is inherently identified to trigger didactic models which guide their instruction around real world data, based on unique learning contexts and delivered with the right information, at the right time, at the right place, at the right way, to the right person[10]. The learning process as ubiquitous learning doesn't restrict to be in the classroom or formal learning environments. On the contrary, the learning involves situating learners in both the real and the virtual world to extend learners' learning experiences[11]. Therefore, ubiquitous learning on smart campus must provide learners with opportunities to learn in their own environment, with another term, selecting and using the proper equipment for their work and with their living experiences, so that it can support an ongoing, contextualized and useful learning to the learner[12][13].

2.2. Use IIMG communication for ubiquitous learning

Internet Messenger (IM) can be used an alternative approach to smart campus based learning environment, IM has a series of attractive features, interaction in IM provides a mobile group communication service among members of the campus community, if a group member send a message to a group, the member sends the message to the IM server and then the server forwards it towards the each of a group members. This interaction makes the IM as an essential component of the learning process. In the u-learning environment, interaction among participants is a crucial concern to lecturer and learner[14]. As a complex and multifaceted component, interaction provides a diverse range of

IOP Conf. Series: Journal of Physics: Conf. Series **801** (2017) 012083 doi:10.1088/1742-6596/801/1/012083

functions in u-learning, and is vital for effectively improving learning[14]. Some researches indicated a positive effect of interaction on learners' academic achievement and performance as well as their satisfaction[15].

2.3. IIMG to improve engagement and behavior

Some literature used term of engagement to indicate readiness for devotes considerable energy to studying, desire and compulsion to actively participate in everyday campus activities involving things like attending to classes, participates actively in leaner organizations, adherence to lecturer's directives in the class like submit a report of progress or assignment[16][17]. Learner engagement includes both the academic and non academic activities and when get more engaged in campus experience, it tends to generate high quality learning[18].

The rapid development of IM like whatsapp, Catfiz, facebook messenger has sparked the creative incorporation of media into current pedagogical application and processes, this technology are rapidly moving beyond their original purpose and significantly increased, even impact on leaner's collaboration and engagement[19], thus contributing to the improvement of learning efficiency and learner engagement when the specific instruction is aimed to encourage learner-centered activities[20].

3. Approach and methodology

In describing the methods used to conduct this research, the model design of u-learning using IIMG, participants, data collection and analysis will be discussed and follows the guidelines of quantitative research technique[21].

3.1. U-Learning model using IIMG

This modeling technique adopted to purpose development of scenarios and conceptual design of the u-learning environment as illustrated in fig. 1, the scenario shows that a learner can learn in the ubiquitous learning environment, Notifications and adaptive learning support messages are sent to learners by IM on the group, like illustrated in fig. 2. The learner can use the smartphone or mobile devices to perform learning tasks quickly. Moreover, the model of learning system can be recommended by lecturers to learner to consult on their learning problems by IM.

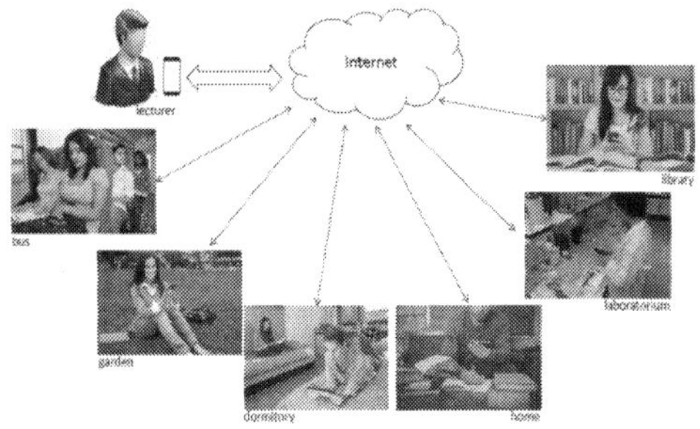

Figure 1. Scenario and conceptual design of the u-learning environment

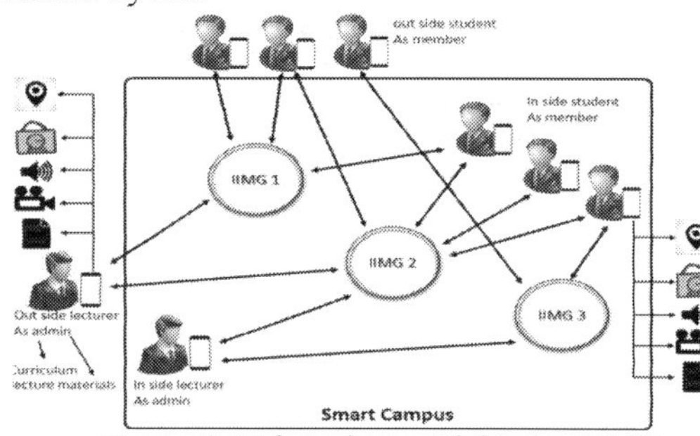

Figure 2. U-learning model using IIMG

In this model we designed a four-mode interactive mechanism for the platform. The first mode is to change a class for certain subjects into groups, lecturer as admin can assemble, organize and coordinate those groups by themselves, and the learner can join to group after registration and approved by the lecturer. The second mode is the scenario to post and circulate information using this platform, for example: lecturer transmit and distribute information like curriculum, syllabus, college contract and learning material. During online course, the learner can control what is distributed by the lecturer, the discussion pace and posting and demonstrate their own understanding or the result of the test or homework via text or visual media. The third mode is to track and record. When interaction exists between lecturer and learner, The IM platform automatically recorded all of feedback on every work from all learners and lecturer. The final mode is to conclude with brief remarks about the learners' knowledge acquirement and achievement via self-report based on the feedback.

3.2. Participant
In 2015/2016, participants in this study were two lecturers of UIN Walisongo Semarang and 147 learners (80 Male, 67 Female) from five classes, they were enrolled form dakwah and communication faculty.

3.3. Data collection and analysis
After the socialization of learning model for students in the mid lecture on December 2015, next, an initial questionnaire was given to the whole class of 147 learners before the start of the experiment to find out if they owned mobile devices and whatsapp. During the pilot program, daily activities of the group were facilitated by the lecturer that as a group administrator, moderator and driver to set the discussion or communication within the group, and researchers join in the group to monitor and determine the extent of use of the IIMG for engagement and collaboration between learner's communities, learner and their lecturer. Just before the end of the semester, the evaluation questionnaire was used to collect data from the students to measure the impact of IIMG on their engagement and behavior. Each participant was provided a questionnaire consisting of 22 items was designed. The survey contained questions regarding learner self-perception of their engagement during IIMG use. The results of attitude survey were converted into a numeric system allowing calculation of mean scores for each question and for each learner.

Measurement of learner attitudes was obtained using an existing attitude survey modified by the researcher. There were a number of questionnaires about learner attitudes toward technology. There is the Computer Attitude Questionnaire (CAQ) a 65-item Likert-scale type instrument for measuring of learners' attitudes was chosen by the researcher. Same item for measuring the importance and enjoyment of computer use were modified into an attitude survey to measure learner perceptions of the impact of the instructional technology being investigated, that is IIMG, and their perceptions about their enjoyment and its importance in groups instruction. This is a free instrument available online at https://iittl.unt.edu/content/computer-attitude-questionnaire-caq developed by The Technology Applications Center for Educator Development. There are 7 subscales measuring various learner attitude components, but only part 1 was used and it was modified by changing the word "computer" to " u-learning model using IIMG.

This instrument has been tested and used extensively by researchers. The reliability and validity data of CAQ are provided as part of the survey packages conducted in 1995 by Knezek and Christensen with a population of five hundred eightylearner and eight seventh and eighth grades in Texas to determine stability of measurement for the instrument. This attitude survey was modified by the researcher for comparative analysis of actual actual learner behavior and their own perception of their attitudes towards the IIMG.

4. Result
This section presents the results obtained from the model defined and implemented in order to develop u-learning model and analyze its impact in real environment, the results of the research based on the data collected and presented. Results are analyzed and presented in various sections below;

- Do you have a smart phone?

This question required for detection of participants to indicate if they had a cell phone and if so which type (for example, smart phone or normal phone). The result from this question, 100% of the participants had a mobile phone and 85% of those devices were smart phone with the capabilities of accessing the internet, sending instant or multimedia messages, and other applications.

- Do you have a whatsapp application?

With our research, it was important to find out the average number of participants that had whatsapp applications on their mobile phone. The result from this question, the data show that 95% of the participants who have a smartphone, they also have a whatsapp application and 5% did not have and did not use whatsapp for personal reasons.

- How often do you use your smart phone to chat with whatsapp application?

This question was directed at finding out how often students used their mobiles to access whatsapp for class-related matters like discussion with peer friends or consult with their lecturer before and after learning models included. It was important to find out the contribution of the learner while accessing whatsapp to ascertain the number of the learner who were active participation, or sometimes or those who were not participating at all. The results are presented in Table 3.

Table 3. Chatting with peer learner or lecture

		Regularly	Sometimes	Never did
Before	With lecturer	5%	11%	84%
	With Peer learner	66%	23%	11%
After	With lecturer	71%	21%	8%
	With Peer learner	76%	17%	7%

- Learns perceptions about benefit using whasapp

The learner was asked to specify their level of agreements or disagreements on a scale to measure of subject attitude. Learner self-perception of personal attitudes of u-learning model using IIMG was collected using the modified CAQ. The results are presented in Table 4.

On the information accessed on the whtasapp as shown in table 4. 74% either agreed or strongly agreed that they are enjoy classroom instruction in the u-learning model using IIMG, while 11% neutral and 15% either disagreed or strongly disagreed, Similarly, 79% of participants agreed or strongly agreed if u-learning model using IIMG can make concentrate better on the lesson and 67% agreed or strongly agreed if they are know how to interaction in IIMG they will be able to get a good job. It is inversely with they are who tired in the u-learning model using IIMG as shown in table 4, 14% either agreed or strongly agreed, while 8% neutral and 78% either disagreed or strongly disagreed.

On the issue of usefulness of u-learning model using IIMG for engagement (Table 4 Item 1, 3 to 11, 13 to 17, 19, 21 and 22) a large number of 78% of the participants believed that the u-learning model using IIMG helped them to engage and be informed about the module in and outside the classroom. 11% neutral position and 11% disagreed.

The results of this observation and investigation showed that u-learning model using IIMG use has a positive effect on the behavior of all learners, thus, on their engagement with the ubiquitous learning environment. Overall, learners were aware of the positive impact of u-learning model using IIMG use on their engagement in ubiquitous classroom instruction. They regard positively, and this was evidenced by their task behavioral improvement. The investigation data showed general improvement in student behavior which translates into improved student engagement

Table 4. U-learning model using IIMG Attitude Survey Results

No	QUIZ	SA	A	N	DA	SDA
1	I enjoy classroom instruction in the u-learning model using IIMG	32%	42%	11%	7%	8%
2	I am tired in the u-learning model using IIMG	7%	7%	8%	46%	32%
3	I will be able to get a good job if I know how to interaction in IIMG	33%	34%	21%	9%	3%
4	I can concentrate better on the lesson in u-learning model using IIMG	38%	41%	12%	7%	2%
5	I would work harder if in u-learning model using IIMG	25%	49%	13%	7%	6%
6	I know that u-learning model using IIMG gives me more opportunities to learn new things	42%	37%	12%	7%	2%

7	I understand the lesson better when the u-learning model using IIMG is used	36%	39%	12%	9%	4%
8	I believe that if more teachers used the u-learning model using IIMG, I would enjoy school more	38%	38%	11%	7%	6%
9	I believe that it is important for me to be able to use technologies such as the u-learning model using IIMG	47%	35%	9%	6%	3%
10	I can learn new things when the u-learning model using IIMG is used	41%	36%	9%	9%	5%
11	I feel comfortable in u-learning model using IIMG	53%	32%	5%	8%	2%
12	I think lessons take longer in u-learning model using IIMG	3%	8%	3%	36%	50%
13	In u-learning model using IIMG does not scare me	43%	35%	10%	8%	4%
14	In u-learning model using IIMG does not make me nervous	51%	31%	9%	6%	3%
15	In u-learning model using IIMG not difficult	38%	46%	8%	5%	3%
16	In u-learning model using IIMG, I want to work and learn whenever I can	53%	31%	5%	8%	3%
17	I work very hard on my schoolwork	33%	34%	21%	8%	4%
18	I do not try hard in school	5%	9%	14%	46%	26%
19	I pay attention in class	27%	47%	13%	8%	5%
20	When I am in class, I just act as if I'm working	4%	20%	46%	21%	9%
21	It is important to do my best in campus	48%	36%	5%	8%	3%
22	I always try to complete my assignments	47%	35%	8%	7%	3%

*The scoring scale for questions 2, 12, 18, and 20 was reversed

5. Conclusion and Recommendations

The result of this research shows that the u-learning model using IIMG emphatically positive and can improve significantly on learner's engagement and behavior. Our analysis indicates that learners and lecturer, highly engaged in the learning process in ways transcended traditional classroom activities. The experimental evidence shows that IIMG can be used as an educational tool for u-learning in smart campus to help learners to engage and collaborate. Therefore, we conclude that, u-learning model using IIMG could be effective for student to engage as well as succeed in their campus activities.

For future work, the opportunities exist for further studies in this area. Corporation between u-learning model with mobile learning or e-learning it is possible to be done to produce models more smart in order to create a smart campus dream, so the learning activities between students and professors increasingly interwoven dynamic and output is expected to have high competitiveness.

References

[1] Tsai I and Yeh C 2015 Using a Novel Model to Measure Smart Campus Service Quality: An Empirical Case *IJEETDM* vol 1 no 4 pp 175–181
[2] Chen G D, Chang C K and Wang C Y 2008 Ubiquitous learning website: Scaffold learners by mobile devices with information-aware techniques *Comput. Educ*, vol. 50 no. 1 pp. 77–90
[3] Smit I 2015 WhatsApp with learning preferences? *Proc. - Front. Educ. Conf. FIE*, vol 2015 Decem
[4] Chuang Y T 2015 SSCLS: A smartphone-supported collaborative learning system *Telemat. Informatics* vol 32 no 3, pp 463–474
[5] Dong X, Kong X, Zhang F, Chen Z and Kang J 2016 OnCampus: a mobile platform towards a smart campus *Springerplus* vol 5 no 1 p 974
[6] Gu R G R, Zhang S Z S and He J H J 2009 Instant Messaging Tool Based Online Question-and-

Answer System: An Effort for a New Group Learning Environment *Int. Conf. Comput. Intell. Softw. Eng.* pp 1–4
[7] Atif Y, Mathew S and Lakas A 2015 Building a smart campus to support ubiquitous learning *J. Ambient Intell. Humaniz. Comput.* vol 6 no 2 pp 223–238
[8] Atif Y and Mathew S 2013 A Social Web of Things Approach to a Smart Campus Model *IEEE Int. Conf. Green Comput. Commun. IEEE Internet Things IEEE Cyber Phys. Soc. Comput*
[9] Xinyou Z and Toshio O 2011 Adaptive multimedia content delivery for context aware u-learning *J.Mobile Learn. Organ.* vol 5 no 1 pp 46–63
[10] Fischer G 2012 Context-Aware Systems: The ‘ Right ’ Information, at the ‘ Right ’ Time, in the ‘ Right ’ Place , in the ‘ Right ’ Way , to the ‘ Right ’ Person, *ACM*
[11] Shih J L, Chu H C, Hwang G J and Kinshuk 2011 An investigation of attitudes of students and teachers about participating in a context-aware ubiquitous learning activity *Br. J. Educ. Technol.* vol 42 no 3 pp 373–394
[12] Wong L H, Milrad M and Specht M 2015 Seamless learning in the age of mobile connectivity *Seamless Learn. Age Mob. Connect.* pp 1–500
[13] Yahya J, Ahmad 2010 The definition and characteristics of ubiquiteous learning *Int. J. Educ. Dev. using Inf. Commun. Technol.* vol 6 no 1 pp 117–127
[14] Bental D S, Jennings E, Papadopoulou E, Taylor N K, Williams M H, Blackmun F R, Ibrahim I S, Lim M Y, Mimtsoudis I and Whyte S W 2015 Smartening Up the Student Learning Experience with Ubiquitous Media *ACM Trans. Multimed. Comput. Commun. Appl.* vol 12 no 1s pp 1–23
[15] Molomer S and Trausan-Matu S 2014 Analysis of interactions in a collaborative learning session of instant messaging *Proc. - RoEduNet IEEE Int. Conf.* pp 5–8
[16] Coates H 2006 student Engagement in Campus-Based and Online Education *Student Engagem. Campus-Based Online Educ. Univ. Connect.* vol. 9780203969 pp 1–212
[17] Ivala E, Gachago D, Condy J and Chigona A 2013 Enhancing student engagement with their studies: A digital storytelling approach *Creat. Educ.* vol 4 no 10 pp 82–89
[18] Nelson K, Clarke J, Stoodley I and Creagh T 2014 *Establishing a framework for transforming student engagement, success and retention in higher education institutions.* http://studentengagementmaturitymodel.net/
[19] Badge J L, Saunders N F W and Cann A J 2012 Beyond marks: New tools to visualise student engagement via social networks *Res. Learn. Technol.* vol 20 no 1 pp 1–14
[20] Kim Y, Jeong S, Ji Y, Lee S, Kwon K H and Jeon J W 2015 Smartphone response system using twitter to enable effective interaction and improve engagement in large classrooms *IEEE Trans. Educ.* vol 58 no 2 pp 98–103
[21] Creswell J W 2014 *Research Design: Qualitative, Quantitative, and Mixed Methods Approaches*, 4 Edition. SAGE Publications Inc. United State of America

Remote Monitoring of Post-eruption Volcano Environment Based-On Wireless Sensor Network (WSN): The Mount Sinabung Case

Soeharwinto[1], Emerson Sinulingga[2] and Baihaqi Siregar[3]

Electrical Engineering Department FT USU[1,2], Information Technology Department FASILKOM-IT USU[3]

E-mail: soeharwinto@usu.ac.id[1], emerson.sinulingga@usu.ac.id[2], baihaqi@usu.ac.id[3]

Abstract. An accurate information can be useful for authorities to make good policies for preventive and mitigation after volcano eruption disaster. Monitoring of environmental parameters of post-eruption volcano provides an important information for authorities. Such monitoring system can be develop using the Wireless Network Sensor technology. Many application has been developed using the Wireless Sensor Network technology, such as floods early warning system, sun radiation mapping, and watershed monitoring. This paper describes the implementation of a remote environment monitoring system of mount Sinabung post-eruption. The system monitor three environmental parameters: soil condition, water quality and air quality (outdoor). Motes equipped with proper sensors, as components of the monitoring system placed in sample locations. The measured value from the sensors periodically sends to data server using 3G/GPRS communication module. The data can be downloaded by the user for further analysis.The measurement and data analysis results generally indicate that the environmental parameters in the range of normal/standard condition. The sample locations are safe for living and suitable for cultivation, but awareness is strictly required due to the uncertainty of Sinabung status.

1. Introduction

Mount Sinabung (3.17°N, 98.39°E) is one of 130 volcanoes in Indonesia, lies on Pacific Ring of Fire — an arc of volcanoes and fault lines encircling the Pacific Basin. On 2010, the stratovolcano type with summit elevation 2460 m (8071 ft) erupted after dormant for more than 400 years. Since then the volcano intermittently erupting. The volcano dormancy more or less same as mount Pinatubo at the Philippines, which has a potentiality for massive magmatic eruption.

The eruption caused an impact in three sectors: economic, social and environmental. Volcanic ashes and hot cloud have ruined the resident houses and farms. The authorities reported until the end of 2014, for the productive economic sector, has lost about 896.64 billion rupiahs. At least 16 lives have dead according to the authorities. An estimation counts from the authorities showed more than 29000 residents from 32 villages had moved to safety locations. The authorities have defined the hazard area with 3, 4, 5 and 7 km radius from Sinabung crater. Unfortunately, the uncertainty of Sinabung condition pushed some of them back to their home and farms, which is unnecessary and disobeyed a warning noticed from authorities. Water resources and environment around Sinabung could be contaminated by hazardous materials produced by eruptive activity, which is not being noted by the residents.

To avoid unnecessary casualties, accurate information about the environmental (soil, water, and air) condition should be available for the authority accordingly as well as for other stakeholders. The information should be easy to access too. Many applications has been built using the WSN technology [1, 2, 3, 4, 5, 6, 7].

This paper describes the implementation of the environmental remote monitoring system in mount Sinabung area. The sensors for monitoring the water quality, soil and gasses in the air were placed at the pilot points. Then, the information gathered by the sensors send to data server using GSM/3G channel communications and presents it in the application web. Finally, authorities and other stakeholders can access the web via the Internet to get the information. The Open source applications has been used in data analysis [8, 9, 10].

2. Methods and Configuration

An Arduino based system was used for processing the signal from sensors. A Waterproof packaging of the mote was required since the sensors will be placed at the outdoor. Smart Water (SW), Smart Agriculture(SA) and Smart Environment(SE) from Libelium Plug&Sense! product has been deployed at sampled locations.

The sensor and motes locations are at Tanjung Merawa, Tiganderket and Perbaji village, Tiganderket residence (Figure 1). Each motes will sense different environment indicators. The Smart Water will monitor the water quality parameters: Oxidation-Reduction Potensial (ORP), water pH, water temperatur, Dissolved Oxygen, and Conductivity. The Smart Agriculture will monitor the parameters related to soil condition, such as temperature (°C),humidity (%RH),atmospheric pressure(kPa), soil temperature (°C), Anemometer (km/hour), Pluviometer (mm/day), Soil moisture (Hz), Vane direction, Leaf wetness(%RH). As the Smart Environment will monitor the presence of certain gases in the air, such as Carbon dioxide (CO_2), Nitrogen dioxide (NO_2), Oxygen (O_2), and Carbon monoxide (CO).

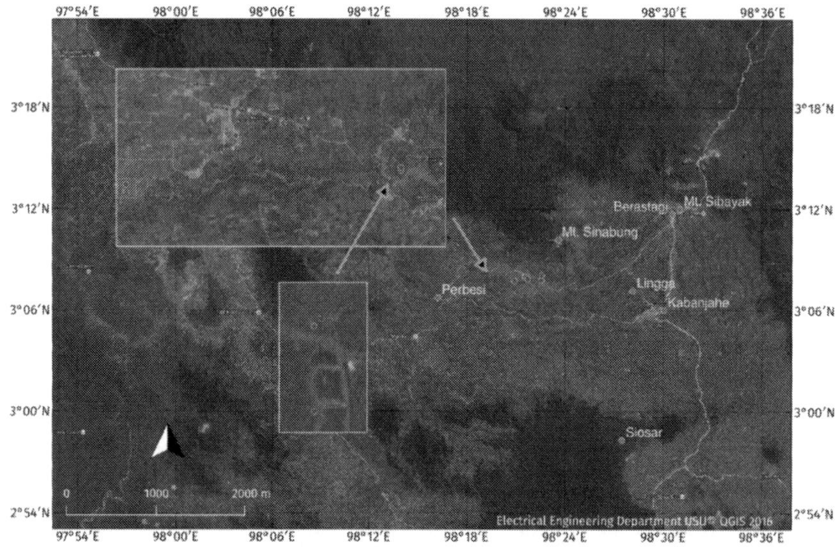

Figure 1: Motes location for remote monitoring environment in mount Sinabung area

Since the number of equipment is limited, the data collection is done successively at pilot locations. The first location for data collection is at Tanjung Merawa and Tiganderket village, and Perbaji village for the second location.

The SW, SA and SW use GSM/3G as channel communication link. Then, the link connecting the SW, SA, and SE to BTS. Through the Internet cloud, the operator BTS forward the data packets to the data server. The topology of the system show on Figure 2. Periodically, motes activate and run a web service on data server accordingly. This web service then triggers the sensor to sampling the current value of the parameters. Afterward, the sampled value send back to the data server. In data server, a simple table has provided to store the information from the sensors. Each mote has one table to store their record sampled value. A web page has been build to present this data online. To analyze data further, the stored data at online web available for download in CSV format.

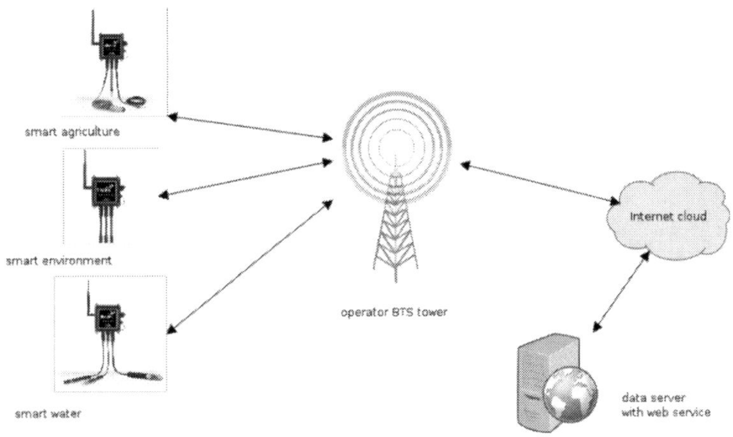

Figure 2: Network topology remote environment monitoring system

3. Result and Discussion

3.1. Soil Condition Monitoring

The Smart Agriculture sensor measured soil moisture in frequency unit (Hz), therefore the conversion is needed to represent it in soil moisture percentage unit. After calibrated the sensor, the maximum frequency F_{max} = 7986 Hz, while the minimum frequency is F_{min} = 51 Hz. This means, for every 1% change in soil moisture, will provide the frequency difference of 39.59 Hz. Soil with high-value moisture percentage represents of high humidity soil. Likewise, soil with low-value moisture percentage represents of dry soil (low humidity soil). The soil moisture and temperature for 25 cm depth sensor shows on Figure 3.

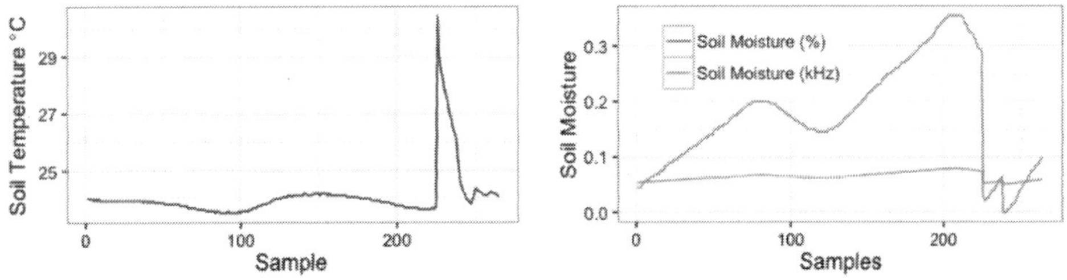

(a) Soil temperature (b) Soil moisture

Figure 3: Monitoring result for soil temperature and moisture with 25 cm sensor depth

Measured soil temperature in the range of 23.50°C – 30.43°C, with mean temperature 23.93°C and median temperature 24.08°C. Maximum soil moisture from sampled data is 0.355%, this result inferred that water fraction in soil is very small at 25 cm depth below ground surface.

Although the anemometer recorded maximum wind speed at 24.8 km/hour, but the mean wind speed was about 5.69 km/hour or equal to 1.58 m/s. This is below the minimum wind speed required to turn wind generator plant to produce electricity, which is about 4 m/s. Vane direction showed the most wind direction to South-West (Figure 4).

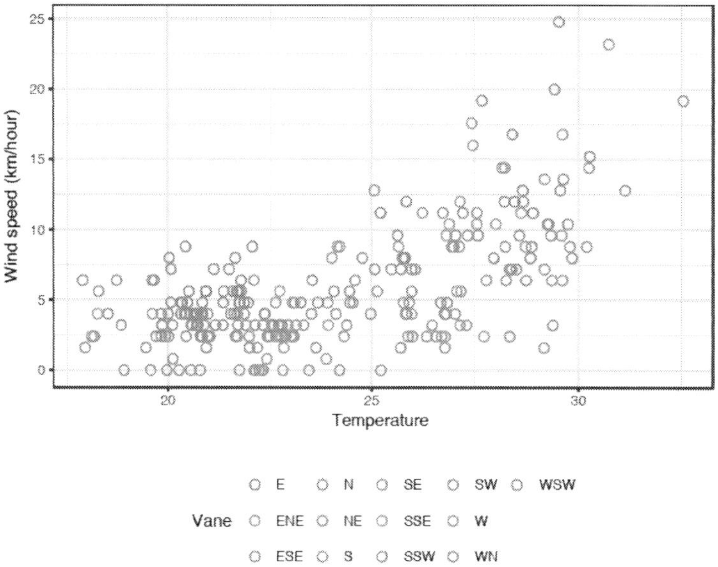

Figure 4: Monitoring result for wind speed and temperature with vane direction. Sensor height about 15m height above ground level

3.2. Water Quality Monitoring

Oxidation-Reduction Potential (ORP) parameter measured in negative number. A negative ORP value indicate that a substance is a reducing agent. The lower the value measured, the more anti-oxidizing it is. ORP measured with ±15 cm sensor depth below water surface. The result in range of -1.90 - -0.60 Volts, with mean value at -1.011 Volts (Figure 5). This inferred that the sample water has big reduction potential. A reference ORP with 0 V is at 25°C for hydrogen reaction $2\,H^+(eq) + 2\,e^- \longrightarrow H_2(g)$. ORP value at 25°C for water (H_2O) with reaction $2\,H_2O\,(l) + 2\,e^- \longrightarrow H_2(g) + 2\,OH^-(eq)$ is -0.83 V.

Another parameter to determine the water quality is Dissolved Oxygen (DO), which refers to oxygen (O_2) gas molecules present in the water. In real, O_2 present as part of a water molecule (H_2O), that can not be used directly by animal and plant in the water. The animal and plant depend on dissolved oxygen for respiration. Oxygen sources come from the surrounding air and as a product of photosynthesis from aquatic plants. In general, high levels of dissolved oxygen are best for a healthy water ecosystem.

An alternative expression of DO is in terms of air saturation percentage in water. A DO with 100% of air saturation in water means that the water is holding as many dissolved gas molecules as it can in equilibrium. The maximum percentage of DO from measured sample is 160%. During the process of photosynthesis, oxygen is produced as a waste product. This

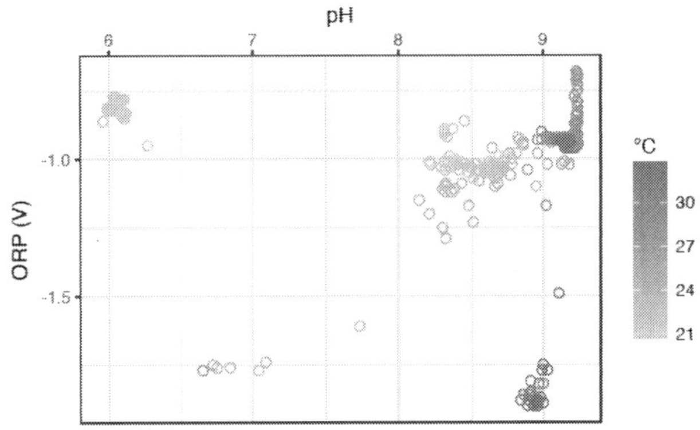

Figure 5: Monitoring result for Oxidation-Reduction Potential. Sensor depth about 15 cm below the water surface

adds to the dissolved oxygen concentration in the water, potentially bringing it above 100% air saturation. The DO percentage with temperature depicted in Figure 6. Mean DO value is 89.37%, which is good for plant and animal in the water.

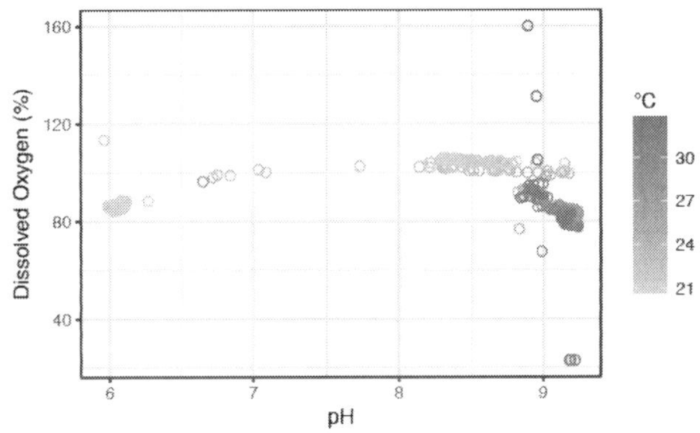

Figure 6: Monitoring result for Dissolved Oxygen. Sensor depth about 15 cm below the water surface

Water conductivity is a measure of the capability of a solution such as water to pass an electric current. This is an indicator of the concentration of dissolved electrolyte ions in the water. It doesn't identify the specific ions in the water. However, significant increases in conductivity may be an indicator that polluting discharges have entered the water.

The basic unit of measurement for water conductivity is micro-mhos per centimeter (μmhos/cm) or micro siemens per centimeter (μS/cm). It is a measure of the inverse of the amount of resistance an electric charge meets in traveling through the water. Distilled water has a conductivity ranging from 0.5 to 3 μS/cm, while most streams range between 50 to 1500 μS/cm. Freshwater streams ideally have a conductivity between 150 to 500 μS/cm to support diverse aquatic life.

Figure 7 depicted water conductivity with pH and temperature from collected data. Data variation of water conductivity quite diverse, ranging from 0.82 μS/cm to 1149.08 μS/cm, with mean data at 509.59 μS/cm. Conductivity data clustered into three range regions: Region

A for values below 150 μS/cm, region B for values around 150 μS/cm and region C for values between 900 μS/cm – 1200 μS/cm. Values of pH in region B less than 6.5 with low temperatures have conductivity about 150 μS/cm. In region C, the water temperature values are about more than 27°C with pH values above 9 have conductivity ranging between 900 — 1200 μS/cm. Conductivity values in region A are around 0.82 μS/cm with pH values ranging between 6.5 — 9.

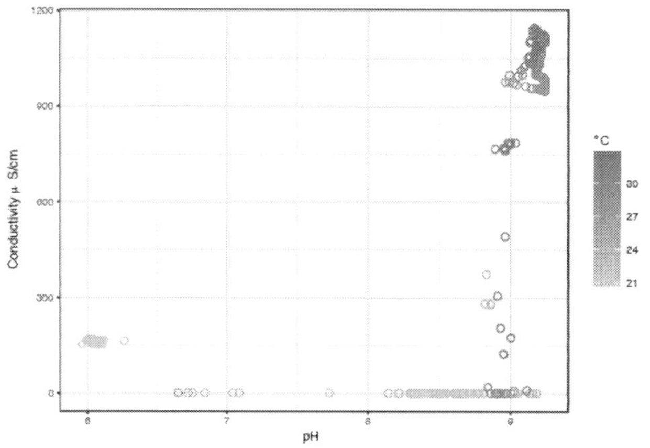

Figure 7: Water conductivity and pH with water temperature. Sensor depth about 15 cm below the surface

3.3. Gasses monitoring in air

The gasses monitored by *Smart Environment* mote are: O_2, N_2O, CO_2 and CO. As default features, the temperature and relative humidity monitored as well. All of the gasses measured in voltage unit, the unit conversion to percentage or parts per million volume unit is required. Although WaspSensorGas_v20.h and WaspSensorGas_v20.cpp provide the function to convert the measured voltage unit into percentage or ppm, but none of the function was implemented. Only the function to read the data from sensors implemented. The conversion to percentage or ppm unit calculated manually. The specific equations [11] are used for conversions.

The mean O_2 from data collected is 17.18%, ranging in 16.51% - 18.40%(red dashed line at Figure 8). Which is below to expected atmospheric concentration i.e: 21%. Like other motes, Smart Environment mote equipped with small solar PV module and battery. The mean battery status of the mote is 83.28%, which is good enough for supplying the mote in normal operation. The Formula 1 used to approximate the percentage concentration of O_2 in air, where V_{read} is the voltage read by the sensor.

$$O_2\ (\%) = \frac{V_{read} - 0.45}{0.318} \tag{1}$$

Two steps needed to convert voltage value read by CO_2 sensor into ppm concentration. First, calculate the ΔEMF (equation 2). Then applying estimated ppm concentration CO_2 using calculated ΔEMF as parameter (equation 3). Others parameter are EMF_1 as baseline voltage reading for 350 ppm CO_2, and EMF_2 as voltage reading from sensor. The GAIN parameter setup at 100. The conversion result depicted on Figure 9, which showed the CO_2 mean concentration at 485.8ppm.

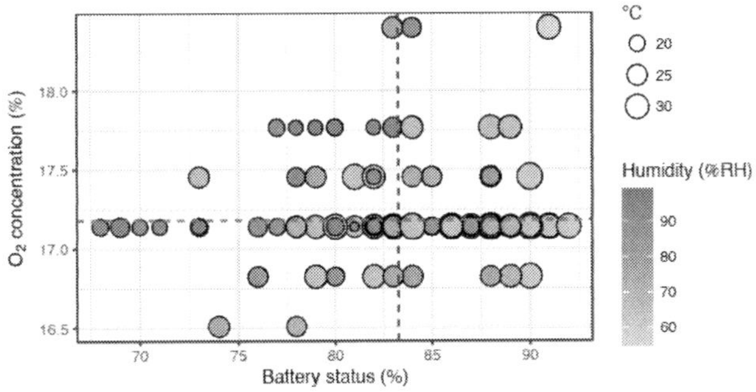

Figure 8: O_2 monitoring result from Smart Environment

$$\Delta EMF(mV) = \frac{EMF_1 - EMF_2}{GAIN} \times 1000 \tag{2}$$

$$CO_2(ppm) = 10^{\left(\frac{\Delta EMF + 163.4}{64.25}\right)} \tag{3}$$

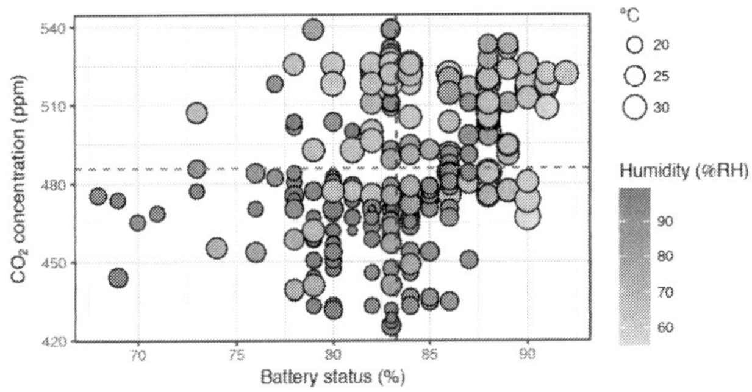

Figure 9: CO_2 monitoring result from Smart Environment

The NO_2 sensor has typical air resistance R_O 2200 Ω with configured power supply V_{cc} at 1.8 volts and 2000 Ω load resistance R_L. To calculate the NO_2 concentration in ppm, first calculate sensor resistance using the equation 4.

$$R_S = \frac{R_L}{V_{cc} - V_{read}} \times V_{read} \tag{4}$$

The approximation of NO_2 concentration in ppm, calculate with equation 5. Figure 10 depicted the NO_2 concentration (ppm) calculation result in mote battery status (%) with temperature and relative humidity. The concentration of NO_2 ranging from 0.02006 - 0.03044 ppm, which is below of significant harm level (US EPA Guidelines) at 2.0 ppm. Mean NO_2 concentration is 0.02465 ppm, while the mean of temperature and humidity are 23.70 °C and 80.48% respectively.

$$NO_2(ppm) = 10^{\left(\frac{\log\left(\frac{R_S}{R_O}\right) - 2.702}{1.865}\right)} \tag{5}$$

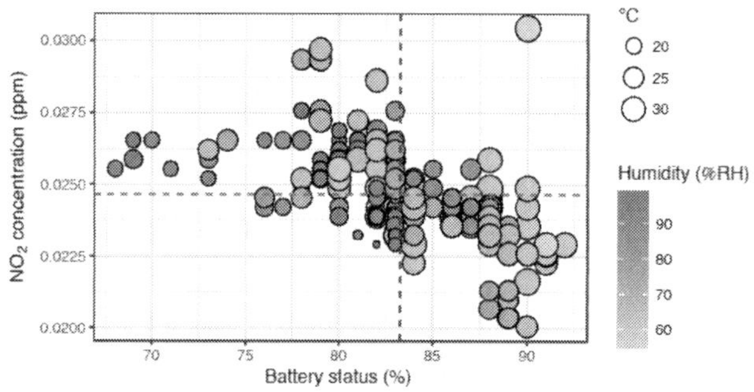

Figure 10: NO_2 monitoring result from Smart Environment

Estimation of CO concentration in ppm, calculate using Equation 6. The CO gas concentration ranging in 11.37 - 24.12 ppm, with mean value at 22.62 ppm.

$$\mathrm{CO}(ppm) = 10^{\left(\frac{0.09721-\log\left(\frac{R_S}{R_O}\right)}{0.6483}\right)} \tag{6}$$

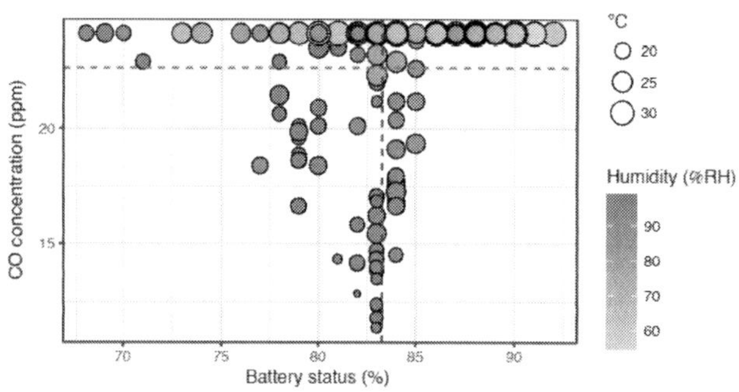

Figure 11: CO monitoring result from Smart Environment

4. Conclusion

The small scale remote environment monitoring have been implemented using motes equipped with small solar PV module and 3G/GPRS module. Three environmental aspects have been measured, i.e: soil condition, water quality, and air quality. The WSN methods provide an alternative technic for measurement of environmental parameters with acceptable accuracy. The availability of 3G/GPRS operator BTS infrastructure helps to implement the motes in the rural area.

The occurring cost in this system mostly for 3G/GPRS SIM card which has to be "top up" regularly in order to keep it working. Another maintenance need is a replacement for battery, roughly once in a year. Hybrid communication between motes and BTS can be studying further to solve the single point of failure issues and building resilience system. Another further study based on WSN such as landslide detection could be done as a continuation of this research.

The measurement results, generally indicate that the environmental aspects after mount Sinabung eruption outside the "red zone" is safe for living. Awareness is strictly required, failure to obeyed authorities warnings could be harmful.

Acknowledgments

This research was funded by Directorate of Community Service and Research, Directorate General Strengthening of Research and Development, Ministry of Research, Technology and Higher education Indonesia. We thank our colleagues at Research Institute (Lembaga Penelitian) University of Sumatera Utara for administration support. We also would like to express our gratitude to colleagues and students for sharing their knowledge with us during this research.

References

[1] Wark T, Swain D, Crossman C, Valencia P, Bishop-Hurley G and Handcock R 2009 *IEEE Pervasive Computing* **8** 30–36
[2] Dinh T L, Hu W, Sikka P, Corke P I, Overs L and Brosnan S 2007 *in Proc. 2nd IEEE Int. Workshop Practical Issues in Building Sensor Netw. Appl*
[3] Mainwaring A, Polastre J, Szewczyk R, Culler D and Anderson J 2002 *WSNA*
[4] Werner-Allen G, Lorincz K, Welsh M, Marcillo O, Johnson J, Ruiz M and Lees J 2006 *IEEE Internet Computing*
[5] Werner-Alle G, Johnson J, Ruiz M, Lees J and Welsh M 2005 *EWSN*
[6] Corke P, Wark T, Jurdak R, Hu W, Valencia P and Moore D 2010 *Proceedings of the IEEE: Invited Paper*
[7] Keshtgari M and Deljoo A 2012 *Scientific Research Wireless Sensor Network*
[8] Team R C 2016 *R: A Language and Environment for Statistical Computing* R Foundation for Statistical Computing Vienna, Austria URL `https://www.R-project.org/`
[9] Team Q D 2009 *QGIS Geographic Information System* Open Source Geospatial Foundation URL `http://qgis.osgeo.org`
[10] Wickham H 2009 *ggplot2: Elegant Graphics for Data Analysis* (Springer-Verlag New York) ISBN 978-0-387-98140-6 URL `http://ggplot2.org`
[11] EIF P 2016 Australia education investment fund wiki URL `https://eif-wiki.feit.uts.edu.au`
[12] Iguchi M, Ishihara K, Surono and Hendrasto M 2011 Learn from 2010 eruptions at merapi and sinabung volcanoes in indonesia Tech. rep. Dissaster Prevention Research Institute Kyoto University
[13] Ong J, You Y, Mills-Beale J, Tan E, Pereles B and Ghee K 2008 *IEEE Sensors*
[14] Buratti C, Conti A, Dardari D and Verdone R 2008 *Journal of Sensors*

A Wi-Fi based Electronic Road Sign for Enhancing the Awareness of Vehicle Driver

A Bhawiyuga, R A Sabriansyah, W Yahya, R E Putra

Faculty of Computer Science, University of Brawijaya, Malang, Indonesia

E-mail: bhawiyuga@ub.ac.id, sabrian@ub.ac.id, widhi.yahya@ub.ac.id, rizki.ek.putra@gmail.com

Abstract. Reducing the road accident rate is one of the city goal in the area of transportation. One of the effort to reach that goal is done by deploying various signs across the road. However, the role of those road signs can be diminished once the vehicle drivers intentionally or unintentionally disobey the rule indicated on those signs. In order to increase the awareness of the driver, we can employ the vehicular network concept in which a vehicle can communicate with another vehicles or with the infrastructure installed along the road. For realizing that idea, we propose the implementation of communication equipped road sign system which consists of two components: Road Side Unit (RSU) module deployed at road sign and On Board Unit (OBU) module deployed at each vehicle. In our proposed scheme, both of the devices communicate each other through the widely-used Wi-Fi protocol (IEEE 802.11n) operating in ad-hoc mode. While a OBU equipped vehicle is moving towards the communication range of RSU, it will make an association to a predefined wireless ad-hoc network. Once it is associated, the OBU can receive message broadcast by the RSU. Upon reception, OBU display alert message indicating that the vehicle is approaching a road sign. From performance testing we observe that the proposed system can give relatively good service the vehicle moving as fast as speed 90km/h with the distance as far as 90m.

1. Introduction

In recent years, there exists an increasing trend of road accident in Indonesia. According to Indonesian Statistics Board (BPS), the road accident rate in Indonesia increases from 108.696 cases in 2011 to 117.949 cases in 2012 [1]. There are several factors that contribute to those increase such as the lack of proper road facilities and the human error factor. As the amount of car and motor in the road rises, this worrying trends is predicted to continue in upcoming years.

One of the effort to reduce the amount of road accident is done by deploying the road sign [2]. There are various kind of deployed road sign including the red-green light signal in junction, crossing pedestrian sign, maximum/minimum speed sign and no turn left/right sign. Those kind of road signal are important for both managing the road traffic and reducing the road accident rates. However, the role of those various road signs may be diminished once the road users disobey the rule indicated to those road sign both intentional and unintentional. Therefore, we think that it is an important problem to improve the awareness of driver on the road towards the existence of deployed road signs.

In literature, there have been several works to cope with aforementioned problem. In [3], the author proposed a traffic sign recognition system through image processing by capturing the sign with camera and then, by utilizing the classification method, it can perceive the meaning of a sign and give an early information to the driver. Since those proposed systems utilize the camera, its performance may highly depends on the amount of lighting in those specific road segment. By realizing the problem in previous work, one of the solution is proposed by incorporating communication between the deployed road signal and the driving entity. In [4, 5], authors proposed the design and prototype of a road sign delivery system using RFID. Since it utilizes the wireless communication, the proposed system does not be affected the lightning problem as in the image processing solution. However, the communication range of the RFID may limit the functionality of proposed system. Therefore, we may require a wireless communication method with longer transmission range and reasonable delay.

Fortunately, in recent years, the researchers are actively involved in the research topic related to the road safety using relatively long range wireless communication between entities on the road called vehicular network. In short, vehicular network is a subset of wireless network that allows a vehicle to communicate and exchange information's with its nearby vehicles called vehicle to-vehicle (V2V) communication or with the deployed road infrastructure called vehicle-to-infrastructure (V2I) communication [6, 7, 8]. By utilizing the exchanged information, each vehicle can detect its surrounding environment and react accordingly while an endangered situation happened [10, 11]. With its ability, we believe that the vehicular network can be one of the promising solution for helping the driver to be aware toward the road signs. In addition, we can see the vehicular network as a part of the whole integrated smart city solution, especially in the area of transportation [9].

In this paper, we propose the implementation of communication equipped road sign system which consists of two components: Road Side Unit (RSU) module deployed at road sign and On Board Unit (OBU) module deployed at each vehicle. The prototype of both modules are built in Raspberry Pi equipped with wireless dongle while the communication between those two entities are performed using IEEE 802.11n protocol. In this case, both the RSU module and its OBU counterpart operates in ad-hoc mode. While a vehicle equipped with OBU is moving towards the communication range of RSU, it will make an association to the predefined wireless ad-hoc network. Once it is associated, then the OBU can receive the road sign information broadcast by the RSU. Upon receiving that information, the OBU module shows up an alarm notification to the driver indicating that the vehicle is approaching a road sign together with a certain signal. Notice that, for carrying a specific information of a road sign with a minimal packet size, we define custom Ethernet frame with customizable payload.

The remainder of this paper is organized as follows. In section 2, we describe architecture of system. Based on this architecture, in section 3 we present the implementation of our proposed system. The feasibility testing results are then discussed in section 4. Finally, we conclude our paper at section 5.

2. System Architecture

Fig. 1(a) illustrates the architecture of the proposed system. Our system consists of two main components: the sender module deployed at road sign and receiver module deployed at each vehicle. Notice that according to vehicular network specification described in Section 1, the sender module acts as Road Side Unit (RSU) while the receiver module acts as the On Board Unit (OBU). Both of OBU and RSU communicate each other using IEEE 802.11n with ad-hoc mode. We choose the ad-hoc mode rather than management mode since this kind of mode does not require additional handshaking time as if we utilize the management mode.

The mechanism of proposed system can be described with block diagram in Fig. 1(b). While a vehicle equipped with OBU is moving towards the communication range of RSU, it will make an association to the wireless ad-hoc network with predefined ESSID. Once it is associated, then the OBU can receive the road sign information broadcast by the RSU. Upon receiving

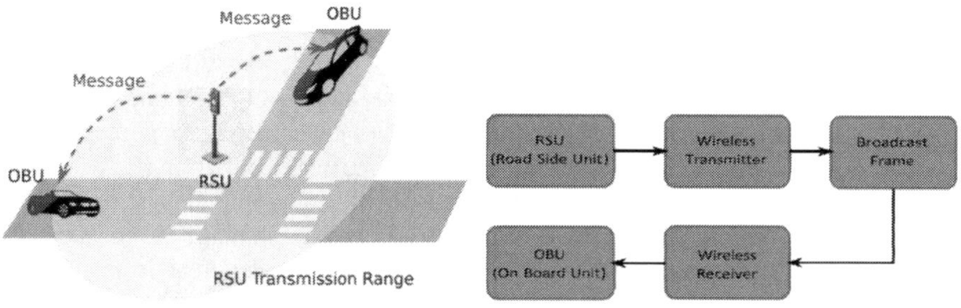

(a)Systemarchitecture (b)Blockdiagram

Figure 1. System architecture and block diagram.

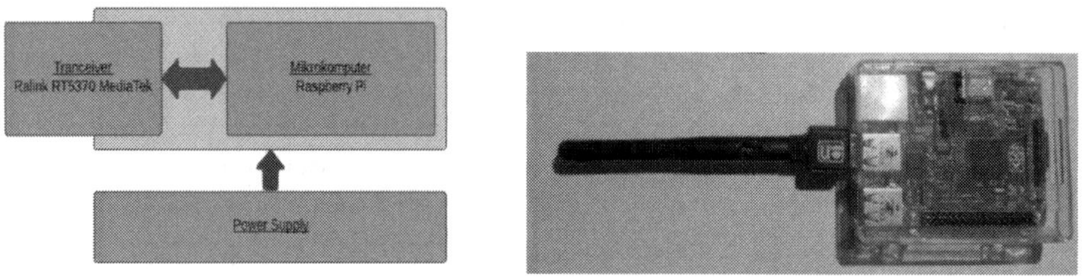

(a) Block diagram of OBU/RSU (b) Hardware implementation

Figure 2. OBU/RSU block diagram and implementation.

that information, the OBU module shows up an alarm notification to the driver indicating that the vehicle is approaching a road sign together with a certain signal. Furthermore, in order to reduce the size of message, instead of using the standard TCP/UDP/IP packet, we utilize the custom Ethernet frame with customizable frame type and payload.

3. Implementation

In this section we present the implementation of both RSU and OBU modules. The implementation is performed in both hardware and software part.

3.1. *Hardware Implementation*

Fig. 2 shows the block diagram of the hardware for both OBU and RSU. In general, the hardware part consists of two main components: processing and communication part. For processing part, we utilize microcomputer entitled Raspberry Pi Model B with 900 Mhz Quad Core CPU, 1 GB RAM, 4 USB ports and 40 GPIO pins. The Raspberry Pi is chosen since it provides enough computation power with the low power consumption. For the RSU part, the power consumption of the hardware can be an important issue since it can be deployed on the road with insufficient source of power. To perform the communication, processing part is equipped by the low cost IEEE 802.11n dongle with 2dBi internal antenna, -70+/-1dBm receive sensitivity, and 15+/-1dBm transmit sensitivity. The assembled hardware consisting of processing and communication part is shown in Fig. xxx.

3.2. Software Implementation

For the software part, we utilize Python based Scapy framework to develop the application for both RSU and OBU part. Scapy itself is a framework for developing a network-based application [12]. With Scapy, we can send and sniff the packet, create a frame with custom payload and decode the incoming packet. Those features are important since we employ Layer 2 broadcasting to reduce the overhead in upper layer. Moreover, since each road sign may have different signal, its payload content can also be different.

```
class Infra (Packet) :
name = "Test"
  fields _desc = [ LongField ("time" , 1) , LongField ("lon" , 2) ,
  LongField ("lat" , 3) , ShortField ("msg" , 0) ]
while (True) :
   curr _time = time . time ()
  lon , lat , max _speed = 129.231212 , 30.1213234 , 60
  frame=Ether (dst="ff:ff:ff:ff:ff:ff " , type=0x8088)/ Infra (
      time=curr _time , lon=lon , lat=lat ,  msg=max _speed)
  sendp(frame , iface="wlan1")
```

Listing 1. Source code of RSU Application

Listing 1 shows the main code of application in RSU part. At first, we define the class containing all information that would be broadcast to all nearby vehicles. For example, we define the message for maximum speed road sign containing several fields including the current time, position of road sign in longitude/latitude format and the allowed maximum speed at that road segment. At next, the Ethernet frame with frame type **0x8088**. Since the RSU should send the frame to all OBUs connected to the similar ad-hoc network, the destination address is set to Layer 2 broadcast address **FF:FF:FF:FF:FF:FF**.

```
def packets (pkt) :
   if pkt.haslayer(Ether) :
    if pkt.type == 0x8088:pack = Infra(pkt.load ) print str(pkt.type )+'
     '+str (pack.time )+' '+str (pack.lon )+' '+str (pack.lat )+' '+str
     (pack.msg)
 sniff ( iface="wlan1" , prn = packets )
```

Listing 2. Source code of OBU Aplication

In the other side, the code of application in OBU part is shown in Listing 2. We first define the callback function which will be called once OBU receives any frame from outside. The callback function is then specified as a parameter of sniffing process performed by the **sniff()** function. The **sniff()** itself is a blocking function which means that its following code will not be executed until a frame received. Upon reception of a frame, it then checks whether the frame is Ethernet with type **0x8088** as predefined in RSU or not. If so, it then prints frame information including: source address, type and message.

4. Performance Testing

In this section, we present the performance testing result of our proposed system. We first present the testing environment and then, we shows the performance result and analysis.

4.1. Testing Environment

During the test, we place the RSU node in the junction and let the vehicle equipped with OBU to go on the road segment near that junction. The OBU node is moved along the road with following scenarios:

(i) The OBU moves along the road with various speed i.e. 10, 20, 30, 40km/h to measure the impact of vehicle speed on the performance of proposed system.

(ii) The OBU stop at various distance from the RSU i.e. 50, 100, 150m to measure the impact of OBU-to-RSU distance on the performance of proposed system.

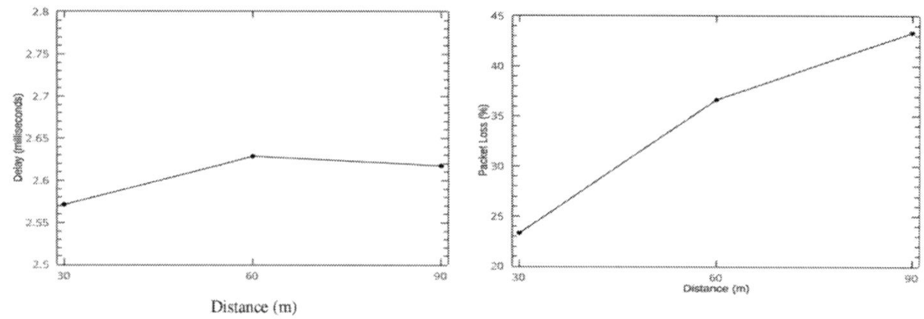

(a) Delay (b) Packet loss

Figure 3. Delay and packet loss from different distance parameter.

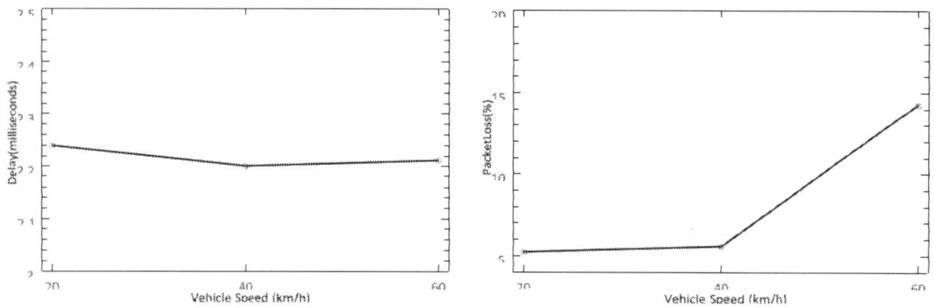

(a) Delay (b) Packet loss

Figure 4. Delay and packet loss from different speed parameter.

For each of the scenario, we measure the performance of proposed system in term of delay (in milliseconds) and packet loss (in percent). In addition we also perform preliminary testing to measure the average communication range from RSU to OBU.

4.2. Testing Result

Fig. 3 (a) and (b) show the delay in milliseconds and packet loss in percent from different distance parameter respectively. From Fig. 3(a) we observe that the delay seems to increase a little bit as the distance increases. However, with average delay 2.61ms, if the vehicle is moving with the speed as fast as 100km/h or 30m/s and the RSU-to-OBU distance is as close as 30m, the message can still be delivered. On the other side, from Fig. 3 (b) we realize that the packet loss tends to significantly increase as the RSU-to-OBU distance increases. This trends can be happened due to the impact of wireless signal attenuation in which the transmission signal decreases as the transmitter-receiver distance increases.

We next examine the impact of vehicle speed to the performance of proposed system in term of delay and packet loss. Fig 4 (a) and (b) show the delay and packet loss from different vehicle speed parameter with fixed starting distance i.e. 60m respectively. From Fig. 4(a) we observe that the vehicle speed seems to have a little impact to the delay. On the other side, the packet loss significantly increases as the vehicle speed increases. This different pattern happened due to the utilization of non-reliable frame transmission in which the loss packet will not be retransmitted if the frame dropped, however, if the frame is successfully received, its delay is relatively small.

In the end, from overall results, we can observe that the proposed system is feasible enough to be implemented in road segment to vehicle signal delivery since the RSU is statically placed and its state shift can be predicted, and therefore, each vehicle may only need to receive the message at least once.

5. Conclusion

In this paper, we presented the implementation of a Wi-Fi-based electronic road sign for enhancing the awareness of vehicle driver. The proposed system consists of two components: The Road Side Unit (RSU) placed at road sign and the On-Board Unit (OBU) placed in the vehicle. While the OBU equipped vehicle is moving toward the RSU, it makes an association with the RSU through a predefined ad-hoc Wi-Fi connection. Once it is associated, OBU can receive the custom Ethernet frame broadcast by RSU containing specific road sign information. Upon reception, the OBU can show an alert message to the driver. From the performance testing we show that the proposed system is feasible enough to be implemented.

References

[1] Biro Pusat Statistik, "Statistik Transportasi (Transportation Statistics),"*Biro Pusat Statistik (Indonesian Statistic Bureau), Indonesia*, 2013.

[2] Department of Transport, "Traffic Signs Policy Paper,"*Department for Transport, Great Minster House, United Kingdom*, Oct. 2011.

[3] Ling W H, and Woo C S, "Traffic sign recognition model on mobile device" *Computers & Informatics (ISCI), 2011 IEEE Symposium on*. IEEE, 2011.

[4] Paul A, et al., "Design and prototype of an in-vehicle road sign delivery system using RFID," *ITS Telecommunications (ITST), 2012 12th International Conference on*. IEEE, 2012.

[5] Rajale, Harshada, Ajit K, and Atul O, "Design of a road sign informing system based on GPS and RFID." *Control, Instrumentation, Communication and Computational Technologies (ICCICCT), 2014 International Conference on*. IEEE, 2014.

[6] Wenshuang L, Zhuorong L, Hongyang Z, Shenling W, and Rongfang B, "Vehicular Ad Hoc Networks: Architectures, Research Issues, Methodologies, Challenges, and Trends," *International Journal of Distributed Sensor Networks*, No. 745303, Nov. 2014.

[7] Kosch T, Schroth C, Strassberger M, and Bechler M, "Automotive Internetworking," *Wiley, New York, NY, USA*, 2012.

[8] Faezipour M, Nourani M, Saeed A, and Addepalli S, "Progress and challenges in intelligent vehicle area networks," {em Communications of the ACM}, vol. 55, no. 2, pp. 90100, 2012.

[9] Barba C T, et al. "Smart city for VANETs using warning messages, traffic statistics and intelligent traffic lights," {em Intelligent Vehicles Symposium (IV), 2012 IEEE}. IEEE, 2012.

[10] ETSITR102638, *Intelligent Transport System (ITS); Vehicular Communications; Basic Set of Applications; Definition*, ETSI Std. ETSI ITS Specification TR 102 638 version 1.1.1, June 2009.

[11] ITS JPO, "Vehicle safety applications," US DOT IntelliDrive(sm) Project - ITS Joint Program Office, Tech. Rep., 2008.

[12] Biondi, P. and Community, *Scapy*. 2010. Available at : http://www.secdev.org/projects/scapy/doc/introduction.html

IOP Conf. Series: Journal of Physics: Conf. Series **801** (2017) 012086 doi:10.1088/1742-6596/801/1/012086

Harnessing the hybrid power supply systems of utility grid and photovoltaic panels at retrofit residential single family building in Medan

A B Pangaribuan[1], R F Rahmat[2], M S Lidya[2] and M Zálešák[1]

[1]Univerzita Tomáše Bati ve Zlíně, Department of Automation and Control Engineering, nám. T. G. Masaryka 5555, 76001 Zlín, Czech Republic

[2]Universitas Sumatera Utara, Department of Information Technology, Jl. Universitas No. 9A, 20155 Medan, Indonesia

Email: artambo@gmail.com

Abstract. The paper describes improvisation mode of energy supply source by collaboration between national utility grid as represented by fossil fuels and PV as independent renewable power resource in order to aim the energy consumptions efficiently in retrofit single family house. In this case, one existing single family house model in Medan, Indonesia was observed for the possibility of future refurbishment. The eco-design version of the house model and prediction analyses regarding nearby potential renewable energy resource (solar system) had been made using Autodesk Revit MEP 2015, Climate Consultant 6.0 and Green Building Studio Analysis. Economical evaluation of using hybrid power supply is discussed as well.

1. Introduction

The escalation of energy demands in digital age are significantly increased due to the using of automation devices that need electric power source perpetually in order to make the integrated automation systems working properly. To fulfil the power demand from time to time, the energy supply no longer can hang on to the fossil fuel source but should migrate into conservation of the renewable resources such as solar, geothermal, wind, etc.

Contribution of renewable energy resources as energy supplies have gradually increased shifting the role of fossil fuels technology around the world and it depends on the type of potential renewable energy that could be found surround the power demand area. Positive impacts to the level of environment quality, i.e. reducing carbon dioxide emissions and economical values for the long term, i.e. lifecycle cost for 50 years are expected to be realized.

Residential buildings have 3rd largest portion in term of consume energy and produce waste energy in all over the world [1, 6]. The solution to reduce this trend is applying the concept of Near Zero Energy Building (NZEB). Recently, the automation building systems are the latest issue that could be brought to make residential buildings become more efficient in term of managing the energy circulation for itself. Moreover, old / existing residential buildings have big possibility to refurbish and become the eco-friendly building.

Medan is located in tropical area and has characteristic hot and humid, by using software Climate Consultant 6.0 analyzed based on the most frequency distribution relative humidity is between 80%-

90% (30% of the year). The annual average temperature is 28 - 29 °C and relatively constant. The hottest month is in March with the average daily max/min temperature is 33 °C/ 23°C and the coldest month is December with average daily max/min temperature is 31 °C/ 22°C.

Sophisticated solar collector technologies are very potential to be installed in the area since the solar resources available throughout the year based on geographical climate analysis. In other hand, the cooling load for indoor air quality will take big part in power demand for reaching thermal comfort of the occupants [6]. However, there is limited research on the quality level of occupants' satisfaction in residential buildings since too many variation of variables must be taken so that the comfort satisfaction standard could be formulized.

2. Methodology

The study established an observation of one existing conventional single family house built on 1983 without any of central electrical/mechanical cooling system, electrical security system, domestic hot water installation, nor renewable energy conversion equipment. The spacious area is 1570 m^2 contains house building with size 300 m^2, an ex-office building 140 m^2 (20m x 7m), one 24m^2 bedroom (6m x 4m) and a warehouse / garage building with size 84 m^2(12m x 7m). The building shape is not exactly rectangle, for the rest of the area is garden and paving block. Front side of the house and the office building in the area is heading South, meanwhile for the warehouse entrance door is heading West.

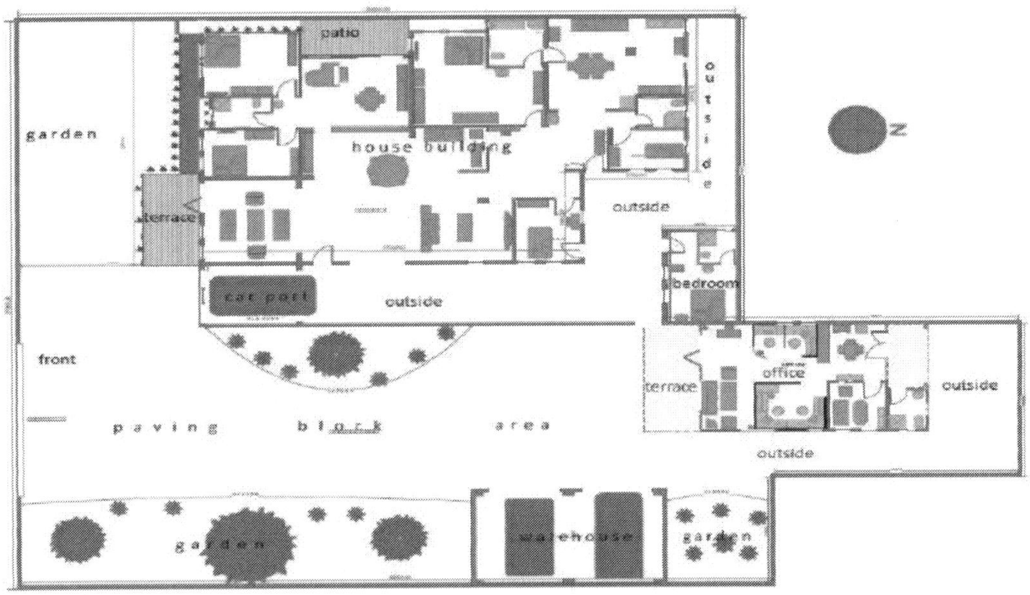

Fig. 1. Spacious Area

The house contains 5 bedrooms, 3 bathrooms, kitchen, living room, terrace, carport and patio. The power load only estimate for the house building and surrounding area, meanwhile the ex-office building and warehouse will not be discussed in this paper.

2.1. *Net Zero Energy Building*

According to National Renewable Energy Laboratory US Government, a net zero building is the building that receive energy supply from at least 50 % renewable resources per annum and reproduce at same quantity. The intent is to achieve aggressive level of energy efficiency (5 kWh/m^2/year) for cooling / heating purpose [3]. Residential Energy Services Network issues index for rating the residential energy system as displayed below:

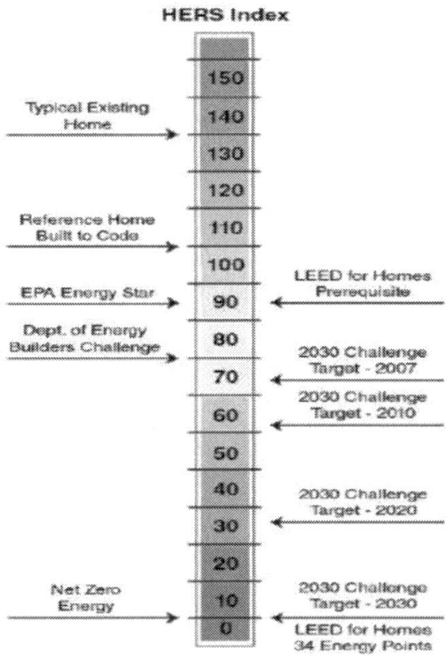

Fig. 2. Home Energy Rating System Index (kWh/m^2per year) [2]

2.2. *Electricity Load*

As modern house, many electronic devices will be installed and used inside the building. Hence, the power consumption considered higher than conventional single family house. The electricity load maybe varies, but approximately load could be determined as shown in Table 1.

Table 1 Electrical Load per Area

Area	Lighting Load(W)	Other Loads (W)
Hall	350	1506
Bathroom 1	35	66
Bathroom 2	35	51
Bathroom 3	35	84
Bedroom 1	75	200
Bedroom 2	75	400
Bedroom 3	75	153
Bedroom 4	75	500
Bedroom 5	75	100
Living Room	135	50
Kitchen	75	340
Patio	30	N/A

Based on the electrical load data, energy simulation analysis had been done using Green Building Studio, the electric demand each month in a year found the highest peak in the middle of March and mid of June around 7.1 kW as shown in the figure 3.

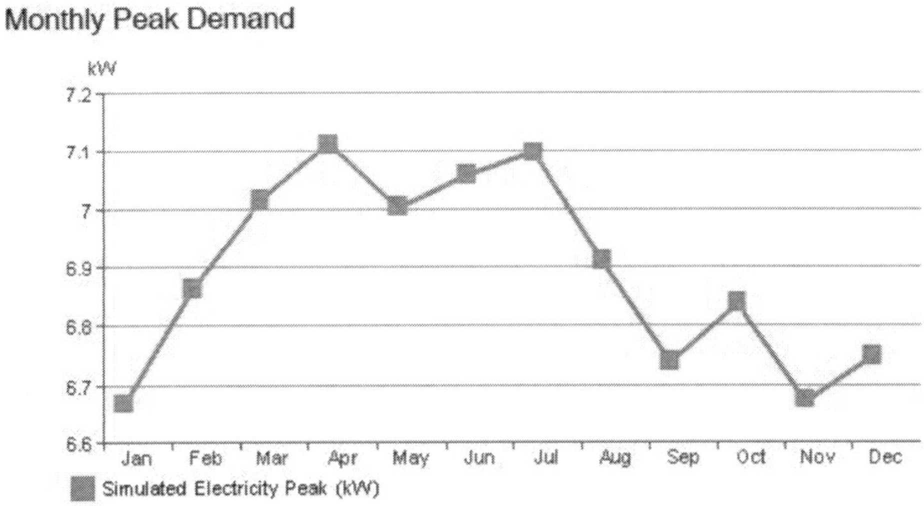

Fig. 3. Simulation Electricity Monthly Peak Demand

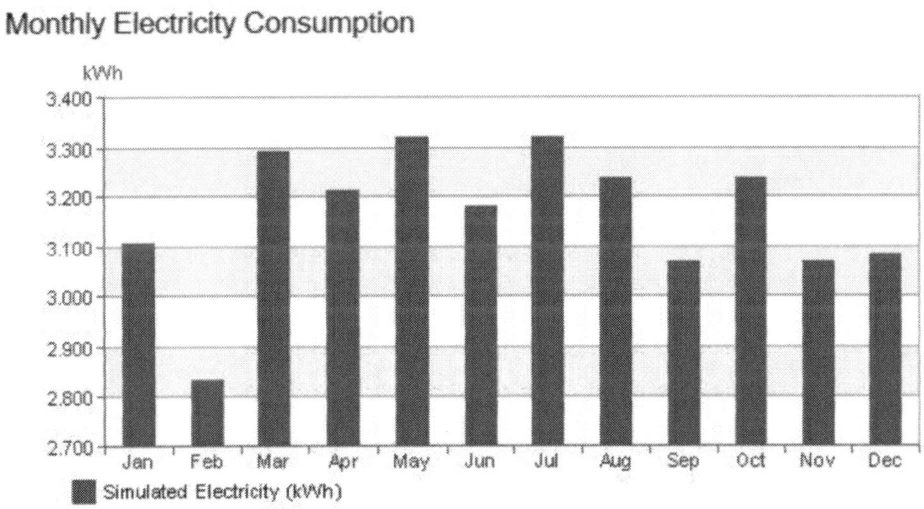

Fig .4. Simulation Electricity Monthly Consumption, 2015

The data shows that the highest consumption is in May and July around 3300 kWh, meanwhile the lowest is in February. This data related to the weather condition in Indonesia in 2015, which May and July are the highest average temperature during the year. Hence, the active cooling is needed longer and more than the other months.

2.3. Electric Utility Service and PV Panel Hybrid System

Recently, there is no policy issued by Indonesian national electric company to the clients who using PV panel regarding intensive cost or reduction price of the electric tariff. There is no standardisation for PV arrays installation had been officially published by the government. Generally, for the residential buildings, the PV panel attached on the roof. The hybrid scheme and PV arrays placement on the roof of the refurbishment house shown in Figure 3 & 4 respectively;

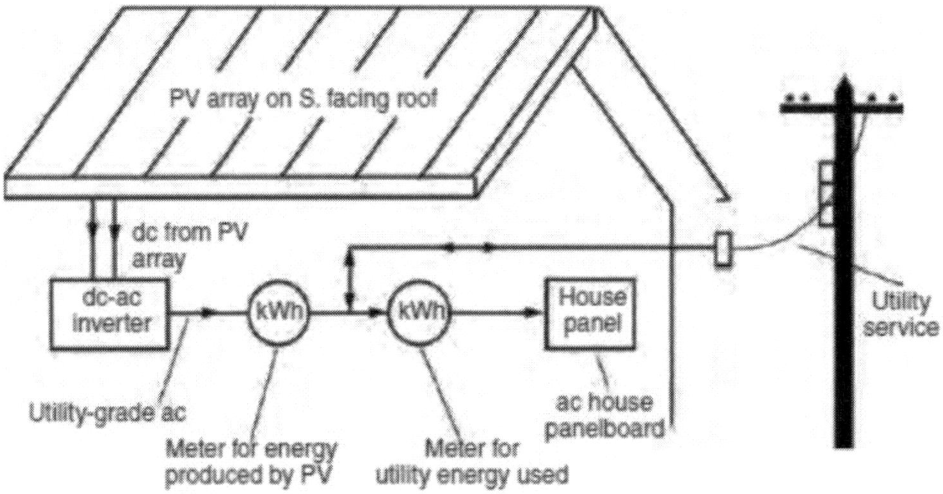

Fig. 5. PV panels and utility service connection scheme [1]

The electrical panel connected to power meter from utility grid and DC to AC inverter as PV panel source, since the loads in the system are all AC. Battery 12 V 100 AH is used with backup capability.

Fig. 6. Mount Roof PV Panel design using Revit MEP 2015

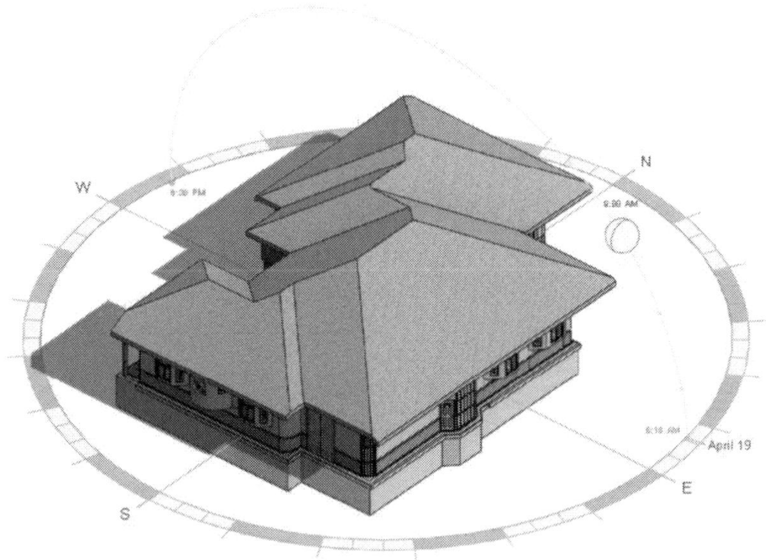

Fig. 8. Sun Position at 9 am, GMT+7

The PV array installed heading east and west using grid connected technique, since the sun exposure mostly bright from 9am-3pm during the year in Indonesia.

Fig. 7. Power meter and DC to AC inverter at retrofit house eco-design using Revit MEP 2015

2.4. *Solar Potential Energy in the Area*

Indonesia has 4.8 kWh /m^2 per day solar energy, equals to 112000 GW [4]. The specification of PV panel used in this plan is Single Crystalline with efficiency 13.8% (102745.6 kWh /year) [5]. By analyse the estimation of PV system energy using software Green Building Studio, can be shown;

Table 2. PV System Energy Estimation per Annum

Renewable Energy Potential

Roof Mounted PV System (Low efficiency):	37,227 kWh / yr
Roof Mounted PV System (Medium efficiency):	74,453 kWh / yr
Roof Mounted PV System (High efficiency):	111,680 kWh / yr
Single 15' Wind Turbine Potential:	147 kWh / yr

*PV efficiencies are assumed to be 5%, 10% and 15% for low, medium and high efficiency systems

The energy efficiency of PV system is calculated based on equation:

$$\eta = \frac{P_m}{E.A_c} \quad (1)$$

Where:
P_m is performance of a PV panel [W]
E is total intensity of solar radiation [W/m^2]
A_c is surface area of PV cell [W m^2]

2.5. *Economic Evaluation*

Based on electric tariff adjustment categories issued by PLN (Perusahaan Listrik Negara), the Indonesian national electric company per June 2015, this house belongs to R-3/TR (residential category, more than 6600 W), that is 1524 IDR/kWh (€0.09/kWh). The prices still subsidize by the Indonesian government.

Table 2. Electric Tariff in Indonesian Rupiah before Adjustment, June 2015

Golongan Tarif Tariff Class	Tegangan Voltage	TTL Tariff	BPP Basic Cost of Electricity Production	Subsidi Subsidy
P.1 s/d 2.200 s/d 5.500 VA	TR	1.018,1	1.473,0	454,9
P.1/> 6.600 s/d 200 kVA	TR	1.337,4	1.473,0	135,6
P.2/> 200 kVA	TM	966,5	1.273,5	307,0
P.3	TR	926,6	1.473,0	546,4
T/> 200 kVA	TM	737,5	1.273,5	536,0
C/> 200 kVA	TM	1.169,7	1.273,5	604,8
L	TM	1.169,7	1.273,5	103,8

The subsidies are calculated from negative difference between the selling prices of electricity on average (IDR/kWh) of each tariff group minus the basic cost of electricity production (IDR/kWh) on the voltage at each tariff group multiplied by sales volume (kWh) for each tariff group. Common

IOP Conf. Series: Journal of Physics: Conf. Series **801** (2017) 012086 doi:10.1088/1742-6596/801/1/012086

residential house in Indonesia using utility grid with 220 V single phase 2 wire. For power supply source, which provide from national electrical grid may be vary based on the price per kWh between 1300VA, 2200VA, 3500VA -5500VA, and more than 6600VA. In this model, the house use 6600VA from utility grid and 1100VA from the photovoltaic panel, hence the total available power is 7700W (35A).

Simple PV payback period of the investment is given by the equation:

$$T_S = \frac{IN}{CF} \qquad (2)$$

Where:
IN is the capital expenditure of the project
CF is the annual benefits of the project

Table 3. PV Payback Period

Panel Type	Installed Panel Cost	Applied Electric Cost	Max Payback Period
Single Crystalline – 13.8%	€5/watt	€0.09/kWh	20 years/ surface

*1€= 15000 IDR

Photovoltaic panel economic calculations start with the initial cost of the PV panel for 1100 W is €5 x 1100 = €5500 with annual return is €275. The payback period is within 20 years; hence the installation of PV panels is still not profitable in this project in the current years. The other consideration is low tariff of utility grid because the subsidy from national government.

3. Conclusion

In the future, the PV price in the world market will decrease in conjunction with government's policy to withdraw the subsidy gradually. Overall result of the paper should can be applied and contributed in Indonesia near future for new residential building as one of the reference or as the second opinion towards the national government's program about saving energy in building and harnessing the Indonesian renewable energy potential.

References

[1] A. G. K. Walter T. Grondzik, Mechanical and electrical equipment for buildings, 12th ed. wiley, 2014.
[2] "HERS index." [Online]. Available: http://www.hersindex.com/understanding. [Accessed:15-May-2015].
[3] P. Torcellini, S. Pless, M. Deru, and D. Crawley, "Zero Energy Buildings: A Critical Look at the Definition," in *ACEEE Summer Study Pacific Grove*, 2006, p. 15.
[4] Meita Rumbayan, Asifujiang Abudureyimu, Ken Nagasaka, "Mapping of solar energy potential in Indonesia using artificial neural network and geographical information system," 2012.
[5] L. Lu , H.X. Yang, Environmental payback time analysis of a roof-mounted building-integrated photovoltaic (BIPV) system in Hong Kong, 2010.
[6] C. S. Clemens Richarz, *Energy efficiency refurbishments: Principles, Details, Case studies*, 2013.
[7] ASHRAE, "ANSI/ASHRAE Standard 62.2-2013, Ventilation and Acceptable Indoor Air Quality in Low- Rise Residential Buildings," 2013.

Monitoring of Solar Radiation Intensity using Wireless Sensor Network for Plant Growing

B Siregar[1], F Fadli[1], U Andayani[1], LA Harahap[1], F Fahmi[2]

[1]Faculty of Computer Science and Information Technology, University of Sumatera Utara, Jl. Dr. Mansur 9 Medan Indonesia
[2]Departement of Electrical Engineering, Faculty of Engineering, University of Sumatera Utara, Jl. Dr. Mansur 9 Medan Indonesia

Email : fahmimn@usu.ac.id

***Abstract*—** Plant growth is highly depending on the sunlight, if the consumption of sunlight is enough, it will grow well. The plant will be green because of its chlorophyll and it can perform photosynthesis at maximum; but if the plants get less sunlight, it will make the plants be yellowing. Radiation is electromagnetic waves that are good for plants, so-called visible light. In the electromagnetic wave spectrum the best wavelength range from 400-700 nm for the plant. A monitoring of sun intensity is needed in order to obtain sufficient solar radiation consumption and provide notification if there is a high radiation. In this study, several sensors and devices were combined such as photosynthetic solar radiation sensors, GSM / GPRS and waspmote as a main board or a microcontroller. The test was carried out on at least three occasions; the system has a stable radiation in the morning with an average of 505.51 micrometers. IN this study, we have successfully developed a monitoring tools for solar radiation intensity applied on plant growth by using wireless sensor network.

1. Introduction

Information of intensity of solar radiation in the greenhouse is needed. If there is too much exposure to sunlight, the growth will be not be optimum, and vice versa if the plants are less exposed to sunlight it cannot perform photosynthesis at its maximum.

Technology is developing very fast. Technology penetrates almost all sectors, especially agricultural sector, to facilitate the farmers in producing a quality crop. One of them is the using of wireless network sensors. Sensor is a device that is able to detect changes in the physical environment or chemical which converts it into electrical signals both electrical current and voltage [1]. In the development of the latest sensor technology using wireless sensor network (WSN), a sensor-based wireless technology that can allow users to access the sensor from a distance without using wires like for robot [2]. The role of these wireless sensor technology on human life as the data retrieval and delivery of information including our surrounding environment [3].

There was a study in 2013 designing a control the intensity of sunlight, temperature and humidity in the greenhouse with PID method. The study consisted of two phases: the design of hardware and software. Hardware used was the sensor SHT11 to measure humidity and temperature, LDR sensors to measure the intensity of light, microcontroller ATmega32 as controlling the overall system, and several actuators as control and maintain stability in the greenhouse such as blower (fan), heater (lights), water sprayer, fertilizer sprayer, ground sprinklers, ventilation, and curtains. Software used was Bascom AVR programming language. [4]

Another study conducted in 2013 focused on the design of a distributed sensor network for monitoring the temperature, humidity and light intensity. Researchers designed a system of distributed sensors to monitor temperature, humidity and light intensity in the greenhouse using the Arduino Uno board. The system consisted of two sensor-actuator nodes and one node controller connected to an Ethernet network using Ethernet Shield board. Sensor-actuator nodes with DHT sensor 11 served to retrieve information environment of temperature, humidity, and light intensity. Results were displayed in a web site [5].

A study about landslide monitoring system was conducted in China using a wireless sensor network. The data collection proved to be consistent with natural phenomena, compared with traditional monitoring methods [6]. The study describes a system for monitoring the temperature in the greenhouse by placing temperature sensors in several positions observations. The proposed system allows a more representative and comprehensive observations to conduct an analysis of changes in temperature using wireless communication. Besides glass house temperature can be monitored optimally [7].

The aim of this study was to develop a system that can monitor level of intensity of solar radiation on the plants in the greenhouse in real time and provides a warning or notification that in the greenhouse deficiency or excess solar radiation. In this study, GSM / GPRS were used for data transmission and for monitoring solar radiation using solar radiation sensor.

2. Material and methods

Solar radiation is a very important element in agriculture. First, light is a source of energy for a green plant through photosynthesis is converted into chemical energy. Second, radiation plays an important role as a source of energy in the process of evaporation that determines crop water needs. [8].

Spectral distribution of sunlight received by the leaves on the surface title (1900 umol m-2s-1) is greater than the leaves in the shade (17.7 umol m-2s-1). In the shade of light that can be utilized for the process of photosynthesis is very little [9].

Solar radiation is comprised of spectra of ultraviolet (wavelengths less than 0:38 microns) which have damaging because the fuel is very high, spectra Photosynthetically Active Radiation (PAR), which acts raised up the process of photosynthesis and spectra of infrared (greater than 0.74 microns) which is temperature-controlled. Spectra PAR radiation can be broken down into bands of spectrum that each has certain characteristics. It turns blue spectrum contributed the most potential in photosynthesis [10].

The general architecture of this study can be seen in Figure 1.

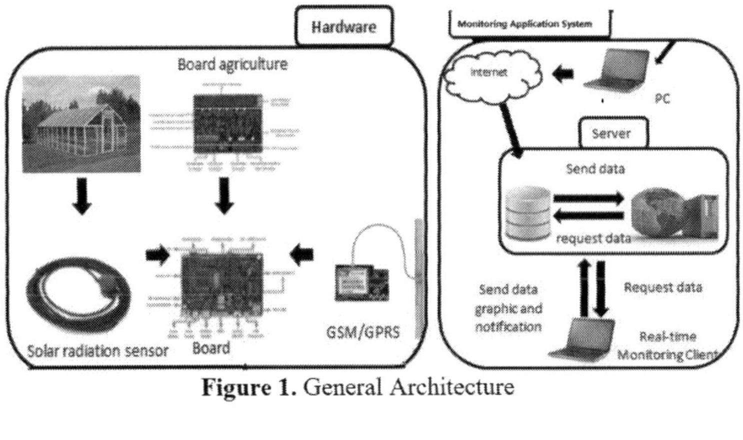

Figure 1. General Architecture

2.1 *Hardware Design*

2.1.1 Photosynthetic Solar Radiation Sensor. Photosynthetic Solar Radiation sensors are placed in the greenhouse, the agriculture board, there are some pins are connected to the Photosynthetic Solar Radiation sensor. Ie pin +, -, GND dam. This pin is used as the sender of the data from the sensor to the main board waspmote after it is stored in SD card, seen in Figure 2.

Figure 2. Photosynthetic Solar Radiation Sensor and Board Agriculture

2.1.2 Waspmote Master Board and GSM/GPRS Module. Waspmote main board is a vital part of the sensor, because on this board program code to be uploaded and also on this board will be put agriculture board, GSM / GPRS, and waspmote 802.15.4, as well as on the main board, there are waspmote on / off, port usb and also a battery which is connected directly. It can be seen in Figure 3.

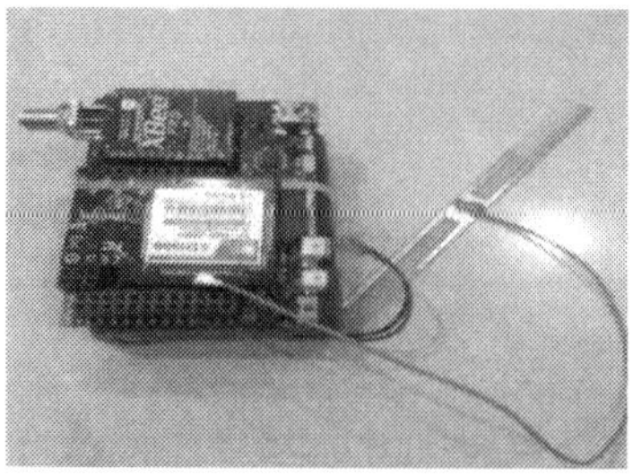

Figure 3. Main Board, *GSM/GPRS,* Waspmote 802.15.4

This board is a data repository that is sent after the waspmote store sensor data into SD card. In the GSM / GPRS module there is a sim card slot that can be used as a signal gprs to send data to the internet or to a server database. Furthermore waspmote Pro 802.15.4, this module is a wireless communication device that sends a signal contained in waspmote sensor. The tool communicates using the 2.4GHz frequency with 12 channels and each channel using a 5MHz Bandwidth.

2.2 *Data Communication and Monitoring system*

2.2.1 Data Communication. Figure 4 describes data communication using the GSM / GPRS, initially solar radiation data captured by the sensor, then the data stored in the SD card contained in the sensor board, from SD card then the data is sent using the gsm module / GPRS. In the gsm module / GPRS card providers were used as signals to transmit data to the database, the data in the three grades who submitted that value, temperature, and battery. Value is the radiation data, temperature is the temperature of the sensor board, and the battery is a battery sensor, and these values are sent to the url and then entered into the database according to the field in the database.

2.2.2. Monitoring. Each incoming data from the sensors will be sent via GSM / GPRS to the url and then entered into the database server. And this data is then sent to the web server and ready to be shown to the client in the form of real-time graph, and the client can see all the information about the state of the sensor and greenhouse conditions were monitored remotely.

2.2.3. *Notification.* The incoming data on the sensor waspmote will be entered into the database, and if it has been entered in the database, the data will always be checked by php. In the solar radiation there is a range of values between 400nm-700nm good for crops as the supply of energy in the process of photosynthesis. The system checked if there was a value more than 800 nm then php will provide notification letting you know that radiation is too high, then a notification is displayed to the client , But if the data is entered into the database less than 800 nm, the program will check back but not sent a notification on the client side. The notification flowchart can be seen in Figure 5.

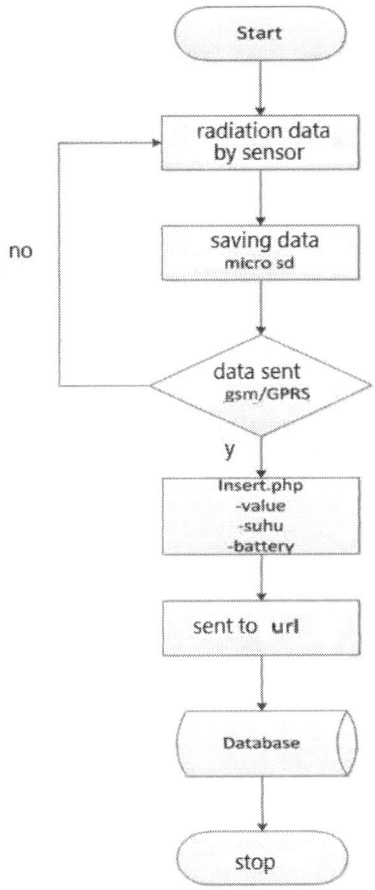

Figure 4. *Flowchart data communication using GSM/GPRS*

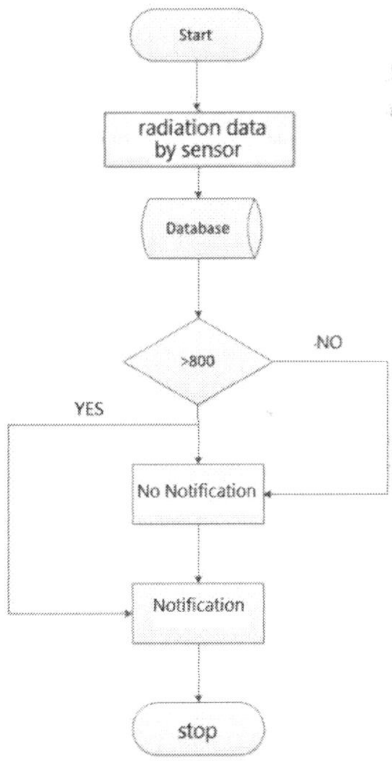

Figure 5. *Flowchart Notification*

When there is high solar radiation it will show a notification popup message on the right top on the web. This notification appears if radiation captured by the sensor was more than 800nm, and the value on the graph will change as well. The page views high radiation can be seen in Figure 6.

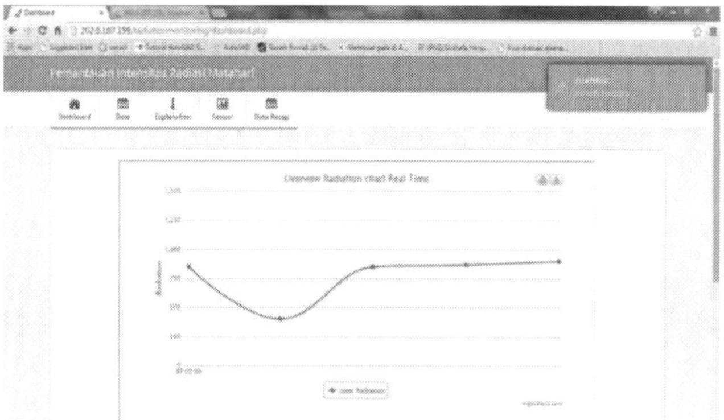

Figure 6. Notification on high radiation

3. Result

System performance testing was conducted to determine the level of intensity of solar radiation at the time was monitored by sensors. In this test we focused on monitoring solar radiation. The monitoring process can be seen at the time of the solar radiation captured by the sensor, after the radiation data obtained and the data is sent to the database server. Then the result was displayed to the user in graphical form. The higher the solar radiation it will be bad for the plant because not all of the radiation can be absorbed by plants. Only the range of 400-700 nm can be absorbed by plants. In performance testing systems to analyze the level of solar radiation we divided in three periods: morning, noon and afternoon. For a morning test observation was held on 8:00 to 10:00 am, noon test at 11:00 am to 02:00 pm, and for the afternoon at 03:00 to 05:00 am. The test system can be seen in Figure 7.

Figure 7. System Testing

Test performance was done to know the average solar radiation in the morning, noon and afternoon. The test was done for three days in a row. The first test was conducted in the morning with the sun's radiation to produce an average 505.31 micrometers, solar radiation produced quite stable in the morning.

The first test showed the radiation generated in the observation in the morning was 505.31 micrometers. The test results are shown in Table 1.

Table 1. Test in morning

No	Radiation (mm)	Time	Delay (s)
1	749	9:49:42	46
2	700	9:50:28	45
3	572	9:51:13	45
4	529	9:51:58	45
5	511	9:52:43	45
6	512	9:53:28	46
7	542	9:54:14	45
8	593	9:54:59	46
9	667	9:55:45	46
10	786	9:56:31	43
radiation	**505,51**	**Avg delay**	**47,03**

Table 2. Test in noon

No	Radiation (mm)	Time	Delay (s)
1	166	13:27:39	47
2	177	13:26:28	45
3	184	13:29:11	45
4	182	13:29:56	47
5	174	13:30:43	45
6	163	13:31:28	51
7	146	13:32:19	45
8	137	13:33:04	52
9	137	13:33:56	43
10	140	13:34:39	46
radiation	**191,55**	**Avg delay**	**46,65**

During the noon, the intensity was decreased compared with the d the morning. The test results are shown in Table 2. The second test is done during the day, at noon produce solar radiation in the range 191.55 micrometers. The third testing done in the afternoon, on the afternoon of 22.34 micrometers generating solar radiation, radiation produced later in the day lower compared with the observations made in the morning and afternoon. The test results are shown in Table 3.

Table 3. Test in afternoon

No	Radiation (mm)	Time	Delay (s)
1	59	17:16:42	48
2	59	17:17:30	46
3	58	17:18:16	46
4	57	17:19:02	50
5	56	17:19:52	48
6	56	17:20:40	48
7	54	17:21:28	49
8	54	17:22:17	46
9	51	17:23:03	46
10	48	17:23:49	45
radiation	**22,34**	**Avg delay**	**45,06**

The test results were then converted into a line chart and a comparison chart can be seen test data on at least three occasions, conducted over three consecutive days. Charts difference can be seen in the morning, noon and afternoon. Solar radiation generated in the afternoon at about 22.34 micrometers, these results are lower compared with the observations made in the morning and afternoon. In the system testing conducted by the authors, the resulting solar radiation in the morning is around 505.31 micrometers. The graph line on the test data delay on the system can be seen in Figure 8.

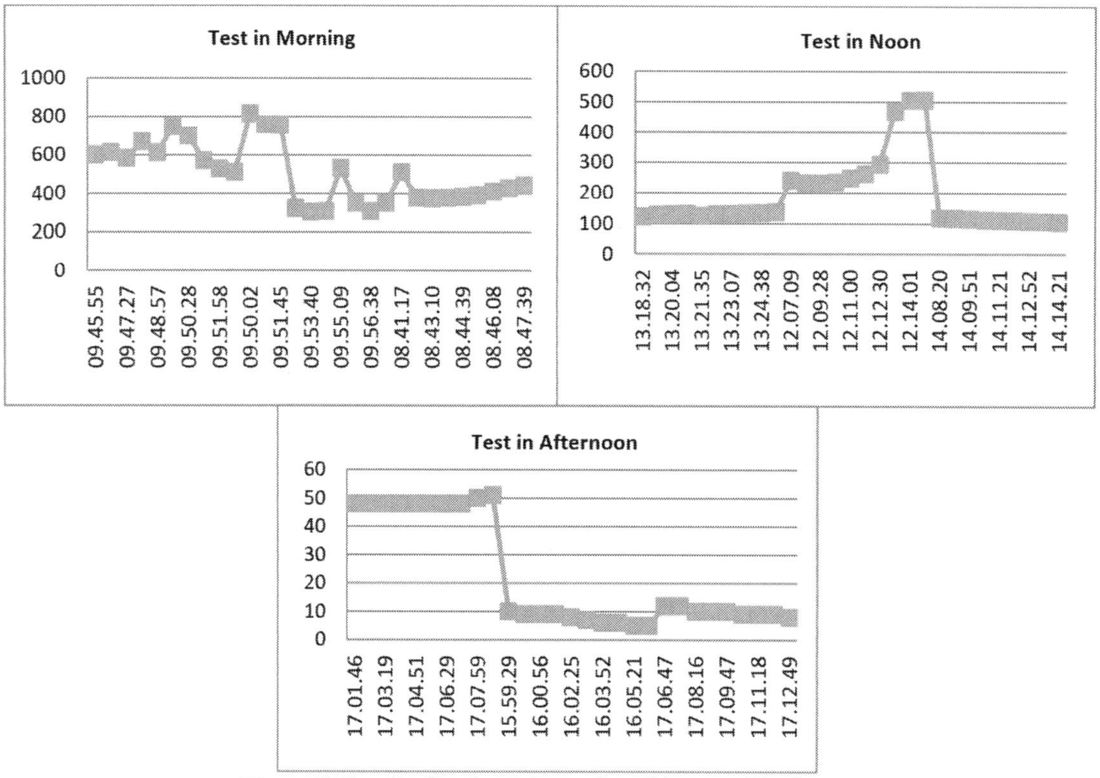

Figure 8. Result of testing (a) morning; (b) noon; (c) afternoon

Based on the comparison chart in Figure 8 it can be seen that the test system to monitor solar radiation on at least three occasions, the morning, afternoon and evening. On the morning of solar radiation has an average 505.31 micrometers, daylight has an average 191.55 micrometers and afternoon had an average of 22.34 micrometers. So in this test showed a stable solar radiation that is in the morning had an average of 505.31 micrometers solar radiation.

And the data transmission system is built has a delay, on the morning had an average delay of 47.03 seconds, 46.65 seconds daytime and evening 45.06 seconds. So at this test has the longest time is in the morning which had an average delay of 47.03 seconds.

4. Discussions

The system we developed used Libelium waspmote for monitoring the intensity of solar radiation and for data communication using the GSM network / GPRS as well as the card provider.

In the process of solar radiation monitoring conducted by the solar radiation sensor, the observations were made at different times three parts ie morning, afternoon and evening. in the morning hours of observations were made at 08:00 to 10:00 pm, during the day is observed on the clock 11:00 to 14:00 pm, and in the afternoon made observations on the clock 15:00 to 17:00 pm. The third part of this time each obtained an average solar radiation, on the morning of the average solar radiation 505.31 micrometers, the average daytime solar radiation 191.55 micrometers, and the afternoon of the average solar radiation 22 34 micrometers.

In the data transmission system is constructed to have a delay, on the morning had an average delay of 47.03 seconds, 46.65 seconds daytime and evening 45.06 seconds. So at this test the longest time was in the morning which had an average delay of 47.03 seconds.

IOP Conf. Series: Journal of Physics: Conf. Series **801** (2017) 012087 doi:10.1088/1742-6596/801/1/012087

The observations made by the author only three days, therefore the authors suggest the observations should be longer to get more data.

5. Conclusions

We have successfully developed a monitoring tools for solar radiation intensity applied for plant growth by using wireless sensor network. Here, a WSN system with integrated sensors simultaneously in real time situation to monitor the intensity of solar radiation. This result will help farmers optimize the growing and encourage the idea of *smart farming*.

References

[1] Kretschmar M and Welsby S 2005 Capacitive and Inductive Displacement Sensors *Newnes Burlington*.
[2] Siregar B, Purba AH, Efendi S and Fahmi F 2016 Fire extinguisher robot using ultrasonic camera and wifi network controlled with android smartphone *Proc. Of Annual Applied Science and Engineering Conference (*AASEC*)Bandung*
[3] Siregar B, Nasution ABA and Fahmi F 2016 Integrated Pollution Monitoring System for Smart City *Proc of Int. Conf on ICT for Smart society (ICISS) Surabaya*
[4] Defriyadi YS 2013 Pengendali Intensitas Cahaya, Suhu, Dan Kelembapan Pada Rumah Kaca Dengan Metode PID *under publication*
[5] Pamungkas BA 2013 Perancangan Jaringan Sensor Terdistribusi Untuk Pengaturan Suhu, Kelembaban, Dan Intensitas Cahaya *under publication*
[6] Yueshun H and Wei Z 2013 The reseach on wireless sensor network for landslide monitoring *IntJ On Smart Sensing and Intellegent Systems* Vol. **6**, No. 3.
[7] Hastriyadi H et al 2014 A Multi Sensor System for Temperature Monitoring in a Greenhouse Using Remote Communication *IntJ of Latest Research in Sci and Tech* Volume **3**, Issue4: Page No.81-87.
[8] Wisnubroto and Soekardi 1981Asas Meteorologi pertanian *Ghalia Indonesia* Jakarta.
[9] Taiz L and Zeiger E 1991 Plant Physiology *The Benyamin/Cumming Publishing Company Inc* Tokyo. p: 219-247.
[10] Koesmaryono and Yonny 1991 *Kapita Selekta Agroklimatologi* IPB. Bogor

A framework to apply ICT for bequeathing the cultural heritage to next generation

R Syah[1], T E Nuradi[2], M K M Nasution[3*]

[1,3]Information Technology, Fasilkom-TI, Universitas Sumatera Utara, Padang Bulan USU Medan 20155, Indonesia.
[2]Perencanaan Wilayah, Universitas Sumatera Utara, Padang Bulan USU Medan 20155, Indonesia.

[*]Email: mahyuddin@usu.ac.id

Abstract. Of course, preservation of cultural heritage is a task given by UNISCO charter, but bequeath the cultural heritage to the next generation is most important. The cultural heritage contains the lost intelligence, displaced by modernization and globalization, but ICT can be a way out although ICT has a negative effect. Based on the component we apply ICT, we derive a framework for applying ICT as core and next generation as consideration scope.

1. Introduction

The cultural heritage is the hidden knowledge in a particular community [1], but modernization and globalization has penetrated into the heart of every culture and change it [2], especially ICT has accelerated the flow of change through the information [3]. The changes prosecute individual by individual in a community be similar between one to another [4, 5].

In the old society or a particular community has existed local wisdom that can deliver a community to next community until their generation today [6]. However, the old culture gradually eroded by a new culture, and old community and their generation are not same [7]. There is the lost intelligence between them. Old community may be heterogeneous, but new communities tend to be homogeneous [8], for example. The other side of ICT is aimed at improving the welfare of the community. It reveal also the hidden knowledge to any people. However, a few of studies have done to implement ICT for bequeathing the cultural heritage of Indonesia to next generation. Therefore, this paper aims to address a framework to develop ICT for bequeathing the cultural heritage to next generation.

2. Basic Concept and Motivation

Cultural heritage is a condition of tangible or intangible knowledge of a community or society that are delivered from past generation, maintained in the present and bestowed for the benefit of future [9]. This condition as an expression of the ways of living that has developed by a community for surviving and existing currently [10]. Arguing that every community has its own challenges, the growing culture with the community itself. In this case, each person has different capabilities, and is able to survive based on that ability [11]. At that time, the competence of each person depends on social activities. However, now everyone pressed to have almost the same competence: They tend to have the same attitude toward both socially and outlook on profits. For example, everyone is not feel perfect life if he/she does not have mobile phone, afterwards get the gadget, and more often alone [12].

All technologies have double-sided blade: negative of positive. Technology as a result of culture will serve to give added value to a community. Meanwhile, technology as a tool tends to ignore the community life but focus on individual life [13]. ICT (Information and Communication Technology) to be a term that includes any communication devices like radio, television, cellular phones, computer and network, hardware and software, satellite systems and others in various services or applications, or it is an extended term for IT (information technology) [14]. ICT is the essence of every life, and provide the main characteristics for layer generations of a community such as z-generation now. Currently, ICT becomes a gateway for the entry of a culture to other cultures. Interacting and build other cultures. ICT however also become the frontline to inform all stakeholders about something [15].

A culture is built from the interaction of environment conditions, the challenges of life, and the approach by human [16]. Therefore, a culture is established by habit, habit comes from the implementation of either by nature or rule force. Approach repeatedly is a way of using force indirectly to change something cultural, such as advertising approach to the consumer via short message service (sms) [17]. So the habit may be generated through the things that are recurring (in frequency of events). An event despite having a single moment, but read many times by a group of peoples (often) cause the community members are forced to change. ICT causing an information about an event can be accessed repeatedly, and then be part of community. Therefore, ICT offers a way to deliver the cultural heritage to the next generation, so that no the loss intelligence from Indonesia.

3. An Approach

To establish a framework for applying ICT to something it is necessary to express the definitions and scope of work. In the scientific culture, the definition is the basis for developing a theory and its application. So the definition of framework is intended to express the basics of the framework and possibilities its application. Therefore, a framework is an abstraction in which something providing generic functionally can be selectively changed by the user based on an approach [18]. In general, a framework for applying ICT has different components as follows [19]:

a. Stakeholder is social actors related to something.
b. Content is all records about something.
c. Processing about converting the raw input into a meaning form.
d. Management is a group that organize all people for those activities.
e. Cost of programs is a plan about something for generating budgets.
f. Facilities and infrastructures are all facilities for succeeding this program.
g. Test and evaluation are an activity for sustaining this program forever.
h. Result or product of this program are a system that bridges the relevant stakeholders.

The arrangement of the components based on the relationship between them can be built as Figure 1. A model proposed for preserving cultural heritage, where all the components are denoted in uppercase (A, B, C, D, E, F, G, H), and their relations in symmetric are (A,B), (A,C), (A,H), (B,D), (C,E), (H,G), and (G,F) [20].

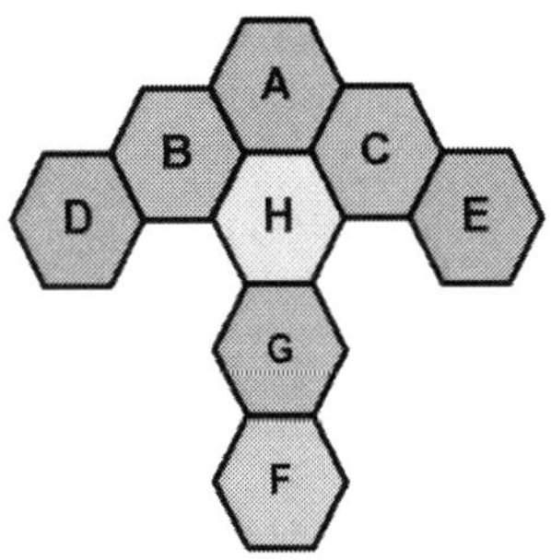

Figure 1. A proposed framework with 8 components

IOP Conf. Series: Journal of Physics: Conf. Series **801** (2017) 012088 doi:10.1088/1742-6596/801/1/012088

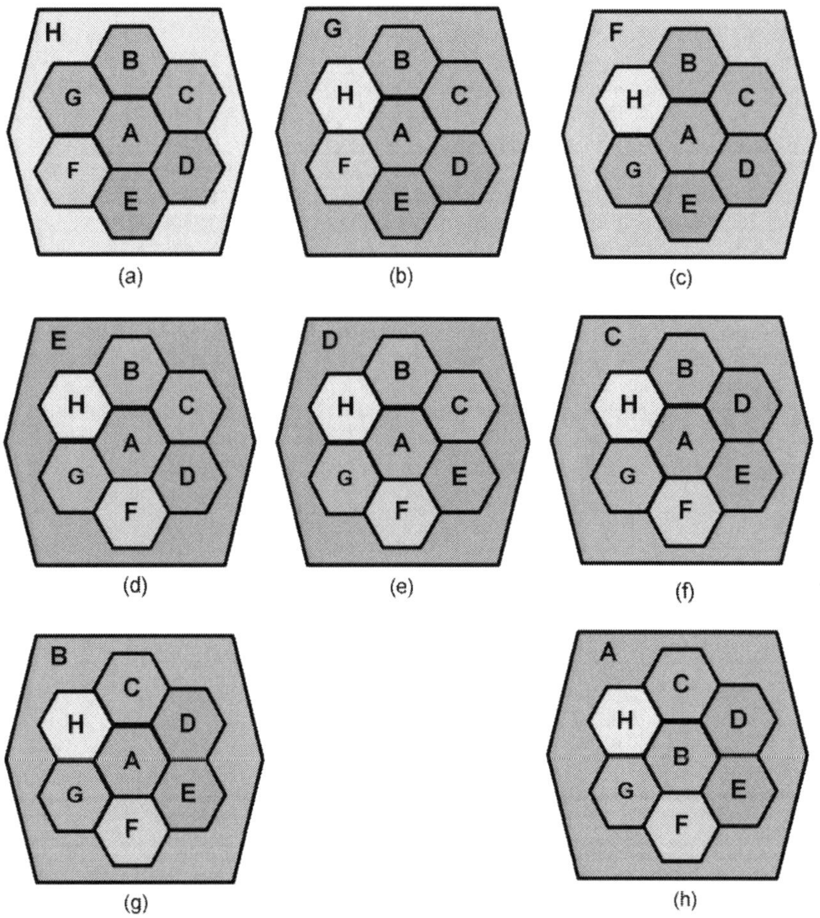

Figure 2*(a)*, 2*(b)*, 2*(c)*, 2*(d)*,2*(e)*, 2*(f)*, 2*(g)*, 2*(h)*. 8 frameworks with 8 components

4. A Discussion about Relation between Components

We discuss first about the cultural heritage of the area. For example, the cultural heritage in the province of Sumatera Utara. The term Sumatera Utara (North Sumatra) is to refer to an area on Sumatra Island that has its own cultural heritage, where there are 8 communities or indigenous tribes (Melayu, Mandailing, Toba, Karo, Angkola, Simalungun, Pakpak, and Nias) and have the culture that is hardly the same. This can be seen from tangible knowledge: buildings, monuments, landscapes, books, works of art, artifacts; intangible knowledge: folklore, traditions, language and science/technologies; and natural heritage i.e. culturally significant landscapes and biodiversity. Because all exist in Sumatera Utara (*Sumut*) we call the framework as *SumutSiana*: The term is derived from two words. The term *Sumut* is abbreviation of North Sumatera, while the term "*Siana*" means as treasure/treasury or scientific wealth. Therefore, *SumutSiana* is owned as cultural or natural wealth that exists in that provincial, implicates the regional assets consisting of 8 indigenous people [19].

The above definition is the basis for the establishment of a framework to bequeath the *SumutSiana* (as cultural heritage) to the next generation. The framework consists of 8 components [19]:

a. Stakeholder is either authors or receivers, either individually or a group, or with categories: individual, student, group, functionary, researcher, lecturer, or professor.
b. Content is all records of: actors or agents of history, local language, art, traditional houses, transportation, lighting and irrigation, customs, religion, environment, flora and fauna.
c. Processing is a transformation raw input into a meaning form.
d. Management is about leading for information flow: text, graphics, audio, and video; about historical information: manuscripts, painting, drawings, narrative, investigative reports; about cultural environment, the history of characters or actors, descent and family; about results of mining like scientific papers, imagery, dialogue, and journey.
e. Budget is cost of this programs.
f. Facilities and infrastructure are all devices: software and hardware about ICT, or information system: *E-SumutSiana* or *M-SumutSiana*.
g. Test and evaluation: any software used first tested prior to launch, and then do an evaluation to enhance a system regularly.
h. Result or product is files about text, video, audio, and database, and then local language dictionary, semantic technology, or media.

In connection with the activities currently undertaken by various levels of society about 8 indigenous tribes, we can see through the frequency of web pages provided by search engines. This information may contain descriptions of components as the cultural heritage of Sumatera Utara [19].

Framework based on the merit of the work focus about 8 components of applying ICT have the combination of components related their relation to be $c(n,k) = n!/(n-k)!k! = 8!/(8-2)!2! = 28$ or for $n = 8$ there are $n(n-1)/2 = 28$ relations between 8 components [21]. If taken one be core and one be the scope of work, then we have 8 arrangements of the components (see Figure 2) where $n = 8$ and $k = 6$. Thus, 8 framework models can be used one of them in accordance with the focus being the development of systems that deliver the information of cultural heritage from stakeholders to system or vice versa. For example, let us focus on ICT like internet and Web, so we consider all components against the facilities and infrastructure: For stakeholder, system should be user friendly to members of stakeholder; About content, the system capable of recording and presenting all information about the content; For processing, the system has ability to process data into information or new meaning; For management, the system can help administrators to manage the entire application about preserving cultural heritage; through the system can be made budget planning for sustaining this application; For test and evaluation, the system is able to tell deficiency, error, and the destruction, and; Statistically the system provide important information on the results or products themselves.

	A	B	C	D	E	F	G	H
A. stakeholder		0.0043	0.0034	0.0008	0.0085	0.0229	0.0095	0.0061
B. content	0.0043		0.2588	0.0192	0.1468	0.4901	0.4183	0.4202
C. process	0.0034	0.2588		0.0061	0.1410	0.4836	0.3676	0.3448
D. management	0.0008	0.0192	0.0061		0.1845	0.4852	0.4966	0.3067
E. budget	0.0085	0.1468	0.1410	0.1845		0.4803	0.3216	0.5994
F. facility	0.0229	0.4901	0.4836	0.4852	0.4803		0.1083	0.0953
G. test	0.0095	0.4183	0.3676	0.4966	0.3216	0.1083		0.2276
H. result	0.0061	0.4202	0.3448	0.3067	0.5994	0.0953	0.2276	

Figure 3. Strength relations between components

Through the information provided in the internet, we can see the relationship of 8 components to disclose any information most widely available and interlinked to develop a system [21, 22, 23], where implementation ICT exists based on this framework, see Figure 3, where frequency of words forms the basis for generating the value of strength relations. Then, the value was sorted from highest to lowest, and for 9 top relations we have a network whereby component F as root (core) and component A as scope.

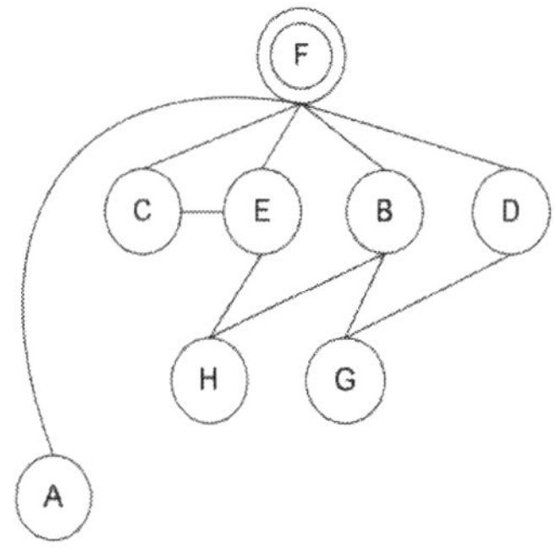

Figure 4. Network of components

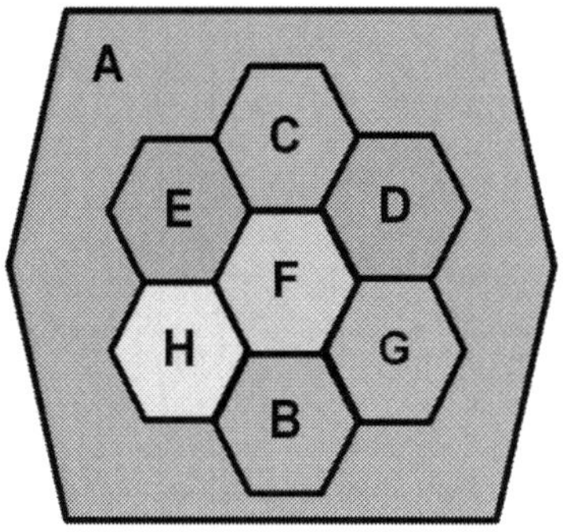

Figure 5. A framework with focus to ICT and scope is stakeholder (generation)

A framework as result of this discussion is a framework that focuses on applying ICT as facilities and infrastructure (F) within the scope of stakeholders. The stakeholder not consist of individual, student, group, functionary, researcher, lecturer, or professor, but stakeholder also refer to next generation. Therefore, a framework to apply ICT for bequeathing the cultural heritage to the next generation is based on Figure 5.

5. Conclusion

The approach involves defining cultural heritage, 8 components and the construction of a model obtained several options about framework to deliver the old cultural to new generations. Based on the resulting to new generations. Based on the resulting selection of some computing framework that focuses on ICT and scope of the new generation. The future work is the developing of information system.

References

[1] I Taksa and J M Flomenbaum 2009 An integrated framework for research on cross-cultural information retrieval *IEEE Sixth International Conference on Information Technology: New Generation*.

[2] G Parayil 2003 Mapping technological trajectories of the green revolution and the gene revolution from modernization to globalization *Research Policy* **32**.

[3] S P Gill 1999 Mediation and communication of information in the cultural interface *AI & Soc* **13**.

[4] K M Nasution, S A M Noah and S. Saad 2011 Social network extraction: Superficial method and information retrieval, *Proceeding of International Conference on Informatics for Development* (ICID'11).
[5] M K M Nasution and S A Noah 2012 Information retrieval model: a social network extraction perspective *IEEE International Conference on Information Retrieval & Knowledge Management*.
[6] A Fuxman, P Tsaparas, K Achan and R Agrawal 2008 Using the wisdom of the crowds for keyword generation *WWW*.
[7] Z A Hasibuan 2011 An overview of integrated approach to digital preservation: case study of Indonesian E-Cultural Heritage and natural history information retrieval system, ICACSIS.
[8] D Lepelley, P Pierron and F Valognes 2000 Scoring rules, Condorcet efficiency and social homogeneity *Theory and Decision* **49**.
[9] D Hamdani, I Inayaturohman, and D Darwanto 2011 State-of-the-aart EMC regulation and conformity of products in Indonesia *IEEE International Conference on Electronical Engineering and Informatics*.
[10] M K M Nasution, M Elveny, R Syah, and S A M Noah 2015 Behaviour of the resources in the growth of social network *IEEE Proceeding of the 5th International Conference on Electrical Engineering and Informatics* Bali.
[11] M K M Nasution and S A Noah 2011 Extraction of academic social network from online database *Proceeding of 2011 International Conference on Semantic Technology and Information Retrieval*, Putrajaya, Malaysia.
[12] R Silverstone 2005 Media, technology and everyday life in Europa: From information to communication *ACM DL*
[13] S A Brown 2008 Household technology adoption, use, and impacts: Past, present, and future *Inf Syst Front* **10.**
[14] P Kleve, R V de Mulder, and K van Noortwijk 2007 Surveillance technology and law: the social impact *Int. J. Intercultural Information Management* **1**.
[15] T Mukhopadhyay, F J Lerch, and V Mangal 1997 Assessing the impact of information technology on labor productivity *Decision Support Systems* **19**.
[16] A da Veiga 2016 A Cybersecurity culture research philosophy and approach to develop a valid and reliable measuring instrument *SAI Computing Conference*.
[17] N Buchbinder, M Feldman, A Ghosh and J Naor 2014 Frequency capping in online advertising *J Sched* **17**.
[18] D G Choi, H Lee, and T-k Sung 2011 Research profiling for 'standardization and innovation' *Scientometrics* **88**.
[19] M K M Nasution, T E Nuradi, and R Syah SumutSiana: A framework for applying ICT to preserve the cultural heritage of Sumatera Utara Indonesia, to appear
[20] M K M Nasution, O S Sitompul, E P Sinulingga, and S A N 2016 An extracted social network mining, *IEEE SAI Computing Conference*, London, UK.
[21] M K M Nasution and S A M Noah 2010 Superficial method for extracting social network for academic using web snippets *Rough Set and Knowledge Technology* LNAI **6401**, Heidelberg, Springer.
[22] M K M Nasution 2012 Simple search engine model: Adaptive properties, arXiv:1212.3906v1 [cs.IR] 17 Dec 2012.
[23] M K M Nasution 2012 Simple search engine model: Adaptive properties for doubleton, arXiv:1212.4702v1 [cs.IR] 19 Dec 2012

Some of Indonesian Cyber Law Problems

D D Machmuddin[1] **and B Pratama**[2]
[1]Langlangbuana University, Bandung
[2]Bina Nusantara University, Jakarta

Email: bpratama@binus.edu / bptama@gmail.com

Abstract. Cyber regulation is very important to control human interaction within the Internet network in cyber space. On the surface, innovation development in science and technology facilitates human activity. But on the inside, innovation was controlled by new business model. In cyber business activities mingle with individual protection. By this condition, the law should keep the balance of the activities. Cyber law problems, were not particular country concern, but its global concern. This is a good opportunity for developing country to catch up with developed country. Beside this opportunity for talented people in law and technology is become necessity. This paper tries to describe cyber law in Indonesia. As a product of a developing country there are some of weakness that can be explained. Terminology and territory of cyber space is become interesting to discuss, because this two problems can give a broad view on cyber law in Indonesia.

1. Introduction

Information and Communication Technology (hereinafter ICT) exponentially has changed human life globally. Amazingly, ICT has various force driven new form of culture in society. Today's society is not only information society, but toward information civilization, in which the world seen as a set of data that can be viewed, accessed, and share in a different ways [1]. Behind the jargon of modern society, Zuboff also warn that the new civilization is a new model of capitalism, and therefore the law should respond to follow the change [2].

Zuboff and Unger message is very clear to develop new paradigm of law in order to create responsive law. This mean, that *jurist* must: think outside the box, holistically and interdisciplinary think, and have strategic thinking. However, in terms of legal thinking, Arief Sidharta [3] reminded that the law is not value-free, because the law has bound to the constitution of a country, and must start from applicable law as a legal basis.

In today's civilization, legal doctrine above reasonably true, because legal rule sometimes become barrier for technology development, especially in ICT. They are saying: the technologist secretly waiting policy maker to sleep, then they can run the technology. If the rule makers awaken, they only become a burden. Law science (jurisprudence) as normology science is bound to the constitution on a country. This is the uniqueness; therefore, the value of Indonesian law is Pancasila as a philosophy of constitution.

In relation of ICT in law perspective, cyber law in Indonesia can be considered as a new study in law. But in practice demanding clear and fast policy to respond the dynamic of innovation. In global perspective, current law issue in developed country also drag and insist developing country to respond, because interconnection of information do not differentiate country level playing field. On the other hand, this is a good opportunity for developing country to catch. But in practice, lack of law expert in cyber become serious problems. The case of data protection is today's global issue in cyber law. In developed country, they already had robust legal framework to protect the data. But in Indonesia, legal standard for data protection and data treatment remain unregulated. This situation made the technologist set the rule self-regulated. In spite of ISO 27001 used as a standard data treatment in practice, but in legal perspective the rule is vague.

Indonesian cyber law is regulate by the Law No. 11/2008 regarding Information and Electronic Transaction (hereinafter ITE Law). Based on practice, there are two fundamental problems in ITE law. First; terminology problems, and second; cyber space problems. This two weakness must be clearly define, if is not, there are more problems occur. Most of cyber law case in Indonesia is defamation cases, in which the case mostly ignored the participant of the crime itself. For simple explanation in defamation case, the criminals can be more than one person, but the convicted always one person.

2. Indonesian Cyber Law Anatomy

Cyber law/ITE law in Indonesia was builds by the convergence of three areas of law that become cyber law pillars, namely: telecommunications law, media law and law of informatics. If the law concept placed in the contestation of existing positive law, Indonesia has not recognized media law and law informatics. Media law and law informatics is associated with the field of intellectual property law and also associated with telecommunication law and press law. Based on the pillars of cyber laws, its principle must be inheritance into ITE law. But it is not in reality. Principle of law that only runs in particulars law is known as sectorial law principle [4].

Thus, it can be said that ITE law epistemologically unaccountable. Apart from legal systematic interpretation, urgency of cyber regulation in 2008 is extremely high, because there is no cyber regulation. But the consequences of this hasty decision, ITE law loose focus on information and electronic transaction. However, if we see ITE law statue, by its name, this law should have more focus in information and electronic transaction, not regulate anything else, like defamation, interception, hate speech, and gamble.

3. Some of Problems in Indonesian Cyber Law

3.1 Cyber Terminology Problems

Before discussing about cyber law, common use of terminology is important to reach an agreement of legal meaning to enhance legal opinion [5]. Cyber terminology in Indonesia is very important to explain, because of disagreement among legal expert. Unfortunately, cyber terminology differences are also reflected in legislation. Referring to Black's Law Dictionary, 'cyber' terminology was uncovered inside the dictionary. Blacks Law only explanation cyber law, not cyber. However, regarding cyber concept, Blacks Law suggested concept of space in it [6]. Space in cyber concept indicate territory, because the law prevail jurisdiction to apply (*ius constitutum*).

Cyber terminology referred to the telecommunications law, in which etymologically derived from French words of *'telematique'* [7]. Others expert said that cyber was formed by the convergence of ICT [8] that began in the early 1990s of electronic commerce or electronic commerce (e-commerce) [9]. Others expert argue that cyber terms also known with another name, that is; Information Technology law (Law of Information Technology), the World Law Maya (Maya World Law) and the Law of Mayantara, which is all the life in virtual (cyber/Internet) [10]. As a result of this disagreement, inconsistency of cyber terms appears not only in ITE Law, but in several legislation, namely: Intelligent Law (Law No. 17/2011) and law of Notaries (Law No. 2/2014).

The Three laws above are still apply until now, although contain terminology errors. This mistake consequently can bring error on substance. Sitompul found that 'cyber' term was not just a term, but it was born based on the concepts of cybernetics. The paradigm of this concept is to see information as an extension of the mind and the eye. Thus forming imagination and reality, including the new world [11]. Cyber terms also have equivalent words in Indonesian, which is '*siber*', not '*maya*' (unrealistic). The reason of 'maya' words is not appropriate because Indonesian dictionary define '*maya*' as imagination or fantasy. Telematics terms also inappropriate because telematics is short for three components: telecommunications, multimedia and informatics. Josua also argue that cyber terms must interpret extensively to get broad meaning. In respond to the situation, Indonesian Constitutional Court judge decided that to understand cyber is the media that use to do an activity that has impacted the lives of people in real world.

Josua arguments have similarity with constitutional court decision to use cyber terminology. Therefore, the next task is to find the constituent elements. In the case of cyber element, according to Kang [12], there are four elements, that is: (1) temporal engagement, (2) communication Initiation, (3) audience scope, and (4) media richness. New type of communication that formed cyber world also have rules in it. Lessig argue in cyber there are two kind of law, namely code is law and legal code [13], which in principle to regulate human interaction not only with computer, but also inter-human communication.

By contrast, terminology disagreement of cyber arises from different view among legal experts. The most vulnerable argument is an argument that takes computer (tools) to inside the argument. In fact, to access the cyber space, the tool was a necessity. Therefore, inserting tools into the argument is disturbing the argument itself. By the proposition of legal expert above, cyberspace is a place/region/territory that generated by Interconnection of computer network. Therefore, fitting legal concept into cyber space is appropriate, and compatible with Cicero adagium *ubi societas, ibi ius.* If we understand cyber law is the law to govern *netizen* (Internet user) to interact with other *netizen,* from the argument above, there are at least three reason to enhanced the argument: (1) cyber is common terminology globally, (2) the words cyber is not apple-to-apple with 'maya' words, because there is legal consequences in cyber, particularly in doing criminal offence (*schuld*), (3) to enter cyber space, require Internet and device connectivity, any action on offline computer considered as non-cyber crime.

The last line of the argument can be explain: if the crime use criminal law, there are two type of crime, namely: the crime using tools and cyber crime. In cyber crime, the use of computer is inevitable and done within Internet network. In the concept of crime using computer the action is not necessarily cybercrime if the action done without Internet connection. For example: a portable computer belonging to a taken by B. The crime committed by B is not the cybercrime. Even if a arguing there is stolen computer data, but the crimes cannot classified as cyber-crime. The compliance of cyber element into B will appear if B stole computer data using computer network. This is the uniqueness of Indonesian cyber law, but in most developed country, the regulation can reach all of computer crime, because they cybercrime regime regulate computer related crime.

3.2 Space in Cyber

A philosophical foundation in freedom of expression has been mandated in 1945 Indonesian Constitution. As a democratic country, one of the indicators is legal guarantee on freedom of expression, freedom of speech and freedom of the press [14]. Indonesian laws regulate freedom of expression by wrapping it inside right to informed (to received, to process, to send, and to use information). So, the idea of freedom of expression is an activity using information. In its relation, individual information is vis-à-vis with another individual information; in which each of individual have the same rights. In this condition, the law turn up to solve individual conflict, by regulating them. In terms of information, they are three type information, which are: public information, private information, and national secrets information. The classification not only based on juridical perspective, but also had similarity with privacy [15]. Solove also explain that in personal information a person have a right to express his/her idea, notion, data and fact [16].

Although, Soloves concept in privacy based on individualism philosophy, which is 'men are created free', but to have strong understanding in informational right, Indonesian law system limited individual right with public interest. In a real world regarding space, we know public and private space to deliver information. The question raised for public and private space is on how the law defined this concept in cyber space. To answer this question, let's take a closer look on communication theory, which was initiated by Edward T. Hall or known as Proxemics theory [17]. Halls theory can be described below.

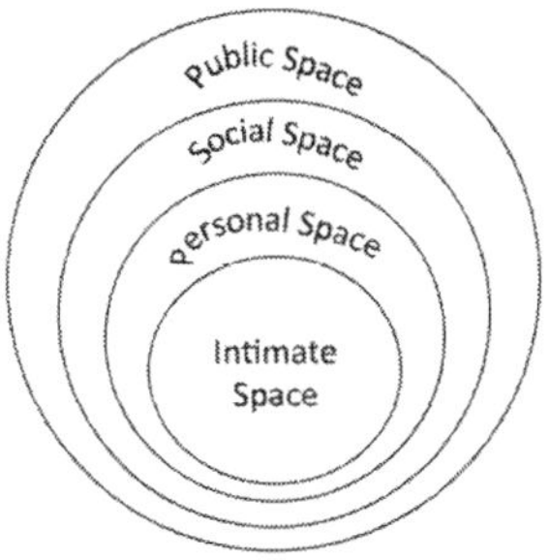

Figure 1. Proxemics Theory

Hall's separate communication by determines the sphere with its relationship. Halls concept, is very useful to classifying legal action clearly. For example: if someone in an intimate sphere humiliate someone else dignity, this action cannot be categorize as defamation. The reason if this arguments because of intimate sphere is not public sphere, and the law prerequisite public sphere in order if defamation offence.

In cyber space territory, defamation offence also must fulfilled public sphere, or done in public space. Ludlow framing this concept by explaining that public space in cyber is a space that's not required invitation or password to enter [18]. However, of course, Ludlow opinion was necessary must be seen case-by-case, because in social media, the situation is far more complicated. Regardless, the important thing from Ludlow is in cyber space the concept of communication sphere also can be used.

Within legal perspective, defamation also regulate in ITE Law. But this norm was taken to judicial review court, because it was feared become highly ruled and harm the democracy spirit. The constitutional court, decided that cyber defamation, basically same with a defamation in non-cyber. But in practice, unfortunately defamation norm mostly used to criminalize by particular person.

To strengthen argumentation, there are at least three cases that can be serve as an example. (1) Prita Mulyasari vs. Omni International Hospitals (2009), (2) satay seller vs. Indonesian President, Joko Widodo (2014), and (3) the National Narcotics Agency (BNN), Indonesian Republic Police (POLRI), Indonesian National Army (TNI) vs. Harris Azhar (The Commission for Disappeared and Victims of Violence/KONTRAS) (2016). These three cases drew public attention, because they are victims. Some of research data claimed they are hundred victims of defamation in, but the exact number was never showed.

The most important things in cyber defamation is the fulfillment of the norms element of public spaces. For example, social media, Tweeter, urged that all of Tweeter users are in public space when they Tweet, that's why Tweeter user must very careful to write in Tweeter. Regarding this concept, Facebook imposed another concept or give option for user to set type of his/her writing in Facebook, whether send in public or private space. So the treatment in Facebook is casuistic, depend on the cases.

4. Conclusion

As a part of the innovation dynamic, cyber phenomenon is hard to catch by the law-maker to formulate robust legal framework. Some of the developed country may see cyber law as a simple space to rule with some adjustment of law concept, but in developing country like Indonesia, the degree of difficulty is several times more difficult. ITE law was an example of law product in Indonesia that trying to catch a developed country. Key important things that can be learned from ITE law are the conceptual of technology as an object of the law must be clearly defined. If the concept was right, then the norms will also be clear to determine.

Reference

[1] Shoshana Zuboff, "Big Others: Surveillance Capitalism and The Prospect of an Information Civilization," *Journal of Information Technology*, p. 77, 2015.

[2] Roberto Mangabeira Unger, *Free Trade Reimagine*. New Jersey: Princeton University Press, 2007.

[3] Bernard Arief Sidharta, *Refleksi tentang Struktur Ilmu Hukum*. Bandung: CV. Mandar Maju, 2000.

[4] Gabriel Hallevy, *A Moern Treatise on The Principle of Legality in Criminal Law*. Berlin: Springer-Verlag, 2010.

[5] Hilman Hadikusuma, *Bahasa Hukum Indonesia*. Bandung: PT. Alumni , 2013.

[6] Bryan A. Garner, Ed., *Black's Law Dictionary*. Dallas, USA: Thomson Reuters, 2009.

[7] Edmon Makarim, *Kompilasi Hukum Telematika*. Jakarta: PT. Raja Grafindo Persada, 2003.

[8] Danrivanto Budhijanto, *Hukum Telekomunikasi, Penyiaran dan Teknologi Informasi, Regulasi dan Konvergensi*. Bandung: Refika Aditama, 2010.

[9] Shinta Dewi, *Cyber Law, Perlindungan Privasi atas Informasi Pribadi dalam E-Commerce Menurut Hukum Internasional*. Bandung: Widya Padjajaran, 2009.

[10] Ahmad M. Ramli, *Pengaruh Teknologi Informasi Terhadap Eksistensi Hak Kekayaan Intelektual dan Urgensi Hukum Siber (cyber law) dalam Sistem Hukum Nasional*. Bandung: Refika Aditama, 2004.

[11] Josua Sitompul, *Cyberspace, Cybercrime, Cyber Law, Tinjauan Aspek Hukum Pidana*. Jakarta: PT. Tata Nusa, 2012.

[12] Jerry Kang, "Cyber-Race," *Harvard Law Review*, pp. 1148-1147, March 2000.

[13] Lawrence Lessig, *Code Version 2.0*. New York, USA: Cambridge Center, 2006.

[14] Amir Efendi Siregar, *Mengawal Demokratisasi Media: Menolak Konsentrasi, membangun Keragaman*. Jakarta, Indonesia: PT. Kompas Media Nusantara, 2014.

[16] Paul M. Schwartz and Daniel J. Solove, *The PII: Privacy and a New Concept of Personality Identifiable Information.*: NYW Law Review, 2011.

[15] Daniel J. Solove, *The Digital Person - Technology and Privacy in the Information Age*. New York: New York University Press, 2004.

[17] Edward T. Hall, *The Hidden Dimension*. New York, USA: Garden City, 1966.

[18] Peter Ludlow, *Crypto Anarchy, Cryberstates, and Pirate Utopias*. USA: Massachusetts Institute of Technology, 2001.

IOP Conf. Series: Journal of Physics: Conf. Series **801** (2017) 012090 doi:10.1088/1742-6596/801/1/012090

Legal Prescription on Virtual Property and Its Rights

B Pratama[1]

[1] *Business Law Department – Bina Nusantara University, Jakarta – Indonesia*

Email: bpratama@binus.edu / bptama@gmail.com

Abstract. Type of property in digital era radically changed. The emersion of cyberspace, ontologically show up legal question: why kind of law applied to answer virtual property right? Whether using new law or adjusting existing law by reconceptualising it to be applied in cyberspace. After the discussion of legal concept, the concept of legal right also raised as the consequences of the legal problems. If the law is simplified, the rule element consists of: legal subject, legal object and legal action. Based on legal premises, formulating legal prescription of virtual property right become interesting to discuss. With clear certainty concept of virtual property right, then the right inherence also become visible.

1. Introduction

Massiveness of Internet user is inevitable by most countries. Parallel with Internet user, the problems also encountered, like security issue, data privacy, freedom or expression, piracy, data theft, and other legal problems. The key of this legal problem in cyberspace is data regulation, because in computer communication, data (in broad) is the key of all. The answer of this question as if been left by legal maker, and is vague under the dynamic of information and technology.

There will be certain condition that data regulation become important, and by the explosion of Internet user. In this condition, the urgency of data regulation is become more important. The clarity right of the data can be narrowed by determining virtual property right as a part type of computer data. In the field of information and communication technology, they are three important factors, namely: hardware, software, and brain ware. From those three factors, this paper focused on software. The reason is because software is the most prone are to have legal dispute, because the object is intangible. Legal right (or property right) in computer software is unique, and also applied intellectual property right concept. The exclusive right in intellectual property rights includes right to give permission or to prohibit someone else to use copyrighted works [1].

Exclusive right arise in copyright, because copyright is made by human creation [2] that needs some skill, the product of the creation is tangible and human sensible. This element is applied to software, that's why copyright defined it as a part of copyright object. The law treats copyright as a property and the right attached to it is legal right. It means the rights that occur in copyright are by law. The problems in the right by law, the object is written by law. This condition made limitation of the law to capture new object of property, for example virtual property right like: information, database, and assets in online game. Based on this logic thinking, this paper tries to discuss and prescribing legal thinking of virtual property rights.

2. Property Rights Concept

The difference among legal system in European Continental and *Anglo-Saxon* was bounded by their history. In *Anglo-Saxon* legal system, the property law concept was developed by the relationship between landlord and the vassal in feudalism era [3]. Regarding this concept, Lock declined the superiority of landlord. To do so, he offer moral rights concept for the vassal to enjoy some of their rights of his work. Lock's doctrine known as labor theory and latter became one of the mainstream concepts in intellectual property law development.

In European Continental property right concept emerged in Roman law codification, the system is known by *numerous clausus*. It said that a person cannot have a right accept the law determine it, the concept also known as closed system [4]. The consequence of this system, the law cannot capture easily new type of property, like virtual property, unless the law regulate it. Closed system principle is one of the characteristic in European Continental law system, but in Spain, they did not adopt this system even their legal system is European Continental. In spite of historical background, the first movement of legal codification in European Law is started from four type of Justianus codification, which are: 1) *Institutiones;* legal definition and legal institution, 2) *Pandecta;* compilation of jurist opinion, 3) *Code;* compilation of law (*leges lex*) by the command of Roman emperor, 4) *noveles;* compilation of comment on *codex* [5].

Nowdays, property concept mostly known as a bundle of rights, but the weakness lies beneath its scope. But at the end, property is a comprehensive terms to describe one or all of legal relation between person with the property [6]. Property right theory in general can be viewed by the concept of ownership, and this concept most of the time used to justified property rights. Type of ownership can be described in table below.

Table 1. Theory of Ownership [7]

Theory	Description
Occupation Theory	A person who have a control of land for a very long time.
Labor Theory	Ownership is a right from his work
Property and Personality	Individual ability to act as free personality and ability to control his property.
Economic Theory	Property right can be said if a person can have economic benefits.

Adapted from: Weir, 2001.

Beside the theory of property, some of classic doctrine envisages ownership as a jurisdiction. Gaus said that this doctrine is a primitive argument, because jurisdiction have individual bound. Even though, jurisdiction can be mean as a territorial of someone rights of goods or its property [8]. Gaus philosophical have two argumentations, which are: 1) *property is about division of resources,* and 2) *the property rights equilibrium is mutually beneficial.* He also said that this two claim are easily to determine. But if we refer to the right of natural resources as a scarce resource that had goal of society prosperity, Gaus argument is become ideal.

Balancing between private properties with society is moral equilibrium, but the idea must be run through the law. In this tangency is a debate between socialist and liberalist regime. Gaus also deny freedom of ownership concept, because to some point individual rights can be defeated by public interest. Even the concept of property right stand in individual concept, but in contemporary US society, public interest has more power then individual rights. A Dutch jurist who done research about common law concept in Indonesia, Vollenhoven found a new terms of legal system, namely Adat law. In legal science, Adat law have similarity from *Anglo-Saxon* law system, that the law is based on common tradition (*der normatived Kraft des Faktischen*). They are two argument Vollenhoven in Adat law, namely: 1) law from west is not better law compared with Adat law, on a contrary Adat law is better then law from west, 2) legal construction of law from west if it applied in Adat law is a disaster [9]. Key point to see a good law that have strong protection of public interest, and balance between public interest with individual interest. Therefore the law had social control in society.

3. Type of Virtual Property Right

Property right and proprietary right are two type of ownership that admitted by law. Property right is ownership right in general, and proprietary right is intellectual property right. Copyright law, already accommodated database as a part of copyright type. In some countries like US, they inserted information as well, but in Indonesia, the law does not set it as copyright. In this paper, I will not discuss about database as copyright object because in clearly regulate and protected under copyright law started in 1995 by the agreement on Trade-Related Aspect of Intellectual Property Rights (TRIPS) [10]. In spite of the copyright regime, asset in online game, especially MMORPG games still unregulated in most of country. Table below is a description of property right and its example.

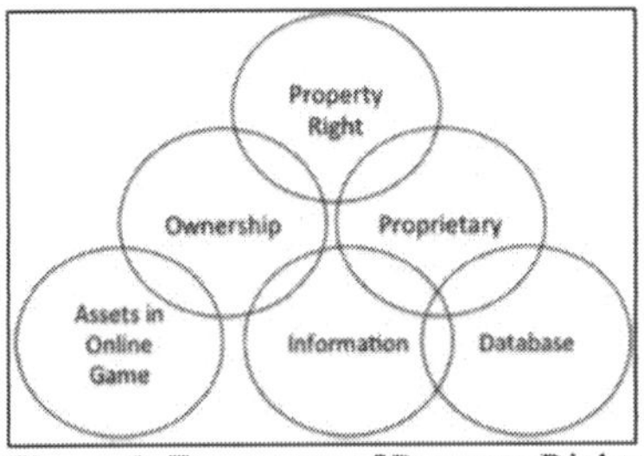

Figure 1. Taxonomy of Property Rights

Even the development of information technology moving fast, but the law development to capture asset in online games is opposite. To regulate the phenomenon of virtual assets in games, the foundation of property concept must have been firmed, whether belong to intellectual property rights regime or property rights in general. In this paper, I argue that virtual property right in games is belong to intellectual property rights regime, because the element of property right is fulfilled, such as: made by human, made by someone who have skill in something (playing games skill), and perceived by human sense. In the next part of this paper I will explain the type of virtual property rights.

3.1 Assets in Online Game

To capture virtual property right, they are two types of law regime, copyright and cyber law regime. In copyright law, the law does not regulate and protect virtual property right inside games, but only protect the game. In this condition we must search an answer in cyber law norms. The law regarding cyber in Indonesia is regulated under the law No. 11/2008 regarding Information and electronic transaction (herein after ITE law). To give holistic description about virtual property in games, the figure below can be described.

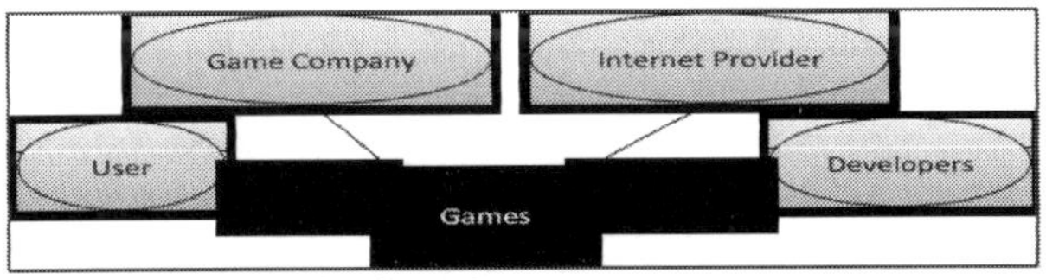

Figure 2. Relationship in Online games

In the figure shows that in an online game there are at least four interactions parties, namely: developers, Internet providers, game company, and gamers (user). Standing position of each party are: (1) developers as game developers, usually always updating and improving when the game if the game damage, security update or others. (2) Internet provider is a provider of Internet bandwidth that are usually subscribed by Internet user or customer. (3) In game company position, there are party who rule as a developer and game company, and also there is separate party between game developer and game company. But in the law perspective, we must separate these two positions. (4) User is a game player.

In online games, usually played by user to have certain hour to have certain game assets. Some user also can pay a voucher for shorten game playtime. Relation between game companies with user can determine as producer and consumer relationship, but this relation also have obligation to game company as producer to protect the customer assets. In reality, game company avoiding this obligation by making standard clause of contract containing not to responsible if there any consumer lost because of server error or system error. Table below is contract comparison from three of game company that includes obligation avoidance.

Table 2. Contract Comparison in *Online Game* [11]

Ketentuan	Gemscool	Lyto	Dasa Game
Obligation to register	Yes	Yes	Yes
Data confidentiality, except to business partner	Yes	Yes	Yes
User have full responsibility to the user ID	Yes	Yes	Yes
User duty to keep and take preventive action	Yes	Yes	Yes
Game company is not responsible for any lost of ID, password because of hacking, keyloger, virus, fraudulence, and/or user negligence.	Yes	Yes	Yes
Prohibition for user to have economy benefit from game facilities.	Yes	Yes	Yes
All game items belong to the game company.	Yes	Yes	Yes
Clossing account if it is idle for some time.	Yes	No	No
Prohibition to distribute, transmition without permission from game company.	Yes	Yes	Yes
Disclaimer to indemnify any consumer lost from system error or natural disasters.	Yes	Yes	Yes
The company's right to shut down the service in anytime without notice.	Yes	Yes	Yes
Oblige user constantly to check the desclimer	No	Yes	Yes

© Pratama, 2012.

From the table above clearly visible that game company-avoiding obligation by making the disclaimer in the contract. There must be solution of indemnify for system error, because the causality not by user, and its unfair of the game company won't give compensation. For user virtual assets, the user doesn't have any receive that the assets is belong to the user. If there any case that user A borrowed the assets from user B, and apparently the asset is sold by user B, then the user A does not have the power to claim it. This is what happens in the case of Qiu in China in 2005, which physically killed his friend because of sold his virtual sword [12]. It is still unfair that user doesn't have any claim of ownership in virtual property right. There is inconsistency concept of copyright, if game publisher want to applied this concept. Figure below is a comparison of inconsistency copyright concept in online game.

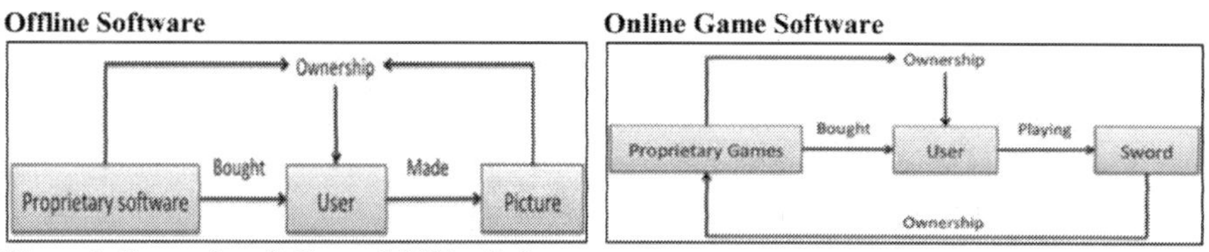

Figure 3. Comparison Online and Offline Software

Figure above describe inconsistency ownership concept between offline and online software. For example if you bough Adobe Photoshop software, then you made a picture with the software. The picture that you've been made is certainly belongs to you as a creator, and you are protected by copyright law once you publish the picture. But if you buy an online game software and playing it online, the asset that you produce inside the game is not belong to you, it's belong to the copyright holder of the game, which is game company.

At this condition, they are inconsistency ownership concept. Why we must use copyright concept? Because in many country, the law regulate software as part of copyright, and the consequence we must applied copyright concept consistently. Inconsistency concept of ownership rise in online game because of the right to control is different between offline software. In classic property right law books it said that right of controlled of the property it will give logical consequence of property right ownership [13]. We cannot change the concept of copyright that applied in offline software, because the concept is internationally applied and have been use more in more than hundred country member of World Intellectual Property Organization (WIPO). The only change is to adjust the concept of ownership in online game to be the same with offline software to have legal coherency. If the concept cannot be modified, then the ownership concept is anomaly and unfair, because the game itself is protected under copyright law, but the asset that produce by user cannot protected by copyright law.

3.2 Information

Some of intellectual legislation classifying information as a part of copyright, for example US copyright law. But some country is not, Indonesia for example. If information is not part of intellectual property law regime, classic property right is the pedestal right. But, sometimes people forgot the limitation concept of freedom of information. In law perspective, they are limitation of individual freedom from another individual. The law rule to keep balanced between individual freedoms, but the common goal is society order. In terms of information freedom, the proposition above must be used to determine which information that can be right of individual, public or state secrets. My research founded that they are 3 type of information's, the classification is based on normative concept in law. These three of information can be described below.

Table 3. Type of Information [14]

No	Type	Description
1	Private Information	Personal information, autonomy, physical space and property [15].
	a. Person	
	b. Legal entity	Unknown information by public in terms of technology, business or others that have economic value.
2	Public Information	Contain state administration in good governance information
3	State Secret Information	Information, goods, personnel, works, activity, that protected by the state.

© Pratama, 2016.

The taxonomy of information is clear measurement to determine which information that belongs to the person, legal entity or state. Point to remember from intellectual property law concept is derivative works, and its concept also applied to information. For example, the law or judicial decision is public information, but if someone compiled the law or judicial decision, then the compilation is copyrighted protected. Even the contain of information is publics, that every person can freely accessed, but the compilation is belong to the author. In terms of state secret information, the state ownership, but the problems is how the state gains the information? The answer is debatable! For example Snowden cases, if the government spying on the citizens then the government was breaking the law to gain information. But once again it's debatable, but I'm pretty sure that government have impunities right to close the information, with the purpose to keep the state secure from the intrusion foreign and domestic.

4. Conclusion

To view virtual property rights using property law, they are three type of it, namely: database, asset in online game, and information. In online game assets, they are inconsistency of copyright ownership between online software with offline software. In this case, online game is classified as online software. Inconsistency occur because of control right of the copyright object is on game publisher, not by user. So the consequences is the ownership is on game publisher not, by user. If we want to use copyright concept consistently then the concept of ownership in online game must be transferred by game publisher to the user. The last type of virtual property right that can be defined is information, and information can be separated into three type, namely: personal information (owned by person, or by legal entity/corporation), public information (regarding state good governance) and state secrets information (regarding state defense information). In terms of information ownership claim, the claimed must be defined clearly, to determine legal right of information. By determining the type of information, the scope of information become clear and the determination of legal right also become clear.

References

[1] Richard A. P , 2005 , "Intellectual Property: The Law and Economic Approach," *Journal of Economic Perspectives*, vol. 19, p. 61

[2] J. Janewa O, 2015, "Corporate "Human Rights" to Intellectual Property Protection?," *Santa Clara Law Review*, vol. 55, p. 61

[3] Yun-Chien C , Henry E S, 2012, "An Economic Analysis of Civil Versus Common Law Property," vol. 88

[4] Habil C U S. Et. All., *Real Property Law and Procedure in The European*. Würzburg: European University Institute (EUI), 2--5.

[5] Rachmadi U,2011, *Hukum Kebendaan*. Jakarta: Sinar Grafika

[6] Robbins. The Common Law and Civil Law Traditions. [Online]. HYPERLINK "https://www.law.berkeley.edu/library/robbins/pdf/CommonLawCivilLawTraditions.pdf https://www.law.berkeley.edu/library/robbins/pdf/CommonLawCivilLawTraditions.pdf

[7] Michael W,2001, "Concept of Property, ," *The National Legal Eagle*, vol. 7, no. I, p. 17

[8] Gerald G,2012, *Property and Ownership*, David Estlund, Ed.: Oxford University Press.

[9] Van V, 1994, *Our Laws Their Lands, Land Laws and Land Use in Moderns Colonial Societies*, Dietmar Rothermunds Jap de Moor, Ed. Hamburrg: Lit Verlag

[10] John G S,2014, *The International law of Property*. United Kingdom: Oxford University Press

[11] Bambang P,2012, "Perlindungan Hukum Digital Property di Indonesia," Business Law Department, Universitas Bina Nusantara, Jakarta

[12] Cao L. China Daily. [Online]. HYPERLINK "www.chinadaily.com.cn/english/doc/2005-06/08/content_449600.htm," www.chinadaily.com.cn/english/doc/2005-06/08/content_449600.htm,

[13] Sri S M S,1977, *Hukum Perdata: Hukum Benda*. Yogyakarta, Indonesia: Liberty, vol. 5.

[14] Bambang P, 2016, "Karakteristik Hak Cipta dalam Sistem Hukum Siber Indonesia sebagai Penunjang Ekonomi Nasional Berbasis Pengetahuan," Postgraduate Law School, Law Disertation Catholic Parahyangan University, Bandung

[15] John L M, 2008, *Privacy, The Lost Righ*. New York: Oxford University Press

We are Different but Alike: A Comparative Analysis between Students' and Teachers' Use of Facebook

Erlin, Susandri, Triyani Arita Fitri

Informatics Engineering, STMIK Amik Riau, Pekanbaru, Indonesia

erlin@stmik-amik-riau.ac.id

Abstract. The purpose of this study is to compare the use of Facebook by students and teachers in higher education. This study is not merely discussed Facebook in intercourse educational environment, or about teaching and learning purposes but more focus to explore and analyze the similarities and differences between students and teachers use of Facebook. This study released an online survey that consists of 17 questions in a questionnaire. A total of 378 respondents from three colleges in Indonesia had given feedback. The finding reveals that neither students nor teacher's positivity agreed with the statements related to the potential of Facebook for education. Students and teachers stated that Facebook could be used to quickly share information, file, news, events, and resource courses. There were differences in perception on question teachers use the chat facility on Facebook to guide students' final project. Almost half of students agreed with this statement while teachers vice versa. Based on the similarities more than the differences indicate that Facebook can be used as a possible educational tool and media assistant for academic communication and academic purposes.

1. Introduction

The rise of Web 2.0, led to the increasing popularity of Social Networking Sites (SNS), in line with social escalation interaction, collaboration, and sharing among internet users. SNS such as Facebook, Twitter, Instagram, Path, Google+, etc. have gained immense popularity in the today's digital society and are being adopted briskly by billions of users. It has entered almost all facets of society. Business, politics, health, trade and education cannot be separated from the role of SNS in it. In 2017, it is estimated that there will be around 2,51 billion social network users around the globe, up from 2,34 billion in 2016 [1]. This number will continue to increase over time as more quickly gained internet access by users around the world.

In spite of various SNS, Facebook is the most popular and the biggest social networking site with the largest number of users. In the third quarter of 2012, the number of active Facebook users had surpassed 1 billion. As of the second quarter of 2016, Facebook had 1,71 billion monthly active users [1]. A huge and vastly growing number of Facebook users, cause a lot of scholars, conducted research on Facebook such as research on Facebook usage [2][3]. Another scholar discussed and analysed the potential Facebook for education [4][5]. Furthermore, several scholars pointed out the effects of using Facebook [6][7], including academic performance [8].

Facebook's user base is dominated by 25 years old or older, but they still have more than 50 million users under 25 years old [9]. Since Facebook has a lot of a number of users under the age of 25 years old, many researchers interested to explored and analysed the use of Facebook for an educational environment. El Bialy and Jalali [10] stated that Facebook is not convenient for education because teachers do not offer a suitable mode of learning. Moreover, Manca and Ranieri [11] argued that more favourable attitude towards personal sharing and connecting with peers in professional networks rather than integrating social media into their teaching practices. In contrast [12] [13], revealed that Facebook could be used to enhance the learning process and promote a mindset of continual learning.

Although many studies conducted in several different countries in understanding how much, why and how students use Facebook as a popular social networking site, but authors have not yet found the articles or empirical studies on the use of Facebook by students and teachers in Indonesia. Therefore, the primary objective of this study was to investigate and compare the use of Facebook by students and teachers at three colleges in Indonesia and also investigate the purpose of the students and teachers using Facebook in educational settings. This study is a continuation previously study that focused on Facebook usage at Riau college students [3].

2. Research Method

2.1. Respondents
This study used students and teachers of college as research subjects. The respondents were contacted through their on-campus email accounts and sent a link to a survey hosted on SurveyMonkey.com, the world's most popular commercial survey online. A total of 135 respondents agreed to participate in the pre-test survey, 92,59% (125) are students and 7,41% (10) are teachers. 93,33% (126) pre-test is valid, including 118 (response rate 94,4%) students and 8 (response rate 80%) teachers. After the approval of three colleges in Riau, Indonesia; STMIK Amik Riau, STIE Pelita Indonesia, and STIKOM Pelita Indonesia, the researchers send the questionnaire that has been validated by SPSS version 2.0 a statistical software package. A total of 400 questionnaires were released, 350 for students and 50 for teachers; and there were 378 valid responses for an overall 94,5% response rate. Of the 378 respondents, 87,04% (329) are students and 12,96% (49) are teachers.

2.2. Measures
Respondents were required to complete an online survey. The first part of survey questions was general information that consists of four questions for students (gender, age, year of study and time using Facebook) and five question for teachers (gender, age, education, department, and time using Facebook). The second part was about Facebook usage that consists of five survey questions; log in to Facebook per day, time spent on Facebook per day, the number of friends on Facebook, the number of group on Facebook, and favourite social networking sites. The last part of survey questions was about the perceptions of students and teachers on Facebook for education. This part consists of seven questions; a group created by teachers, teachers used Facebook group to share information, academic staff used Facebook group to share information, teachers used Facebook chat to guide students on final project, Facebook conveys information quickly, Facebook is convenience in sharing file, multimedia, and other resources among fellow academicians, and Facebook is better for reviewing and searching through past article on a certain topic. This part used 5-point Likert scale; 1=strongly disagree, 5=strongly agree.

2.3. Procedures
This study uses a survey online in the form of a questionnaire to gather opinions and perceptions of students and teachers towards the use of Facebook. Before the questionnaire distributed to the respondents, the questionnaire was prior tested to a small number of respondents to determine the validity and reliability of the survey. The validity of the survey's items was tested using the Pearson correlation method ($p < 0.05$) and the reliability was tested using Cronbach α with Cronbach's α of 0.83. After reaching the standard value, the questionnaire is ready to be shared with the students and teachers of three colleges in Indonesia.

The technical on filling out the questionnaire by respondent was designed using nonparametric statistical which are closed. Analysis obtained from this technique is descriptive and explanatory, and the data were expected valid and filled with actual by the respondent to describe how the real data available in the field.

3. Results and Discussion

3.1. Students and Teachers Demographic Profile

Table 1 shows the students demographic profile that consisted of four items; gender, age, year of study and time using Facebook. In term of gender, 69% (227) of respondents identified as male, while 31% (102) of respondents are female. In age, 27,66% (91) of respondents under 20 years of age, whereas 67,17% (221) of respondents were 20 to 25 years old, and 5,17% (17) of respondents above 25 years old. In the year of study, 26,44% (87) of respondents at first year, 20,67% (68) at second years, 31% (102) at third years, 19,45% (64) at the fourth years; whereas only 2,43% (8) of respondents at more than four years. Finally, in term of time using Facebook, almost all respondents were long term Facebook users, with the vast majority (95,13%, 323 respondents) indicating they had used Facebook for more than 2 years, 4,56% (15) for 1 to 2 years and only one student had used Facebook less than a year.

Table 1. Students Demographic Profile

Demographic	n (329)	n (%)
Gender		
Male	227	69,00
Female	102	31,00
Age		
< 20 years old	91	27,66
20 – 25 years old	221	67,17
> 25 years old	17	5,17
Year of Study		
1st year	87	26,44
2nd year	68	20,67
3rd year	102	31,00
4th year	64	19,45
>4th year	8	2,43
Time Using Facebook		
<1 year	1	0,30
1-2 years	15	4,56
3-4 years	92	27,96
5-6 years	129	39,21
>6 years	92	27,96

Table 2. Teachers Demographic Profile

Demographic	n (49)	n (%)
Gender		
Male	30	61,22
Female	19	38,78
Age		
< 30 years old	27	55,10
30 – 40 years old	11	22,45
> 40 years old	11	22,45
Education		
Master Degree	48	97,96
Doctorate Degree	1	2,04
Department		
Informatics Engineering	31	63,27
Information Systems	18	36,73
Time Using Facebook		
< 1 year	0	0
1-2 years	0	0
3-4 years	2	4,08
5-6 years	14	28,57
> 6 years	33	67,35

In spite table 1 shows the student's demographic profile, table 2 presents the teacher's demographic profile that consisted of five items; gender, age, education, department and time using Facebook. In gender term, 61,22% (30) of respondents identified as male, while 38,78% (19) of respondents are female. In age, 55,10% (27) of respondents under 30 years of age, whereas 22,45% (11) of respondents were 30 to 40 years old, and 22,45% (11) of respondents above 40 years old. In term of education, only 1 (2,04%) of respondent had a doctorate degree; the rest are master degree (97,96%). Majority of respondents from informatics engineering department (63,27%-31), while 36,73% (18) from information systems department. Similarly, in term of time using Facebook, most of the respondents (67,35%) used Facebook for more than 6 years, 28,57% (14) for 5 to 6 years, 4,08% (2) for 3 to 4 years; no one indicated they had been using Facebook for less than a year as well as for 1 to 2 years.

3.2. Facebook Usage

Comparison of students and teachers on Facebook usage as shown in table 3. On the first question related to the login to Facebook per day, there is a difference between students and teachers. Most of the students (38,60% - 127) log in to Facebook three to six times per day, while most of the teachers (53,06%-26) login to Facebook less than three times per day. However, in term of time spent on Facebook, neither students nor teachers spent their time on Facebook around 30 to 60 minutes per day. It means that, although teachers did not log in as much as students but time spent on Facebook alike. Means, once logged into Facebook the teachers spending more time than the students.

In a number of friend's term, about half of respondents for both of students and teachers claim to have the number of friends of more than 1000 shows that both of students and teachers receiving quite a lot of friends and inviting someone to be a friend on Facebook. From the student's side, some of other respondents (23,71%-78) have a number of friends between 500 to 1000, and 90 respondents (27,36%) have a number of friends less than 500 friends. From the teacher's side, 18 respondents (36,73%), have a number of friends between 500 to 1000 friends, and the remaining, 8 respondents (16,33%) have a number of friends less than 500 friends.

Table 3. Comparison of student and teacher on Facebook usage

Question Items	Indicator Question	Student (N=329)		Teacher (N=49)	
		N	n %	n	n %
Log in to Facebook per day	< 3 times	83	25,23	26	53,06
	3 – 6 times	127	38,60	20	40,82
	> 6 times	119	36,17	3	6,12
Time spent on Facebook per day	< 30 minutes	71	21,58	12	24,49
	30 – 60 minutes	150	45,59	27	55,10
	> 60 minutes	108	32,83	10	20,41
Number of Friends on Facebook	< 500	90	27,36	8	16,33
	500 – 1000	78	23,71	18	36,73
	> 1000	161	48,94	23	46,94
Number of Group on Facebook	< 5	84	25,53	11	22,45
	5 – 10	135	41,03	26	53,06
	> 10	110	33,44	12	24,49
Favorite Social Networking Sites	Facebook	256	80,82	38	77,55
	Google+	138	43,40	13	26,53
	Plurk	13	4,09	0	0
	Twitter	181	56,92	17	34,69
	Instagram	191	60,06	29	59,80
	LinkedIn	23	7,32	4	8,16
	Pinterest	14	4,40	0	0
	Path	92	28,93	13	26,53
	Others	59	18,55	7	14,29

Furthermore, the questions relating to the item the number of the Facebook group that owned by respondents. Most of respondents, both students (41.04%-135) and teachers (53.06%-26) have a Facebook group between 5 to10 groups, shows that both of them are equally open towards the formation of the group on Facebook related to their profession as a student and a teacher. From the student's views, (33,44%-110) of respondents stated that they have a Facebook group are more than 10 groups, and the remaining, 84 respondents (25,53%) have a Facebook group are less than 5 group. Instead of the teachers, 12 respondents (24,49%) claimed to have a Facebook group of more than 10 groups and the remaining 11 respondents (22,45%) have a Facebook group of less than 5 groups.

The last questions related to the social networking site which is available and be a favourite for the respondent by right now. Although there are numerous social media sites available, Facebook is a significantly popular site. The overwhelming majority of respondents, either students (256-80,82%), nor teachers (77,55%-38) stated that Facebook was the most used and preferred SNS by students and teachers at the college compared to others sites such as Google +, Twitter, Instagram, LinkedIn, Path, Plurk, Pinterest and others (Line, BBM, WA, etc). It means that Facebook becomes a popular choice for students and teachers as social media to interact each other. Both types of respondents, students, and teachers alike agree that Instagram becomes another popular social media after Facebook, followed by Twitter, Google+, Path, Others (Line, BBM, WA, etc), LinkedIn, Pinterest and the last Plurk. Although the order of the favorite social media in the same order, but the difference lies in social media Plurk and Pinterest, where is no one teacher stated that they have the account on Plurk and Pinterest.

3.3. Facebook For Education

Table 4 and table 5 depicts the perceptions of students and teachers on Facebook for education. In this study, the perception of respondents which is categorized as strongly agree and agree would be classified as total agree, whereas perception of respondents which is categorized as strongly disagree and disagree would be classified as total disagree.

The results showed that in general, students and teachers were more likely a group created by teachers. One of feature that is owned by Facebook is the ease of forming the group consisting of similar cohorts with similar objectives. Overall agreement by students (strongly agree=41,56% + agree=44,38%, total agree = 85,94%), and overall agreement by teachers (strongly agree=44,90% + agree=46,94%, total agree = 91,84%). Regarding the teachers used Facebook group to share information, most of the respondents (students = 86,16%, teachers = 93,86%) indicated that they were more likely Facebook group as a tool to share information with teachers. Teachers feel more comfortable to share information with the member of the group in their class that they hace created before.

Due to the teachers found that it easy to share information through Facebook, it is not surprising that there are the number of teachers who agree the academic staff use Facebook to share information is also very high (91.84%) of the respondents. It is also directly proportional to the student's perception that also prefers to academic staff use Facebook to share information (76,08%).

Regarding teachers used Facebook chat to guide students on the final project, the result reported that less than half (46,11%) of students agree to use chat feature of Facebook as a tool to guide them in the final project. On the other hand, (81,63%) of teachers, disagree about that. Teachers have a certain view, not unimpeded for guiding the final project through the chat tool on Facebook.

Furthermore, teachers experience when using the Facebook group, where rapid in conveying information to members of the group, providing a view of the 87,76% of teachers agree that Facebook can deliver information quickly, while 73,38% of students have the same perception of the teachers. Only a small of respondents who disagree with this statement.

The capabilities of Facebook in sharing files, multimedia, and other learning resources easily become rationale 93,88% of teachers, and 71,96% of students agree with this statement. Share files in multi-media format have become another advantage of Facebook in attracting the attention of respondents. Moreover, during follow and become a member of the Facebook group, the respondents confirmed that Facebook is very easily used as a tool to share files and other resources for their communities that they follow.

When a user uploads a file in various formats, whether text, images, video or other sources, users quickly browse and looking back the files in question. A total of 72,82% of students and 87,76% of teachers agreed to this statement. It proves that Facebook is very easy to use and has a search facility which is a requirement for social networking application.

Table 4. The perceptions of students on Facebook for education

Questions	strongly agree	agree	neutral	disagree	strongly disagree	skipped	total
A group created by teachers	41.56% 133	44.38% 142	11.88% 38	1.56% 5	0.63% 2	9	320
Teachers used Facebook group to share information	40.25% 128	45.91% 146	11.64% 37	0.63% 2	1.57% 5	11	318
Academic staff used Facebook group to share information, news, and event	34.47% 111	41.61% 134	18.01% 58	5.28% 17	0.62% 2	7	322
Teachers used Facebook chat to guide students on final project	17.45% 56	28.66% 92	24.92% 80	23.05% 74	5.92% 19	8	321
Facebook convey information quickly	34.06% 110	39.32% 127	19.81% 64	5.88% 19	0.93% 3	6	323
Convenience in sharing file, multimedia or textual educational resources	24.92% 80	47.04% 151	23.99% 77	3.74% 12	0.31% 1	8	321
Facebook is better for reviewing or searching through past article	28.44% 91	44.38% 142	22.81% 73	4.06% 13	0.31% 1	9	320

Table 5. The perceptions of teachers on Facebook for education

Questions	strongly agree	agree	neutral	disagree	strongly disagree	skipped	total
A group created by teachers	44,90% 22	46,94% 23	4,08% 2	2,04% 1	2,04% 1	0	49
Teachers used Facebook group to share information	51,02% 25	42,86% 21	4,08% 2	0,00% 0	2,04% 1	0	49
Academic staff used Facebook to share information, news, and event	44,90% 22	46,94% 23	0,00% 0	4,08% 2	4,08% 2	0	49
Teachers used Facebook chat to guide students on final project	10,20% 5	6,12% 3	2,04% 1	30,61% 15	51,02% 25	0	49
Facebook convey information quickly	46,94% 23	40,82% 20	4,08% 2	4,08% 2	4,08% 2	0	49
Convenience in sharing file, multimedia or textual educational resources	48,98% 24	44,90% 22	2,04% 1	2,04% 1	2,04% 1	0	49
Facebook is better for reviewing or searching through past article	40,82% 20	46,94% 23	4,08% 2	4,08% 2	4,08% 2	0	49

4. Conclusion

Facebook is one of the most popular social networking sites that has been widely used by students and teachers in their daily life activity. Facebook use became more ingrained into their life, and as a consequence, has the potential to become valuable resources to support academic communication and academic purposes.

Based on the results, there are some similarities compared to differences between students and teachers use of Facebook. Students and teachers have the same perception on the four of five questions related to the Facebook usage. They only differ on the first item about the matter how many times a user logged into Facebook per day. Although there is a different login level, however, the time spent on Facebook per day has the same time, 30 – 60 minutes per day.

Furthermore, on the part of Facebook for education, students and teachers agree on almost all questions. The difference is only found on the item in question number 4 (four) that teachers use the chat facility on Facebook to guide a student in the final project. Nearly half the students agreed with this statement, while most teachers precisely do not agree. The percentage of the teachers agree greater than the students in 9 (nine) of the 10 (ten) item in question. For the student, the whole question agrees more dominant compared to strongly agree, while for teachers, 5 of the 7 questions, strongly agree more dominant than agreed, which means that the teachers strongly agree Facebook used as a tool for educational purposes.

Based on this study, indicates that Facebook as a good medium not only for communication but also for engagement in the process of learning, especially when teachers and students use the facebook group. Facebook has the potential to enhance teaching and learning process and support formal learning. Hence, finding of this study support the point of view that social networking site, particularly Facebook, can be used as a part of improving and arousing students' and teachers' experience and learning.

Acknowledgement
The researchers would like to thank the Ministry of Research, Technology and Higher Education (Kemenristekdikti), Indonesia, for funding this research. We also thanks to teachers and our students for their assistance and the survey filling.

References

[1] Statista.com, *Number of worldwide social network users*. Retrieved Juni 2016.
[2] L. Ahern, J. Feller, and T. Nagle, "Social media as a support for learning in universities: an empirical study of Facebook Groups," *J. Decis. Syst.*, vol. 25, no. sup1, pp. 35–49, 2016.
[3] E. Erlin, A. F. Triyani, and S. Susandri, "Using Social Networks: Facebook Usage at the Riau College Students," in *Procedia Computer Science*, 2015, vol. 59, pp. 559–566.
[4] S. Kumar, A. Joshi, and H. Sharma, "A multi-analytical approach to predict the Facebook usage in higher education," *Comput. Human Behav.*, vol. 55, pp. 340–353, 2016.
[5] D. Lambić, "Correlation between Facebook use for educational purposes and academic performance of students," *Comput. Human Behav.*, vol. 61, pp. 313–320, 2016.
[6] J. H. Tang, M. C. Chen, C. Y. Yang, T. Y. Chung, and Y. A. Lee, "Personality traits, interpersonal relationships, online social support, and Facebook addiction," *Telemat. Informatics*, vol. 33, no. 1, pp. 102–108, 2016.
[7] N. Leelathakul and K. Chaipah, "Quantitative effects of using facebook as a learning tool on students' performance," in *10th International Joint Conference on Computer Science and Software Engineering (JCSSE) 2013*, pp. 87–92, 2013
[8] R. Junco, "Student class standing, Facebook use, and academic performance," *J. Appl. Dev. Psychol.*, vol. 36, pp. 18–29, 2015.
[9] Socialbakers.com, "Facebook statistics." Retrieved Juni 2016.
[10] S. El Bialy and A. Jalali, "Go Where the Students Are: A Comparison of the Use of Social Networking Sites Between Medical Students and Medical Educators," *JMIR Med. Educ.*, vol. 1, no. 2, p. e7, 2015.
[11] S. Manca and M. Ranieri, "Is it a tool suitable for learning? A critical review of the literature on Facebook as a technology-enhanced learning environment," *J. Comput. Assist. Learn.*, vol. 29, no. 6, pp. 487–504, 2013.
[12] O. Ozyurt and H. Ozyurt, "Using Facebook to enhance learning experiences of students in computer programming at Introduction to Programming and Algorithm course," *Comput. Appl. Eng. Educ.*, vol. 24, no. 4, pp. 1–9, 2016.
[13] Z. A. Weber and A. H. Vincent, "Facebook as a method to promote a mindset of continual learning in an ambulatory care pharmacy elective course," *Curr. Pharm. Teach. Learn.*, vol. 6, no. 4, pp. 478–482, 2014.

IOP Conf. Series: Journal of Physics: Conf. Series **801** (2017) 012092 doi:10.1088/1742-6596/801/1/012092

The Use of Absorptive Capacity in Improving the New Product Development (NPD

W Gunawan[1], P Gerardus[12], B J Tji[12], and K Richard[12]

[1]Faculty member, Master in Information Systems Management,
[2]Senior Faculty member, DRM Programmes, Bina Nusantara University.

Abstract. The term Absorptive Capacity (AC) refers to maximizing the external knowledge transfer into the organization in order to improve its performance. Since its introduction in year 1990, AC has been applied widely in many fields such as: economy, business, KM, HR, intellectual capital, IT, operation management, marketing, etc. Due to its wide application, nevertheless, The AC application in both Indonesian industry and higher education institutions (HEIs) are still rare to find. The Indonesian Directorate General of Higher Education (DGHE) has encouraged creating effective collaboration model that enables to link the HEIs with the industries in order to improve knowledge creation in improving product development that can be used by the firms. For this reason, the article examines the effective AC model that enables to assist in improving new product development (NPD) process in the polytechnic perspectives.

1. Introduction

Researchers have found that firms that use many other external sources of knowledge such as competitors, suppliers, and customers, private institute, trade fairs and associations etc. also tend to use the higher education research more intensively (1) (2). Prior knowledge of collaboration allows the higher education institutions (HEIs) to assimilate and exploit new knowledge, and enhance its learning capabilities and AC (3). Li emphasizes that global research and development (R&D) literature portrays international R&D conducted by multinational corporations (MNCs) have begun to draw on new sources of innovation such as local HEIs for knowledge creation and new product development (NPD). Li addressed that many HEIs are shifting from long-term research to near-term research and commercialization of research output.

The HEIs have traditionally provided know-how (skills and capability), and know-why (general principles and laws), and they need partner that can focus on commercializing the knowledge (4). The DGHE enables the HEIs to take role in offering the consultancy services, and entering into industry collaborative relationship expansion into acquiring know-what (facts) and know-who (establishing the relationships).

The national government states the important role of the HEIs in the national master plan (5) to support the growth of SMEs and large enterprises (5). The leading HEIs are subject to the beneficial reputation effects either in local and global. Inter-organisational ties are effective means to diffuse and transfer the complex knowledge. It enables to create a mutual trust, embeddedness, and social cohesion between partners (6). Petruzelli addressed that firm relationship enables to provide opportunities in:

easy access to wide range of industries; learn the different knowledge from many industries; and link knowledge across industries and sectors. The HEIs have the ability to recombine and integrate the external knowledge and act as knowledge brokers that span multiple markets and technology domains and transfer the knowledge for any company that needs it. These effects lead to the improvement of innovation performance of the polytechnics as measured for instance, by patent applications and registrations (4).

The article examines the needs of applying the absorptive capacity (AC) term, to acquire the firm (external) knowledge and internalize it into product development in the state-owned polytechnics. The state-owned polytechnics are selected due to their strategic position in the national economic development plan (5).

2. Theoretical Framework

2.1. Application Absorptive Capacity (AC) in Higher Education Institutions (HEIs).

Improving the AC in the HEIs is not an easy task (7). Huggins et.al. (4) warn that the collaboration of higher education and firms contains high complexity and can easily ends up in failures. The article addresses the major problems faced by state-owned polytechnics from the AC perspectives such as: (a) major polytechnics and firms have few clues on what they expect from each other. The firms have very little clue on acquiring the academic knowledge owned by the HEIs; on the other hand, the HEIs have difficulty to transform the academic knowledge into practical knowledge (8) that can be commercialized by the firms. Both parties may experience many difficulties on maximizing the knowledge transfer in NPD (1) (6); (b) Lack of collaboration model has led to the difficulty of assessing the knowledge transfer from firms to polytechnic and vice versa (9).

Broring et.al. (10) addressed the important of all firms to collaborate with the HEIs in creating value in their NPD process. Broring et.al. encourage the firms to search academic partnerships in: patent development, literature analyses, conference attendance and close contact with the leading universities or polytechnics. They argue that firms should be aware of their competence gaps, and needs to improve their AC in order to improve front-end of the NPD process.

The idea of improving AC in supporting industry's NPD has limited application for the state-owned polytechnics in Indonesia, where there are wide disparities amongst those polytechnics on maximizing and internalizing the knowledge transfer with the firms. Studies conducted by the U.S. National Science Foundation (NSF) (11), show that most of collaboration programmes end up in failures. NSF noted that most HEI faculties gained experience in competing for and managing federal research projects, however they have not learned how to develop and manage cooperation research framework. Edmondson addressed the collaboration problems lies on the gap of understanding framework, experience and skills required in the collaboration.

The active role of stakeholders determines the successful of HEIs collaboration with the firms. There are several important factors affect to the collaboration such as (11): understanding the nature of polytechnic and firms; multidisciplinary and cross-cultural approach to research and learning; shared vision and develop a clear strategy, is important to determine the successful of collaboration, and to be supported with the appropriate AC framework; rethinking the roles of traditional Technology Transfer Office (TTO), as the main mission to support the activity of research and community/industry service; formalization of the partnerships; other important factors are not discussed in the article due to its low relevance to the AC framework, such as entrepreneurship skills, time devotion and full commitment from top management of both higher education institutions, etc.

The concept of AC (12) can be applied in all these steps through emphasizing the collaboration with the firms as important external sources with following activities such as (13): (a) NPD is a business process that is highly knowledge-intensive. It needs the involvement of the firms (customers) as key business actor for creating new organizational knowledge; (b) NPD projects are under pressure to accelerate development cycles and decrease development costs, while increasing designing quality and flexibility; (c) From learning perspective for an organisation, NPD is the context from which the organization is most likely to transfer methods (resources and capabilities) to other areas of the organization; (d) In NPD, customer knowledge-related activities play a critical role, and thus provide excellent leverage points for support system and AC model enhancement; (e) NPD project are increasingly using customer sources and other external resources to overcome the learning curves related to new market and new technologies; In order to measure the final NPD success, the article examines the indicators such as: time, economy, market, and quality (14).

Zahra and George (15) addressed the importance of AC comprised of two main domains: potential absorptive capacity (PAC), and realised absorptive capacity (RAC). The article examines the PAC and RAC (16), and relate them with the determinant factors such as: ideation, prior-customer related knowledge, and motivation (17), as important steps to produce the new product advantage (NPA) (14).

2.2. *Ideation (Id)*

Ideation process is summarized into: (a) ideation relates to the PAC (Idpac), refers to getting important external (firm) idea and insights of the product design; (b) ideation relates to the RAC (Idrac), refers to internalising the idea into creating production methods/procedures (18).

2.3. *Prior Customer-related Knowledge (Pkmpac).*

The prior customer-related knowledge (Pkmpac) generally differs from one customer to the other, and it determines the experience of using the product (19). Hausberg (20) suggests that the more firms try to absorb the knowledge from HEIs, the easier they will approach and absorb knowledge from them.

2.4. *Motivation (Mot)*

Individuals with high motivation will likely be able to increase invention experience to select the creative ideas from groups that is at the frontier of knowledge and solution development. Motivation factor relates with PAC (motpac), refers to the firm (external) knowledge acquisition process, while the motivation factors relates with RAC (motrac), to relates with the transformation of external knowledge into creating effective production methods and creating new product advantage (NPA) (15).

2.5. *Support System (SS)*

Support System (SS) refers to the use of IT tools to support the entire AC process. It can be done through intra/inter organisational networks, such as: intranet and extranet. The role of SS (21) (22) in PAC (sspac) refers to gathering firm's (customer) knowledge in order to improve the collaboration firm-polytechnic. The role SS in RAC (ssrac), refers to setting up collaboration networks with external source that enables open collaboration networks through internet technologies.

2.6. *Project Complexity and Risk (Pro)*

Scholars agree that project structure plays important role in both acquiring the firm knowledge (PAC) and creating NPA (23). It depends highly on the structure (phase) of the project in dealing with complexity and risk (20). The NPD framework comprised of the following steps (adapted from (10)): project visualization (actions, interactions, commitments and customizes product definition); project planning (making commitments, choosing activities, structuring the process, estimating, optimizing, and improving key variables (time, cost, etc.); allocating resource; and project execution and control (monitoring commitments, assessing progress, re-directing and re-planning); and project development (continuous improvement, organisational learning and knowledge management, training, setting up metrics and compliance).

2.7. *Environmental Turbulences (etrac)*

The nature of NPD is generally affected by environmental turbulence in the stage of the realising the external knowledge in product development (RAC) such as (6): (a) high uncertainty, refers to the nature of business competition marked by the dynamic changing of customer's needs and competition; (b) information asymmetries, refers to the gap of information takes place amongst the firms, customers and polytechnics; (c) transactional costs. The factors of high uncertainty and asymmetric information enable to escalate the transactional (hidden) costs of the collaboration; (d) appropriability hazards. In case of high competition, market information processing is thought to be crucial for quick and adequate reaction in order to move and adapt. Similarly, the processing of superior market insights is important in highly turbulence markets, where it requires firms to continuously uncover changing customer preferences and quickly adjust product offerings to meet these most current needs (14).

2.8. New Product Advantage (NPA)

The term NPA is used as an indicator to measure the successful of the collaboration with the firm (24). Customers are more likely to purchase a new product/service when it offers superior features and unique benefits that cannot found in products that already exist in the market (25). Product/service advantage appears as the most important product characteristic in explaining the adoption and success of the new product (26). Langerak et.al. (27) further addressed the importance of creating market-oriented culture to ensure that new products help customers achieve their production objectives. It can be done through attaching the customer collaboration as the essential factor of creating market-oriented culture.

3. Research Model

The formulation of hypotheses of AC model are summarized as follows: (See figure 1)

- H1: Acquiring the prior firm (customer) related knowledge (Pkm-PAC) is important to encourage the creativity process in inventing product ideation in the collaboration process (Id-PAC).
- H2: The prior firm (customer) related knowledge (Pkm-PAC) delivers the strong roles in the improving the Potential Absorptive Capacity (PAC).
- H3: The prior firm (customer) related knowledge (Pkm-PAC) is important to increase the motivation in collaborating with the (firms) customer (Motpac).
- H4: Acquiring the product ideation from firm (customer) (Id-PAC) is important to encourage creating important idea to select appropriate production method (Id-RAC).
- H5: Motivation that involves in acquiring the firm knowledge (Motpac) is important to improve potential absorptive capacity (PAC).
- H6: The firm collaboration motivation (Motpac) is important to increase the motivation in the entire production process (Motrac).
- H7: The Potential Absorptive Capacity (PAC) delivers strong impact in increasing the Realised Absorptive Capacity (RAC).
- H8: Motivation that involves in the production is important to improve the Realised Absorptive Capacity (RAC).
- H9: The Realised Absorptive Capacity (RAC) delivers the strong impact to creating new product advantage (NPA).

4. Data Collection Method

The questionnaires are designed with the Likert scale of 5, and focused to the staffs that have actively involved in collaboration with the firms. They are ranging from staffs, supervisors, head of programmes, and directors. The population are selected from the 31 national and 17 local state owned polytechnics that are recommended by the DGHE through their reputations and experiences in the firm-polytechnics collaboration. The questionnaires later are processed by using SEM method, with SmartPLS software in order to forecast their results (28) (29) (30).

There are 1000 questionnaires are distributed through e-mails, phone calls, internet forums, seminars, and interviews. From 1,000 questionnaires had been distributed, only 113 had returned and validated. The respondents comprised of 18 (out of 31) national state-owned polytechnics, and 9 (out of 17) local state-owned (*Pemda*) polytechnics, with 79 respondents from national, and 34 respondents from local polytechnics. See Table 1 and 2 for the distribution of respondents.

Table 1. Distribution of the respondents.

		Polytechnics		
No	Title	National	Local	Total
1	Vice and Director (Top Managerial Position)	14	8	22
2	Head of Programme (Middle Management)	25	12	37

		Polytechnics		
No	Title	National	Local	Total
3	Lecturer and Operational Staff (Operational Staff)	40	14	54
	Total Respondents	79	34	113

Table 2. Distribution respondents of the institutions.

National Polytechnics (verified)	Local Polytechnics (verified)	Respondents in National Polytechnics	Respondents in Local Polytechnics
18 (58%)	9 (53%)	79	34

5. Findings

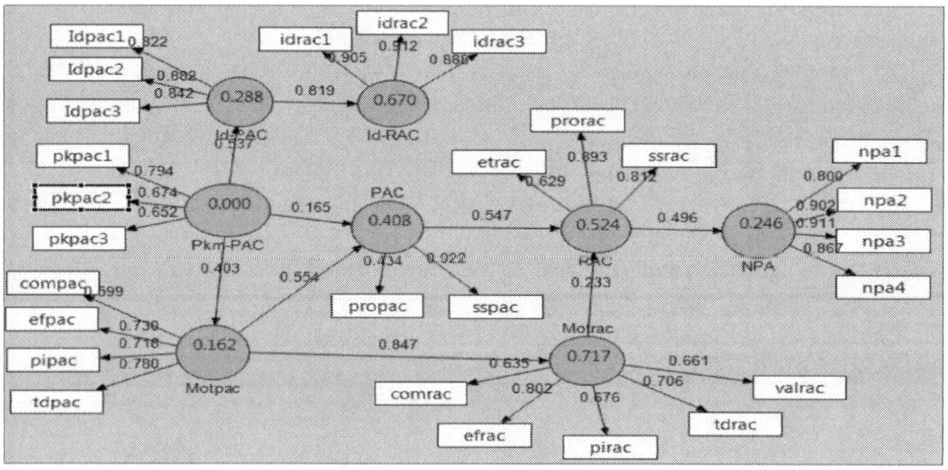

Figure 1. Absorptive Capacity (AC) model in the New Product Development (NPD). (Adapted from (14), (15) (31) (32))

Table 3. Summary validity and reliability Test.

Latent Variable	Convergent Validity (AVE)	Composite Reliability (CR)	Explained Variance (R^2)	Communality	Predictive Relevance (Q^2)	Status
Id-PAC	0.72	0.88	0.32	0.72	0.22	valid
Id-RAC	0.77	0.91	0.64	0.77	0.43	valid
Motpac	0.56	0.83	0.14	0.56	0.06	valid
Motrac	0.57	0.84	0.71	0.52	0.35	valid
NPA	0.68	0.9	0.22	0.68	0.12	valid
PAC	0.56	0.71	0.39	0.56	0.18	valid
Pkm-PAC	0.51	0.76		0.51	0.06	valid

RAC	0.64	0.84	0.53	0.64	0.3	valid
GoF = 0.51 (Large Effect) (33) (34)						

Table 3 to 8 show that model is tested valid against several measurements, and eligible for further analysis. Table 7 and 8 Outer Loading shows path coefficients > 0.5, eligible to be used as further prediction to the AC model. Table 9 shows three path coefficients<0.5 or having potential weak effects: (a) H2: the prior firm (customer) related knowledge (Pkm-PAC) delivers the weak roles in the improving the Potential Absorptive Capacity (PAC); (b) H3: the prior firm (customer) related knowledge (Pkm-PAC) does not show important roles to increase the motivation in collaborating with the (firms) customer (Motpac). (c) H6: the firm collaboration motivation (Motpac) does not show important roles to increase the motivation in production process (Motrac). It can be understood, generally, the transfer knowledge mechanism into polytechnics is still relative new, where the firms depend more on the principal products, rather than innovation capability that comes from collaboration scheme with the polytechnics in new product development.

Table 4. Discriminant Validity (Matrix Correlations)

Latent Variable	AVE	CR	Commu-nality	Q^2	Status
Id-PAC	0.72	0.88	0.72	0.22	valid
Id-RAC	0.77	0.91	0.77	0.43	valid
Motpac	0.56	0.83	0.56	0.06	valid
Motrac	0.57	0.84	0.52	0.35	valid
NPA	0.68	0.9	0.68	0.12	valid
PAC	0.56	0.71	0.56	0.18	valid
Pkm-PAC	0.51	0.76	0.51	0.06	valid
RAC	0.64	0.84	0.64	0.3	valid

Note: **Correlation is significant at the 0.01 level (2-tailed) *Correlation is significant at the 0.05 level (2-tailed).

Table 5. Cross-Loadings (Discriminant Validity)

Indicators	Path Coeffi-cient	T-distribu-tion (***)	Status
ldpac1	0.84	25.91	valid
ldpac2	0.89	53.7	valid
ldpac3	0.82	24.77	valid
Compact	0.58	10.29	valid
Comrac	0.66	12.71	valid
Efpac	0.77	16.18	valid
Efrac	0.82	33.5	valid
Etrac	0.66	8.94	valid
idrac1	0.89	39.83	valid
idrac2	0.89	53.53	valid
idrac3	0.86	32.52	valid
npa1	0.76	18.91	valid
npa2	0.88	33.28	valid
npa3	0.85	27.84	valid

Table 6. Cross-Loadings (Discriminant Validity) (continued)

Indicators	Path Coeffi-cient	T-distribu-tion (***)	Status
npa4	0.81	16.8	valid
pipac	0.81	17.3	valid
pirac	0.72	15.48	valid
Pkmpac1	0.8	18.93	valid
Pkmpac2	0.63	8	valid
Pkmpac3	0.7	16.16	valid
propac	0.57	5.96	valid
prorac	0.89	59.45	valid
sspac	0.89	36.55	valid
ssrac	0.82	33.83	valid
tdpac	0.8	25.19	valid
tdrac	0.74	22	valid
valrac	0.64	9.79	valid

Note: (***) significant with p<0.01, t>2.58 (bootstrapping with case 200 and samples 1000).

Table 7. Outer Loadings (Unidimensionality Validity Type)

Latent Variable	Indicator	Path Coeffi-cient	t-distribu-tion (***)	R Square
Potential Absorptive Capacity (PAC)	propac -> PAC	0.57	3.79	0.39
	sspac -> PAC	0.89	24.77	
Ideation (PAC)	Idpac1 -> Id-PAC	0.84	18.56	0.72
	Idpac2 -> Id-PAC	0.89	36.66	
	Idpac3 -> Id-PAC	0.82	18.56	
Prior Customer Related Knowledge (PAC)	Pkmpac1 -> Pkm-PAC	0.8	12.44	0.39
	Pkmpac2 -> Pkm-PAC	0.63	5.78	
	Pkmpac3 -> Pkm-PAC	0.7	10.98	
Motivation (PAC)	pirac -> PAC	0.58	7.09	0.14
	efpac -> PAC	0.77	17.11	
	pipac -> PAC	0.81	10.37	
	tdpac -> PAC	0.8	11.9	

Table 8. Outer Loadings (Unidimensionality Validity Type) (continued)

Latent Variable	Indicator	Path Coeffi-cient	t-distribu-tion (***)	R Square
Realised Absorptive Capacity (RAC)	prorac -> RAC	0.89	22.79	0.53
	ssrac -> RAC	0.82	37.32	
	etrac -> RAC	0.66	6.44	
Ideation (RAC)	idrac1 -> Id-RAC	0.89	27.63	0.64
	idrac2 -> Id-RAC	0.89	36.26	
	idrac3 -> Id-RAC	0.86	22.24	
Motivation (RAC)	comrac -> RAC	0.66	8.74	0.71
	pirac -> RAC	0.72	10.48	
	tdrac -> RAC	0.74	14.55	
	valrac -> RAC	0.64	6.87	
	efrac -> RAC	0.82	21.32	
New Product Advantage (NPA)	npa1 -> NPA	0.76	12.62	**0.22* (Mode-rate High)**
	npa2 -> NPA	0.88	18.51	
	npa3 -> NPA	0.85	15.65	
	npa4 -> NPA	0.81	9.88	

(***) significant with $p< 0.01$, $t>2.58$; Bootstrapping with case = 200, samples = 1000.
(*) $R^2 > 0.2$, is considered high in disciplines such as consumer behavior especially in explaining customer satisfaction or loyality (34) (35) (36)

Table 9. Summary of Hypotheses Test.

Hypothesis Test	**Confirm with the findings**	**Coefficient Path**	**T- Dist. (***)**	**Impact**
H1	Yes	0.57	7.47	Moderate
H2	No	0.22	2.97	Not Supported
H3	No	0.37	4.11	Not Supported
H4	Yes	0.8	16.31	Significant
H5	Yes	0.84	22.9	Significant
H6	No	0.26	2.78	Not Supported
H7	Yes	0.52	4.75	Moderate
H8	Yes	0.47	6.65	Moderate
H9	**Yes**	**0.82**	**14.16**	**High***

Note: ***significant with $p < 0.01$, $t > 2.58$; Bootstrapping with case = 200, samples = 1000; *moderate for exploration research (range from 0.4 to 0.6).

6. Discussion and Summary

The article is a preliminary study to the use of absorptive capacity (AC) concepts in the polytechnics. The findings infer that more efforts need to be done in order to improve the collaboration mechanism to maximize internalization external knowledge into product development. The findings show the ideation process runs smooth from the firms to creating production methods, and high motivation of polytechnics stakeholders to acquire external (firm) knowledge. The area that has potential improvements such as: using the firm knowledge to improve creativity process; flows from potential knowledge into realized knowledge through creating effective production methods; more incentive involves in acquiring firm knowledge and creating production methods; and maximizing knowledge in creating NPA.

The article also emphasizes the roles of support system, project complexity and risk, and managing environmental turbulence as important roles in acquiring the firm knowledge (PAC) and internalize it into creating effective production methods (RAC) domains. The outcome of the article is expected to deliver several objectives: (a) for manager of the polytechnics and firms, The model provides guideline and reference to maximize the knowledge transfer from/to the polytechnics and firms; (b) for the DGHE. The model enables to create an effective model for the polytechnics to collaborate with the firms.

Acknowledgements:
Special thanks are given to the SP4 task force of the Directorate General of Higher Education (DIKTI) Indonesia for their endorsements.

References

1. Li J. Global R and D alliances in China: Collaborations with universities and research institutes. *IEEE Trans Eng Manag*. 2010;57(1):78–87.

2. Abreu M, Grinevich V, Hughes A, Kitson M, Ternouth P. Universities, *Business And Knowledge Exchange*. 2008;

3. Laursen K, Salter A. Searching high and low: What types of firms use universities as a source of innovation? *Res Policy*. 2004;33(8):1201–15.

4. Huggins R, Prokop D. University-Industry Networks : Interactions With Large R & D Performers. Business. 2010;44(0):0–33.

5. Kuncoro M. 2013 ECONOMIC GEOGRAPHY OF INDONESIA : CAN MP3EI REDUCE INTER-REGIONAL. 2013;2(2):8–9.

6. Petruzzelli A M. University-industry R&D collaborations: A joint-patents analysis. DRUID CBS Copenhagen Bus Sch. 2009;

7. Cuddy N, Leney T, Adams J M, Bousquet S. Vocational education and training in the United Kingdom Short description. *Cedefop Panor Ser Luxemb Off Off Publ Eur Communities*. 2005;

8. Janicot C, Mignon S. Knowledge codification in audit and consulting firms: a conceptual and empirical approach. *Knowl Manag Res Pract*. 2012;10(1):4–15.

9. Tijssen R J W, van Leeuwen T N, van Wijk E. Benchmarking university–industry research cooperation worldwide: performance measurements and indicators based on co-authorship data for the world's largest universities. Res Eval [Internet]. 2009;18(1):13–24. Available from: http://openurl.ingenta.com/content/xref?genre=article&issn=0958-2029&volume=18&issue=1&spage=13

10. Bröring S, Cloutier L M, Leker J. The front end of innovation in an era of industry convergence: Evidence from nutraceuticals and functional foods. *R D Manag*. 2006;36(5):487–98.

11. Edmondson G, Valigra L, Kenward M, Hudson R L, Belfield H. Making Industry-University Partnerships Work. Lessons from Successful Collaborations. *Sci Bus Innov Board*. 2012;

12. Cohen W M, Levinthal D A. Absorptive capacity: A new perspective on learning and innovation. *Adm Sci Q*. 1990;128–52.

13. Carlsson S. Enabling and enhancing potential absorptive capacity through the use of ICT. In: *Proceedings of the Fifth European Conference on Organizational Knowledge, Learning, and Capabilities*. 2004.

14. Kuester S, Homburg C, Hildesheim A. The Specific Role of Sales Force Integration in New Product Development. 2011;

15. Zahra S A, George G. Absorptive capacity: A review, reconceptualization, and extension. Acad Manag Rev. 2002;27(2):185–203.

16. Van Den Bosch F A J, Volberda H W, De Boer M. Coevolution of firm absorptive capacity and knowledge environment: Organizational forms and combinative capabilities. Organ Sci. 1999;10(5):551–68.

17. Foss NJ, Lyles MA, Volberda HW. Absorbing the concept of absorptive capacity: how to realize its potential in the organization field. 2009;

18. Kallio A. Enhancing absorptive capacity in a non-research and development. An action research approach to converting individual observations into organisational awareness. Acta Univ Lappeenrantaensis. 2012;

19. Ertl B, Kopp B, Mandl H. Effects of an individual's prior knowledge on collaborative knowledge construction and individual learning outcomes in videoconferencing. In: Proceedings of th 2005 conference on Computer support for collaborative learning: learning 2005: the next 10 years! 2005. p. 145–54.

20. Hausberg J P. Departmental Cross-Functional Absorptive Capacity. DRUID Soc CBS Copenhagen, Denmark. 2012;

21. Ming X G, Yan J Q, Lu W F, Ma D Z. Technology solutions for collaborative product lifecycle management--status review and future trend. Concurr Eng. 2005;13(4):311–9.

22. Bhattacharya P J, Seddon P B, Scheepers R. Enterprise systems for innovation in products and processes: Beyond operational efficiency. In: ACIS 2012: *Proceedings of the 23rd Australasian Conference on Information Systems 2012*. 2012. p. 1–11.

23. Mandal P, Gunasekaran A. Issues in implementing ERP: A case study. *Eur J Oper Res.* 2003;146(2):274–83.

24. Browning TR, Fricke E, Negele H. Key concepts in modeling product development processes. Syst Eng. 2006;9(2):104–28.

25. Björk J, Di Vincenzo F, Magnusson M, Mascia D. The impact of social capital on ideation. Ind Innov. 2011;18(6):631–47.

26. Heising W. The integration of ideation and project portfolio management□A key factor for sustainable success. Int J Proj Manag. 2012;30(5):582–95.

27. Langerak F, Hultink EJ, Griffin A. Exploring mediating and moderating influences on the links among cycle time, proficiency in entry timing, and new product profitability. J Prod Innov Manag. 2008;25(4):370–85.

28. William D R, Goldstein M. Multivariate Analysis. Methods and Applications. Probab Math Stat Wiley, New York. 1984;

29. Akter S, D'Ambra J, Ray P. An evaluation of PLS based complex models: the roles of power analysis, predictive relevance and GoF index. 2011;

30. Chin WW. The partial least squares approach to structural equation modeling. Mod methods Bus Res. 1998;295(2):295–336.

31. Salojärvi H, Sainio L-M. Applying absorptive capacity construct to customer-related knowledge processing. 2006;

32. Jensen MB, Johnson B, Lorenz E, Lundvall BÅ. Forms of knowledge and modes of innovation. Res Policy. 2007;36(5):680–93.

33. Jarvis CB, MacKenzie SB, Podsakoff PM. A critical review of construct indicators and measurement model misspecification in marketing and consumer research. J Consum Res. 2003;30(2):199–218.

34. Fornell C, Larcker DF. Evaluating structural equation models with unobservable variables and measurement error. J Mark Res. 1981;39–50.

35. Henseler J, Ringle CM, Sinkovics RR. The use of partial least squares path modeling in international marketing. Adv Int Mark. 2009;20(1):277–319.

36. HULT GTM, RINGLE C, SARSTEDT M. A Primer on Partial Least Squares Structural Equation Modeling (PLS-SEM). SAGE Publications, Incorporated, Thousand Oaks, CA; 2013.

Development of software for estimating clear sky solar radiation in Indonesia

H Ambarita

Sustainable Energy Research Centre, Faculty of Engineering University of Sumatera Utara, Jl. Almamater Kampus USU Medan 20155, Indonesia

*Email: himsar@usu.ac.id

Abstract. Research on solar energy applications in Indonesia has come under scrutiny in recent years. Solar radiation is harvested by solar collector or solar cell and convert the energy into useful energy such as heat and or electricity. In order to provide a better configuration of a solar collector or a solar cell, clear sky radiation should be estimated properly. In this study, an in-house software for estimating clear sky radiation is developed. The governing equations are solved simultaneously. The software is tested in Medan city by performing a solar radiation measurements. For clear sky radiation, the results of the software and measurements ones show a good agreement. However, for the cloudy sky condition it cannot predict the solar radiation. This software can be used to estimate the clear sky radiation in Indonesia.

1. Introduction

Solar energy is a potential renewable energy resource for the globe. The annual solar radiation reaching the earth surface is approximately 3,400,000 Exajoule. The annual energy consumption of human being is approximately 450 Exajoule. This fact reveals that only 1 hour 10 minutes of solar radiation can fill the global annual energy consumption. Solar energy resource will be crucial for future energy resources. In Indonesia, the potency of solar energy is very significant. According to Ministry of Mineral and Energy Resources of the Government of Indonesia, the potency of solar energy in Indonesian islands is 4.8 kWh/m^2/day in average and the duration of radiation is 10 – 12 hours per day. This is a big potency to be collected. Research on solar energy applications in Indonesia has come under scrutiny in recent years. Solar radiation is harvested by solar collector or solar cell and convert the energy into useful energy such as heat and or electricity.

Several studies on solar energy application can be found in literature. Thirugnanasambandam et al [1] reported a literature review of solar thermal technologies. According to the study all solar thermal technologies can be divided into eight groups. They are solar water heater [2], solar cooker [3], solar driers [4], solar ponds [5], solar architecture [6], solar air-conditioning [7], solar chimneys [8], solar power plants [9], and solar water purification [10]. Those studies only several examples that show that research on the use of solar energy in recent years has reached a remarkable edge. The main driving forces for this are the perceiving scarcity of fossil fuel and the global warming catastrophe. In Indonesia, there is a strong commitment from the Government of Indonesia to accelerate the research on the use of solar energy. Thus, number of reported studies on solar energy applications in Indonesia are increasing. Dina et al [11] reported a study on effectiveness of continuous solar drier integrated with desiccant thermal storage for drying cocoa beans. In this study, in day time solar energy is used as energy for drying and to heat thermal energy storage. The thermal energy storage is then used in the night time to provide continuous drying. Ambarita and Kawai [12] reported an experimental study on solar-powered adsorption refrigeration cycle with activated alumina and activated carbon as adsorbent. Here, solar radiation is used to heat generator of the adsorption cycle. Recently, Ambarita [13] reported a study on the performance of natural vacuum desalination system using low grade heat source.

The above literatures show that the solar energy from the sky is used as energy input to the system. Mainly, the solar radiation is used to calculate the performance of the system. However, to the best knowledge of the author, there is no reported study focuses on the estimation of the solar radiation for Indonesia case found in literature. In this work, in order to estimate solar radiation on the flat surface of a solar collector, an in-house software is developed. The developed software will be validated by performing a solar radiation measurements in Medan city of Indonesia. The results are expected to supply the necessary information to support the Government of Indonesia in development of the solar collector technologies.

2. Method

In this study a software for estimating clear sky radiation is developed. The results of the developed software are validated by using measurement data in Medan city of Indonesia. Thus, governing equations used in the software and the experimental method are explained in the following sections.

2.1. Governing equations

The total clear sky radiation is the sum of beam radiation (G_{beam}) and diffuse radiation (G_{difuse}). The following equation is used to calculate the clear sky radiation.

$$G_{total} = G_{beam} + G_{difuse} \tag{1}$$

The beam and diffuse radiations are calculated using the following equations:

$$G_{beam} = G_{on}\tau_b \cos\theta_z \tag{2}$$

$$G_{difure} = G_{on} \cos\theta_z \left[0,271 - 0,294\tau_b\right] \tag{3}$$

where G_{on} and θ_z are extra-terrestrial radiation and zenith angle, respectively. While, τ_b is the atmospheric transmittance for beam radiation. The zenith angle is calculated by the following equation:

$$\cos\theta_z = \cos\phi\cos\delta\cos\omega + \sin\phi\sin\delta \tag{4}$$

where $\phi = 3°34'$ is the latitude in at Medan city, δ is declination angle depend on the day measured, and ω is hour angle. The declination angle is calculated by the following equation.

$$\delta = 23,45\sin\left(360\frac{284+n}{365}\right) \tag{5}$$

where n is the day of the year. The hour angle is calculated by:

$$\omega = 15(STD-12) + (ST-STD)\times\frac{15}{60} \tag{6}$$

where STD and ST are represent the standard time and solar time, respectively. The solar time is calculated by using the following equation:

$$ST = STD - 4(L_{st} - L_{loc}) + E \tag{7}$$

where L_{st} is the standard meridian for the local time zone and L_{loc} is longitude of the location. The parameter E is the equation of time and calculated by the following equation.

$$E = 229,2(0,000075 + 0,001868\cos B - 0,032077\sin B - 0,014615\cos 2B - 0,04089\sin 2B \tag{8}$$

and B is given by

$$B = (n-1)\frac{360}{365} \tag{9}$$

The atmospheric transmittance for beam radiation is calculated by:

$$\tau_b = a_0 + a_1 \exp\left(\frac{-k}{\cos\theta_z}\right) \tag{10}$$

The parameters a_0, a_1 and k in equation (10) are calculated using the following equations, respectively:

$$a_0 = r_0\left(0,4237 - 0,00821(6-A)^2\right) \tag{11}$$

$$a_1 = r_1\left(0,5055 + 0,00595(6,5-A)^2\right) \tag{12}$$

$$k = r_k\left(0,2711 + 0,01858(2,5-A)^2\right) \tag{13}$$

As suggested by Hottel [13], for Medan city the parameters in the above equations are $r_0 = 0,95$ $r_1 = 0,98$, and $r_k = 1,02$. The altitude of the present experimental apparatus is $A = 0,02$ km.

The above governing equations are solved simultaneously. Here, a Fortran code program is written to solve those governing equations. The Fortran, stand for Formula Translation, is a general purpose, imperative programming language that is especially suited to numeric computation and scientific computing.

2.2. Solar radiation measurements

In order to validate the developed software, a solar radiation measurement apparatus has been designed and fabricated and shown in Figure 1.

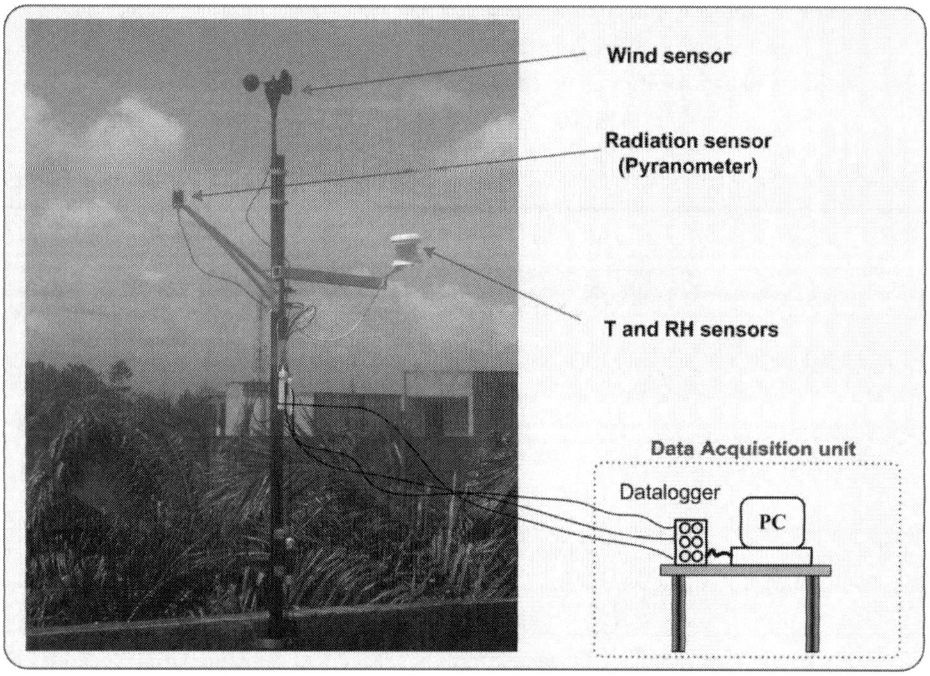

Figure 1. Solar radiation measurement apparatus

The apparatus is made of A HOBO micro station data logger which consists of temperature/RH smart sensor, solar radiation sensor, and wind velocity smart sensors. The solar radiation smart sensor is a light sensor (silicon pyranometer) with a measurement range of 0 to 1280 W/m^2 over a spectral range of 300 to 1100 nm. The measurement resolution and accuracy are 1.25W/m^2 and $\pm 5\%$, respectively.

The temperature/RH smart sensor has temperature measurement range of -40°C to 75°C. The temperature measurement and accuracy are 0.02°C (at 25°C) and ±0.21°C (from 0 to 50°C). Measurement range of the wind velocity sensor is 0 to 76 m/sec. The measurement accuracy and resolution are ± 1.1 m/s and 0.5 m/s, respectively. By using the above data acquisition unit and the smart sensors solar radiation measurements are carried out. The interval of measurement is 2 minutes.

3. Results and Discussions

Solar radiation resulted from the developed software and measurement by experimental apparatus will be presented and discussed here. The measurements take place on top roof of "Gedung S2 Teknik Mesin" of University of Sumatera Utara in Medan city of Indonesia. Several measurements have been carried out. All measurements can be divided into clear sky condition and cloudy sky condition. Thus, the results and discussions will be divided into clear sky and cloudy sky conditions as shown in the following sub sections.

3.1. Clear sky condition

For the clear sky radiation condition, the measurement and software results are presented in Figure 2. The measurements were taken from July 28 of 2015. In the figure, the software result is shown by red solid line and measurement results shown by blue circle markers. The vertical axis of the graphic shows solar radiation and the horizontal axis shows time in local time of Medan (WIB). Both measurement and software results show the same trend. In the morning time, as the sun rises solar radiation increases as time increase. In this section, the measurement starts at 7.00 WIB in the morning. The software and measurement results show almost the same value of radiation. The software and the measurement results are 66.77 W/m^2 and 68.10 W/m^2, respectively. For the measurement results, a maximum value of solar radiation of 846 W/m^2 is captured at 12.50 WIB. While from the software results, the maximum value is 916.8 W/m^2 at 12.50 WIB. These facts reveal that measurement and software show a different maximum value of radiation. There is a discrepancy of 8.3%. However, the software and measurement show the same time when the maximum radiation occurs. After the maximum value is reaching, the software result and measurement show the same trend. The solar radiation decrease as time increase. In the figure, some measurement show lower values. This is because of the presence of cloud in the sky. But, the number of lower value is not significant. Based on these results, a preliminary conclusion can be drawn here. When the sky is clear, the software can predict the solar radiation. The solar radiation trend and the time of maximum value can be predicted perfectly. The maximum value shows a discrepancy only 8% or less than 10%.

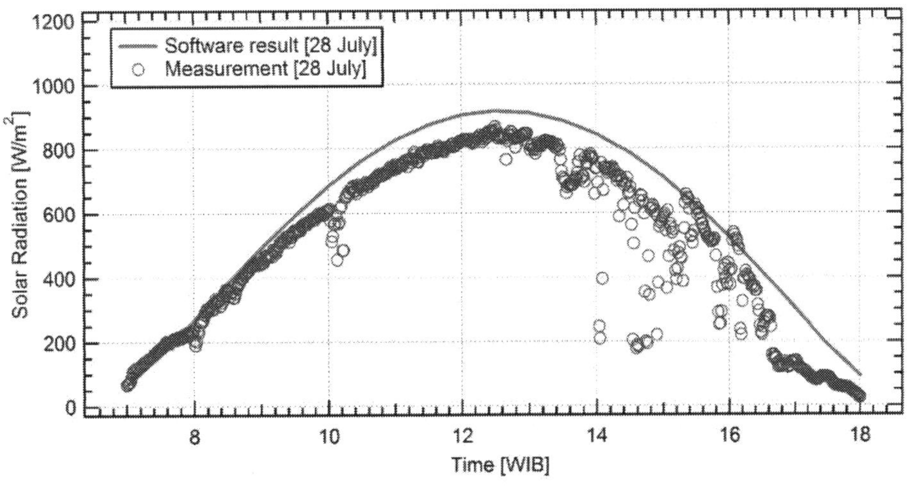

Figure 2. Software and Measurement results for clear sky condition

3.2 Cloudy sky condition

In the measurement, there are several days are categorized cloudy sky. One of the measurement is shown in Figure 3. The figure shows the solar radiation resulted by software and by measurement on January 1st of 2015. In the figure the measurement results shown by blue circle markers and the software result shown by solid red line. As expected, the software result shows the same trend with the previous clear sky radiation. In the morning, the radiation increase as time increase. After reaching a maximum value the solar radiation decrease as time increase. The measurement shows a different trend with the previous clear sky radiation. It is very unpredictable. It can be seen clearly that measurement and software results show different value in almost all times.

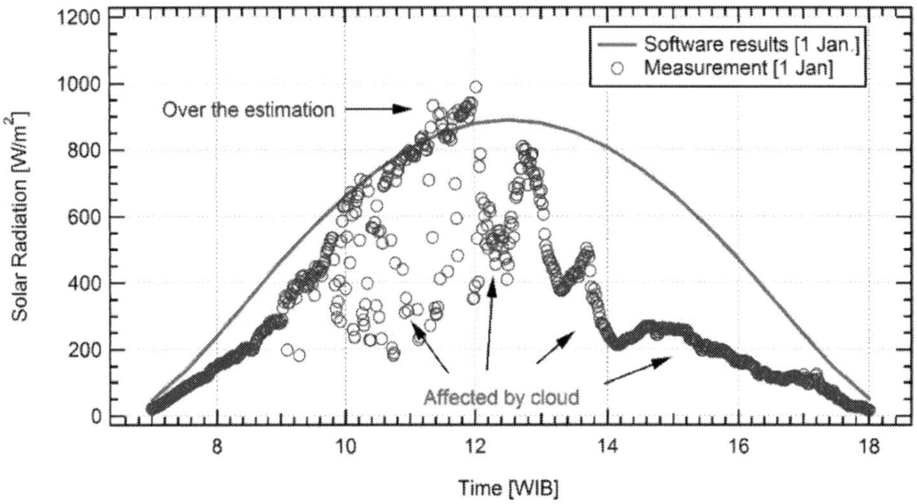

Figure 3. Software and Measurement results for cloudy sky condition on January 1st, 2015

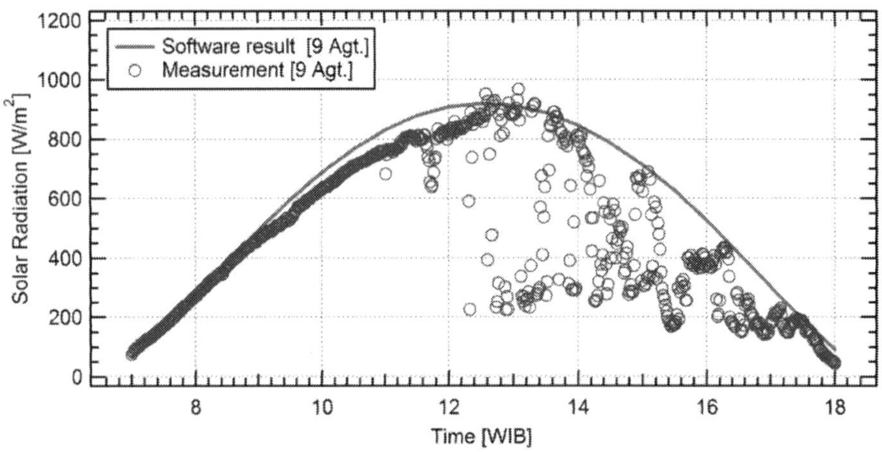

Figure 4. Software and Measurement results for combination clear and cloudy sky conditions on August 9th, 2015

A combination of clear sky and cloudy sky conditions is shown in Figure 4. The measurement was taken on August 9th, 2015. The figure shows that in the morning the software and measurement show the same results. However, after the maximum value is reaching, the software and measurement show a significant different. This is because the solar radiation from the sky is blocked by the cloud.

4. Conclusions

In this study a software for estimating clear sky radiation has been developed. In the software, a set of governing equations is solved simultaneously by using a FORTRAN code. In order to validate the software an experimental apparatus has been developed and several measurements in Medan city of Indonesia have been carried. The software results and measurements have been compared. The conclusions of this study are as follows. For the clear sky condition, the software can predict the solar radiation with acceptable value. The discrepancy less than 10%. However, for the cloudy sky condition the solar radiation is very unpredictable. It is suggested that this software can be used to predict solar radiation in order to design solar collector.

References

[1] Thirugnanasambandam M, Iniyan S, Goic R 2010 *Renewable and Sutainable Energy Review* **14**, 312-322.
[2] Mathioulakis E, Belessiotis V 2002 *Solar Energy* **72**(1) 13-20.
[3] Schwarzer K, Vieira da Silva ME, 2008 *Solar Energy* **82** 157-63.
[4] Mahapatra AK, Imre L 1997 *Solar Energy* **59** (4-6) 227-31.
[5] Velmurugan V, Srithar K 2008 *Renewable and Sustainable Energy Reviews* **12** 2253-63.
[6] Kischkoweit-Lopin M 2002 *Solar Energy* **73**(2) 77-82.
[7] Elsafty A, Al-Daini AJ 2002 *Renewable Energy* **5** 569-83.
[8] Pretorius JP, Kroger DG 2006 *Solar Energy* **80** 534-44.
[9] Quaschning V 2004 *Solar Energy* **77** 171-8.
[10] Tripathi R, Tiwari GN 2006 *Solar Energy* **80** 956-67.
[11] Dina SF, Ambarita H, Napitupulu FH, Kawai H 2015 *Cases Studies in Thermal Engineering* **5** 32-40.
[12] Ambarita H, Kawai H 2016 *Cases Studies in Thermal Engineering* **7** 36-46.
[13] Ambarita H 2016 *Cases Studies in Thermal Engineering* **8** 346-58.
[14] Hottel HC. 1976 *Solar Energy* **18** 129.

Computational Analysis of Suspended Particles in the Irrigation Channel

Tulus[1*], Suriati[2] and M Situmorang[1]

[1]Department of Mathematics, University of Sumatera Utara, Jl. dr. Mansur Kampus USU Medan 20155, Indonesia
[2]Department of Informatics, Sekolah Tinggi Teknik Harapan Medan, Jl. HM Joni No.70 Medan 20218, Indonesia

[*]Email: tulus@usu.ac.id

Abstract. Irrigation channel that is used for bringing water to land in order to make plants grow is one of important thing in tropical country, especially in Indonesia. It has been also a central feature of agriculture for over 5,000 years and is the method in which water is supplied to plants at regular intervals for agriculture. This paper is to analyse the sedimentation process in the irrigation channel in District Deli Serdang, Sumatera Utara Province. The analysis is based on the mixture model in turbulent flow. Computation using Finite Element Method is performed in the domain in two dimensional. The values of dispersed phase density parameter vary from 1100 kg/m^3 to 1250 kg/m^3. The results have shown that the structure of mixture velocity distribution and the dispersed phase volume fraction.

1. Introduction

North Sumatra Province is one of province in Indonesia that the plantation agriculture is one of the priority economic sectors. The Ministry of Agriculture of the Republic of Indonesia has published statistical data, i.e. land area of rice in 2015 reached 781.769Ha (5.54% of the rice land area of the entire Indonesian) with a growth rate last reached 8.98%. Especially, the production of rice in 2015 from the province of North Sumatra reached 4.044.829Ton (5.37% of rice production throughout Indonesia) with the last growth reached 11.40% [1]. This rate of growth of land area is still high, even though it is still low compared to several provinces in Java.

Researches on sediment transport are mostly considered to derive the prediction equations for solid discharge and to analyze the grain kinematics and dynamics [2]. The smooth bed configuration can give insight into how the turbulent flow affects the sediment motion [3]. On the other hand, an experimental configuration filters out the interaction of moving particles with the bed roughness and with other moving grains [4]. Building settling basin is traditionally built based on retention time [5]. The evolutions of flow pattern and sediment transportation at a 90° open-channel confluence with different discharge ratios of the tributary flow to the total flow have been studied experimentally by Liu et. al [6].

Analysis using computation on a physical phenomenon that occur in a real situation can be performed using computer-aided technology considering some governing equation and boundary conditions. The process of sedimentation in a building settling basin can be analyzed to obtain the optimal parameters, so it can produce more efficient performance. The application of characteristics of water flow and sediment transport based simulations to predict sediment transport and water in the river system has been studied by Sil, B.S. and Choudhury [7]. Some authors have also derived the profile shape model of the river, and sedimentation models on a tube and have also conducted experimental studies some basic changes in a channel and formulate conditions channel geometry and characterize the effects of geometry and flow [8, 9].

A laboratory study of open-channel confluences using a 3D has been performed by Bradbrook et al. Some elliptic solution of the Reynolds-averaged Navier-Stokes equations have been analyzed, including a method for approximating the effects of water surface elevation patterns [10]. There are some other researchers studied the flow pattern and sediment transportation at channel confluence with respect to the momentum equations, energy equations, and potential theory [11, 12].

2. Sedimentation Phenomenon

The derivation of the phenomenological theory for settling of sedimentation with compression was conducted by Concha and Bustos [13]. The hydrodynamic can be modeled as a Reynolds Averaged Navier-Stokes (RANS) equations and on the standard *k-ε* model for turbulence processes. The sediment transport can be derived as a mass balance equations for the re-suspended sediment [14].

A suspension is a mixture of solid particles and a liquid. In this theory, a suspension was modeled as a mixture of two superimposed continuous media. The dynamics of a suspension can be modelled by a momentum transport equation for the mixture, a continuity equation, and a transport equation for the solid phase volume fraction. The Mixture Model, Laminar Flow interface automatically sets up these equations. It uses the following equation to model the momentum transport:

$$\rho\frac{\partial \mathrm{u}}{\partial t}+\rho(\mathrm{u}\cdot\nabla)\mathrm{u}=-\nabla p-\nabla\cdot\left(\rho c_s(1-c_s)\mathrm{u}_{slip}\mathrm{u}_{slip}\right)+\nabla\cdot\left[\eta\left(\nabla\mathrm{u}+\nabla\mathrm{u}^T\right)\right]+\rho\mathrm{g} \tag{1}$$

where $\mathbf{u}$ is the mass averaged mixture velocity (m/s), p denotes the pressure (Pa), $\mathbf{g}$ refers to the acceleration of gravity (m/s^2), c_s is the dimensionless particle mass fraction, and $\mathbf{u}_{slip}$ gives the relative velocity between the solid and the liquid phases (m/s).

3. Finite element analysis for Sedimentation

Sediment deposition problems on the channel need to be handled appropriately. The deposition of sediments in these channels will cause problems in the development of agricultural production and the cultivation of paddy fields. Based on the above ideas, it will be evaluated the building settling basin. The analysis in this paper considers the 2D of the channel. To analyze the sedimentation process, it is considered the equation of momentum transport from the equations (1) to give the dynamic of the problem. So, the following governing equations are considered to the problem of mixture model in turbulent flow.

$$\rho\frac{\partial \mathrm{u}}{\partial t}+\rho(\mathrm{u}\cdot\nabla)\mathrm{u}=\nabla\cdot\left[-pI+(\mu+\mu_T)\left(\nabla\mathrm{u}+\nabla\mathrm{u}^T\right)\right]-\nabla\cdot\left\{\rho c_d(1-c_d)\mathrm{u}_{slip}\mathrm{u}_{slip}\right\}+\rho\mathrm{g}+\mathrm{F} \tag{2}$$

$$(\rho_c-\rho_d)\left\{\nabla\cdot\left[\phi_d(1-c_d)\mathrm{u}_{slip}-D_{md}\nabla\phi_d\right]+\frac{m_{dc}}{\rho_c}\right\}+\rho_c(\nabla\cdot\mathrm{u})=0,\quad \frac{\partial\phi_d}{\partial t}+\nabla\cdot N_{\phi_d}=-\frac{m_{dc}}{\rho_c}, \tag{3}$$

$$\phi_d=phid,\quad N_{\phi d}=\phi_d u_d,\qquad u_d=u+\phi_d(1-c_d)u_{slip}-\frac{D_{md}}{\phi_d}\nabla\phi_d \tag{4}$$

$$\rho=\phi_c\rho_c+\phi_d\rho_d,\quad c_d=\frac{\phi_d\rho_d}{\rho},\qquad D_{md}=\frac{\mu_T}{\rho\sigma_T} \tag{5}$$

The boundary conditions are considered with respect to the real condition of the model, that are wall, slip wall, inlet, and outlet.

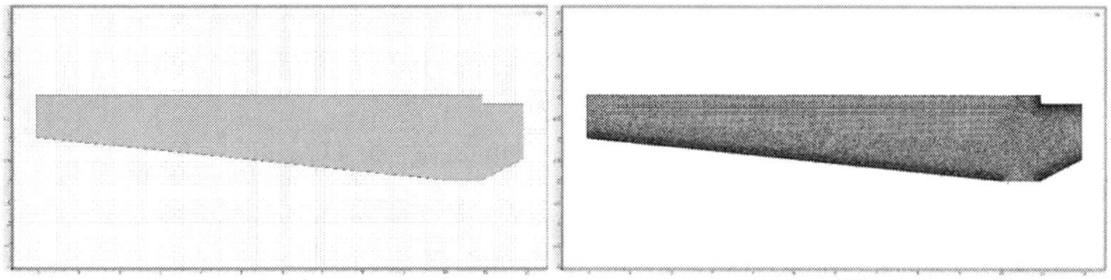

(a) (b)

Figure 1. Model: (a) Geometry and (b) Triangle mesh of the geometry

The 2D geometric model and the boundary conditions are designed as depicted in Figure 1 (a). The water inlet is applied at the left side, and the water outlet is applied at the right side and the right bottom side. Wall slip is applied at top side, and the rest are set as the wall. By using Finite Element Method, the mesh is generated using triangle elements as depicted in Figure 1 (b). An initial-boundary value problem was derived by introducing constitutive assumption, simplifying the balance, and restricting to one space dimension. The problem was of the form of a second order partial differential equation for volumetric solid concentration. The mixed type nature corresponds to the interface between the compression zone, where the solid effective stress varies, and the hindered settling zone.

Table 1. Specification of the material properties

No	Parameter	Symbol	Value	Unit
1	Continuous phase density	ρ_c	1000	kg/m³
2	Continuous phase viscosity	μ_c	0.001	Pa·s
3	Dispersed phase density	ρ_d	1100-1300	kg/m³
4	Dispersed phase particle diameter	D_d	2E-4	m

The parameters used in the computation are considered as a description for the material properties of water and the averaged diameter of particle dispersed in water as described in Table 1. On the other hand, some values are given in initial values of the computation with deal to the condition in the channel as listed in the Table 2.

Table 2. Some variables that used in the computation

No	Variables	Symbol	Value	Unit
1	Inlet velocity	v_{in}	1.25*step function	m/s
2	Outlet velocity	v_{out}	1.25*step function	m/s
3	Inlet dispersed phase volume fraction	$\phi_{d\,in}$	0.003	-
4	Dispersed phase mass-out flux	$q_{d\,out}$	$2\pi x\phi_{d\,in}$	kg/(m·s)

4. Results and Discussions

This section will show the distribution of velocity and volume fraction of the dispersed phase in the channel in some units of time. Figure 2 shows the distribution of velocity in the channel time by time, i.e. from 0.0s to 0.4s. The computational observation is performed to the distribution of velocity. There are some ununiformly velocity that can cause a turbulence in the channel. The distribution of the mixture velocity varies time by time. Figure 2 shows also that the maximum velocity occurs at the surface of the channel. The initial velocity in the computation was set to be 0m/s at $t = 0$. Table 3 shows the maximum mixture velocity in the channel at some time of computation.

Table 3. The maximum mixture velocity in the channel

	Maximum Velocity (m/s)		
Time (sec)	$\rho_d = 1100$ kg/m^3	$\rho_d = 1200$ kg/m^3	$\rho_d = 1300$ kg/m^3
0.00	0.0000	0.0000	0.0000
0.10	1.1983×10^{-5}	2.7546×10^{-5}	1.5645×10^{-4}
0.20	4.4709×10^{-5}	1.1765×10^{-4}	5.0363×10^{-4}
0.30	1.0702×10^{-4}	3.1265×10^{-4}	9.9879×10^{-4}
0.40	1.4391×10^{-4}	7.8201×10^{-4}	1.8859×10^{-3}

Figure 3 shows the graph of maximum velocity versus time with respect to the dispersed phase densities. Higher dispersed phase density, higher the maximum velocity. Compared with the Figure 2, it appears that higher in speed changes causes the velocity distribution is not consistent. This can lead to turbulence in that area. The streamlines of the mixture dispersed phase volume fraction at several times of computation. From the Figure 2, the distribution of volume fraction is suitable with the velocity field. Opposing effects of gravity settling and turbulence-induced particle dispersion produce volume-fraction gradients in the interior. The magnitude of the mixture strain rate (and hence the turbulence production) decreases with the distance from the inlet side.

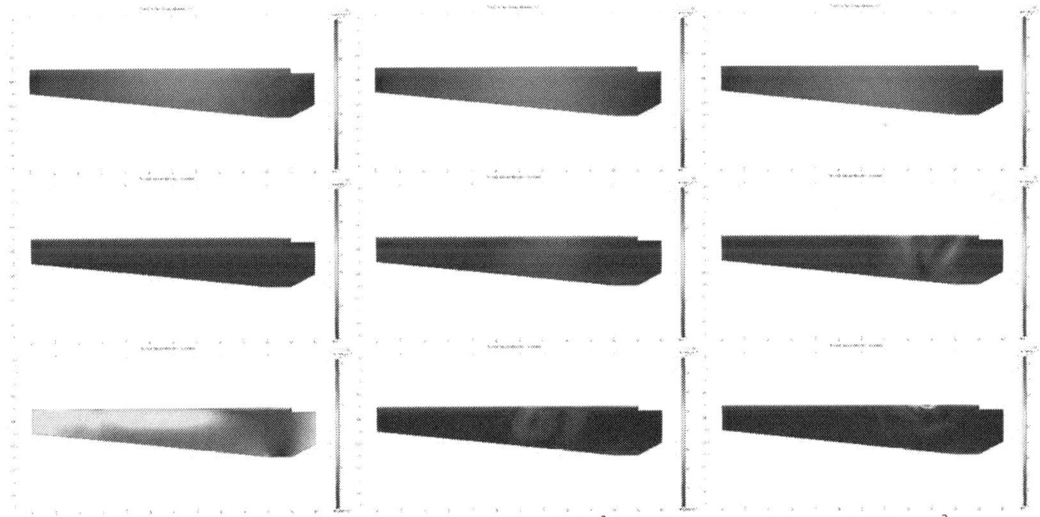

(a) ρ_d=1100 kg/m^3 (b) ρ_d=1200 kg/m^3 (c) ρ_d=1300 kg/m^3

Figure 2. Velocity distribution of the mixture from time 0s until 0.4s

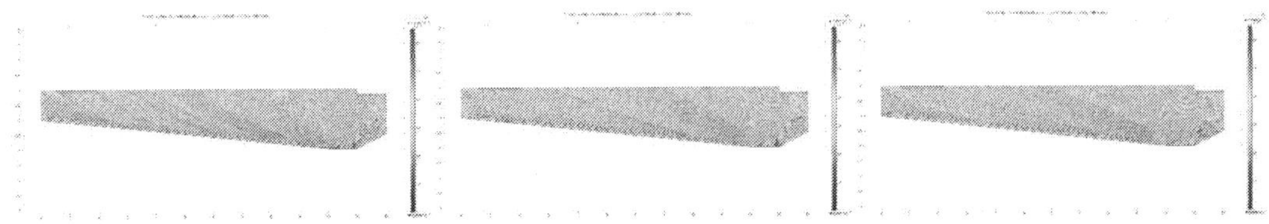

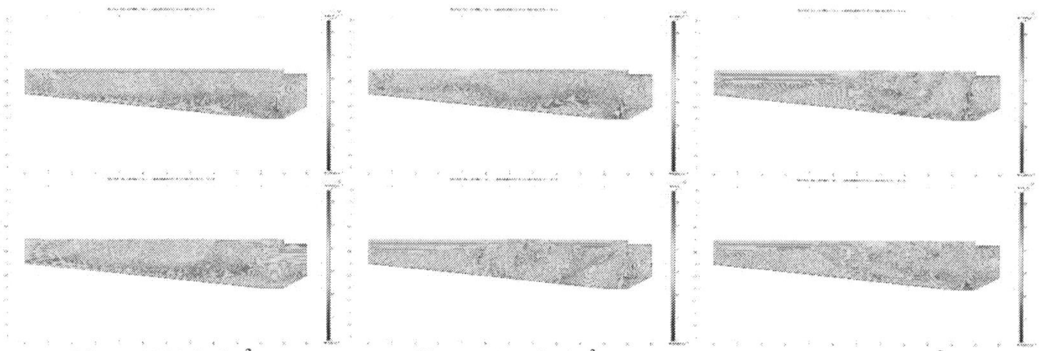

(a) ρ_d=1100 kg/m^3 (b) ρ_d=1200 kg/m^3 (c) ρ_d=1300 kg/m^3

Figure 3. Volume fraction distribution of the dispersed phase

The streamlines show the differences of mixture fraction in some areas. At time 0.0, the form of the streamline is relatively uniform. The streamlines change from time to time. In times of 0.1 s to 0.4 s, the solid fall to channel bed with respect to the form of the streamlines. The different values of dispersed phase density affect the structure of streamline of the mixture fraction. The value 1100 kg/m^3 of dispersed phase density is more suitable compared with the other higher values.

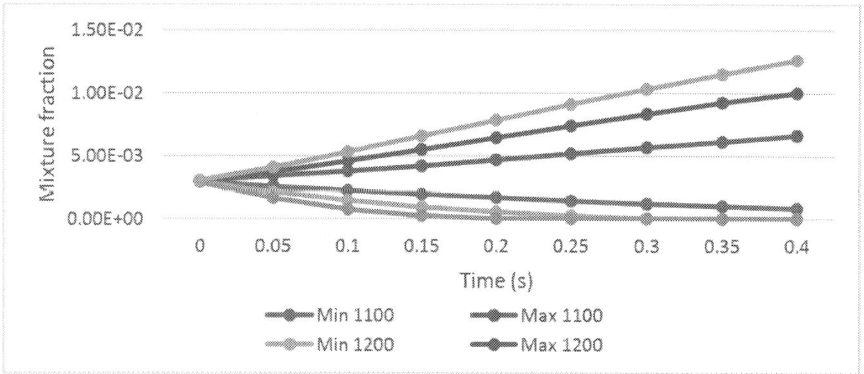

Figure 4. Minimum and maximum volume fraction versus time

The minimum and maximum of mixture fraction change with respect to time computation. From the Figure 4, the minimum mixture fractions decrease, with respect to dispersed phase densities 1100 kg/m3, 1200 kg/m3 and 1300 kg/m3, from 2.9793 x 10^{-3} to 7.802 x 10^{-4}, from 2.9537 x 10^{-3} to 0, and from 2.9235 x 10^{-3} to 0, respectively. On the other hand, the maximum mixture fraction tends up, from 3.0185 x 10^{-3} to 6.6335 x 10^{-3}, from 3.0327 x 10^{-3} to 1.0101 x 10^{-2}, and from 3.0443 x 10^{-3} to 1.26 x 10^{-2}, respectively. It means that the mixture fraction in the channel firstly almost uniform, and the range of mixture fraction tends to big. The higher value of dispersed phase density, then faster the particles tend down. Physically, it means that in some areas of the channel, the solid in mixture fall to ground of the channel.

5. Conclusions

A 2D numerical computation based on momentum transport equation for the mixture is presented to search the relationship between the sediment movement, and the pattern of volume fraction of the dispersed phase in the channel. From a computational point of view, results have shown the importance streamlines of the mixture velocity and the dispersed phase volume fraction. From a physical point of view, results highlighted the contour of the volume fraction according to the water flow with deal to the inlet and outlet position.

Acknowledgments
The authors gratefully acknowledge that the present research is supported by Ministry of Research and Technology and Higher Education Republic of Indonesia. The support is under the research grant BP-PTN USU of Year 2016 Contract Number 98/UN5.2.3.1/PPM/SP/2016.

References

[1] Minister of Agriculture Republic of Indonesia 2016 Sub Sector Food Crops [http://www.pertanian.go.id/ap_pages/mod/datatp] Downloaded 22 June 2016

[2] Ancey C, Davison A C, Böhm T, Jodeau M and Frey P 2008 Entrainment and motion of coarse particles in a shallow water stream down a steep slope *Journal of Fluid Mechanics* 595 83–114

[3] Campagnol J, Ballio F, Sadabadi, S A H, and Sazadul H 2014 Particle motion of bed-load sediment moving over a smooth bed. River Flow (2014). *In Proceedings of VII International Conference on Fluvial Hydraulics*, Lausanne, Switzerland

[4] Ballio F and Radice A 2015 Fluctuations and time scales for bed-load sediment motion over a smooth bed *Intl Journal of Sediment Research* 30 (2015) 321–327

[5] Gary F 2001 Settling Basin Design *WRAC Publication No.106 06-2001* 1-6 [http://aqua.ucdavis.edu/DatabaseRoot/pdf/WRAC-106.PDF] Downloaded 22 July 2016

[6] Liu T, Chen L, & Fan, B 2012 Experimental study on flow pattern and sediment transportation at a 90° open-channel confluence *International Journal of Sediment Research* 27 178-187

[7] Sil B S and Choudhury P 2016 Muskingum equation based downstream sediment flow simulation models for a river system *Intl. Journal of Sediment Research* 31 139–148

[8] Tulus 2013 Mathematical Modeling of Sedimentation on Thickening Processes *Proc. of The 2nd Intl. Seminar on Operational Research* (InteriOR) 2013 241-244

[9] Nazari-Giglou A, Jabbari-Sahebari A, Shakibaeinia A and Borghei S M 2016 An experimental study of sediment transport in channel confluences *Intl. Journal of Sediment Research* 31 87–96.

[10] Bradbrook K F, Lame S N, Richards K S, Biron P M and Roy A G 2001 Role of bed discordance at asymmetrical river confluences *Journal of Hydraulic Engineering* 127 No. 5 pp 351–368

[11] Ze-yu M, Sheng-wei Z and Sheng L 2005 Study on the separation zone in open channel confluence *Advances in Water Science* **6** 1 pp 7–12

[12] Liu T,. Wei G and Zhan Lei 2009 Experimental study of the velocity profile at 90° open channel confluence *Advances in Water Science* **20** 4 pp 25–29

[13] Concha F and Bustos M C 1985 Theory of sedimentation of flocculated fine particles In B M Moudgil, P Somasundaran (Eds.) *Proc. of the EF Conf. on Flocculation, Sedimentation and Consolidation* Sea Island GA AIChE, New York, pp 275-284

[14] Ji S, Ouahsine A, Smaoui H and Sergent P 2014 3D Modeling of sediment movement by ships-generated wakes in confined shipping channel *Intl. Journal of Sediment Research* 29 pp 49-58

[15] Cao Z, Pender G, Wallis S, and Carling P 2004 Computational dam-break hydraulics over erodible sediment bed *Journal of Hydraulic Engineering* **130** 7 pp 689–703

A numerical study on combustion process in a small compression ignition engine run dual-fuel mode (diesel-biogas)

H Ambarita, T.I Widodo, and D.M Nasution

Sustainable Energy Research Centre, Faculty of Engineering University of Sumatera Utara, Jl. Almamater Kampus USU Medan 20155, Indonesia

Email: himsar@usu.ac.id

Abstract. In order to reduce the consumption of fossil fuel of a compression ignition (CI) engines which is usually used in transportation and heavy machineries, it can be operated in dual-fuel mode (diesel-biogas). However, the literature reviews show that the thermal efficiency is lower due to incomplete combustion process. In order to increase the efficiency, the combustion process in the combustion chamber need to be explored. Here, a commercial CFD code is used to explore the combustion process of a small CI engine run on dual fuel mode (diesel-biogas). The turbulent governing equations are solved based on finite volume method. A simulation of compression and expansions strokes at an engine speed and load of 1000 rpm and 2500W, respectively has been carried out. The pressure and temperature distributions and streamlines are plotted. The simulation results show that at engine power of 732.27 Watt the thermal efficiency is 9.05%. The experiment and simulation results show a good agreement. The method developed in this study can be used to investigate the combustion process of CI engine run on dual-fuel mode.

1. Introduction

Green Houses Gases (GHGs) emissions can cause global warming and it can make the earth in dangerous. The main sources of GHGs emissions are deforestation and burning fossil fuel. Indonesia is one of the main GHGs emitter in the world. The Government of Indonesia (GoI) has released its target on reduction GHGs emissions [1]. The burning of fossil fuel includes activities that using engines to produce power. As a note, the main consumption of fossil fuel is diesel which is used in Compression Ignition (CI) engines. Thus, reducing diesel oil consumption will reduce GHGs emissions significantly. Due to thermodynamic principle of combustion engines and the Carnot cycle limit, the emission target can only be achieved with fuel change or blending with biofuel. Here biogas is promoted to replace the biodiesel. Biogas is produced from anaerobic biodegradation of organic material. It consists of methane typically ranges from 40 - 70% and its low heating value is between 15 and 30 MJ/Nm^3. There are two methodologies that can be used to run CI engines with biogas. The first methodology is converting the CI engine into pure biogas engine and the second one is dual-fuel mode (diesel-biogas). In the CI engine with dual-fuel mode, after compression of the charge comprised of biogas and air, a small amount of diesel, called the pilot is injected. This injected pilot fuel gets self-ignited and then becomes the ignition source for the inducted biogas. The main advantage of dual-fuel CI engines is that they can work with a wide variety of gaseous fuels without engine modifications [2]. This study focuses on CI engine run on dual-fuel mode.

Several studies on CI engine run on dual-fuel mode have been found in literature. Bedoya et al [3] reported a study on the effect of mixing system and pilot fuel quality on diesel-biogas dual-fuel engine performance. Cacua et al [4] studied experimentally the effects of oxygen enriched air on the operation and performance of a diesel-biogas dual-fuel engine. Tippayong et al [5] carried out a study on electricity production for on-farm using a small CI dual-fuel diesel-biogas. Their main objective was to evaluate the effect of long-term operation on performance and wear of the dual-fuel engine. The CI engine was tested for 2000 hours of operation. Makareviciene et al [6] explored the impacts of CH_4 compositions in a big CI four stoke and four-cylinders engine when operated under dual-fuel biogas-diesel mode. The impact of exhaust gas recirculation (EGR) was also explored. Tonkunya and Wongwuttanasatian [7] reported a study on the utilization of biogas-diesel mixture as fuel in a fertilizer pelletizing machine for reduction of GHG emission in small farms. The results showed that by using biogas as dual-fuel mode in the CI engine, a reduction in diesel fuel of 63% was achieved. This result was equivalent to 13 ton CO_2eq/year/farm in Thailand case. Nathan et al [8] performed an experimental study on the biogas-biodiesel HCCI mode of engine operation. The objective was to investigate the potential of the HCCI concept to utilize biogas effectively. Recently, Ambarita et al [9] reported an experimental study on the performance of small CI engine run on raw biogas. Mainly the studies of CI engine run on dual-fuel uses experimental method and only several studies use numerical method [10,11]. Yousefi et al [11] used CFD method to investigate effect of premixed fuel on the performance and emissions of dual-fuel pilot diesel engine.

The above literatures show that study on CI engine run on dual-fuel mode in order to decrease the fossil fuel has come under scrutiny. However, when it is operated on dual-fuel (diesel-biogas), thermal efficiency is relatively lower. In order to improve the thermal efficiency, several modifications have been proposed by researchers. Numerical simulation can be used to explore the combustion process on diesel engine run on dual-fuel mode. The better understanding of the combustion process in the CI engine run on dual-fuel mode is needed to improve the thermal efficiency. In this study, a numerical method is developed to investigate the combustion process in the CI engine run on dual-fuel mode. The main objective is to provide an alternative method in exploring the combustion process in the CI engine run on dual-fuel mode. The results are expected to supply the necessary information on development of alternative solutions for increasing thermal efficiency of CI engine run on dual-fuel mode.

2. Method

In the simulation, the governing equations are solved numerically. In order to simulate the combustion process in the chamber a set of governing equations have been developed. The governing equations consist of Navier-Stokes and energy equations, turbulent equations, and chemical kinetics equations [12]. The Navier-Stokes and energy equations are explained below.

$$\frac{\partial \rho}{\partial t}+\left(\nabla \cdot \rho \mathrm{U}\right)=0 \tag{1}$$

$$\frac{\partial(\rho U_i)}{Dt}=\rho g_i-\frac{\partial P}{\partial x_i}+\frac{\partial}{\partial x_j}\left[\mu\left(\frac{\partial U_i}{\partial x_j}+\frac{\partial U_j}{\partial x_i}-\frac{2}{3}\frac{\partial U_k}{\partial x_k}\delta_{ij}\right)-\rho\overline{u_i' u_j'}\right]=0 \tag{2}$$

$$\frac{D(\rho H)}{Dt}=\rho \dot{q}_g+\frac{\partial P}{\partial t}+\frac{\partial}{\partial x_i}(U_j\tau_{ij})+\frac{\partial}{\partial x_j}\left(\lambda\frac{\partial T}{\partial x_j}\right) \tag{3}$$

$$\frac{D(\rho C)}{Dt}=\rho\dot{r}+\frac{\partial}{\partial x_j}\left(D_{if}\frac{\partial C}{\partial x_j}-\rho\overline{cu_i}\right) \tag{4}$$

Where U (m/s), ρ [kg/m^3], P [Pa], μ [kg/ms], H[J/kg], λ [W/mK], C[kg], and D_{if} [m^2/s] are velocity, density, pressure, viscosity, specific enthalpy, conductivity, mass concentration of species, and diffusivity coefficient.

In order to simulate the turbulent flow in the combustion chamber, the two-equations of k-epsilon model is employed. The governing equations and the source terms are calculated using the following equations:

$$\frac{D(\rho k)}{Dt}=P+G-\varepsilon+\frac{\partial}{\partial x_j}\left(\left(\mu+\frac{\mu_t}{\sigma_k}\right)\frac{\partial k}{\partial x_j}\right) \tag{5}$$

$$\frac{D(\rho\varepsilon)}{Dt}=\left(C_{\varepsilon 1}P+C_{\varepsilon 3}G+C_{\varepsilon 4}k\frac{\partial U_k}{\partial x_k}-C_{\varepsilon 2}\varepsilon\right)\frac{\varepsilon}{k}+\frac{\partial}{\partial x_j}\left(\frac{\mu_t}{\sigma_\varepsilon}\frac{\partial \varepsilon}{\partial x_j}\right) \tag{6}$$

$$P=-2\mu_t S:S-\frac{2}{3}\left[\mu_t(trS)+k\right](trS) \tag{7}$$

$$G=-\frac{\mu_t}{\rho\sigma_p}\nabla P \tag{8}$$

$$\mu_t=C_\mu \rho\frac{k^2}{\varepsilon} \tag{9}$$

Where k [m/s^2], ε [kg/ms^3], and $\dot{r}$ [kg/s] are turbulent kinetic energy, weighted turbulence dissipation rate, and species source production. To determine the necessary sources terms in the momentum and energy equations in each time step, the following general form of reaction is used.

$$\sum_{i=1}^{N_s} v'_{k,i}S_i+TB \square \sum_{i=1}^{N_s} v'_{k,i}S_i+TB \qquad k=1,2,\ldots N_r \tag{10}$$

The reaction rate (RR) will be calculated using

$$RR_k=\left(\sum_{i=1}^{N_s}a_{k,i}[TB_i]\right)\left(K_{f,k}\prod_{i=1}^{N_s}[S_i]^{v'_{k,i}}-K_{b,k}\prod_{i=1}^{N_s}[S_i]^{v''_{k,i}}\right) \tag{11}$$

In this equation, K is the reaction rate which is calculated using Arrhenius equation formulated below.

$$K=A_{Ar}T^{\beta}\exp\left[\frac{-Ea}{R_u T}\right] \tag{12}$$

In each time step production or destruction rate ($\dot{\omega}_i$) of each species is determined by using:

$$\dot{\omega}_i=\sum_{i=1}^{N_R}RR_k\left(v'_{k,i}-v''_{k,i}\right) \tag{13}$$

All of those governings equations are converted into linier equations by employing finite volume method. The system of liner equations for all fields are coupled using SIMPLE algorith. The commercial code of ANSYS FLUENT is used to carry out the simulation. In the simulation the used biogas has methane percentage of volume is 60%. In this study a CI engine which is originally used in a small tractor for agricultural will be tested in the experimental apparatus. The specifications of the CI engine are presented in Table 1.

Table 1. Specification of the CI engine

No	Parameter	Value
1	Commercial name/model	Tiger Diesel Engine R175 AN
2	Number of cylinder/stroke	Single-cylinder/4 strokes and Horizontal
4	Bore × Stroke	75 mm × 80 mm
5	Maximum output	4.86 kW

In order to validate the numerical results, experiments are performed. An experimental apparatus has been designed and developed. It consists of a unit of CI engine, generator, series of lamps, biogas tank, gas mixer, and measurements apparatus. In single fuel mode (pure diesel oil only) the CI engine will be tested without modification. In dual-fuel mode, a gas mixer has been designed and developed in order to mix the fresh air with biogas. The mixture of the fresh air and biogas will be injected into the CI engine. To simulate the load, the CI engine will be coupled with a single phase synchronous generator using pulley.

3. Results and Discussions

3.1. The model

In this simulation, three-dimensional model of the combustion chamber was developed using CAD software. The developed CAD model and mesh of the combustion chamber are shown in Figure 1. The model is divided into several sub-models in order to make the analysis.

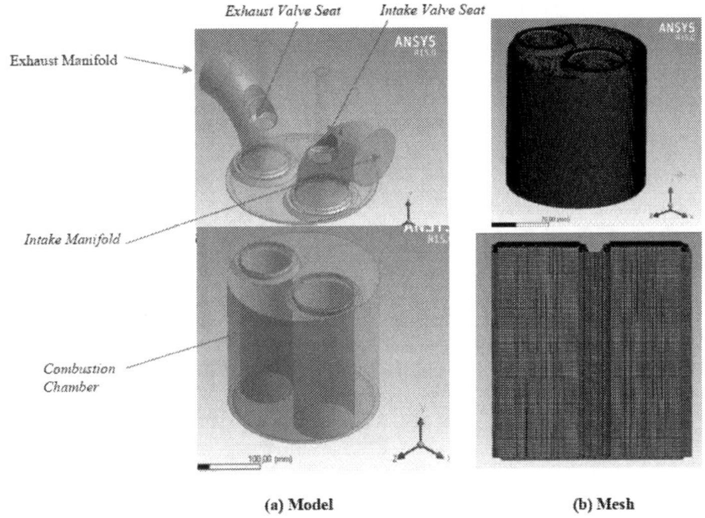

Figure 1. The developed CAD model (a) and mesh of the combustion chamber

3.2. Numerical Validation

In order to validate the numerical method, a numerical validation has been carried out by comparing the experimental results and numerical one. The comparisons were made for engine speed of 1000 rpm and biogas injection rate of 2L/minute. The engine was loaded with a load of 2500 Watt. The output power, torque and thermal efficiency are compared as shown in Table 2. The table shows that the experimental and numerical results do agree well. Based on these validation, the numerical method will be used to simulate the combustion process in combustion chamber.

Table 2. Numerical and Experimental results at load 2500 W and 1000 rpm

Parameter	Experimental result	Numerical result	Difference [%]
Power [W]	736.44	732.27	0.55
Torque [Nm]	7.04	6.996	0.65
Efficiency [%]	9.10	9.05	0.54

3.3. Pressure distribution

Pressure distributions inside the combustion chamber are shown in Figure 2. The figure shows pressure distributions at 270°, 360°, 450°, and 540°, respectively. In the figure, the position of piston at the mentioned crank angle is also shown. It can be seen clearly that at angle of 270°, 360°, and 450° pressure distributions inside the combustion chamber are almost homogenous. The values of the pressure are 6 MPa, 49.89 MPa, 1.172 MPa, respectively. On the other hand, when the crank angle at position 540° pressure difference is present in the combustion chamber. Even though, the difference is very small. It varies between 1.72 MPa and 1.73 MPa. This is because, the volume of the combustion chamber is in the maximum value.

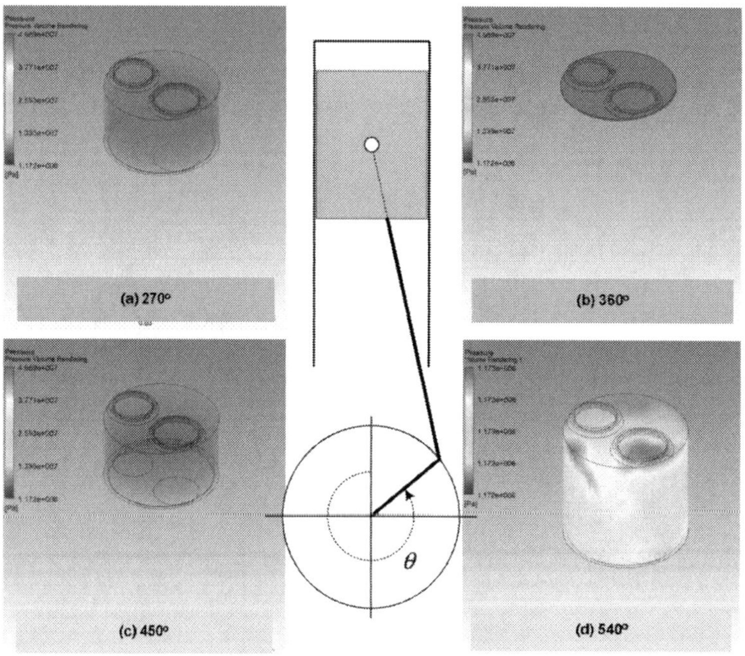

Figure 2. Pressure distributions in combustion chamber

3.4. Averaged pressure and Temperature

Averaged pressure and temperature inside the combustion chamber are shown in Figure 3a and 3b, respectively. The pressure history shows that in the compression stage (from 270° to 360°) the pressure increases as angle increases. When the combustion occurs (from 326° to 360°) the pressure is constant at 49.89 MPa. This suggest that the combustion is an isobaric process. The temperature history is shown in Figure 3b.

3.5. Streamlines

In this simulation, the streamlines in the combustion chamber are shown in Figure 4. The figure clearly shows the streamlines when the compression and expansion stages occur in the combustion chamber. It can be seen in the figure that more lines were captured at crank angle of 540°. As expected, this is because volume of the chamber is at maximum value. It can be said that the simulation can describe the streamline clearly.

IOP Conf. Series: Journal of Physics: Conf. Series **801** (2017) 012095 doi:10.1088/1742-6596/801/1/012095

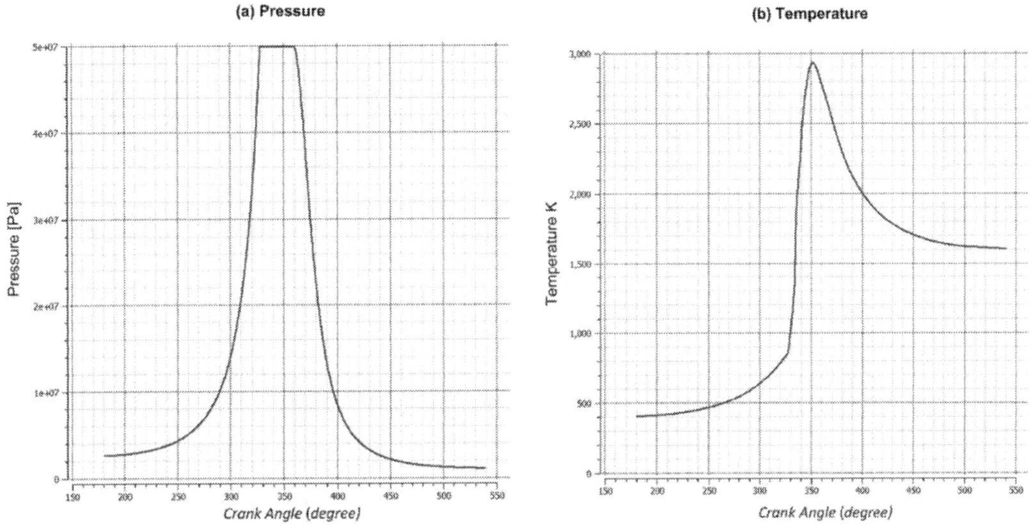

Figure 3. Averaged Pressure and Temperature

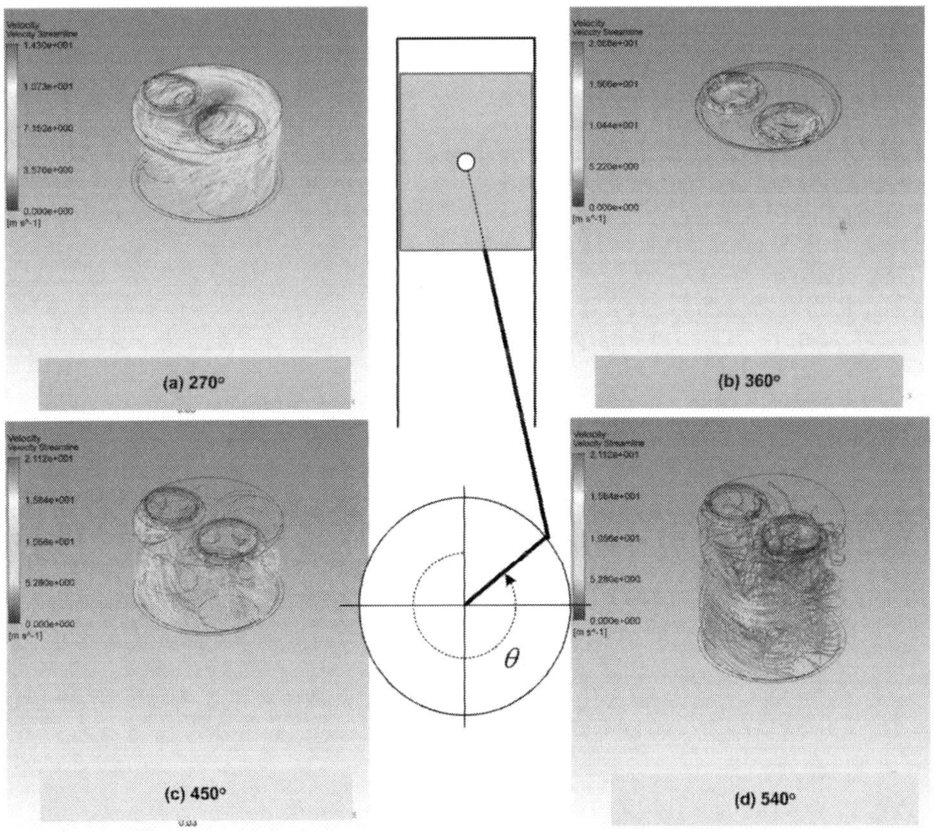

Figure 4. Streamlines in the combustion chamber

4. Conclusions

In this study, the combustion process in the combustion chamber of a small compression ignition engine run on dual-fuel mode (diesel-biogas) has been studied numerically. Three sets of governing equations (Navier-Stokes, Turbulent equations, and Chemical Kinetic) have been solved using commercial code ANSYS FLUENT. At the tested fuels and engine, simulation and numerical results show a good agreement. The simulation method can be used to show the temperature and pressure distributions, averaged temperature and pressure history, vector velocities, and streamlines. Based on the results it can be concluded that the present numerical method can be used to provide an alternative method in exploring the combustion process in the CI engine run on dual-fuel mode in order to increase the efficiency.

Acknowledgments

The authors gratefully acknowledge that the present research is supported by Ministry of Research and Technology and Higher Education Republic of Indonesia. The support is under the research grant BP-PTN USU of Year 2016.

References

[1] Government of Indonesia Presidential Decree No No 61 Year 2011, National Action Plan For Reducing Greenhouse Gas Emissions.
[2] Henham A, Makkar MK, 1998 *Energy Conversion Management* **39**, 2001-2009.
[3] Bedoya ID, Saxena S, Cadavid FJ, Dibble RW, Wissink M 2012 *Applied Energy* 618-629.
[4] Cacua K, Amell A, Cadavid F 2012 *Biomass and Bioenergy* **45** (2) 159-167.
[5] Tippayawong N, Promwungkwa P, Rerkkriangkrai 2007 *Biosystem Engineering* **98** 26-32.
[6] Makareviciene V, Sendzikiene E., Pukalskas S, Rimkus A, Vegneris R 2013 *Energy Conversion and Management* **75** 224 – 233.
[7] Tonkunya N, Wongwuttanasatian T 2013 *Energy for Sustainable Development* **17** 240 – 244.
[8] Nathan SS, Mallikarjuna JM, Ramesh A 2010 *Energy Conversion and Management* **51** 1347 – 1353.
[9] Ambarita H, Sinulingga EP, Nasution MKM, Kawai H 2016 *IOP Journal Conference series* (Accepted for pubilcation)
[10] Kong SC 2007 *Fuel* **86** 1483 – 1489.
[11] Yousefi A, Birouk M, Lawler B, Gharehghani A 2015 *Energy Conversion Management* **106** 322 – 336
[12] Kou KK 1986 Principles of combustion. 1st ed. John Wiley & Sons Inc.

Transient natural convection and conduction heat transfers on hot box of a coke drum in Pre-heating stage

A S Siahaan[1], H Ambarita[2], H Kawai[1], and M Daimaruya[1]

[1] Department of Mechanical Systems of Engineering, Muroran Institute of Technology 27-1 Mizumoto-cho, Muroran 050-8585, Japan
[2] Sustainable Energy Research Centre, Faculty of Engineering, University of Sumatera Utara, Jl. Almamater Kampus USU Medan 20155, Indonesia

E-mail: himsar@gmail.com

Abstract. In an oil refinery unit, coke drum is subjected cyclic thermal stress and mechanical loads due to cyclic heating and cooling loads. Thus, the useful life of a coke drum is much shorter than other equipment. One of the most severe locations due to thermal stress is shell to skirt junction. Here, a hot box is proposed. In this study effectiveness of a hot box will be analyzed numerically. The addition of hot box (triangular cavity) was expected to generate natural convection, which will enhance heat transfer. As for the result show that heat flux conduction and natural convection have the same trend. The peak of conduction heat flux is 122 W/m^2 and for natural convection is 12 W/m^2. In the heating stage of coke drum cycle it found that the natural convection only provide approximately 10 % of heat transfer compare to conduction heat transfer. In this study it was proved that in the heating stage, the addition of triangular enclosure is less effective to enhance the heat transfer than previously thought.

1. Introduction

The aforementioned studies reveal that the function of a hot box is debateable. Thus, further analysis is extremely needed. In this study, the transient numerical analysis of convection and conduction heat transfer in a hot box of a coke drum is carried out. Since the operation time of a coke drum is very long, only the preheating stage will be analysed.

As for the data provided by S.H.I Examination and Inspection, four identical coke drums is taken into consideration. Which can we said as coke drum A, B, C, and D. the shell and skirt of coke drums made by low alloy steel SA-387 Gr.11 CL.2. The cladding inside the coke drum is made of 3 mm thick SA 240TP. 410S material. Thickness of shells vary within the range 28.5 mm – 34 mm and the skirt thickness are 24 mm. Ratio between diameter and wall thickness of the coke drum varied from 125 to 150 mm/mm. For the coke drums aforementioned above, dimension, and thermocouple positions is shown in Fig.1. The measurements were carried out in the point which shown in the Fig.1, where the thermocouple and strain gauges is located. Temperatures and strains were taken continuously every minute for around 6 months. The cycle recorded for each coke drum A, B, C, D were 52, 53, 53, and 54 cycles respectively.

The cyclic duration of this four coke drums is around 48 hours which are preheating stage from 0 to 460 min, filling stage from 460 to 2000 min, cooling stage from 2000 to 2400 min and cutting from 2400 min to 2800 min. For preheating stage, it divided again as preheating and switching stage, which is for coke drums consider in this paper, the switching time is around 80 minutes, start from approximately 380 min.

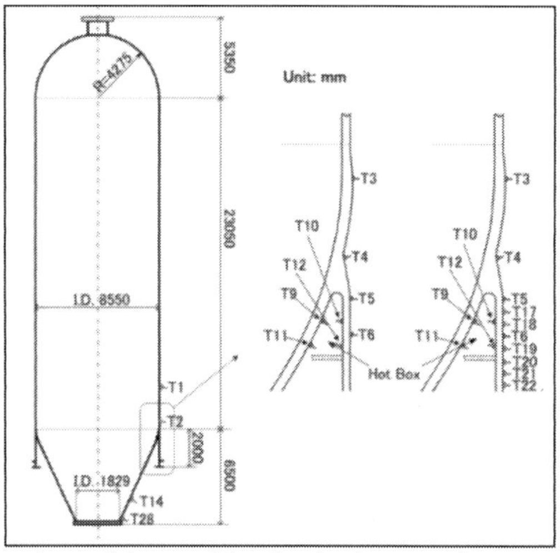

Fig. 1 Dimensions of the considered coke drums and measured location.

In preheating stage of coke drum, it is heated to around 350°C, and increasing rapidly in switching temperature to desirable injection temperature around 440°C to 460°C. Because coke drum is heated from the inside, the temperature in shell part of the coke drum increase more rapid than in skirt part, thus the temperatures at points T1-T4 is larger than T5-T8. The typical stages of the coke drum shown in Fig.2 below.

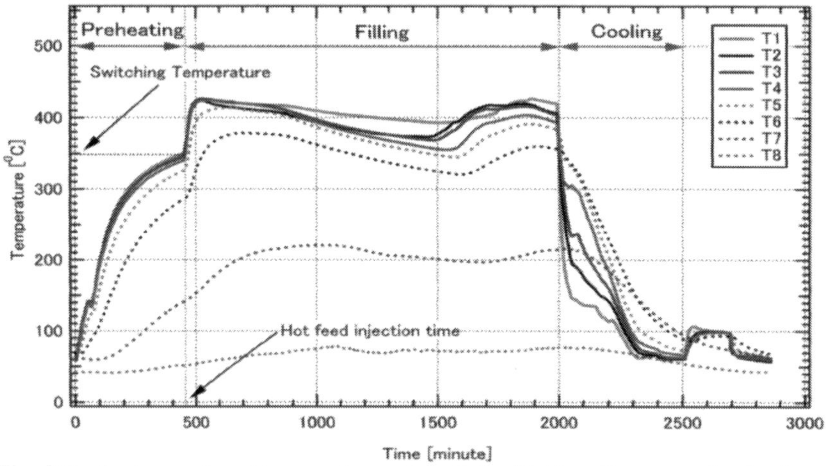

Fig. 2 Typical operational temperatures at shell-to-skirt junction area of coke drum.

Nowadays natural convection in enclosure is a topic of contemporary importance, because enclosure filled with fluid are central components in long list of engineering and geophysical system. Natural convection in an enclosure is the result of the complex interaction between a finite sized fluid system in thermal communication with all the walls that confine it. The complexity occurs for the diversity of flow that exists inside enclosures [1]. Some studies had been done in the field of natural convection inside triangular enclosure, whether with experimental method or in computational fluid dynamics to further understanding in the flow and heat transfer inside triangular enclosure.

In coke drum, the shell-to-skirt junction tends to be the most severe part [2] because of the thermal stress due to temperature difference. Hot box is a triangular cavity in the shell-to-skirt portion, which be added in hope to generate natural convection to enhance heat transfer inside shell-to-skirt junction. Thus, the main objective of this present study is to observe the transient natural convection and conduction heat transfers on hot box in pre-heating stage, and compare these two heat transfer, therefore we can concluded whether the hot box is effective or not.

2. Method

Some of the cycle in coke drum was observed, and we found that coke drum C have the most exemplary cycles compared with cycles from coke drum A, B, and D, thus we choose coke drum C cycle as the representative to be analyzed with numerical analyses, to see the effectiveness of hot box in the shell-to-skirt junction part of coke drum.

The purpose of this study was to analyses thermal distribution in the model geometry at different times during cycle. In this present study only preheating and heating stages will observed. For the model that we analyzed with numerical analyses we propose shell-to-skirt junction of coke drum model which shown in Fig.3. The numerical FEM were performed by ANSYS Fluent, using a two-dimensional axisymmetric model. The model was developed based on fabricated drawing of the coke drum.

For numerical analyses, temperature boundary condition was treated as follow; the temperature of inner shell part of coke drum is modeled by UDF, imitate measured temperature from the real cycle of coke drum C cycle number 3, which start from minute 1 at 62°C and end in minute 597 at 422°C. According to Antalffy et.al [3], even though the feed is around 500°C the maximum drum skin temperatures that recorded during operation rarely exceed 440°C. Apparently, an insulating skin of coke deposits onto the cooler drum shell and insulates the drum shell from the 500°C process fluid temperature within the drum.

Because the effect of insulating layer not taken into consideration in this paper, thus for the modeled UDF, for the temperature in inner shell we only set it 20°C higher than the measured temperature from real cycle; 82°C (min 1) and 442°C (min 597) respectively. As for surrounding area we used stream temperature of 30°C, and in the under skirt part it was applied stagnant air of 90°C to enhance heat transfer in the lower part of triangular enclosure due to convection heat transfer.

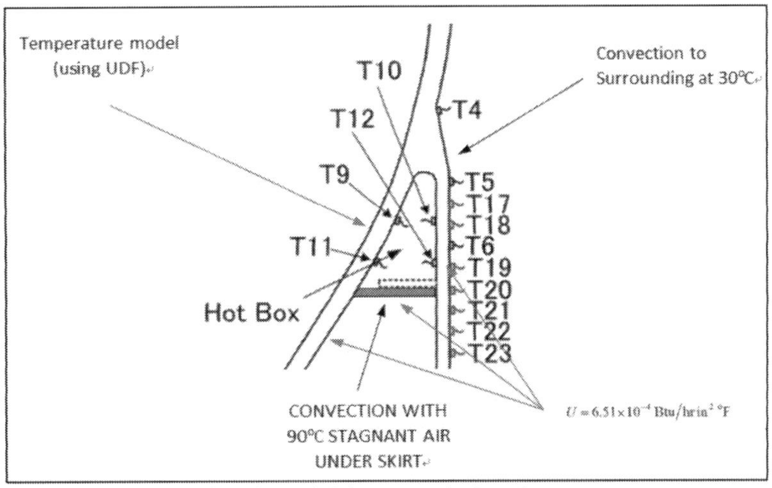

Fig. 3 Model for numerical analyses

To ensure the accuracy of the numerical result, some nodes were generated. There are three nodes numbers were proposed in the numerical analyses, which are 1263, 4657, and 10321 respectively. Important parameters such as temperatures in point T9, T10, T11, T12, and T4 were compared for each nodes numbers.

All nodes number show a small difference of temperature plot in selected point of T9, T10, T11, T12, and T4. However, to make a better compatibility with measured result, the model with 10321 nodes was carried for all analyses. The coke drum material density is 7850 kg/m3 and 40.6 W/mK for thermal conductivity. For the insulator we used rockwool with thermal conductivity of 0.04 W/mK. In this analyses the wind, piping, and seismic overturning moments are excluded.

3. Result and Discussion

Fig. 4 shows temperatures plots in some points in this considered model. The numerical analyses results and thermocouple measurement results shows similar trend even though there occurred some differences from numerical and measured temperatures plot. Mostly, problem happened in peak temperature of numerical model and measured model. In numerical model the peak temperature (in T4) reached 440°C, but in the measured temperatures the peak temperature is below the numerical result. These phenomena occurred due to the insulated skin deposit in shell wall, which not taken into consideration in this paper. Thus, this problem also occurred in all temperature points. Set aside these differences between numerical and measured one, the results is still acceptable and further analyses will be taken to affirm it.

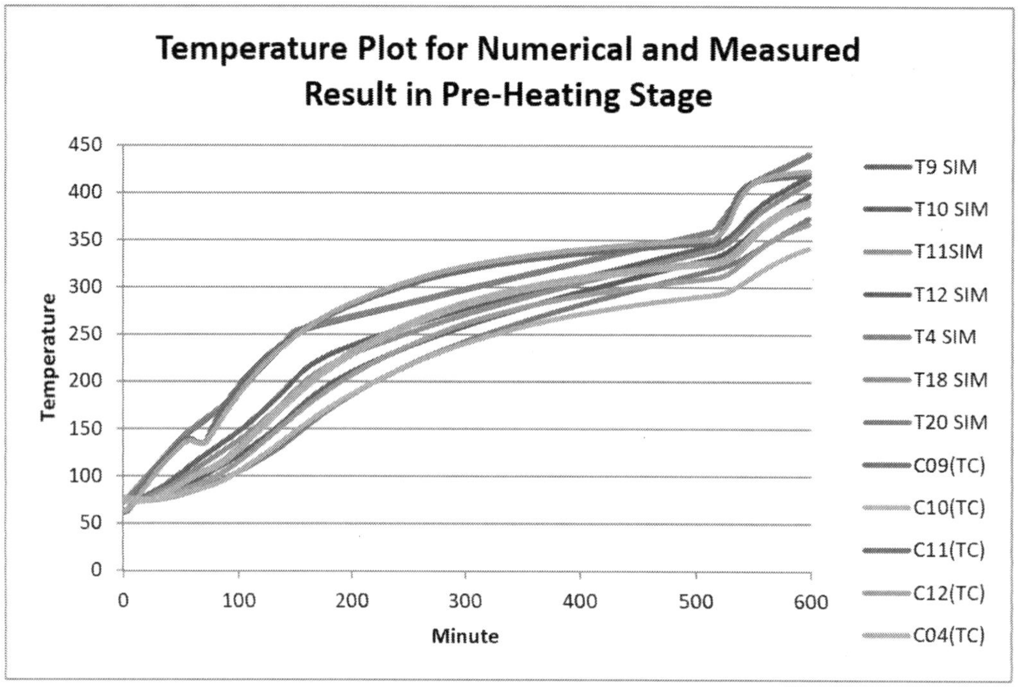

Fig.4 Temperature Plot For Measured and Numerical Result

In the beginning of the cycle because of the stagnant air under skirt, the triangular enclosure was heated from below. Fig.5 shows the temperature (a, c, e) and velocity (b, d, f) profiles in triangular cavity. From these figures it shown that the contour of temperature and velocity was developed first in lower part of enclosure, thus the purpose of stagnant air is to enhance the heat transfer in lower part of triangular enclosure. From fig. 5 (d) and (f), contour of velocity shows low velocity rate in the lower part of triangular enclosure, thus it concluded that stagnant air become less effective as the cycle goes by. Fig.5 (c) and (e) shows that temperature distribution in skirt part cannot keep up the increment rate of temperature in shell part, even though the hot box (triangular enclosure) already attached. For initial conclusion it can be said that the triangular enclosure is not very effective to enhance heat transfer in shell-to-skirt part. To ensure this statement, further analyses has been done in this paper.

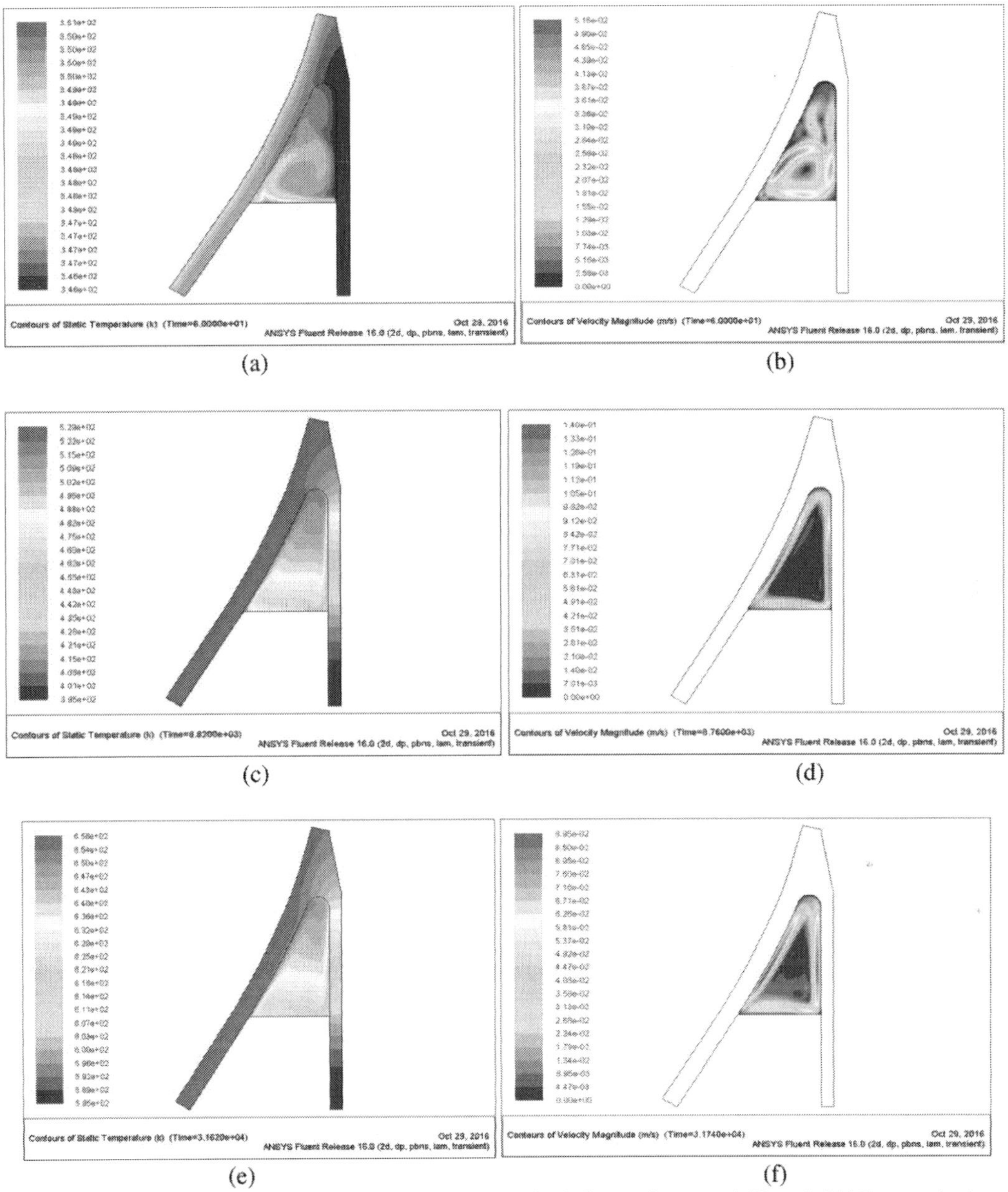

Fig.5 Temperature (a, c, e) and Velocity Contours (b, d, f) at Minute 1, 147, and 544 Respectively

Temperature differences in 4 points in triangular enclosure (T9 and T10) and (T11 and T12) both for numerical results and measured thermocouples were compared. T9 and T10 have the same elevation (upper part of triangular enclosure) in Y-coordinate where T9 is located in shell part, and T10 is located in skirt part and so the temperature difference of these points will be measured, thus for T11 and T12 which is located in lower part of triangular enclosure respectively. Because the main purpose of the addition of the hot box (triangular enclosure) is to enhance the heat transfer, this four points was often be used to measure the effectiveness of hot box.

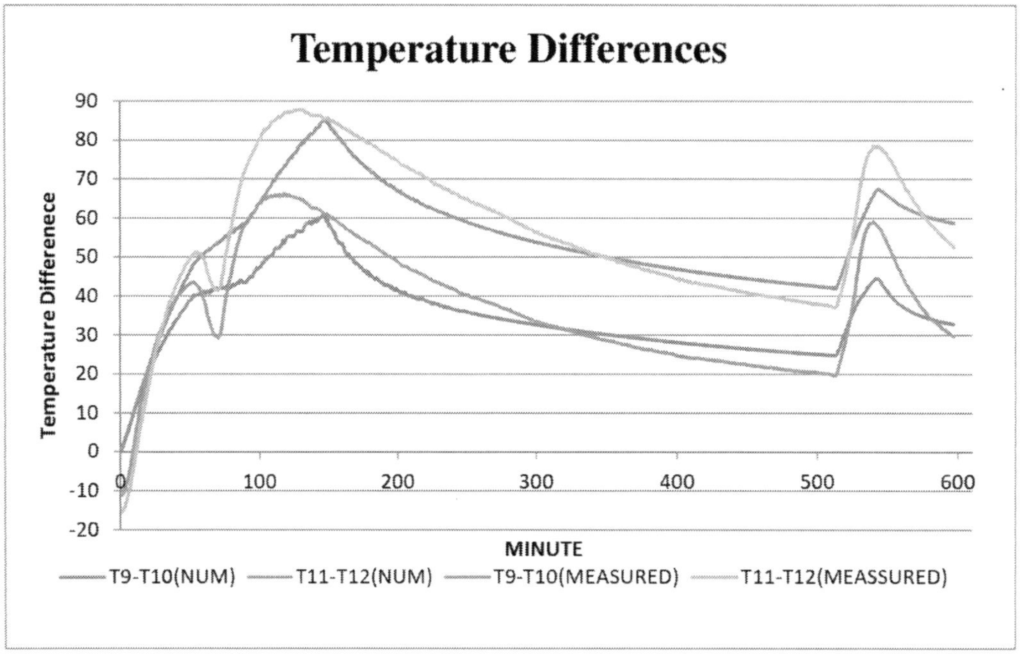

Fig.6 Temperature Differences for Numerical and Measured Result

Fig 6, the result of temperature differences show a good agreement both for numerical analyses and measured result. The temperature difference have the same trend both for numerical and measured one, even though there are some slightly difference because some parameters which occurred in real cycle not taken into consideration in the numerical analyses.

The peak of temperature difference (T11 and T12) in this cycle occurred not in in the switching temperature, but in the beginning of pre-heating stage which was appeared in minutes 147. From the numerical (87°C) and measured thermocouples (85°C) we found the slightly difference of 2°C, which is acceptable. The problem occurred in switching stage, that the difference of temperature peak is 11°C, which exists in minutes 544.

Because of the peak of temperature difference (T11 and T12) in the numerical results are lower than the measured one, for heat flux from numerical result we will multiply by scale-up factor, because heat flux is the function of temperature. The scale–up factor is defined as comparison of maximum temperature difference from measurement and numerical one. For cycle C3 the maximum temperature difference in switching stage from measurement and numerical (shown in fig.6) are 78°C and 67°C respectively. Thus the scale up factor will be 1.17. Thus the calculated heat flux both for conduction and natural convection will be calibrated multiplying by 1.17 (fig 7).

From fig. 7 for heat flux result of transient conduction and natural convection in triangular cavity we can see that both have the same trend. In the beginning stage of preheating, heat flux conduction gradually increased up to 70 W/m^2 and 5 W/m^2 for heat flux in triangular enclosure. This high rate heat flux increment happened due to the hot steam which started to enter in the beginning of pre-heating stage. Heat flux tends to rise until hot steam spread evenly inside of the coke drum. Started from minute 147 heat flux fell to around 53W/m^2 and became stagnant until the beginning of switching stage. This statement was validated due to the temperature plot from fig.4 which shown significant temperature increasing in the beginning of preheating stage and decreases toward the end of preheating stage.

In switching stage, because of the rapid temperature increasing in a short time, the heat flux conduction increasing rapidly up to 122 W /m^2, and for heat flux in triangular cavity was increased up to 12 W/m^2. Heat flux increasing rapidly from the beginning of switching stage and decreases toward the end of switching stage, the same reason as the preheating stage that in the temperature rising significantly in the beginning of switching stage and the end of the switching stage temperature gradient becomes low.

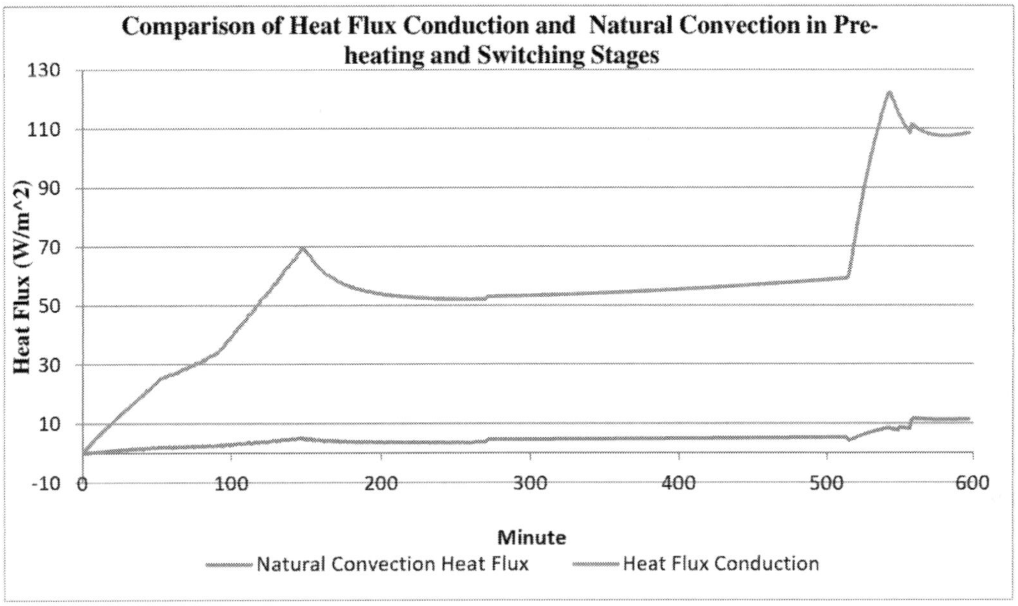

Fig. 7 Heat Flux Conduction and Natural Convection

The result shown, hot box merely have a little effect to enhance heat transfer in shell-to–skirt junction. The natural convection inside the triangular cavity of shell-to-skirt junction approximately only took apart for 10% of the heat transfer. This result has a good agreement with the previous research for the same coke drum. According to M.Oka et.al [4], a significant temperature difference exist between T9 and T10; 120°C (preheating stage) and 150°C for T11 and T12 (preheating stage) respectively. These temperatures indicate very significant differences, thus it was concluded that hot box (triangular enclosure) is less effective to enhance heat transfer in shell-to-skirt junction part unlike previously thought. Therefore the new method to enhance the heat transfer in the shell-to-skirt junction should be considered in the future research, to reduce the thermal stress and increase the life span of coke drum.

4. Conclusion

From this research it can be concluded that the numerical analysis can be used as an affordable method to predicted transient heat transfer in shell-to-skirt junction. Although some slightly difference occurred between measured and numerical result is undeniable due to some boundary conditions which occurred in measured system such as complex insulating layer in inner shell which was not taken into consideration in this paper, but it still acceptable with using scale-up factor which offered.

The addition of triangular cavity (hot box) to generated natural convection in shell-to-skirt junction merely took a part to enhance the heat transfer unlike previously thought. The natural convection inside the triangular cavity of shell-to-skirt junction approximately only took apart for 10% of the heat transfer. This result has a good agreement with the previous research about the same coke drum.

For the development of the coke drum fatigue life can be done further research in any ways that can increase shelf life, especially in the shell-to-skirt junction portion, which is the most severe part of coke drum, for example the provision of oil or nano-fluid instead of air to enhance the heat transfer (natural convection) inside of the triangular cavity of the shell-to-skirt-junction.

Acknowledgement
This research was supported by data from Sumitomo Heavy Industries Examination and Inspection, Ltd., 1501 Imazaike, Saijo, Ehime 799-1393, Japan.

References
[1] A Bejan 1984 *Convection Heat Transfer* (A Wiley Interscience Publication) p 159
[2] Ramos A, Rios C, C Vargas, J Tahara, and T Hasagewa 1997 Mechanical Integrity Evaluation of Delayed Coke Drums *ASME PVP* 359 pp 291-298.
[3] Antalffy L P, Bardia K L, Baxter J E, Malek D W, and Taagepera J 1995 Analyses of Alternate Skirt Attachment to Coke Drums *ASME PVP* 315 pp 437-436
[4] Oka A, Ambarita H, Daimaruya M, Fujiki H 2011 Study on the effects of switching temperature on the thermal fatigue life of the shell to skirt junction of coke drum *ASME Pressure Vessel Technology* 133

Mechanical Response of Open Channel Cover Made of Concrete Foam Due to External Loadings

BustamiSyam[1], Alexander Sebayang[2], Indra Gunawan[3], Maraghi Muttaqin[2] , Herry Darmadi[2], Basuki WS[4], M. Sabri[1], and S. Abda[5]

[1] Impact and Fracture Research Center, Dept. of Mechanical Engineering, Faculty of Engineering, University of Sumatera Utara (USU), E-mail: bstsyam@gmail.com
[2] Graduate Student, MechEngr Study Program, USU
[3] Undergraduate Student, Dept of MechEngr Study, USU
[4] Dept. of Chemistry, Faculty of Science, University of Sumatera Utara
[5]Graduate student, Dept. ofMech. Engr, Faculty of Engineering, University of Sumatera Utara

Abstract. Open channel conduit is commonly constructed on side roadway, car park area, etc., with the aims to collect rainwater or seapage water. To avoid garbages following the waterflow, the conduit is equipped with adequate cover. In addition, the cover can also be functioned for temporary car park area. Thus, the conduit cover should strong enough to sustain external loading. This paper discusses the design and response of the conduit cover using finite element software ANSYS MECHANICAL version 17.5.
Keywords: Conduit cover, parking bumpers, foam concrete, impact strength, Fiber Oil Palm Empty Fruit Bunch (EFB).

1.Introduction

Open channel conduit is one of ways to reduce excess water, whether from rain, seepage, or excess irrigation water in an area, drainage is included in one of the important components of urban infrastructure in tackling the problem of flooding and waterlogging. On the roadway [1] as shown in Figure 1 and 2, open channel conduits should function the same; however conduit covers are needed.

This paper discusses the design and response of the conduit cover using finite element software ANSYS MECHANICAL version 17.5. Two types of conduit covers as shown in Figure 3 with a size of 920 × 200 × 150 mm with the use of concrete foam-fiber reinforced empty oil palm bunches (EFB) were designed, manufactured, and tested. The conduit covers were subjected to impact load using a free-fall apparatus. Results were also compared with numerical simulation using a commercial software ANSYS Mechanical version 17.5. We consider also the impact test is with the assumption that the covers are also to functioned as parking bumper at the moment vehicles bump on kerbline and jump on to the conduit cover.

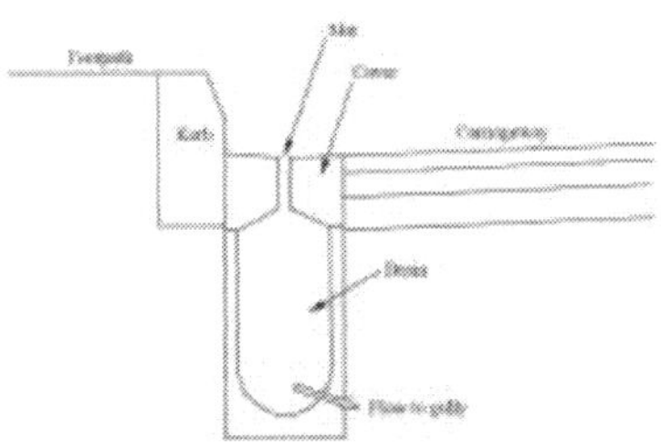

Figure 1. Slot drain

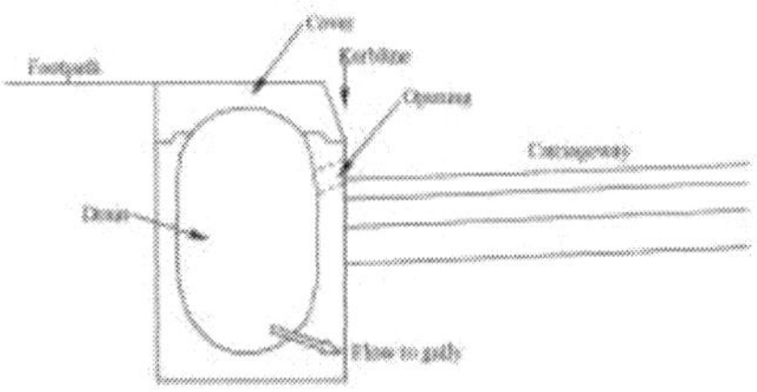

Figure 2. Kerb drain

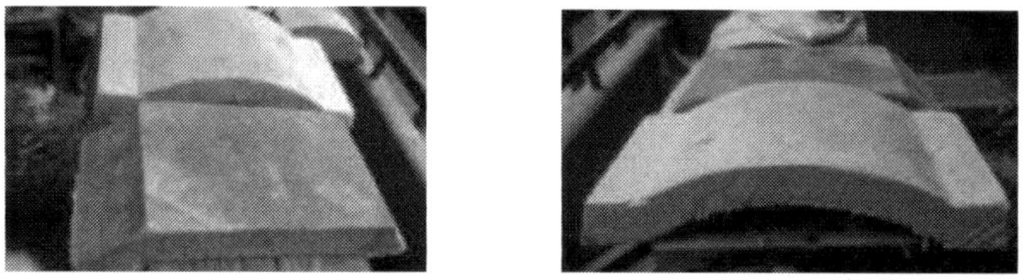

(a) (b)

Figure 3. Conduit covers
(a) flat cover (type 1) (b) curvy cover (type 2)

2. Materials and Methods

2.1Material

As shown in Figure 3, two types of conduit cover are used to function as walk way as well as parking bumper. Both covers should meet technical specifications, e.g. strong enough to sustain loadings from vehicles passing on it. In addition, the covers should look good and do not endanger pedestrians walking or crossing over it. To achieve the technical specs, we select the newly developed material that is lightweight and strong enough to withstand a static load and impact, i.e. concrete foam (confoam). There are several classes of materials confoam [3]. In this study we choose the type B4 [2] in which the physical and mechanical properties of the material shown in Table 1. The material has been extensively used of some light structures, such as tiles [3] and speed bump [4].

Table1.Confoam mechanical properties

Specimen	Age (day)	Berat (g)	Compressive Strength (MPa)	Tensile strength (MPa)	Modulus Elastisitas (MPa)
B4	28	4.85	5.49	0.025	43.824

2.2. Geometry and Dimensions

Geometry and dimensions of the conduit covers are shown in Figures 4 and 5, respectively.

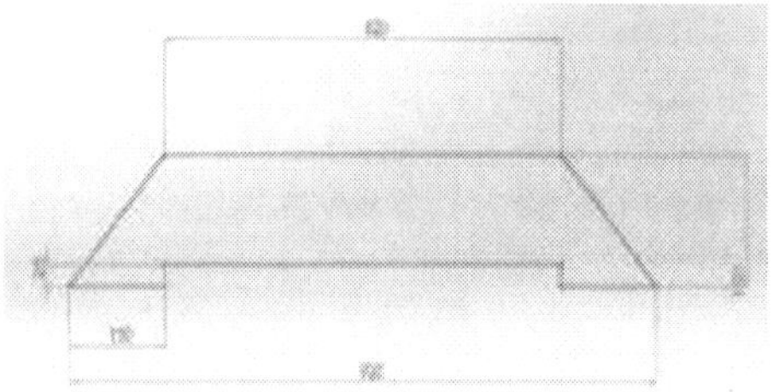

Figure 4. Flat (type 1)

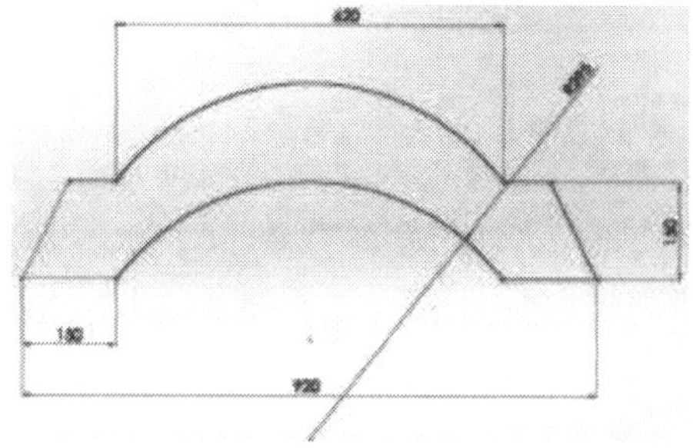

Figure 5. Curvy (type 2)

3. Numerical Simulation

3.1. Load Model

The conduit covers geometry and dimensions were modelled as shown Figure 6 and 7. The model is subjected static load of 3,290 Newton per contact area of 2.000 mm2. The impact load, applied to the parking bumber in lateral direction is focused on the calculation of the stress distribution in the x, y, and priciple direction using commercial FEM software, with a 3-D element.model; we also calculate the principle stress (σ_1). Let us observe the stress contour of the models one-by-one.

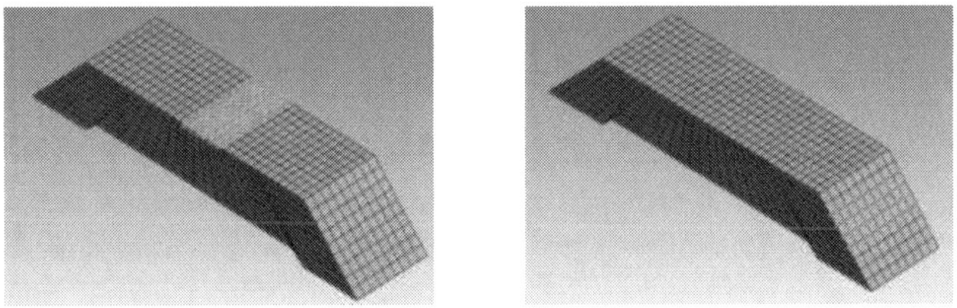

Figure 6. Cover Type 1

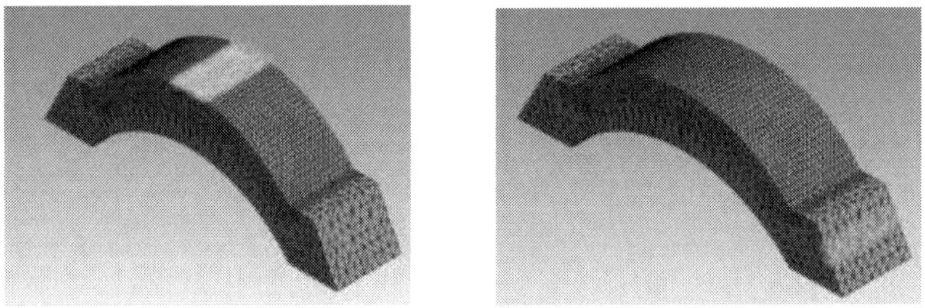

Figure 7. Cover Type 2

3.2. Stress contour

The stress distribution is presented in the x, y, and direction of the principles in the parking bumper models. All calculations are shown in Table 2. As shown, the radial load from vehicles from radial direction still cannot be accepted safely. However the covers are able to withstand impact load from lateral direction. Static and Dynamic side impact for Type 1 and 2 is shown in Figures 8 and 9, respectively. Simulation of static and dynamic in radial direction for type 1 and 2 is shown in Figures 10 and 11, respectively.

Tabel 2. Max stress in x, y, and principle direction (in MPa)

	σ_x	σ_y	σ_1
Dynamic Simulation Type 1 lateral	0.076	0.122	0.180
Static Simulation Type 1 lateral	0.082	0.020	0.118
Dynamic Simulation Type 2 lateral	0.026	0.052	0.251
Static Simulation Type 2 lateral	0.035	0.047	0.100
Dynamic Simulation Type 1 radial	0.121	0.021	0.001
Static Simulation Type 1 radial	5.867	1.872	5.869
Dynamic Simulation Type 2 radial	0.055	0.004	0.057

Static Simulation Type 2 radial	3.521	0.388	3.521

Side Impact Type 1 (dynamic)

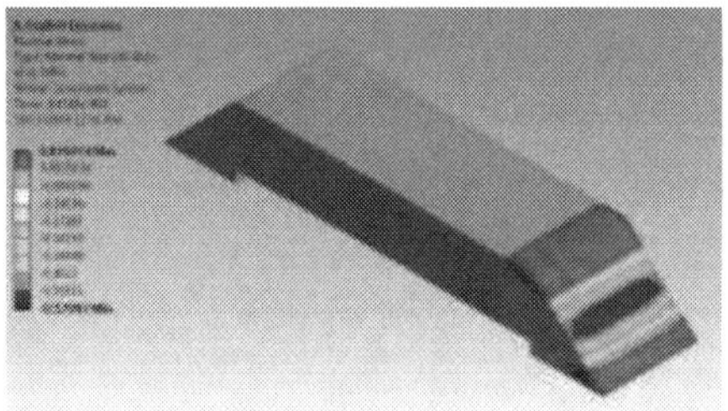

σ_x

Side Impact Type 1 (static)

σ_x

Side Impact Type 1 (dynamic)

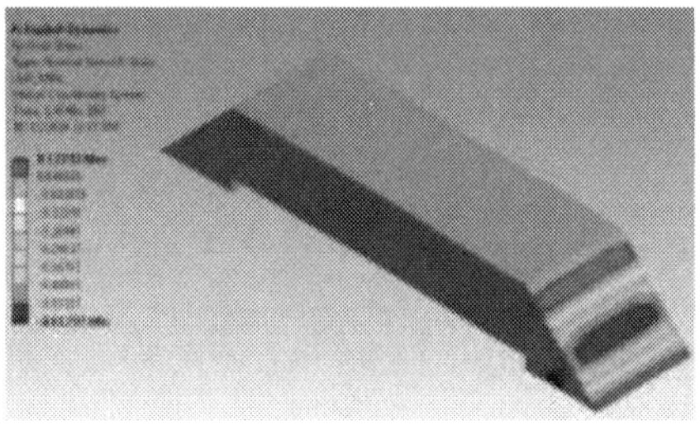

σ_y

Side Impact Type 1 (static)

σ_y

Side Impact Type 1 (dynamic)

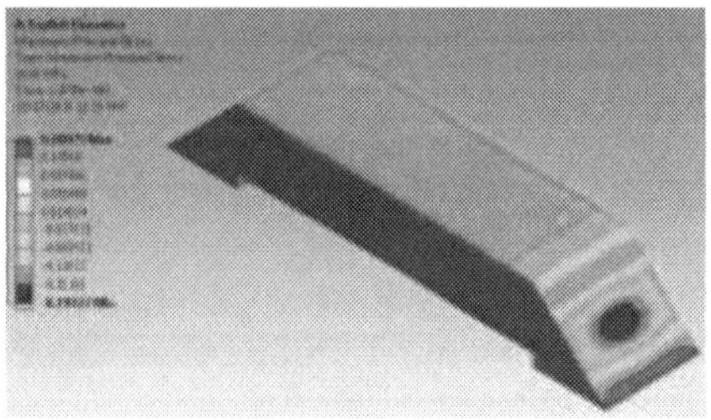

σ_1

Side Impact Type 1 (static)

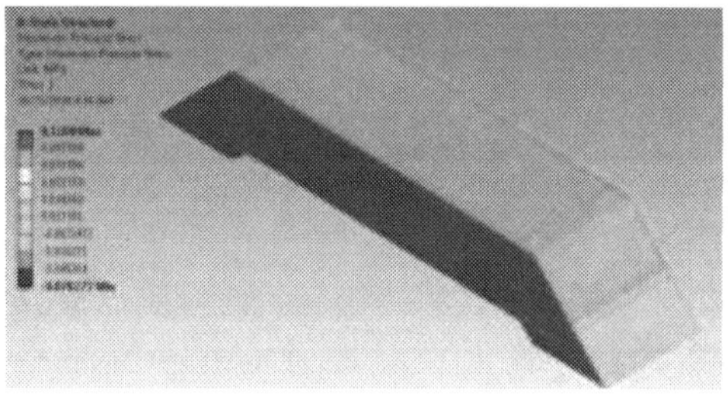

σ_1

Figure 8. Static and dynamic lateral impact

Side Impact Type 2 (dynamic)

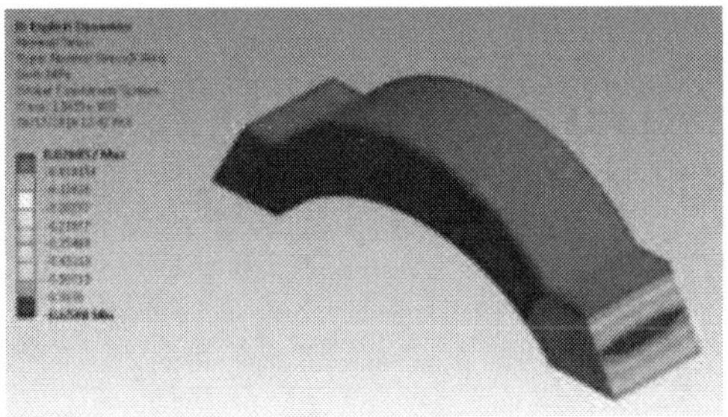

σ_x

Side Impact Type 2 (static)

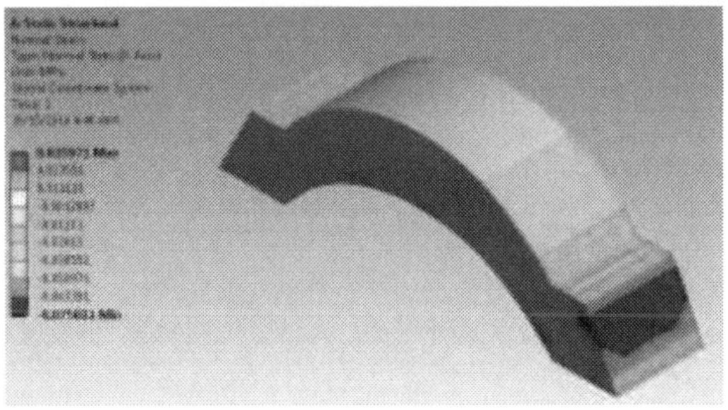

σ_x

Side Impact Type 2 (dynamic)

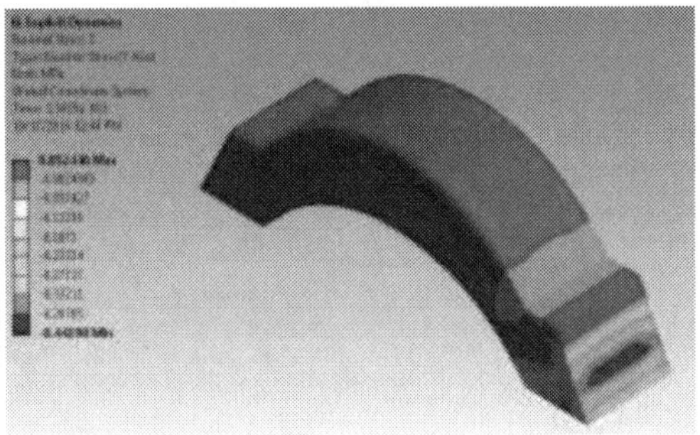

σ_y

Side Impact Type 2 (static)

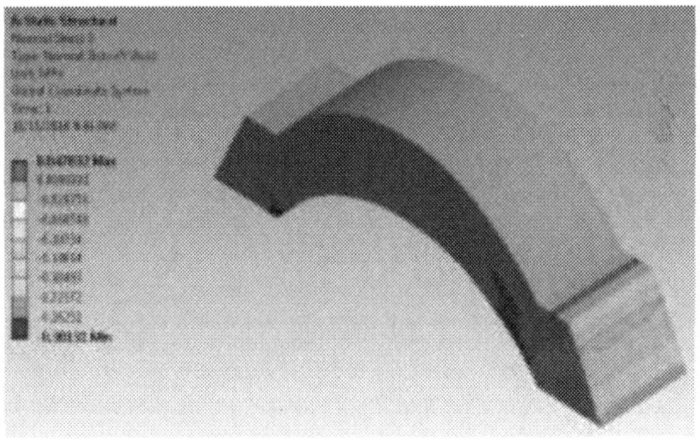

σ_y

Side Impact Type 2 (dynamic)

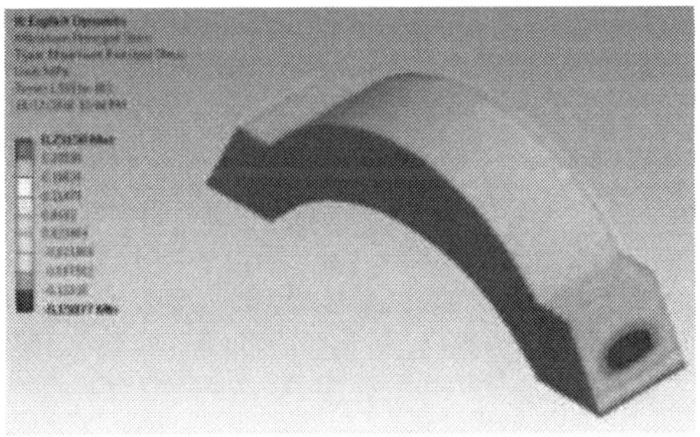

σ_1

Side Impact Type 2 (static)

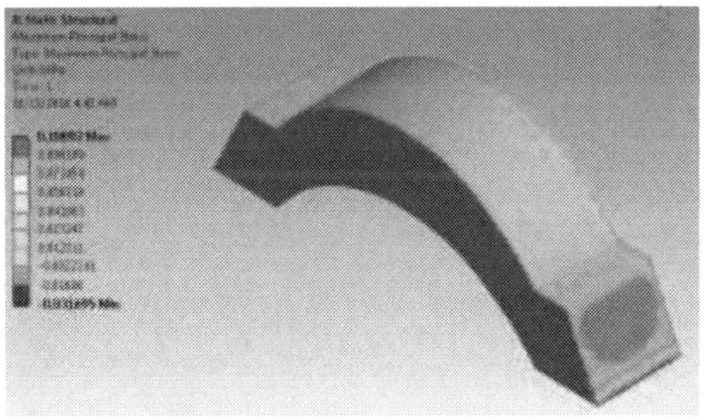

σ_1

Figure 9. Static and dynamic lateral impact

Radial Impact Type 1 (dynamic)

σ_x

Radial Impact Type 1 (static)

σ_x

Radial Impact Type 1 (dynamic)

σ_y

Radial Impact Type 1 (static)

σ_y

Radial Impact Type 1 (dynamic)

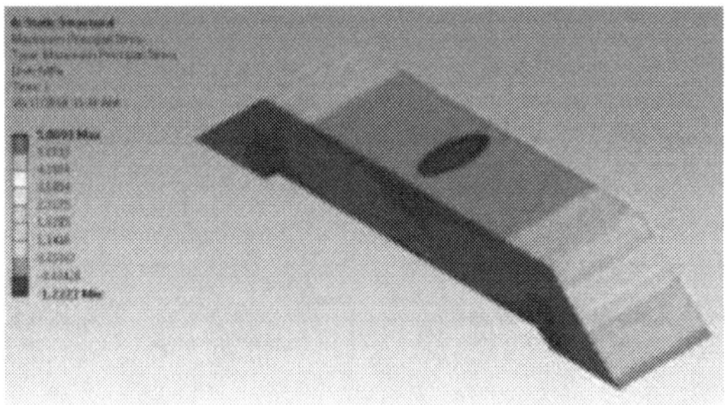

σ_1

Radial Impact Type 1 (static)

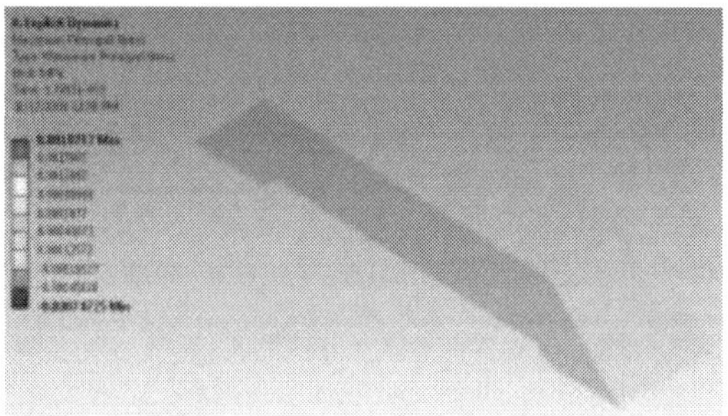

σ_1

Figure 10. Static and dynamic radial impact

Radial Impact Type 2 (static)

Radial Impact Type 2 (dynamic)

IOP Conf. Series: Journal of Physics: Conf. Series **801** (2017) 012097 doi:10.1088/1742-6596/801/1/012097

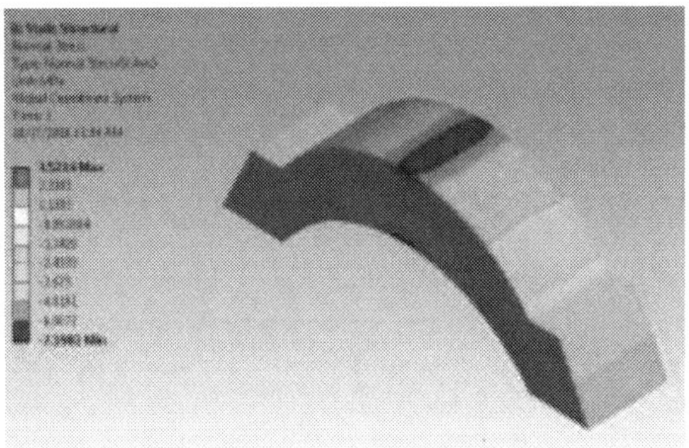

σ_x

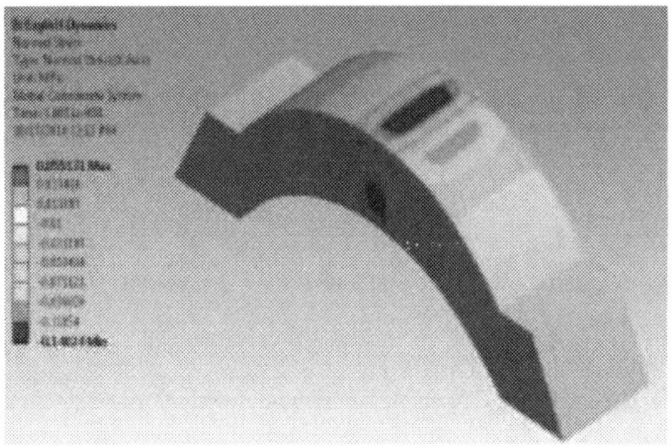

σ_x

Radial Impact Type 2 (static)

Radial Impact Type 2 (dynamic)

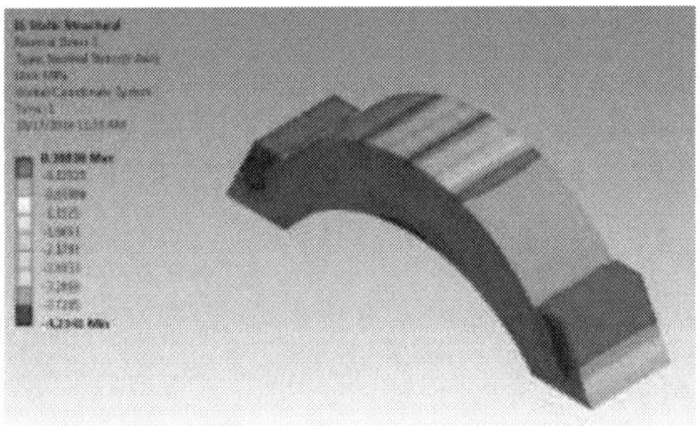

σ_y

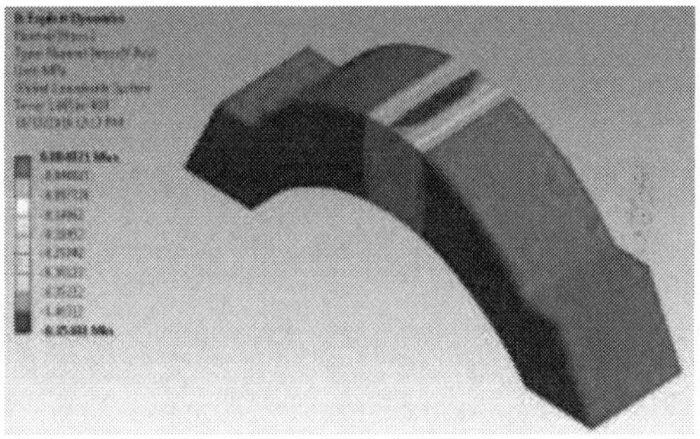

σ_y

Radial Impact Type 2 (static)

Radial Impact Type 2 (dynamic)

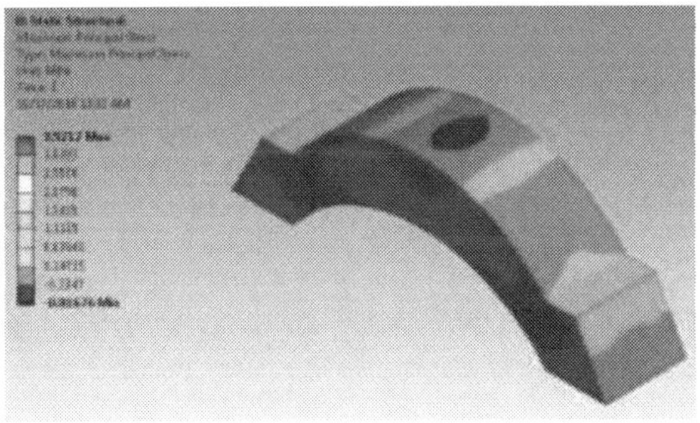

σ_1

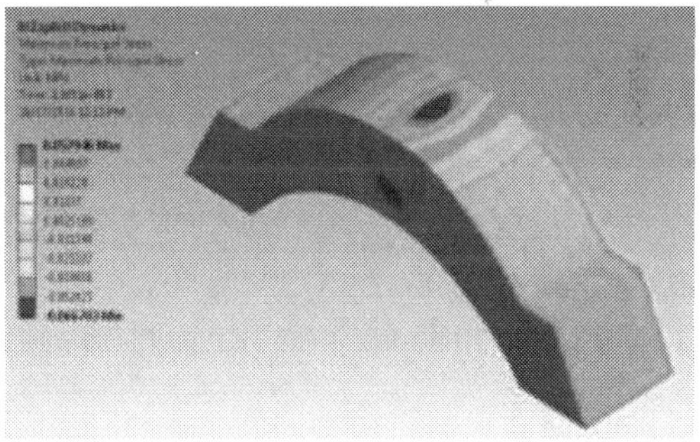

σ_1

Figure 11. Static and dynamic radial impact

4. Conclusions

Two models of covers of the open channel conduit have been designed, manufactured, tested, and simulated in our laboratory. A newly developed light weight material so called confoam was used to build the structure. The cover was sought for its structure integrity, i.e. its response when they are subjected to both static and dynamic external loading in radial and lateral directions, respectively. The conclusions that can be drawn are that the two models function well when they are subjected to lateral loading. Thus, the covers may be functioned as parking bumper. However, under radial loading both types failed to withstand the external loading. Here, redesign of the geometry and dimensions as well selecting other types of confoam materials are needed.

References

[1]. Guidance Notes on Road Pavement Drainage Design, RD/GN/035, May 2010

[2]. Syam, Bustami, B. Wirjosentono, dan S. Rizal, "Pengembangan Bahan Komposit Busa Polimer (Polymeric Foam Composite) Diperkuat Serat Tandan Kosong Kelapa Sawit (TKKS) untuk Pembuatan Berbagai Produk Industri Transportasi dan Olah Raga", Laporan Akhir PENPRINAS MP3EI, DIKTI, 2013.

[3]. Ade Irwan, Bustami Syam, dan Basuki W.S, Analisa Respon Mekanik Genteng dari Bahan Confoam Diperkuat Serat TKKS Akibat Beban Impak Hujan Es, Prosiding Seminar DIES USU 2016

[4]. Syam, Bustami, Marari Muttaqin, Dino Hastrino, Alexander. Sebayang, Basuki WS, M. Sabri, and Syahrul Abda, Analysis of Power Generating Speed Bumps Made of Concrete Foam Composite, International Journal IOP, 2016 (in progress).

ACKNOWLEDGEMENTS

The authors acknowledge the research fund obtained from Kemenristekdikti under BPPTN scheme 2016.

Pseudo Order Kinetics Model to Predict the Adsorption Interaction of Corn-Stalk Adsorbent Surface with Metal Ion Adsorbate Cu (II) and Fe (II)

B Haryanto[1], W B Singh[2], E S Barus[3], A Ridho[1] and M R Rawa[1]

[1]Department of Chemical Engineering, University of Sumatera Utara, Medan 20155, Indonesia
[2]Department of Chemistry, Manipur University, Canchipur, Imphal, Manipur-795003, India
[3]Department of Technical Informatics, Institute of Management Informatics and Computer, Neumann Christian, Medan, 20123, Indonesia

E-mail: haryanto_bode@yahoo.com

Abstract. The adsorption process using cornstalk as adsorbent was used to remove the single metal ion such as copper ion and cadmium ion. The adsorption kinetics of each contaminant then used to predict the interaction type of metal ion on surface of corn stalk by calculating pseudo order 1st and 2nd. The identification type as chemically or physically interaction was predicted from the quality of r^2 by plotting the adsorption capacity (q) and time (t). The r^2 were 0.01 and 0.99 for pseudo order 1st and 2nd respectively for Fe (II). The r^2 were 0.26 and 0.999 for pseudo order 1st and 2nd respectively for Cu (II). The result of adsorption interaction of metal ion and surface function of corn stalk is chemical type. SEM/EDX confirmed the Cu^{2+} presence on cornstalk surface.

1. Introduction

The corn crop is one of Indonesian main plant. Corn is a primary food in some parts of the region in Indonesia. It is grown in all regions of Indonesia (AR Nabilla et al., 2014). The side product of corn plantation such as corncob and cornstalk is available abundantly and usually ignored or burned. The characteristic of the cornstalk with porous structure has potential to use as natural adsorbent (Wang et al., 2016).

Heavy metals cause problems to the environment because it can induce high toxicity to ecosystems and human (PB Tchounwou et al., 2012). Adsorption is a process that occurs when a fluid has an existing substance dissolved in solution, bound to a solid (adsorbent) caused by chemical-physical force between the substances and the absorbent. Heavy metal as adsorbate substances adsorbs by adsorbents generally studied using a batch system. Several factors affect the adsorption process are such as pH, temperature, concentration and contact time (S Vafakhah, et al., 2014; M Karnib, et al., 2014; B Haryanto and CH Chang, 2014).

Equation 1 is to calculate the amount of adsorbed metal per unit mass of adsorbent at equilibrium. Equation 2 is used to calculate the amount of adsorbed metal per unit mass of adsorbent at time t. Equation 3 is used to calculate the percentage of metal ions removed at time t. The equations are as follows:

$$q_e = \frac{(C_0 - C_e)V}{m_{ads}} \quad (1)$$

$$q_t = \frac{(C_0 - C_t)V}{m_{ads}} \quad (2)$$

$$R\% = \frac{(C_0 - C_e).100\%}{C_0} \quad (3)$$

In which qe and qt is the adsorption ability at equilibrium and at time (t) respectively (mg g-1); R% is the removal percentage of metal ions (%). C_0 is the initial metal concentration (mg L-1), Ct is the metal concentration at time t (mg L^{-1}) and Ce is the metal concentration at equilibrium (mg L-1). V is the volume of solution (L) and mads is the mass of adsorbent (g) (L Wen, et al., 2013; L Haibin, et al., 2013; B Haryanto, et al., 2014).

Measurement of the adsorption ability (qt) is by increasing contact time (t) during adsorption process which produces the adsorption kinetic data. Haibin, et al., 2013 have used the adsorption kinetic data empirically to describe and evaluate adsorption type using pseudo first order (equation 4) and second order (equation 5) models. The models were used to identify the interaction of adsorbate on absorbent during the adsorption process. Correlation coefficient (r^2) qualities have been widely used to identify the adsorption interaction either chemically or physically. Pseudo 1st order is related to physical interaction and pseudo order 2nd is related to chemical interaction (Chen Suhong, et al., 2011; Liu Haibin, et al., 2013; M Arshadi, et al., 2014). The pseudo first order and second order shown as following equations:

$$\frac{1}{q_t} = \frac{k_t}{(q_e)t} + \frac{1}{q_e} \quad (4)$$

$$\frac{t}{q_t} = \frac{t}{q_e} + \frac{1}{(k_2 q_e^2)} \quad (5)$$

The idea of this study to apply the pseudo order 1st and order 2nd equations to identify the type of adsorption interaction of metal ion Fe^{2+} and Cu^{2+} on cornstalk surface base on correlation coefficient (r^2). The empirical data of adsorption process on single substance of Fe^{2+} and Cu^{2+} as adsorbates and cornstalk as adsorbent with batch system were used.

2. Method

For this study, the method applied has tree steps. The first step was the adsorption experiment with the batch system. The adsorption kinetic used equation (2) and (3) to get the ability of corn stalk (qt) and percentage of removal metal ion (R%) by adding the adsorption time until reach equilibrium condition. The calculation base on initial concentration of metal ions (Co =50 ppm) dissolved in liquid phase and the concentration at certain time (Ct) of metal ion adsorbed on adsorbent surface; with the liquid phase pH 4.5 and shacked in Erlenmeyer with 200 rpm.

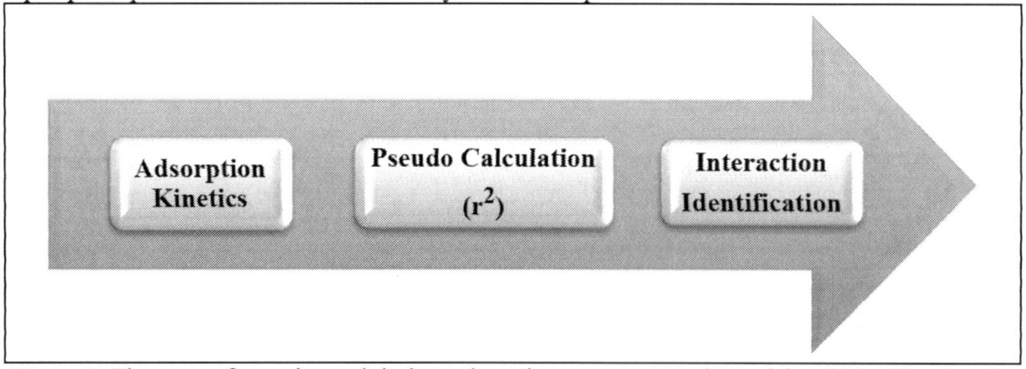

Figure 1. The steps of experimental design: adsorption process, pseudo model and identification the interaction

The second step was applying equation (4) to plot the 1/qt versus 1/t to get information of the relationship (r^2) for pseudo order 1st. The equation (5) was used to plot the t/qt versus t to get information of the relationship (r^2) for pseudo order 2nd. The correlation coefficient (r^2) was then used to identify the interaction type.

The correlation coefficient (r^2) quality of first order kinetic is used to indicate the physical interaction of adsorbate and adsorbent. The correlation coefficient (r^2) quality of second order kinetic is used to indicate the chemical interaction. The quality of correlation coefficient (r^2) is near to one indicate the interaction type on each model. The Figure 1 is the steps flow that has been used in this study. Confirmation in the presence of metal ion was applied by Scanning Electron Microscope-Energy Dispersive Spectroscopy (SEM/EDS) Bruker. Metal ion concentration was measured by Atomic absorption spectroscopy (AAS), AA-700 Series, Shimadzu Corporation, Japan.

3.Result

The Table 1 shows the result of adsorption capacity (qt) and removal percentage (R%) of Fe and Cu by series of time. With the same result then used to get the R% and then the data were plot in Figure 2 for Fe and 3 for Cu. The concentration of metal ion was analyzed by using AAS. From the data at Table 1 and Figure 2 then searched the adsorption capacity which tends to reach equilibrium for Fe (II) at 80 to 120 min. It was similar to Cu (II) which was about 100 to 120 minutes with higher capacity. The adsorption capacity of Fe (II) was about 60% and Cu (II) was about 85%.

Chemical	**Fe (II)**		**Cu (II)**	
t (time)	**qt**	**R%**	**qt**	**R%**
0	0.000	0.000	0.000	0.000
10	2.881	57.615	4.606	90.788
20	3.750	75.065	4.521	90.958
30	3.319	66.375	4.541	90.918
40	3.766	75.315	4.539	90.922
50	3.384	67.685	4.557	90.887
60	3.003	60.055	4.574	90.852
70	2.915	58.305	4.571	90.859
80	2.828	56.555	4.567	90.866
90	2.992	59.838	4.551	90.897
100	3.156	63.120	4.536	90.928
110	3.028	60.555	4.512	90.979
120	2.900	57.990	4.519	90.962

Figure C shows the same drift kinetic base on the adsorption capacity (qt). The both results then used to measure the information of relationship (r^2) for pseudo order 2nd. The correlation coefficient (r^2) was then used to identify the interaction type by using equation 4 and 5.

First and second orders of pseudo equation were used to plot the data of adsorption kinetic. Adsorptions kinetic of Fe (II) are as shown in Figure 3 (A) for pseudo order 1st and 3 (B) for order 2nd. The quality of correlation coefficient (r^2) was significantly different. The 1st order is 0.0089 and the 2nd order is 0.9909. The correlation for Fe (II) which interacts with functional group on cornstalk is chemically (Chen Suhong, et al., 2011; Liu Haibin, et al., 2013; M Arshadi, et al., 2014).

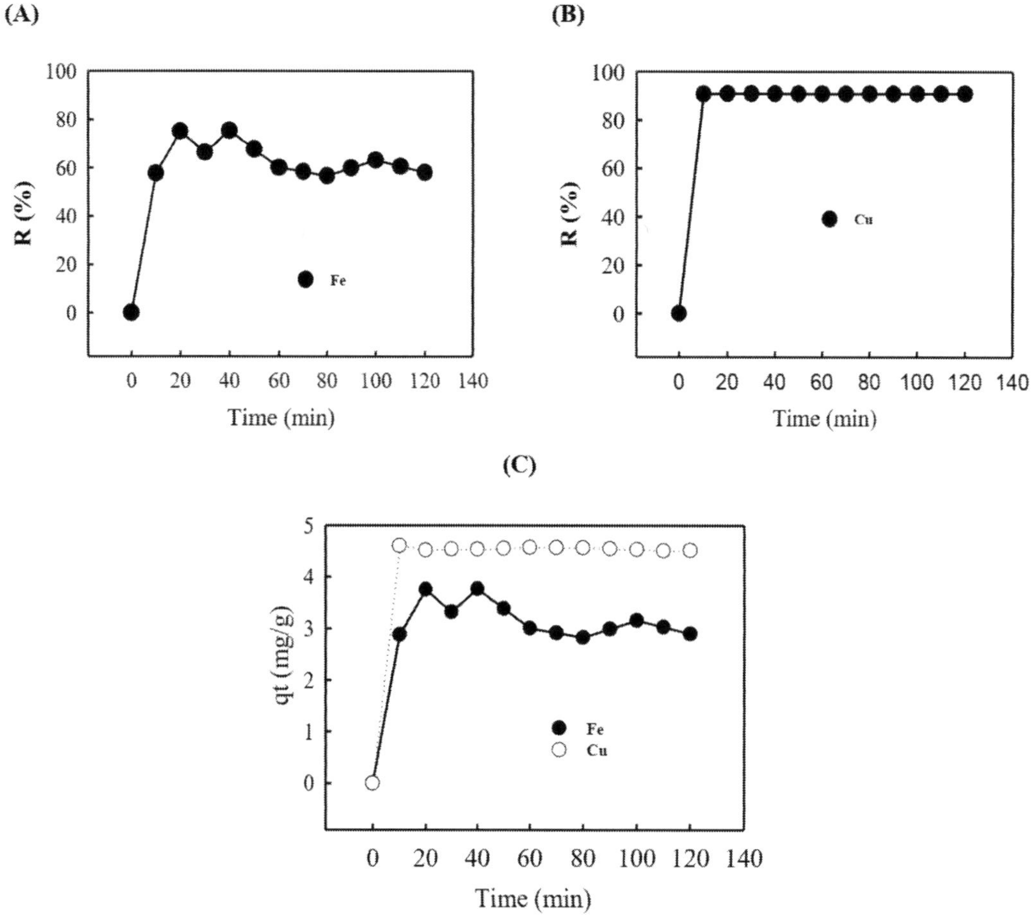

Figure 2. Adsorption kinetic of (A) Ion Fe^{2+}; (B) Cu^{2+} base on R% and (C) base on adsorption capacity

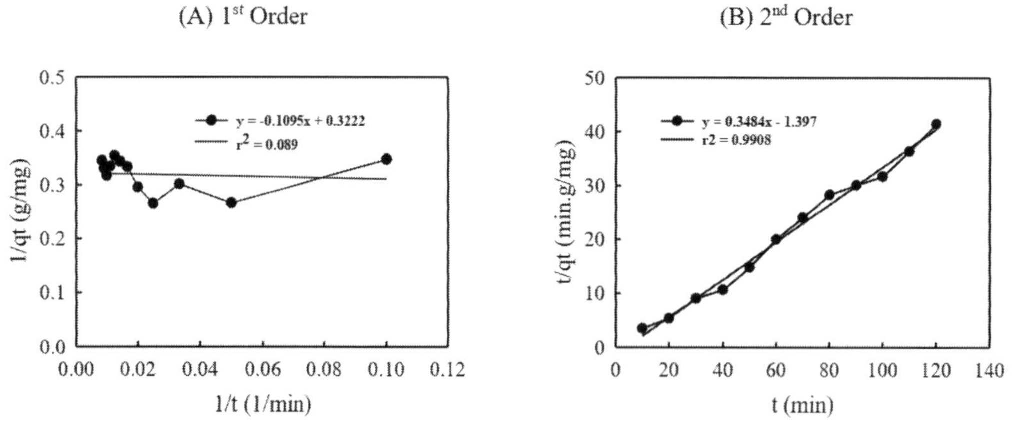

Figure 3. Pseudo order kinetic for Fe (II)

For Cu (II) shows similar trend of the quality of correlation coefficient (r^2). First and second order of pseudo equation were found that 0.2609 and 0.9999 respectively. As shown in Figure 4 (A) for pseudo order 1st and 4 (B) for order 2nd, the quality of correlation coefficient (r^2) was also significantly different. For both of chemical compound, Fe (II) and Cu (II) have chemically interaction on cornstalk (Chen Suhong, et al., 2011; Liu Haibin, et al., 2013; M Arshadi, et al., 2014).

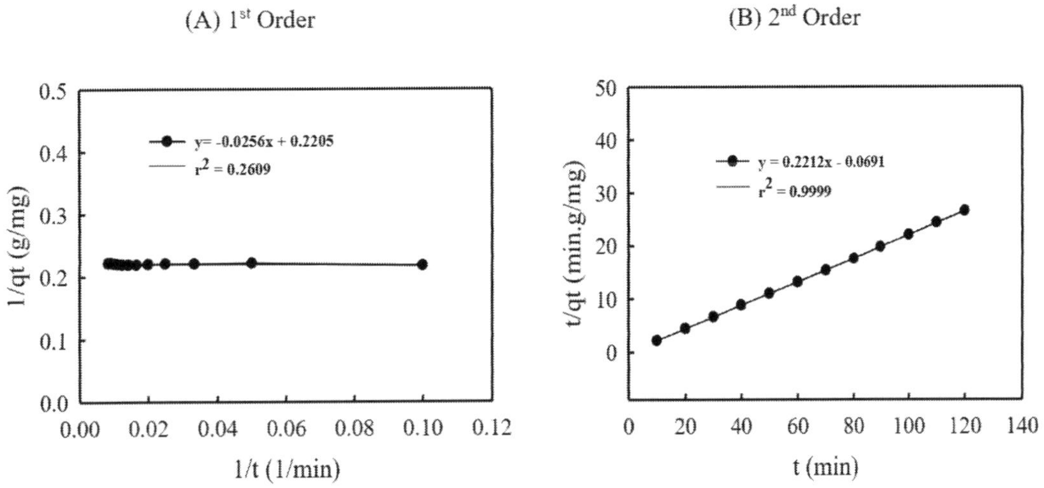

Figure 4. Pseudo order kinetic for Cu (II)

Figure 5. SEM fresh cornstalk 50 mesh

The physical structure of cornstalk adsorbent was taken by SEM/EDX. Figure 5 shows the profile of the 70 mesh cornstalk. It is the SEM result with 500 x magnitude. The porous characteristic of cornstalk is an advantage to use as adsorbent. The higher the surface area an adsorbent will increase the potential to adsorb the metal ions. The EDX confirmed the presence of metal ion contaminant represent by analyzing Cu (II) on cornstalk adsorbent. Table 2 is the confirmation of the chemical compound on cornstalk surface. It was compared to Table 3 as the result of chemical compound after contaminated by metal ion Cu (II). The result is indicating the possible interaction of copper ion chemically on sand surface as predicted by calculating the adsorption kinetic by using pseudo order 1st and 2nd.

Table 2 The EDX chemical compound result before used as adsorbent to adsorb Cu (II).

Spectrum: FRESH+CLEAN=70MESH

El	AN	Series	unn. C [wt.%]	norm. C [wt.%]	Atom. C [at.%]	Error (1 Sigma) [wt.%]	K fact.	Z corr.	A corr.	F corr.
C	6	K-series	57.44	57.44	68.51	7.69	0.872	0.659	1.000	1.000
O	8	K-series	34.14	34.14	30.57	5.30	0.281	1.216	1.000	1.000
Au	79	L-series	7.73	7.73	0.56	0.44	0.034	1.923	1.000	1.185
Si	14	K-series	0.70	0.70	0.35	0.07	0.002	4.350	1.000	1.007
		Total:	100.00	100.00	100.00					

Table 3 The EDX chemical compound result after used as adsorbent to adsorb Cu (II).

Spectrum: SAMPEL 2

El	AN	Series	unn. C [wt.%]	norm. C [wt.%]	Atom. C [at.%]	Error (1 Sigma) [wt.%]	K fact.	Z corr.	A corr.	F corr.
C	6	K-series	9.73	49.04	64.91	1.67	0.103	4.780	1.000	1.000
O	8	K-series	6.05	30.48	30.29	1.18	0.039	7.911	1.000	1.000
Au	79	L-series	2.17	10.94	0.88	0.17	0.120	0.803	1.000	1.133
Si	14	K-series	0.96	4.83	2.73	0.08	0.025	1.886	1.000	1.007
Cu	29	K-series	0.94	4.72	1.18	0.08	0.024	1.637	1.000	1.223
		Total:	19.84	100.00	100.00					

4.Conclusions

The r^2 of data adsorption kinetics for Fe (II) were 0.01 and 0.99 for pseudo order 1st and 2nd respectively. For Cu (II), the r^2 were 0.26 and 0.999 for pseudo order 1st and 2nd respectively. The r^2 of both contaminants have quality for pseudo order 2nd. The interaction type of metal ion on surface on corn stalk with higher quality r^2 pseudo 2nd was identified. From the result was found that the interaction of metal ions Cu^{2+} and Fe^{2+} with the functional group on surface of corn stalk is chemically. SEM/EDX confirmed the Cu^{2+} presence on cornstalk surface. The result of adsorption interaction of metal ion and surface function of corn stalk is chemically adsorption of interaction type.

References

[1] Amanda Rizka Nabilla, Rahmanta Ginting and Sinar Indra Kesuma 2014 Faktor-faktor yang Mempengaruhi Produksi dan Pendapatan Petani, Studi Kasus: Desa Lau Bekeri, Kecamatan Kuta Limbaru, Kabupaten Deli Serdang

[2] Yu Wang, Feng Shen and Xinhua Qi 2016 A corn stalk-derived porous carbonaceous adsorbent for adsorption of ionic liquids from aqueous solution, RSC Adv., 6, 32505-32513

[3] Paul B Tchounwou, Clement G Yedjou, Anita K Patlolla, and Dwayne J Sutton 2012 Heavy Metals Toxicity and the Environment, EXS. 101, 133–164

[4] Vafakhah S, M E Bahrololoom, R. Bazarganlari dan M Saeedikhani 2014 Removal of copper ions from electroplating effluent solutions with native corn cob and corn stalk and chemically modified corn stalk, Journal of Environment Chemical Engineering, 2, 356-361

[5] Mona Karnib, Ahmad Kabbani, Hanafy Holail and Zakia Olama 2014 Heavy Metals Removal Using Activated Carbon, Silica and Silica Activated Carbon Composite, Energy Procedia, 50,113-120

[6] Haryanto Bode and Chang Chien-Hsiang 2014 Foam-enhanced removal of adsorbed metal ions from packed sands with biosurfactant solution flushing. Journal of the Taiwan Institute of Chemical Engineers. 45, 5, 2170-2175

[7] Wen Liu, Ting Wang, Alistair G L Borthwick, Yanqi Wang, Xiaochen Yin, Xuezhao Li and Jinren Ni 2013 Adsorption of Pb^{2+}, Cd^{2+}, Cu^{2+} and Cr^{2+} onto titanate nanotubes: Competition and effect of inorganic ions. Journal Science of The Total Environment. 456, 171-180

[8] Liu Haibin,Wang Cuiping, Liu Jingting, Wang Baolin and Sun Hongwen 2013 Competitive adsorption of Cd(II), Zn(II), and Ni(II) from their binari and ternary acidic systems using tourmaline. Journal of Environmental Management. 128, 727-734

[9] Bode Haryanto, Jo-Shu Chang and Chien-Hsiang Chang 2014 Application of Biosurfactant Surfactin on Copper Ion Removal from Sand Surfaces with Continuous Flushing Technique. 51, 5, 1-8

[10] Chen Suhong, Yue Qinyan, Gao Baoyu, Li Qian and Xu Xing 2011 Removal of Cr(VI) using modified corn stalks. Journal Chemical Engineering. 168, 909-917

[11] Arshadi M., Amiri M.J. and Mousavi Sajjad 2014 Kinetic, equilibrium and thermodynamic investigations of Ni(II), Cd(II), Cu(II) and Co(II) adsorption on barley straw ash. Water Resources and Industry. 6, 1-17